Geophysical Monograph Series
American Geophysical Union

Geophysical Monograph Series

A. F. Spilhaus, Jr., Managing Editor

1 **Antarctica in the International Geophysical Year,** A. P. Crary, L. M. Gould, E. O. Hulburt, Hugh Odishaw, and Waldo E. Smith (editors)

2 **Geophysics and the IGY,** Hugh Odishaw and Stanley Ruttenburg (editors)

3 **Atmospheric Chemistry of Chlorine and Sulfur Compounds,** James P. Lodge, Jr. (editor)

4 **Contemporary Geodesy,** Charles A. Whitten and Kenneth H. Drummond (editors)

5 **Physics of Precipitation,** Helmut Wieckmann (editor)

6 **The Crust of the Pacific Basin,** Gordon A. Macdonald and Hisashi Kuno (editors)

7 **Antarctic Research: The Matthew Fontaine Maury Memorial Symposium,** H. Wexler, M. J. Rubin, and J. E. Caskey, Jr. (editors)

8 **Terrestrial Heat Flow,** William H. K. Lee (editor)

9 **Gravity Anomalies: Unsurveyed Areas,** Hyman Orlin (editor)

10 **The Earth Beneath the Continents: A Volume in Honor of Merle A. Tuve,** John S. Steinhart and T. Jefferson Smith (editors)

11 **Isotope Techniques in the Hydrologic Cycle,** Glenn E. Stout (editor)

12 **The Crust and Upper Mantle of the Pacific Area,** Leon Knopoff, Charles L. Drake, and Pembroke J. Hart (editors)

13 **The Earth's Crust and Upper Mantle,** Pembroke J. Hart (editor)

14 **The Structure and Physical Properties of the Earth's Crust,** John G. Heacock (editor)

15 **The Use of Artificial Satellites for Geodesy,** Soren W. Henriksen, Armando Mancini, and Bernard H. Chovitz (editors)

16 **Flow and Fracture of Rocks,** H. C. Heard, I. Y. Borg, N. L. Carter, and C. B. Raleigh (editors)

17 **Man-Made Lakes: Their Problems and Environmental Effects,** William C. Ackermann, Gilbert F. White, and E. B. Worthington (editors)

18 **The Upper Atmosphere in Motion: A Selection of Papers With Annotation,** C. O. Hines and Colleagues

19 **The Geophysics of the Pacific Ocean Basin and Its Margin: A Volume in Honor of George P. Woollard,** George H. Sutton, Murli H. Manghnani, and Ralph Moberly (editors)

20 **The Earth's Crust,** John G. Heacock (editor)

The Earth's Crust

geophysical monograph 20

The Earth's Crust

Its Nature and Physical Properties

JOHN G. HEACOCK
editor
GEORGE V. KELLER
JACK E. OLIVER
GENE SIMMONS
associate editors

American Geophysical Union
Washington, D. C.
1977

Published under the aegis of the AGU Geophysical Monograph Board; Bruce Bolt, Chairman; Thomas E. Graedel, Rolland L. Hardy, Pearn P. Niiler, Barry E. Parsons, George R. Tilton, and William R. Winkler, members.

International Standard Book Number 0-87590-020-8

Copyright © 1977 by the American Geophysical Union
1909 K Street, N.W.
Washington, D. C. 20006

Library of Congress Card Number 77-083153

EDWARDS BROTHERS INC., ANN ARBOR, MICHIGAN

FOREWORD

The papers in this monograph are based on research contributions presented at the Symposium on 'the Nature and Physical Properties of the Earth's Crust' held August 2 to 6, 1976, at Vail, Colorado. This symposium was cosponsored by the Office of Naval Research (ONR) and the Colorado School of Mines.

The year 1976 seemed to be a good one in which to hold this symposium. It is memorable as the two-hundredth anniversary of the independence of the United States of America. By coincidence, it was also the fiftieth anniversary of the Department of Geophysics of the Colorado School of Mines and the thirtieth anniversary of the Office of Naval Research.

The year 1976 also seemed to be a good time to evaluate our knowledge of the earth's crust. Great strides had been made through worldwide cooperation, initially during the International Geophysical Year (1957 to 1959), then during the International Upper Mantle Project (1962 to 1970), and, more recently, during the International Geodynamics Project (IGP) (1971 to 1979). Strong patterns of international cooperation have been established, and significant progress has been made in the understanding of earth phenomena on a broad scale. Progress has been made in relating structural configurations, geochemical sequences, paleontological records, and geophysical phenomena to each other through their dependence on geodynamic processes.

It had been 6 years since an earlier ONR/University of Colorado symposium was held which reviewed the state of the art of evaluating crustal properties on a multidisciplinary basis and which resulted in the publication by the American Geophysical Union of <u>The Structure and Physical Properties of the Earth's Crust</u>, volume 14 of this series.

Also, in 1976 the international earth science community began to evaluate seriously the question of where it should focus its attention following the termination of the IGP at the end of 1979. In the United States, considerable weight of opinion currently appears to favor a broad study of crustal dynamics. Progress in the understanding of the earth based on geodynamic phenomena and our present knowledge of the earth's interior provides both an academic challenge and a platform from which to launch a study of the fundamental processes active in the development of the crust. Additional impetus for expanding our knowledge of the crust is derived from the need to meet the increasing demands of society for natural resources.

The purpose of this symposium was to assemble from a broad range of the fundamental geological and geophysical disciplines up-to-date knowledge of crustal properties and processes by reporting the latest field research findings and to establish relationships among the physical properties of crustal materials from laboratory studies. Using these data on a multidisciplinary basis, we should be able to improve our ability to infer deep crustal properties from surface geophysical observations. Ultimately, it may be possible to compare laboratory and field data in order to infer such deep crustal properties and physical parameters as the lithology, strength, state of stress, temperature, porosity, electrical and thermal conductivities, permeability, and the fluid content of crustal materials. We may then apply these results to the evaluation of such important physical aspects as the lifetime and energy content of geothermal reservoirs, earthquake stress concentrations and associated risks, the utility of portions of the earth for waste disposal, and the environment, location, and availability of deep mineral and fossil fuel resources.

Many problems remain to be solved before we can unravel the origin and history of development of the earth's crust. Yet a solution of these problems must necessarily be made in terms of understanding relationships among the elements of the crust and the upper mantle and in terms of understanding the basic processes which have been active or are currently active in the crust. To do this requires, among other things, a development of the basic research tools needed to probe the crust from the surface in order to measure its physical properties and to interpret them in geological terms. We hope that this volume, which deals with the current state of our knowledge of the crust and techniques for measuring crustal properties, will contribute in a fundamental way toward achieving these goals.

John G. Heacock
Editor

ACKNOWLEDGEMENTS

The job of obtaining two professional reviews of each manuscript in this volume and of providing technical evaluation for each paper was the **responsibility** of the associate editors, G. V. Keller, J. E. Oliver, and G. Simmons. Many of the authors served as readers for other papers in the volume and provided careful reviews. I sincerely appreciate these important contributions which were an integral and invaluable part of the editorial process.

The responsibility for administering the organization of the Symposium on the Nature and Physical Properties of the Earth's Crust was shouldered by George V. Keller and his very able assistant, Catherine Skokan of the Colorado School of Mines. The smooth functioning of the symposium attests to their conscientious and pleasant handling of these details.

I appreciate the expertise and reliability of Lottie M. Dobrai, my secretary, who handled the heavy editorial correspondence, and of Mary J. Scroggins, the AGU copy editor who was responsible for the production of this book.

Funding was provided by the Earth Physics Program of the Office of Naval Research (except where company or contract support was available to some individuals).

Finally, I wish to express my gratitude for the responsive and enthusiastic contributions made by the authors and meeting participants to the creation of this volume.

John G. Heacock

CONTENTS

Foreword vii

Acknowledgements ix

1. GEOLOGY

   The Evolution of the Earth's Crust and Sedimentary Basin
      Development . . . . A. R. Green   1
   Aspects of the Deep Crustal Evolution Beneath South Central
      New Mexico . . . . . . Elaine R. Padovani and James L. Carter   19
   Evolution of Silicic Magma Chambers and Their Relationship to
      Basaltic Volcanism . . . John C. Eichelberger and R. Gooley   57
   Significance of Fe-Mg Cordierite Stability Relations on
      Temperature, Pressure, and Water Pressure in Cordierite
      Granulites . . . . Sang Man Lee and M. J. Holdaway   79

2. LABORATORY STUDIES

   Physical Properties/Microcracks

   Microcracks in Crustal Igneous Rocks: Physical Properties
         Michael Feves, Gene Simmons, and Robert W. Siegfried   95
   Characterization of Modulus-Pressure Systematics of Rocks:
      Dependence on Microstructure . . . . Nick Warren   119

   Geology/Microcracks

   Microcracks in Crustal Igneous Rocks: Microscopy
                        Dorothy Richter and Gene Simmons   149

   Electrical Study

   Electrical Characteristics of Igneous Precambrian Basement
      Rocks of Central North America
                        R. M. Housley and J. R. Oliver   181

   Seismic Studies

   Internal Friction Measurements and Their Implications in
      Seismic Q Structure Models of the Crust . . . B. R. Tittmann   197
   In Situ and Laboratory Measurements of Velocity and
      Permeability
         H. R. Pratt, H. S. Swolfs, R. Lingle, and R. R. Nielsen   215

   Geothermal Study

   Geothermal Systems: Rocks, Fluids, Fractures
                        Michael L. Batzle and Gene Simmons   233

3. SEISMOLOGY

   Field Reflection Studies

   Complexities of the Deep Basement From Seismic Reflection
      Profiling . . . . .Jack Oliver and Sidney Kaufman   243
   Seismic Velocity, Reflections, and Structure of the
      Crystalline Crust
         Scott B. Smithson, Peter N. Shive, and Stanley K. Brown   254
   Crustal Velocities From Marine Common Depth Point Seismic
      Reflection Data
         Joel S. Watkins, Richard T. Buffler, Mark H. Houston,   271
         John W. Ladd, Thomas H. Shipley, F. Jeanne Shaub,
               John B. Sinton, and J. Lamar Worzel

   Field Refraction Studies

   A New Model of the Continental Crust. . . . . Stephan Mueller   289
   The Nature of the Earth's Crust in Canada
                              M. J. Berry and J. A. Mair   319
   The Structure of the Crust-Mantle Boundary Beneath North
      America and Europe as Derived From Explosion Seismology
                                             Claus Prodehl   349
   Inversion of Oceanic Seismic Refraction Data
         John A. Orcutt, Leroy M. Dorman, Paul K. P. Spudich   371

   Physical Property Evaluation

   Geophysical Evidence for a Magma Body in the Crust in the
      Vicinity of Socorro, New Mexico
         A. R. Sanford, R. P. Mott,Jr., P. J. Shuleski,   385
         E. J. Rinehart, F. J. Caravella, R. M. Ward,
                                    and T. C. Wallace
   A Summary of Seismic Surface Wave Attenuation and Its
      Regional Variation Across Continents and Oceans
      Brian J. Mitchell, Nazieh K. Yacoub, and Antoni M. Correig   405

   Model Study

   Interpretation of Crustal Velocity Gradients and Q Structure
      Using Amplitude-Corrected Seismic Refraction Profiles
                                       Lawrence W. Braile   427

4. ELECTRICAL RESISTIVITY

   Determining the Resistivity of a Resistant Layer in the Crust
                     George V. Keller and Robert B. Furgerson   440
   Electrical Studies of the Deep Crust in Various Tectonic
      Provinces of Southern Africa. . . . . .Jan S. V. van Zijl   470
   Flambeau Anomaly: A High-Conductivity Anomaly in the Southern
      Extension of the Canadian Shield
                           Ben K. Sternberg and C. S. Clay   501

High Electrical Conductivities in the Lower Crust of the Northwestern Basin and Range: An Application of Inverse Theory to a Controlled-Source Deep-Magnetic-Sounding Experiment. . . . . . Barry R. Lienert and David J. Bennett ... 531

Crust and Upper Mantle Near the Western Edge of the Great Plains. . . . . W. Chaipayungpun and M. Landisman ... 553

5. STRESS

   Field Mechanical Study

   Crustal Stress in the Continental United States as Derived From Hydrofracturing Tests. . . . . . . .Bezalel C. Haimson ... 576

   Field Electrical Study

   High-Accuracy Determination of Temporal Variations of Crustal Resistivity
   H. Frank Morrison, Robert F. Corwin, and Mark Chang ... 593

   Laboratory Study

   Kinematics and Dynamics of Sliding in Saw-cut Rock
   H. M. J. Illfelder and H. F. Wang ... 615

6. GEOTHERMAL STUDIES

   Broad Scale

   Heat Flow in the United States and the Thermal Regime of the Crust . . . . .Arthur H. Lachenbruch and J. H. Sass ... 626

   Area

   Characteristics of Selected Geothermal Systems in Idaho
   James K. Applegate and Paul R. Donaldson ... 676

   Convection, Conduction, Hydrology

   Fluid Circulation in the Earth's Crust. . . . . .Denis Norton ... 693
   Numerical Solutions for Transient Heating and Withdrawal of Fluid in a Liquid-Dominated Geothermal Reservoir
   Ping Cheng and Lall Teckchandani ... 705
   Numerical Calculation of Two-Temperature Thermal Convection in a Porous Layer with Application to the Steamboat Springs Thermal System, Nevada
   D. L. Turcotte, R. J. Ribando and K. E. Torrance ... 722

7. APPENDIX. Recommendations for Future Crustal Research ... 737

8. INDEX ... 751

THE EVOLUTION OF THE EARTH'S CRUST AND SEDIMENTARY BASIN DEVELOPMENT

A. R. Green

Exxon Production Research Company, Houston, Texas 77001

Abstract. The earth's crust is a thin dynamic shell that changes in both thickness and composition through time. It is postulated that the changes that alter the crust are primarily the result of subcrustal processes and lateral interactions of crustal plates. The type of crust that underlies a sedimentary basin determines the physical framework, stability, structural style, and conditions of sedimentation and environment throughout the evolution of the basin. Thus as sediments are deposited, they record the tectonic history of the basin. These stratigraphic data can be used to develop conceptual genetic models that put the evolution of oceanic, continental, and transitional types of crust into perspective. An orderly cycle of crustal evolution is proposed which suggests that oceanic crust is thickened and continental crust is thickened and thinned by a number of natural processes. The resulting transitional crustal types, which represent intermediate steps in the continuum, occupy a realm between thin basic oceanic crust and thick acidic continental crust. These transitional, somewhat unstable crustal types host most of the world's sedimentary basins. An attempt is made to step back from the detail of the complex interplay involved in the process of crustal genesis and sedimentary basin development and look at the overall natural system. A set of hypothetical genetic models depicting the proposed origin of various crustal types is presented so that geophysical data can be compared and tested against them.

The Crustal Setting of Sedimentary Basins

Sedimentary deposits greater than 2000 m thick accumulate in depressions within the major continents or adjacent to their margins (Figure 1). Basins within the major continents contain a large percentage of Precambrian, Paleozoic, and Mesozoic sediment, which furnishes a record of crustal evolution during these earlier periods of the earth's development. Some of these older rocks have been metamorphosed to such a degree that they can no longer be considered part of the sedimentary section but rather are more closely related to the underlying crystalline crust. Basins on the marginal shelves, slopes, and rises of the Atlantic and Indian oceans contain sediments of Mesozoic and Tertiary age and form some of the thickest sediment accumulations in the world. Most of the sedimentary basins around the margins of the Pacific Ocean have formed more recently and furnish valuable clues concerning the Tertiary crustal history of that part of the world. Indeed, many of the Pacific margins are presently in an active stage of basin formation and provide modern analogs to ancient mechanisms of crustal evolution.

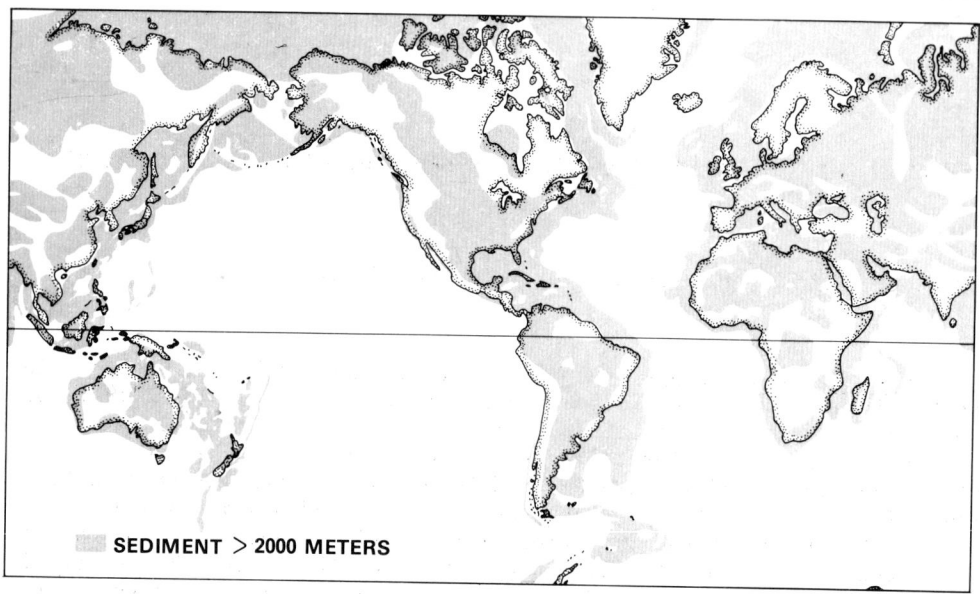

Fig. 1. Generalized sedimentary basins of the world (after H. R. Hopkins, personal communication, 1976).

Most of the world's sedimentary basins overlie crustal types that are neither purely continental nor purely oceanic in terms of thickness and composition. They generally develop upon thinned continental crust or upon accreted constructed continental crust (Figure 2). Most of the sediment that has accumulated upon thickened oceanic and purely oceanic crust occurs at the edges of large continental masses. The sediment-covered continental rises, deep-sea sedimentary cones, and the distal portions of deltas make up a large percentage of the sediment volume found upon oceanic crust.

Cycle of Crustal Evolution

Gutenberg's [1959] definition of the earth's crust, as that part of the earth's outer shell which is above the Mohorovicic discontinuity is utilized. For the purpose of this discussion, crustal type is defined on the basis of both thickness and composition.

If crustal thickness [Cummings and Shiller, 1971] is empirically plotted against composition (acidic to basic), a cycle of crustal evolution can be envisioned, and the origin of various types of sedimentary basins can be set into a crustal framework to show the interaction of tectonics and stratigraphy (Figures 3 and 4). It is important to establish this relationship, because it is the layers of sedimentary rocks that record the pulse of tectonic activity in the underlying crust.

In the plot of crustal thickness versus composition in Figure 3, most of the earth's crustal types occur within the white corridor that expands from the upper left to the lower right. The shading depicts the proposed cycle of evolution from thin basic oceanic crust to thick

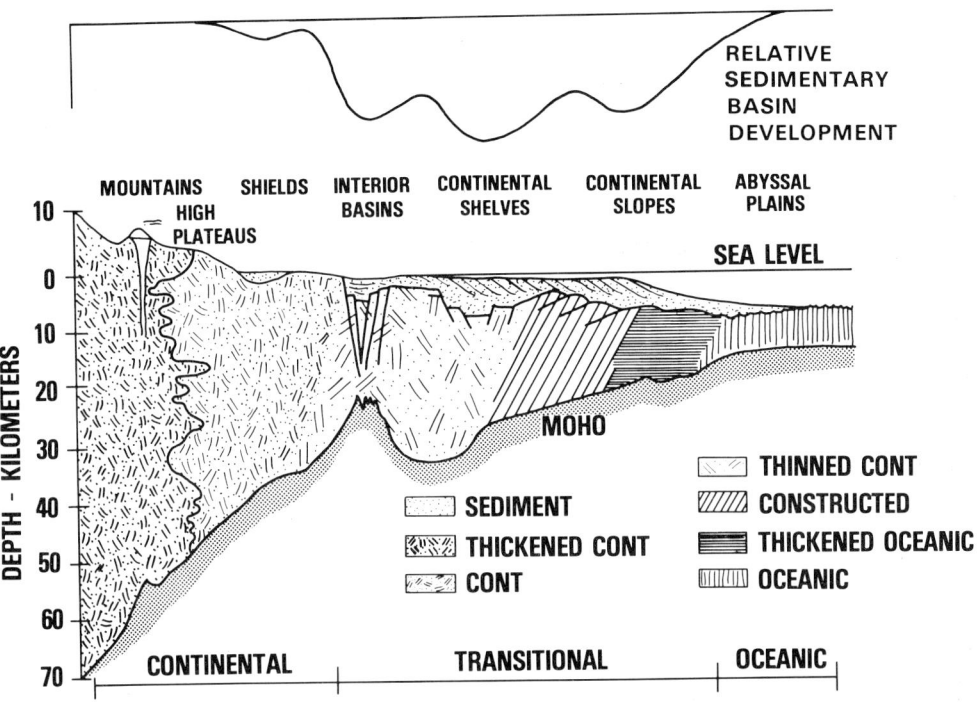

Fig. 2. Primary crustal types related to sedimentary basin development.

acidic orogenic belts. The continental crust is then subjected to rifting and transverse shearing and injected with basic rocks and thus follows a path back toward a thinner and more basic type of crust. The schematic models shown in Figure 4 represent stages in what is actually a continuum of crustal evolution.

Models of Oceanic Crustal Thickening

Oceanic crust, as referred to in this discussion, is typically 4-9 km thick, but it can be thickened in a number of ways. Mantle plumes, or hot spots, that penetrate oceanic crust and discharge significant volumes of basaltic flows and ash deposits may thicken the crust to more than 20 km (Figure 4a). The comparatively thick crust that underlies Iceland and the Faeroe ridge, for instance, has been confused with continental crust, even though it is primarily basaltic in nature [Casten, 1973] and is therefore neither continental nor oceanic but rather transitional crust.

Another example is found in the Azores, in the north central Atlantic. This area is underlain by 'oceanic' crust that is about 60% thicker than the surrounding oceanic crust, and the upper mantle seismic velocities are anomalously low [Searle, 1976]. The trilobate domed morphology of this part of the mid-Atlantic ridge, combined with geophysical and rare-earth evidence, is consistent with a model of an

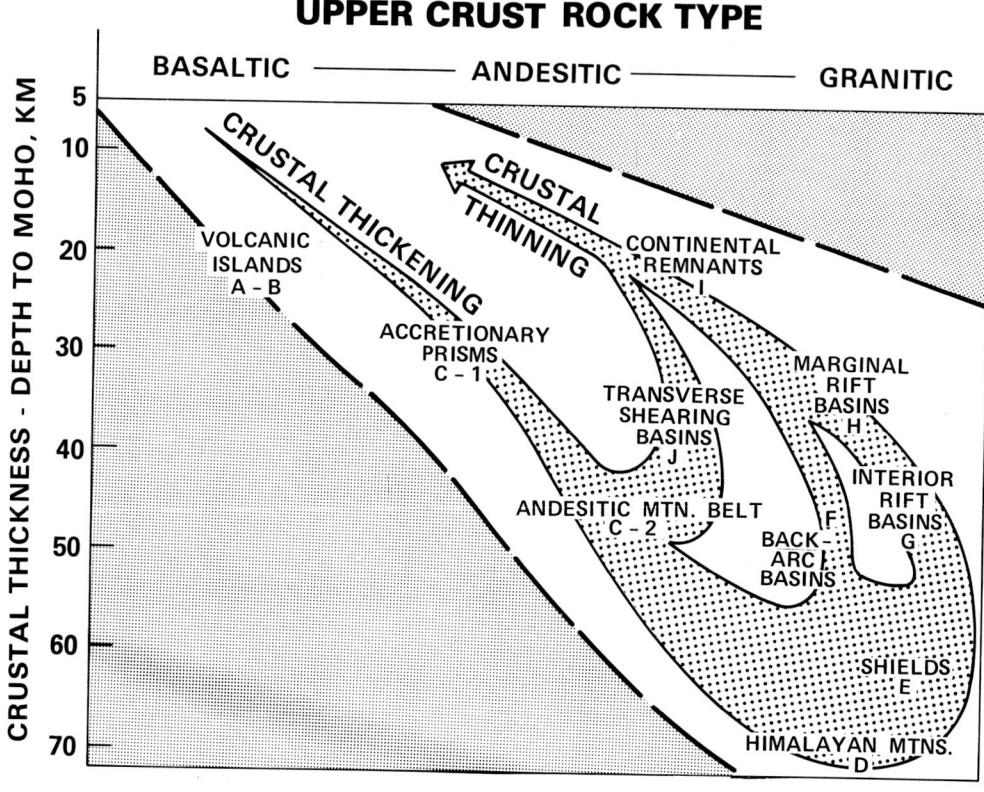

Fig. 3. Hypothetical cycle of crustal evolution. (Letters refer to models in Figure 4.)

upwelling of igneous material (plume) under the Azores plateau [Schilling, 1975].

Many such hot spots are associated with regional topographic uplift. Vogt [1972] suggests that this uplift reflects the existence of hot low-density regions in the asthenosphere derived from plume activity.

The Hawaiian-Emperor seamount chain is a thickened welt of extruded basaltic material. Jackson et al. [1972] believe that this welt formed over a melting spot, roughly 300 km in diameter, as the Pacific plate passed over it. Others have suggested that the volcanic zones are associated with a zone of fractures (D. L. Turcotte, personal communication, 1976).

MacDonald [1972] has used the term 'platillo' crust to describe all such undeformed volcanic masses which contain great thicknesses of basaltic rocks overlying otherwise normal oceanic crust.

The aseismic ridges in the equatorial Atlantic near Saint Paul's Rocks may represent the formation of thickened welts of oceanic crust by 'leaky' transforms [Thompson and Melson, 1972]. Some north-south extension may have occurred along the east-west transform faults, resulting in the extrusion of diapiric peridotites and basalts with a strong alkaline character. These extrusions form aseismic submarine ridges

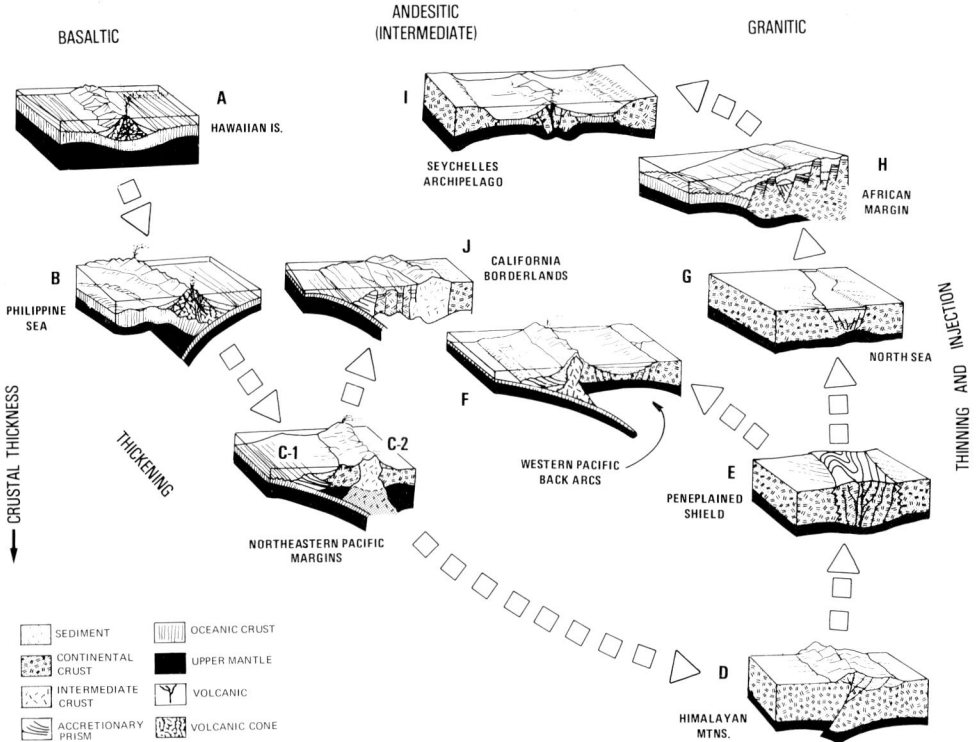

Fig. 4. Schematic cycle of crustal evolution.

that are thicker than normal oceanic crust. Thus even in the process of generating oceanic crust, atypical or transitional types are developed.

Descent of one oceanic segment of crust under another oceanic segment (oceanic/oceanic subduction) results in the formation of a trench and a thickened volcanic arc (Figure 4b). At depths of 110-200 km the descending slab undergoes partial melting, and the resulting magma erupts as submarine volcanos. Continued eruptions and volcanic upbuilding form an island arc which is initially basaltic in composition but becomes more andesitic with crustal thickening and magma chamber differentiation [Mitchell and Reading, 1971]. The arc is therefore more acidic and thicker than 'normal' oceanic crust and is termed 'transitional.'

The volcanic arcs of the Philippine Sea [Uyeda and Ben-Avraham, 1972] and eastern Panama may be examples of such an in situ origin of thickened oceanic crust from oceanic/ oceanic subduction. Case [1974] indicates that the thickening of the eastern isthmus of Panama has been accompanied by an uplift of at least 6 km since Late Cretaceous time. Thus an area that was once part of an ocean basin has been thickened and uplifted and is now part of the Central American landmass.

Sedimentary basins may develop along the flanks of these volcanic islands, or piles. The basin fill in such areas is characteristically composed of ashfalls and base surge deposits that are generated during

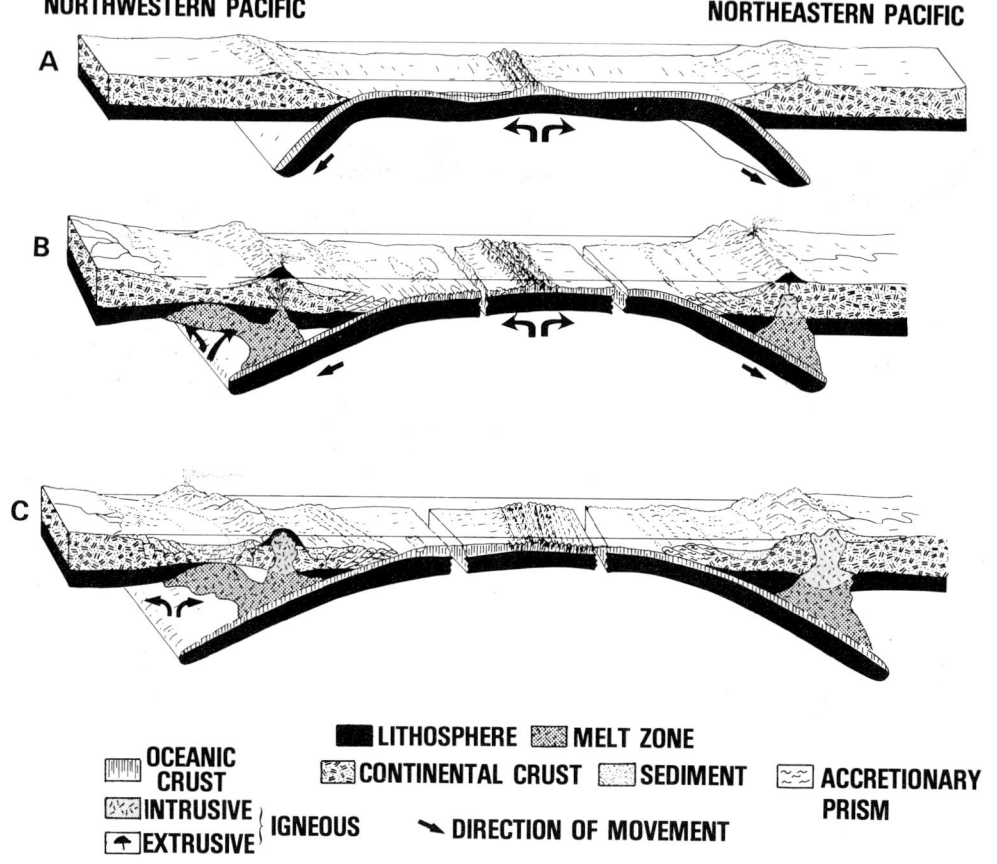

Fig. 5. Schematic evolution of trench margins of the northwestern and northeastern Pacific.

volcanic eruption and subaerial erosion of the volcanic islands [Waters and Fisher, 1971]. Reef carbonates and planktonic accumulations may also contribute to the sedimentary section and thus increase the depth to the Moho substantially. Jamaica appears to have had such an origin [Horsfield and Roobol, 1974] and is thus another type of transitional crust.

Models for the Development of 'Continental' Crust

When basaltic oceanic crust is subducted beneath continental crust (oceanic/continental subduction), two transitional types of crust are formed (Figure 4c): an accretionary wedge of greywacke sediment and continental crustal thickening by plutonic injection.

As subduction proceeds, the sediment cover on the downgoing oceanic plate and some of the igneous crust appear to be stripped off the plate at the landward side of the trench. The compression dewaters the sedi-

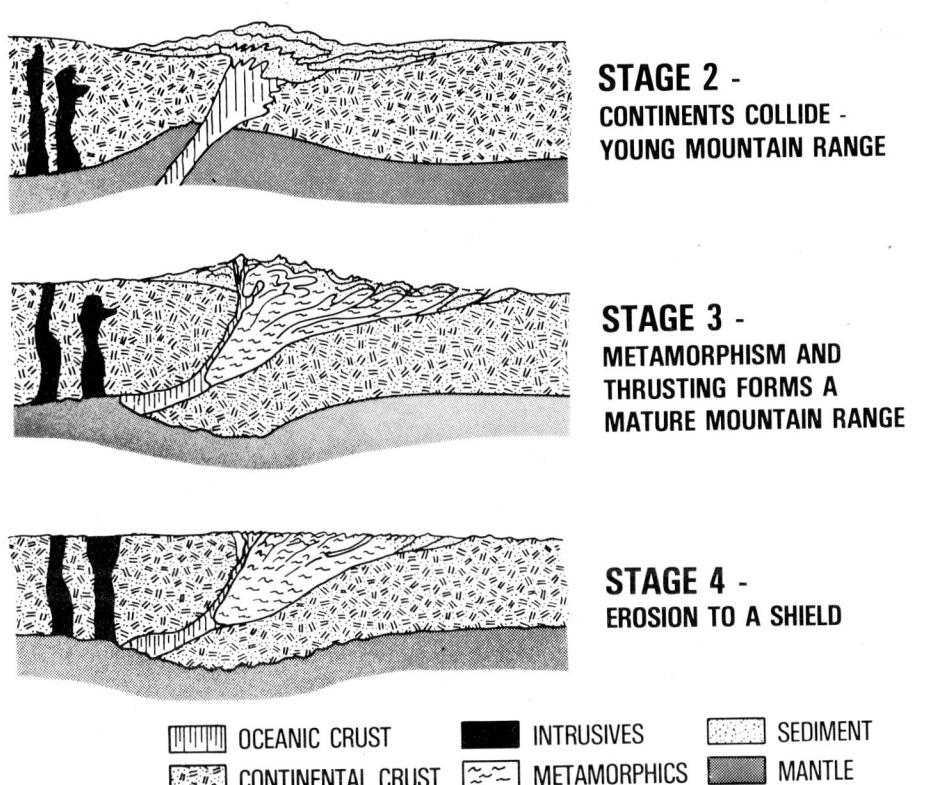

Fig. 6. Evolution of Himalayan type orogenic belt [after Dewey and Bird, 1970].

ment, and thrust sheets develop in a snowplow effect (Figure 5b). The thrust slices move upslope as subsequent thrusts develop seaward. Thus an accretionary prism is formed at the edge of the continent [Karig and Sharman, 1975]. The prism forms a tectonic dam, and forearc sedimentary basins develop between it and the continent (Figure 5c). If the process continues, the continent builds seaward, as it has along the northwest coast of the United States [Silver, 1972]. Thus a sedimentary basin develops as the result of ponding against the tectonic dam rather than as the result of isostatic subsidence of the crust.

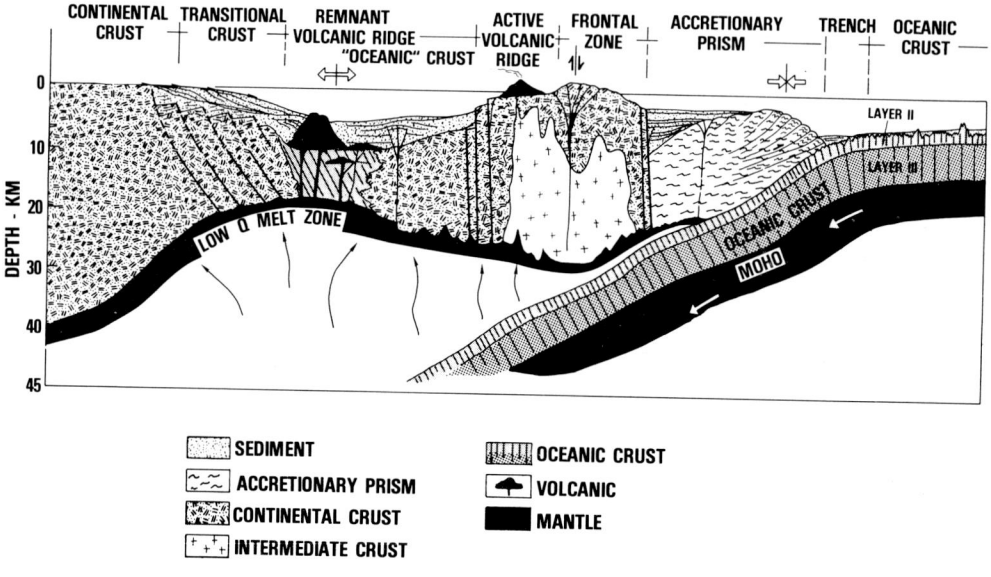

Fig. 7. Diagrammatic cross section of the western Pacific island arc system [after Karig, 1971].

Both the thickness and the composition of the original continental crust adjacent to the trench margin are changed by the injection of plutonic material of intermediate composition. As some of the descending oceanic slab undergoes partial melting, the resulting magma rises through continental crust to form a mountain chain of intermediate or andesitic composition (Figure 5b). The Sierra Nevada range of California is thought to be the result of such processes [Dickinson, 1974].

The thickest and most acidic crust forms when two continents collide (Figure 4d), as is thought to be the case in the Himalayas (Figure 6) [Dewey and Bird, 1970]. From seismic and gravimetric data the crust has been estimated to be up to 80 km thick south of the high range and to average 60 km thick under the Tibetan plateau [LeFort, 1975].

When such areas are eroded and peneplained, the roots of these and other types of orogenic belts become parts of shield complexes. Thus the thinning phase of the crustal cycle begins (Figure 4e).

Models of Crustal Thinning

Back-arc basin development. The back-arc basins of the western Pacific margin are thought to have developed by crustal extension [Karig, 1971]. The thinning of the crust is postulated to have been caused by the upwelling and intrusion of partially molten mantle material, associated with the subducted lithospheric plate (Figure 7). During the Tertiary the island arcs of the western Pacific have moved away from the Asian continent, and new basaltic oceanic crust has developed in the resulting small ocean basins [Horváth et al., 1975].

Attenuation of high-frequency seismic compressional waves above the

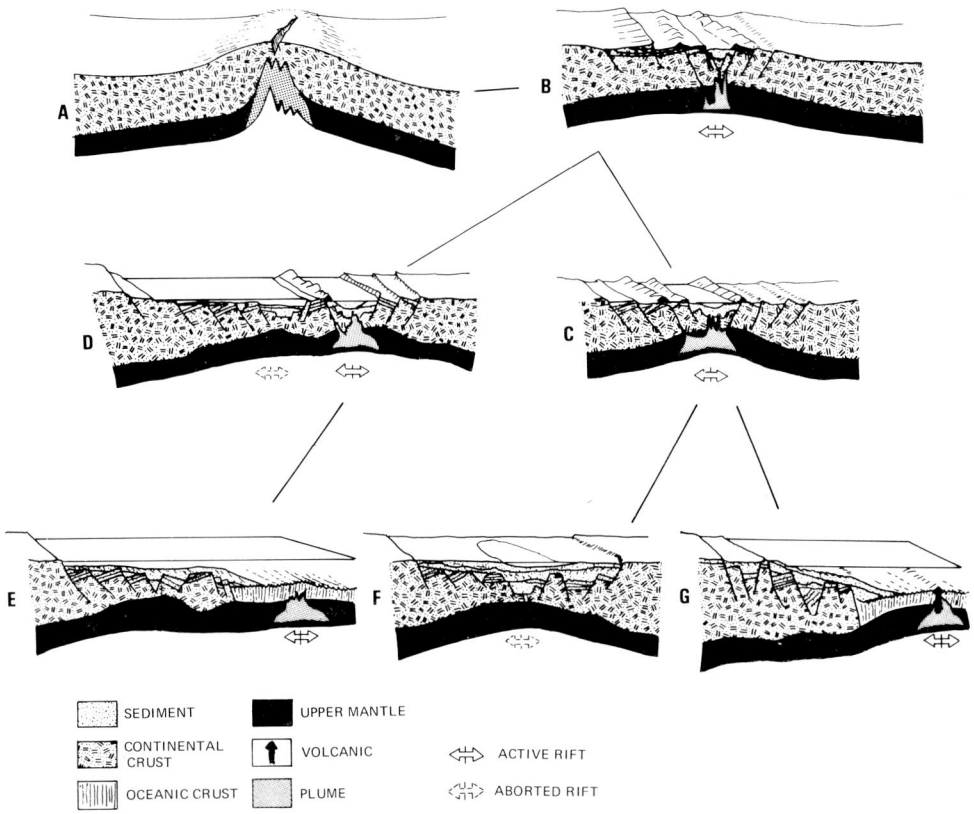

Fig. 8. Schematic models of rift basin development.

downgoing oceanic slab and the high heat flow conditions present in back-arc basins reinforce the idea that the subsiding basin may be underlain by a zone of partially melted material [Barazangi et al., 1975]. Normal faulting further attenuates the crust on the landward side of the basin and results in additional subsidence by normal block faulting (Figure 7). As these basins rapidly subside, the gradient of rivers entering the basins increases. As a result, some of the world's largest rivers and associated delta systems have developed adjacent to the back-arc basins of the western Pacific. This sediment loading from the major river systems causes further subsidence, thus perpetuating basin development.

Rift basin development. Rift basins are generally described as topographic depressions formed by subsidence along faults with roughly parallel strike and associated with volcanic and seismic activity [Dennis, 1967]. However, the physical manifestation of a rift, that is, the subsidence of fault blocks, may result from the complex and variable interaction of a number of crustal processes that fracture and thin the crust and/or alter its composition. The mechanisms for attenuating the crust are problematical. Both subcrustal processes and lateral movement (distension) are probably involved in most cases (Figure 8).

Differential lithospheric cooling, phase changes, ductile flowage, subcrustal erosion, surface erosion following thermal uplift, injection and stoping action of dense material, and related magma chamber collapse have all been considered as mechanisms of continental crustal thinning and subsidence [Fischer, 1975].

The evolution of the interior basins of Africa appears to follow a pattern of domal uplift and erosion, which is in turn followed by subsidence [Burke and Whiteman, 1973]. The uplift is interpreted as isostatic response to mass deficiencies produced by partial melting in the lithosphere. Uplift and injection of dense material into the continental crust are normally followed by alkaline volcanicity and trilobate crustal rift formation (Figure 8a). Subsequent cooling results in subsidence (Figures 8b and 8c). If crustal alteration aborts at this stage, an amoeboid-shaped intracratonic basin may result (Figures 4g and 8f).

Haxby et al. [1976] suggest that the injection of the mantle diapir caused the transformation of the lower crustal gabbroic rocks to eclogite (phase change), and as the dense intruded body cooled, the basin subsided under the load of the eclogite.

Drake [1976] has postulated that the Michigan basin was formed by doming of continental lithosphere over a mantle plume and subsequent subsidence as the plume activity ceased, an intracratonic basin thus being created.

If tectonic extension and attenuation are not aborted at this stage, a complete disruption of the continental lithosphere results, and oceanic crust is generated between the segmented continental blocks (Figures 4h and 8g). Once new oceanic crust has developed adjacent to the young continental margin, a number of tectonic processes become active.

Marinelli [1971] and Morton and Black [1975] suggest that crustal attenuation in the Afar of Ethiopia is enhanced by faulting and block tilting in the upper crust, while ductile flowage occurs at depth. In this case the block faults do not initiate the crustal thinning, but once it has started, the normal fault blocks further thin the crust.

After continental crustal separation, subsidence of the shelf area may occur as isostatic response to crustal thinning caused by hot creep of lower continental crustal material toward the oceanic mantle [Bott, 1971].

Sleep [1971] indicates that the subsidence of Atlantic type continental margins is the result of cooling and thermal contraction of the crust and upper mantle as the rifted continental blocks drift away from the spreading ridge. The observed subsidence rate on the Atlantic coast of the United States has declined exponentially with a time constant of about 50 m.y.

Sedimentary basins along rifted margins may therefore result from a series of events. The sequence may be (1) injection of dense material and resultant uplift of continental crust, (2) erosion of the uplifted surface, (3) cooling and collapse, (4) attenuation of the crust by faulting, (5) ductile flowage of the lower crust toward newly formed oceanic crust, and (6) sediment and water loading of the thinned and injected unstable continental crust, producing additional subsidence. The continental margin basins of west Africa are thought to have formed in such a manner. Thus the crust underlying such basins is transitional and is the product of extreme crustal thinning and injection.

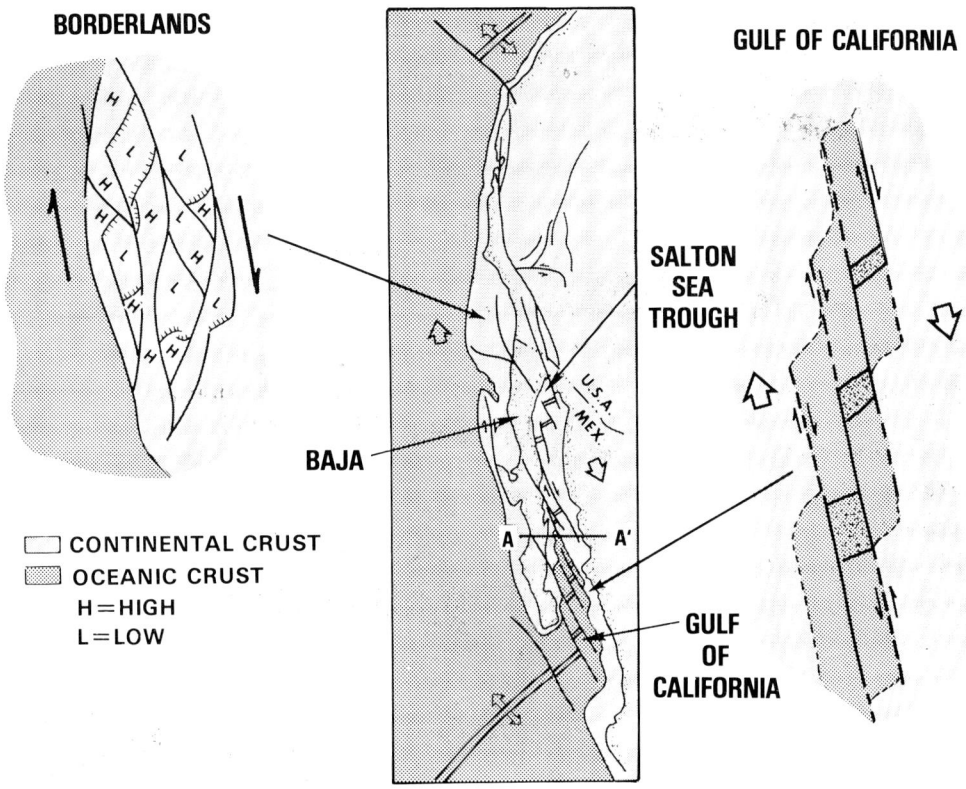

Fig. 9. Transverse shear basin development [after Crowell, 1974].

If the zone of igneous injection moves from one position to another (Figure 8d), a wide complex rifted margin may develop (Figure 8e).

<u>Transverse shear basin development</u>. Strike-slip extensional basins develop along transverse plate boundaries where the crust has been shattered and attenuated (Figure 9). At the present time the Gulf of California and the Salton Sea trough are widening as the continental Baja block moves obliquely to the northwest away from the main North American block [Crowell, 1974]. The subsiding basins in the northern part of the Gulf of California lie above attenuated and fragmented continental crust which has been intruded by igneous masses (Figure 10). In the southern Gulf of California, young oceanic crust has formed as complete continental separation has occurred.

In the California borderlands, rhomboid and lens-shaped basins are developed with similarly shaped ridges. The fragmented crust moves differentially along strike-slip faults that converge and diverge in plan view. Where right lateral faults diverge, the area is extended, and the rhomb-shaped terrane subsides [Crowell, 1974]. Thus the sedimentary basin in the northern Gulf of California has formed as one block of continental crust moves obliquely away from another, whereas the sedimentary basins of the borderlands result from differential shearing of continental crust (Figure 4j).

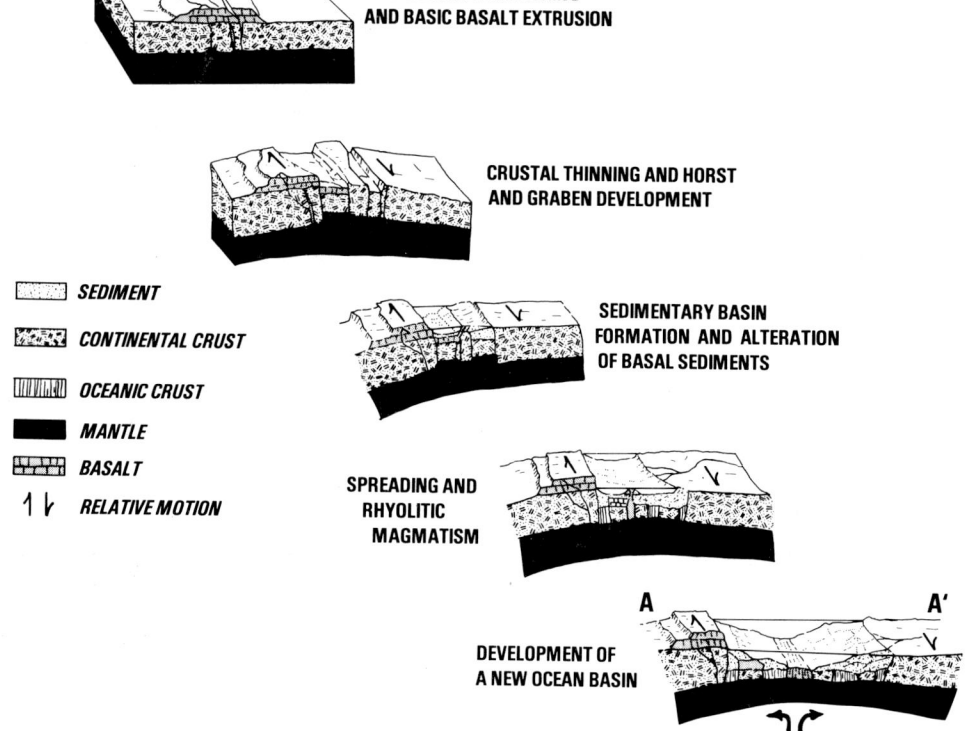

Fig. 10. Evolution of transverse shear basins. (See Figure 9 for location of cross section.)

Holcombe et al. [1973] have documented another example of crustal shearing subsidence and creation of new oceanic crust. The Cayman trough rift in the Yucatan basin has developed in response to differential relative motion between the Americas and Caribbean plates. The gap that has resulted from the eastward movement of the Caribbean plate away from the Americas plate is being filled with new oceanic crust, the Cayman trough thus being formed. A similar tectonic situation is operative within continental crust where the Dead Sea rift basin is developing.

Examples of Transitional Crustal Types

In the process of crustal evolution, oceanic crust is thickened, and continental crust is thickened and thinned, the result being a wide variety of transitional crustal types which are intermediate in both composition and thickness. The resulting thin and often unstable crust is the host for most of the world's sedimentary basins.

Some of the processes involved in the cycle of oceanic crustal evolution are summarized in Figures 11 and 12. Oceanic crust is from 4 to

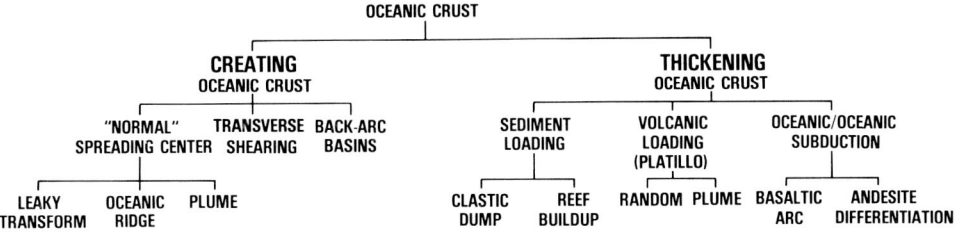

Fig. 11. Transitional crustal types developed from oceanic crust.

9 km thick, and the upper portion is basaltic in composition. Oceanic crust may form as continental crust is rifted apart and a midocean spreading ridge is developed [Talwani et al., 1965] in back-arc basins or along oblique shearing margins.

Oceanic crust may evolve to transitional crust by a number of processes, some of which are illustrated in Figure 12 and discussed below:

1. Volcanic flows may cover older oceanic crust and increase its thickness even though its composition remains basic. This may occur over an igneous plume, such as Iceland and the Hawaiian Islands (Figure 12, profile 6), or as random volcanic flows, such as those in the western Atlantic.

2. One oceanic plate subducting under another oceanic plate (oceanic/oceanic subduction) may result in uplift and volcanism, which causes the formation of thickened volcanic arcs, such as those of the Philippine Sea (Figure 12, profile 7).

3. As oceanic crust is thickened by volcanic extrusion, magma chambers may develop in the thickened section, and differentiation to andesites may occur, as is the case in the Tonga Island area of the South Pacific (Figure 12, profile 10).

4. Massive reef buildups form on thickened oceanic crust and further thicken the total crustal section, as they have in Jamaica (Figure 12, profile 5). Extensive influx of detrital material may also thicken the section (Figure 12, profile 9).

5. At subduction margins the sediment cover on the downgoing oceanic plate is stripped off, and thrust sheets develop in a snowplow effect, an accretionary wedge (Figure 12, profile 8) of transitional crust thus being formed.

The processes involved in the cycle of continental crustal evolution are summarized in Figures 13 and 14. Continental crust ranges from 25 to 70+ km in thickness, averages about 36 km thick, and is granitic in composition in its upper portion.

Continental crust may evolve to even thicker crust by orogenic activity:

1. If a continental margin collides with another continent (continental/continental collision), a Himalayan type mountain belt results, and very thick crust is formed [Dewey and Bird, 1970] (Figure 14, profile 24).

2. The Mérida Andes of Venezuela appear to be a well developed

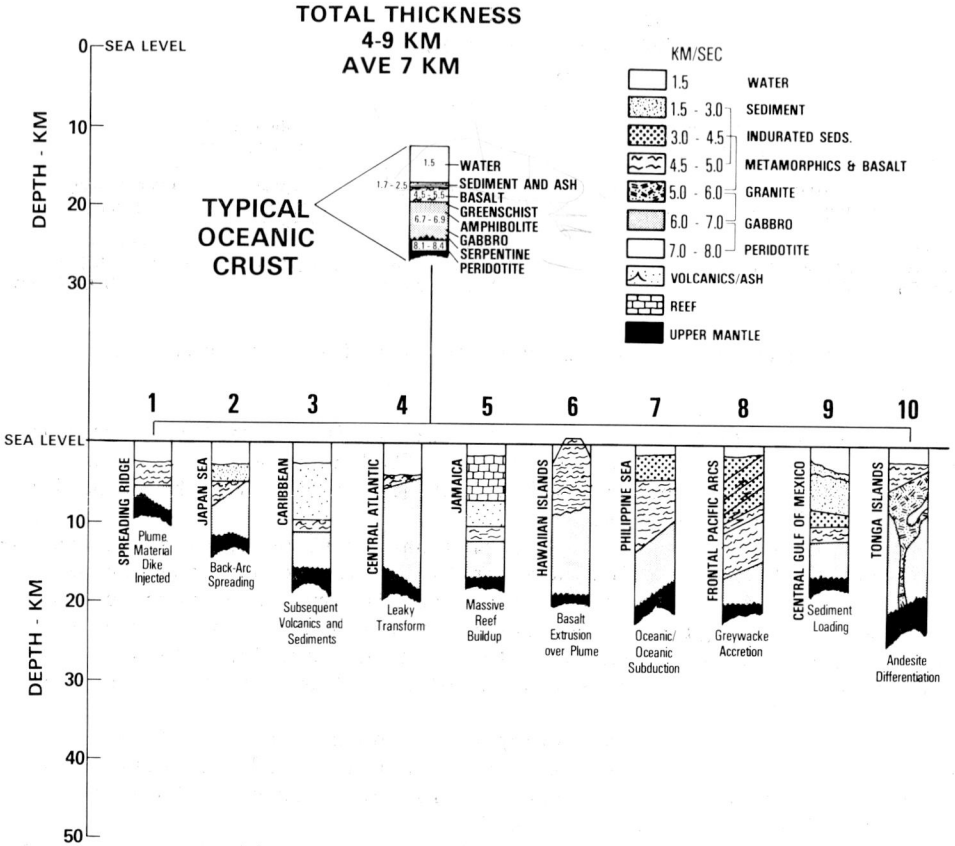

Fig. 12. Oceanic crust and models of thickening.

Mesozoic graben system that experienced compression and inversion during the Miocene and formed a mountain range with thickened crust. Thus the process is another type of continental/continental collision (Figure 14, profile 22).

3. If oceanic crust is subducted under continental crust, a mountain belt develops as calc-alkaline magmas are generated and inject and thicken the older continental crust (Figure 14, profile 23).

Continental crust may be thinned to transitional crust by a number of processes:

1. Injection of dense material into the upper crust can effectively thin the crust. With subsequent cooling, isostatic subsidence results (Figure 14, profile 11).

2. According to the phase transition hypothesis, cooling at depth causes the conversion of gabbro into eclogite with contraction of the deep rocks and resultant subsidence at the surface [Wyllie, 1971]. Alvarez [1972] has suggested that the lithosphere, and subsequently the continental crust, can be eroded by a convection cell in the upper mantle. When convection ceases, thinned continental crust or new

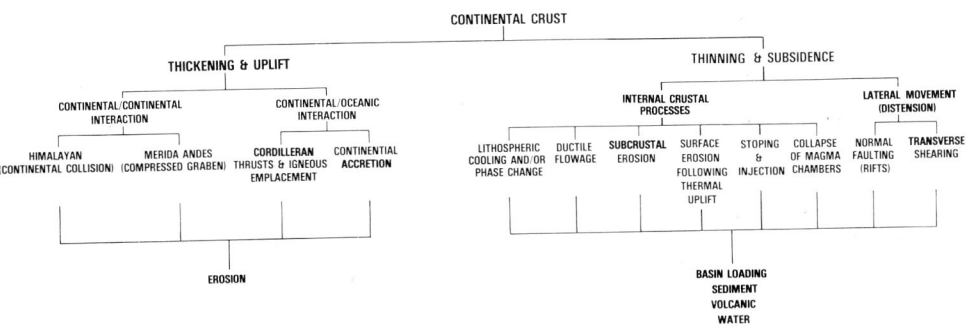

Fig. 13. Transitional crustal types developed from continental crust.

oceanic crust would theoretically subside to its normal isostatic level, a basin thus being formed.

3. Rifting of continental crust by extension or transverse shearing forms interior continental basins if the process is incomplete (Figure 14, profile 18) and continental margin basins if new oceanic crust develops between the rifted continental blocks (Figure 14, profile 16).

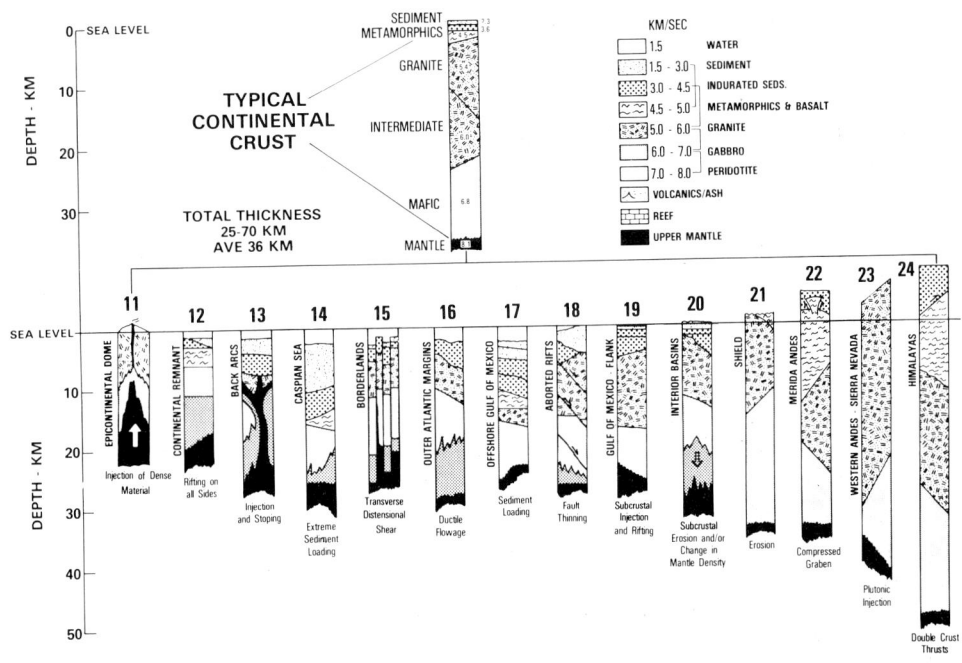

Fig. 14. Continental crust and models of thickening and thinning.

4. Back-arc basins in the western Pacific have developed as the arcs have moved away from the Asian continent (Figure 14, profile 13).

5. After any of the crustal thinning processes have been active and rifting is initiated, a number of secondary processes may become important. As the area subsides and breaks up, fault thinning may occur (Figure 14, profile 18), and ductile flowage of the lower crust may further increase the subsidence. As the area subsides, sediments load the unstable thinned crust and promote further subsidence (Figure 14, profile 14).

In the cyclic process of crustal evolution many types of transitional crust have developed. Sedimentary basins owe their very existence as well as many of their internal characteristics to the dynamic nature of the earth's crust which hosts them.

References

Alvarez, W., Uncoupled convection and subcrustal current ripples in the western Mediterranean, Studies in Earth and Space Science, Geol. Soc. Amer. Mem., 132, 119-132, 1972.

Barazangi, M., W. Pennington, and B. Isacks, Global study of seismic wave attenuation in the upper mantle behind island arcs using pP waves, J. Geophys. Res., 80, 1079-1092, 1975.

Bott, M. H. P., Evolution of young continental margins and formation of shelf basins, Tectonophysics, 11, 319-327, 1971.

Burke, K., and A. J. Whiteman, Uplift, rifting and break-up of Africa, in Implications of Continental Drift to the Earth Sciences, vol. 2, edited by D. H. Tarling and S. K. Runcorn, p. 735, Academic, New York, 1973.

Case, J. E., Oceanic crust forms basement of eastern Panama, Geol. Soc. Amer. Bull., 85, 645-652, 1974.

Casten, U., The crust beneath the Faeroe Islands, Nature Phys. Sci., 241, 83-84, 1973.

Crowell, J. C., Origin of late Cenozoic basins in southern California, Soc. Econ. Paleontol. Mineral. Spec. Publ., 22, 190, 1974.

Cummings, D., and G. I. Shiller, Isopach map of the earth's crust, Earth Sci. Rev., 7, 97-125, 1971.

Dennis, J. G. (Ed.), International Tectonic Dictionary, English Terminology Mem. 7, American Association of Petroleum Geologists, Tulsa, Okla., 1967.

Dewey, J. F., and J. M. Bird, Mountain belts and the new global tectonics, J. Geophys. Res., 75, 2625-2647, 1970.

Dickinson, W. R., Plate tectonics and sedimentation, Soc. Econ. Paleontol. Mineral. Spec. Publ., 22, 1-27, 1974.

Drake, B., Saginaw Bay graben and its implications for the origin of the Michigan basin and Pleistocene glaciation (abstract), Eos Trans. AGU, 57, 760, 1976.

Fischer, A. G., Origin and growth of basins, in Petroleum and Global Tectonics, pp. 47-79, Princeton University Press, Princeton, N. J., 1975.

Gutenberg, B., Physics of the Earth's Interior, 2nd ed., p. 240, Academic, New York, 1959.

Haxby, W. F., D. L. Turcotte, and J. M. Bird, Thermal and mechanical evolution of the Michigan basin, Tectonophysics, 36, 57-75, 1976.

Holcombe, T. L., P. R. Vogt, J. E. Matthews, and R. R. Murchison, Evidence for sea-floor spreading on the Cayman trough, Earth Planet. Sci. Lett., 20, 357-371, 1973.

Horsfield, W. T., and M. J. Roobol, A tectonic model for the evolution of Jamaica, J. Geol. Soc. Jam., 14, 31-38, 1974.

Horváth, F., L. Stegena, and B. Géczy, Ensimatic and ensialic interarc basins: Comments on 'Neogene Carpathian arc: A continental arc displaying the features of an island arc' by M. D. Bleahu, B. Boccaletti, P. Manetti, and S. Peltz, J. Geophys. Res., 80, 281-283, 1975.

Jackson, E. D., E. I. Silver, and G. B. Dalrymple, Hawaiian-Emperor chain and its relation to Cenozoic circumpacific tectonics, Geol. Soc. Amer. Bull., 83, 601-618, 1972.

Karig, D. E., Origin and development of marginal basins in the western Pacific, J. Geophys. Res., 76, 2542, 1971.

Karig, D. E., and G. F. Sharman III, Subduction and accretion in trenches, Geol. Soc. Amer. Bull., 86, 377-389, 1975.

LeFort, P., Himalayas: The collided range, Present knowledge of the continental arc, Amer. J. Sci., 275-A, 1-44, 1975.

MacDonald, W. D., Continental crust, crustal evolution, and the Caribbean, Geol. Soc. Amer. Mem., 132, 351-362, 1972.

Marinelli, G., La province géothermique de la dépression Dankali, Ann. Mines, 123-134, 1971.

Mitchell, A. H., and H. G. Reading, Evolution of island arcs, J. Geol., 79, 253-284, 1971.

Morton, W. H., and R. Black, Crustal attenuation in Afar, Afar Depression of Ethiopia, vol. 1, Inter-Union Comm. on Geodynamics Sci. Rep. 14, pp. 55-65, E. Schweizerbart'sche Verlagsbuchhandlung (Nägele u. Obermiller), Stuttgart, 1975.

Schilling, J. G., Azores mantle blob: Rare-earth evidence, Earth Planet. Sci. Lett., 25, 103-115, 1975.

Searle, R. C., Lithospheric structure of the Azores plateau from Rayleigh-wave dispersion, Geophys. J. Roy. Astron. Soc., 44, 537-546, 1976.

Silver, E. A., Pleistocene tectonic accretion of the continental slope off Washington, Mar. Geol., 13, 239-249, 1972.

Sleep, N. H., Thermal effects of the formation of Atlantic continental margins by continental break up, Geophys. J. Roy. Astron. Soc., 24, 325-350, 1971.

Talwani, M., X. Le Pichon, and M. Ewing, Crustal structure of the midocean ridges, 2, Computed model from gravity and seismic refraction data, J. Geophys. Res., 70, 341-352, 1965.

Thompson, G., and W. G. Melson, The petrology of oceanic crust across fracture zones in the Atlantic Ocean: Evidence of a new kind of sea-floor spreading, J. Geol., 80, 526-538, 1972.

Uyeda, S., and Z. Ben-Avraham, Origin and development of the Philippine Sea, Nature Phys. Sci., 240, 176-178, 1972.

Vogt, P. R., Evidence for global synchronism in mantle plume convection, and possible significance for geology, Nature, 240, 338-342, 1972.

Waters, A. C., and R. V. Fisher, Base surges and their deposits: Capelinhos and Taal volcanoes, J. Geophys. Res., 76, 5596-5614, 1971.

Wyllie, P. J., The Dynamic Earth: Textbook in Geosciences, p. 233, John Wiley, New York, 1971.

ASPECTS OF THE DEEP CRUSTAL EVOLUTION BENEATH SOUTH CENTRAL NEW MEXICO

Elaine R. Padovani[1] and James L. Carter

The University of Texas at Dallas, Program for Geosciences,
Richardson, Texas 75080

Abstract. The abundant quartzofeldspathic and mafic granulite facies xenoliths which occur at Kilbourne Hole maar in New Mexico provide an opportunity to study samples of the earth's deep crust which have been directly sampled by and incorporated into ascending magmas during explosive volcanic eruptions. Sillimanite-bearing garnet granulite, garnet orthopyroxenite, two-pyroxene granulite, charnockite, and anorthosite are the rock types represented in the deep crustal xenolith population. Equilibration temperatures of 750°-1000°C and pressures of 6-9 kbar are derived from the observed chemistries and phase assemblages of the garnet granulites. The chemistry of the more mafic charnockites and two-pyroxene granulites suggests equilibration in a similar temperature-pressure regime. Mineral equilibria in the suite of crustal xenoliths have been applied to estimate the geothermal gradient at their depth of origin prior to their incorporation into the eruptive process which formed Kilbourne Hole maar. The geotherm thus defined has a slope of about 30°/km at depths between 22 and 28 km. This slope agrees with the value predicted from heat flow interpretations in the southern Rio Grande Rift. The whole-rock chemistry and anhydrous nature of the garnet granulites suggest further that they may represent residues from the partial melting of a pelitic sediment and extraction of a water-rich acidic magma, which probably crystallized in the intermediate crust. The association of these rock types suggest that a granulite facies metamorphic complex, perhaps related to minor anorthosite, is representative of the lower crust in this region.

Introduction

The crust of the earth is defined as the outer part of the lithosphere where compressional seismic velocities are less than or equal to 7.7 km/s [Wyllie, 1971]. Major ideas about the nature of the crust-mantle boundary have centered around three hypotheses: (1) that it is an isochemical phase change from basalt to eclogite [e.g., Lovering, 1958; Ito and Kennedy, 1971]; (2) that it is a chemical discontinuity from granulite facies rocks of intermediate composition to peridotite [e.g., Ringwood and Green, 1966; Green and Ringwood, 1967, 1972]; or (3) that it is a hydration reaction from peridotite

---

[1] present address: Department of Earth and Planetary Sciences, Massachusetts Institute of Technology, Cambridge, Massachusetts 02139

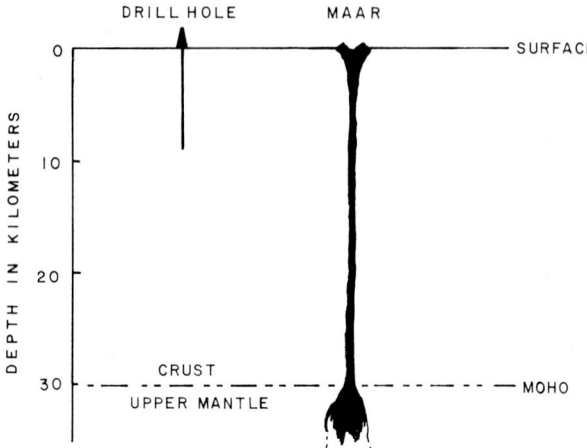

Fig. 1. Diagram illustrating relative depth of the earth's crust sampled by the world's deepest drill hole, 9.6 km, Lone Star Producing Borden 1, Beckam County, Oklahoma [U.G.M.S., 1976], compared to the depth potentially sampled by a maar volcano.

in the mantle to serpentine in the crust, especially the oceanic crust [Hess, 1962].

Since no hole has yet been drilled to the mantle (the deepest drill holes are less than 10 km [Utah Geological and Mineral Survey, 1976]) to obtain samples directly (Figure 1), geophysical and experimental petrological techniques have been extensively used to supply physical and chemical information about the nature of the crust-mantle region. Geophysical techniques can supply information on certain properties of the lower crust, such as average density, thermal and electrical conductivity, and seismic velocity, while experimental petrology can define a range of possible mineral assemblages under various PT conditions, using simple and complex compositional systems. However, neither of these techniques can define a unique composition for the crust at any given depth. Thus it is important to pursue alternative techniques to determine the chemical, mineralogical, and textural relationships of deep crustal materials, which will allow a better understanding of the processes that have shaped the earth's crust.

Direct sampling of the rock types which characterize the earth's lithosphere is limited to exposed outcrops, drill cores, and fragments (xenoliths) brought to the earth's surface by erupting magmas. Surface exposures and drill cores of discontinuous segments of the crustal column can give only relative stratigraphic, structural, and paragenetic information, particularly with regard to the lower crust (Figure 1). Moreover, since the deepest segments of the earth's crust thus exposed have been greatly displaced, they have often been highly altered by tectonic processes. In contrast, the deep crustal materials found as xenoliths in the Kilbourne Hole maar ejecta are essentially pristine with respect to their primary phase assemblages [Padovani and Carter, 1973]. Thus the only direct means presently available for possibly determining the earth's entire column at one location is the study of crustal assemblages brought to the surface as xenoliths by volcanic processes [e.g., McGetchin and Silver, 1970; McGetchin and Ullrich, 1973; Padovani and Carter, 1973, 1974, 1975, 1976a, b, c].

Xenoliths of a wide variety of rock types are characteristic components of the ejecta associated with the eruptive phases of maar development [e.g., Reeves and De Hon, 1965; Lorenz, 1970]. They also occur in diatremes, kimberlite pipes, and other basaltic hosts [e.g., Ross et al., 1954; Nixon et al., 1963; Carter, 1965a, b, 1970; Forbes and Kuno, 1967; MacGregor, 1968; Chronic et al., 1969]. These authors interpret such xenoliths as having been accidentally caught up in erupting magmas and consider them to represent equilibrium assemblages that reflect pressures ranging from those in the crust to those in the upper mantle.

Studies focusing on ultramafic xenoliths with the objective of understanding the mineralogy, chemistry, and composition of the upper mantle have for the most part neglected those xenoliths which may represent portions of the deep crust in the vicinity of the crust-mantle interface. The existence of both mafic and ultramafic xenoliths in the Kilbourne Hole maar ejecta implies a subcrustal origin for the magma which was responsible for the formation of this volcanic feature. If the magma was generated in the upper mantle, it had the potential of sampling the entire crustal column.

Previous field studies have focused on the geomorphology and stratigraphy of this maar and its related volcanic features [Lee, 1907; Reiche, 1940; Reeves and De Hon, 1965; Hoffer, 1969a, b, 1971, 1975, 1976]. A detailed chemical and mineralogical study of the mafic and ultramafic xenoliths from the Kilbourne Hole maar ejecta resulted in a partial fusion-partial crystallization model for the upper mantle beneath the maar [Carter, 1965a, b, 1966, 1969, 1970]. However, no systematic study has been made of the mineralogy, chemistry, and texture of the feldspar-rich xenoliths. Until the abstracts of Padovani and Carter [1973, 1974, 1975, 1976a, b, c] on the garnet- and pyroxene-bearing granulites from Kilbourne Hole maar, only brief mention had been made in the geologic literature of the feldspar-rich rock types which dominate the deep crustal xenolithic assemblages in the Kilbourne Hole maar ejecta [Reiche, 1940; Carter, 1965a]. The deep crustal xenolithic assemblages consist of various types of feldspar-rich granulites, including garnet- and sillimanite-bearing varieties, two-pyroxene granulites, charnockite, and anorthosite.

The granulite facies xenoliths are the main focus of this paper, since no rocks of such high metamorphic grade occur either in outcrop or in wells drilled into the Precambrian basement of the area [Denison et al., 1970]. The garnet-bearing granulites are of particular interest, because they constitute a major component of the ejecta and their mineral equilibria can be used to characterize some aspects of the compositional and physical properties of the deep crust in south central New Mexico.

Geologic Setting

The Kilbourne Hole maar is one of three structurally controlled volcanic depressions formed along the trace of the Fitzgerald-Robledo

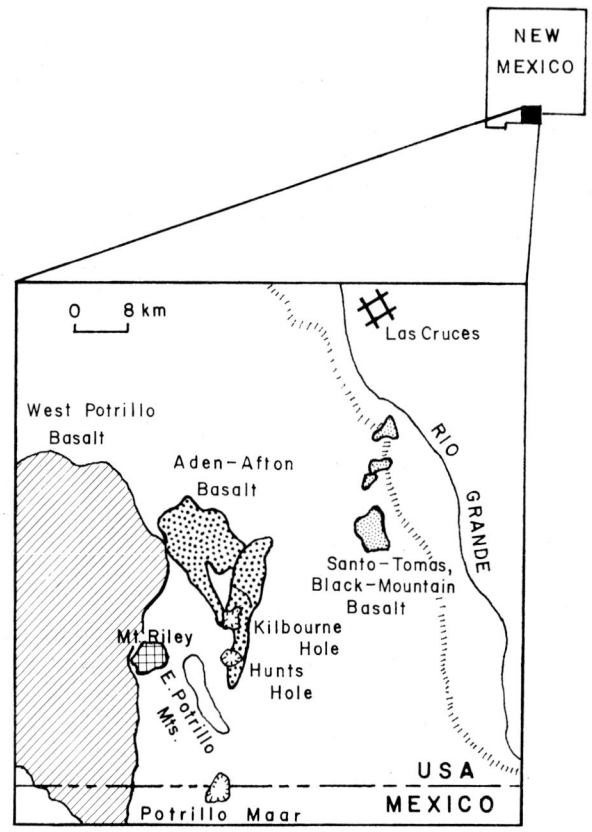

Fig. 2. Index map of Potrillo volcanic field, Dona Ana County, south central New Mexico [after Hoffer, 1975]. Stippled area represents extent of basalt flows, while crosshatched area represents rhyolite flows.

fault system on the mid-Pleistocene basin-filled La Mesa surface in south central New Mexico and northernmost Mexico [Reeves and De Hon, 1965; Hawley and Kottlowski, 1969a]. These maars form part of the Potrillo volcanic field (Figure 2), which can be divided into three regions: (1) a region to the east composed of the Santo Tomas-Black Mountain basalt field; (2) a central region containing several maars (Kilbourne Hole, Hunts Hole, and Potrillo); and (3) a region to the west composed of the East Potrillo Mountains (an uplifted fault block of rocks of Cretaceous age) and the West Potrillo Mountains (a series of spatter and cinder cones along with their related flows and maars) [Hoffer, 1971, 1975, 1976]. This volcanic field lies within the Rio Grande Depression and came into existence after the formation of the La Mesa surface [Hawley and Kottlowski, 1969b]. The Kilbourne Hole, Hunts Hole, and Potrillo maars were formed during late Pleistocene times by explosive steam blast eruptions, as evidenced by the base surge deposits which form elevated rims around the craters [Hawley and Kottlowski, 1969b].

The Rio Grande Rift dominates the regional geology and tectonic setting and is characterized by a series of basins or north trending grabens which are arranged in an en echelon fashion along the course of the Rio Grande [Chapin, 1971]. Rifting began approximately 18 m.y. ago, and evidence from fault scarps on alluvial fans suggests that movement is continuing at present [Kelley, 1952]. Chapin [1971] extends the southern limit of the Rio Grande Rift to the international border with Mexico and considers the East Potrillo Mountains to be an intrarift horst. These mountains are part of a belt of fault block mountain ranges which characterize the southern part of the rift.

Method of Study

Both petrographic and chemical studies were made of each granulite facies xenolith type. Thin sections were made of representative xenolith types for preliminary optical examination, and doubly polished thin sections were made for detailed optical study and electron microprobe analysis. The major and minor element chemistries of the phases from each representative xenolith type were performed by using both energy dispersive and wavelength dispersive electron microprobe X ray techniques [Padovani, 1977].

The major and minor elements in coexisting phases, such as garnet, sillimanite, feldspars, pyroxenes, oxides, and symplectites (intergrowths of glass, orthopyroxene, and spinel), were determined in order to examine the homogeneity of mineral phases and to study possible diffusion of elements across grain boundaries [Padovani and Carter, 1973; Padovani, 1977] (Tables 1 and 2). The electron microprobe data were combined with those from modal analyses to approximate whole-rock compositions (Table 3). This approach is thought to be more meaningful than whole-rock analyses per se because of various degrees of both partial fusion and possible contamination of the feldspar-rich rock types through interaction with the host magma, which would indicate the average chemical results of such reactions rather than the original chemistry of the unaltered rocks.

Observations

Garnet Granulites

The garnet granulite xenoliths range in size from about 3 to 35 cm in maximum dimensions and can have a partial or complete coating of alkali olivine basalt which forms a chilled margin from several millimeters to 1 cm in thickness. The garnet granulite xenoliths can be divided into two main groups: those which are foliated and contain potassium feldspar (sanidine) but not orthopyroxene (group 1) and those which are equigranular and do not contain sanidine but do contain orthopyroxene (group 2). Group 1 garnet granulites have the following possible mineralogies: (1) garnet (almandine), quartz, plagioclase, and sanidine; or (2) almandine, quartz, plagioclase, sanidine, and sillimanite. Group 2 garnet granulites consist of almandine, plagioclase, and orthopyroxene with or without quartz. In addition to these basic mineralogies, group 1 garnet granulites have the following accessories: (1) opaque minerals consisting of graphite,

TABLE 1. Representative Electron Microprobe Analyses of Garnet, Sillimanite, Orthopyroxene, and Opaque Phases From Group 1 and Group 2 Garnet Granulites

Oxide	Garnet 4-1	Garnet 9-8	Garnet 18-3	Garnet 54-4	Garnet 45-1	Sillim 4-4	Sillim 18-3	Sillim 54-4	Orthpx 200-1	Ilmen 33-Host	Ulvosp 33-Lamella
SiO_2	40.42	39.51	39.81	37.77	38.80	37.92	37.78	37.95	50.13	nd	nd
TiO_2	nd	0.13	0.13	0.01	0.21	0.06	0.04	0.04	0.31	49.44	25.66
Al_2O_3	19.13	21.98	21.91	21.83	22.50	61.26	60.06	61.90	4.07	1.04	7.26
Cr_2O_3	nd	nd	nd	nd	nd	0.24	nd	0.09	nd	0.07	0.05
FeO	29.14	24.71	28.28	30.62	24.11	0.49	1.08	0.84	23.71	48.78	66.27
MnO	0.45	0.66	0.45	0.51	0.58	0.16	nd	0.01	0.35	0.25	0.28
MgO	9.29	11.02	8.93	8.39	9.56	nd	0.01	nd	21.65	2.07	2.47
CaO	1.57	2.05	1.38	1.19	4.26	nd	nd	0.06	0.71	nd	nd
Total	99.99	100.07	100.87	100.32	100.06	100.29	99.04	100.91	100.93	101.65	99.99

Values are in weight percent; nd indicates not determined. Total iron is calculated as FeO.

Samples 4, 18, and 54 are group 1 garnet granulites with garnet, plagioclase, potassium feldspar, and sillimanite. Sample 9 is a group 1 garnet granulite without sillimanite. Samples 33, 45, and 200 are group 2 garnet granulites with garnet, plagioclase, and orthopyroxene. Quartz is present in sample 33 but not in samples 45 and 200. Sillim, sillimanite; Orthpx, orthopyroxene; Ilmen, ilmenite; and Ulvosp, ulvospinel.

TABLE 2. Representative Electron Microprobe Analyses of Pyroxene, Mica, Opaques, and Glass From Mafic Granulites and Anorthosite

Oxide	Clinpx 6-1	Orthpx 6-1	Glass 6-3	Clinpx 22-1	Orthpx 22-3	Amp 22-3	Glass 20-1	Olivine 20-1	Ilmen 6-8	Spinel 6-6	Spinel 37-1	Mica 1-1
SiO_2	49.56	51.26	48.76	50.13	50.05	40.21	48.73	37.38	0.24	0.37	0.18	37.67
TiO_2	0.61	0.17	3.54	0.89	0.28	4.67	2.20	0.04	28.55	14.38	12.65	6.87
Al_2O_3	5.34	3.53	22.57	6.30	4.64	13.97	14.21	0.03	0.84	4.19	9.55	13.51
Cr_2O_3	nd	nd	0.02	nd	nd	nd	nd	nd	0.76	1.84	nd	nd
FeO	11.57	21.57	12.68	12.53	22.81	14.79	13.95	28.27	62.57	71.50	73.44	17.75
MnO	nd	nd	0.66	0.29	0.28	0.08	0.26	0.51	nd	0.02	0.26	0.11
MgO	12.42	21.66	4.64	12.43	21.43	10.01	2.58	34.87	2.31	5.49	4.30	12.66
CaO	19.56	1.14	2.34	17.54	1.02	11.67	9.68	0.40	1.70	3.56	0.15	0.21
Na_2O	nd	nd	1.64	0.08	nd	0.66	3.31	nd	nd	nd	nd	0.35
K_2O	nd	nd	2.78	nd	nd	2.35	1.83	nd	nd	nd	nd	9.15
P_2O_5	nd	nd	10.55	nd	nd	nd	5.06	0.82	nd	nd	nd	nd
Total	99.06	99.33	100.19	100.21	100.55	98.41	101.79	102.32	97.00	101.42	100.52	98.27

Values are in weight percent; nd indicates not determined. Total iron is calculated as FeO.

Samples 22 and 37 are two-pyroxene granulite, sample 6 is charnockite, and sample 20 is anorthosite. Clinpx, clinopyroxene; Orthpx, orthopyroxene; Ilmen, ilmenite; and Amp, amphibole.

TABLE 3. Calculated Bulk Compositions for Group 1 and Group 2 Garnet Granulites From Point Counts (1000 Points Minimum)

Oxide	101	54	9	51	60	18	33*	45*	200*
SiO_2	53.69	40.05	57.02	47.18	52.36	62.98	50.59	37.68	42.72
TiO_2	1.83	3.77	nd	1.89	1.4	1.32	4.25	5.12	1.89
Al_2O_3	25.55	27.12	23.03	32.52	31.20	21.15	14.84	20.35	21.40
FeO	9.39	20.48	7.28	13.03	11.4	6.88	22.43	22.99	18.81
MnO	0.15	nd	nd	nd	nd	nd	nd	0.46	0.58
MgO	1.81	4.74	2.72	2.34	2.20	1.98	6.25	7.39	8.70
CaO	0.54	1.18	4.39	0.84	0.73	0.86	1.9	4.09	5.07
Na_2O	nd	1.03	3.17	0.14	0.07	1.06	0.05	0.75	1.49
K_2O	6.72	1.06	2.3	2.08	0.69	3.74	0.01	0.01	0.19
Total	99.68	99.43	99.91	100.02	100.05	99.97	100.32	98.84	100.85

Values are in weight percent; nd indicates not determined.
*Group 2 garnet granulite.

ilmenite, hercynite, or ulvospinel; (2) rutile, ranging in color from yellow orange to deep blue; and (3) zircon. Group 2 garnet granulites have only ilmenite and hercynite as accessory minerals.

The garnets within each xenolith in both groups are homogeneous in composition and display no observable zonation with respect to major or minor elements [Padovani and Carter, 1973]. This homogeneity is of great interest because chemical reactions involving garnet in the crust tend to be sluggish [Chinner, 1962; Evans, 1965]. Thus such uniformity in composition implies that sufficient time was available for these reactions to go to completion. The garnets from group 1 garnet granulites contain less than 5 mol % grossularite and less than 1 mol % spessartite components, while the garnets from group 2 garnet granulites have from 11 to 14 mol % grossularite and similar spessartite contents (Figure 3). The pyrope content of group 1 garnets ranges from 25 to 45 mol %, while that of group 2 garnets ranges between 35 and 40 mol % (Figure 3). Inclusions of rodlike plagioclase and rutile and of rounded quartz are typical of group 1 garnets, while group 2 garnets rarely have inclusions. Representative electron microprobe analyses from both groups are presented in Table 1.

Two feldspars, plagioclase of andesine composition and perthitic potassium feldspar (sanidine with exsolved plagioclase), are characteristic of group 1 garnet granulites. In contrast, group 2 garnet granulites contain only andesine plagioclase. Perthite is confined to the cores of the grains in group 1 garnet granulites, and in some grains, perthite is absent. Perthitic textures include patch, string, bead, and hair perthite. Where individual plagioclase lamellae are large enough to analyze, their composition is close to

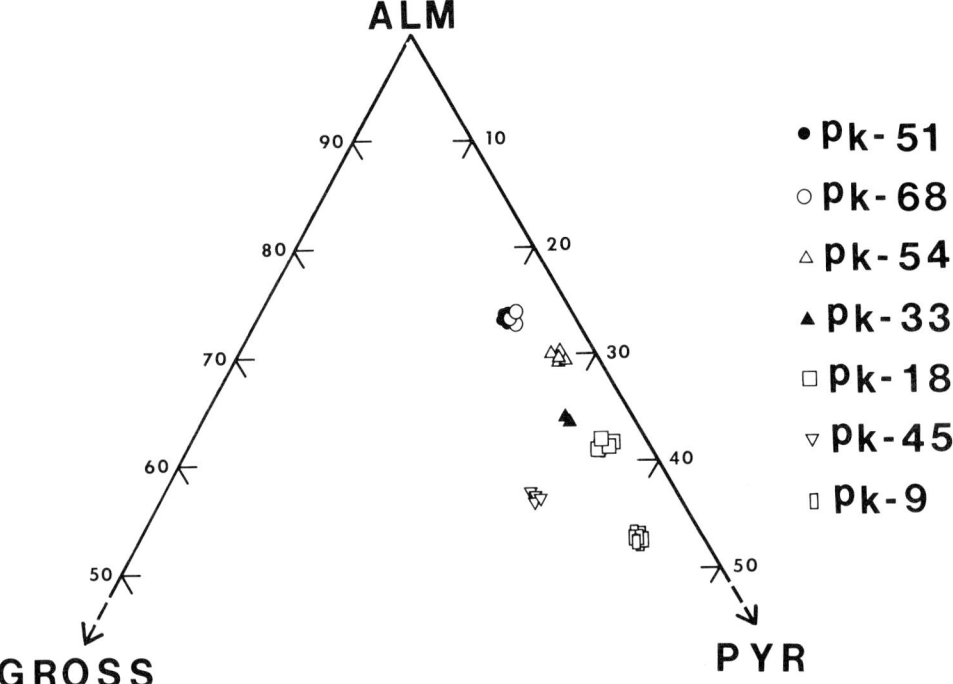

Fig. 3. Electron microprobe analyses of garnets in group 1 and 2 granulites, plotted in terms of mole percent end-members: Gross, grossularite ($Ca_3Al_2Si_3O_{12}$); Alm, almandine ($Fe_3Al_2Si_3O_{12}$); and Pyr, pyrope ($Mg_3Al_2Si_3O_{12}$). Spessartine ($Mn_3Al_2Si_3O_{12}$) accounts for less than 1 mol % in both groups of garnets. Group 1 garnets: pk-51, 68, 54, 18, and 9. Group 2 garnets: pk-33 and 45.

or slightly more calcic than that of the coexisting plagioclase grains. The plagioclase is andesine and ranges in composition from An_{30} to An_{50}. Electron microprobe analyses of the feldspars are plotted in Figures 4 and 5.

Colorless to pale blue sillimanite is ubiquitous in group 1 garnet granulites. It occurs as subhedral platy grains which range up to 5 mm in length. Samples containing blue sillimanite also tend to contain reddish purple to blue rutile as an accessory, while those containing colorless sillimanite have straw yellow to reddish orange rutile as an accessory. Doubly polished single-grain mounts of blue sillimanite have revealed that the blue color can be attributed to inclusions of blue rutile rods and blades which are oriented in the (101) direction parallel to, and often lying in, cleavage traces. Rods and blades of yellow brown rutile have also been observed in the colorless sillimanite. The rutile inclusions are about 1 μm in the shortest direction and represent an edge-on trace of rods and blades of rutile. In general, the blue variety is depleted in FeO, relative to the colorless variety (Figure 6). Representative electron microprobe analyses of sillimanite are found in Table 1.

Rutile (Table 1) ranges from straw yellow to deep blue in color and

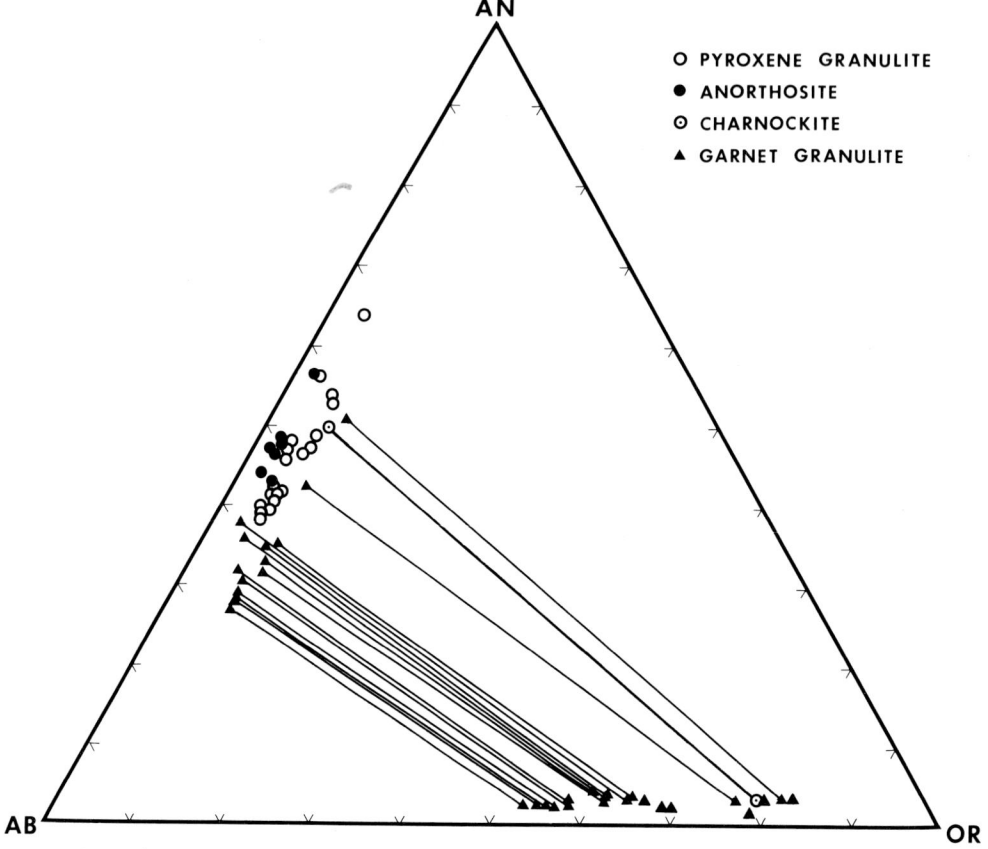

Fig. 4. Electron microprobe analyses of feldspars from group 1 and 2 garnet granulites, anorthosite, charnockite, and two-pyroxene granulites, plotted in terms of mole percent end-members: Ab, albite ($NaAlSi_3O_8$); An, anorthite ($CaAl_2Si_2O_8$); and Or, orthoclase ($KAlSi_3O_8$).

is restricted in occurrence to group 1 garnet granulites [Padovani and Carter, 1974]. Blue color in rutile can be directly correlated with the presence of blue color in sillimanite. Graphite, when it is present, always occurs in garnet granulites containing blue rutile. Ilmenite (Table 1) is present in both group 1 and group 2 garnet granulites and can have an exsolution relationship with ulvospinel, which forms either parallel lamellae within the centers of ilmenite grains or a discontinuous heterogeneous rim in grains which have undergone partial fusion.

Interstitial glass, representing former silicate melt and its quench products, is extensively developed around garnet, although glass is also developed in regions rich in sillimanite, feldspar, and quartz in highly fused xenoliths (equal to or greater than 50%). The glass is quite variable in color and composition (Figure 7) and frequently contains quenchlike crystals of aluminous iron-rich orthopyroxene and cruciform hercynite spinel [Padovani and Carter, 1975]. The glass

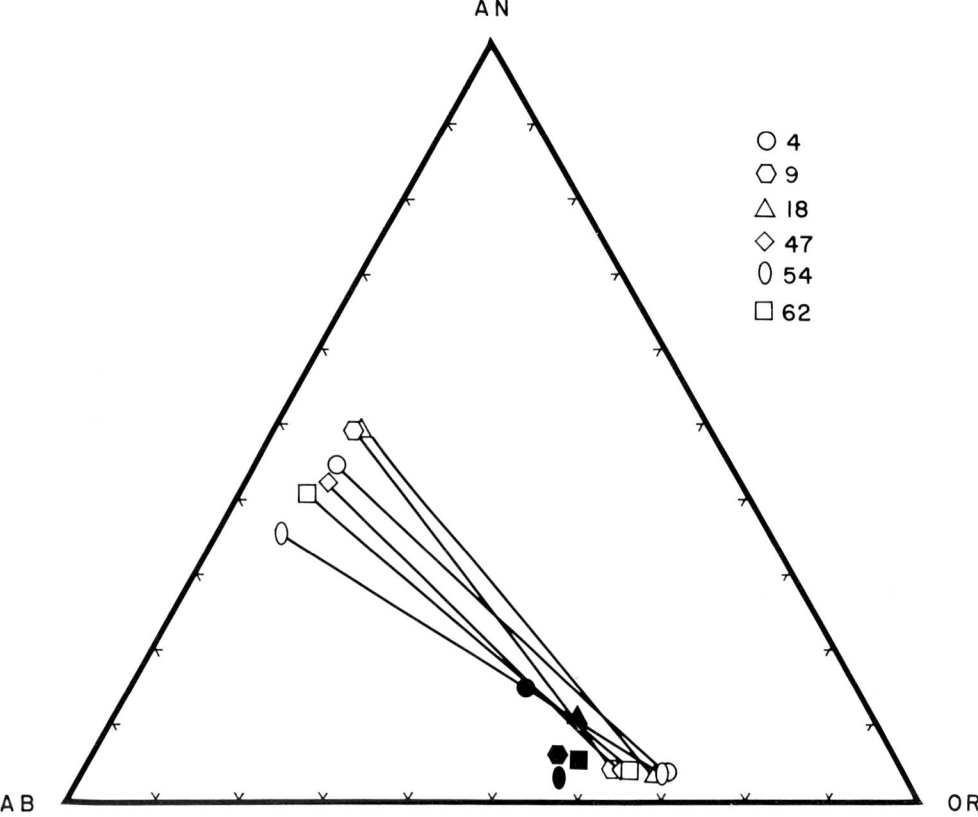

Fig. 5. Electron microprobe analyses of feldspars from group 1 garnet granulites used in geothermometry. Analyses are plotted in terms of mole percent end-members: Ab, albite ($NaAlSi_3O_8$); An, anorthite ($CaAl_2Si_2O_8$); and Or, orthoclase ($KAlSi_3O_8$).

varies in color from colorless through yellow brown to purple. Glass in contact with feldspar is colorless, glass in contact with garnet is yellow brown, and glass in contact with ilmenite is purple. These differences are accentuated where the silicate melts represented by these glasses flowed together but were not thoroughly mixed. The composition of the glass can vary from silica undersaturated to silica oversaturated over a distance of millimeters within a single thin section and appears to reflect the adjacent mineralogies (Figure 7).

Mafic Granulites: Two-Pyroxene Granulite, Charnockite, and Anorthosite

Two-pyroxene granulites are a second major constituent of the feldspar-rich crustal xenolith population at the Kilbourne Hole maar, while charnockite and anorthosite are less commonly observed. The presence of charnockite and anorthosite with two-pyroxene granulites

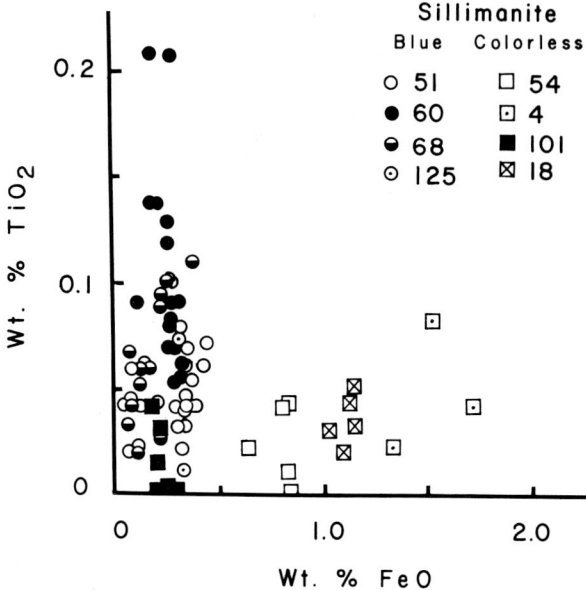

Fig. 6. FeO versus TiO$_2$ relationship in sillimanite from group 1 garnet granulites. Circles represent samples containing blue sillimanite, and squares represent samples containing colorless sillimanite. Fe is calculated as FeO.

is of interest because these rock types frequently occur together in Precambrian shields in association with anorthosite complexes [De Waard, 1967, 1968; De Waard and Romey, 1968]. These xenoliths have an average size larger than that of the garnet granulites and range from 80 to 350 mm in maximum dimension. An alkali olivine basalt crust may totally or partially envelope the xenoliths and, as was the case with the garnet granulites, form a sharp contact with the xenolithic mineral assemblage.

The two-pyroxene granulites exhibit layering which can be coarse, having mineral segregations up to 80 mm in thickness, or fine, having segregations 1-2 mm in thickness. The segregations of mafic minerals are composed of orthopyroxene and clinopyroxene, ilmenite being the primary accessory. Amphibole and mica are uncommon but, when they are present, appear to have formed at the expense of pyroxene. The segregations of light-colored minerals are composed of plagioclase feldspar.

The charnockite is composed of orthopyroxene, clinopyroxene, potassium feldspar, and plagioclase, ilmenite and ulvospinel being accessories. The potassium feldspar is sanidine and is characterized by coarse drop perthite of calcic andesine up to 0.1 mm in maximum dimension. These rocks are generally equigranular and are gray in color. The grain size in charnockites is most commonly in the 1- to 2-mm range, but one sample of charnockite with an average grain size of 5 mm was collected.

Anorthosite is the least abundant of the three types of mafic

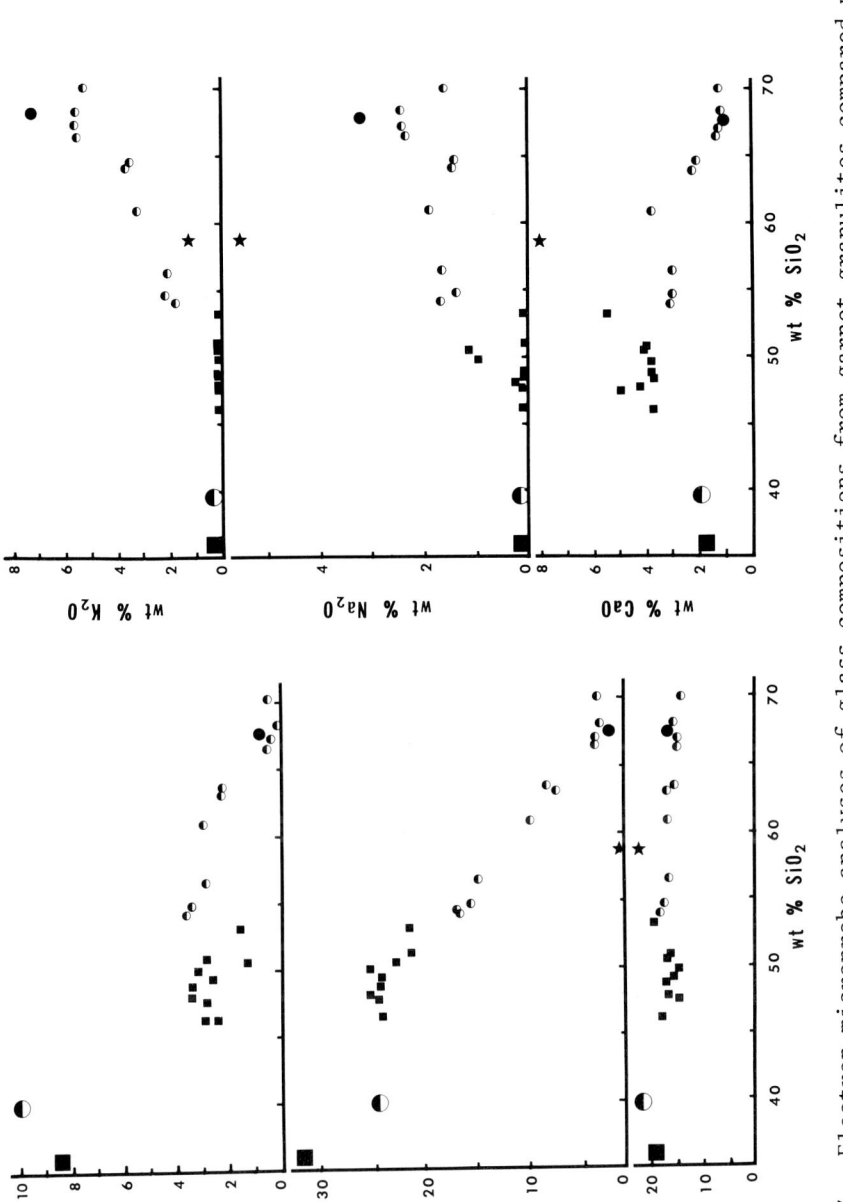

Fig. 7. Electron microprobe analyses of glass compositions from garnet granulites compared with analyses of major phases. Small squares and half-filled circles represent the range of compositions of glasses from two xenoliths representing group 2 (sample pk-33) and group 1 (sample pk-9) garnet granulites, respectively. The large squares and half-filled circles represent the average composition of garnet from the same samples. The stars and the dots represent matrix plagioclase and sanidine compositions, respectively.

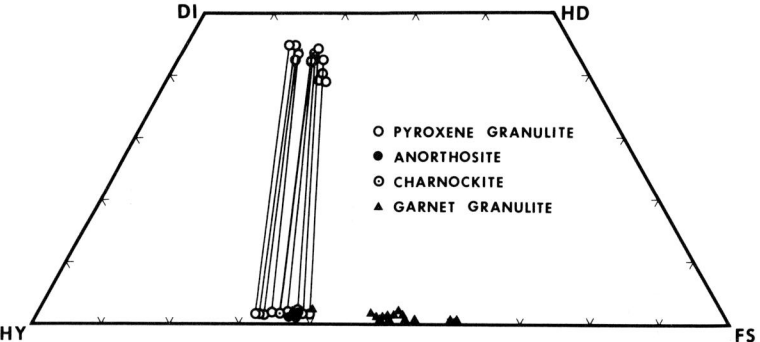

Fig. 8. Electron microprobe analyses of pyroxenes from garnet granulite, anorthosite, charnockite, and two-pyroxene granulite, plotted in terms of mole percent end-members: En, enstatite ($Mg_2Si_2O_6$); Fs, ferrosilite ($Fe_2Si_2O_6$); Wo, wollastonite ($Ca_2Si_2O_6$); Di, diopside ($MgCaSi_2O_6$); and Hd, hedenbergite ($FeCaSi_2O_6$).

feldspar-rich xenoliths. It occurs as small 40- to 80-mm xenoliths which are dark gray in color and with can easily be mistaken for pyroxenite in the field. The anorthosites have been plastically deformed and extensively recrystallized. Orthopyroxene, apatite, and ilmenite are minor phases.

Clinopyroxene from the two-pyroxene granulites and charnockites ranges from 0.5 mm to 1 cm in maximum dimension and varies in composition from $Wo_{45}En_{41}Fs_{14}$ to $Wo_{42}En_{38}Fs_{20}$ (Table 2 and Figure 8). Rare exsolution lamellae of orthopyroxene occur in clinopyroxene in coarse-grained xenoliths with thick accumulations of mafic minerals. Figure 8 illustrates the compositions of coexisting pyroxenes in both two-pyroxene granulites and charnockites, using the pyroxene quadrilateral.

Orthopyroxene occurs in all types of mafic granulites and ranges in size from 0.5 mm to 1 cm in maximum dimension. The variation in composition is from $Wo_2En_{70}Fs_{28}$ to $Wo_2En_{60}Fs_{38}$ (Table 2 and Figure 8). It is strongly pleochroic from green to pink or reddish pink. Exsolution lamellae of ilmenite have been observed in orthopyroxene in all three types of mafic granulites. Rare exsolution lamellae of clinopyroxene occur in orthopyroxene in coarse-grained xenoliths with thick accumulations of mafic minerals. Representative analyses of both orthopyroxenes and clinopyroxenes are found in Table 2.

Plagioclase occurs in all three types of mafic granulites. In the two-pyroxene granulites the plagioclase ranges in composition from An_{40} to An_{58} (Figure 6). The composition of plagioclase in charnockite falls within the upper part of this range. Plagioclase in anorthosite has the composition of calcic andesine and ranges from An_{45} to An_{48}. Inclusions of apatite and rutile occur in plagioclase from anorthosite and charnockite. Both plagioclase and sanidine occur in charnockite. This type of perthite, which frequently exhibits albite twinning, is characteristic of charnockitic rocks from anorthositic terrains in Precambrian shields [Smith, 1974]. In anorthosite, large grains tend to be deformed, while finer-grained recrystallized aggregates occur in elongate zones and patches along its margins. The recrystallized plagioclase in anorthosite tends to be 2-3 mol % lower in An and higher in Ab than the larger-grained type.

Phlogopite and dark brown amphibole occur infrequently in mafic granulites and appear to have formed at the expense of pyroxene. The hydrous minerals compose less than 2% of the minerals when they are present and are commonly enriched in TiO_2 (Table 2). Apatite occurs as a minor phase in anorthosite. Ilmenite, hercynite, apatite, and rutile are the principal accessory minerals.

Partial fusion has affected the mafic granulites to a much lesser extent than it has the garnet granulites. Amphibole, ilmenite, and apatite are the phases most affected. Electron microprobe analyses of glass in anorthosite and charnockite show these glasses to be enriched in P_2O_5 and TiO_2 and depleted in Al_2O_3 when they are compared to glasses from garnet granulites (Table 2). These differences reflect the influence of apatite, ilmenite, and amphibole (and/or pyroxene) on the melt composition. Melt crystallization products are uncommon in these glasses, except in anorthosite, where heteroepitaxial quench crystals of olivine have formed in glass around the margin of orthopyroxene which has been partially fused (Table 2). No more than 10% melt has been observed in the two-pyroxene granulites.

Discussion

Point counts were made of samples characteristic of group 1 and 2 garnet granulites in order to estimate the bulk composition of the rock. Calculation of the total garnet contribution was made by assuming that the symplectite has the same bulk composition as the garnet, which is the phase most affected by the partial fusion process resulting from heating by the enclosing host basaltic magma [Padovani and Carter, 1973]. This partial fusion can be minor, affecting only the margins of the garnets and sometimes the opaque phases and affecting the other phases little, or this fusion can be so extensive as to consume the garnets entirely, leaving in their place melt crystallization products consisting of glass, orthopyroxene, and spinel (symplectites). With this assumption, not taking into account the partial fusion of the other phases would lower the total estimated for the elements contributed by those phases but would not do so sufficiently to change the paragenetic interpretation unless a phase were totally fused and not taken into account (Table 3).

The results of the estimation of the major element chemistry plus the textures of the garnet granulites suggest that they were residues from the partial fusion of sediments before their incorporation into the basaltic magma. The high Al_2O_3 contents combined with generally low amounts of Na_2O and K_2O are characteristic of metamorphosed pelites [Jackson, 1970; Blatt et al., 1972]. Estimates of the mineralogy of an average unmetamorphosed garnet granulite (Table 4) suggest that the rock was composed of approximately 60% kaolinite and was a pelitic sediment. Estimates of the chemistry of the parent before partial fusion, using the method of Bowen [1968], support this inference and suggest further that it was a shale. The extent of partial fusion would be about 30% if granodiorite were the liquid extracted and 40% if quartz monzonite were the liquid extracted [Padovani, 1977]. This partial fusion event apparently occurred at

TABLE 4. Calculated Bulk Composition of Unmetamorphosed Pelite K60

Oxide	Kaolinite	Chlorite	Quartz	Hematite	Others*	Whole Rock
SiO_2	32.98	3.32	15.19	0.0	0.0	51.49
Al_2O_3	27.95	2.77	0.0	0.0	0.0	30.72
TiO_2	0.0	0.0	0.0	0.0	1.40	1.40
FeO	0.0	3.34	0.0	9.59	0.8	12.93
MgO	0.0	2.20	0.0	0.0	0.0	2.20
CaO	0.0	0.0	0.0	0.0	0.73	0.73
K_2O	0.0	0.0	0.0	0.0	0.69	0.69
Na_2O	0.0	0.0	0.0	0.0	0.07	0.07
Total	60.93	11.63	15.19	9.59	2.89	100.23

Values are in weight percent. Mineral compositions were assumed to be stoichiometric.

*Other minerals are present in minor amounts. CaO would be present as calcite, TiO_2 as rutile, and K_2O and Na_2O as sericite.

K60 is a group 1 garnet granulite with garnet, sillimanite, sanidine, quartz, rutile, and ilmenite.

least 1 b.y. ago. Preliminary whole-rock lead-lead ages on a garnet granulite and on a two-pyroxene granulite from the Kilbourne Hole maar gave an apparent average age in excess of 1 b.y. (W. I. Manton and T. J. O'Donell, personal communication, 1974). This age falls within the range of ages determined for Precambrian basement rocks in southern New Mexico [Denison et al., 1970].

The major element chemistry and chemical homogeneity of phases within each xenolith imply that the garnet granulites were equilibrium assemblages at depth. These data can be applied to experimentally derived geobarometers and geothermometers for the appropriate mineral equilibria to estimate the ambient PT conditions that prevailed before the xenolithic materials were incorporated into the pipe. The assumption of equilibrium must be carefully applied to xenoliths which have been extensively fused (e.g., equal to or greater than 50%), because in addition to the extensive fusion of garnet, certain other phases, such as ilmenite, sanidine, and plagioclase, have partially melted and because diffusion gradients involving major cations have been observed near their margins. Thus electron microprobe analyses of the centers of such grains may or may not reflect chemical equilibria. For this reason, few xenoliths lend themselves indiscriminately to the application of geothermometry and geobarometry, because either (1) they lack one of the necessary phases or (2) fusion processes have affected the centers of the grains and caused diffusion of major cations. Six samples were specially selected for detailed analysis for the purpose of calculating equilibrium PT conditions (Figure 6) and estimating possible oxygen fugacities.

Since graphite and blue rutile occur together only in samples which contain blue sillimanite, one can place limits on the oxygen fugacity at the time of transport to the surface. The presence of blue color in rutile has been determined experimentally to be limited by the oxygen buffer assemblages nickel-nickel oxide and wustite-magnetite (D. H. Eggler and S. E. Kesson, personal communication, 1974). For the Kilbourne Hole garnet granulites with blue sillimanite the oxygen fugacity, which is temperature dependent, has an implied range of 10^{-12}–10^{-16} atm [Euster and Wones, 1962; Buddington and Lindsley, 1964]. In addition, the parallel exsolution lamellae within the centers of ilmenite grains in both group 1 and group 2 garnet granulites have reached equilibrium and thus can be used to indicate the temperature and oxygen fugacity of the xenolith under equilibrium conditions before its incorporation into the eruptive process [Buddington and Lindsley, 1964]. A temperature of about 1000°C and an oxygen fugacity of 10^{-12} atm are indicated by these data. The oxygen fugacity range indicated by the presence of blue rutile includes the value indicated by the ilmenite-ulvospinel equilibria and suggests that both groups of garnet granulites equilibrated under reducing conditions. Since the blue rutile does not occur in all group 1 garnet granulites, the granulites with straw yellow to orange rutile probably equilibrated under relatively more oxidizing conditions.

The Fe-Mg cordierite geobarometer of Holdaway and Lee [1977], the two-feldspar geothermometer of Stormer [1975], and the plagioclase-garnet-aluminum silicate geothermometer-geobarometer of Ghent [1976] can be applied to the Kilbourne Hole garnet granulite data to indicate the limits of pressure and temperature which the garnet granulites may have experienced at depth. The breakdown of Fe-Mg cordierite plus biotite to form almandine, potassium feldspar, aluminum silicate, and quartz is of particular interest because it is pressure dependent and thus has potential as a geobarometer [Kerrick, 1972; Holdaway and Lee, 1977]. These breakdown products are characteristic of the mineral assemblages present in group 1 garnet granulites. Since cordierite and biotite do not occur in the group 1 assemblages, it can be assumed that the xenoliths crystallized at pressures above the breakdown of cordierite and at temperatures above the dehydration temperature of biotite. Iron-magnesium ratios in the garnets range from 55 to 73 mol %, which corresponds to K_D values of Fe_{20}–Fe_{34} in cordierite, if it was present. The minimum PT conditions which result from the application of these data indicate P_{total} to be greater than or equal to 5.4 kbar, T to be greater than or equal to 725°C, and X_{H_2O} to be less than 0.2 ($P_{H_2O} < 0.2\ P_{total}$).

The two-feldspar geothermometer of Stormer [1975] and the plagioclase-garnet-aluminum silicate-quartz geothermometer-geobarometer of Ghent [1976] can be applied to the Kilbourne Hole data to indicate further limits of pressure and temperature which the granulites may have experienced at depth. The Stormer geothermometer is based on the partitioning of albite between plagioclase and potassium feldspar. Using an expression which relates temperature and pressure over a wide range of compositions, Stormer plotted a series of isotherms relating feldspar compositions at different pressures. When pressure can be estimated, temperatures can be determined with an accuracy of ±30°C. Since pressure has been taken into account in the generation of these isotherms, their position will vary depending on the estimated pressure (Figures 9-11). The method lacks accuracy for feldspars in which plagioclase has high potassium contents and in which potassium feldspar has high calcium contents. The albite contents in plagioclases from the Kilbourne Hole garnet granulites fall within the range of precision of the method for pressures greater than 5 kbar.

The application of the Stormer geothermometer to the Kilbourne Hole data is dependent on two major assumptions about the compositions of the feldspars. The first assumption is that the rims of the sanidine grains containing perthitic cores are in equilibrium with the coexisting plagioclase. This assumption is based on the fact that sanidine of similar composition without perthite, occurring in the same xenolith or in other xenoliths, appears to be in equilibrium with the coexisting plagioclase. The second assumption is that the composition of the host sanidine in the centers of grains which contain perthite approximates rim compositions. This assumption is based on electron microprobe analyses of several sanidine cores in which patch perthite was coarse enough to be distinguished from the host (e.g., samples 54 and 62).

The bulk composition of the perthitic cores is richer in sodium and calcium than is that of their respective rims. If the perthitic cores represent exsolution after homogenization during a heating event, the rims must represent either overgrowth or migration of potassium from the cores. The composition of the host potassium feldspar in the cores is almost identical to that of the rim feldspar, the core feldspar being relatively depleted in potassium. This small difference may result from (1) incomplete unmixing of the feldspar components on cooling, (2) overgrowth, or (3) difficulty in analyzing the very small andesine exsolution phase. In many xenoliths which are rich in garnet, sillimanite, and quartz, only sanidine is present. This observation suggests that compositional variations are significant in the xenoliths and that in cases in which sanidine is the only feldspar present, there may have been insufficient sodium and calcium present for plagioclase to have formed in the rocks.

Isotherms based on the analytical formula of Stormer [1975, p. 670] were plotted with the aid of a PDP-15 computer and a CALCOMP plotter. Isotherms were generated for pressures of 5000, 7000, and 9000 bars in order to determine the range of temperatures at these pressures (Figures 9-11). These pressures were chosen because they span a range of pressures between a minimum suggested by the experimental results of Holdaway and Lee [1977] and a maximum suggested by the base of the crust in the Rio Grande Rift [Decker and Smithson, 1975]. The data base for these plots is given in Tables 5 and 6. In Figures 9, 10, and 11, rim compositions for potassium feldspars are represented by open symbols and are the result of spot analyses. Core compositions are represented by solid symbols and are the result of broad beam scans over the perthite-rich regions. The latter analyses should reflect the bulk composition of the perthite.

The temperatures determined by this method range from 750° to 900°C for the rims and from 950° to 1100°C for the perthitic cores at 5000 bars (Figure 9). At 9000 bars the temperatures of the rims tend to

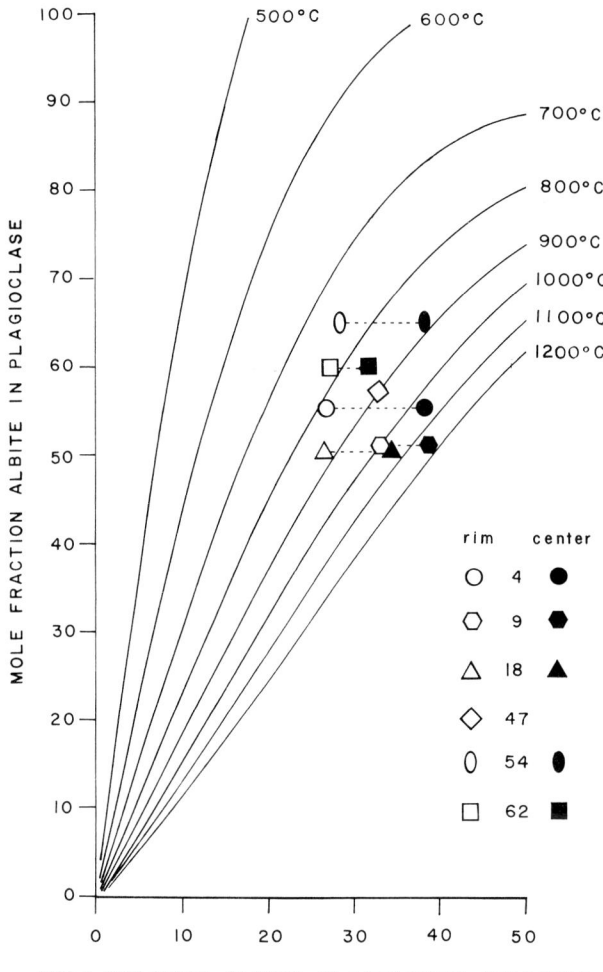

Fig. 9. Isotherms calculated according to the method of Stormer [1975] at 5000 bars. Electron microprobe analyses of coexisting feldspars from group 1 garnet granulites are plotted in terms of mole fraction of albite in plagioclase and in alkali feldspar.

increase by an average of 100°C (Figures 9 and 11). Two samples, 54 and 62, show a smaller range of temperature between core and rim than do samples such as 18 and 4. Sample 9 exhibits anomalously high temperatures. The sanidine in this sample is rarely perthitic and is less common in occurrence than plagioclase. This sample lacks sillimanite and its garnet composition is the most pyrope rich of any determined (45 mol %). Less than 3% fusion in the form of glass along the margins of garnet and ilmenite were observed. The high albite content determined by electron microprobe analyses is perhaps due to (1) the small grain size of sanidine, resulting in the possible inter-

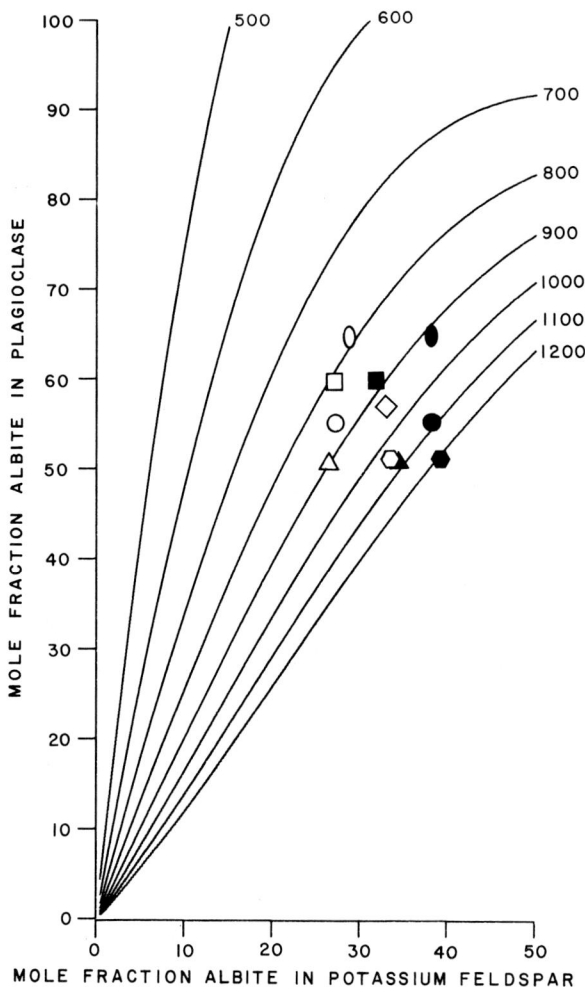

Fig. 10. Isotherms calculated according to the method of Stormer [1975] at 7000 bars. Electron microprobe analyses of coexisting feldspars from group 1 garnet granulites are plotted in terms of mole fraction of albite in plagioclase and in alkali feldspar.

action of the electron beam with adjacent plagioclase, or (2) the homogenization of the potassium feldspar by a previous heating event without subsequent separation of the two phases on cooling. The first possibility seems the less likely, because the sanidine compositions have a narrow range. Since the mineral assemblage appears to be depleted with respect to water, it can be treated essentially as a dry system. The calculated melting temperature at 5 kbar for a dry system is about 1100°C [Luth, 1968]. Thus if the temperature fell sufficiently rapidly, exsolution could have been suppressed [Smith, 1974]. On the other hand, even slow cooling to temperatures above the sanidine high-albite solvus would inhibit exsolution unless tempera-

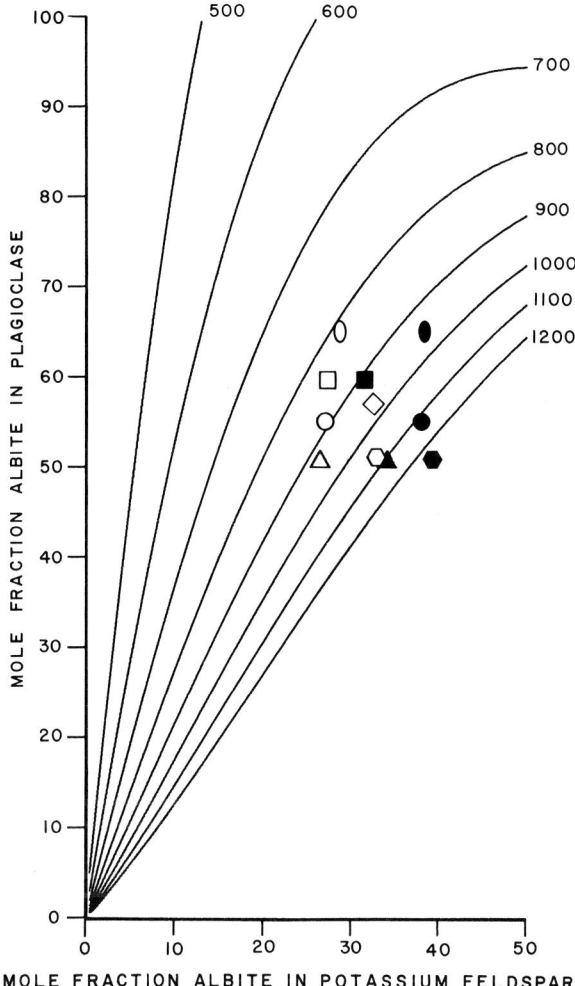

Fig. 11. Isotherms calculated according to the method of Stormer [1975] at 9000 bars. Electron microprobe analyses of coexisting feldspars from group 1 garnet granulites are plotted in terms of mole fraction of albite in plagioclase and in alkali feldspar.

tures fell below about 650°C at 5 kbar or 750°C at 10 kbar for the composition involved [Luth, 1968]. If the above conditions were obtained, the determined temperatures would be real and the presence or absence of perthite would be a function of ambient pressure, temperature, and bulk composition.

The plagioclase-garnet-quartz-aluminum silicate geobarometer of Ghent [1976] can be used to calculate the maximum pressure at which group 1 garnet granulites could have existed at any given temperature. It is based on the distribution of CaO between garent and anorthite. When sillimanite is the aluminum silicate polymorph, the equilibrium relationship can be described by

TABLE 5. Electron Microprobe Analyses Used for Geothermometry and Geobarometry

Oxide	Garnet KH4	Plag KH4	Kspar KH4 (Rim)	Kspar KH4 (Center)	Garnet KH9	Plag KH9	Kspar KH9 (Rim)	Kspar KH9 (Center)	Garnet KH18	Plag KH18	Kspar KH18 (Rim)	Kspar KH18 (Rim)	Kspar KH18 (Center)
SiO_2	40.42 (6.261)	59.09 (7.934)	65.17 (9.032)	62.05 (8.560)	39.37 (6.001)	54.74 (7.803)	66.64 (8.853)	64.82 (8.841)	37.66 (5.936)	58.52 (7.807)	63.20 (8.870)	64.04 (8.896)	63.21 (8.861)
TiO_2	nd	nd	nd	nd	0.17 (0.02)	nd	nd	nd	0.12 (0.014)	nd	nd	nd	nd
Al_2O_3	19.13 (3.493)	25.69 (4.066)	18.00 (2.941)	21.16 (3.441)	22.24 (3.995)	26.35 (4.198)	20.01 (3.147)	20.05 (3.224)	21.62 (4.015)	26.67 (4.194)	19.05 (3.152)	18.96 (3.104)	20.60 (3.340)
FeO	29.14 (3.775)	nd	nd	nd	24.07 (3.068)	nd	nd	nd	27.27 (3.726)	nd	0.09 (0.011)	nd	nd
MnO	0.45 (0.059)	nd	nd	nd	0.67 (0.086)	nd	nd	nd	0.46 (0.061)	nd	nd	nd	nd
MgO	9.29 (2.145)	nd	nd	nd	10.95 (2.448)	nd	nd	nd	8.74 (2.054)	nd	nd	nd	nd
CaO	1.57 (0.260)	7.41 (1.066)	0.76 (0.112)	2.98 (0.441)	1.94 (0.316)	8.27 (1.198)	1.03 (0.175)	1.20 (0.175)	1.37 (0.232)	8.35 (1.194)	0.68 (0.102)	0.70 (1.104)	2.31 (0.340)
Na_2O	nd	6.35 (1.652)	2.95 (0.793)	4.30 (1.149)	nd	5.83 (1.528)	3.92 (1.009)	4.11 (1.088)	nd	5.85 (1.514)	2.88 (0.785)	3.31 (0.864)	3.89 (1.033)
K_2O	nd	1.65 (0.282)	11.4 (0.015)	8.01 (1.410)	nd	1.58 (0.272)	10.88 (1.843)	8.68 (1.510)	nd	1.70 (0.290)	11.46 (2.051)	11.47 (2.032)	9.31 (1.627)
Total	99.99	100.18	98.25	98.49	99.45	99.77	102.56	98.94	98.23	101.09	97.37	98.37	99.40
X_{Na}	nd	55.1	27.2	38.3	nd	51.0	33.6	39.2	nd	50.5	26.7	28.8	34.4
X_{Ca}	0.042	0.355	nd	nd	0.054	0.509	nd	nd	0.037	0.398	nd	nd	nd

Values are in weight percent and (in parentheses) mole percent; nd indicates not determined. Total iron is calculated as FeO. Plag, plagioclase; and Kspar, potassium feldspar. X_{Na} is the mole fraction of albite component in plagioclase or potassium feldspar. X_{Ca} is the mole fraction of anorthite component in plagioclase or the mole fraction of grossular component in garnet.

TABLE 6. Electron Microprobe Analyses Used for Geothermometry and Geobarometry

Oxide	Garnet KH47	Plag KH47	Kspar KH47 (Rim)	Kspar KH47 (Center)	Garnet KH54	Plag KH54	Kspar KH54 (Rim)	Kspar KH54 (Center)	Garnet KH62	Plag KH62	Kspar KH62 (Rim)	Kspar KH62 (Center)
SiO_2	39.63 (6.117)	59.96 (7.990)	62.23 (8.863)	nd	38.43 (5.933)	62.04 (8.167)	64.26 (8.928)	66.01 (9.111)	39.27 (6.142)	59.54 (8.031)	64.58 (8.890)	64.45 (8.890)
TiO_2	nd	nd	nd	nd	nd	nd	nd	nd	nd	nd	nd	nd
Al_2O_3	20.81 (3.785)	25.44 (4.000)	18.99 (3.137)	nd	22.01 (3.942)	24.70 (3.832)	18.76 (3.072)	17.51 (3.802)	20.86 (3.969)	24.97 (3.110)	19.16 (3.110)	19.56 (3.162)
FeO	29.09 (3.755)	nd	nd	nd	30.48 (3.947)	nd	nd	nd	30.43 (3.935)	nd	nd	nd
MnO	0.81 (0.206)	nd	nd	nd	0.44 (0.058)	nd	nd	nd	0.58 (0.076)	nd	nd	nd
MgO	8.63 (1.978)	nd	nd	nd	8.68 (2.008)	nd	nd	nd	7.65 (1.763)	nd	nd	nd
CaO	1.45 (0.241)	7.00 (1.00)	0.91 (0.137)	nd	1.23 (0.198)	5.90 (0.832)	0.48 (0.072)	0.49 (0.073)	1.44 (0.238)	6.71 (0.969)	0.74 (0.110)	1.10 (0.162)
Na_2O	nd	6.66 (1.722)	3.64 (0.990)	nd	nd	7.62 (1.945)	3.19 (0.859)	4.39 (1.175)	nd	6.84 (1.789)	3.10 (0.828)	3.56 (0.948)
K_2O	nd	1.64 (0.278)	10.48 (1.873)	nd	nd	1.34 (0.224)	11.67 (2.069)	9.56 (1.683)	nd	1.41 (0.242)	11.74 (2.062)	10.80 (1.890)
Total	100.42	100.68	97.29	nd	101.08	101.59	98.36	98.00	100.66	99.47	99.32	99.48
X_{Na}	nd	57.4	33.0	nd	nd	64.8	28.6	38.4	nd	59.63	27.6	31.6
X_{Ca}	0.040	0.333	nd	nd	0.319	0.277	nd	nd	0.039	0.323	nd	nd

Values are in weight percent and (in parentheses) mole percent; nd indicates not determined. Total iron is calculated as FeO. Plag, plagioclase; and Kspar, potassium feldspar. X_{Na} is the mole fraction of albite component in plagioclase or in potassium feldspar. X_{Ca} is the mole fraction of anorthite component in plagioclase or the mole fraction of grossular component in garnet.

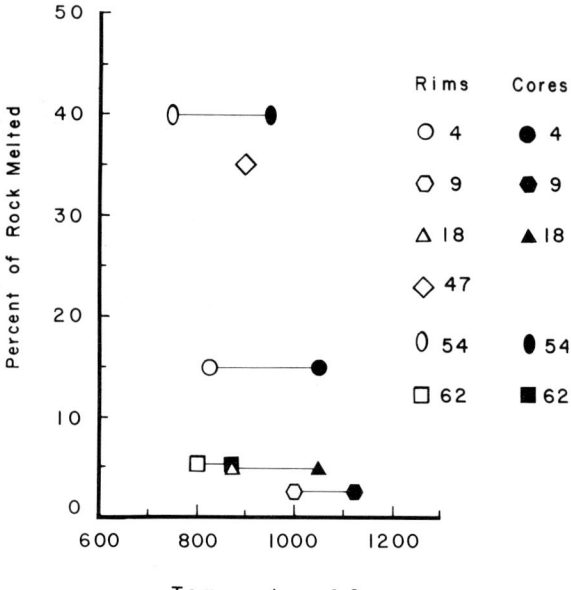

Fig. 12. Percent of rock melted, as established by point counting, versus its temperature range calculated by using the method of Stormer [1975].

$$0 = \frac{-2551.4}{T} + 7.1711 - \frac{0.2842(p - 1)}{T} + \log K_D \quad (1)$$

where p is in bars and T is in degrees Kelvin. Log K_D is defined by

$$\log K_D = 3 \log X_{Gr} - 3 \log X_{An} \quad (2)$$

where X_{Gr} and X_{An} are the mole fractions of grossular and anorthite, respectively. For any given metamorphic grade it has been observed that the equilibrium constant K_D is separate from that for any other metamorphic grade. Equation (1) can be used to calculate PT curves for log K_D = -2.00 and -3.00. These curves are plotted in Figure 12 to show the relative positions of the log K_D values for group 1 garnet granulites.

Pressures were calculated from temperatures determined from Figures 9-11 and plotted in Figures 13 and 14 by using the method of Ghent [1976]. The data base for these calculations is found in Tables 5-7. Figure 13 is a temperature-depth plot showing the range of temperatures as a function of depth for the Kilbourne Hole group 1 garnet granulites, the geothermometer of Stormer [1975] at 5000 bars being used. It was assumed that the temperatures derived from the rim or host compositions of sanidine were the equilibrium temperatures, and these were used for pressure calculations. Temperatures determined from the core compositions were assumed to reflect a prior heating event and were plotted on the assumption that a pressure change did not occur, at the same pressure as that calculated for the rims.

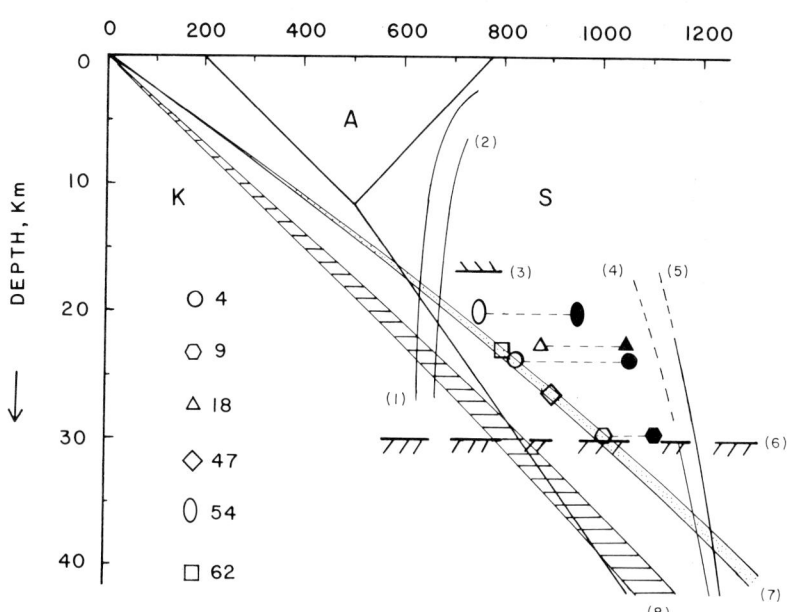

Fig. 13. Temperature-depth plot showing the range of temperature as a function of depth calculated for the Kilbourne Hole garnet, sillimanite granulite xenoliths using the geothermometer of Stormer [1975] and geothermometer-geobarometer of Ghent [1976]. Open symbols represent temperatures calculated from perthite-free sanidine margins, while closed symbols represent temperatures calculated from perthitic sanidine cores. The stability relations of the Al_2SiO_5 polymorphs are from the results of Holdaway [1971]. K, A, and S indicate the stability fields of kyanite, andalusite, and sillimanite, respectively. Curves 1 and 2 represent the melting of granite at $X_{H_2O} = 1$ and $X_{H_2O} = 0.5$, respectively [Kerrick, 1972]. The hatched line (3) indicates the minimum pressure for the xenoliths as predicted by the experimental results of Holdaway and Lee [1977]. Curves 4 and 5 represent eutectic reactions in the binary systems sanidine-silica and albite-silica under dry conditions [Luth, 1968]. The dry melting curve for granite would fall on the low-temperature side of these curves. Hatched line (6) indicates the base of the crust in the southern Rio Grande Rift [Decker and Smithson, 1975]. Curves 7 and 8 represent the range of temperatures as a function of depth for surface fluxes (q_s) of 2.4 HFU (μcal cm^{-2} s^{-1}) for the southern Rio Grande Rift and of 2.0 HFU for the Basin and Range Province, respectively [Decker and Smithson, 1975].

Figure 14 supports these inferences. It contains a plot of the calculated range of possible temperatures for a pressure range of 5-9 kbar. This range reflects the upper and lower limits required by the data of Holdaway and Lee [1977] and by the estimated depth to the mantle [Decker and Smithson, 1975]. If an average value of tempera-

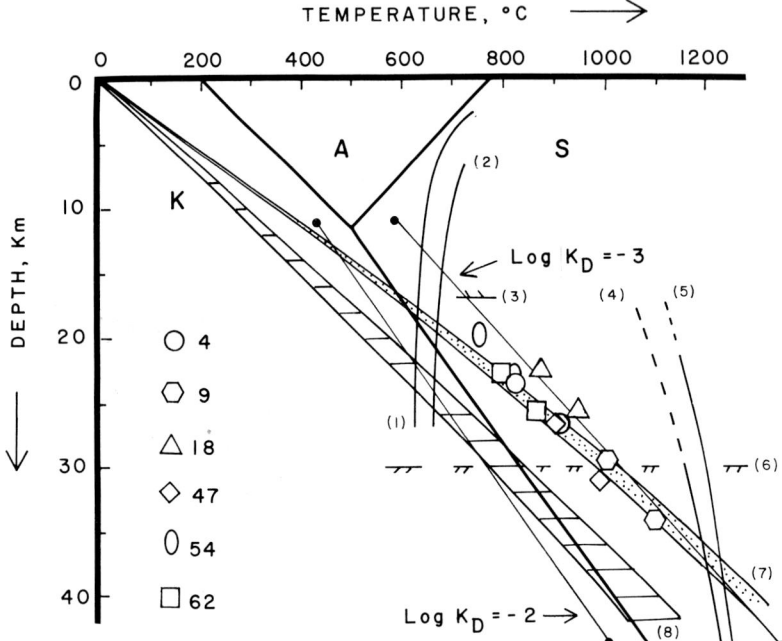

Fig. 14. Temperature-depth plot showing the aluminum silicate stability field with respect to hypothetical K_D values of -3 and -2 for coexisting garnet and plagioclase, according to the method of Ghent [1976]. The positions of group 1 garnet granulites have been calculated by using the methods of Stormer [1975] and Ghent [1976]. See Tables 5-7 for data points. Symbols are the same as those in Figure 13.

ture within the range of each sample is assumed, the depth varies between about 22 and 33 km. If sample 9 is discounted for reasons discussed below, the depth varies between about 22 and 28 km. Sample 9 differs from other group 1 garnet granulites in that sillimanite has not been observed as a constituent of the mineral assemblage. For this reason the Ghent geobarometer is not applicable, in principle, and thus the PT conditions calculated for it may be too high. Alternatively, the anomalously high temperature determination from feldspar geothermometry may reflect the proximity of this rock type at depth to a magma body. If the point is moved to lower pressures at constant T, it approaches the dry melting curve for granite.

The facts that the rim compositions plot on the calculated temperature-depth curve for the geotherm in the southern Rio Grande Rift at a slope of about 30°/km [Decker and Smithson, 1975] and that the core compositions plot close to the dry melting curve for granite [Luth, 1968] are of great interest. They imply that the garnet granulites were formed under equilibrium conditions and that their mineral assemblages reflect the ambient geothermal gradient in the deep crust under the Kilbourne Hole maar. They also imply that the core temperatures probably reflect the dry melting curve for granulite facies rocks in the lower crust beneath the southern Rio Grande Rift and that the lower crust in this area is anhydrous as a result of partial fusion and incorporation of the volatile-rich phases into the magma. The implication of a negative slope to the melting curve for water-saturated rocks and minerals (Figures 13 and 14) [e.g., Yoder, 1965; Luth, 1968; Eggler, 1973; Burnham and Davis, 1974; Holdaway and Lee, 1977] is that movement of a water-rich magma into the intermediate crust would probably result in its crystallization there before it could reach the earth's surface. The net result of this process over geologic time would be the dehydration of the lower crust and the subsequent enrichment of the intermediate crust in water and other volatiles and in low melting constituents [e.g., Lambert and Heier, 1967].

Application of geothermometry and geobarometry to the mafic granulites is also possible because of the overall homogeneity of their major phases. The assumption of equilibrium is supported by electron microprobe analyses which revealed no major inhomogeneities except in highly fused grains in which partial fusion had produced large diffusion gradients.

A pyroxene geothermometer based on composition-temperature relationships in coexisting pyroxene in the 0- to 5-kbar range can be used to estimate a minimum temperature of crystallization for the mafic granulites from Kilbourne Hole [Ross and Huebner, 1975]. Temperatures of 850°C $\pm$ 100°C result from the application of this geothermometer. The recent experimental results of Lindsley and Dixon [1976] show the effect of higher pressures on the pyroxene solvus and can be used to place an upper limit on temperature. From these data, temperatures of 850°-1000°C are indicated in the 10- to 15-kbar range.

The results of Irving [1974] from experimental studies on the subsolidus phase relationships of two-pyroxene granulites from the Delegate Pipe (Australia) place further limits on the possible range of pressure for the two-pyroxene granulites from the Kilbourne Hole maar. The compositions studied by Irving are similar to those at Kilbourne Hole, except for the presence of minor scapolite in the Delegate samples. These results indicate that the subsolidus field of the two-pyroxene granulite assemblage lies between 6 and 10 kbar at 1000°C and, by extrapolation, between 5 and 9 kbar at 900°C.

Application of the Stormer [1975] two-feldspar geothermometer to the charnockites yields temperatures of 800°C for sample 6 and 900°C for sample 42. The same variation occurs when the host sanidine and drop perthite analyses are used. These results support the hypothesis of Smith [1974] that drop perthite is a primary growth feature due to epitaxial nucleation of plagioclase on potassium feldspar. Electron microprobe analyses of charnockitic rocks from Norway show a similar relationship in composition between drop perthite, the host potassium feldspar, and separate grains of plagioclase [Smith, 1974].

Studies of granulite facies rocks from Precambrian terrains have resulted in similar temperature estimates for two-pyroxene granulites and charnockite of the type seen as xenoliths at Kilbourne Hole maar. For example, temperature estimates for pyroxenes from Broken Hill granulites (Australia) and from the Wilmington gneiss complex (Delaware) are between 860° and 900°C [Hewins, 1975]. Sighinolfi and

TABLE 7. Calculated Temperatures as a Function of Assumed Pressure and Their Effect on Geobarometry and Geothermometry Using the Methods of Stormer [1975] and Ghent [1976]

T, °C	Assumed P, bars	T, °K	log K_D	Calculated P, bars	Depth, km	T_{max}, °C
			Sample 4			
825	5,000	1098	-2.793	7,939	23.8	1050
850	6,000	1123	-2.793	8,322	25.0	1075
860	7,000	1133	-2.793	8,476	25.4	1090
875	8,000	1148	-2.793	8,707	26.1	1100
890	9,000	1163	-2.793	8,939	26.8	1125
900	10,000	1173	-2.793	9,093	27.3	1150
			Sample 9			
1000	5,000	1273	-2.935	9,981	30.0	1100
1025	6,000	1298	-2.935	10,369	31.1	1200
1050	7,000	1323	-2.935	10,742	32.2	1210
1075	8,000	1348	-2.935	11,115	33.3	1225
1075	9,000	1348	-2.935	11,115	33.3	1250
			Sample 18			
875	5,000	1148	-3.081	7,545	22.6	1050
875	6,000	1148	-3.081	7,545	22.6	1050
900	7,000	1173	-3.081	7,904	23.7	1075
920	8,000	1193	-3.081	8,192	24.5	1090
930	9,000	1203	-3.081	8,336	25.0	1120
950	10,000	1223	-3.081	8,623	25.9	1130
			Sample 47			
900	5,000	1173	-2.777	9,026	27.1	nd
925	6,000	1198	-2.777	9,545	28.6	nd
930	7,000	1203	-2.777	9,622	28.9	nd
960	8,000	1233	-2.777	10,086	30.3	nd
975	9,000	1248	-2.777	10,318	30.9	nd
990	10,000	1263	-2.777	10,550	31.7	nd
			Sample 54			
750	5,000	1023	-2.816	6,700	20.1	875
775	6,000	1048	-2.816	7,082	21.2	900
780	7,000	1053	-2.816	7,159	21.5	925
800	8,000	1073	-2.816	7,465	22.4	930
815	9,000	1088	-2.816	7,695	23.1	960
830	10,000	1103	-2.816	7,925	23.8	975
			Sample 62			
800	5,000	1073	-2.758	7,685	23.1	875
800	6,000	1073	-2.758	7,685	23.1	875
825	7,000	1098	-2.758	8,072	24.2	880
830	8,000	1103	-2.758	8,150	24.5	890
850	9,000	1123	-2.758	8,461	25.4	915
860	10,000	1133	-2.758	8,616	25.8	939

TABLE 7. (cont'd)

Abbreviation nd indicates not determined.
T (°C) (P, bars, assumed) is the range of possible temperatures for group 1 garnet granulites obtained by using the rim compositions of sanidine and coexisting plagioclase. Pressure range of 5,000-10,000 bars was used to calculate T (°C) because of the boundary conditions imposed by the experimental data of Holdaway and Lee [1977] and by the geophysical data of Decker and Smithson [1975]. Calculated P (bars) is the calculated range of pressures according to the method of Ghent [1976] using temperatures calculated according to the method of Stormer [1975]. T_{max} (°C) is the maximum temperature derived by using centers of sanidine grains (host plus perthite).

Gorgoni [1975], from their studies of small anorthosite bodies in high-grade metamorphic terrains, consider that temperatures of up to 900°C and pressures of 6-10 kbar are normal for such terrains. The authors further note that anorthosite is often sporadically present within rock associations such as the mangerite-charnockite suite. This point is relevant to the presence of charnockite and anorthosite in minor amounts in the Kilbourne Hole xenolith population and implies that the minor occurrence of anorthosite may reflect sporadic occurrence of this rock type at depth.

A plot of Al_2O_3 versus TiO_2 for pyroxenes from mafic granulites from Kilbourne Hole is presented in Figure 12. The apparent interrelationship between clinopyroxene and orthopyroxene can be seen and

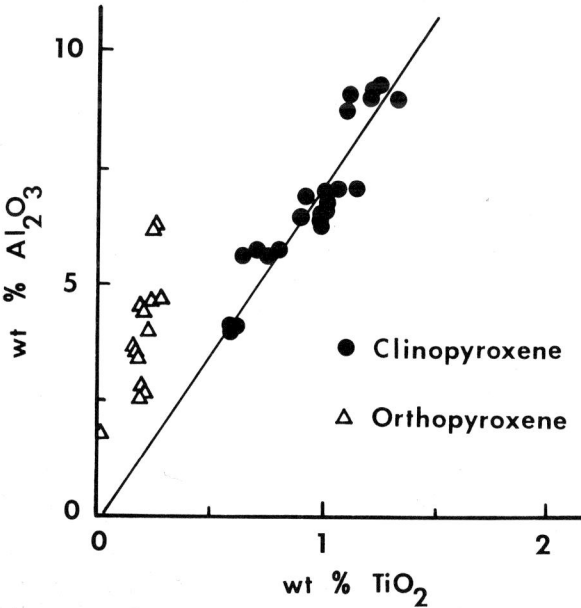

Fig. 15. Al_2O_3 versus TiO_2 relationships in pyroxenes from anorthosite, charnockite, and two-pyroxene granulites.

reflects a direct correlation between Al and Ti. This correlation suggests that crystallization occurred under similar PT conditions. A least squares fit of the clinopyroxene data results in a line with a slope of about 7. Orthopyroxene appears to have a constant ratio of about 20 for these elements as well, but perturbations of this value may exist because of exsolution of ilmenite and rutile.

The data from the Kilbourne Hole mafic granulites imply that two-pyroxene granulite, charnockite, and anorthosite xenoliths are related and appear to have formed under similar PT conditions. The major element chemistry, homogeneity of major phases, and occurrence of ilmenite exsolution in orthopyroxene are important features that the three types of mafic granulites have in common. The Al/Ti values in clinopyroxene form a constant trend, and such a trend may occur in orthopyroxene. The composition of clinopyroxene in the mafic granulites follows a crystallization path characteristic of pyroxene crystallization in the granulite facies pressure-temperature regime [Hewins, 1975]. This is supportive evidence for similar conditions of formation for all the mafic granulites. Furthermore, the results have much internal consistency. The range of PT conditions estimated for the mafic granulites coincides with the trend for the garnet granulites and has a slope close to that of the calculated geotherm for the southern Rio Grande Rift [Decker and Smithson, 1975]. Thus these rock types may be closely associated in space, possibly in a regionally metamorphosed complex in the deep crust.

A similar association of anorthosite and mafic granulites from the San Francisco volcanics (Arizona) has been studied by Stoeser [1973]. The major element chemistry of these granulites is very similar to that of the Kilbourne Hole granulites, the pyroxene and feldspar compositions from the two localities being indistinguishable from one another. Stoeser [1973] considers the two-pyroxene granulites and charnockites from the San Francisco volcanics to represent wall rocks around a layered cumulus intrusion in the deep crust.

An important aspect of this study of deep crustal xenoliths from the Kilbourne Hole maar is the presence of healed and sealed microcracks. Most of the healed microcracks contain glass. The glasses appear to be trapped silicate melts of variable composition and could be the result of (1) partial fusion of the host grain, (2) injection of partial melt products of adjacent grains, or (3) melt formed from material which had previously occupied the crack.

The importance of microcracks in igneous and metamorphic rocks has been demonstrated by Richter and Simmons [1977a], Feves et al. [1977], and Batzle and Simmons [1977], who show that the nature and history of fracturing and crack-sealing events and fracture porosity within a given rock type can have great influence on the physical properties of the rock at depth (e.g., the electrical resistivity and the velocity of compressional and shear waves). The presence of microcracks in the Kilbourne Hole deep crustal xenoliths is of particular interest, since Richter and Simmons [1977a, b] have shown that most microcracks in shallow basement igneous rocks have been closed by lithostatic pressure at depths ranging from 2.5 to 4 km. The significance of healed and sealed microcracks in the deep crustal material from the Kilbourne Hole maar is that they could have served as reaction surfaces and paths for fluid conduction under the PT conditions of the deep crust from 22 to 28 km. Such cracks either formed and were simultaneously filled with nucleating material (e.g., cracks sealed with apatite in plagioclase) or formed and remained open, permitting the injection of silicate melt during a heating event such as that which caused the partial melting in the xenoliths. Such a heating event could also have caused a sufficient increase of vapor pressure locally (from dissolved CO, CO_2, or NH_4) to allow the cracks to remain open until they were sealed with silicate melt. The formation of a melt phase may also have resulted directly in the formation and filling of cracks due to volume expansion on melting [Rosenfield and Chase, 1961].

The percent of rock melted (established by point counts) was plotted versus the range of temperature to establish whether or not there is a correlation between calculated temperature and amount of fusion within the xenoliths (Figure 15). As can be seen in this figure, the lowest temperature (sample 54) correlates with the most extensive fusion, so there appears to be no direct relationship between visible fusion and calculated temperature. No more than 10% melt has been observed in the two-pyroxene granulites, compared with over 50% in some garnet granulites. This fact may have a bearing on the proximity and length of exposure of each rock type to the heat source which induced the partial fusion as well as on the detailed PT relationships between the liquidus and solidus surface for each rock type.

From evidence provided by xenoliths and from rocks in Precambrian terrains in deeply eroded shield regions the lower crust appears to be heterogeneous in nature and not to be dominated by rocks of gabbroic composition. The presence or absence of water in the lower crust is also variable (for example, a dry crust is suggested by the xenolith population at Kilbourne Hole, while a wet crust appears to be the case beneath the Colorado plateau [McGetchin and Silver, 1970]). Results of seismic velocity studies on granulite facies rocks demonstrate that such rock types have a range of velocities comparable to those reported from the lower continental crust [Christensen and Fountain, 1975]. Since neither eclogite nor serpentine has been found in the xenolith population at Kilbourne Hole, the model of Green and Ringwood [1966] seems to be best supported by our present data. Eclogite may exist in some areas of the crust on a restricted basis but, from the available data on crustal xenoliths, not to any great extent. Since the lower crust has undergone a polymetamorphic history and reflects the effects of multiple tectonic and thermal events, one would expect a mixture of many rock types with complex relationships. The lower crust beneath areas of active rifting which are characterized by high heat flow and crustal thinning may be more dehydrated at shallower depths than the crust at similar depths in other tectonic environments, such as shield areas.

Thus the importance of studying deep crustal xenoliths is that they can directly yield important information about the nature and physical properties of the lower crust and about the physical and chemical processes which operate therein. Knowledge of properties such as chemical composition, water content, grain size, mineralogy, and oxygen fugacity is important for the refinement and calibration of physical parameters such as seismic velocity, temperature, electrical conductivity, and heat flow. For example, heat flow measurements at the surface can give only a relative estimate of the temperature distribution at depth, while the xenoliths may be used to specify the exact temperature distribution throughout a given depth range in the crust.

Conclusions and Inferences

The following conclusions and inferences are drawn from this study of the feldspar-bearing granulite xenoliths from the Kilbourne Hole marr in New Mexico.

1. The garnet granulites appear to be equilibrium assemblages whose major element chemistry, in general, has been relatively unaffected by heating in the basaltic host magma during transport to the earth's surface. However, this heating event produced the interstitial silicate liquids which subsequently quenched at the earth's surface to form symplectites of glass, skeletal orthopyroxene, and cruciform hercynitic spinel.

2. The mineral assemblage present in group 1 garnet granulites and the composition of the garnets indicate that these granulites formed at pressures greater than those required for the breakdown of cordierite. Furthermore, the absence of the hydrous phase biotite indicates that the equilibrium temperature had to be greater than that required for its breakdown. The minimum PT conditions suggested by these data are 5.4 kbar and 700°C.

3. Equilibrium temperatures of 750°-1000°C are indicated from feldspar geothermometry. Application of geobarometry to the garnet-plagioclase equilibria yields a pressure range corresponding to depths of 22-28 km. This range is equivalent to a geothermal gradient of approximately 30°/km in the depth range of 22-28 km beneath the Rio Grande Rift.

4. The garnet granulite xenoliths plot along the geotherm derived from heat flow measurements in the southern Rio Grande Rift, a result which implies that the mineral assemblages reflect the ambient geothermal gradient in the deep crust as estimated from heat flow measurements. These data also indicate that temperatures and pressures derived from the cores of sanidine grains may define the dry melting curve for the deep crust.

5. Anhydrous garnet-bearing granulite facies rocks may be constituents of the deep crust in tectonic environments similar to the tectonic environment in the southern Rio Grande Rift. This implies that anhydrous garnet granulites are residues from partial melting and extraction of acidic magmas in the lower crust. These volatile-rich magmas, as a result of a negative slope of the melting curve, would probably crystallize before reaching the earth's surface, thus enriching the intermediate crust in water and other volatiles.

6. The ilmenite-ulvospinel equilibria and the occurrence of blue rutile suggest that an oxygen fugacity of $10^{-12}-10^{-16}$ atm was prevalent. The presence of graphite in xenoliths containing blue rutile provides additional evidence for reducing conditions.

7. The mafic granulites (two-pyroxene, charnockite, and anorthosite) have many features in common, particularly the constant trend of Al/Ti values in orthopyroxene and clinopyroxene, the similarity in composi-

tion of pyroxenes and feldspars in different rock types, and the occurrence of ilmenite exsolution in more than one rock type.

8. The application of pyroxene and feldspar geothermometry and geobarometry to the mafic granulite data gives results which are internally consistent. The resulting temperatures for the Kilbourne Hole mafic granulites are between 800° and 1000°C within a pressure range of 6-10 kbar.

9. This range of estimated PT conditions agrees with that of the garnet-bearing granulites and implies that the mafic granulites also reflect the ambient geothermal gradient at depth.

10. This similar PT relationship suggests that a granulite facies complex, perhaps related to minor anorthosite, is representative of the lower crust in the southern Rio Grande Rift.

11. The presence of healed and sealed microcracks in all types of granulite facies xenoliths suggests the possibility that open microcracks are available as reaction surfaces and paths for fluid conduction under the PT conditions of the lower crust.

Appendix: Glossary

Because this is a multidisciplinary publication assembled for the purpose of advancing our understanding of the major characteristics of the crust of the earth and is therefore addressed to specialists in many disciplines and subdisciplines, this glossary of terms is included for the convenience of the nonspecialist who may wish to read this article in depth.

anorthosite:	plutonic rock composed of greater than 95% plagioclase.
charnockite:	quartzofeldspathic granulite or gneiss with hypersthene.
drop perthite:	intergrown plagioclase and potassium feldspar, where the latter is the host for large droplike patches of the former.
granulite facies:	rocks produced by high-grade regional metamorphism at elevated temperatures and pressures.
K_D:	equilibrium constant.
oxygen fugacity:	effective partial pressure of oxygen.
perthite:	intergrowth of potassium feldspar and plagioclase.
symplectite:	melt crystallization products consisting of intergrown orthopyroxene, spinel, and glass.

Acknowledgments. This work was supported by NSF grant DES 74-14532. The suggestions of M. J. Holdaway and R. Kay have contributed to the improvement of the manuscript. Contribution 321 of the Program for Geosciences at the University of Texas at Dallas.

References

Batzle, M. L., and G. Simmons, Geothermal systems: Rocks, fluids, fractures, in: The Earth's Crust, Geophys. Monogr. Ser., vol. 20, edited by J. G. Heacock, AGU, Washington, D. C., this volume, 1977.

Blatt, H., G. Middleton, and R. Murray, Origin of Sedimentary Rocks, 634 pp., Prentice-Hall, Englewood Cliffs, N. J., 1972.

Bowen, R. W., Mineral Distribution Program, No. A417, Computer Center Division, Menlo Park Computation Branch, U.S.G.S., Aug. 19, 1968.

Buddington, A. F., and D. H. Lindsley, Iron-titanium oxide minerals and synthetic equivalents, J. Petrology, 5(2), 310-357, 1964.

Burnham, C. W., and N. F. Davis, The role of H_2O in silicate melts, II, Thermodynamics and phase relations in the system $NaAlSi_3O_8$ - H_2O to 10 kilobars, 700° to 1100°C, Amer. J. Sci., 274, 902-940, 1974.

Carter, J. L., The origin of olivine bombs and related inclusions in basalts, Ph.D. thesis, 264 pp., Rice Univ., Houston, Tex., 1965a.

Carter, J. L., A geochemical investigation of ultrabasic and basic inclusions in the Kilbourne Hole, New Mexico, basalt, (abstract), EOS Trans. AGU, 46(1), 186-187, 1965b.

Carter, J. L., Chemical composition of the primitive upper mantle, (abstract), Geol. Soc. Amer. Spec. Pap., 101, 35-36, 1966.

Carter, J. L., Mineralogy and chemistry of the earth's upper mantle based on the partial fusion-partial crystallization model, Geol. Soc. Amer. Bull., 81, 2021-2034, 1970.

Chapin, C. E., The Rio Grande Rift, 1, Modifications and additions, Guidebook to the San Luis Basin, N. Mex. Geol. Soc. Field Conf. Guideb., 22, 191-201, 1971.

Chinner, G. A., Almandine in thermal aureoles, J. Petrology, 3(2), 316-340, 1962.

Christensen, N. I., and D. M. Fountain, Constitution of the lower continental crust based on experimental studies of seismic velocities in granulite, Geol. Soc. Amer. Bull., 86, 227-236, 1975.

Chronic, J., M. E. McCallum, C. S. Ferris, Jr., and D. H. Eggler, Lower Paleozoic rocks in diatremes, southern Wyoming and northern Colorado, Geol. Soc. Amer. Bull., 80, 149-156, 1969.

Decker, E. R., and S. B. Smithson, Heat flow and gravity interpretation across the Rio Grande Rift in southern New Mexico and west Texas, J. Geophys. Res., 80(17), 2542-2552, 1975.

Denison, R. E., W. H. Burke, Jr., E. A. Hetherington, and J. B. Otto, Basement rock framework of parts of Texas, southern New Mexico and northern Mexico, in: Symposium in Honor of Ronald K. DeFord, pp. 3-14, West Texas Geological Society, 1970.

De Waard, D., The occurrence of garnet in the granulite-facies terrain of the Adirondak Highlands and elsewhere, an amplification and a reply, J. Petrology, 8(2), 210-232, 1967.

De Waard, D., The anorthosite problem: The problem of the anorthosite-charnockite suite of rocks, in: Origin of Anorthosite and Related Rocks, Mem. 18, pp. 71-92, edited by Y. W. Isaachsen, University of the State of New York, 1968.

De Waard, D., and W. D. Romey, Petrogenic relationships in the anorthosite-charnockite series of Snowy Mountain Dome, south-central Adirondaks, in: Origin of Anorthosite and Related Rocks, Mem. 18, pp. 307-316, edited by Y. W. Isaachsen, University of the State of New York, 1968.

Eggler, D. H., Role of CO_2 in melting processes in the mantle. Carnegie Inst. Wash. Yearb., 72, 457-467, 1973.

Euster, H. P., and D. R. Wones, Stability relations of the ferruginous biotite, annite, J. Petrology, 3(1), 82-125, 1962.

Evans, B., Pyrope garnet-pieometer or thermometer?, Geol. Soc. Amer. Bull., 76, 1295-1300, 1965.

Feves, M., G. Simmons, and R. Siegfried, Microcracks in crustal igneous rocks: Physical properties, in: The Earth's Crust, Geophys. Monogr. Ser., vol. 20, edited by J. G. Heacock, AGU, Washington, D. C., this volume, 1977.

Forbes, R. B., and H. Kuno, Peridotite inclusions and basaltic host rocks, in: Ultramafic and Related Rocks, pp. 337-349, edited by P. J. Wyllie, John Wiley, New York, 1967.

Ghent, E. D., Plagioclase-garnet-Al_2SiO_5-quartz: A potential geobarometer-geothermometer, Amer. Mineral., 61, 710-714, 1976.

Green, D., and A. E. Ringwood, An experimental investigation of the gabbro to eclogite transformation and its petrological applications, Geochim. Cosmochim. Acta, 31, 767-833, 1967.

Green, D., and A. E. Ringwood, A comparison of recent experimental data on the gabbro-garnet granulite-eclogite transition, J. Geol., 80, 277-288, 1972.

Hawley, J. W., and F. E. Kottlowski, The Sante Fe Group in the south-central New Mexico border region, Border Stratigraphy Symposium, Circ. 104, pp. 52-67, Mex. State Bur. of Mines and Mining Res., 1969a.

Hawley, J. W., and F. E. Kottlowski, Quaternary geology of the south-central New Mexico border region, Border Stratigraphy Symposium, Circ. 104, pp. 89-104, Mex. State Bur. of Mines and Mining Res., 1969b.

Hess, H., History of ocean basins, in: Petrologic Studies: A Volume to Honor A. F. Buddington, edited by A. E. J. Engel, H. L. James, and B. F. Leonard, Geological Society of America, New York, 1962.

Hewins, R. H., Pyroxene geothermometry of some granulite facies rocks, Contrib. Mineral. Petrol., 50, 205-209, 1975.

Hoffer, J. M., Preliminary note on the Black Mountain basalts of the Potrillo Field, south-central New Mexico, Border Straigraphy Symposium, Circ. 104, pp. 116-121, N. Mex. State Bur. of Mines and Mining Res., 1969a.

Hoffer, J. M., Volcanic history of the Black Mountain-Santo Tomas basalts, Potrillo volcanics, Dana Ana County, New Mexico, N. Mex. Geol. Soc. Field Conf. Guideb. 20, 108-115, 1969b.

Hoffer, J. M., Mineralogy and petrology of the Santo Tomas-Black Mountain basalt field, Potrillo volcanics, south-central New Mexico, Geol. Soc. Amer. Bull., 82, 603-612, 1971.

Hoffer, J. M., The Aden-Afton basalt, Potrillo volcanics, south-central New Mexico, Tex. J. Sci., 26(3-4), 380-390, 1975.

Hoffer, J. M., The Potrillo basalt field, south-central New Mexico, N. Mex. Geol. Soc. Spec. Publ., 5, 89-92, 1976.

Holdaway, M. J., Stability of andalusite and the aluminum silicate phase diagram, Amer. J. Sci., 271, 97-131, 1971.

Holdaway, M. J., and S. M. Lee, Significance of Fe-Mg cordierite stability relations on temperature, pressure and water pressure in cordierite granulites, in: The Earth's Crust, Geophys. Monogr. Ser., vol. 20, edited by J. G. Heacoc, AGU, Washington, D. C., this volume, 1977.

Irving, A. J., Geochemical and high pressure experimental studies of garnet pyroxenite and pyroxene granulite xenoliths from the Delegate basaltic pipes, Australia, J. Petrology, 15(1), 1-40, 1974.

Ito, K., and G. C. Kennedy, An experimental study of the basalt-garnet granulite-eclogite transition, in: The Structure and Physical Properties of the Earth's Crust, Geophys. Monogr. Ser., vol. 14,

edited by J. G. Heacock, pp. 303-314, AGU, Washington, D. C., 1971.
Jackson, K. C., Textbook of Lithology, 552 pp., McGraw-Hill, New York, 1970.
Kelley, V., Tectonics of the Rio Grande depression of central New Mexico, N. Mex. Geol. Soc. Field Conf. Guideb., 93-105, 1952.
Kerrick, D. M., Experimental determination of muscovite and quartz stability with $P_{H_2O} < P_{total}$, Amer. J. Sci., 72, 946-958, 1972.
Lambert, I. B., and K. S. Heier, The vertical distribution of uranium, thorium and potassium in the continental crust, Geochim. Cosmochim. Acta, 31, 377-390, 1967.
Lee, W. T., Afton craters of southern New Mexico, Geol. Soc. Amer. Bull., 18, 211-220, 1907.
Lindsley, D. H., and S. Dixon, Diopside-enstatite equilibria at 850°-1400°C, 3-35 kbar, Amer. J. Sci., 276, 1285-1301, 1976.
Lorenz, V., Some aspects of the eruption mechanism of the Big Hole Maar, central Oregon, Geol. Soc. Amer. Bull., 81, 1823-1830, 1970.
Lovering, J. G., The nature of the Mohorovicic discontinuity, EOS Trans. AGU, 39, 947-955, 1958.
Luth, W. C., The influence of pressure on the composition of eutectic liquids in the binary systems sanidine-silica and albite-silica, Carnegie Inst. Wash. Yearb., 66, 480-484, 1968.
MacGregor, I. D., Mafic and ultramafic inclusions as indicators of the depth of origin of basaltic magmas, J. Geophys. Res., 73, 3737-3745, 1968.
Matthews, V., III, and C. E. Anderson, Yellowstone convection plume and break-up of the western United States, Nature, 243, 158-159, 1975.
McGetchin, T. R., and L. T. Silver, Compositional relations in minerals from kimberlite and related rocks in the Moses Rock Dike, San Juan County, Utah, Amer. Mineral., 55, 1738-1771, 1970.
McGetchin, T. R., and G. W. Ulrich, Xenoliths in maars and diatremes with inferences for the moon, Mars, and Venus, J. Geophys. Res., 78, 1833-1853, 1973.
Nixon, P. H., D. von Knorring, and J. M. Rooke, Kimberlites from and associated inclusions of Basutoland: A mineralogical and chemical study, Amer. Mineral., 48, 1090-1132, 1963.
Ollier, C. D., Phreatic eruptions and maars, in: Physical Volcanology, edited by L. Civetta, P. Gasparini, G. Luongo, and A. Rapolla, pp. 289-311, Elsevier, New York, 1974.
Padovani, E. R., Granulite facies xenoliths from Kilbourne Hole maar, New Mexico and their bearing on deep crustal evolution, Ph.D. dissertation, Univ. of Tex. at Dallas, Richardson, 1977.
Padovani, E. R., and J. L. Carter, Mineralogy and mineral chemistry of a suite of anhydrous, quartzo-feldspathic, garnet-bearing granulites from Kilbourne Hole maar, New Mexico, Geol. Soc. Amer. Abstr. Programs, 5, 761-762, 1973.
Padovani, E. R., and J. L. Carter, Blue sillimanite in garnet granulite xenoliths from Kilbourne Hole, New Mexico, (abstract), EOS Trans. AGU, 55(4), 482, 1974.
Padovani, E. R., and J. L. Carter, Aluminous, iron-rich orthopyroxene and iron-rich spinel in garnet granulite xenoliths from Kilbourne Hole maar, New Mexico, (abstract), EOS Trans AGU, 56(6), 465, 1975.
Padovani, E. R., and J. L. Carter, Labradorite anorthosite from south

central New Mexico, (abstract), EOS Trans. AGU, 57(4), 338, 1976a.

Padovani, E. R., and J. L. Carter, Aspects of the deep crustal evolution beneath south central New Mexico, paper presented at Symposium on the Nature and Physical Properties of the Earth's Crust, Office of Nav. Res., 1976b.

Padovani, E. R., and J. L. Carter, Non-equilibrium fusion due to decompression effects in crustal and mantle xenoliths, (abstract), in: Chapman Conference on Partial Melting in the Earth's Upper Mantle, p. 31, 1976c.

Reeves, C. C., Jr., and R. De Hon, Geology of Potrillo maar, New Mexico, and northern Chihuahua, Mexico, Amer. J. Sci., 263, 401-409, 1965.

Reiche, P., The origin of Kilbourne Hole, New Mexico, Amer. J. Sci., 238, 212-225, 1940.

Richter, D., and G. Simmons, Microcracks in crustal igneous rocks: Microscopy, in: The Earth's Crust, Geophys. Monogr. Ser., vol. 20, edited by J. G. Heacock, AGU, Washington, D. C., this volume, 1977.

Ringwood, A. E., and D. H. Green, An experimental investigation of the gabbro-eclogite transformation and some geophysical implications, Tectonophysics, 3(5), 383-427, 1966.

Rosenfeld, J. L., and A. B. Chase, Pressure and temperature of crystallization from elastic effects around solid inclusions in minerals, Amer. J. Sci., 259(7), 519-541, 1961.

Ross, C. S., M. D. Foster, and A. T. Meyers, Origin of dunites and of olivine-rich inclusions in basaltic rocks, Amer. Mineral., 39, 693-737, 1954.

Ross, M., and J. S. Huebner, A pyroxene geothermometer based on composition-temperature relationships of naturally occurring orthopyroxene, pigeonite and augite, paper presented at International Conference on Geothermometry and Geobarometry, 1975.

Sighinolfi, G. P., and C. Gorgoni, Genesis of massif-type anorthosites: The role of high-grade metamorphism, Contrib. Mineral. Petrol., 51, 119-126, 1975.

Smith, J. V., Feldspar Minerals, vols. 1, 2, Springer, New York, 1974.

Stoeser, D. B., Mafic and ultramafic xenoliths of cumulus origin, San Francisco volcanic field, Arizona, Ph.D. dissertation, Univ. of Oreg., 1973.

Stormer, J. C., A practical two-feldspar geothermometer, Amer. Mineral., 60, 667-674, 1975.

Utah Geological and Mineral Survey, Deepest well in the world, Survey Notes, 10(2), 5, 1976.

Wyllie, P. J., The Dynamic Earth: Textbook in Geosciences, John Wiley, New York, 1971.

Yoder, H. S., Jr., Diopside-anorthite-water at five and ten kilobars and its bearing on explosive volcanism, Carnegie Inst. Wash. Yearb., 64, 82-89, 1965.

EVOLUTION OF SILICIC MAGMA CHAMBERS AND THEIR RELATIONSHIP TO BASALTIC VOLCANISM

John C. Eichelberger and R. Gooley

Geological Research Group, University of California, Los Alamos Scientific Laboratory, Los Alamos, New Mexico 87545

Abstract. Andesitic and rhyolitic volcanism is commonly preceded by the eruption of basalt. Similarly, the earliest phases of granitic plutonic complexes are often gabbro. Thus basaltic magma apparently plays a role in the initiation of a large silicic magma system. Three lines of evidence suggest that basaltic magma also enters silicic chambers and influences their further evolution: (1) Contemporaneous basalt vents flank silicic volcanic centers. (2) Thermal models of silicic magma bodies suggest that their heat must be replenished to maintain them in the upper crust for their observed life-span. (3) Petrologic data indicate that 'cognate' mafic clots and xenoliths common in granodiorite and andesite represent basaltic magma quenched within active silicic magma chambers. The presence of abundant basaltic material in intermediate volcanic and plutonic rocks, phase assemblages in volcanic rocks, and the bulk composition of volcanic and plutonic rocks of the basalt-andesite-rhyolite and gabbro-granodiorite-granite suites show that these rock associations form by mixing of basaltic and rhyolitic magmas. It is proposed that basaltic magma from the upper mantle provides heat for generating rhyolitic melt in the lower crust and that the resulting rhyolitic magma body continues to receive injections of basaltic magma as it rises through the crust. Implications of this model are that (1) the development of a silicic system depends on the intensity of basaltic volcanism and the ability of the lower crust to produce a rhyolitic melt and (2) the relative volume of intermediate versus basaltic and rhyolitic end products depends on the extent of mixing. It appears that the influence of tectonic setting on crustal residence time of silicic chambers controls to what degree the system is homogenized.

Introduction

Spatial and temporal association between basalt and rocks of higher silica content within both plutonic complexes and volcanic fields implies a genetic relationship. This paper discusses field evidence of this relationship and presents new petrologic data indicating that it is the interaction of basaltic magma with initially rhyolitic magma bodies which produces much of the compositional diversity within the basalt-andesite-rhyolite and gabbro-granodiorite-granite suites. On the basis of this evidence from diverse complexes a general model is proposed to explain silicic igneous activity in terms of stage of evolution and tectonic setting. Recent discussions of petrologic evidence of mixing have been presented by Eichelberger [1975] and Anderson [1976]. Earlier papers on mixing include those by Bunsen [1851], Fenner [1926], Larsen et al. [1938], and Wilcox [1944].

Relationship of Basaltic Volcanism to Silicic Magma Chambers

Most rocks of the compositional range andesite to rhyolite or granodiorite to granite are closely associated in the field with basalt or gabbro. In general, intrusions of basaltic magma precede, continue during, and sometimes postdate the development of more silica-rich magmas. This observation applies to both igneous complexes where intermediate (andesite or granodiorite) compositions predominate and those where extreme (basalt-rhyolite or gabbro-granite) compositions predominate. For example, the earliest phases of the predominantly granodioritic Snoqualmie batholith are gabbros [Erikson, 1969], and the northern Cascade batholiths of this type were intruded by basaltic dikes late in their history [Erikson, 1969; Tabor and Crowder, 1969]. Mayo [1941] referred to the earliest intrusives of the Sierra Nevada batholith as mafic 'forerunners' and there, too, mafic magmas invaded the later silicic bodies. Ring dike complexes commonly begin with emplacement of gabbros, followed by a copious variety of granites [e.g., Moorbath and Bell, 1965; Chapman, 1942; Modell, 1936]. The intermingling of gabbroic and granitic rocks within some of the granite bodies suggests that mafic magmas were still present when the granites were intruded.

Essentially all volcanic fields in the western United States contain some basalt. Basalt has erupted throughout the Cascade Range, where andesite is an important rock type, and predominates volumetrically in parts of it [McBirney et al., 1974]. The more bimodal Snake River plain-Yellowstone plateau field erupted basalt before and after major caldera-forming rhyolite eruptions [Eaton et al., 1975]. The relative volume of early basalt or gabbro in a complex may be underestimated because it is concealed by later lava or stoped downward by later intrusives.

In both plutonic and volcanic complexes, mafic bodies are typically satellitic to centers of silicic activity. Joplin [1959] stated this general observation for discordant granitic batholiths. It has been mentioned more recently in reference to the Snoqualmie batholith [Erikson, 1969] and the Boulder batholith [Tilling, 1973]. The equivalent spatial arrangement in volcanic fields is represented by silicic volcanic centers with older to contemporaneous flanking basalt vents. These include the major volcanos of the Cascade Range (the relationship is especially clear in the southern Cascades), Long Valley caldera (California), San Francisco Peaks (Arizona), Mount Taylor (New Mexico), and the Valles caldera (New Mexico).

Eaton et al. [1975] and Bailey et al. [1976] have used the distribution of basalt vents to infer the position of active silicic magma bodies by assuming that dense mafic liquid cannot pass upward through such a body. Field observations at Yellowstone and Long Valley indicate that the roof of a silicic chamber, as defined by a caldera, lacks basalt vents during the period when the caldera is active. As silicic activity related to the caldera declines, basalt vents encroach on the caldera region, presumably because the underlying body is crystallizing inward. Similar observations can be made for other centers lacking calderas, where large accumulations of silicic lava imply the presence of a large chamber.

Thus it appears that silicic magma chambers typically lie within a region of upward flux of basaltic magma. Lachenbruch et al. [1976] have shown that such a model could account for the discrepancy between the observed life-span of the Long Valley magma chamber and the shorter life-span calculated for thermal models if heat loss by conduction alone is assumed. Trapping of basaltic magma provides an efficient means of keeping the silicic magma hot.

These observations have important petrogenetic implications. The existence of mafic forerunners to batholiths suggests that basaltic magma may provide heat for development of silicic magmas by partial melting in the lower crust. Evidence that the upward flux of basalt continues after an active silicic pluton has formed suggests the possibility of direct interaction. Mass as well as heat will be added to a low-density magma which traps basalt, a change in composition as well as temperature thus being produced.

Petrologic Evidence of Interaction of Basalt and Rhyolite Magmas

Although silicic volcanic centers lack basalt vents while they are active, lavas of intermediate composition within these centers commonly contain abundant xenoliths of basaltic composition, which range in size from 1-m globules to 1-mm crystal clots to individual crystals derived from these clots. These xenoliths have no exact counterparts among the country rocks. Texturally similar xenoliths occur in granodiorites of plutonic complexes. Textural variations are common among or even within xenoliths [Tabor and Crowder, 1969], but the mineralogy remains the same (dominantly plagioclase plus pyroxene or plagioclase plus hornblende). The most common textural types are fine-grained porphyritic and medium-grained nonporphyritic. Where these textures are found in the same xenolith the former type makes a rind on the latter (Figure 1) [Didier, 1973]. In the plutonic case the compositions of phases within the xenoliths match those of the host, and only the proportions differ; but in volcanic rocks, there is striking evidence of disequilibrium between xenolith and host.

Basaltic xenoliths are the subject of a long and lively literature under such names as mafic microgranular enclaves, mafic segregations, secretions, cognate xenoliths, restites, and so forth. Probably the most common interpretation is that they are cumulates or equivalent products of fractional crystallization [e.g., Williams, 1931; I. A. Nicholls, 1971]. Others [Piwinskii, 1973; Presnall and Bateman, 1973] have suggested that they are refractory residuum, 'restites,' from partial melting of the lower crust. Joplin [1959] suggested that they are engulfed mafic forerunners. The compositionally equivalent but genetically different interpretation is that they are 'basaltic pillows' [Blake et al., 1965; Zeck, 1970], basalt chilled and crystallized in silicic magma.

We have studied samples of these xenoliths and their host lavas from a number of volcanic centers in Ecuador, the Lesser Antilles, and the

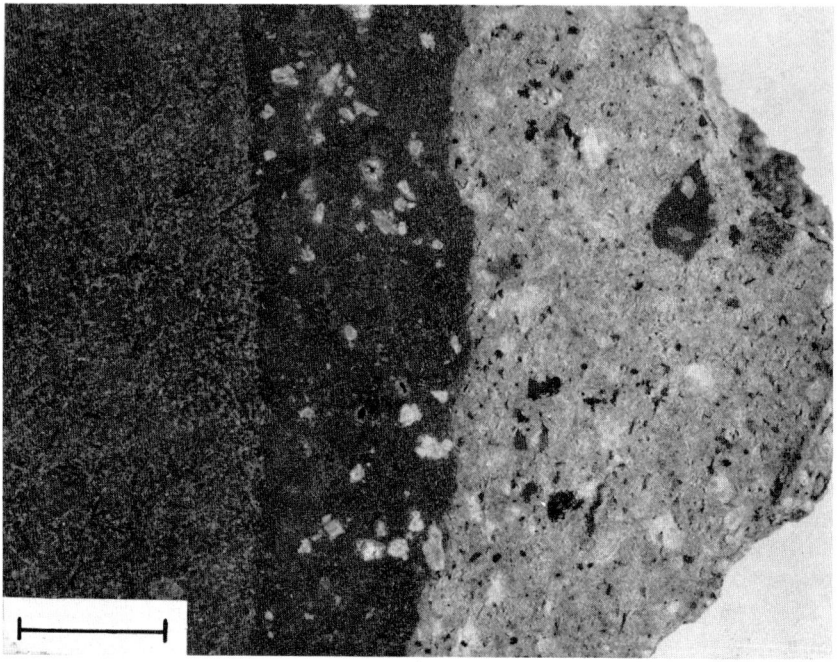

Fig. 1. Portion of large zoned plagioclase-hornblende xenolith (dark gray) and dacite host (light gray) from Chaos Crags, Lassen Volcanic National Park, California. From left to right are the medium-grained core, porphyritic fine-grained rind (Table 1, column 4), and dacite host. Note the small fragments of both zones of the xenolith in the dacite at the far right. For the plutonic equivalent, see Didier [1973, p. 228]. Bar length is 2 cm.

western United States, including a suite of Lassen Volcanic National Park lavas spanning the compositional range of basaltic andesite to rhyodacite. Several lines of evidence indicate that the xenoliths represent basaltic magma injected into and crystallized within silicic magma chambers, that is, that they represent the basaltic magma which does not reach the surface above these chambers:

1. The xenoliths are globular in shape, sometimes with a finely contorted border.
2. Many of the larger xenoliths show chilling against the host. In xenoliths in Lassen dacite, average grain size decreases toward the border, and hornblende changes from subequant to highly elongate.
3. Many of the larger xenoliths are surrounded by a mixed rind containing partially melted or reacted phenocrysts from the host. The matrix is similar to but finer grained and more glass rich than the interior of the xenolith (Figure 1).
4. Most of the xenoliths contain interstitial vesicular glass. Since the crystals are subhedral to euhedral, the texture is clearly one of advanced crystallization rather than of beginning of melting.
5. In bulk composition the xenoliths approach the composition of

TABLE 1. Whole Rock and Phenocryst Composition of Some Basaltic Xenoliths in Andesites and Dacites

	1	2	3	4	5
SiO_2	51.4	54.7	53.4	58.5	50
TiO_2	1.0	0.9		0.6	
Al_2O_3	18.7	18.5	19.2	17.8	17
FeO	8.6	7.0	7.5	5.9	
MgO	5.3	5.8	4.9	3.5	
CaO	10.7	8.7	9.8	7.2	
Na_2O	3.2	3.0	2.9	3.4	
K_2O	0.5	1.0	1.0	1.5	
Olivine	Fo70-80	Fo80	Fo80	Fo80	Fo70
Plagioclase	An80	An70-80	An90	An30, An90	An80-85

Column 1 represents the average of xenoliths in the modern series, Santorini volcano, Greece [I. A. Nicholls, 1971]; 2 a xenolith in dacite, Medicine Lake highland, California [Anderson, 1941; Eichelberger, 1975]; 3 a xenolith in dacite, Lassen Volcanic National Park, California [Williams, 1931]; 4 the porphyritic rind on xenolith in dacite (Figure 1), Lassen Volcanic National Park, California (analyst was John Husler, University of New Mexico); and 5 a microxenolith in andesite, Mount Baker, Washington. The silica and alumina content are estimated from mode, and bytownite occurs as cores of some plagioclase grains.

associated basalt and contain primary phenocryst phases which match phenocrysts in associated basalts (Table 1). The most common of these are highly calcic plagioclase and magnesium-rich olivine.
6. Metamorphic textures are completely absent.

Evidence that basaltic material has been added to an igneous rock implies that before this event, there existed a magma body higher in silica content than the present silica content of the rock. Products of disaggregation of the basaltic material are so abundant on a microscopic scale in rocks of intermediate composition that the parent magma body must have been much more silica rich in composition, rhyolitic, before mixing. For example, subtraction of basalt-derived material from the mode of Lassen dacite gives an estimate of silica content before mixing of 75 wt %. Additional evidence for the involvement of rhyolitic magma is provided by the presence of partially reacted or resorbed phenocrysts, appropriate for magma of rhyolitic composition, in virtually all intermediate lavas. Further, rhyolite-derived phenocrysts in andesites from Lassen (Figure 2), Glacier Peak, San Francisco Peaks, Ecuador, and the Lesser Antilles contain inclusions of rhyolite glass in their unmelted cores.

The most common rhyolite-derived phenocrysts in andesite and dacite are sodic plagioclase and quartz. In response to injection of basalt into the rhyolite magma chamber, sodic plagioclase partially dissolves leaving a skeletal framework of more calcic plagioclase. This process

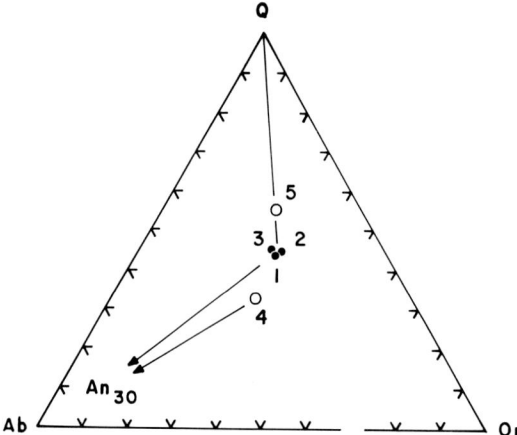

Fig. 2. Average normative composition of glass inclusions in andesine and quartz phenocrysts in dacite and andesite from Lassen Volcanic National Park based on complete electron microprobe analyses. Locations of glasses are: 1, the core of unmelted plagioclase in Chaos Crags dacite (eight analyses); 2, the unmelted core of partially melted plagioclase in the porphyritic rind of a basaltic xenolith (Figure 1), Chaos Crags dacite (three analyses); 3, the core of quartz in Chaos Crags dacite (five analyses); 4, the unmelted core of partially melted plagioclase in Prospect Peak andesite (three analyses); and 5, the core of quartz in Prospect Peak andesite (two analyses). The position of the rhyolite glasses in the andesite with respect to the clustered analyses for the dacite is apparently due to expansion of the liquid stability field in the hotter andesite hybrid. Phenocrysts representative of locations 3, 4, and 5 are shown in Figure 3.

results in a cloudy zone consisting of calcic plagioclase, glass, pyroxene, and opaques surrounding an unmelted sodic core (Figures 3 and 4). When thermal equilibrium is reached and crystallization resumes, a normally zoned calcic overgrowth forms on the crystal. Depending on the proportion of basalt to rhyolite, quartz is either resorbed (dacite) or reacts with the liquid to form pyroxene (andesite). If the rhyolite is at or above its liquidus when mixing occurs, no rhyolite-derived phenocrysts will be present in the hybrid [e.g., Eichelberger, 1975]. The phase assemblage of the rhyolite at the instant of mixing is shown clearly by the assemblage of rhyolite-derived phenocrysts in the mixed rind of basaltic xenoliths (in Lassen dacite, quartz plus andesine plus biotite).

Textural relationships thus indicate that basalt magma breaks into globules as it intrudes rhyolitic magma. Mixing occurs at the interfaces, and the products of interaction are then strewn throughout the magma body by convective stirring. The basalt globules are vesicular because they are cooled to near solidus, vapor saturation of the residual liquid thus resulting. Direct mixing of mafic and silicic liquids is minor if the proportion of introduced basalt is small because the globules are quickly crystallized. In Lassen dacite, this limited mixing is represented by the thin rinds on the xenoliths (Figure 1).

Conversely, when the proportion of basalt is large, direct mixing of liquids is more thorough. The initial mixture is still heterogeneous, however, and when thermal equilibrium is reached, the more basaltic portions are more crystal rich and are strewn through the magma as clots and microxenoliths. Figure 3 illustrates the products of basalt and rhyolite interaction for dacite and andesite from Lassen Volcanic National Park. Note that the thermal effects on rhyolite-derived phenocrysts increase with an increase in the proportion of basalt in the mixture (Figure 3). The partial melt zone on sodic plagioclase is wider in the hotter andesitic hybrid than in the dacite, and quartz reacts rather than resorbs. The higher temperature in the andesite hybrid results in formation of pyroxene rather than hornblende in the xenoliths. Figure 4 shows similar features in dacite from Ecuador and andesite from Mount Baker. It should be emphasized that these features are present in abundance in every thin section and that these samples are typical of rocks from their respective regions. Similar features have been described in the lavas of Mount Rainier [Fiske et al., 1963] and Glacier Peak [Tabor and Crowder, 1969]. The basaltic microxenoliths in these lavas correspond to similar xenoliths in intermediate parts of associated Miocene batholiths. The similarity in texture of the latter to xenoliths in Lassen dacite is noted in Figure 3. It is interesting to note that a large negative gravity anomaly [Pakiser, 1964], which can be interpreted as an active batholith in the upper crust, encompasses the hybrid vents of Lassen Volcanic National Park but excludes the recent basaltic vents. The areal extent of this anomaly is comparable to the Miocene batholiths of the northern Cascade Range, such as Snoqualmie and Cloudy Pass. Erikson [1969] suggested that most of the Snoqualmie batholith, from granite to gabbro, was in a molten state at the same time. Thus the analogy among Lassen andesites and dacites, northern Cascade andesites and dacites, and northern Cascade granodiorites appears to hold. Production of the modern volcanic hybrids takes place in sizable chambers as represented by the exhumed Miocene batholiths.

A test of this model in terms of major element composition of these rocks is presented in Table 2. Average compositions of Snoqualmie gabbro (51% SiO_2) and granite (75% SiO_2) are taken as being representative of the parent magmas for Snoqualmie granodiorite. For Glacier Peak, Quaternary basalt (51% SiO_2, White Chuck cinder cone) of the area and average granite of the associated Miocene batholith (73% SiO_2) are used for parent compositions. Similar data have been presented previously for Lassen [Eichelberger, 1975]. Church [1976a] reported isotopic data from Glacier Peak, Mount Baker, and Mount Shasta indicating mixing of lead from two sources and concluded that the upper mantle and lower crust are the probable sources. Mixing of magmas derived from the upper mantle and lower crust provides a mechanism for accomplishing the mixing of lead.

It has been suggested that quartz [J. Nicholls et al., 1971] and sodic plagioclase [Nash, 1973] in andesite are products of crystallization at high pressure. The presence of rhyolitic glass inclusions in the unmelted cores of quartz and sodic plagioclase in andesite from the Lesser Antilles, Ecuador, Lassen, and San Francisco Peaks clearly rules out this possibility for these lavas. Further, Drake [1976] demonstrated that there is no significant

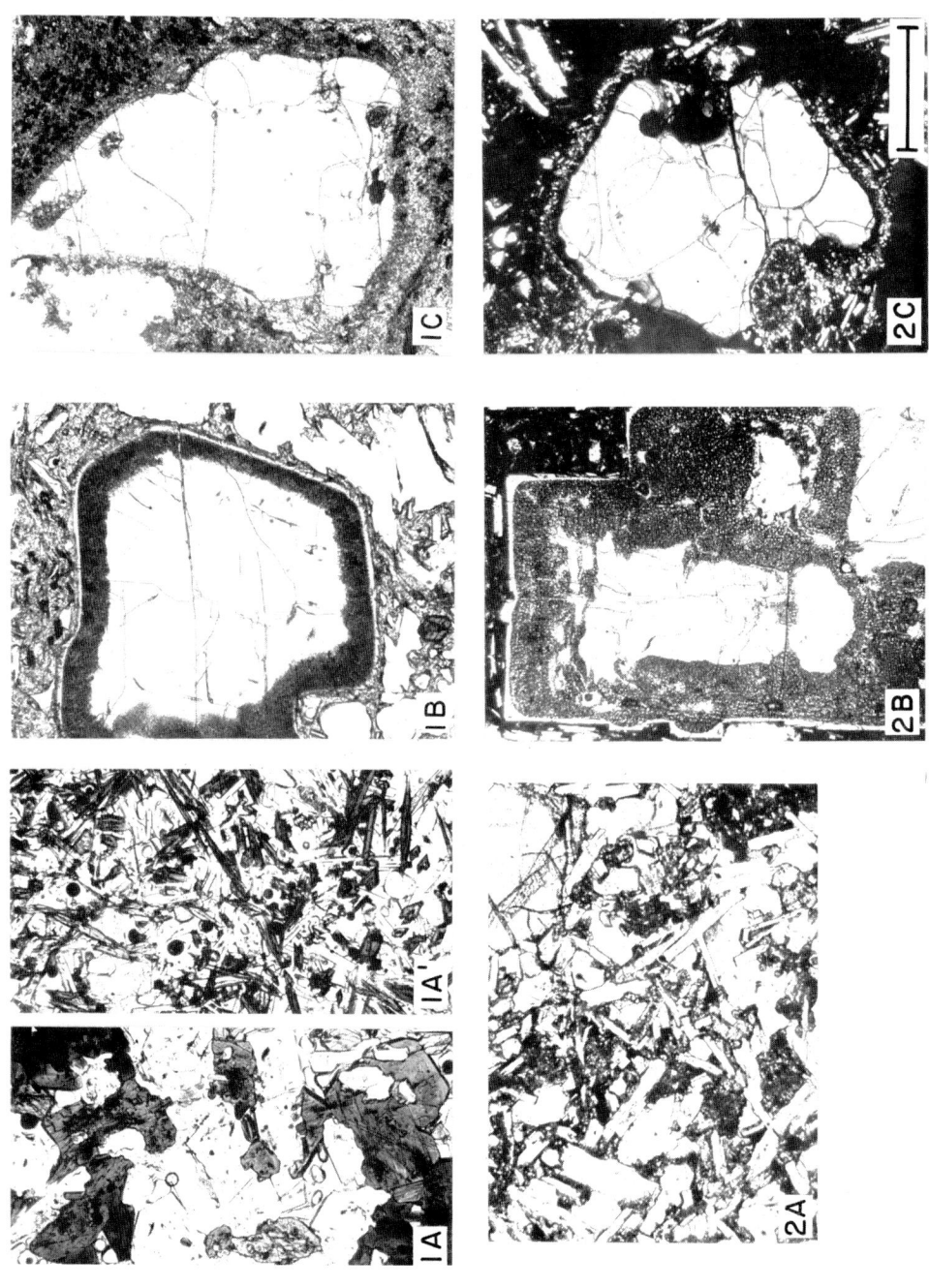

pressure effect to 10 kbar on plagioclase melt equilibria in dry systems of basalt to rhyolite composition. The effect of water at high pressures would shift plagioclase toward a more calcic composition [Yoder, 1969], and ascent of a hydrous magma would cause crystallization, not melting. Doe et al. [1969] found that partially melted sodic plagioclase in an andesite from the San Juan Mountains is in isotopic disequilibrium with the matrix, an indication that the crystals did not grow from a liquid represented by the matrix.

Stewart [1975] has proposed that plagioclase-pyroxene crystal clots in andesites are breakdown products of high-Al amphiboles. In Lassen rocks it is clear that the plagioclase-pyroxene or hornblende clots are simply fragments of the basaltic xenoliths, since a continuous size spectrum exists. The presence of glass, calcic plagioclase, and magnesian olivine in the clots cannot be reconciled with the breakdown product hypothesis. Stewart [1975] makes a distinction between plagioclase plus pyroxene plus glass clots as xenoliths and plagioclase plus pyroxene clots as amphibole products. However, Fo70 olivine and An85 plagioclase are present in both glass-present and glass-absent clots at Mount Baker, and the Lassen xenoliths suggest that the amount of glass is caused by minor differences in the mixing and cooling history of injected basalt. If as the petrologic evidence suggests, northern Cascade andesite forms in a manner similar to that of southern Cascade andesite, then the predominance of andesite at Mount Baker indicates that the parent magma body is well stirred. Thus it is not surprising to find the basalt represented by smaller xenoliths.

Thoroughness of Mixing and Evidence of Partially Mixed Magma Bodies

Examples such as Mount Baker and Glacier Peak show that mixing of basaltic and rhyolitic magmas can be so thorough that products of intermediate composition predominate and little or no rhyolitic magma reaches the surface. The basalt-rhyolite association of the Yellowstone plateau represents the opposite extreme, where mixed lavas are negligible in volume. Other volcanic and plutonic complexes such as

Fig. 3. (Opposite) Comparison of basaltic xenoliths and rhyolite-derived phenocrysts in (top) dacite and (bottom) andesite from Lassen Volcanic National Park. Column A is basaltic xenoliths, B is partially melted plagioclase, and C is quartz. The dacite is from Chaos Crags and represents approximately four parts rhyolite to one part basalt by weight. The andesite is from Prospect Peak and represents approximately two parts rhyolite to three parts basalt by weight.
(1A) Basaltic xenoliths are plagioclase plus hornblende plus glass. (1A') Chilled margin [cf. Tabor and Crowder, 1969, p. 13]. (1B) Plagioclase has an An28-40 core, a thin partial melt zone (dark), and a normally zoned An75-50 overgrowth. Plagioclase which did not encounter hot basalt globules has an An28-40 core, no partial melt zone and An50 rims (not shown). (1C) Quartz is resorbed. (2A) Basaltic xenoliths are plagioclase plus pyroxene plus glass. (2B) Plagioclase has an An28-40 core, a wide partial melt zone, and a normally zoned An70-55 overgrowth. All sodic plagioclase in the andesite has a partial melt zone. (2C) Quartz has a pyroxene reaction rim. Bar length is 0.5 mm.

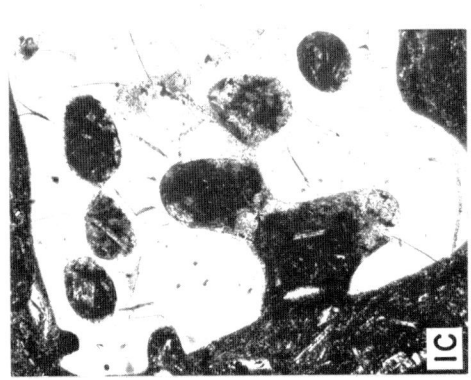

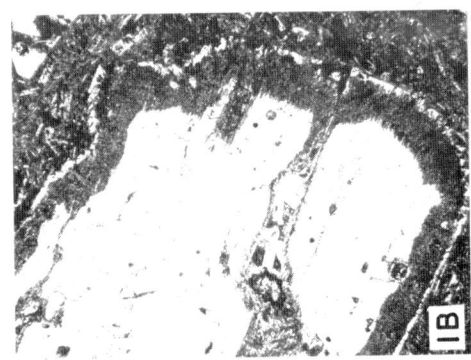
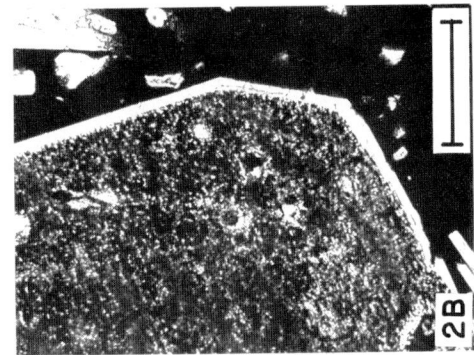

TABLE 2. Comparison of Observed Bulk Compositions with Compositions Calculated for Mixtures of Rhyolite and Basalt

	Snoqualmie		Glacier Peak			
	1	2	3	4	5	6
SiO_2	65.6	65.3	57.0	57.4	63.4	63.9
TiO_2	0.5	0.6	0.8	0.9	0.7	0.6
Al_2O_3	15.7	15.0	18.0	17.1	16.9	15.9
FeO	4.5	5.0	5.9	6.1	4.4	4.3
MgO	2.3	2.5	4.9	5.2	2.9	3.2
CaO	4.5	4.9	7.0	7.2	4.8	4.9
Na_2O	3.4	3.1	3.6	3.4	4.1	3.7
K_2O	2.3	2.0	1.2	1.2	1.8	2.0

Column 1 represents the average granodiorite (33 analyses), 2 a mixture of three parts rhyolite (9 analyses) to two parts basalt (6 analyses), 3 andesite of Gamma Ridge (1 analysis), 4 a mixture of three parts rhyolite (3 analyses) to seven parts basalt (2 analyses), 5 average dacite (11 analyses), and 6 a mixture of three parts rhyolite to two parts basalt. All data are from Erikson [1969] and Tabor and Crowder [1969]. See the text for a discussion of selected basalt and rhyolite parent compositions.

Lassen Volcanic National Park and the Snoqualmie batholith represent intermediate cases, where mixing is substantial, but basaltic and rhyolitic end-members are still present. Bimodal (unmixed) volcanic fields tend to occur in strongly block faulted terrane. This relationship has been noted previously on a large scale and correlated with plate tectonic setting [Lipman et al., 1972; Christiansen and Lipman, 1972; Martin and Piwinskii, 1972], and it holds on a smaller scale as well, within the Cascade Range. Fields which erupt the entire spectrum of composition such as Lassen, Medicine Lake, and Newberry Volcano lie in areas of prominent faulting, while intermediate centers occur nearby, where faulting is more subdued. Christiansen and Lipman [1972] have shown that a switch from intermediate to bimodal compositions has occurred at many centers in the western United States and that this change took place when block faulting began. This suggests that the tectonic regime which produces normal faulting in some way inhibits mixing. A likely explanation is that in an extensional

Fig. 4. (Opposite) Features analogous to those of Figure 3 in (top) dacite from the vicinity of Nevado Antizana, Ecuador, and (bottom) andesite from Mount Baker, Washington. (1A) Basaltic microxenoliths are plagioclase plus hornblende. (1B) Plagioclase has an An35 core, a partial melt zone, and a normally zoned An65-50 overgrowth. (1C) Quartz is resorbed. (2A) Basaltic microxenoliths are plagioclase plus pyroxene plus glass. (2B) Plagioclase is partially melted throughout with a normally zoned An75-50 overgrowth. Bar length is 0.5 mm.

tectonic environment, rhyolite diapirs rise more rapidly to the upper crust or surface, where they crystallize or erupt before thorough mixing with basaltic magma can take place.

In complexes where mixing is not thorough, large plutons or volcanic units may be expected to show the effects of partial mixing. The simple case in which basaltic magma enters the bottom of a rhyolite magma chamber has been described by Rice and Eichelberger [1976]. Initially, two separate convection systems will be present because of the density difference between the magmas, and a mixed (andesitic) crystal-rich layer will grow at the interface. Compositionally zoned tuff sheets in which crystal-rich dacite or andesite overlies crystal-poor rhyolite may represent the partial emptying of such a stratified chamber. Although such sheets have been interpreted as products of fractional crystallization [Hedge and Noble, 1976], several features are not in accord with this hypothesis:

1. The sheets are isotopically as well as compositionally zoned. Phenocrysts are in isotopic disequilibrium with the matrix in the dacitic or andesitic portions [Noble and Hedge, 1969].

2. Compositional variations in the tuff sheets are abrupt. Most of the variation in bulk composition is due to a change in composition of the matrix rather than to the higher proportion of phenocrysts in the andesitic portions [Lipman, 1967; Lipman, et al., 1966].

Aso caldera tuff sheets show an upward increase in partially melted plagioclase and appearance of Fo80-85 olivine and An80 plagioclase in some of the andesitic upper units and have abundant crystal-rich mafic xenoliths [Lipman, 1967]. Noble and Hedge [1969] report that sanidine in the quartz latite top of a zoned tuff in Nevada isotopically matches the rhyolite bottom of the sheet. These data are better explained as the product of a partially mixed stratified magma body in which radiogenic rhyolitic magma overlies less radiogenic basaltic magma.

The sequence of eruption at other partially mixed centers appears to match these tuffs. Some San Juan calderas which erupted rhyolitic tuff subsequently filled with dacitic or andesitic lava [Steven and Lipman, 1975]. The Long Valley caldera has erupted dacitic and andesitic lavas following the Bishop rhyolitic tuff [Bailey et al., 1976]. Centers which erupted andesite following extrusion of rhyolitic domes or flows include Mount Taylor [Baker and Ridley, 1970] and the Summer Coon volcano [Lipman, 1968].

The latter examples may reflect the same mafic downward zonation of magmas as the tuffs or merely an increasingly mafic mixture with time because the rhyolitic magma in the chamber was not replenished. The most common compositionally equivalent zonation in plutonic bodies is mafic outward. Mafic xenolith-bearing granodiorites surround granite cores in some Sierra Nevada plutons [Bateman and Wahrhaftig, 1966]. These may represent a convection configuration in which the hybrid is swept up the outer portion of the chamber [A. Rice, personal communication, 1976]. Another possibility is that the upward flux of basaltic magma ceased between hybridization of the granodiorite and emplacement of the last rhyolitic magma diapir.

Statement of Model

Field and petrologic evidence discussed above implies that basaltic magma is the heat source for the production of rhyolitic melt, that continued upward flux of basaltic magma results in mixing of basaltic and rhyolitic magmas, and that the volume distribution of rock types in the resulting igneous complex depends on the amount of mixing. This model is depicted in Figure 5. The process is simply one by which upward migration of a hot melt, if there is enough of it, promotes melting out of a lower melting fraction at higher levels. Under some conditions the cooler melt retains its identity because its physical properties differ greatly from the hot melt. In the model the lower crust is chosen as a likely source for rhyolitic melt because it is the deepest region, and therefore has the highest initial temperature, in which rocks of composition capable of producing significant amounts of rhyolite are likely to occur. Thermal calculations show that production of rhyolitic melt in the lower crust in response to emplacement of basaltic sills can be an efficient process [Lachenbruch, 1976; Younker and Vogel, 1976; U. Nitsan, personal communication, 1976]. Formation of relatively homogeneous hybrids by injection of basaltic magma into rhyolitic magma bodies is apparently accomplished through rapid convection. Rice and Eichelberger [1976] calculate that a rhyolitic magma body need be only 10 m deep to convect and estimate convection rates of the order of 1 cm/s within sizable (1 km) rhyolitic plutons. The size constraint on convection, the large volumes of homogeneous hybrids, and the presence in hybrids of rhyolite-derived phenocrysts and globules of quenched basaltic magma require that much of the mixing occur within magma chambers. Generally, the rhyolitic melt has coalesced, separated from its source region, and begun to cool when infusion of fresh basaltic magma results in hybridization. Only limited mixing can occur between basaltic magma and dispersed rhyolitic melt within the lower crust.

The model implies that development of a silicic igneous complex is dependent on intensity of the basalt flux and ability of the lower crust to produce a rhyolitic melt. Continental crust is not required [Helz, 1976], although significantly less rhyolitic melt is produced by partial melting of oceanic crust. Some continental basaltic fields may have insufficient rate or volume of basalt intrusion or fail to trap enough of the basalt in the lower crust to generate a silicic system.

Although the most compelling petrologic evidence for the model comes from volcanic fields, it is consistent with many kinds of data on batholiths. In the case of the Sierra Nevada batholith the model provides a mechanism for mixing of components from upper mantle and lower crustal sources, preferred by Kistler and Peterman [1973] and Doe and Delevaux [1973] on the basis of isotopic data. It also provides an explanation for the mafic xenoliths in the granodiorite [Bateman and Wahrhaftig, 1966] and the high liquidus temperatures for the granodiorite observed experimentally by Piwinskii [1973].

Isotopic Data and Inferences Concerning the Rhyolite Source Region

Mixing of two relatively consistent parent magmas results in linear variation of the whole rock composition of the products. Obviously, this holds for isotopic as well as elemental abundances. A linear array of whole rock isotopic data from an igneous suite (Figure 6 and

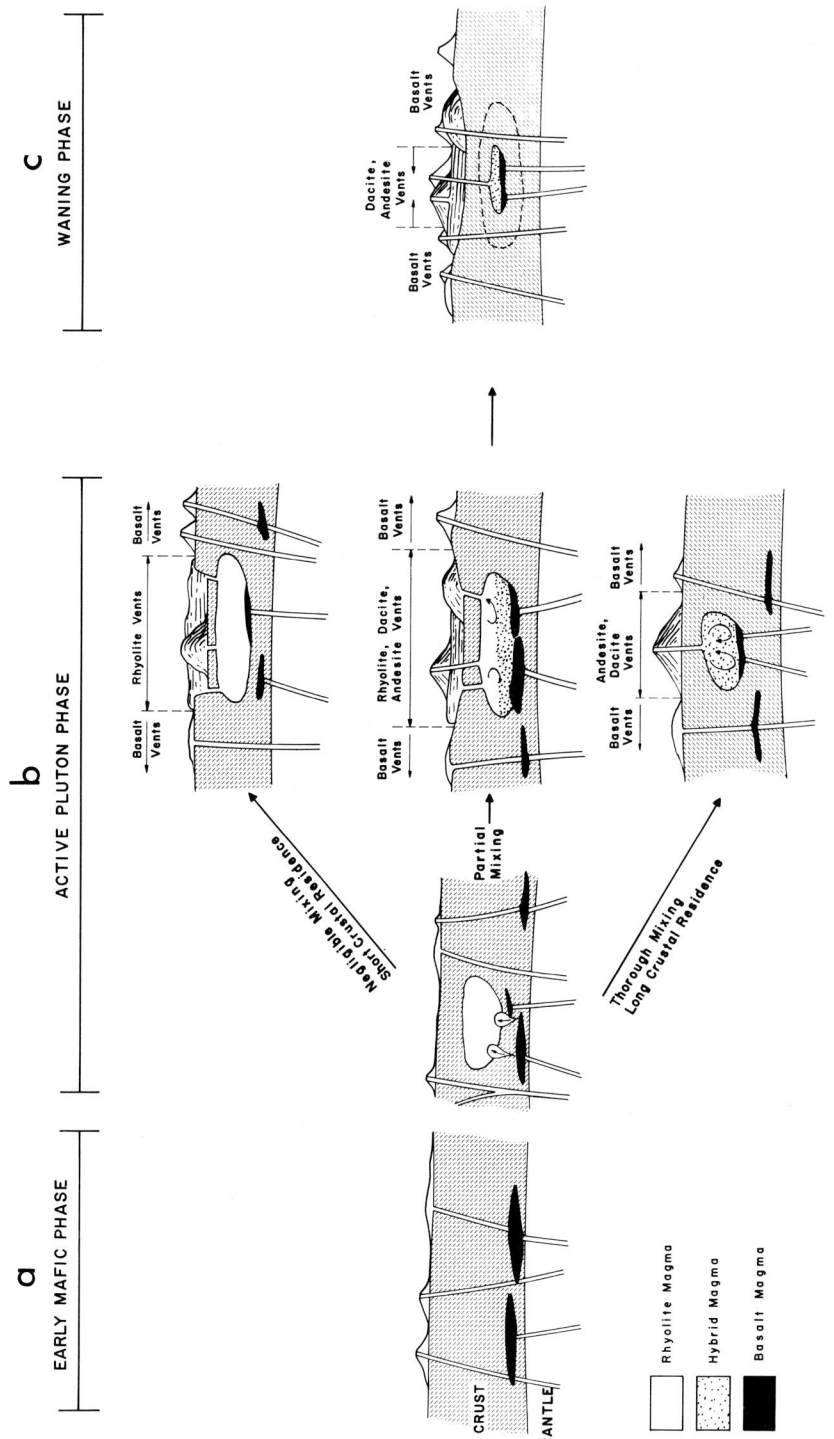

Table 3) can represent an isochron, mixing line, or both. For $^{87}Sr/^{86}Sr$ versus $^{87}Rb/^{86}Sr$, products of a fractionation (melt-liquid segregation) event will plot on a line of zero slope at the time of fractionation. This is because fractionation changes $^{87}Rb/^{86}Sr$ but not $^{87}Sr/^{86}Sr$. If the products of fractionation remain as closed systems with respect to Rb and Sr after fractionation, the data will continue to lie along a straight line whose slope increases nearly linearly with time owing to decay of ^{87}Rb to ^{87}Sr. The slope of the line gives the time since fractionation, and the y intercept gives the initial composition of the fractionation products and their source at the time of fractionation [Faure and Powell, 1972]. A suite of fresh volcanic rocks which represents magmas related to a single source by fractionation should plot as a horizontal line (time equals zero, since magmatic events are short in comparison to the half-life of ^{87}Rb). An older suite should plot as a line with a slope representing the age of the suite, i.e., its age-corrected (initial) compositions should plot as a horizontal line. Yet whole rock initial $^{87}Sr/^{86}Sr$ commonly shows a positive linear correlation with $^{87}Rb/^{86}Sr$ (Figure 6) [Brooks et al 1976]. Clearly, a process other than or in addition to fractionation of magma from a single source is involved. Because fractionation changes $^{87}Rb/^{86}Sr$, this linear correlation can have meaning as an isochron only in special cases: (1) the observed igneous complex was produced by whole melting ($^{87}Rb/^{86}Sr$ is unchanged) of an older igneous complex, (2) the complex formed through bulk assimilation of crustal material by basaltic magma, and (3) a magma differentiated and then remained in its chamber in a molten state, without convection, for a period of time before the eruptive event which is the observed age of the complex.

In case 1 the isochron gives the age of the source complex, in case 2 the approximate age of the contaminant, and in case 3 the age of the differentiation event. Case 1 is chemically unreasonable, because it requires that no fractionation can occur during melting. Case 2 requires unreasonable amounts of contamination to produce silicic igneous rocks. Figure 6 and Table 3 show that the data give ages too young for cases 1 and 2 and too old for case 3 and are therefore pseudoisochrons.

In the model proposed in this paper, basaltic magma forms by

Fig. 5. (Opposite) A model for evolution of silicic igneous complexes: (a) Basalt from the upper mantle heats the crust. (b) Partial melting in the lower crust produces rhyolite liquid which gathers into diapirs and ascends. In a strongly block faulted region (top) the diapirs reach the upper crust rapidly, and little mixing occurs (e.g., Yellowstone plateau). In areas lacking extensional tectonics (bottom), mixing is thorough (e.g., Mount Baker). The case in the middle is intermediate; mixing occurs, but some unmixed rhyolite reaches the surface (e.g., Lassen Volcanic National Park). Large chambers in these centers are compositionally zoned as shown (e.g., San Juan Volcanic Field). Activity may cease with eruption or emplacement of silicic bodies if the basalt flux declines. (c) If rhyolite is not replenished due to depletion of the lower crust but basaltic activity continues, magma in the chamber becomes more mafic, and basalt vents encroach on the silicic center (e.g., Long Valley).

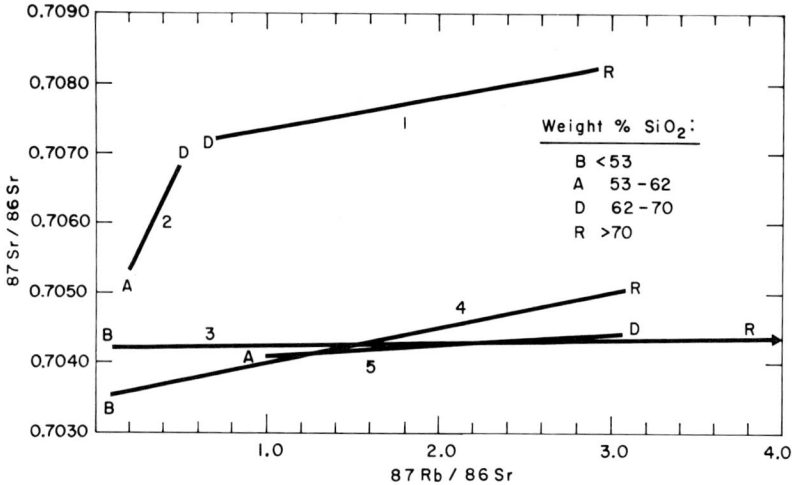

Fig. 6. Whole rock initial $^{87}Sr/^{86}Sr$ versus $^{87}Rb/^{86}Sr$ for several volcanic suites. Descriptive data are given in Table 3. Letters denote composition of samples at extremes of each data set (1-5) described in Table 3: B is basalt, A is andesite, D is dacite, and R is rhyolite. Others examples have been compiled by Brooks et al. [1976].

partial melting in the upper mantle, rhyolitic magma forms by partial melting in the lower crust, and intermediate magmas are mixtures of basaltic and rhyolitic magmas. Thus we interpret pseudoisochrons as mixing lines which contain information about the lower crust and upper mantle. A true isochron whose slope represents the time since the rhyolite source (lower crust) and basalt source (upper mantle) were last in isotopic equilibrium (i.e., two-stage model age) passes through points representing the rhyolite and the basalt sources. Data on partitioning of Rb and Sr between crystals and melts [Philpotts and Schnetzler, 1970] indicate that under most circumstances the Rb/Sr value of a partial melt will be higher than that of its source. Therefore a true isochron lies to the left (lower $^{87}Rb/^{86}Sr$) of a pseudoisochron in Figure 6. Because the $^{87}Rb/^{86}Sr$ value of basalt is very low, such a line must have approximately the same y intercept as the pseudoisochron but a steeper slope. Thus the pseudoisochron gives a lower limit for the two-stage model age of the lower crustal rhyolite source (time since the lower crust gained a higher Rb/Sr value than that of the upper mantle).

For U-Pb, linear variation of whole rock $^{207}Pb/^{204}Pb$ with $^{206}Pb/^{204}Pb$ in an igneous suite can be both a two-stage isochron and a mixing line, because fractionation does not affect either variable. It does not matter whether the data points on the line represent melts from different sources which are related to each other by some previous magmatic event or are generated by mixing of end-member parent magmas. The secondary isochron will thus give the age of the lower crust [Church, 1976b] in cases where plate tectonic motion has not moved the deep and shallow magmatic sources

TABLE 3. Descriptive Data for Whole Rock Initial $^{87}Sr/^{86}Sr$ Versus $^{87}Rb/^{86}Sr$ for Several Volcanic Suites Shown in Figure 6

Data Set	Suite	Location	Pseudoisochron 'Age,' m.y.	Analyses	Reference
1	Doggie Spring member of Superstition tuff (compositionally zoned tuff)	Arizona	33	3	Stuckless and O'Neil [1973]
2	Barroso volcanics	Peru	290	8	James et al. [1976]
3	Fantale volcano	Ethiopia	3	12	Dickinson and Gibson [1972]
4	Aden volcano, main cone series	Yemen	35	8	Carter and Norry [1976]
5	Aden volcano, Shamson caldera series	Yemen	12	5	Carter and Norry [1976]

independently. Since a knowledge of this age would closely constrain the position of a true isochron in Figure 6, the lead and strontium data together give an indication of Rb/Sr in the lower crust and the change in Rb/Sr during partial melting. The greater the difference between the apparent lead and strontium ages, the greater the difference between Rb/Sr of the rhyolitic magma and its source. This in turn places constraints on the conditions of melting in the lower crust.

At the present there appears to be no complete set of Sr and Pb isotope data on a compositionally diverse suite of rocks closely related in time and space. Lacking Pb data on the age of the previous igneous event for a two-stage model, we can assume that the age of the crystalline basement under a volcanic complex is the age of the lower crust. If the rhyolite source of the Aden volcano (Figure 6 and Table 3) separated from the basalt source 600 m.y. ago, then Rb/Sr of the lower crust in this region is between 0.06 and 0.10. By applying the data of Philpotts and Schnetzler [1970] the change to Rb/Sr = 1.0 of the rhyolitic magma could be accomplished by the order of 10% melting if the residuum is plagioclase and hornblende or by a larger amount of melting if clinopyroxene is in the residuum.

Conclusions

The model proposed here explains the origin of a variety of igneous complexes and is consistent with a large body of data, including the phase assemblage, elemental and isotopic composition, texture, and field relationships of volcanic and plutonic rocks. The model suggests that rift and subduction zone igneous activity are manifestations of similar parent magmas produced in response to a heat source and that the variety of end products is controlled by tectonics. Although this contradicts the widely accepted relationship between andesite genesis and subduction, the presence of voluminous andesite in such places as Colorado demonstrates that subduction is not a requirement. Since the model applies to island arcs such as the Lesser Antilles, it implies that continental material is being generated by partial melting of oceanic crust. Finally, by identifying the parent liquids the model provides a means of exploring their source regions.

Acknowledgments. Many of the ideas presented here evolved during discussions with our colleagues at Los Alamos, especially B. Crowe, F. Koch, U. Nitsan, and A. Rice. We thank T. Gregory, G. Heiken, T. McGetchin, D. Mann, C. Nelson, J. Rowley, and L. Smith for their help and/or encouragement. Samples from Ecuador were collected with the guidance of M. Hall, Escuela Politecnica National, Quito. This work was performed under the auspices of the Division of Physical Research, U. S. Energy Research and Development Administration.

REFERENCES

Anderson, A. T., Magma mixing: Petrological process and volcanological tool, J. Volcanol. Geotherm. Res., 1, 3-33, 1976.

Anderson, C. A., Volcanoes of the Medicine Lake Highland, California, Calif. Univ. Publ. Geol. Sci., 25, 347-422, 1941.

Bailey, R. A., G. B. Dalrymple, and M. A. Lanphere, Volcanism, structure, and geochronology of Long Valley caldera, Mono County, California, J. Geophys. Res., 81, 725-744, 1976.

Baker, I., and W. I. Ridley, Field evidence and K, Rb, Sr data bearing on the origin of the Mt. Taylor volcanic field, New Mexico, USA, Earth Planet. Sci. Lett., 10, 106-114, 1970.

Bateman, P. C., and C. Wahrhaftig, Geology of the Sierra Nevada, Geology of Northern California, Calif. Div. Mines Geol. Bull., 190, 107-172, 1966.

Blake, D. H., R. W. D. Elwell, I. L. Gibson, R. R. Skelhorn, and G. P. L. Walker, Some relationships resulting from the intimate association of acid and basic magmas, Quart. J. Geol. Soc. London, 121, 31-50, 1965.

Brooks, C., D. E. James, and S. R. Hart, Ancient lithosphere: Its role in young continental volcanism, Science, 193, 1086-1094, 1976.

Bunsen, R., Über die Prozesse der volkanischen Gesteinbildungen Islands, Ann. Phys. Chem., 83, 197-272, 1851.

Carter, S. R., and M. J. Norry, Genetic implications of Sr isotopic data from the Aden volcano, south Arabia, Earth Planet. Sci. Lett., 18, 161-166, 1976.

Chapman, R. W., Ring structures of the Pliny region, New Hampshire, Geol. Soc. Amer. Bull., 53, 1533-1568, 1942.

Christiansen, R. L., and P. W. Lipman, Cenozoic volcanism and plate-tectonic evolution of the western United States, II, Late Cenozoic, Phil. Trans. Roy. Soc. London, 271, 249-284, 1972.

Church, S. E., The Cascade Mountains revisited: A re-evaluation in light of new lead isotopic data, Earth Planet. Sci. Lett., 29, 175-188, 1976a.

Church, S. E., Lead isotopic results from rocks from the central Andes: A comparison with data from other orogenic volcanic rocks in crustal arcs, (abstract), EOS Trans. AGU, 57 (12), 1023, 1976b.

Dickinson, D. R., and I. L. Gibson, Feldspar fractionation and anomalous Sr^{87}/Sr^{86} ratios in a suite of peralkaline silicic rocks, Geol. Soc. Amer. Bull., 83, 231-239, 1972.

Didier, J., Granites and their Enclaves, 393 pp., Elsevier, Amsterdam, 1973.

Doe, B. R., and M. H. Delevaux, Variations in lead-isotopic compositions in Mesozoic granitic rocks of California: A preliminary investigation, Geol. Soc. Amer. Bull., 84, 3513-3526, 1973.

Doe, B. R., P. W. Lipman, C. E. Hedge, and H. Kurasawa, Primitive and contaminated basalts from the southern Rocky Mountains, U. S. A., Contrib. Mineral. Petrol., 21, 142-156, 1969.

Drake, M. J., Plagioclase-melt equilibria, Geochim. Cosmochim. Acta, 40, 457-465, 1976.

Eaton, G. P., R. L. Christiansen, H. M. Iyer, A. M. Pitt, D. R. Mabey, H. R. Blank, Jr., I. Zietz, and M. E. Gettings, Magma beneath Yellowstone National Park, Science, 188, 787-796, 1975.

Eichelberger, J. C., Origin of andesite and dacite: Evidence of mixing at Glass Mountain in California and at other circum-Pacific volcanoes, Geol. Soc. Amer. Bull., 86, 1381-1391, 1975.

Erikson, E. H., Jr., Petrology of the composite Snoqualmie batholith, central Cascade Mountains, Washington, Geol. Soc. Amer. Bull., 80, 2213-2236, 1969.

Faure, G., and J. L. Powell, Strontium Isotope Geology, 188 pp., Springer, New York, 1972.

Fenner, C. N., The Katmai magmatic province, J. Geol., 34, 673-772, 1926.

Fiske, R. S., C. A. Hopson, and A. C. Waters, Geology of Mt. Rainier National Park, Washington, U. S. Geol. Surv. Prof. Pap., 444, 93, 1963.

Hedge, C. E., and D. C. Noble, Isotopic studies of ash-flow tuffs, Geol. Soc. Amer. Abstr. Programs, 8 (5), 591, 1976.

Helz, R. T., Phase relations of basalts in their melting ranges at P_{H_2O} = 5Kb, II, Melt compositions, J. Petrology, 17, 139-193, 1976.

James, D. E., C. Brooks, and A. Cuyubamba, Andean Cenozoic volcanism: Magma genesis in the light of strontium isotopic composition and trace-element geochemistry, Geol. Soc. Amer. Bull., 87, 592-600, 1976.

Joplin, G. A., On the origin and occurrence of basic bodies associated with discordant bathyliths, Geol. Mag., 96, 361-373, 1959.

Kistler, R. W., and Z. E. Peterman, Variations in Sr, Rb, K, Na, and initial Sr^{87}/Sr^{86} in Mesozoic granitic rocks and intruded wall rocks in central California, Geol. Soc. Amer. Bull., 84, 3489-3512, 1973.

Lachenbruch, A. H., J. H. Sass, R. J. Munroe, and T. H. Moses, Jr., Geothermal setting and simple heat conduction models for the Long Valley caldera, J. Geophys. Res., 81, 769-784, 1976.

Larsen, E. S., J. Irving, F. A. Gronyer, and E. S. Larsen, III, Petrologic results of a study of the minerals from the Tertiary volcanic rocks of the San Juan region, Colorado, Amer. Mineral., 23, 227-257, 1938.

Lipman, P. W., Mineral and chemical variations within an ash-flow sheet from Aso caldera, southwestern Japan, Contrib. Mineral. Petrol., 16, 300-327, 1967.

Lipman, P. W., Geology of Summer Coon volcanic center, eastern San Juan Mountains, Colorado, Quart. Colo. Sch. Mines, 63, 211-236, 1968.

Lipman, P. W., R. L. Christiansen, and J. T. O'Connor, A compositionally zoned ash-flow sheet in southern Nevada, Geol. Surv. Prof. Pap., 524-F, 1-47, 1966.

Lipman, P. W., H. J. Prostka, and R. L. Christiansen, Cenozoic volcanism and plate-tectonic evolution of the western United States, I, Early and Middle Cenozoic, Phil. Trans. Roy. Soc. London, 271, 217-248, 1972.

Martin, R. F., and A. J. Piwinskii, Magmatism and tectonic settings, J. Geophys. Res., 77, 4966-4975, 1972.

Mayo, E. B., Deformation in the interval Mt. Lyell-Mt. Whitney, California, Geol. Soc. Amer. Bull., 52, 1001-1084, 1941.

McBirney, A. R., J. F. Sutter, H. R. Naslund, K. G. Sutton, and C. M. White, Episodic volcanism in the central Oregon Cascade Range, Geology, 2, 585-589, 1974.

Modell, D., Ring-dike complex of the Belknap Mountains, New Hampshire, Geol. Soc. Amer. Bull., 47, 1885-1932, 1936.

Moorbath, S., and J. D. Bell, Strontium isotope abundance studies and rubidium-strontium age determinations on Tertiary igneous rocks from the Isle of Skye, northwest Scotland, J. Petrology, 6, 37-66, 1965.

Nash, W. P., Plagioclase resorption phenomena and geobarometry in basic lavas, (abstract), EOS Trans. AGU, 54 (4), 507, 1973.

Nicholls, I. A., Petrology of Santorini volcano, Cyclades, Greece, J. Petrology, 12, 67-119, 1971.

Nicholls, J., I. S. E. Carmichael, and J. C. Stormer, Jr., Silica activity and P-total in igneous rocks, Contrib. Mineral. Petrol., 33, 1-20, 1971.

Noble, D. C. and C. E. Hedge, Sr^{87}/Sr^{86} variations within individual ash-flow sheets, U. S. Geol. Surv. Prof. Pap., 650-C, 133-139, 1969.

Pakiser, L. C., Gravity, volcanism, and crustal structure in the southern Cascade Range, California, Geol. Soc. Amer. Bull., 75, 611-620, 1964.

Philpotts, J. A., and C. C. Schnetzler, Phenocryst-matrix partition coefficients for K, Rb, Sr, and Ba, with applications to anorthosite and basalt genesis, Geochim. Cosmochim. Acta, 34, 307-322, 1970.

Piwinskii, A. J., Experimental studies of igneous rock series, central Sierra Nevada batholith, California, II, Neues Jahrb. Mineral. Monatsh., 5, 193-215, 1973.

Presnall, D. C., and P. C. Bateman, Fusion relations in the system $NaAłSi_3O_8-CaAł_2Si_2O_8-KAłSi_3O_8-SiO_2-H_2O$ and generation of granitic magmas in the Sierra Nevada batholith, Geol. Soc. Amer. Bull., 84, 3181-3202, 1973.

Rice, A., and J. C. Eichelberger, Convection in rhyolite magma, (abstract), EOS Trans. AGU, 57 (12), 1024, 1976.

Steven, T. A., and P. W. Lipman, Calderas of the San Juan volcanic field, southwestern Colorado, U. S. Geol. Surv. Prof. Pap., 958, 1-35, 1975.

Stewart, D. C., Crystal clots in calc-alkaline andesites as breakdown products of high-Ał amphiboles, Contrib. Mineral. Petrol., 53, 195 204, 1975.

Stuckless, J. S., and J. R. O'Neil, Petrogenesis of the Superstition-Superior volcanic area as inferred from strontium- and oxygen-isotope studies, Geol. Soc. Amer. Bull., 84, 1987-1998, 1973.

Tabor, R. W., and D. F. Crowder, On batholiths and volcanoes: Intrusion and eruption of late Cenozoic magmas in the Glacier Peak area, North Cascades, Washington, U. S. Geol. Surv. Prof. Pap., 604, 1-67, 1969.

Tilling, R. I., Boulder batholith, Montana: A product of two contemporaneous but chemically distinct magma series, Geol. Soc. Amer. Bull., 84, 3879-3900, 1973.

Wilcox, R. E., Rhyolite-basalt complex on Gardiner River, Yellowstone Park, Wyoming, Geol. Soc. Amer. Bull., 55, 1047-1080, 1944.

Williams, H., The dacites of Lassen Peak and vicinity, California, and their basic inclusions, Amer. Sci., 222, 385-403, 1931.

Yoder, H. S., Jr., Calcalkalic andesites: Experimental data bearing on the origin of their assumed characteristics, Proceedings of the Andesite Conference, Ore. Dept. Geol. Miner. Ind. Bull., 65, 77-89, 1969.

Younker, L. W., and T. A. Vogel, Plutonism and plate dynamics: The origin of circum-Pacific batholiths, Can. Mineral., 14, 238-244, 1976.

Zeck, H. P., An erupted migmatite from Cerro del Hoyazo, S. E. Spain, Contrib. Mineral. Petrol., 26, 226-246, 1970.

SIGNIFICANCE OF FE-MG CORDIERITE STABILITY RELATIONS ON TEMPERATURE,
PRESSURE, AND WATER PRESSURE IN CORDIERITE GRANULITES

Sang Man Lee[1] and M.J. Holdaway

Department of Geological Sciences, Southern Methodist University
Dallas, Texas 75275

Abstract. The assemblage cordierite-garnet has been proposed as a geothermometer and geobarometer applicable to pelitic granulites. Experimental work and calculated phase diagrams for the Fe-Mg reactions of cordierite-K feldspar to sillimanite-biotite and cordierite to sillimanite-garnet (Figure 3) show that the cordierite-garnet-K feldspar assemblage is restricted to a low-water environment and a narrow temperature range near 700°C. Within this field, pressure, temperature, and ratio of water pressure to total pressure may be estimated from cordierite composition in rocks containing the assemblage cordierite-garnet-K feldspar-sillimanite-biotite-plagioclase-quartz. Typical granulites containing this assemblage crystallized at temperatures between 690° and 710°C, at pressures between 4.0 and 5.7 kbar, and at a ratio of water pressure to total pressure between 0.2 and 0.5. Necessary geothermal gradients in the overlying rocks were about 40°C/km, perhaps explaining the common occurrence of cordierite-garnet granulites among Precambrian rocks and their less common occurrence in Paleozoic rocks.

Several lines of evidence suggest that rocks of the granulite facies of regional metamorphism are important constituents of the middle and lower crust: (1) Granulite facies rocks are common among the deepseated crystalline rocks of Precambrian shield areas [Turner, 1968]. (2) Experimental evidence, summarized by Ringwood [1975, pp. 23-27] indicates that at least basic granulites should occur in the middle and lower crust. (3) Granulitic rocks of both basic and pelitic compositions are common crustal inclusions in maars and breccia pipes [Lovering and White, 1964; McGetchin and Silver, 1970; Padovani and Carter, 1977].

Most granulites fall into two compositional categories: basic granulites (including hornblende-orthopyroxene, two-pyroxene, and garnet-clinopyroxene granulites) and pelitic granulites (including cordierite-garnet and garnet-sillimanite granulites). Our report deals with the conditions of formation (P_{tot}, X_{H_2O}, and T; notation list) of pelitic granulites, especially cordierite-garnet granulites. De Waard [1966] has suggested that cordierite-garnet granulites crystallize at lower

---

[1] Present address: Department of Geology, College of Natural Sciences, Seoul National Univeristy, Seoul, Korea.

pressures than garnet-sillimanite granulites. Cordierite-garnet granulites are commonly seen in the Precambrian shields and occur less often at the highest metamorphic grades in Paleozoic metamorphic belts [e.g., Hess, 1971]. In addition to cordierite and garnet these granulites contain K feldspar and quartz and may contain sillimanite, biotite, and/or sodic plagioclase (oligoclase or andesine). In general, these rocks have been derived from surface-deposited pelitic sediments (shales rich in Al, Mg, Fe, and K) metamorphosed to high temperatures ($680°-800°C$), moderate pressures (4-6 kbar), and low water pressures.

During much of the metamorphic process, water dominates the fluid phase as it is continually being produced by dehydration reactions. (The fluid phase is a supercritical fluid with a density between that of liquid and gas, which is a homogeneous mixture of fluid components (H_2O, CO_2, CH_4, etc.) and occupies pore space in the rock. In the absence of a fluid phase the rock may still contain molecular layers of the above components (along grain boundaries). Fluid phase and silicate melt are only partly miscible.) However, the high-grade granulites have long been thought of as rocks which crystallized under conditions of low P_{H_2O} [Turner, 1968; Winkler, 1974]. Low P_{H_2O} is required in pelitic granulites because the quartz-K feldspar-plagioclase combination common in such rocks would melt under conditions of $P_{H_2O} = P_{tot}$ at about $650°C$ [Kerrick, 1972], and only rocks deficient in one of these minerals or water would remain. Low P_{H_2O} could result from the absence of a fluid phase in the rock or from a fluid phase rich in other components. The main sources of other fluid components are graphite, which reacts with water to produce CO_2 and CH_4, impure marbles, which release CO_2 during metamorphism, and possibly CO_2 from the upper mantle.

The first reduction of water content in the fluid during progressive metamorphism of pelitic rocks probably occurs as granitic melt begins to form by the decomposition-melting reaction of muscovite-quartz-plagioclase along a line from $650°C$, 4 kbar to $670°C$, 7 kbar [Kerrick, 1972]. As granitic melt forms, water is concentrated in the melt, while CO_2 and probably CH_4 concentrate in the fluid remaining behind in the solid portions of the rocks [Mysen et al., 1976]. Thus a fluid phase rich in CO_2 and CH_4 and relatively low in H_2O may persist to higher temperatures in pelitic rocks. The presence of such fluids has been demonstrated by fluid inclusion studies in basic granulites [Touret, 1971]. Harris [1976] has cited evidence that at least some cordierite-garnet granulites are the solid remnants (restites) from an episode of granite melting. Any remaining fluid in such rocks should be low in H_2O. In this paper we will assume the existence of a fluid phase during granulite facies metamorphism and designate its water content as $X_{H_2O} = P_{H_2O}/P_{tot}$. If for some rocks no fluid phase existed during metamorphism, X_{H_2O} is still a valid expression as defined above, but the sum of the various partial pressures was less than P_{tot} (notation list).

Cordierite-Garnet Geothermometers and Geobarometers

Previous Results

In recent years a number of schemes have been devised to estimate T and P_{tot} in cordierite-garnet rocks. The cordierite-garnet geothermometer requires only a chemical analysis of the two minerals in the rock

TABLE 1. Minerals Used in This Paper

Mineral	Abbreviation	Formula	Characteristics and Significance
Cordierite	Cd	$(Fe, Mg)_2Al_4Si_5O_{18} \cdot nH_2O$*	Low pressure, pelitic, metamorphic
Garnet	Ga	$(Fe, Mg)_3Al_2Si_3O_{12}$	Solid solution end-members: almandine Fe (Alm), pyrope Mg (Py), spessartite Mn, and grossularite Ca.
Biotite	Bio	$K_{0.84}(Fe, Mg)_{2.52}Al_{1.72}Si_{2.74}O_{10}(OH)_2$+	Dark mica
Muscovite	Mus	$KAl_3Si_3O_{10}(OH)_2$	White mica
K feldspar	Ksp	$KAlSi_3O_8$	Orthoclase, perhaps having albite solid solution ($NaAlSi_3O_8$)
Oligoclase-Andesine	An_{30}	$Na_{0.7}Ca_{0.3}Al_{1.3}Si_{2.7}O_8$	Plagioclase solid solution in pelitic granulite
Sillimanite	Sil	Al_2SiO_5	High temperature, pelitic, metamorphic
Al silicate	Als	Al_2SiO_5	Sillimanite, kyanite, or andalusite
Quartz	Q	SiO_2	
Hercynite	Hc	$FeAl_2O_4$	Member of spinel group
Water	V	H_2O	Main fluid component

*Water content is given by Schreyer and Yoder [1964] as a function of P_{H_2O} and T.
+From Guidotti et al. [1975].

and is independent of presence or absence of other minerals. An exchange reaction may be written between the end-members of these two solid solution series:

Mg cordierite + Fe garnet $\rightleftharpoons$ Fe cordierite + Mg garnet (1)

(All mineral formulae are given in Table 1). The equilibrium constant for an exchange reaction, if ideal solid solution is assumed, is the distribution coefficient K_D:

$$K_D = \frac{X_{Fe}^{Cd} X_{Mg}^{Ga}}{X_{Mg}^{Cd} X_{Fe}^{Ga}} \qquad (2)$$

For this exchange reaction, ΔV is small [Holdaway and Lee, 1977] so K_D is mainly a function of temperature. In short, K_D measures how similar the garnet and cordierite are in Fe/Mg ratio, and this ratio in turn is a function of temperature.

A cordierite-garnet geobarometer depends on the coexistence of these minerals with sillimanite and quartz. The reaction

3 Fe cordierite $\rightleftharpoons$ 2 Fe garnet + 4 sillimanite + 5 quartz + $3nH_2O$ (3)

has been determined experimentally by Richardson [1968] and by Holdaway and Lee [1977]. The stability curve is highly pressure dependent, and the products appear on the high-pressure side (Figure 1). Addition of known amounts of Mg to the cordierite raises the equilibrium pressure and concentrates Mg in the cordierite in relation to the garnet. Thus in theory, knowledge of cordierite and garnet composition in a cordierite-garnet-sillimanite-quartz rock allows unique determination of both pressure and temperature.

Currie [1971] has experimentally calibrated the cordierite-garnet geothermometer and geobarometer at $P_{H_2O} = P_{tot}$. His results suggest that K_D, as defined here, decreases with increasing temperature. His data, applied to a typical granulite terrain in Ontario, give T = $600°$-$730°$C at P_{tot} = 5.3-6.3 kbar. We believe that the $600°$C value is too low as we will discuss subsequently.

Hutcheon et al., [1974] used a theoretical approach to estimate the metamorphic conditions for the Daly Bay complex northwest of Hudson Bay as being T = $610°$-$770°$C and P_{tot} = 5.3-6.5 kbar.

Experimental results of Hensen and Green [1973] in a dry system give pressures about 1.5 kbar higher than those of Currie. Newton's [1972] experiments on Mg cordierite in a wet and a dry system suggest that in a dry system, cordierite breaks down at lower pressures than in a wet system. Newton's results on the wet system are consistent with the extrapolation of Currie's [1971] pressures to pure Mg cordierite.

Hensen and Green's [1973] K_D values increase with temperature. Thompson [1976] has plotted calibrated natural occurrences and finds that the K_D values agree with those of Hensen and Green. Holdaway and Lee [1977] using a different approach on natural occurrences, have made slight adjustments to the Thompson curve but are still in agreement with the Hensen and Green K_D values.

In summary, it is possible that the experimental results of Currie [1971] are best for pressure because they agree with Newton's results, while at the same time, the K_D values of Hensen and Green [1973] are best in that they appear to agree with natural observations. This apparent contradiction is possible because K_D values depend on the system's reaching total compositional equilibrium, which is more likely to be achieved at the higher temperatures of Hensen and Green's experiments. On the other hand, the piston-cylinder apparatus used by Hensen and Green is more likely to have pressure calibration errors than the gas pressure apparatus used by Currie [1971] (see also Thompson [1976] and Holdaway and Lee [1977]).

Present Approach

Most of these studies on cordierite-garnet have failed to take into account two important observations: (1) X_{H_2O} has an effect on cordierite decomposition reactions because cordierite is actually a hydrate [Newton, 1972; Schreyer and Yoder, 1964], (2) With decreasing temperature the assemblage cordierite-garnet-K feldspar typical of many pelitic granulites reacts to sillimanite-biotite-quartz. The temperature of this reaction at $X_{H_2O} = 0.4$ for average cordierite compositions is about 700°C [Holdaway and Lee, 1977], strongly suggesting that previously derived temperatures near 600°C are incorrect for cordierite-garnet granulites.

We have used a different approach to try to determine P_{tot}, X_{H_2O}, and T of granulites which contain the assemblage cordierite-garnet-sillimanite-biotite-K feldspar-oligoclase (or andesine)-quartz. This assemblage is unusually common among cordierite-garnet granulites and will be called the seven-phase cordierite-garnet assemblage (the assemblage without plagioclase is the six-phase assemblage). With knowledge of crystallization conditions of the seven-phase assemblage, something can then be said regarding metamorphic conditions of common subassemblages (garnet-absent, biotite-absent, and plagioclase-absent).

Experimental and Calculated Results

Our experimental procedure and methods of calculation are described elsewhere [Holdaway and Lee, 1977], and only a brief summary will be provided here. Experiments were conducted by using synthesized minerals, $X_{H_2O} = 1$, and a quartz-fayalite-magnetite oxygen buffer to maintain Fe in the ferrous state and duplicate natural conditions.

Stability relations for two Fe end-member reactions were determined (Figure 1). At lower temperatures in the presence of K feldspar, Fe cordierite breaks down with increasing pressure according to the reaction

$$1.26 \text{ Fe cordierite} + 0.84 \text{ K feldspar} + (1-1.26n)H_2O \rightleftharpoons$$

$$1 \text{ Fe biotite} + 2.08 \text{ sillimanite} + 4 \text{ quartz} \qquad (4)$$

At higher temperatures, Fe cordierite breaks down according to the reaction in (3) instead. The phases involved in (4) begin to melt at 715°C, but if we ignore melting, the reactions in (3) and (4) meet at a

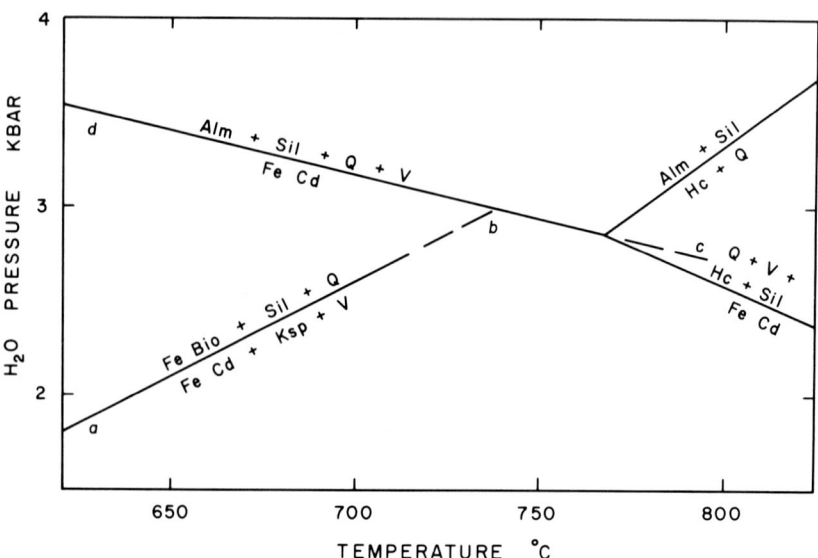

Fig. 1. Experimentally determined curves for the Fe end-member reactions in (3) and (4). Abbreviations are as given in Table 1. Metastable segments are dashed. Calculations for Figures 2-4 are based on segments ab and bc. In the presence of K feldspar, segment db is metastable. Addition of Mg diminishes the field of hercynite-sillimanite-quartz. The figure is based on experimental work of Holdaway and Lee [1977].

hypothetical invariant point at 740°C, 3 kbar (Figure 1). For calculation of Fe-Mg equilibria, the hercynite-sillimanite field of Figure 1 may be ignored, since Mg greatly reduces the hercynite stability field.

Fe-Mg P-X loops were determined for the reaction in (4) at 675°C (Figure 2) and for that in (3) up to 4 kbar [Holdaway and Lee, 1977, Figure 5] to show the effect of Mg on each reaction. In much the same way as occurs in the melting diagram of a solid solution mineral like plagioclase, the lower curve of Figure 2 gives the cordierite composition, enriched in Mg, and the upper curve gives the biotite composition, which may coexist with cordierite, sillimanite, and quartz. At 675°C the pressure may be uniquely determined if the cordierite composition is known and the other minerals are present. The garnet-cordierite behavior is similar, but the pressure effect is greater for a given change in cordierite composition.

From these experimental results, P_{tot}-T diagrams were calculated for various values of X_{H_2O} (Figure 3). Calculations were carried out isothermally in a two-step process [Holdaway and Lee, 1977]. The first step involved calculation of a complete P_{tot}-T diagram for $X_{H_2O} = 1$, and the second step involved calculation of diagrams for other values of X_{H_2O}. During the second step, corrections were made for the average composition of K feldspar (20 mol % albite) and garnet impurities (grossularite + spessartite = 7 mol %) in cordierite-garnet granulites.

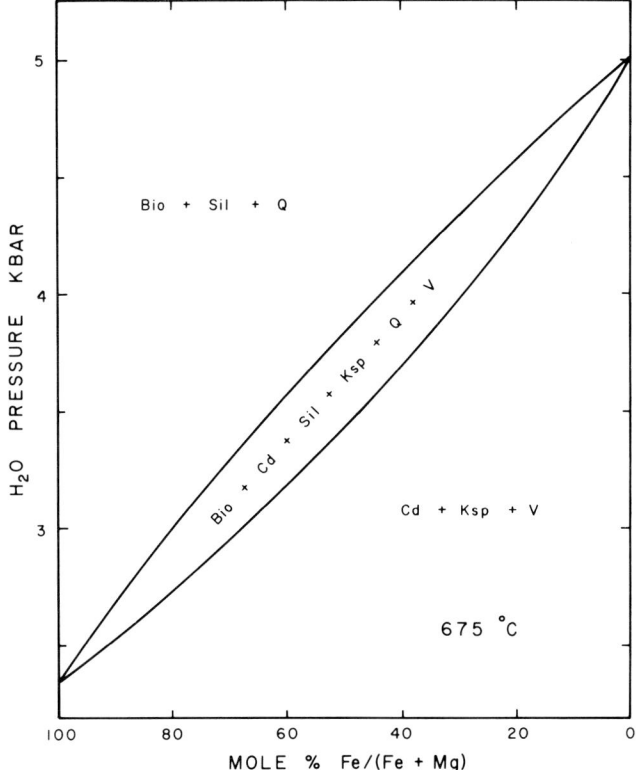

Fig. 2. P-X diagram for the Fe-Mg reaction in (4) at 675°C and X_{H_2O} = 1. Abbreviations are as given in Table 1. The lower curve gives the maximum Fe/(Fe + Mg) ratio of cordierite for any given pressure, and the upper curve gives the minimum Fe/(Fe + Mg) ratio of biotite. The two curves fix the composition of biotite and cordierite in the assemblage biotite-cordierite-sillimanite-K feldspar-quartz at any given pressure and 675°C. The lower curve is based on experimental results, and the upper curve is based mainly on calculations assuming a K_D of 0.53 for cordierite-biotite [Holdaway and Lee, 1977].

The general form of the equation for the calculations is

$$\Delta P_{tot} \Delta V_s + \Delta P_{H_2O} \Delta V_{H_2O} = -RT \ln K \qquad (5)$$

where volume change of the reaction in (3) or (4) is split into ΔV of solids and ΔV of water, K is the equilibrium constant which corrects for changes in mineral compositions between the initial and the final point of the calculation, R is the gas constant (notation list), and T is in degrees Kelvin. In addition to the experimental results, molar volumes of the minerals, water content of cordierite, and K_D for cordierite-almandine and for cordierite-biotite as a function of temperature must

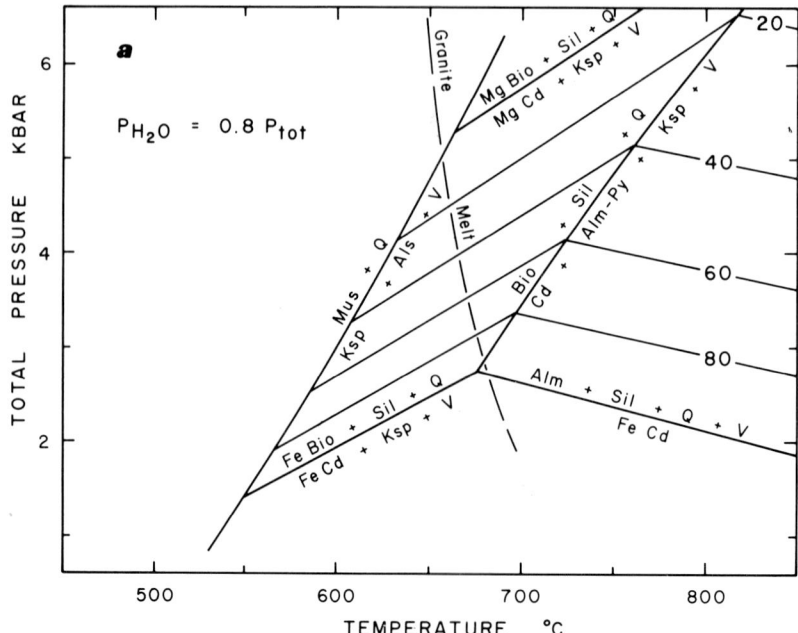

Fig. 3. Calculated P-T curves for the Fe-Mg reactions in (3) and (4) and the reaction producing cordierite, garnet, and K feldspar, for (a) $X_{H_2O} = 0.8$, (b) $X_{H_2O} = 0.6$, and (c) $X_{H_2O} = 0.4$. Garnet has 7 mol % spessartite plus grossularite, and K feldspar has 20 mol % albite. The numbers refer to the percent of the Fe/(Fe + Mg) ratio in cordierite stable with breakdown products (see also Figure 2). Abbreviations are as given in Table 1. Cordierite is possible throughout the contoured area depending on the rock composition. Without the breakdown products, cordierite may be more Mg-rich than the indicated compositions (Figure 2), but with breakdown products present the cordierite Fe/(Fe + Mg) ratio is a function only of X_{H_2O}, P_{tot}, and T. The granite melt curves from Kerrick [1972] and the muscovite curves from Chatterjee and Johannes [1974] were modified by using data of Kerrick [1972] and Thompson [1974].

be known. The total effect of the calculations is to convert a point on the equilibrium curve for pure Fe cordierite at $X_{H_2O} = 1$ to a point on an equilibrium curve for another cordierite composition at a new value of X_{H_2O}, all at one given temperature.

From the diagrams (Figure 3) it can be seen that at low temperatures, cordierite breaks down first by reaction with K feldspar. The Fe/(Fe + Mg) ratio of cordierite with biotite, sillimanite, K feldspar, and quartz is a function mainly of pressure and is indicated in percent on the diagram. At higher temperatures, cordierite breaks down alone according to the reaction in (3). The intersections of lines of constant cordierite composition in Figure 3 for the two reactions determine the following reaction involving the six-phase cordierite-garnet assemblage:

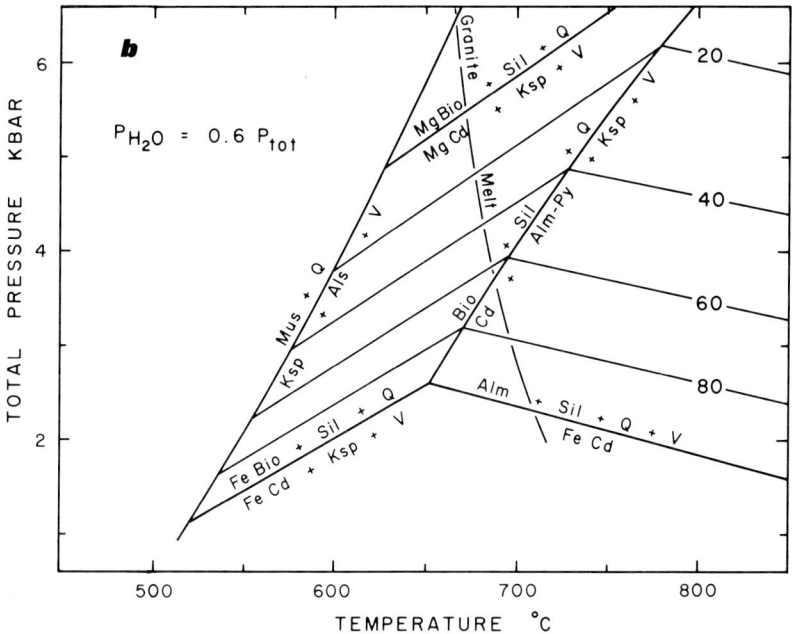

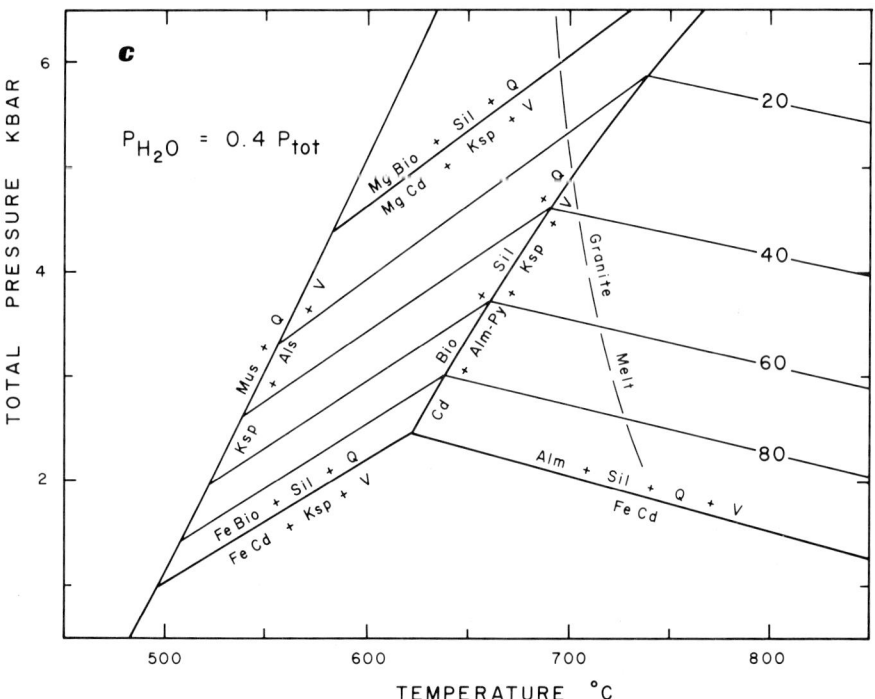

Fig. 3. (continued)

$$\text{biotite + sillimanite + quartz} \rightleftharpoons \text{cordierite +}$$

$$\text{almandine + K feldspar + H}_2\text{O} \qquad (6)$$

At any known value of X_{H_2O} the cordierite composition present in the six-phase cordierite-garnet assemblage uniquely determines both P_{tot} and T.

On each of the X_{H_2O} diagrams (Figure 3) the breakdown reaction of muscovite with quartz and the melting reaction of granite is shown. These reactions do not involve Fe or Mg, so cordierite, biotite, and almandine are not directly involved in them. The muscovite and granite melt curves do affect the other minerals of the reactions in (3) and (4). That in (4), for example, must change to a new reaction involving muscovite instead of K feldspar and having a positive P-T slope [Holdaway and Lee, 1977].

Application to Nature

General Considerations

As we noted above, in a rock containing the six-phase cordierite-garnet assemblage a measurement of the Fe content of cordierite allows the determination of P_{tot} and T provided that X_{H_2O} can be independently estimated. If plagioclase is also present in the rock, we can set limits on X_{H_2O} because the quartz-plagioclase-K feldspar combination melts at the granite melting curve. If significant melting has occurred, it can normally be recognized texturally or by the removal of the melted minerals from the rock as the melt migrates toward the surface. In order not to have melted, the rock must have crystallized on or below the granite melting curve for the X_{H_2O} which existed at the time. Thus for rocks containing the seven-phase cordierite-garnet assemblage, crystallization must have occurred along the six-phase line (reaction in (6) and Figure 3) but on or below the granite melt curve.

Figure 4 shows the six-phase cordierite-garnet curves and the granite melting curves for various X_{H_2O} taken from Figure 3. Also shown are the cordierite compositions stable with the six- or seven-phase assemblage. The heavy dashed line, determined by the intersections of respective six-phase curves and granite melting curves, is a biotite-sillimanite-K feldspar-quartz-plagioclase melting curve. If the P_{H_2O} is not externally controlled by the environment but is allowed to drop as small amounts of granite melt form and absorb water, cordierite, almandine, K feldspar, biotite, sillimanite, plagioclase, and quartz may all coexist with the melt. A measurement of cordierite composition uniquely determines P_{tot}, T, and X_{H_2O}.

In nature it may be difficult to recognize the former presence of small amounts of melt. However, the above condition of seven phases plus minor melt is probably quite common. This is because of the buffering effect of the assemblage on X_{H_2O}. As the reaction in (6) proceeds, it gives off water and increases X_{H_2O}, a result that means that successively higher temperatures are required to produce more cordierite and garnet (compare the six-phase curves for X_{H_2O} = 0.4, 0.6, and 0.8 in Figure 4). As X_{H_2O} increases, the granite melt curve drops in temperature. Eventually, X_{H_2O} and T have increased to the point where the water produced by sillimanite-biotite breakdown goes directly into granite melt; X_{H_2O} is thus prevented from increasing any further. Thus the intersection of the cordierite composition lines with the heavy dashed line determines P_{tot}, T, and X_{H_2O} for rocks containing the seven-phase cordierite-garnet assemblage and minor melt. If the absence of granite melt can be established, the estimated T, P_{tot}, and X_{H_2O} are all upper limits, as can be seen from Figure 4.

Examples

We shall now consider some examples from high-grade pelitic rocks. We assume conditions of incipient granite melting.

Three analyzed specimens of the Daly Bay complex [Hutcheon et al., 1974] contain the seven-phase assemblage with cordierites between Fe_{19} and Fe_{23}. These rocks crystallized at about P_{tot} = 5.5 kbar, T = 710°C, and X_{H_2O} = 0.3 (Table 2 and Figure 4). Of two additional rocks from the area, both are missing biotite, and one is missing plagioclase. These rocks crystallized above the heavy dashed curve (Figure 4) probably at higher temperatures and lower X_{H_2O}.

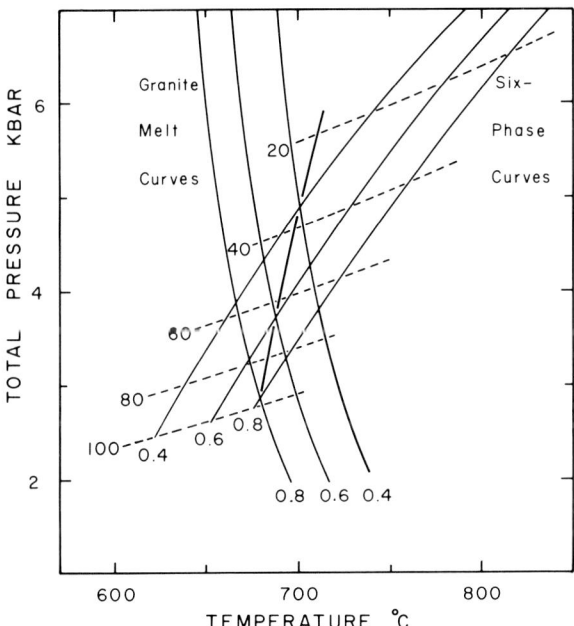

Fig. 4. Synopsis of six-phase cordierite-garnet curves for the reaction in (6) and granite melt curves from Figure 3. The heavy dashed line shows vapor-absent melting of sillimanite-biotite-K feldspar-plagioclase-quartz, and the light dashed lines show the percent of the Fe/(Fe + Mg) ratio in cordierite stable with the six-phase assemblage at various X_{H_2O}. X_{H_2O} for each solid line is given at the base of the line. During progressive metamorphism a rock containing the seven-phase cordierite-garnet assemblage crystallized near the intersection of its respective cordierite composition line and the heavy dashed line. Table 2 gives examples.

TABLE 2. Metamorphic Conditions in Localities Containing the Seven-Phase Cordierite-Garnet Assemblage of High-Grade Pelitic Granulites

Cordierite Fe	P_{tot}, kbar	T, °C	X_{H_2O}
Rocks From Daly Bay, N. W. T., Hutcheon et al. [1974]			
19	5.7	711	0.29
22	5.6	709	0.30
23	5.5	708	0.31
Rocks From Opinicon Lake, Ontario, Currie [1971]			
20	5.7	710	0.29
23	5.5	708	0.31
30	5.2	704	0.36
37	4.8	699	0.42
43	4.5	696	0.46
Rocks From Steinach, Bavaria, Okrusch [1971]			
50	4.3	694	0.50
52	4.2	693	0.52
54	4.1	692	0.53
54	4.1	692	0.53
54	4.1	692	0.53

Rocks of the Opinicon Lake region, Ontario, studied by Currie [1971] also contain the seven-phase assemblage with cordierite composition from Fe_{20} to Fe_{43}. The conditions of metamorphism average about P_{tot} = 5 kbar, T = 700°C, and X_{H_2O} = 0.35. Rocks to the southeast studied by Reinhardt [1968] are entirely above the sillimanite-biotite reaction and presumably crystallized at higher temperatures and lower X_{H_2O}.

From the Steinach contact metamorphic aureole in Bavaria, Okrusch [1971] has analyzed cordierite from rocks containing the seven-phase assemblage. In the highest-grade rocks, cordierite compositions are Fe_{50}-Fe_{54}, and the deduced conditions are P_{tot} = 4 kbar, T = 690°C, and X_{H_2O} = 0.5. The same assemblage exists at somewhat lower grades near the point where muscovite disappears, but the garnet in these rocks contains 14-42 mol % spessartite and grossularite, which stabilizes the garnet-cordierite-K feldspar assemblage to lower temperatures. The present calculated results are only directly applicable to garnet containing 7 ± 2 mol % impurities in addition to the almandine and pyrope components.

Padovani and Carter [1977] have studied crustal inclusions from Killbourne Hole Maar, New Mexico. Pelitic granulites of deep crustal origin here do not contain cordierite or biotite but do contain garnet, sillimanite, K feldspar, plagioclase, and quartz. Feldspar geothermometry indicates crystallization temperatures of about 750°C. In the garnets, the Fe/(Fe + Mg) ratio ranges from 55 to 73 mol %. K_D values [Holdaway and Lee, 1977] indicate that if cordierites were present, they would be Fe_{20}-Fe_{34}. The fact that none of the inclusions contain any cordierite suggests that all the inclusions crystallized at pressures above the line for cordierite Fe_{20}, although some of the more Fe-rich garnets could have fortuitously crystallized at somewhat lower pressures. Extrapolation of the lines for cordierite Fe_{20} in Figures 3b and 3c and the melting curves in Figure 4 indicate that at 750°C, P_{tot} was above 5.4 kbar, and X_{H_2O} was below 0.2. This is consistent with the total absence of hydrous minerals in these rocks.

Stoeser [1973] has found cordierite-hypersthene granulites as xenoliths from one volcanic center in the San Francisco volcanic field in the Colorado Plateau in Arizona. Composition of the cordierite was not determined. Correlation of laboratory seismic velocities of rocks with seismic velocities in the area suggests to Stoeser that these xenoliths along with other granulitic xenoliths come from the upper part of the crust above a seismic discontinuity which occurs at about 20-25 km. Such depths are consistent with the present experimental results.

Conclusions

1. Whereas the cordierite-garnet pair has been proposed as a geothermometer and geobarometer crystallizing over a wide temperature range, our experimental work shows that many cordierite-garnet rocks cyrstallized over a narrow temperature range near 700°C. On the low temperature side, garnet, cordierite, and K feldspar are replaced by biotite and sillimanite, while on the high-temperature side, granite melting occurs. Exceptions are (1) low-K and/or low-H_2O rocks which would crystallize over a wider range, (2) plagioclase-free rocks which would persist to higher temperatures without melting, and (3) rocks in which garnet is stabilized to lower temperatures and pressures by high spessartite (Mn) or grossularite (Ca) content. All cordierite-garnet geothermometry is subject to large errors, and some of the K_D versus T calibrations are incorrect.

2. Typical cordierite-garnet granulites containing the seven-phase assemblage have cordierite Fe_{20}-Fe_{50} and crystallized at temperatures between 690° and 710°C at pressures between 4.0 and 5.7 kbar and at X_{H_2O} of 0.2-0.5. Garnet-sillimanite granulites crystallized at higher pressures.

3. The common occurrence of assemblages on the sillimanite-biotite reaction results in part from the added variance of the reaction produced by garnet and K feldspar impurities and in part from the buffering effect of this reaction and the granite melting reaction on water. For a given depth (P_{tot}) and temperature the reaction buffers X_{H_2O}. As temperature increases, more reaction takes place, and X_{H_2O} increases, now being buffered at a higher level (Figure 4). Eventually, X_{H_2O} is high enough to promote granite melting, which sets an upper limit to X_{H_2O}. The melting of part of the rock requires heat and tends to buffer the temperature. If sufficient heat is added from below, the sillimanite-biotite reaction may go to completion, and temperature may continue to climb. Melting decreases or stops as X_{H_2O} drops, as plagioclase becomes more calcic, or as plagioclase, quartz, or K feldspar disappears. Thus rocks which contain cordierite-almandine-K feldspar-plagioclase-quartz and either sillimanite or biotite may well be restites left over from granite melting. This is especially likely if quartz, K feldspar, or plagioclase is depleted or missing [Harris, 1976]. Granitic melts produced in such a way are Al-rich [Green, 1976] and crystallize cordierite and perhaps garnet when they are intruded at shallower levels. Such granites have been described by Flood and Shaw [1975] from New South Wales.

4. If we assume that 1 kbar of pressure corresponds to 3.5 km of

depth, the average geothermal gradient in rocks above cordierite-garnet granulites at the time of metamorphism was 40°C/km. Garnet-sillimanite granulites probably experienced somewhat lower gradients. The cordierite-garnet granulites crystallized at depths of 16-20 km, while garnet-sillimanite granulites required greater depths. The apparent abundance of cordierite-garnet granulites in Precambrian rocks may relate to higher geothermal gradients in such rocks. However, common occurrence of such rocks does not require average geothermal gradients of 40°C/km in Precambrian rocks. The granulites may still result from anomalous gradients which would be expected from the rise of granitic magmas formed at greater depths. The problem needs further exploration.

5. Present-day temperatures in the lower crust may often be much lower than the temperatures under which lower crustal rocks crystallized. In stable areas, average geothermal gradients of 15°C/km may be expected in comparison to the 25°-40°C/km required to produce granulites. Some granulites, produced at 700°C or more, exist now at 300°-450°C and are probably not at equilibrium. The high-temperature assemblages may have been preserved over long time periods by (1) consumption of fluid components by minor rehydration, (2) decreasing reaction rates at lower temperatures, and (3) lack of penetrative deformation [Turner, 1968]. In all crustal rocks it is important to distinguish between conditions of origin and present conditions.

Notation

For further information concerning symbols and terms the reader is referred to any standard physical chemistry or petrology text.

P_f fluid pressure in kilobars.

P_{H_2O}, P_{CO_2}, P_{CH_4} partial pressures of fluid components. If a fluid phase exists in the rock, $P_{H_2O} + P_{CO_2} + P_{CH_4} + \ldots = P_f$.

P_{tot} total rock or overburden pressure. For average rock densities, 1 kbar pressure is produced for every 3½ km of depth. If the rock contains a fluid phase, $P_{tot} \approx P_f$ during metamorphism [Turner, 1968]. If no fluid phase exists, $P_{tot} > P_{H_2O} + P_{CO_2} + P_{CH_4} + \ldots$.

X_{H_2O} P_{H_2O}/P_{tot}.

X_{Fe}^{Cd} mole fraction Fe/(Fe + Mg) in cordierite.

Fe_{30} designation for composition of a solid solution mineral, Fe/(Fe + Mg) = 30%.

K_D distribution coefficient, equilibrium constant for an exchange reaction, which expresses the partitioning of two solid solution elements between two minerals.

K equilibrium constant for a reaction.

ΔV_s volume change of solids for a reaction.

ΔV_{H_2O} volume change of water for a reaction.

R gas constant.

T temperature in degrees Kelvin or Centigrade.

Acknowledgment. This project was supported by National Science Foundation grant GA-35644.

References

Chatterjee, N. D., and W. Johannes, Thermal stability and standard thermodynamic properties of synthetic $2M_1$-muscovite $KAl_2(AlSi_3O_{10}(OH)_2)$, Contrib. Mineral. Petrol., 48, 89-114, 1974.

Currie, K. L., The reaction 3 cordierite = 2 garnet + 4 sillimanite + 5 quartz as a geological thermometer in the Opinicon Lake region, Ontario, Contrib. Mineral. Petrol., 33, 215-226, 1971.

De Waard, D., The biotite-cordierite-almandine subfacies of the hornblende-granulite facies, Can. Mineral., 8, 481-492, 1966.

Flood, R. H., and S. E. Shaw, A cordierite-bearing granite suite from the New England batholith, N. S. W., Australia, Contrib. Mineral. Petrol., 52, 157-164, 1975.

Green, T. H., Experimental generation of cordierite- or garnet-bearing granitic liquids from a pelitic composition, Geology, 4, 85-88, 1976.

Guidotti, C. V., J. F. Cheney, and P. D. Conatore, Interrelationship between Mg/Fe ratio and octahedral Al content in biotite, Amer. Mineral. 60, 849-853, 1975.

Harris, N. B. W., The significance of garnet and cordierite from the Sioux Lookout region, English River gneiss belt, northern Ontario, Contrib. Mineral. Petrol., 55, 91-104, 1976.

Hensen, B. J., and D. H. Green, Experimental study of the stability of cordierite and garnet in pelitic compositions at high pressures and temperatures, III, Synthesis of experimental data and geological applications, Contrib. Mineral. Petrol., 38, 151-166, 1973.

Hess, P. C., Prograde and retrograde equilibria in garnet-cordierite gneisses in south-central Massachusetts, Contrib. Mineral. Petrol., 30, 177-195, 1971.

Holdaway, M. J., and S. M. Lee, Fe-Mg cordierite stability in high-grade pelitic rocks based on experimental, theoretical, and natural observations, Contrib. Mineral. Petrol., in press, 1977.

Hutcheon, I., F. Froese, and T. M. Gordon, The assemblage quartz-sillimanite-garnet-cordierite as an indicator of metamorphic conditions in the Daly Bay complex, N. W. T., Contrib. Mineral. Petrol., 44, 29-34, 1974.

Kerrick, D. M., Experimental determination of muscovite + quartz stability with $P_{H_2O} < P_{tot}$, Amer. J. Sci., 272, 946-958, 1972.

Lovering, J. F., and J. R. White, The significance of primary scapolite in granulitic inclusions from deep-seated pipes, J. Petrology, 5, 195-218, 1964.

McGetchin, T. R., and L. T. Silver, Compositional relations in minerals from kimberlite and related rocks in the Moses Rock dike, San Juan county, Utah, Amer. Mineral., 55, 1738-1771, 1970.

Mysen, B. O., D. H. Eggler, M. G. Seitz, and J. R. Holloway, Carbon dixide in silicate melts and crystals, I. Solubility measurements, Amer. J. Sci., 276, 455-479, 1976.

Newton, R. C., An experimental determination of the high-pressure stability limits of magnesian cordierite under wet and dry conditions, J. Geol., 80, 398-420, 1972.

Okrusch, M., Garnet-cordierite-biotite equilibria in the Steinach aureole, Bavaria, Contrib. Mineral. Petrol., 32, 1-23, 1971.

Padovani, E., and J. L. Carter, Aspects of the deep crustal evolution beneath south-central New Mexico, in *The Earth's Crust*, Geophys.

Monogr. Ser., vol. 20, edited by J. G. Heacock, AGU, Washington, D.C., this volume, 1977.

Reinhardt, E. W., Phase relations in cordierite-bearing gneisses from Gananoque area, Ontario, Can. J. Earth Sci., 5, 455-482, 1968.

Richardson, S. W., Staurolite stability in a part of the system Fe-Al-Si-O-H, J. Petrology, 9, 467-488, 1968.

Ringwood, A. E., Composition and Petrology of the Earth's Mantle, McGraw-Hill, New York, 1975.

Schreyer, W., and H. S. Yoder, Jr., The system Mg-cordierite-H_2O and related rocks, Neues Jahrb. Mineral. Abh., 101, 271-342, 1964.

Stoeser, D. B., Mafic and ultramafic xenoliths of cumulus origin, San Francisco volcanic field, Arizona, Ph.D dissertation, Univ. of Oreg., Eugene, 1973.

Thompson, A. B., Calculation of muscovite-paragonite-alkali feldspar phase relations, Contrib. Mineral. Petrol., 44, 173-194, 1974.

Thompson, A. B., Mineral reactions in pelitic rocks, II, Calculation of some P-T-X (Fe-Mg) phase relations, Amer. J. Sci., 276, 285-308, 1976.

Touret, J., Le facies granulite en Norvège méridionale, II, Les inclusions fluides, Lithos, 4, 423-436, 1971.

Turner, F. J., Metamorphic Petrology, McGraw-Hill, New York, 1968.

Winkler, H. G. F., Petrogenesis of metamorphic rocks, 3rd ed., Springer-Verlag, New York, 1974.

MICROCRACKS IN CRUSTAL IGNEOUS ROCKS: PHYSICAL PROPERTIES

Michael Feves, Gene Simmons, and
Robert W. Siegfried[1]

Department of Earth and Planetary Sciences
Massachusetts Institute of Technology
Cambridge, Massachusetts 02139

Abstract. Microcracks in crustal igneous rocks in situ and in the laboratory control such physical properties as velocity of compressional waves V_p, velocity of shear waves V_s, compressibility β, and electrical resistivity ρ. In situ electrical resistivity as a function of V_p, V_s, and depth is determined from laboratory measurements of crack porosity, V_p, and V_s as functions of pressure. Our data on suites of rocks from central Wisconsin and southeast Missouri show that (1) electrical resistivity in these Precambrian basement rocks attains values of at least 10^6 Ωm at depths of only 2-3 km and (2) elastic properties of these rocks (V_p, V_s, and β) approach intrinsic values at depths of 2-3 km. We suggest that field measurements of shear velocity can be used to determine resistivity profiles in situ.

Introduction

Natural microcracks in igneous and metamorphic rocks are extremely common. They control the values of many physical properties of rock in situ to depths of at least 5-15 km. In this paper we report new data that relate certain microcrack characteristics to various physical properties and therefore provide an improved basis for using laboratory data to estimate properties of rock in situ. We also provide empirical relations between pairs of various physical properties which give an improved basis for interpreting field data on one property in terms of another property (e.g., measured shear or compressional velocities can yield electrical resistivity).

Physical properties are affected significantly by microcracks in rocks, even if the crack porosity is only 0.1%. At pressure below 2 kbars, corresponding to depths in the earth's crust of less than 6-8 km, the effects of microcracks are dramatic. Compressibility [Birch and Bancroft, 1938; Brace, 1965; Todd et al., 1972; Simmons et al., 1974], electrical resistivity [Brace and Orange, 1968; Parkhomenko, 1967], the velocity of both compressional and shear waves [Birch, 1960, 1961; Hughes and Cross, 1951; Simmons, 1964a, b; Nur and Simmons, 1969a, b], thermal conductivity [Walsh and Decker,

---
[1]Present address: Corning Glass Works, Corning, New York 14830

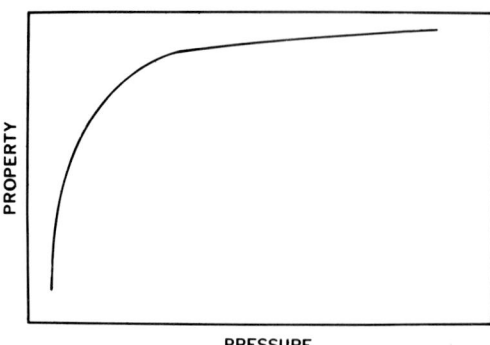

Fig. 1. Variation of physical properties with pressure (schematic).

1966], and so on all exhibit the behavior with pressure that is shown schematically in Figure 1. Theoretical discussions of the effects of cracks on various physical properties have been given in a series of papers by Walsh and co-workers [Walsh, 1965, 1973; Walsh and Brace, 1966; Walsh and Decker, 1966] and, more recently, by Budiansky and O'Connell [Budiansky and O'Connell, 1976; O'Connell and Budiansky, 1974] and by Warren and Nashner [1976].

A knowledge of microcracks and of their relationship to physical properties has great practical and scientific importance. For example, consider the following:

1. The determination of suitable underground sites for radioactive waste disposal depends on the ability to predict physical properties at depth. In particular, the presence of microcracks is undesirable at disposal sites because of dispersal of radioactive material by groundwater circulation through open microcracks.

2. Electromagnetic wave propagation above the ground is affected by the electrical properties of the subsurface rock. Regional variations in conductivity can cause significant errors in the position of aircraft determined with loran-type systems.

3. The possibility of electromagnetic wave propagation for significant distances through the lithosphere, a long-sought goal for hardened communication systems [Institute of Electrical and Electronic Engineers, 1963; Tsao, 1975; de Bettencourt, 1966], depends on the physical properties of the crust. The electrical resistivity must be high at shallow depth and the wave guide must be continuous over large distances [Keller et al., 1966].

4. Many oil, gas, and mineral prospecting techniques depend upon measurements of in situ physical properties. The deposition of oil, gas, and minerals may also be controlled by microcracks.

5. Exploration for, and development of, geothermal systems are based on rock properties that are controlled by the microcracks. Flow of geothermal fluids may reduce permeability by sealing fractures and microcracks [Batzle and Simmons, 1976].

An important goal of our laboratory studies on microcracks and their effects on physical properties is that of predicting the physical properties of crustal rocks throughout large regions. Our set of hypotheses is the following: (1) Microcracks have been pro-

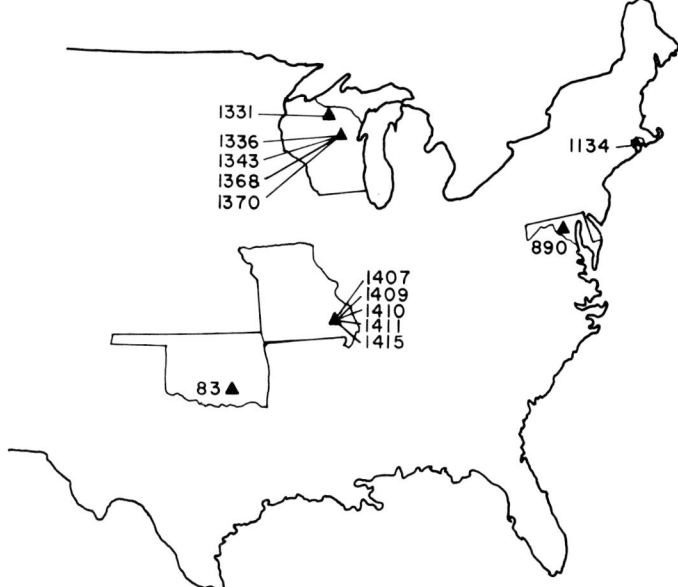

Fig. 2. Sample location map. See Table 1 for key to sample names and numbers.

duced (and also modified) by several geologic processes. (2) The geologic processes that produce cracks in rocks most likely extend over large geographic regions. (3) Samples from a few regions are representative of large areas. (4) Physical properties can be related to measurable crack characteristics. From the complete characterization of microcracks we can determine the processes which produced the cracks and estimate reliably the physical properties of rocks in situ as functions of depth over large geographic regions.

Microcracks can be characterized both microscopically [Simmons and Richter, 1976; Simmons et al., 1975] and with precise measurements of certain physical properties. In this paper we report work on a suite of rocks in which we use differential strain analysis [Simmons et al., 1974] to characterize open microcracks. In a companion paper, Richter and Simmons [1977] report microscopic observations on a subset of our rocks. We used the measured crack characteristics and an empirical relation between crack porosity and electrical resistivity to determine electrical resistivity at depth in situ in central Wisconsin and southeast Missouri.

Experimental Details

The Samples

The rocks examined in this study consist of Precambrian igneous rocks from central Wisconsin and southeast Missouri. Data for a few younger granites and diabases are also included for comparison. Sample locations are shown in Figure 2, and precise locations are

TABLE 1. Samples Examined in this Study

MIT Sample	Name
83	Troy (Okla.) granite
890	Frederick (Md.) diabase
1134	Westerly (R. I.) granite
1331	Mellen (Wis.) gabbro
1336	Wausau (Wis.) granite (Prehn quarry)
1343	Wausau (Wis.) granite (Linden quarry)
1368	Tigerton (Wis.) gabbro
1370	Red River (Wis.) quartz monzonite
1407	Middlebrook (Mo.) felsite
1409	Graniteville (Mo.) granite
1410	Graniteville (Mo.) granite
1411	Stouts Creek (Mo.) rhyolite
1415	Skrainka (Mo.) diabase

reported by Richter and Simmons [1977]. Modal compositions of the samples are determined by point counting, and chemical compositions of phases are determined with an electron microprobe. Massachusetts Institute of Technology (MIT) sample numbers and names are listed in Table 1 and compositions in Table 2.

Differential Strain Analysis

Differential strain analysis (DSA) is a technique for measuring strain in materials with very high precision. It provides information on the cracks present in the material. The very high precision, ± 2 or 3×10^{-6}, is obtained with a simple trick: the measured strain of a reference sample of fused silica is subtracted from the measured strain of a rock, measured at the same time and under identical conditions. A graphical representation of a DSA compression curve is shown in Figure 3. In Figure 3a the dashed line represents the linear strain measured as a function of pressure for a fused silica reference, and the solid line represents the strain of a typical cracked solid. If we subtract the strain of the fused silica from the strain of the cracked solid, we obtain the curve in Figure 3b which is the DSA differential strain curve.

If we assume that the compressibility of fused silica is constant for pressures below 2 kbars, we have

$$\hat{\varepsilon} = \varepsilon + P\beta_{FS} \qquad (1)$$

where $\hat{\varepsilon}$ is the differential strain, ε is the measured rock strain at pressure P, and β_{FS} is the fused silica compressibility and we use the convention that compressive strains are negative. In our previous work [Simmons et al., 1974; Simmons et al., 1975] we assumed constant β_{FS}; however, in the present work we have corrected for the fused silica compressibility pressure dependence, using the data of Peselnick et al. [1967]. Their measurements show that the linear compressibility of fused silica varies from 0.913×10^{-6} bar^{-1} at P = 0 to 0.945×10^{-6} bar^{-1} at 2 kbars. The actual sample strain is calculated by adding to the differential strain the fused silica strain calculated from Peselnick's data. Then, in order to expand the strain scale on plots and to facilitate comparison with previous differential strain data, a linear strain curve with a slope equal to the zero pressure fused silica compressibility determined by Peselnick et al. is subtracted from the actual strain curve. Thus the differential strains in this paper can be converted to true strains with the following formula:

$$\varepsilon = \hat{\varepsilon} - P\beta \qquad (2)$$

where $\beta = 0.913 \times 10^{-6}$ bar^{-1}.

The DSA compression curve is related mathematically to the cracks present in the sample. In a crack-free solid the curve is a straight line with a slope given by the difference in compressibilities of the sample and fused silica. But for a sample containing cracks the strain at pressure P = 0 due to cracks which are completely closed by pressure P_1, denoted by $\zeta(P_1)$, is given by the expression

$$\zeta(P_1) = P_1 \left. \frac{d\hat{\varepsilon}}{dP} \right|_{P_1} - \hat{\varepsilon}(P_1) \qquad (3)$$

where $\hat{\varepsilon}$ is the linear differential strain. Equation (3) has a simple geometrical interpretation. From Figure 3, $\zeta(P_1)$ is seen to be the zero pressure intercept of a line tangent to the strain curve (and also the differential strain curve) at P_1. The linear crack porosity due to all cracks that remain open at P_1 is also shown and may be expressed as

$$\eta(P_1) = \zeta(P_{max}) - P_1 \left. \frac{d\hat{\varepsilon}}{dP} \right|_{P_{max}} + \hat{\varepsilon}(P_1) \qquad (4)$$

where $\hat{\varepsilon}(P_1)$ is the value of differential strain at P_1. If the strains of Figure 3 are volumetric strain, then we obtain $\zeta_v(P)$, which is the volumetric strain at P = 0 due to all cracks that close at $P \leq P_1$ and $\eta_v(P_1)$, which is the (usual) volumetric crack porosity at P_1. Note that $\eta_v(0)$ is the crack porosity that has been used previously by other authors [e.g., Walsh, 1965; Todd et al., 1973].

In addition to crack porosity as a function of pressure, DSA allows us to obtain the distribution of crack porosity (at P = 0) with respect to the crack closure pressure, termed crack spectrum, and the effective orientation of the cracks associated with each value of crack closure pressure. Morlier [1971], on the basis of the work of Walsh [1965], showed that compression curves can be interpreted in terms of the distribution of crack aspect ratios. Siegfried and Simmons [1977] replaced the penny-shaped or elliptical crack models of Morlier and Walsh with the more general assumption that strain is linear over any pressure range in which no cracks close completely. They obtained the following:

TABLE 2.

Mineral	83	890	1134	1331	1336	1343
Plagioclase	43.2	47.3	39.2	53.7	23.8	10.0
K Feldspar	41.4[a]		30.7[b]		44.8[a]	49.6[a]
Pyroxene		46.9		30.1		
Quartz	14.2		22.5		28.9	39.1
Olivine		0.4		1.1		
Biotite	0.8	0.8	5.0			tr
Epidote	0.4					tr
Opaque	tr	2.5	0.7	3.1	1.0	tr
Secondary		2.2	0.4	12.2	1.5	0.8
Matrix[c]						
Others			1.5[f]		tr[d]	0.3[d]
Total	100.0	100.1	100.0	100.2	100.0	99.8
Number of counts	2000	2413	1000	1000	1000	1200

The abbreviation tr represents trace.
[a]Perthite.
[b]Microcline.
[c]Fine-grained quartz and feldspar.
[d]Fluorite with trace of sphene.

$$\nu(P) = P\left(\frac{d^2\hat{\varepsilon}}{dP^2}\right) \tag{5}$$

where $\nu(P)\, dP$ is the strain at zero pressure due to the presence of cracks closing between P and dP. We therefore can determine crack spectra by twice differentiating strain curves. Without the high precision data obtained with DSA it would not be possible to obtain meaningful results from these differentiations. Complete details of the DSA technique and its theoretical basis can be found in the work of Simmons et al., [1974], Feves and Simmons [1976], and Siegfried and Simmons [1977].

Velocities of Elastic Waves

The velocities of both compressional and shear waves were measured as functions of pressure for both dry and water-saturated conditions. Velocities were determined from measurements of the transit time of ultrasonic pulses through rock cores 2 1/2 cm in diameter and 7 to 14 cm long. We used techniques and equipment similar to that described by Birch [1960] for compressional waves and by Simmons [1964b] for shear waves. For saturated conditions we used techniques similar to those of Nur and Simmons [1969b] and Takeuchi and Simmons [1973]. The ends of each velocity core were ground parallel and flat to 25 μ with a surface grinder. All cores were then dried in a vacuum of 10^{-5} torr at room temperature for 3-5 weeks. After V_p and V_s were measured for dry conditions, the

Modal Analysis

1368	1370	1407	1409	1410	1411	1415
52.0	33.7	0.3	29.3	26.9	16.1	66.9
	25.7[b]		30.4[a]	39.3[a]		
24.3	0.9					8.8
	28.5	6.7	39.8	32.8	8.3	
0.3						14.2
18.0	9.5					2.3
	0.6	12.1				
4.1	0.7		tr	tr	0.4	4.7
1.1	0.3	1.0	tr	0.5	2.5	3.1
		77.5			72.7	
0.2[e]	0.1[f]	2.4[g]	0.5[h]	tr[h]		
100.0	100.0	100.0	100.0	99.5	100.0	100.0
1000	1000	1000	1000	1200	1000	1000

[e]Apatite.
[f]Apatite and muscovite.
[g]Sphene.
[h]Fluorite and muscovite.

cores were again placed in vacuum. After 2 days the vacuum chamber was vented with distilled water in order to saturate the cores. Saturated samples were jacketed with distilled water, sand, and tygon tubing in order to prevent the sample from drying out and to provide a reservoir to prevent pore pressure buildup as confining pressure was increased. See Figure 4.

Electrical Resistivity

The laboratory experiment of measuring electrical resistivity of saturated rocks as a function of confining pressure, with pore pressure held at 1 bar, is not a valid analogue for the electrical resistivity of rock in situ. Although the cracks close mechanically with confining pressure (as they would in the earth with depth), they do not 'close electrically.' Surface conduction occurs along the crack walls even after the cracks close mechanically in the laboratory. Surface conduction is probably absent in rocks in situ, if the cracks have been closed mechanically for tens or thousands of years, because the surfaces of the cracks disappear. We suggest then that the electrical resistivity of rocks in situ is best estimated indirectly from crack characteristics and an empirical relation between crack characteristics and resistivity rather than by 'direct' laboratory measurement of electrical resistivity as a function of confining pressure.

Therefore we measured resistivity on a suite of rocks that differ only in crack porosity. We used Frederick (Maryland) diabase, a rock remarkable for its extremely low crack porosity ($<5 \times 10^{-6}$). Various crack porosities were induced into each sample (cores of 12 mm in diameter x 50 mm in length) in the laboratory by thermal

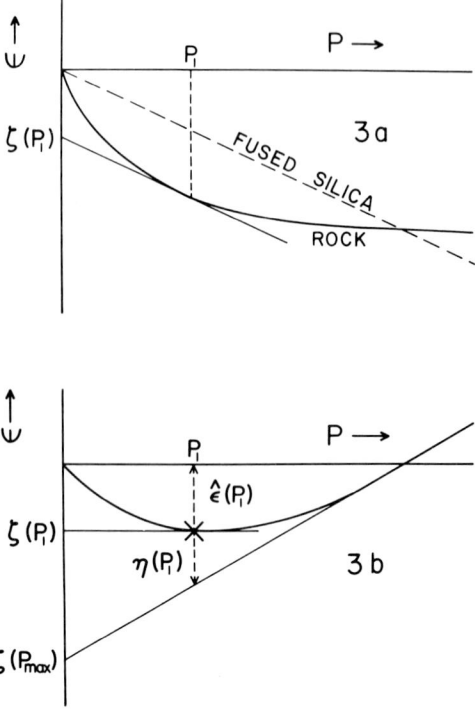

Fig. 3. Differential strain analysis. In Figure 3a, the measured strain of fused silica (dashed line) is subtracted from the measured strain of a rock (solid line) to yield the DSA compression curve in Figure 3b. The zero pressure intercept of a line tangent to the strain curves at any pressure P_1 is the strain at zero pressure associated with cracks which have completely closed by pressure P_1. Here $\eta(P_1)$ is the total open crack porosity at P_1. (See text for details.)

cycling each core to a different maximum temperature [Richter and Simmons, 1974]. Electrical resistivity was measured on the thermally cycled samples fully saturated with 0.1 N NaCl electrolyte. The resistivity of 0.1 N NaCl electrolyte at 20°C is 1.04 Ωm. These resistivity and equivalent salinity are typical for many groundwaters [Keller and Frischknecht, 1970]. Total crack porosity of each sample was determined by DSA.

Our preliminary relation between resistivity ρ and crack porosity $\eta_v(0)$ is shown in Figure 5. A least squares fit through the data in Figure 5 yields:

$$\log \rho = 11.97 - 3.11 \log [\eta_v(0) \times 10^6] \qquad (6)$$

Note that this relation includes contributions from the rock matrix, the electrolyte present in each open microcrack, and from the two surfaces of each open microcrack. Because the resistivity of most silicate rocks is from 10^{10} to 10^{15} times greater than the resisti-

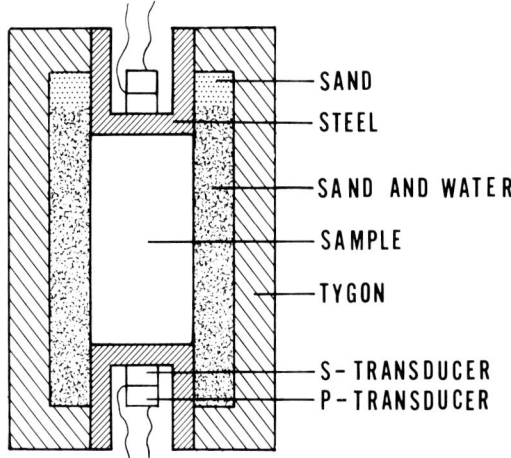

Fig. 4. Diagram of sample configuration used for measuring V_p and V_s of water-saturated samples. The pore space in the sand above the water provides a reservoir to which the water in the rock can move when the confining pressure is increased. It prevents buildup of pore pressure and keeps the sample completely saturated.

vity of the electrolyte, conduction occurs primarily in open microcracks, and we can ignore the effects of mineral conduction (in silicate rocks). Since no mechanically closed but nonhealed microcracks are present in our samples, no conduction occurs along surfaces of closed cracks.

Results

The data of this study consist of several dozen DSA compression curves, measurements of V_p and V_s as functions of confining pressure and saturation condition, and measurements of electrical resistivity, modal and mineral compositions, and density. From these laboratory measurements we are able to calculate the elastic constants of the rocks, predict electrical resistivity in situ, and determine empirical relationships between various pairs of physical properties.

Differential Strain Analysis

With DSA we can determine the distribution of strain due to cracks which completely close at a given pressure $\nu(P_c)$. Examples of crack spectra for several samples are shown in Figure 6. Crack spectra for most crustal igneous rocks examined show that total crack porosity is primarily due to cracks with low (<1 kbar) closure pressure. Therefore most cracks will be closed at depths of 3-5 km.

One useful by-product of DSA is the determination of compressibility as a function of pressure. If we assume $P = Dgh$, where D is the average crustal density (2.67 g/cm^3), g is gravitational constant, and h is depth, we obtain compressibility as a function of

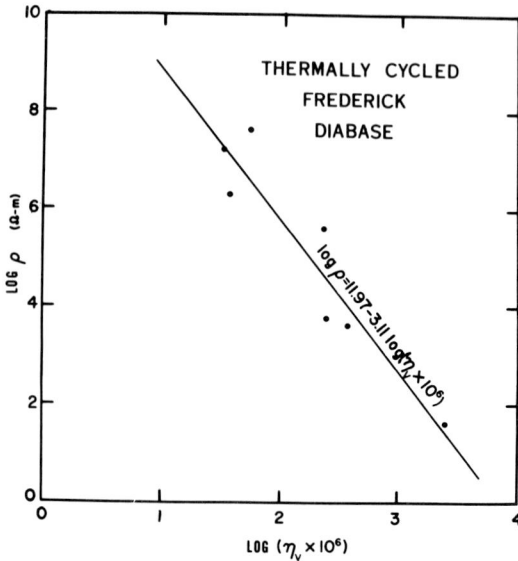

Fig. 5. Electrical resistivity ρ as a function of crack porosity η for Frederick (Maryland) diabase with thermally induced cracks. Each point was measured on a separate sample. All cracks present in each sample were completely saturated with 0.1 N NaCl solution.

depth. In Figure 7 we show two plots of compressibility as a function of depth, plots that are typical for our suite of rocks. If pore pressure is nonzero, then the plots in Figure 7 represent minimum values of compressibility.

The compressibilities obtained in DSA can be used to determine whether all cracks are closed at the maximum pressure used in a DSA experiment. We compare DSA-measured values β_{DSA} with upper and lower bounds on the intrinsic values [Hill, 1952]. The upper bound β_R (Reuss average) is given by

$$\beta_R = \sum_{i=1}^{N} A_i \beta_i \tag{7}$$

where A_i is the fractional volume and β_i is the compressibility of the ith mineral. The β_i are given in Table 3. The lower bound β_V (Voigt average) is given by

$$\beta_V^{-1} = \sum_{i=1}^{N} A_i \beta_i^{-1} \tag{8}$$

In Table 4 we compare β_R, β_V, and the β_{DSA} measured for several samples. The β_{DSA} at 2 kbars approaches, but does not equal, the upper limit of the intrinsic values. We attribute the difference between calculated and observed compressibilities to cracks with closure pressure greater than 2 kbars or to pores which are compressible and not considered in calculating β_R and β_V.

Another by-product of DSA is the crack porosity as a function of depth. It can be obtained from (4) and the relation between pressure and depth (P = Dgh). Several examples of open crack porosity profiles are shown in Figure 8, and total crack porosity at zero pressure $\eta_v(0)$ for all samples is listed in Table 4.

Velocities of Elastic Waves

Compressional and shear wave velocities for dry and saturated samples are given in Tables 5 and 6. These data were obtained by fitting a smooth curve to measured values and taking the value of the curve at the given pressure as the velocity. A typical set of velocity curves and measurements is shown in Figure 9.

Electrical Resistivity

Our technique to obtain resistivity of rock in situ is (1) obtain η_v as function of depth from DSA, Figure 7, and (2) convert η_v to ρ using (6). The results are shown in Figure 10.

Error Analysis

The accuracy of our results depends upon several measurements: electrical resistance of foil strain gages, confining pressure, sample length, delay time of ultrasonic pulses in the sample, electric current through the sample, and voltage across the sample. The accuracy of the gage factor for the foil strain gages is ±1%. With the DSA technique the precision of our strain measurements is ±3 x 10^{-6}. In differentiating DSA curves to obtain crack spectra, five or seven points are fit with a quadratic. This curve fitting tends to smooth the data and produce end effects at zero and 2 kbars, since end points will not be fit by all five or seven points. Errors for pressures below 500 bars are indicated on crack spectra by error bars. At higher pressure, data points are fit over larger pressure ranges. This results in reduction of error and resolution. The accuracy of the Heise gages used for measurements of confining pressure is ±0.5%. The length of samples used for velocity measurements is known to ±0.001 in. No corrections for changes in sample length due to confining pressure are made. These corrections would amount to at most 0.03% at 2 kbars. The largest error in the velocity measurements is due to the determination of the time delay of the ultrasonic pulse. The measurement of the time delay depends upon the ability to pick the first arrival in the wave train and relate this arrival to the time when the input pulse was generated. The initial arrival is gradual and can only be determined to about 2-5% in relation to the input pulse. However, by consistently choosing the onset of the initial pulse a precision of about 0.5% is obtained. The accuracy of electrical resistivity measurements is about 10%. Current and voltage are measured to ±1%, and length is measured to about ±5%. Error is also introduced by electrode polarization and by conduction along the outside surface of the core.

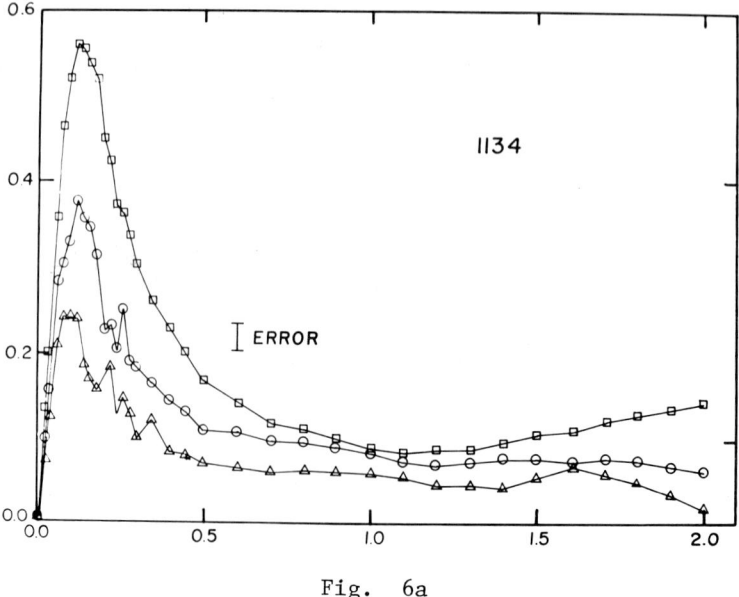

Fig. 6a

Fig. 6. Crack closure spectra in three orthogonal directions for several rocks. The abscissa is the crack closure pressure and the ordinate is $\nu(P) = P(d^2\hat{e}/dP^2)$ in units per megabar. Note that $\nu(P)\, dP$ is the porosity at zero pressure due to cracks closing between pressure P and P + dP. All samples contain many cracks with closure pressure less than 1 kbar, and several samples may contain a few cracks with closure pressure greater than 2 kbars.

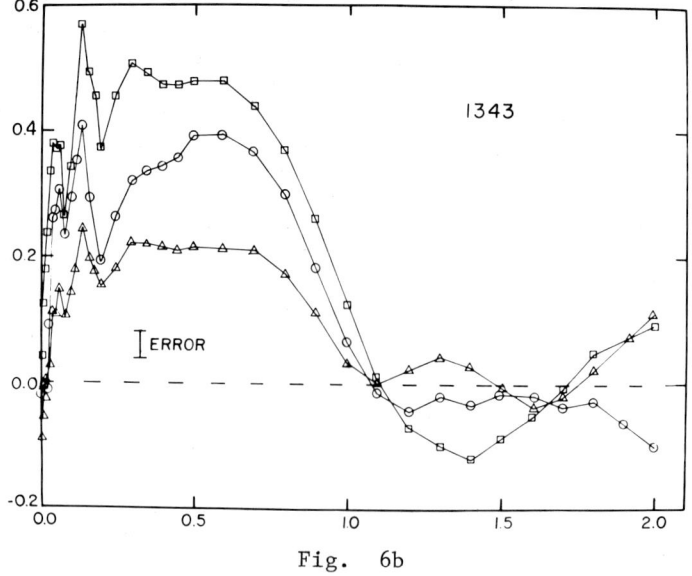

Fig. 6b

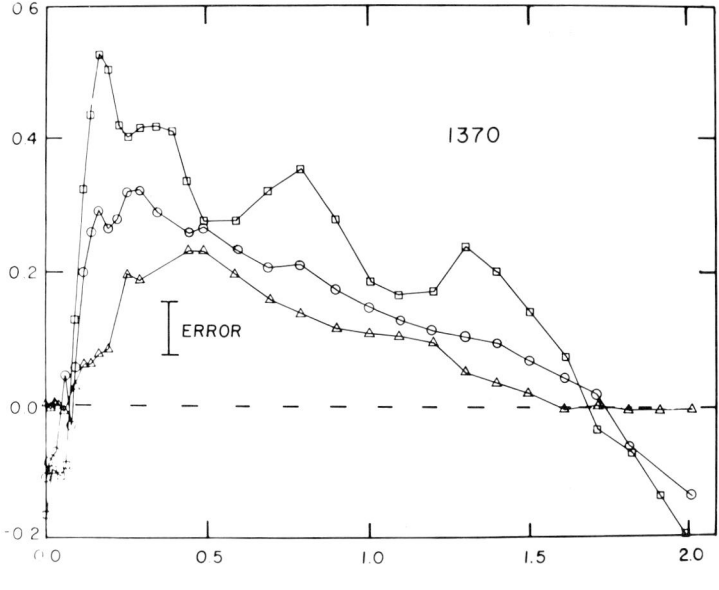

Fig. 6c

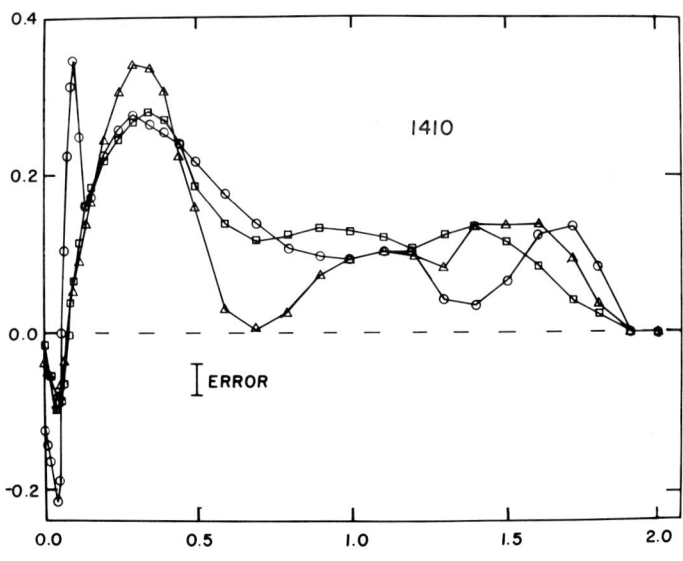

Fig. 6d

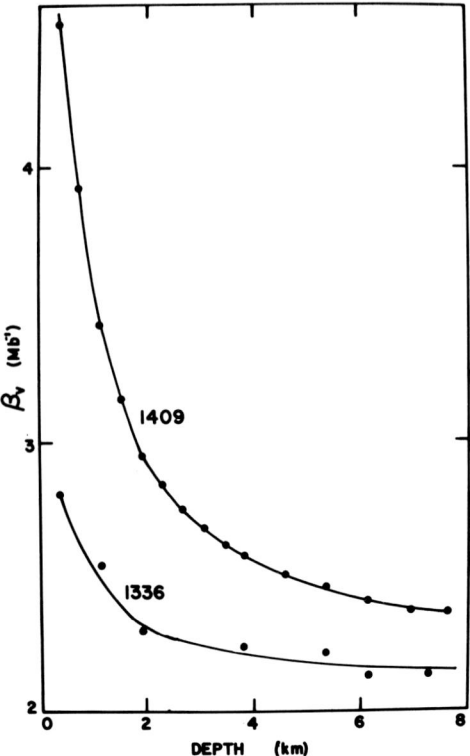

Fig. 7. Examples of compressibility as a function of depth. These profiles were obtained from DSA curves on the basis of an average rock density of 2.67 g/cm³.

TABLE 3. Mineral Compressibilities Used in Calculations

Sample	β, Mbar^{-1}
Quartz	2.63
Augite	1.05
Microcline	1.84
Plagioclase	
An$_9$	1.75
An$_{24}$	1.63
An$_{41}$	1.51
An$_{50}$	1.44
Labradorite	1.33
Olivine	0.76
Biotite	2.05
Muscovite	1.98
Epidote	0.94
Magnetite	0.62

Source: Simmons and Wang [1971].

TABLE 4. Total Crack Porosity and Rock Compressibility

Sample	$\eta_v \times 10^6$*	β_{DSA}*	β_R	β_V
83	746			
890	<5	1.36	1.18	1.15
1134	706	2.06	1.96	1.88
1331	73	1.35	1.31	1.24
1336	370			
1343	859			
1368	60			
1370	854			
1407	156			
1409	883	2.37	2.13	2.06
1410	613	2.21	2.06	2.04
1411	604			
1415	<5	1.45	1.20	1.13

Values are in units per megabar.
*Value at P = 2 kbars.

Discussion

Resistivity

We have determined relations between crack porosity and electrical resistivity. Our data on the suite of thermally cycled Frederick diabase samples support the conclusion of previous workers [e.g.,

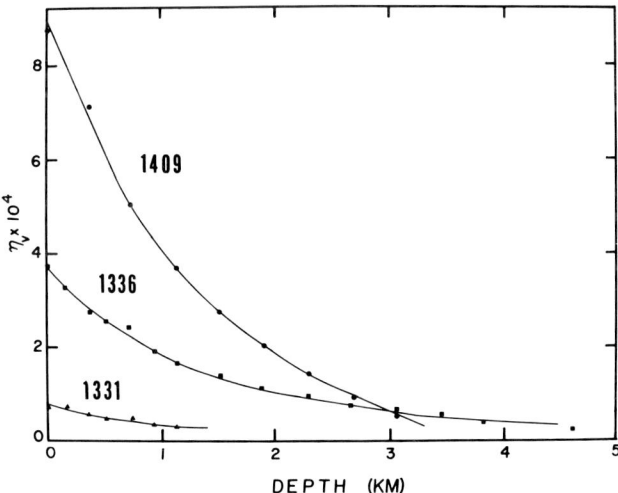

Fig. 8. Examples of crack porosity as a function of depth. These profiles were obtained from DSA curves on the basis of an average rock density of 2.67 g/cm^3.

TABLE 5. Velocities for Dry Conditions

MIT Sample	Density g/cm³	Mode	Confining Pressure, kbars						
			0.0	0.1	0.3	0.5	1.0	1.5	2.0
83	2.62	P	5.11	5.71	6.16	6.35	6.52	6.59	6.60
		S	2.71	2.94	3.17	3.26	3.31	3.34	3.35
890	2.98	P	6.97	7.01	7.03	7.05	7.08	7.11	7.14
1134	2.64	P	4.61	5.12	5.70	5.88	6.08	6.17	6.22
		S	2.11	2.34	2.48	2.53	2.57	2.59	2.60
1331	3.02	P	6.57	6.62	6.68	6.73	6.80	6.84	6.86
		S	4.31	4.34	4.38	4.43	4.47		
1336	2.62	P	5.91	6.03	6.19	6.29	6.43	6.47	6.50
		S	3.39	3.59	3.70	3.75	3.80	3.81	3.82
1370	2.66	P	5.11	5.48	5.81	5.95	6.14	6.22	6.23
		S	4.44						4.51
1409	2.58	P	4.80	5.37	5.60	5.71	5.89	6.01	6.08
		S	3.84	4.03	4.21	4.30	4.40	4.47	4.51
1411	2.61	P	5.61	5.77	5.96	6.06	6.15	6.19	6.22
		S				2.94	2.99	3.01	3.03

Velocities are in units of kilometers per second.

Brace et al., 1965; Greenberg and Brace, 1969] that the electrical resistivity of rocks is primarily controlled by microcracks. Our values differ significantly from their values at high pressure, however.

By combining our ρ-n_v relation (Equation (6)) with data on crack porosity as a function of pressure (depth) we have shown that resistivity may attain high values (>10^6 Ωm) at depths of only 2-3 km in igneous basement rocks. This result is probably valid for most

TABLE 6. Velocities for Rocks Saturated with Distilled Water

MIT Sample	Density g/cm³	Mode	Confining Pressure, kbars						
			0.0	0.1	0.3	0.5	1.0	1.5	2.0
1134	2.64	P	5.74	5.93	5.98	6.00	6.03	6.04	6.04
1331	3.02	P	6.68	6.72	6.79	6.86	6.94	6.97	7.00
1336	2.62	P	6.25	6.30	6.35	6.26	6.39	6.40	6.40
		S	3.39	3.43	3.45	3.46	3.49		
1370	2.66	P	5.84	5.88	5.91	5.95	5.99	6.01	6.03
		S	4.35	4.37	4.41	4.45	4.50	4.54	4.56
1409	2.58	P	5.85	5.94	5.98	6.00	6.04	6.06	6.07
		S	3.97	4.05	4.17	4.28	4.44	4.50	4.54
1411	2.61	P	5.76	5.83	5.88	5.90	5.94	5.96	5.98
		S	2.90	2.93	2.96	2.98	2.99	2.99	3.00

Velocities are in units of kilometers per second.

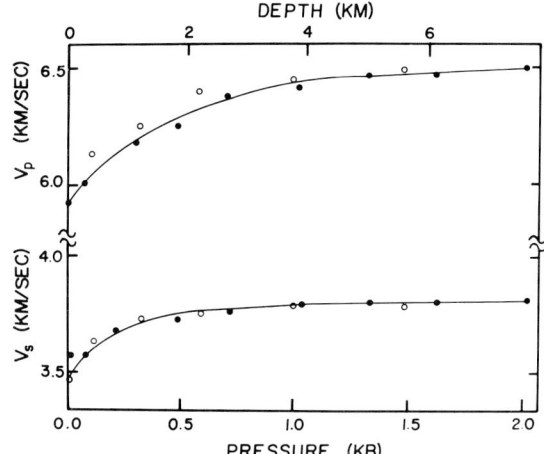

Fig. 9. Example of compressional and shear wave velocity as a function of pressure and depth (sample 1336). For converting pressure to depth we used an average rock density of 2.67 g/cm³. The data in Tables 5 and 6 are obtained from the smooth curve through the measured values.

igneous crustal rocks in the eastern half of the United States, since the samples were obtained from several widely separated locations (Figure 2).

The high value of resistivity which we obtain does not agree with the direct laboratory measurements of resistivity as a function of confining pressure. Brace [1971] found that at no depth should resistivity be greater than 10^6 Ωm. We attribute this discrepancy to effects of surface conduction. Cracks may be closed mechanically but not electrically by confining pressure in the laboratory. Therefore the direct laboratory measurement of resistivity yields values which are too low. Richter and Simmons [1977] have shown that most cracks in Precambrian basement rocks from central Wisconsin and southeast Missouri are partially or totally healed. Cracks in these basement rocks in situ are probably closed both mechanically and electrically.

In determining resistivity profiles we have neglected the effects of pore pressure, temperature, and possible variations in salinity. What is the magnitude of these effects? We believe it is small for our suite of rocks. Consider pore pressure P_p. Its variation with depth is unknown. If the cracks are interconnected with the earth's surface, then P_p is approximately the hydrostatic head, and the correct relation of our laboratory pressures to depths would be $P = 1.67$ gh. If the cracks are not connected with the surface, then we have no method of estimating P_p. An increase in pore pressure reduces the effective confining pressure which results in lower resistivity at a given depth. Consider temperature. It affects the resistivity of minerals and electrolytes. The geotherms calculated by Roy et al. [1968] indicate a temperature of about 100°C at a depth of 7 km in the eastern half of the United States. This tem-

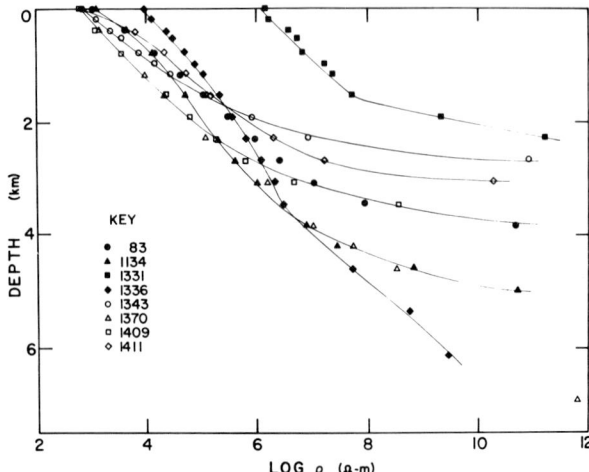

Fig. 10. Electrical resistivity of several crustal igneous rocks in situ. The data were derived from DSA measurements of crack porosity and the Frederick diabase calibration curve. Details of the technique are described in the text.

perature has negligible effect on mineral resistivities [Parkhomenko, 1967]. The electrical resistivity of 0.1 N NaCl solution decreases from about 1.7 Ωm at 0°C to 0.5 Ωm at 100°C [Keller and Frischknecht, 1970]. However, decreasing electrolyte resistivity from 50 Ωm (NaCl solution) to 0.25 Ωm (tap water) results in less than an order of magnitude decrease in electrical resistivity of rocks [Brace, 1971]. Therefore the effects of temperature and salinity variations with depth are within our experimental error (10% for electrical resistivity measurements).

Reliable field measurements of electrical resistivity greater than 10^5 Ωm are difficult to make. We suggest that reliable results may be obtained by measuring electrical resistivity indirectly through measurements of other physical properties. We have chosen to relate measurements of compressional and shear wave velocity to electrical resistivity. Velocity, unlike resistivity, is dependent upon both microcrack and mineral properties. To eliminate effects of mineral variations among samples, we have taken the difference ΔV between measured velocity at 2 kbars pressure and measured velocities at various pressures below 2 kbars. Since most cracks are closed by 1 kbar, the measured velocity at 2 kbars approximates the intrinsic velocity for the rock matrix. By combining relations between ΔV and pressure and ρ and pressure (Figure 10) we obtain curves relating electrical resistivity to compressional and shear wave velocity. Resistivity-velocity curves for several samples are shown in Figure 11. Two general relations are apparent: one for compressional wave velocity ΔV_p and one for shear wave velocity ΔV_s.

The difference between the curves for ΔV_p and ΔV_s is due to the difference in dependence of compressional and shear wave velocity on crack porosity. From DSA data of crack porosity as a function of

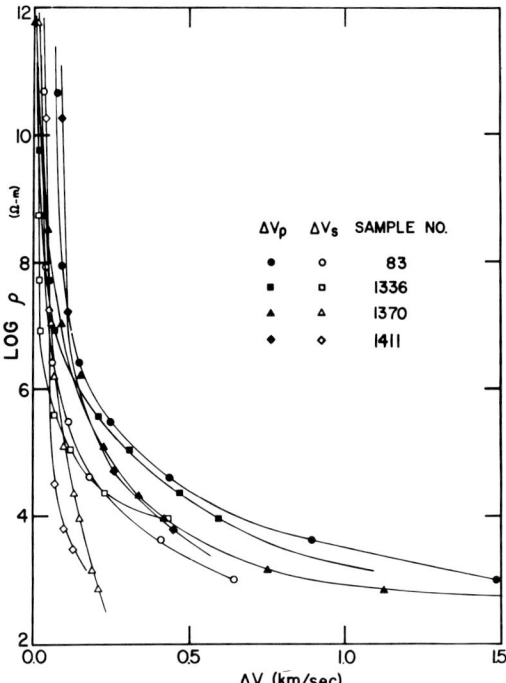

Fig. 11. Relation between velocity and electrical resistivity for several rocks from the eastern United States. The ΔV is the difference between velocity measured at 2 kbars (intrinsic velocity) and velocities measured at various pressures below 2 kbars. Velocity data are for unsaturated samples. These data were obtained by combining velocity-depth profiles (Figure 9) with electrical resistivity-depth profiles (Figure 10).

pressure and measurements of velocity as a function of pressure we obtain relations between velocity and crack porosity. The data in Figure 12 indicate that for dry rocks, compressional wave velocity is more dependent on microcrack porosity than is shear wave velocity.

Saturation conditions also affect elastic wave velocity. Our data are in agreement with those of Nur and Simmons [1969b] and others who have shown that in water-saturated rocks at atmospheric pressure, compressional wave velocity is 30-50% higher than in dry rocks and shear wave velocity is unaffected by saturation condition. The degree of water saturation in shallow crustal rocks below the water table is probably 100%, but above the water table the degree of water saturation is often unknown. Therefore field measurements of shear wave velocity may provide an alternate reliable method of determining in situ electrical resistivity.

Conclusions

Microcracks control physical and electrical properties of rocks. From laboratory measurements of strain, compressional wave velocity,

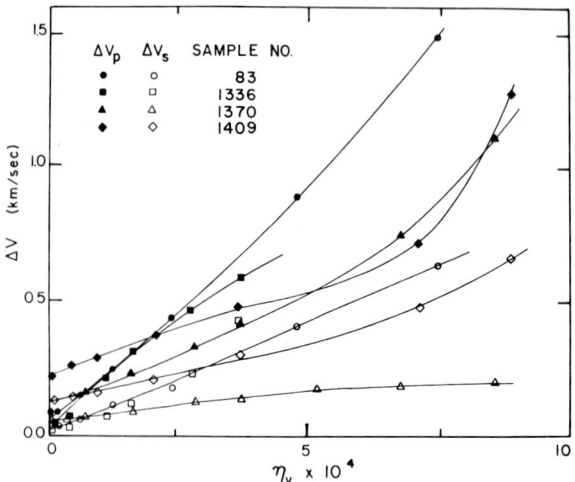

Fig. 12. Examples of compressional and shear wave velocity as functions of crack porosity. The ΔV is the difference between velocity measured at 2 kbars (intrinsic velocity) and velocity measured at various pressures below 2 kbars. Microcracks affect compressional velocity more than shear velocity. However, shear wave velocities are more useful for determining in situ crack properties when the degree of fluid saturation is unknown, since shear velocity is unaffected by saturation condition (see Tables 5 and 6).

shear wave velocity, and electrical resistivity we have characterized cracks in crustal igneous rocks of the eastern half of the United States and have developed relations between several pairs of physical and electrical properties.

Our results show that electrical resistivity is related to crack porosity. Differential strain analysis indicates that the crack porosity of the igneous rocks studied is low at shallow depth. Therefore in situ electrical resistivity is high ($>10^6$ Ωm) at depths of only 2-3 km in basement rocks of the eastern United States. Relations between elastic wave velocities and electrical resistivity are determined from laboratory measurements. These relations can be used to predict in situ resistivity (velocity) given in situ velocity (resistivity). Shear wave velocities are unaffected by saturation condition and therefore provide a reliable means of estimating in situ electrical resistivity when saturation conditions are unknown.

Acknowledgements. We benefited from discussions with Herman Cooper and Dorothy Richter. Michael Batzle, Herman Cooper, Mike Fehler, and Ann Harlow helped obtain the DSA data. Ann Harlow and Lucille Foley typed the manuscript. This work was supported by ONR contract N00014-76-C-0478.

References

Batzle, M. L., and G. Simmons, Microfractures in rocks from two geothermal areas, Earth Planet. Sci. Lett., 30, 71-93, 1976.

Birch, F., The velocity of compressional waves in rocks to 10 kilobars, 1, J. Geophys. Res., 65, 1083-1102, 1960.

Birch, F., The velocity of compressional waves in rocks to 10 kilobars, 2, J. Geophys. Res., 66, 2199-2224, 1961.

Birch, F., and D. Bancroft, The effect of pressure on the rigidity of rocks, II, J. Geol., 46, 113-141, 1938.

Brace, W. F., Some new measurements of linear compressibility of rocks, J. Geophys. Res., 70, 391-398, 1965.

Brace, W. F., Resistivity of saturated crustal rocks to 40 km based on laboratory measurements, in The Structure and Physical Properties of the Earth's Crust, Geophys. Monogr. Ser., vol. 14, edited by J. G. Heacock, pp. 243-255, AGU, Washington, D. C., 1971.

Brace, W. F., and A. S. Orange, Further studies of the effects of pressure on electrical resistivity of rocks, J. Geophys. Res., 73, 5407-5420, 1968.

Brace, W. F., A. S. Orange, and T. R. Madden, The effect of pressure on the electrical resistivity of water-saturated crystalline rocks, J. Geophys. Res., 70, 5669-5678, 1965.

Budiansky, B., and R. J. O'Connell, Elastic moduli of a cracked solid, Int. J. Solids Struct., 12, 81-97, 1976.

De Bettencourt, J. T., Review of radio propagation below the earth's surface, paper presented at Sessions of Commission II, XVth General Assembly, Int. Union of Geod. Geophys., Munich, Germany, June 1966.

Feves, M., and G. Simmons, Effects of stress on cracks in Westerly granite, Bull. Seismol. Soc. Amer., 66, 1755-1765, 1976.

Greenberg, R. J., and W. F. Brace, Archie's law for rocks modeled by simple networks, J. Geophys. Res., 74, 2099-2102, 1969.

Hill, R., The elastic behavior of a crystalline aggregate, Proc. Phys. Soc. London Sect. A, 65, 349-354, 1952.

Hughes, D. S., and J. H. Cross, Elastic wave velocities in rocks at high pressures and temperatures, Geophysics, 16, 577-593, 1951.

Institute of Electrical and Electronics Engineers, Transactions on antennas and propagation, Special Issue on Electromagnetic Waves in the Earth, AP-11, New York, May 1963.

Keller, G. V., and F. C. Frischknecht, Electrical Methods in Geophysical Prospecting, Int. Ser. Monogr. Electromagn. Waves, vol. 10, pp. 1-33, Pergamon, New York, 1970.

Keller, G. V., L. A. Anderson, and J. I. Pritchard, Geological survey investigations of the electrical properties of the crust and upper mantle, Geophysics, 31, 1078-1087, 1966.

Morlier, P., Description de l'etat de fissuration d'une roche a partir d'essais non-destructifs simples, Rock Mech., 3, 125-138, 1971.

Nur, A., and G. Simmons, Stress-induced velocity anisotropy in rock: An experimental study, J. Geophys. Res., 74, 6667-6674, 1969a.

Nur, A., and G. Simmons, The effect of saturation on velocity in low porosity rocks, Earth Planet. Sci. Lett., 7, 183-193, 1969b.

O'Connell, R. J., and B. Budiansky, Seismic velocities in dry and

saturated cracked solids, J. Geophys. Res., 79, 5412-5426, 1974.
Parkhomenko, E. I., Electrical Properties of Rocks, 314 pp., Plenum, New York, 1967.
Peselnick, L., R. Meister, and W. H. Wilson, Pressure derivatives of elastic moduli of fused quartz to 10 kb, J. Phys. Chem. Solids, 28, 635-639, 1967.
Richter, D., and G. Simmons, Thermal expansion behavior of igneous rocks, Int. J. Rock Mech. Min. Sci., 11, 403-411, 1974.
Richter, D., and G. Simmons, Microcracks in crustal igneous rocks: Microscopy, in The Earth's Crust: Its Nature and Physical Properties, Geophys. Monogr. Ser., vol. 20, edited by J. G. Heacock, AGU, Washington, D. C., this volume, 1977.
Roy, R. F., D. D. Blackwell, and F. Birch, Heat generation of plutonic rocks and continental heat flow provinces, Earth Planet. Sci. Lett., 5, 1-12, 1968.
Siegfried, R. W., and G. Simmons, Characterization of oriented cracks with differential strain analysis, submitted to J. Geophys. Res., 1977.
Simmons, G., Velocity of compressional waves in various minerals at pressures to 10 kilobars, J. Geophys. Res., 69, 1117-1121, 1964a.
Simmons, G., Velocity of shear waves in rocks to 10 kilobars, 1, J. Geophys. Res., 69, 1123-1130, 1964b.
Simmons, G., and D. Richter, Microcracks in rocks, in The Physics and Chemistry of Minerals and Rocks, edited by R. G. J. Strens, pp. 105-137, John Wiley, New York, 1976.
Simmons, G., and H. Wang, Single Crystal Elastic Constants and Calculated Aggregate Properties, 370 pp., MIT Press, Cambridge, Mass., 1971.
Simmons, G., R. W. Siegfried, and M. Feves, Differential strain analysis: A new method for examining cracks in rocks, J. Geophys. Res., 79, 4383-4385, 1974.
Simmons, G., R. Siegfried, and D. Richter, Characteristics of microcracks in lunar samples, Proc. Lunar Sci. Conf., 6th, 3, 3227-3254, 1975.
Takeuchi, S., and G. Simmons, Elasticity of water-saturated rocks as a function of temperature and pressure, J. Geophys. Res., 78, 3310-3320, 1973.
Todd, T., H. Wang, W. S. Baldridge, and G. Simmons, Elastic properties of Apollo 14 and 15 rocks, Proc. Lunar Sci. Conf. 3rd, 3, 2577-2586, 1972.
Todd, T., D. A. Richter, G. Simmons, and H. Wang, Unique characterization of lunar samples by physical properties, Proc. Lunar Sci. Conf. 4th, 3, 2639-2662, 1973.
Tsao, C. K. H., Radio propagation via subsurface rock strata, final report for U.S. Air Force Cambridge Research Laboratory, contract AF19(628)-2363, Raytheon, Norwood, Mass., April 1965.
Walsh, J. B., The effect of cracks on the compressibility of rock, J. Geophys. Res., 70, 381-389, 1965.
Walsh, J. B., Theoretical bounds for thermal expansion, specific heat, and strain energy due to internal stress, J. Geophys. Res., 78, 7637-7646, 1973.
Walsh, J. B., and W. F. Brace, Elasticity of rock: A review of some recent theoretical studies, Rock. Mech. Eng. Geol., 4, 283-297, 1966.

Walsh, J. B., and E. R. Decker, Effect of pressure and saturating
 fluid on the thermal conductivity of compact rock, J. Geophys.
 Res., 71, 3053-3061, 1966.
Warren, N., and R. Nashner, Theoretical calculation of compliances of
 a porous medium, in The Physics and Chemistry of Minerals and
 Rocks, edited by R. G. J. Strens, pp. 197-216, John Wiley, New
 York, 1976.

CHARACTERIZATION OF MODULUS-PRESSURE SYSTEMATICS OF ROCKS: DEPENDENCE ON MICROSTRUCTURE

Nick Warren

Institute of Geophysics and Planetary Physics
University of California
Los Angeles, California 90024

Abstract. Dynamic modulus-pressure (M-P) data for rocks show important systematics and characteristics which are related to microstructure. Diagnostics characterizing the functional form of log M versus log P data include average slope, elbows (sharp changes in slope), local inflection regions, hysteresis, and similarities and differences of characteristics within up and down portions of a hysteresis cycle. Characteristics are strikingly similar for rocks with similar petrographic or structural texture descriptions. Many characteristics can be modeled by using theoretical expressions giving the dependence of the modulus on porosity and pore strains. The typical form of log M versus log P data for microcracked rock can be modeled by assuming minimum pore-pore interaction and pore strains which are linearly dependent on pressure. For very strong dependence of pore strains on pore concentration and/or on pressure, power laws are predicted. Theoretically, overcracking a rock yields power law behavior typical of soils. Discrete distributions of crack-closing pressures generate inflection regions in M versus P curves. Conclusions are that M versus P data exhibit systematics dependent on microstructure, that the effects of microstructure on moduli override mineralogical effects even at low porosities, and that velocity-pressure data can be inverted to give detailed characterization of volume-averaged microstructure of rock.

Introduction

 A central and general research problem in rock physics is that of developing quantitative theories relating rock microstructure to physical properties. In this paper, attention is focused on two parts of this problem: the experimental characterization of elastic modulus versus pressure data and the theoretical modeling of the functional forms of the data in terms of structural variables.
 Characteristics of plots of modulus versus pressure and/or velocity versus pressure are to be sought which are related to internal microstructures. Interpretation of curve characteristics will then be considered in terms of a mathematical modeling. Accordingly, the paper contains two main sections. In the first section, dynamic modulus data are presented in order to demonstrate that there are systematics and groupings in such data. Data used here are compiled from a number of laboratories. In the second section the characteristics of the functional forms of these data are discussed by using a modeling theory involving porosity and pore strains as parameters.

Two main points are to be made. First, dynamic modulus versus pressure (M-P) (henceforth the hyphen is used to separate the functions plotted) data for rocks show characteristics which are diagnostic of structure. Second, the general forms of diagnostic characteristics are predicted by modeling theory.

A consequence of the first point is that velocity-pressure data can be 'read' to indicate underlying structure. A consequence of the second point is that the experimental differences between dynamic and static moduli appear to be compatible with time relaxation of pore strains.

Characterizing of Velocity-Pressure Functions

Systematics of the pressure dependence of elastic properties are sought by considering the functional forms of velocity-pressure (V-P) data and of log modulus-log pressure (log M - log P) data. Functional properties are easily brought out in plots of the data.

Log M-log P plots accentuate the form of the power law dependence of moduli on pressure. They show slope changes at pressures which are characteristic of elastic properties of pores and cracks. Log-log data curves are bounded by flat pressure-independent curves at one extreme and by steep power law curves, typical of soils, at the other extreme. Velocity-pressure plots are also used, as they accentuate local inflections in curve shapes which are not as well resolved in log-log plots.

Besides making direct visual inspection of curve shapes, one can also fit least squares polynomials to moduli and/or velocity data, and coefficients can then be used to be fairly direct, but because of the unevenness in the spacing of data, such polynomials must be checked to eliminate mathematical artifacts (unwanted waves) generated between data points. Graphic display of the data curves brings out the salient features of the functional forms and is used here.

If modulus characteristics are related to underlying microstructure, the question arises as to whether static or dynamic moduli are more diagnostic. A priori, it is not clear whether static or dynamic data are better for characterizing the microstructure. However, if both static and dynamic data are controlled by the underlying microstructure, then the dynamic compressional and shear modulus data and the static strain data should be related.

In this article, compressional velocity (V_p) and the compressional velocity modulus C are used, since measurements of compressional velocity have the greatest precision. The modulus C is defined by

$$C = V_p^2 \cdot \rho = K + (4G/3)$$

where K is the dynamic bulk modulus, G is the shear modulus, and ρ is density. Hence C has units of pressure (we will use megabars), and C/ρ has units of Mbar g^{-1} cm^3 (or $10^2 \times km^2/s^2$).

Although shear velocity (V_s) data show characteristics which are very similar to those for V_p data [Warren and Trice, 1975], the V_s data show greater numbers of anomalies and interlaboratory differences. It is hard to judge whether these differences represent intrinsic rock properties or differences in measurements.

Static data, in general, are simply less available and often more

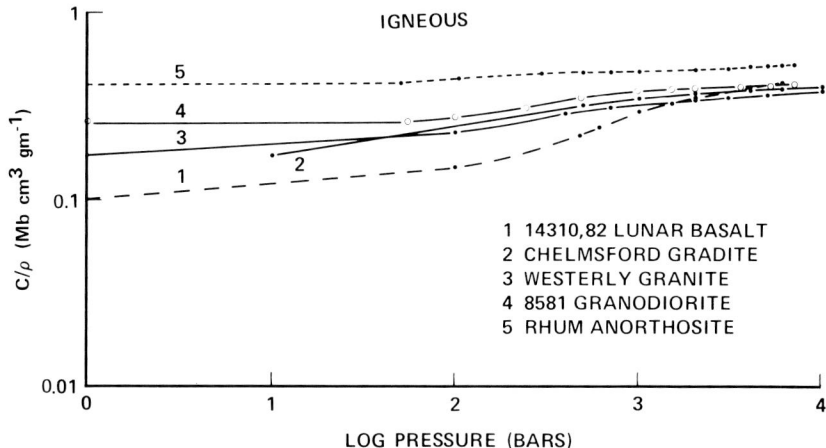

Fig. 1. Log-log plot of C/ρ versus P data for igneous rock. References for curve 1 are the appendix and Trice, et al., [1974]; curve 2, Birch [1960]; curve 3, Simmons and Brace [1965]; curve 4, the appendix; and curve 5, the appendix.

coarsely spaced than V_p data (with the exception of those obtained by differential strain analysis [Simmons et al., 1975]). The relations between static and dynamic moduli, however, must bear on the problem of inverting elastic data into microstructure parameters. This matter is discussed in the section on theory. Next, examples of functional forms of data are presented without interpretation.

Groups

Figures 1 through 6 illustrate functional forms of log C/ρ-log P data for samples ranging from the very competent rocks to soils. Plots of log C/ρ-log P show essentially the same character as plots of log C-log P, since C changes much more with pressure than density does. This is even essentially true of soil data [Warren and Trice, 1975].

The data are characterized both by the general, or smoothed, functional forms of the curves and by local inflections, or variations in the curve forms about the mean trend. Summary descriptions of certain of the samples are given in the appendix, and velocities are given in Table A^1 on microfiche.

Very competent but apparently microcracked rocks such as granites exhibit curves which range in shape from nearly flat to more S-shaped (Figure 1). Such S-shaped curves are nearly flat at low pressure, become concave upward, usually in or near the 100-bar range, and at higher pressure have negative curvature with values that approach zero with increasing pressure, the curve becoming nearly flat again at pressures in the kilobar range.

---

[1] Table A is available with entire paper on microfiche. Order from American Geophysical Union, 1909 K Street, N.W., Washington, D. C. 20006. Document G20-001; $1.00. Payment must accompany order.

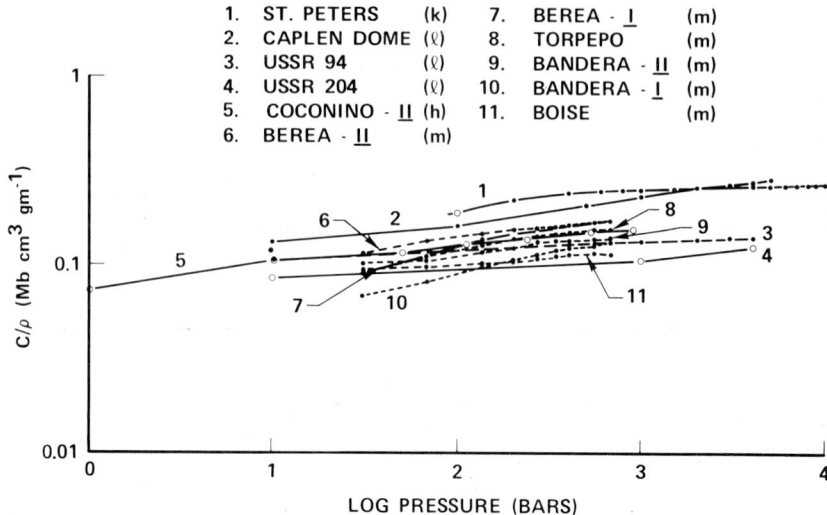

Fig. 2. Log-log plot of C/ρ versus P data for sandstones. The k is from Schock et al. [1974]; 1, Anderson and Liebermann [1968]; h, the appendix; and m, King [1966].

Data for a number of quartz sandstones are shown in Figure 2. Porosities range from 5% to 20%. Curve shapes for sandstone data grade from S-shape to broad arcuate curves with a negative curvature (convex) over most of the measured pressure range. The more S-shaped curves are found for graywackes and low-porosity quartzites. In the pressure range of tens of bars to hundreds of bars, shapes of sandstone curves have curvatures similar to those from the data of competent igneous rock in the high hundreds through the 1000-bar range (Figure 1). In the kilobar range, although the curves are flat, the absolute values are usually well below the value of C/ρ for pure quartz of 0.365 Mbar g^{-1} cm^3.

Figure 3 shows curves for a suite of samples made by compacting sorted quartz grains into 'artificial' sandstone without grain overgrowths. Above 100 bars, all the samples essentially show slopes of 1/3. Only one sample (sample 18) shows a definite flattened curve at pressure below 40 bars. All of the samples were formed by similar processes and have in that respect essentially equivalent histories of formation.

Samples for which data are plotted in Figure 4 are shocked breccias which have certain petrographic descriptive features in common. Lunar sample 15498 is an unrecrystallized breccia with a glassy matrix and no visible porosity, although it has a dense network of cracks. The mineralogy is plagioclase and pyroxene. Apparently, there are no published petrographic descriptions of sample 14318 other than general characterization of it as a shock breccia. Microscopy observations we have made, however, show this sample to have a very similar structure to sample 15498. The Coconino class 2 sample is almost pure quartz, has no porosity other than cracks, and contains up to 10% glass.

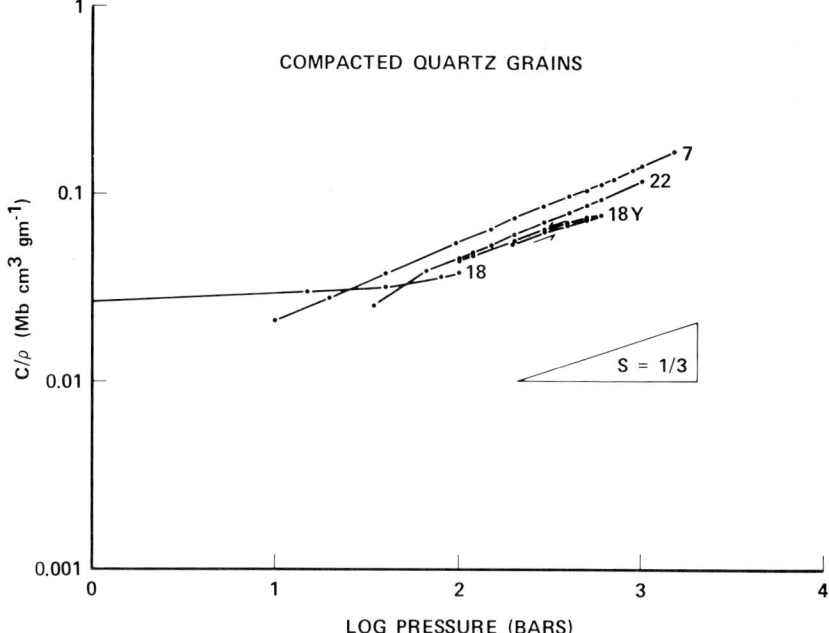

Fig. 3. C/ρ versus P data for a suite of artificially compacted quartz grain samples. See the appendix for description and velocity data.

The pressure dependence of the Coconino class 2 shocked sandstone on the up-pressure run is very similar to that of the other two samples. However, the down-pressure run curve crosses over the up-pressure curve and retains essentially a power law form to the lowest pressure at which V_p was measured. Above 200 bars, all these curves have slopes of 1/3.

Data for two predominantly quartz impactites and one lunar microbreccia (pyroxene-plagioclase) are shown in Figure 5. Both impactites are described as class 1 rocks and were formed by missles impacting a quartz sand region of the White Sands test ground (S. W. Keiffer, personal communication, 1976). Sample WW1 is a 'brown' rock and therefore is apparently less microfractured than sample WW2, which is a 'white' rock. There is unfortunately little published characterization of sample 10065 other than that it is a fine-grained microbreccia of significant porosity.

Figure 6 shows soil data which have power laws of 2/3 to 3/4 at low pressure and approximately 1/3 power laws at pressures above 100 bars.

Trends

The two sets of data shown in Figures 7 and 8 are for sequences of shocked rocks. The Ries Crater samples are a sequence of shocked low-porosity igneous rocks, shock level increasing from 934 (unshocked) to 931 [Todd et al., 1973]. The Coconino sequence (Figure 8) includes data for unshocked Coconino sandstone, the class 2 shocked sample (shown also in Figure 4), two class 1b samples, and, for comparison,

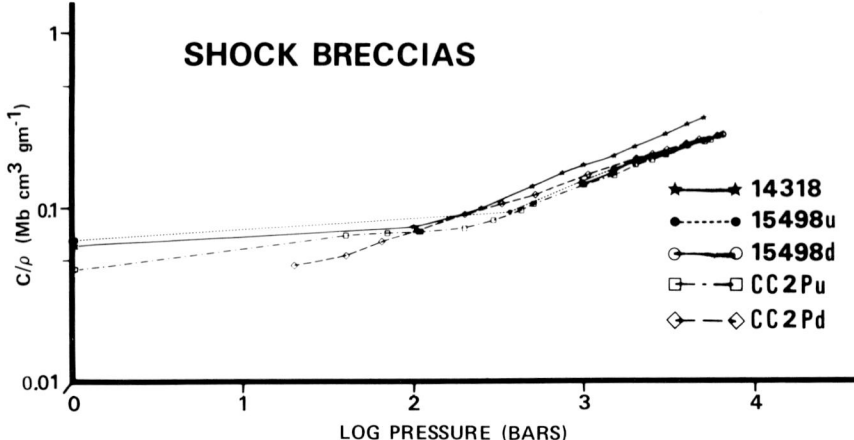

Fig. 4. C/ρ versus P data for two lunar unrecrystallized breccias (samples 14318 and 15498) and a shocked sandstone (sample CC2P). Up- and down-pressure data are denoted by u and d, respectively. See the appendix for lithologies and velocity data for the Coconino sandstone, the appendix and Warren et al., [1973] for sample 15498, and Todd et al., [1973] for data for sample 14318.

an impactite 'dynamite rock' formed by a dynamite explosion during field seismic work in the Imperial Valley, California (the up-pressure curve).

The curves for the class 1b samples and the dynamite rock show average 1/3 slopes, as well as a series of inflections, or kinks, about the mean trend.

Kinks and crossovers can be more apparent in V-P data than in log-log display. Figure 9 shows four sets of data with crossovers. The curves also show a number of local inflections, some of which are reversible, appearing in both up- and down-pressure runs. Reversible inflections are clearly seen superimposed on the broad hysteresis in the Harkless quartzite data at a pressure near 1 kbar (Figure 10).

Characterizing Parameters

The functional forms of the presented data clearly show systematic features. The features may be treated broadly as examples of characteristics which may be used to classify data. General characterizing features, not all of which are treated in this paper, include the following: (1) slope of log M-log P data (average or characteristic slope over large pressure ranges); (2) general curvature of log M-log P data (curvature κ over large ranges of pressure); (3) elbows in log M-log P data (marks changes in average slope below and above the elbow pressure (P_e)); (4) inflection regions in either log M-log P or in V-P data (local alteration of slope, either increasing and then decreasing or decreasing and then increasing, varying around the general trend or slope of the data); (5) comparatives (functional form of data for up- versus down-pressure runs, comparative forms over repeated

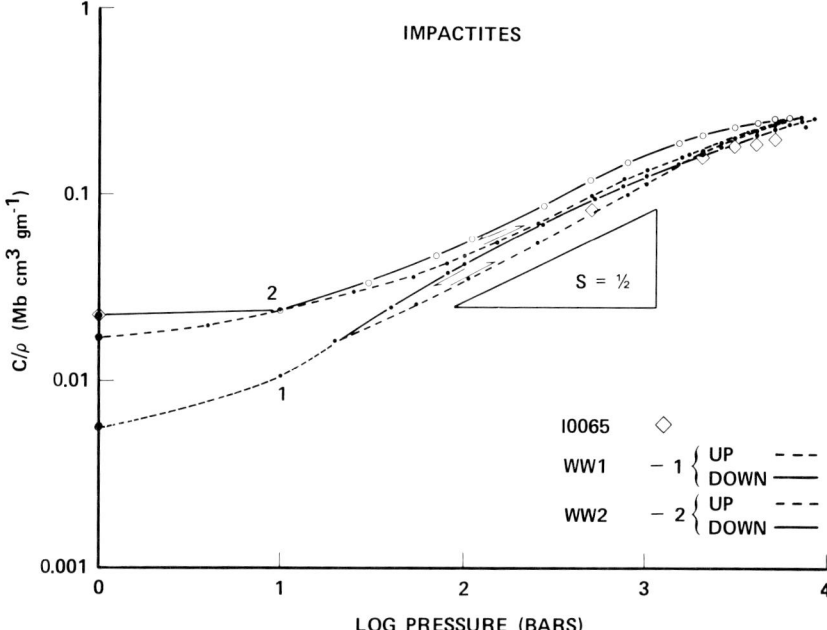

Fig. 5. C/ρ versus P data for impactites from White Sands, New Mexico (samples WW1 and WW2), and for one lunar microbreccia (sample 10065). Velocity data for sample 10065 are from Kanamori et al., [1970]. Samples WW1 and WW2 are listed in the appendix.

cycling, and comparative forms of compressional versus shear and static strain data); and (6) absolute value of data.

The data for igneous rock (Figure 1) show a broad inflection region in the high hundreds of bars, giving the curves their general S shape.

The sandstone data in Figure 2 are characterized by small values of slopes which are positive and decreasing in the 100-bar range (negative curvature), no elbows, and an absence or minimum of inflection regions.

The data for the shocked Coconino sandstone sample CC2P on the up-pressure run (sample CC2PU, Figure 4) are characterized by a slope of 1/3 above an elbow pressure P_e = 200 bars; there is no dominant curvature either above or below the elbow and no inflection regions above P_e. The data for the down-pressure run (CC2Pd) cross over that for the up-pressure run at $P \leq P_e$. The compacted quartz grain data (Figure 3) have a dominant slope of 1/3, no inflection regions, no curvature, and a 'soft' elbow in the data of sample 18.

The curve for the Imperial Valley dynamite rock (Figure 8) is characterized by a general slope of 1/3, large local inflections, and no marked overall curvature.

If in a local inflection region the sense of the alternation is decreasing and then increasing, then the minimum slope S in the region may have a value $S > 0$, $S \simeq 0$, or $S < 0$. Crossover occurs at the highest pressure inflection with minimum slopes $S \simeq 0$. This behavior is true of the most reliable crossovers, in which hysteresis is larger than the uncertainties in accuracy (Figure 9).

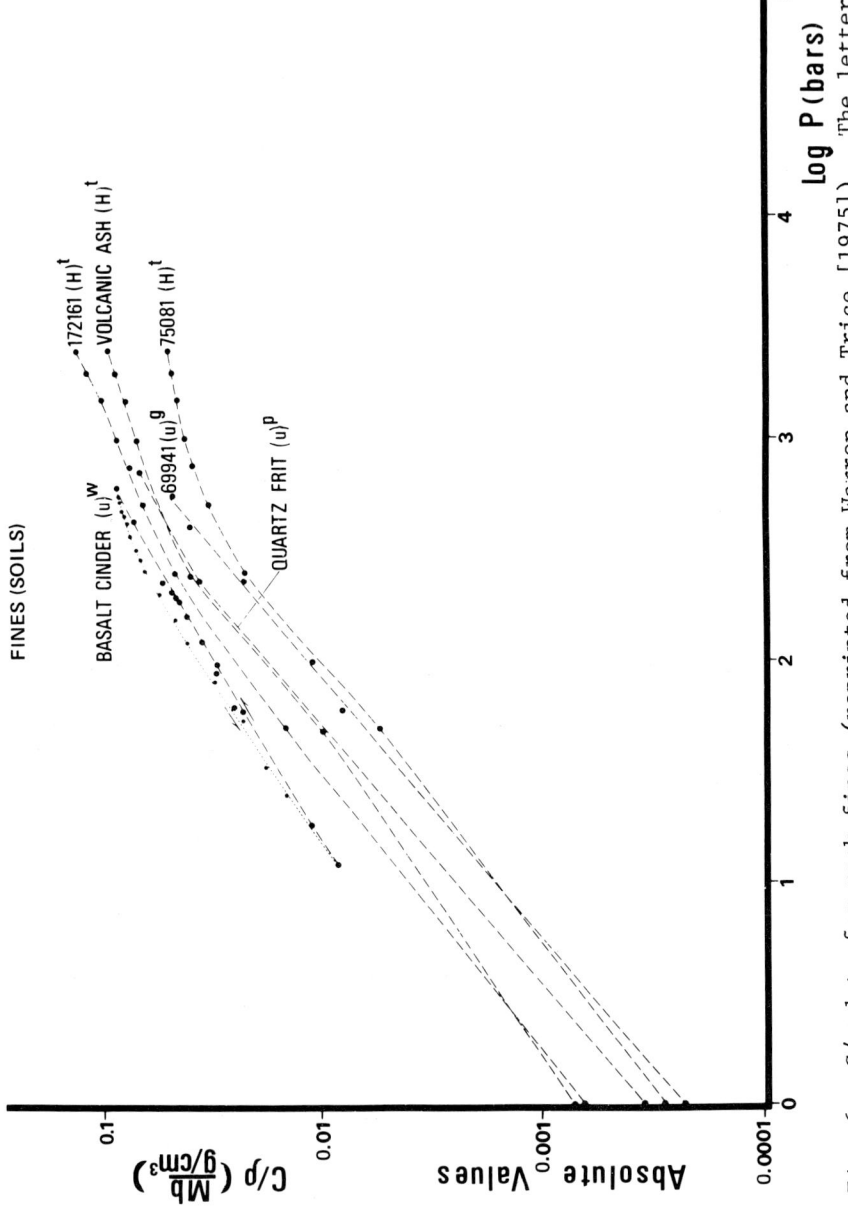

Fig. 6. C/η data for rock fines (reprinted from Warren and Trice [1975]). The letters u and H refer to uniaxial and hydrostatic, respectively. References for w are from Warren and Anderson [1973] and the appendix; t, Talwani et al. [1974]; g, Warren et al. [1973]; and p, Warren [1975].

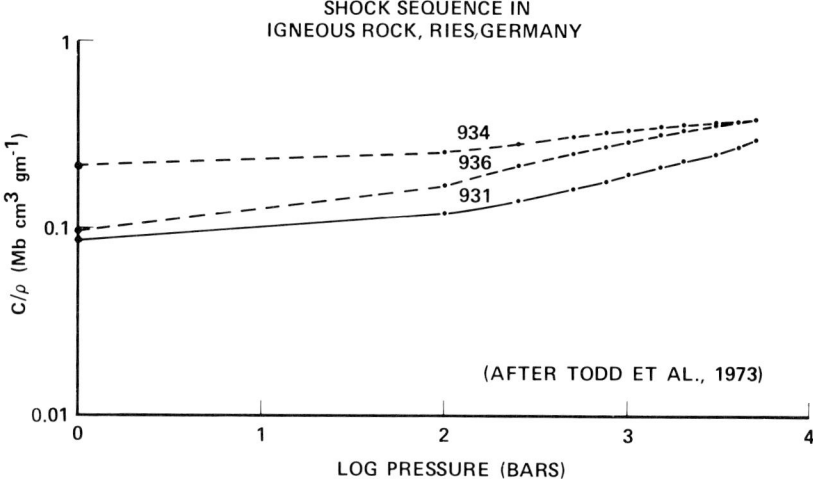

Fig. 7. C/ρ versus P data for shock igneous rock from Ries Crater, Germany.

Theory and Interpretation

Theoretical Functions

Qualitatively, the systematics and characteristics of velocity versus pressure data and of log C/ρ-log P data suggest that data curves can be read in terms of diagnostic features which relate to quantitative microstructural variables. To establish this requires the interpretation of the characteristic functional forms within a framework of a rock-modeling theory. Conversely, rock-modeling theories must be capable of miming the range of observed characteristics, such as S shapeness, power laws, and inflections. In this section, functional forms of solutions to a particular expression for elastic compliances are presented.

A compliance is defined by a ratio of strain to stress and is therefore the inverse of a stiffness modulus. In general tensor notation the compliance tensor $\underline{S}$ is given by

$$\varepsilon_{ij} = S_{ijkl}\sigma_{kl}$$

where ε_{ij} is the strain and σ_{kl} is the stress.

Expressions for compliances can be written in terms of porosities and pore strain. General expressions of this form are derived by Warren and Nashner [1976]. For the case of zero pore pressure, all the expressions take the form

$$\phi/\phi_0 = 1 + \sum_j {_j}\eta(1 - {_t}\eta)^{-1}({_j}\varepsilon_h/\varepsilon_0) \tag{1}$$

where ϕ is a general compliance (ratio of some strain to some stress) of a porous material and ϕ_0 is the corresponding compliance of a solid zero-porosity reference sample made of the same material as the

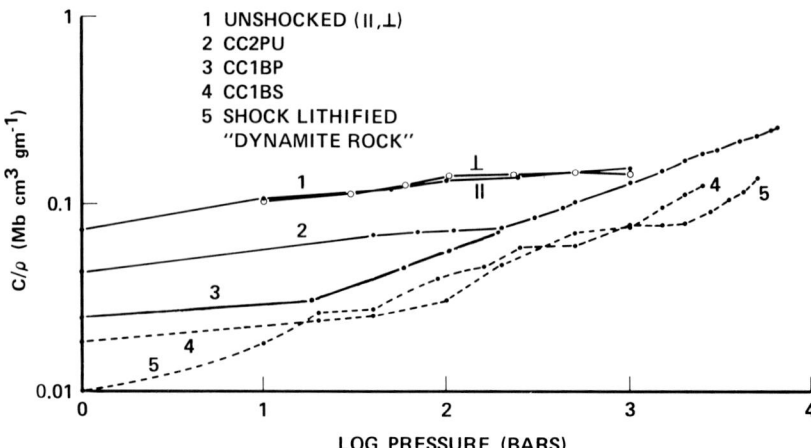

Fig. 8. C/ρ versus P data for a sequence of Coconino sandstones from a meteorcrater in Arizona (curves 1 through 4), and an instant rock (curve 5). Curve 1 is unshocked sandstone perpendicular ($\perp$) and parallel ($\parallel$) to the bedding. CC2PU's are up-pressure data for sample CC2P, and CC1BP and CC1BS are data from class 1b samples. See the appendix for descriptions. The instant rock, plotted for comparison, is also described in the appendix.

porous sample. The term $_j\eta$ denotes the porosity η of the jth type of pore, where each type of pore is parameterized by, say, the geometry. The term $_t\eta$ denotes the total porosity

$$_t\eta = \sum_j {_j\eta}$$

The quantities ($_j\varepsilon_n/\varepsilon_o$) are the ratios of pore strains to reference strains of the reference material. The numerator $_j\varepsilon_h$ denotes the mean pore strain (the pore is denoted by subscript h) of the jth type of pore. The quantity ε_o denotes the equivalent strain component in the solid reference sample under identical conditions of applied stress. Equation (1) is a slight simplification of the general expression by Warren and Nashner [1976] and cannot be applied as it is with complete generality. Furthermore, it is restricted by the assumption that the matrix material is continuously welded. Nevertheless, (1) is fully general enough for application here.

The pressure dependence of (1) depends on the functional relations of pore strains and porosity on pressure. The relations are dependent on pore geometry, such as whether the pores are flat cracks or perhaps intergranular voids due to grain packing. Cracks which can be loosely defined as flat flaws with subparallel faces which close together at finite hydrostatic pressure are often modeled by oblate ellipsoids which have a linear dependence of strain on hydrostatic pressure. On the other hand, strains due to deformations of asperities are nonlinear with pressure. The theory of Hertz [1895] used in soil mechanics to

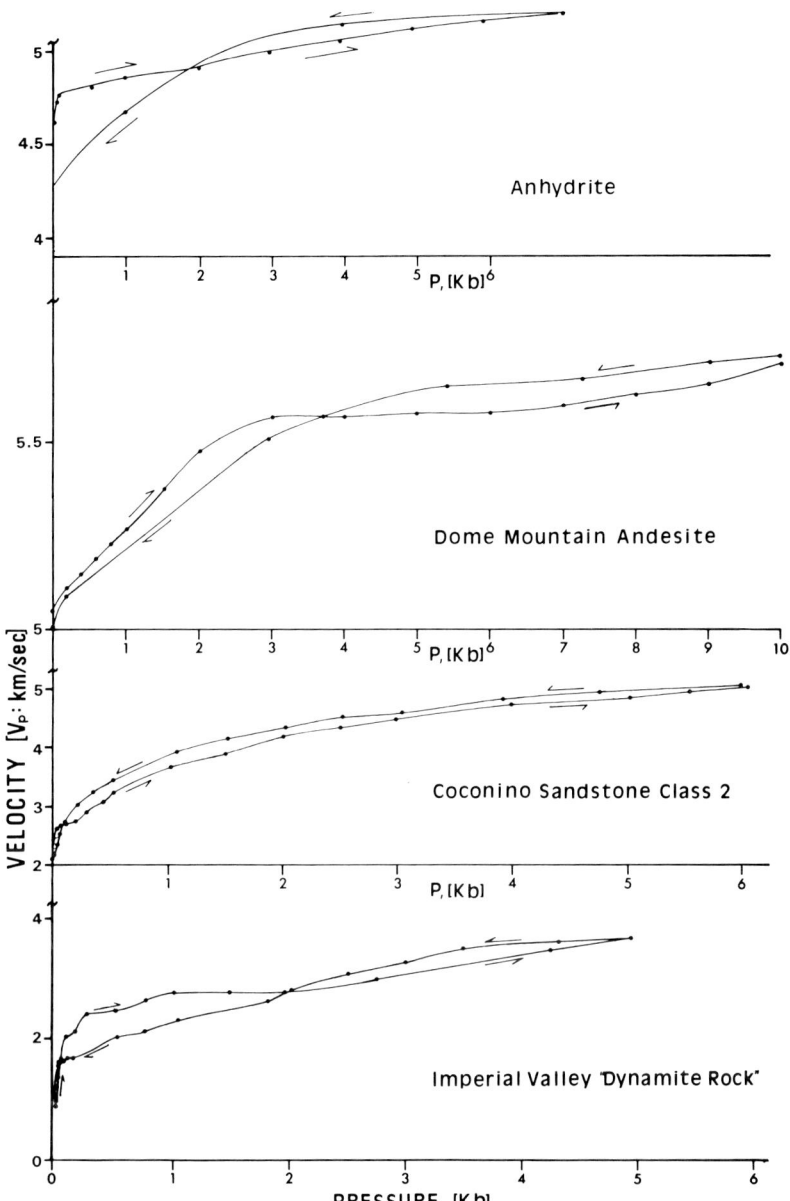

Fig. 9. Velocity-pressure data showing crossing over of up-pressure run data by down-pressure run data. See Nur and Simmons [1969] for an additional example. References for anhydrite are from Warren et al., [1975] (also see the appendix); Dome Mountain andesite, Schock et al., [1974] (down-pressure run data courtesy of B. Bonner, Lawrence Livermore Laboratory); Coconino CC2P, the appendix; and Imperial Valley 'dynamite rock', the appendix.

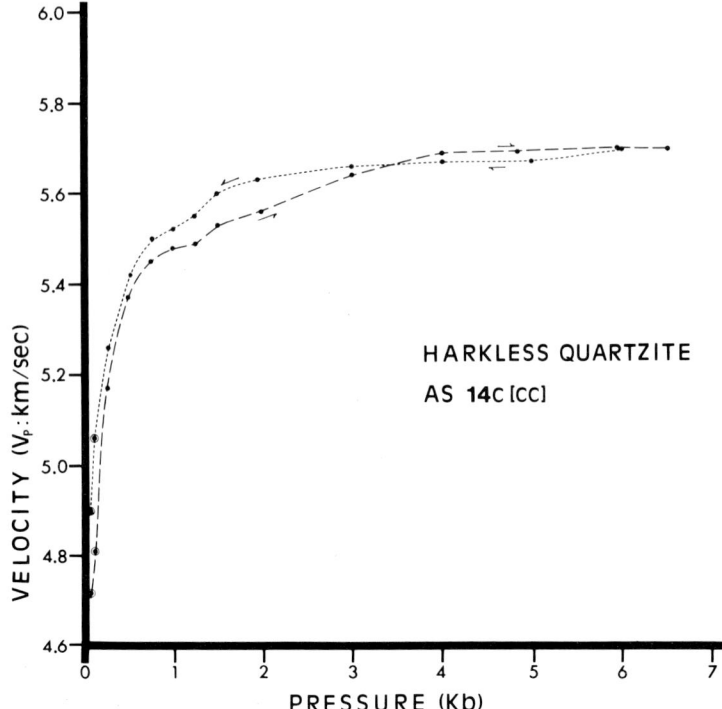

Fig. 10. Velocity-pressure data showing reversible inflection in data at about 1-kbar pressure. Bracketted data have the largest uncertainties in measurement. See the appendix for description.

model grain contacts yields the result that compression between two grains varies as the 2/3 power of the force across the contact.

In addition to the pore strain dependence on pressure, the solutions to (1) depend on assumptions of pore-pore interactions, that is, on the perturbing effects on the strain of one pore caused by the presence of neighboring pores. Two extreme assumptions are typically used. In the first case, pore interactions are completely neglected, and any one pore strains as if it were imbedded in a solid matrix which has moduli equal to those of the solid reference sample. These are nonself-consistent (NSC) solutions. In the other case, pore interactions are assumed to be such that any one pore strains as if it were imbedded in a solid matrix which has elastic moduli equal to the average moduli of the porous material. These are self-consistent (SC) solutions.

For reasonable matrix materials, NSC compliances have finite values for porosities of less than 1.0. Self-consistent compliances can become infinite at very low values of porosity (the material falls apart). In this limit (the 'break apart' value of porosity) the stiffness moduli are always equal to or greater than zero, SC solutions may exist as functions of pressure even for porosities exceeding that required to imply a 'broken' rock.

From the above, we can now look at the functional forms of solutions for elastic moduli such as C, assuming various pore-strain laws. The form of a normalized modulus such as C/C_0 is given by $[\phi/\phi_0]^{-1}$ from (1), where C_0 is the C modulus of the solid reference material. In the following equation the normalized quantities are denoted by asterisks. That is,

$$C/C_0 \equiv C*$$

When solving (1) for the pressure dependence of $C*$, the coefficients involving porosity ($_j\eta/(1 - _t\eta)$) decrease with increasing cumulative pore volume strain.

The porosity $_j\eta(P)$ of the jth type of pore at some pressure P given by

$$_j\eta(P) \approx {_j\eta}(P = 0)[1 - (\Delta V_h/V_h)]$$

where $\Delta V_h/V_h$ is the cumulative pore strain, V_h is the pore volume, and ΔV_h is the total volume change.

The reference solid sample strain ε_0 is given by a linear dependence on stress. The pore strain need not be linearly dependent on pressure; however, the cumulative pore strain must satisfy some function of pressure $f(P)$ in such a way that $f(P) = 0$ at $P = 0$ and that $f(P) = -1$ at the closure pressure P_c of the pore.

Consider a dependence of pore strain ε_h on the rth power of pressure $r \leq 1$. At this point we may drop the subscript j without loss of general argument:

$$\varepsilon_h = -(P/P_c)^r \equiv -(P*)^r$$

For brevity we write $P* = P/P_c$.

For a single crack type, $j = 1$, equation (1) can be differentiated to give

$$\frac{d \log C*}{d \log P*} = -C* P* \frac{d}{dP*} \frac{\eta}{(1 - \eta)} \frac{\varepsilon_h}{\varepsilon_0} \quad (2a)$$

and

$$\frac{d \log C*}{d \log P*} = P* \left[1 + \frac{\gamma \eta_0 (1 - P*^r) P*^{r-1}}{(1 - \eta_0 + \eta_0 P*^r)} \right]^{-1} A \quad (2b)$$

where

$$A = \frac{\gamma \eta_0^2 r(1 - P*^r) P*^{2(r-1)}}{(1 - \eta_0 + \eta_0 P*^r)^2} + \frac{\gamma \eta_0 r P*^{2(r-1)}}{(1 - \eta_0 + \eta_0 P*^r)} - \frac{\gamma \eta_0 (r - 1)(1 - P*^r) P*^{r-2}}{(1 - \eta_0 + \eta_0 P*^r)}$$

The quantity n_o is shorthand for $n(P = 0)$, and γ is a constant arising from expressing $\varepsilon_h/\varepsilon_o$ in terms of the ratio P/P_c. The form of the derivative of the velocity function can similarly be expressed as

$$\frac{dV^*}{dP^*} = \frac{1}{2}\left(\frac{C^*}{\rho^*}\right)^{-1/2}\left(\frac{1}{\rho^*}\frac{dC^*}{dP^*} - \frac{C^*}{\rho^{*2}}\frac{d\rho^*}{dP^*}\right) \tag{3a}$$

where

$$V^* = \frac{V \text{ of the test sample}}{V \text{ of the reference sample}}$$

and the density ρ^* is similarly defined. The expanded form of (3a) is long. However, the basic behavior can be seen by considering the dominant control of the velocity derivative by C^*.

$$\frac{dV^*}{dP^*} \sim \frac{dC^*}{dP^*}^{1/2} = \frac{1}{2}\left[A\left(1 + \frac{\gamma n_o (1 - P^{*r})P^{*r-1}}{1 - n_o + n_o P^{*r}}\right)^{-3/2}\right] \tag{3b}$$

where A is given following (2b).

The general characteristics of the solution for various r can be obtained by assuming small porosity n in such a way that the term $1 - n$ can everywhere be approximated by 1 and the density ρ^* by 1.

If pore strains are linearly elastic, $r = 1$ so that the stress ratio $\varepsilon_h/\varepsilon_o$ is a constant (γ), and for oblate spheroidal pores it is of the form $\gamma \sim a/b$ [e.g., Warren, 1973], where a/b is the ratio of the equator diameter to the pole diameter (i.e., the inverse of the aspect ratio α). Equation (2) becomes

$$\frac{d \log C^*}{d \log P^*} = P^*\left(\frac{1 + n_o \gamma}{n_o \gamma} - P^*\right)^{-1} \quad r = 1 \quad P^* \leq 1 \tag{4}$$

which goes to zero as $P \to 0$ and increases to a maximum as $P \to P_c$; the form of $\log C^*$-$\log P^*$ is flat at small pressures, becomes concave upward as P approaches P_c, and is flat for $P \geq P_c$ (Figure 11). The character of the solution $r = 1$ is therefore similar to S-shaped data curves except in the region of $P \simeq P_c$, where theory predicts a sharp corner. From (3a) and (3b) the derivative of velocity is

$$\frac{dV^*}{dP^*} \sim \frac{1}{2}\left[(\gamma n_o)^{-1/2}\left(\frac{1 + \gamma n_o}{\gamma n_o} - P^*\right)^{-3/2}\right] \tag{5}$$

which implies $V^*(P^*)$ is concave upward over the region $P = 0$ to $P = P_c$. Large-velocity gradients at $P \to 0$ and convex V-P functional forms are not predicted.

If pore strains depend on pressures to some power r, $r = 1 - \xi$, where ξ is small but finite, then

$$\frac{d \log C^*}{d \log P^*} \to \xi \quad P \to 0$$

and $d \log C^*$-$d \log P^*$ is similar to (4) as $P \to P_c$. Although the basic

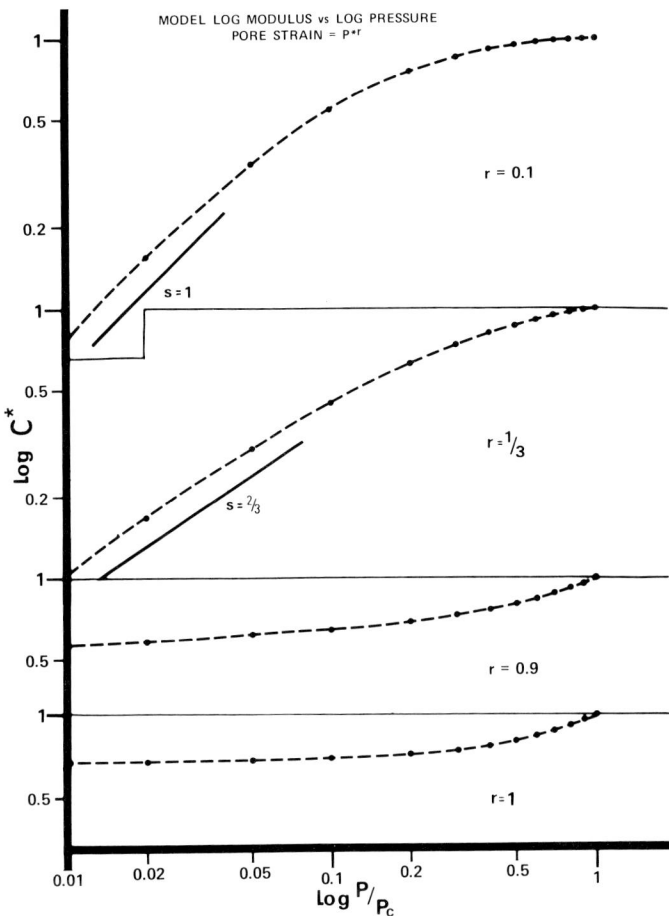

Fig. 11. Model log C*-log P curves for r = 1.0, 0.9, 0.33, and 0.1, where C* is the normalized modulus $V_p^2 \rho$ discussed in the section on theoretical functions.

form of the function log C*-log P* for $r = 1 - \xi$ is similar to the case where r = 1, the dominant term in the velocity derivative goes to approximately p^{-1} as P* → 0, and so the velocity derivative becomes infinite. The limiting value for P → P_c is the same as that for r = 1. The velocity curve $[C*(P)]^{1/2}$ has positive and decreasing slope for pressure sufficiently less than P_c and becomes similar to the velocity profile for r = 1 as P → P_c.

A value of r < 1 of course results in a stiffness modulus C* and velocity V* going to zero at P* = 0, as is true in soil theories. The extreme case of r = χ, where χ is finite and small, may arise if pore strains are very sensitive to pressure in some pressure ranges. In such cases, as P* → 0,

$$\frac{d \log C*}{d \log P*} \to 1$$

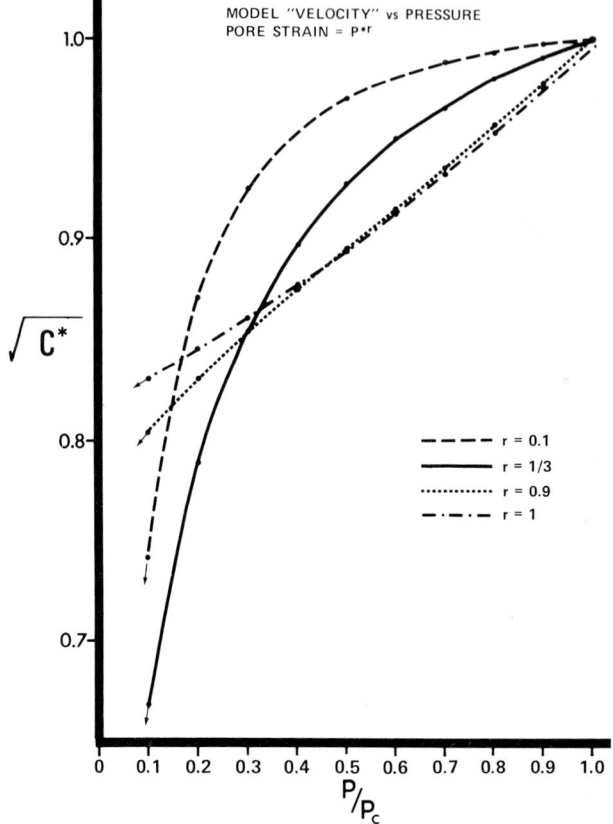

Fig. 12. Model velocity-pressure curves for r - 1.0, 0.9, 0.33, and 0.1.

dV^*/dP^* becomes infinite, and

$$\frac{d \log C^*}{d \log P^*} \to \gamma \eta_o \chi \qquad P \to P_c$$

If $\gamma \eta_o \chi < 1$, log C^*-log P^* has negative and increasing curvature. Typical curve shapes are illustrated in Figures 11 and 12. The actual curves plotted are for log C^* and $(C^*)^{1/2}$, given from $C^* = [1 + 0.5(1 - P^*r)P^{*r-1}]^{-1}$.

This equation for C^* comes directly from (1) if we assume a single pore population, low porosity $(1 - {}_t\eta \simeq 1)$, and a pore strain ratio given by γP^{r-1}, where the value of γ is such that $\gamma \eta_o \simeq 0.5$. The expression $(C^*)^{1/2}$ goes as velocity if we assume normalized density $\rho^* \simeq 1$.

Comparison of the forms of these theoretical curves to data plots such as those in Figures 1 through 8 suggests that data curve characteristics such as slope, curvature, and inflection regions may be interpretable by models for elastic moduli which are controlled by the pressure dependence of the pore strains of the microstructure. In doing so, however, uncertainty arises concerning the need to take

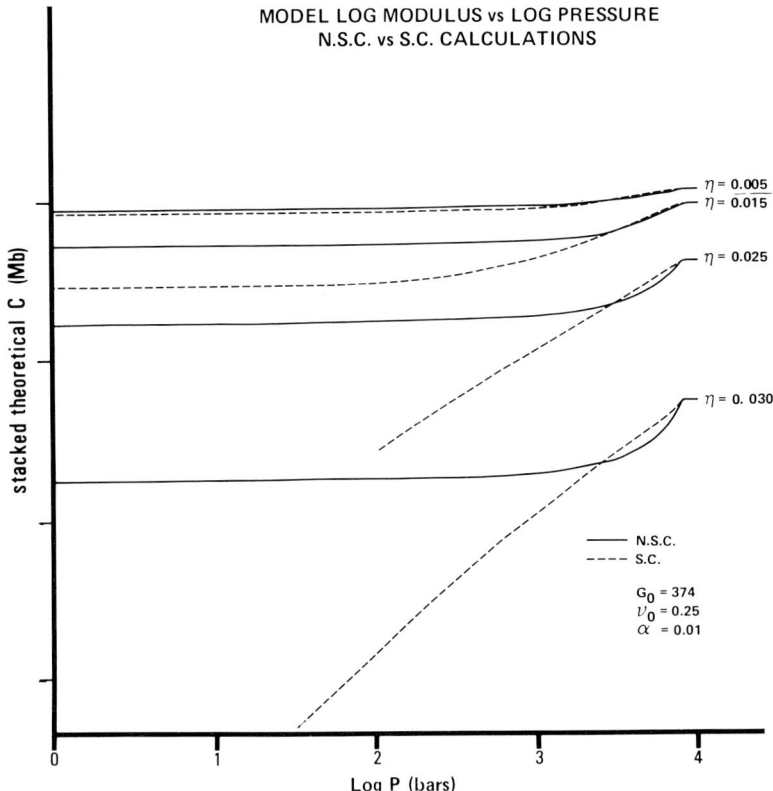

Fig. 13. Self-consistent and nonself-consistent model curves for a material with zero-porosity shear modulus $G_0 = 347$, Poisson's ratio $\nu_0 = 0.25$, and cracks of aspect ratio $\alpha = 0.01$. Porosity at zero pressure is denoted by η.

pore-pore interaction into account in these modulus calculations. Self-consistent calculations are argued for by Budiansky and O'Connell [1976] for the case of cracks. At the other extreme in soil theories, pore-pore interactions are not considered. Simple soil theory is framed in terms of the deformation of isolated pairs of grain contacts.

How do self-consistent results differ from the results discussed above? Figure 13 shows calculated self-consistent and nonself-consistent curves for increasing porosity at constant aspect ratio α. The calculation is for $r = 1$, and the pores are oblate spheroids with $\alpha = 0.01$. For low porosity or weak pore-pore interactions, SC solutions approximate the NSC solutions, the deviations being largest at low pressure. At higher porosities a value is reached where $C^* \to 0$ as $P^* \to 0$. As porosity increases through the break apart value given by $\eta_0 \gtrsim 2b/a$, no convergent solutions for C^* are generated by our program for pressure near zero. However, at higher pressures, convergence is fairly rapid, and the SC solution is found to be essentially a power law in pressure.

In Figure 13 the values of SC moduli exceed the values of NSC moduli for pressures near P_c because in the SC case the cumulated pore strains calculated self-consistently are greater than the cumulated strains found by assuming a linear dependence on pressure as is done in the NSC calculations. Calculated values of NSC and SC of compressional and shear moduli and Poisson's ratio are given in Table 1 for $n_o = 0.015$ and $n_o = 0.025$. Even though $r = 1$ in these calculations, the power law generated by the SC model at high porosities (e.g., $n_o = 0.025$) is somewhat similar in character to the NSC solution for extremely nonlinear pore closure ($0 < r < 1$). This may be expected, since in both cases the pressure dependence of the strain ratio $\varepsilon_h/\varepsilon_o$ approaches P^{-1}. The fact that a critical minimum porosity is required for the existence of a power law solution for the moduli can be demonstrated, at least in a limited case. If it is assumed that a stiffness modulus goes as a power law in pressure $P*$, then the equation for compliance (1) must satisfy

$$n\varepsilon_h/\varepsilon_o = P*^{-f} - 1 \qquad (6)$$

Now, since self-consistency can be expressed as a dependence of pore strain on porosity and both strain and porosity are functions of pressure, a solution for the pore strains satisfying (6) can be written as a function of pressure. A solution for (6) is

$$\varepsilon_h = -\frac{1}{2} - \left[\frac{1}{4} + \frac{\varepsilon_o}{n_o}(P*^{-f} - 1)\right]^{1/2} \qquad (7a)$$

For the case of the bulk modulus we can set $n_o = -P/K_o$, and then for $f = 1$, (7a) becomes

$$|\varepsilon_h| = \frac{1}{2} + \left[\frac{1}{4} - \frac{1}{n_o K_o}(P_c - P)\right]^{1/2} \qquad (7b)$$

This solution is possible only if the radical is positive; this requires that the porosity n_o exceed a critical value which can be cast in terms of material and crack parameters [Walsh, 1965a]

$$P_c \sim E_o \alpha$$

where E_o is the Young's modulus of the reference material and α is the crack aspect ratio. From (7b), n_o must satisfy the relation

$$n_o/\alpha \geq 4E_o/K_o$$

For a Poisson's ratio of 0.25, if n_o is greater than 2.6α, the modulus can show a power dependence on pressure, consistent with the modeling when the material is cracked beyond the break apart limit. A power law solution is compatible with that expected for a fully comminuted material.

Interpretation and Modeling

Equation (1) gives rise to functional forms consistent with the forms of the characteristics of real data curves discussed in the section on the characterizing of velocity-pressure functions. S-shaped

TABLE 1. Self-Consistent Versus Nonself-Consistent Moduli
(in kilobars)

	$\eta_o = 0.015$					$\eta_o = 0.025$				
	NSC		SC			NSC		SC		
P	$C^§$	$G^†$	$C^§$	$G^†$	$\nu^{††}$	$C^§$	$G^†$	$C^§$	$G^†$	$\nu^{††}$
0	613	248	329	132	0.092	475	202			
0.1	617	250	354	157	0.098	478	203	81	39	0.029
0.4	628	253	414	181	0.113	490	208	188	88	0.057
0.6	635	255	449	193	0.121	498	210	240	111	0.070
0.8	643	258	480	204	0.128	506	213	285	130	0.081
1	651	260	508	214	0.135	514	216	326	146	0.091
2	692	273	627	254	0.161	557	231	487	207	0.130
3	740	286	723	283	0.182	608	247	617	250	0.159
4	795	301	819	308	0.199	670	266	733	284	0.183
5	859	317	904	328	0.215	747	288	841	313	0.203
6	935	335	984	345	0.228	845	314	943	337	0.222
7	1030	356	1061	362	0.241	977	345	1043	358	0.238
8	1122	374	1122	374	0.250	1122	374	1122	374	0.250

$G_o = 374$, $\nu_o = 0.25$, and $\alpha = 0.01$.
§C indicates compressional modulus in kilobars
†G indicates shear modulus in kilobars
††ν indicates Poisson's ratio

curves (typical of data for igneous rock) can be generated by letting r = 1 for pressure below P_c and by letting r be less than 1 for pressures greater than or equal to P_c. This amounts to assuming that pore strains are almost ideally elastic at pressures of $P < P_c$ and that they are controlled by contact points at $P \simeq P_c$.

This solution, however, implies concave upward V-P curves in the pressure region in which pore strains are elastic. In Figure 9 the upward scalloped form of some of the data curves (e.g., Dome Mountain andesite up-pressure run, where P < 1 kbar) do indeed suggest elastic piecewise closure of crack networks. In addition, Figure 14 shows V-P data over the first 100 bars for two igneous and two sedimentary rocks (a sandstone and a compacted granular coherent rock). The igneous rock data are concave upward as was predicted. Indeed, most experimentalists simply note that rock velocities change very little in the first 100 bars for 'good' rocks (usually referring to competent igneous rocks, quartzites, etc.)

It follows from the information above that inflections in log C-log P or in V-P plots may be modeled by a distribution of cracks with discrete closing pressures. Such predicted inflection regions should be elastic (i.e., reversible, occurring on both up and down runs). A good example is the 'scallop' at ~ 1 kbar seen in the Harkless quartzite data (Figure 10).

Inflection of the type associated with crossovers (Figure 9) clearly indicate irreversible changes in rock structure. For example, in the case of Coconino sandstone sample CC2P the curve through the down-

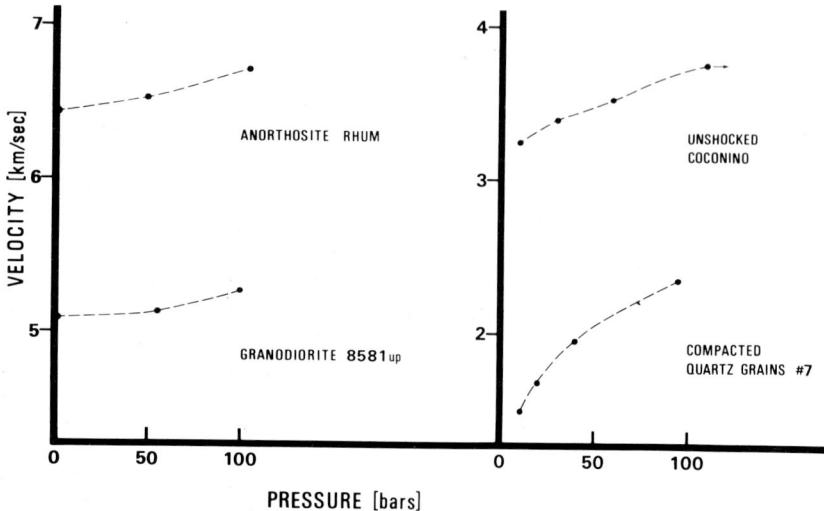

Fig. 14. Velocity-pressure data over first 100 bars of pressure for two igneous rocks, a sandstone and a compacted competent granular rock. See the appendix for descriptions of samples.

pressure run data (CC2Pd) crosses the up-pressure run data (CC2Pu) at about 200 bars, indicating that some structural failure occurred near 200 bars. Although the rock was still competent after the pressure run, some stiffness component (welding component) probably failed and caused contact strains (r < 1) to hold at pressures below 200 bars on the down-pressure run.

For $r \leq 1$ the slope of log C*-logP* at $P < P_c$ is equal to $r - 1$. The CC2Pd data give a slope of 1/3, implying $r = 2/3$ (see Figure 4). Data for a number of other rocks, including competent rocks (such as the lunar microbreccias, Figure 4), also show similar power law behavior. A value of $r = 2/3$ is consistent with pore strains which are controlled by the deformation of grain contacts in a fully compacted material (no loose grains).

In soil mechanics, powers greater than 1/3 are modeled by assuming that the number of grain contacts carrying the stress is pressure dependent [Ko and Scott, 1967; Warren and Anderson, 1973]. In this paper, (1) assumes a continuously welded matrix, so in a model based on (1) the density of grain contacts is not an independent function of pressure. However, the steeper slopes of log C/ρ-log P data for, say, the soils are generated by values of $r < 2/3$. Indeed, the form of the model curve given by $r = 1/3$ (Figure 11) is in excellent agreement with the basalt cinder data in Figure 6. An explanation for this agreement may follow from the earlier argument that indicates that pore-pore interactions couple pore strains to porosities and that this coupling leads to the same power law dependence of the moduli on pressure as that for noninteracting pores with nonlinear pore strains.

The sandstone data are compatible with solutions involving $r < 1$. The positive decreasing curvature of the log-log plots is interpretable as being due to contact structures or to deformable minerals at grain contacts which yield with pressure, the large central pore regions not collapsing.

Data such as that for the dynamite rock (Figure 8 and the appendix), which show numerous inflections about a mean trend with a slope of approximately 1/3, are effects of cracks and collapsing structures on top of a dominant control by grain contacts. Elbows, such as those shown in Figure 4, apparently mark essentially abrupt changes in deformation mechanism. Although the cause and the significance of the elbow pressure P_e is not understood, it appears that P_e increases with increasing induration of compacted materials (compare, for example, Figure 4 and Figure 3). A sharp elbow transition between a flat curve and a power law curve, such as is shown in Figure 4, seems to indicate that deformation has gone from being elastic with $r \sim 1$ (crack controlled) for $P < P_c$ to being dominantly controlled by grain contacts ($r \sim 2/3$).

The more rounded upward concavity of the low-pressure end of S-shaped curves is related to the closing of cracks.

It is apparent that functional forms such as those of log C/ρ-log P data are diagnostic of microstructure and further that log-log plots provide a means of 'reading' not only the dominant mechanism but also the quantitative characteristics of pore structures such as closing pressures of cracks (and hence aspect ratios) and approximate crack porosities. Approximate closure pressures can be obtained from the pressure of the two inflections marking the S. The crack porosity controls the magnitude of the S.

These interpretations are quantified by rock modeling. Comparison of the form of the theoretical S-shaped curves to dynamic modulus data from velocity measurements (Figure 1) suggests that in the igneous samples, pore-pore interactions are minimal or that crack porosities are very low. NSC modeling of granodiorite sample 8581 as a cracked rock gives very good agreement between the theoretical and the measured dynamic moduli (Table 2). Modeling input parameters are zero-porosity shear modulus, Poisson's ratio (from estimates based on mineralogy or from high-pressure data), and crack populations (oblate spheroids) with given porosities (η_o) and aspect ratios.

Since high and low estimates of the moduli are given by NSC and SC calculations, respectively, Warren and Trice [1975] suggested that they should approximate or bound dynamic and static data.

The fit of dynamic data such as, for example, those from sample 8581, should therefore predict approximate values or trends in static data if the observed differences are related to pore interactions in the samples.

Figures 15 through 17 strongly support this suggestion. Figure 15 shows two early models of lunar feldspathic basalt sample 14310 based on measured porosities and published mineralogies.

Figure 16 and Table 3 give recent model calculations and comparisons of moduli to Westerly granite. The NSC and SC models agree well with the dynamic and static moduli. Direct petrographic observation of porosity and aspect ratios for unstressed granite are given by Hadley [1975, Figure 5-7]. Figure 17 shows a comparison of her figure (giving volume-averaged percent porosity versus crack aspect ratio) to the parameters used in the model presented here. The model here was derived primarily from the functional form of the log C/ρ-log P data.

TABLE 2. Granodiorite 8581 Data Versus Model 8581-71576-IV

Compressional Modulus				Shear Modulus			
Pressure	$C^§$	NSC	Δ, %	Pressure	$G^§$	NSC	Δ, %
0.0	697	677	-2.9	0.0	262	269	+2.7
0.055	708	700	-1.1	0.060	273	277	+1.6
0.11	747	724	-3.1	0.11	273	282	-0.3
0.24	841	838	-0.3	0.27	311	313	+0.6
0.49	953	943	-1.0	0.54	330	338	+2.4
0.98	1025	996	-2.8	0.775	337	343	+1.6
1.5	1048	1033	-1.4	1.02	343	349	+1.7
2.0	1058	1072	+1.3	2.0	361	364	+0.9
3.5	1072	1082	+1.0	3.0	364	365	+0.1
5.0	1089	1092	+0.3	4.97	364	368	+0.9
7.0	1099	1105	+0.5	5.8	370	369	-0.3

Pressure and moduli are in kilobars.
Model parameters are (aspect ratio, porosity): (1.0,0.010), (0.011,0.001), (0.002,0.0007), (0.00053,0.0004).
G_0 = 380 kilobars and ν_0 = 0.25.
§These data are from Table A on microfiche.

The significance of the comparison in Figure 17 is that these two independent attempts to characterize crack populations in Westerly granite suggest that there are three main discrete crack populations which can be parameterized by aspect ratios $\alpha \sim 4 \times 10^{-2}$ to 4×10^{-3}, $\alpha \sim 2 \times 10^{-3}$ to 8×10^{-4}, $\alpha \sim 6 \times 10^{-4}$ to 4×10^{-4}. There is reasonable agreement between the identification of these distributions, although in the region of $\alpha \sim 10^{-2}$ the model aspect ratios at $\alpha = 8 \times 10^{-3}$ and $\alpha = 5 \times 10^{-3}$ do not fall on top of the local peaks in the observed distribution. However, the methods used are not exact, and there are definite uncertainties both in modeling and in observational measuring aspect ratios. The general argument, however, seems to imply that these distributions realistically represent the microcrack structure of Westerly granite. Brace [1965], using static strain data, determined that a dominant crack population could be characterized by an aspect ratio of $\alpha = 10^{-3}$, and Warren [1973] attempted to fit a model to the zero-pressure data of Brace and found a need to add a population with aspect ratios $\alpha = 10^{-3}$ to 10^{-4} ($\alpha = 2 \times 10^{-4}$ was used); Hadley's [1975] observation gives the three major peak regions plus a small population with $\alpha = 2 \times 10^{-4}$ to 1×10^{-4}.

In the model given in Table 3 the peaks at $\alpha = 1.2 \times 10^{-3}$ and $\alpha = 5 \times 10^{-4}$ control the shape of the modulus-pressure curve at pressures below about 1.5 kbar. The model fit to data is not strongly sensitive to the exact distribution of pores with $\alpha \geq 10^{-2}$, as these pores mainly affect just the slope of the data at pressures above a few kilobars. The effect of adding a small porosity with an aspect ratio $\alpha \sim 2 \times 10^{-4}$ to the model given here would be to generate a better fit to the zero-pressure data.

In summary, although neither the modeling nor the observations can be taken as exact, it is interesting and perhaps significant that both

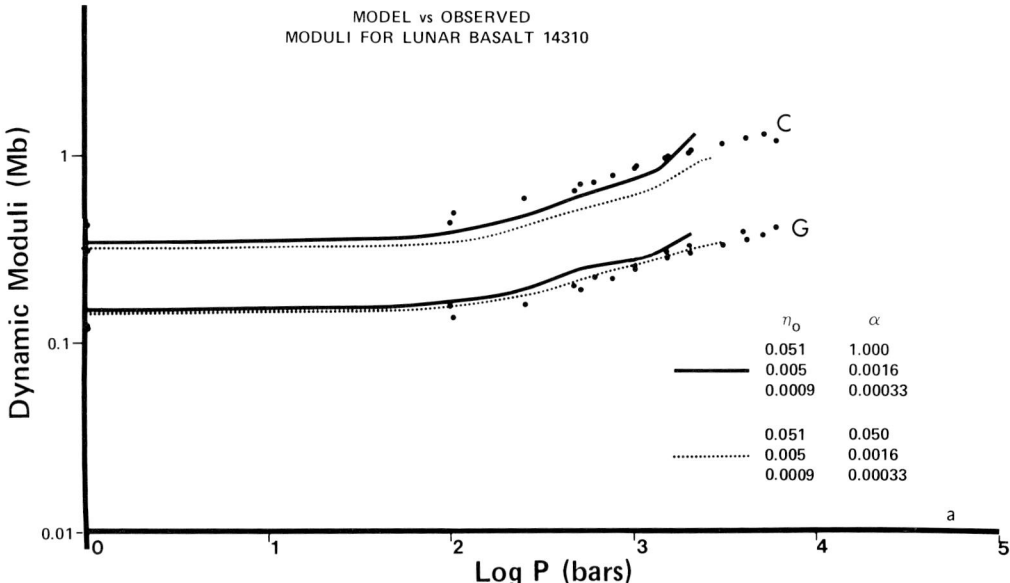

Fig. 15a. Model comparisons to lunar sample 14310. The figure is after Warren and Trice [1975] (with the addition of new static strain data). $C = V_p^2 \rho$ and $G = V_s^2 \rho$. Data (dots) are from both Todd et al., [1973] and Trice et al., [1974]. Model curves are NSC calculations. The η denotes zero-pressure model porosities, and α denotes aspect ratio.

independent methods suggest a similar distribution of cracks parameterized by aspect ratios.

Several consequences follow if such distributions are real and do relate the static moduli to dynamic moduli through a time dependent strain relaxation (pore-pore interactions).

At present, the mechanism of such a pore-strain relaxation mechanism cannot be prescribed. It could be due to the relaxation of stress concentrations in the matrix around a pore or due to friction between crack surfaces as suggested by Walsh [1965b]. However, at low pressures where dynamic and static modulus differences are the greatest, most fractures are open. Therefore if the mechanism involves relaxation of stress concentrations caused by stress interactions between pores (all of which may be open), then although neither the static nor the dynamic data may show hysteresis in themselves, they may differ from each other. This is indeed true of data for good (elastic) rock. In addition, even in the case of nonlinear pore strains the SC moduli will have lower values then the NSC moduli, implying that the static moduli of compacted granular materials should have lower values than those of the dynamic data; this also is observed.

Conclusions

The functional forms of velocity-pressure and dynamic modulus-pressure data are diagnostic of underlying microstructures. Characteristics of this functional form can be represented in terms of a fairly general

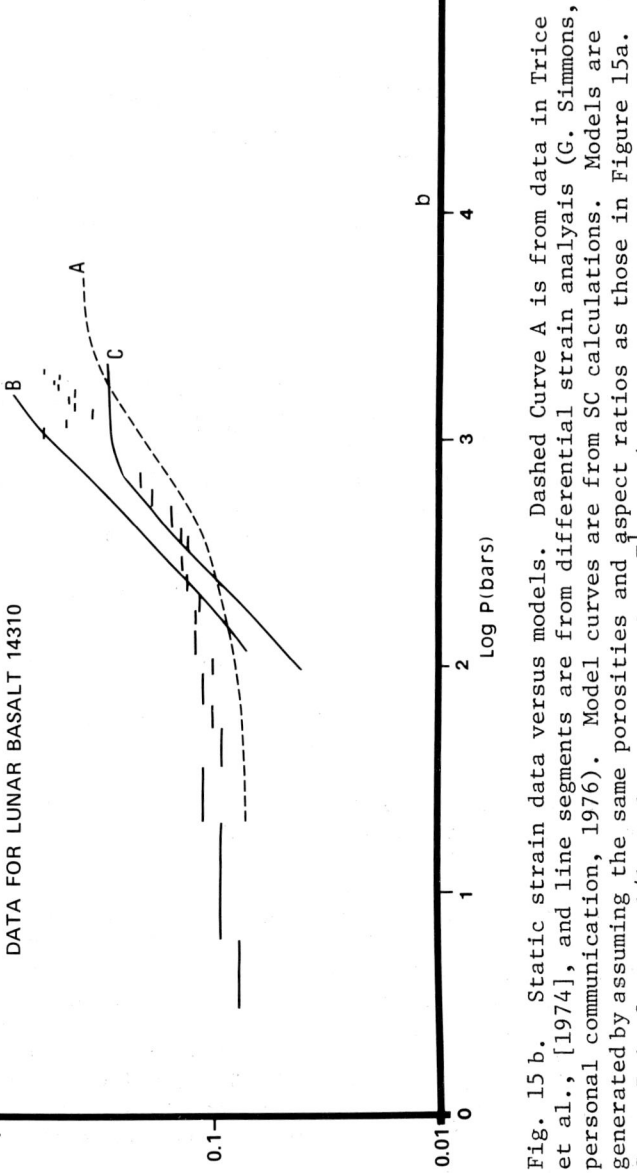

Fig. 15 b. Static strain data versus models. Dashed Curve A is from data in Trice et al., [1974], and line segments are from differential strain analysis (G. Simmons, personal communication, 1976). Model curves are from SC calculations. Models are generated by assuming the same porosities and aspect ratios as those in Figure 15a. Curve B is for $\alpha = 1/1$, and curve C is for $\alpha^{-1} = 20/1$.

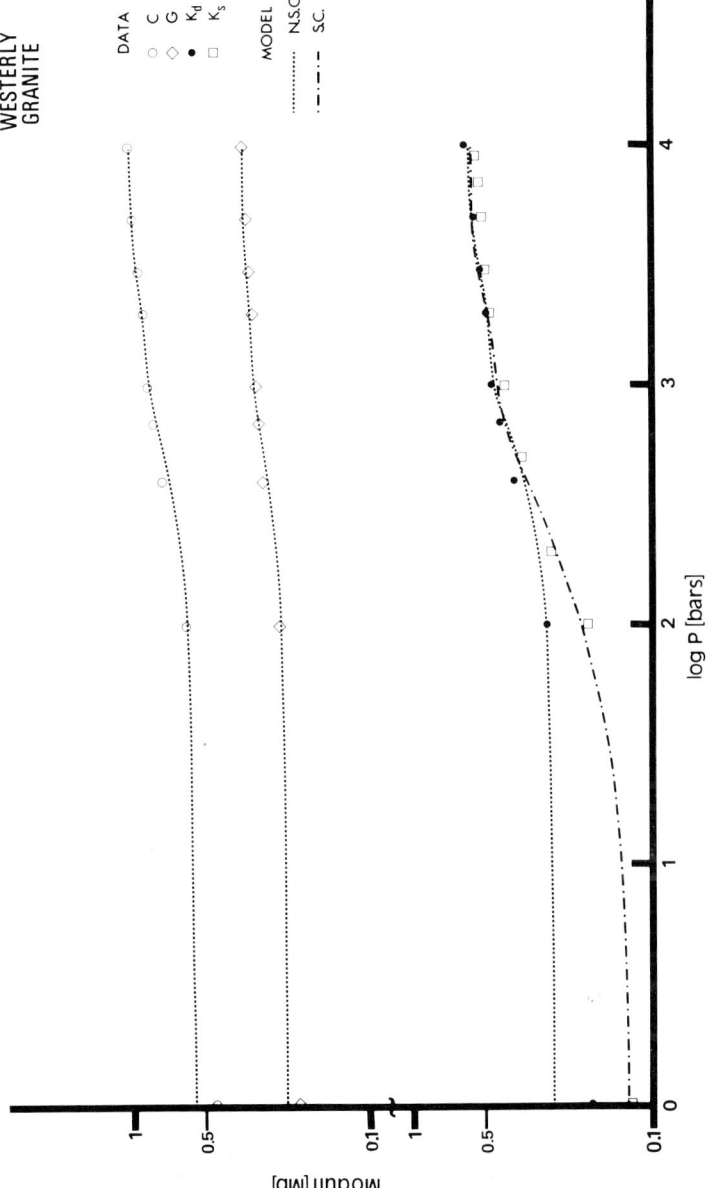

Fig. 16. Model comparisons to Westerly granite data. $C = V_p^2 \rho$ and $G = V_s^2 \rho$, K_s is static bulk modulus, and K_d is dynamic bulk modulus. Data and references are given in Table 2.

TABLE 3. Westerly Granite Data Versus Model WG 71476-I

Pressure	Compressional and Shear Moduli						Pressure	Static Bulk Modulus		
	C§	NSC	Δ, %	G§	NSC	Δ, %		K_s†	SC	Δ, %
0	443	562	+26.7	200	228	+14.0	0	120	124	+3.3
0.1	596	590	-1.0	240	273	+13.7	0.1	185	195	+5.4
0.4	749	692	-7.6	280	266	-5.0	0.2	263	249	-5.3
0.7	809	789	-2.5	290	291	+0.3	0.5	346	356	+2.9
1	859	852	-0.8	300	306	+2.0	1	407	434	+6.6
2	894	898	+0.4	310	316	+1.9	2	463	477	+3.0
3	940	950	+1.0	320	327	+2.2	3	485	519	+0.7
5	983	997	+1.4	333	336	+0.9	5	503	550	+9.3
10	1038	1007	-3.0	340	337	-0.9	7	518	557	+7.5
							9	535	557	+4.1

Pressure and moduli are in kilobars.
Model parameters are (aspect ratio, porosity): (1,0.006), (0.0083,0.0009), (0.005,0.0016), (0.0012,0.001), (0.0005,0.0001). G_o = 341.65, ν_o = 0.248
§Data are from Simmons and Brace [1965].
†Data are from Brace [1965].

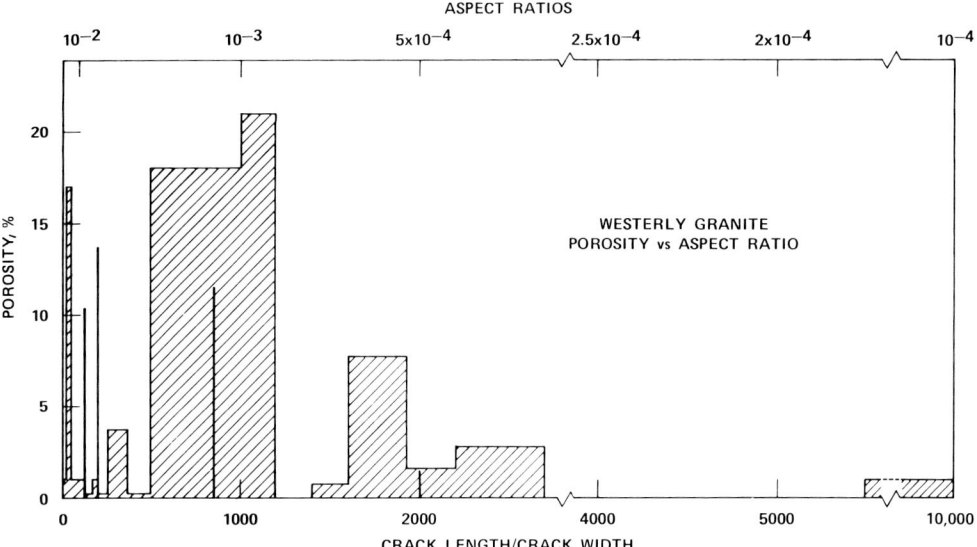

Fig. 17 Comparison of model porosity versus ratios to observed porosity and aspect ratios after Hadley,[1975]. Percent porosity is percent of total porosity. Spherical pore porosities ($\alpha = 1$) are not plotted. Shaded areas are from Hadley, and lines are from the model given in Table 3.

theoretical model, provided we are allowed to vary the pressure dependence of porosity according to plausible microscopic mechanisms. Plots of good data, taken at a sufficient density of pressure points, therefore may provide qualitative and indeed quantitative characterization of the average structures (cracks, pores, and grain contacts) which control velocity. Log modulus-log pressure functions appear to be good diagnostic tools, since they display characteristics controlled by the form of the dependence of pore strains on pressure.

Dynamic data are apparently sufficient to characterize microstructure, and this result lends weight to the suggestion that static and dynamic moduli are related by a time relaxation phenomenon, or dispersion, based on time-dependent pore-pore interactions.

Appendix: Summary of Descriptions

Anhydrite from Canada (courtesy of D. D. Thompson)
  Description by Warren et al., [1975]
    Grains massive and anhedral to well-defined laths;
    bounded by pinacoidal cleavages; grain (lath) sizes
    350µ × 30µ. Two samples designated 1 and 3.
Anorthosite from Rhum, Scotland (courtesy of S. K. Runcorn)
  Description by Wager and Brown [1967, chapters 10, 11, 12] from
  allivalite Rhum unit 13
    Olivine (Fo_{80+}) 25% and plagioclase (bytownite) 75%;
    grain size about 0.5-2 mm. One sample.

Sample 8581 grandiorite from dry well GT-2 at Fenton Hill, Valles
  Caldera, New Mexico (obtained from the Environmental Research and
  Development Administration)
 Description by Trice and Warren (unpublished manuscript)
  Biotite quartz monzonite, grain size 1-2 mm, typical texture,
  moderately fractured in thin section; heavily altered feldspars.
Sample 14310 (lunar feldspathic basalt (courtesy of NASA))
 Description by Simmons et al., [1975]; Brown et al., [1972]; and
  Trice et al., [1974]
  Intersertal texture with plagioclase consisting of prismatic needles;
  grain size of the order of 0.1 mm; (transgranular, cracks in
  pyroxene (0.6µ wide), and thin (0.1µ wide) cracks in plagioclase).
  (One sample designated 14310,82.)
Harkless flat quartzite from Inyo, California (courtesy of J. Nichols
  and Sylvester)
 Description by Sylvester and Christie [1968]
  Quartzite (99% quartz, 1% chlorite, and trace apatite and hematite);
  two sets of cracks, perpendicular and parallel rock a-a direction.
  (One sample designated As-14c (cc).)
Compacted quartz grains (courtesy of D. D. Thompson)
  Dense packing of grains formed by hydrostatic compaction to about
  1 kbar with some consolidation by brine. Below are the dominant
  grain sizes.

Sample	0.5 mm	0.05 mm
7	85%	15%
22	50%	50%
18 (18Y)		100%

Impactites, White Sands, New Mexico (courtesy of S. W. Kieffer)
 Description by Kieffer [1975]
  Formed by missle impacts in quartz sand;
  Two samples designated WW1A and WW2A (WW1A, crumbly reddish brown,
  and WW2A, white sandstone with intragranular fracturing and higher
  shock level than sample WW1A). (Both samples are class 1 shock
  breccias as classified by S. W. Kieffer.)
Impactites from Imperial Valley, California (courtesy of D. D. Jackson)
  Formed by dynamite blast in shallow sand below caliche layer. Clay-
  free alluvium with wide-grain size distribution from very fine to
  few mm. (One sample designated 'dynamite rock'.)
Coconino sandstone (courtesy of S. W. Kieffer)
 Description by Kieffer [1971] and Dumas and Warren (unpublished
 manuscript)
  Unshocked
   95% quartz
   massive porosity about 10%
   well-rounded grains, and grain size 0.09-3.9 mm
   (two samples designated CC perpendicular and CC parallel)
   (perpendicular to bedding) respectively.
  Class 2
   80-95% quartz, 2-5% coesite, and 3-10% glass; well indurated,

grains fractured throughout, and no observable (pore) porosity.
(Two samples designated CC2P and CC2PS.)
Class 1b
40% grains fractured; blocky fractures; disoriented shards form
matrix. (Two samples designated CC1BP and CC1BS.)
Sample 15498 (lunar breccia (courtesy of NASA))
 Description by Christie et al., [1973] (class A unrecrystallized
 breccia in classification of Christie et al.)
  Little or no porosity in reflected light, 50% of matrix may be glass,
  and clasts are fractures. (One sample designated 15498,23.)
Fines
 Quartz frit, 0.05-mm grain size; basalt cinder described by Warren
 and Anderson [1973], grain size of the order of 1 mm.

Acknowledgments. Work was done under Environmental Research and
Development Administration contract E(04-3)34pA224 and support from
Chevron Oil Field Research Company. I especially wish to thank
J. Christie, D. D. Thompson, E. Litov, and H. Kanamori for relevant
discussions, R. Phinney, M. Feves, and B. Bruner for review comments,
and J. Heacock for editorial comments, and Gwen Hummel for preparation
of the final manuscript. This is contribution 1603 of the Institute of
Geophysics and Planetary Physics of the University of California at
Los Angeles.

References

Anderson, O. L., and R. C. Liebermann, Sound velocities in rocks
 and minerals, in *Physical Acoustics*, vol. 4, part B, edited by
 W. P. Mason, p. 330, Academic, New York, 1968.
Birch, F., The velocity of compressional waves in rocks to 10 kbars,
 1, J. Geophys. Res., 65(4), 1083-1102, 1960.
Brace, W. F., Some new measurements of linear compressibility of
 rocks, J. Geophys. Res., 70(2), 391-398, 1965.
Brown, G. M., C. M. Emelius, J. G. Molland, A. Peckett, and R. Phillips,
 Mineral-chemical variations in Apollo 14 and Apollo 15 basalts and
 granitic fractions, Proc. Lunar Sci. Conf. 3rd, suppl. 3, Geochim.
 Cosmochim. Acta, 1, 141-157, 1972.
Budiansky, B., and R. J. O'Connell, Elastic moduli of a cracked solid,
 Int. J. Solids Struct., 12, 81-97, 1976.
Christie, J. M., D. T. Griggs, A. M. Heuer, G. L. Nord, Jr.,
 S. V. Radcliffe, J. S. Lally, and R. M. Fisher, Electron petrography
 of Apollo 14 and 15 breccias and shock produced analogs, Proc.
 Lunar Sci. Conf. 4th, suppl. 4, Geochim. Cosmochim. Acta, 1,
 365-382, 1973.
Hadley, K. Dilatancy: Further studies in crystalline rock, Ph.D. thesis,
 Mass. Inst. of Technol., Cambridge, 1975.
Hertz, H., J. math. (Crelle's Journal) 92, 1881, reprinted in
 Gesammelte Werke, 1, 155, 1895.
Kanamori, H., A. Nur, S. H. Chung, and G. Simmons, Elastic wave
 velocities of lunar samples at high pressure and their geophysical
 implications, Proc. Apollo 11 Lunar Sci. Conf., suppl. 1, Geochim.
 et Cosmochim. Acta, 3, 2289-2293, 1970.
Kieffer, S. W. Shock metamorphism of the Coconino sandstone at Meteor

Crater, Arizona, J. Geophys. Res., 76(23), 5449-5473, 1971.

Kieffer, S. W., From regolith to rock by shock, Moon, 3(1-3), 301-320, 1975.

King, M. S., Wave velocities in rocks as a function of changes in overburden pressure and pore fluid saturants, Geophysics, 31, 50-73, 1966.

Ko, J.-Y., and R. F. Scott, Deformations of sand in hydrostatic compression, J. Soil Mech. Found. Div. Amer. Soc. Civil Eng., 3, 137-156, 1967.

Nur, A., and G. Simmons, The effect of saturation on velocity in low porosity rocks, Earth Planet. Sci. Lett., 1, 183-193, 1969.

Schock, R. N., B. P. Bonner, and H. Louis, Collection of ultrasonic velocity data as a function of pressure for polycrystalline solids, report, Lawrence Livermore Lab., Univ. of Calif., Livermore, April 10, 1974.

Simmons, G., and W. F. Brace, Comparison of static and dynamic measurements of compressibility of rocks, J. Geophys. Res., 70(22), 5649-5656, 1965.

Simmons, G., R. Siegfried, and D. Richter, Characteristics of microcracks in lunar samples, Proc. Lunar Sci. Conf. 6th, suppl. 6, Geochim. Cosmochim. Acta, 3, 3227-3254, 1975.

Sylvester, A. G., and J. M. Christie, The origin of crossed-girdle orientations of optic axes in deformed quantities, J. Geology, 76(5), 571-580, 1968.

Talwani, P., A. Nur, and R. L. Kovach, Implication of elastic wave velocities of Apollo 17 rock powders, Proc. Lunar Sci. Conf. 5th, suppl. 5, Geochim. Cosmochim. Acta, 3, 2919-2926, 1974.

Todd, T., D. A. Richter, G. Simmons, and H. Wang, Unique characterization of lunar samples by physical properties, Proc. Lunar Sci. Conf. 4th, suppl. 4, Geochim. Cosmochim. Acta, 3, 2639-2662, 1973.

Trice, R., N. Warren, and O. L. Anderson, Rock elastic properties and near-surface structure at Taurus-Littnow, Proc. Lunar Sci. Conf. 5th, suppl. 5, Geochim. Cosmochim. Acta, 3, 2903-2911, 1974.

Walsh, J. B., The effect of cracks on the compressibility of rock, J. Geophys. Res., 70(2), 381-389, 1965a.

Walsh, J. G., The effect of cracks on the uniaxial elastic compression of rocks, J. Geophys. Res., 70(2), 399-411, 1965b.

Warren, N., Theoretical calculation of the compressibility of porous media, J. Geophys. Res., 78(2), 352-362, 1973.

Warren, N. and O. L. Anderson, Elastic properties of granular materials under uniaxial compaction cycles, J. Geophys. Res., 78(29), 6911-6925, 1973.

Warren, N., and R. Nashner, Theoretical calculation of compliances of a porous medium, in The Physics and Chemistry of Minerals and Rocks, pp. 197-216, John Wiley, New York, 1976.

Warren N., and R. Trice, Correlation of elastic moduli systematics with texture in lunar materials, Proc. Lunar Sci. Conf. 6th, suppl. 6, Geochim. Cosmochim. Acta, 3, 3255-3268, 1975.

Warren, N., R. Trice, N. Soga, and O. L. Anderson, Rock physics properties of some lunar samples, Proc. Lunar Sci. Conf. 4th, suppl. 4, Geochim. Cosmochim. Acta, 3, 2611-2629, 1973.

Warren, N., R. Trice, and C. Tosaya, Elastic moduli of polycrystalline anhydrite, report, Chevron Oil Field Res. Co., La Habra, Calif., 1975.

MICROCRACKS IN CRUSTAL IGNEOUS ROCKS: MICROSCOPY

Dorothy Richter[1] and Gene Simmons

Department of Earth and Planetary Sciences
Massachusetts Institute of Technology
Cambridge, Massachusetts 02139

Abstract. Precambrian igneous rocks in north-central Wisconsin and the St. Francois Mountains in Missouri contain very few open microcracks but many formerly open microcracks that are now completely healed or which have been sealed with secondary minerals. The common sealing minerals in these exposed basement rocks are quartz, chlorite, rutile, and magnetite. By inductive reasoning from observations made with the petrographic and scanning electron microscopes on these exposed basement rocks we believe that the buried Precambrian crustal igneous rocks in the central United States probably contain very few 'open' microcracks nor do they contain cracks which are merely closed by the lithostatic pressure.

Introduction

The central portion of the North American continent is underlain by an extensive terrain of Precambrian rocks. These rocks represent an extensive terrain of continental sedimentary, igneous, and metamorphic rocks which have been relatively stable for at least 600 m.y. Precambrian terrain is exposed in the Canadian shield, the St. Francois Mountains of Missouri, the Arbuckle Mountains of Oklahoma, and the Llano uplift of Texas. Igneous rocks from these regions are probably representative of a significant portion of the upper continental crust.

Physical properties of igneous rocks measured as a function of pressure in the laboratory may model the variation of physical properties with depth in the crust. If the laboratory experiments are to be analogs for the rocks in situ, then the characteristics of the microcracks in the rocks in situ and in the laboratory specimens must be identical.

In this paper we discuss in detail the microcrack textures present in a suite of Precambrian rocks from central Wisconsin and the St. Francois Mountains of southeast Missouri. In a companion paper, Feves et al. [1977] report on laboratory measurements of the physical properties of the same rocks. Physical properties of igneous rocks that contain microcracks--and most of them do--are strongly dependent on the microcracks present in them. In general, the rocks contain few open microcracks, but there is abundant evidence of healing and sealing of microcracks.

---

[1]Present address: Rock of Ages Corporation, Barre, Vermont 05641

Our terminology follows that of Simmons and Richter [1976] and Simmons et al. [1975]. A healed microcrack occurs where a formerly open microcrack has annealed. Bubble planes (i.e., fluid inclusions), planes of solid inclusions, and offset twin lamellae are the most common evidence of healed cracks. Sealed cracks are formerly open microcracks which have been sealed by secondary or remobilized minerals.

The Samples

The samples we studied are briefly described in Table 1. Modal analyses of our samples are reported by Feves et al. [1977] in their Table 2.

Most of the Wisconsin rocks studied are unmetamorphosed shallow crustal intrusives which were formed in a nonorogenic environment. The Wausau granite, part of an alkaline igneous complex, is interpreted by Lockwood [1970] and Emmons [1953] as an intrusion into the Precambrian basement of Wisconsin. The Mellen gabbro is from a sill-like layered mafic igneous complex near Mellen, Wisconsin [Tabet, 1974]. The Red River quartz monzonite and Tigerton gabbro were collected near Tigerton, Wisconsin, and are part of the epizonal Wolf River batholith [Medaris et al., 1973].

The Missouri samples are from the St. Francois Mountains, a 900 km^2 exposure of late Precambrian epizonal igneous rocks [Tolman and Robertson, 1969; Amos and Desborough, 1970; Kisvarsanyi, 1972]. The Middlebrook felsite and Stouts Creek rhyolite are apparently older than the Graniteville granite and Skrainka diabase. The rhyolites and felsites represent the initial stages of igneous activity in the St. Francois petrogenetic province [Tolman and Robertson, 1969]. The Graniteville granite is one of several related granites which are intrusive into the volcanic rocks. The Skrainka diabase is poorly exposed, but Amos and Desborough [1970] interpreted it as a sill intrusive into the granites and volcanic rocks of the region.

Each specimen was collected carefully and without hammering so that no new cracks would be introduced. Where samples had to be collected at quarries or roadcuts, extreme care was used to take them from as far away from shot points as possible. Preparation of the crack sections for microscopic examination, without producing new cracks, was done as follows: we cut blocks or cores (about 2 x 2 x 4 cm or 1.8 diam x 2 cm) with standard rock saws or coring equipment from rock adjacent to the differential strain analysis (DSA) blocks of Feves et al. [1977]. We then removed at least 6 mm of our sample with a Buehler isomet saw rotating at a very slow speed (perhaps 30-50 rpm) to remove damaged rock. The isomet-cut face was ground by hand on glass plates to remove an additional 1-2 mm, polished to 0.05 µ abrasive on a microcloth, and was mounted on a glass slide with a room temperature curing epoxy. An isomet was used to cut the remainder of the mounted slice to 2 mm. The section was then ground by hand to 100 µ on glass plates and polished to 0.05 µ abrasive on a microcloth. Approximately 10-15 µ of material were removed from each section by ion bombardment, and the samples were coated with ~500 Å of gold. The ion milling removes polishing damage, produces slight topography among the minerals, and also produces the elliptical mounds evident in the figures. Because crack sections are 3 times as thick as standard thin

TABLE 1. Sample Descriptions

Name	Sample Number	Locality	Petrographic Description*	Radiometric Age	References
Wausau granite	1343	Linden quarry, Wausau, SE 1/4, sec. 34, T30N, R8E, Marathon Co., Wisconsin	Coarse-grained (1-3 mm) even textured 'dry' granite composed of quartz, perthite, plagioclase; very few mafic minerals	~1100 m.y.	age: Chaudhuri et al. [1969] geology: Lockwood [1970]
Red River quartz monzonite	1370	Road cut on Wis. highway 29, 0.3 mi. E of Five Corners; NW 1/4, sec. 22, T27N, R12E, Shawano Co., Wisconsin	Weakly foliated, porphyritic (feldspar phenocrysts 0.5-1.5 cm) biotite quartz monzonite; matrix even-textured quartz, plagioclase, microcline, biotite	1450 ± 30 m.y.	age: Van Schmus [1973] geology: Medaris et al. [1973]
Mellen gabbro	1331	Wis. highway 13, 2.1 mi S of intersection with Wis. highway 77, NW 1/4, sec. 6, T43W, R2E, Ashland Co., Wisconsin	Fine-grained (0.2-0.5 mm) massive ophitic olivine gabbro	1640 ± 40 m.y.	age: Van Schmus [1973] geology: Tabet [1974]
Tigerton gabbro	1368	Outcrop on Shawano Co. road J, 1/2 mi. S of intersection with Wis. rt. 29; SW 1/4, sec. 23, T27N, R12E, Shawano Co., Wisconsin	Medium-grained (0.5-1.5 mm) massive gabbro; minor olivine; some alteration of pyroxene to hornblende; some plagioclase highly altered		geology: Medaris et al. [1973]

TABLE 1. (continued)

Name	Sample Number	Locality	Petrographic Description*	Radiometric Age	References
Graniteville granite	1410	Abandoned quarries 30 m E of Mo. highway 21; NE 1/4, sec. 10, T34N, R3E, Iron Co., Missouri	Coarse-grained (1-4 mm) massive granite with few mafic minerals	1273 ± 92 m.y.	age: Bickford and Mose [1975] geology: Tolman and Robertson [1969]
Skrainka diabase	1415	Stream bed on Mo. highway 72, NE 1/4, sec. 10, T33N, R6E, Iron Co., Missouri	Massive coarse-grained (1-2 mm) anorthositic troctolite		geology: Amos and Desborough [1970]
Middlebrook felsite	1407	Road cut on Mo. highway 21, 0.6 mi. S of intersection of Iron Co. road W; NW 1/4, sec. 24, T34N, R3E, Iron Co., Missouri	Porphyritic with aggregates of epidote-chlorite-sphene-quartz as replacements of pyroxene phenocrysts; matrix is very fine-grained (<0.1 mm) quartz and feldspar	1416 ± 32 m.y.	age: Anderson et al. [1969] geology: Anderson [1970]
Stouts Creek rhyolite	1411	Stouts Creek Shut Ins, Mo. highway 72, NE 1/4, sec. 3, T33N, R4E, Iron Co., Missouri (type locality)	Porphyritic with partially resorbed phenocrysts (1-2 mm) of quartz and plagioclase; matrix is very fine (<0.9 mm) quartz and feldspar	1327 ± 27 m.y.	age: Bickford and Mose [1975] geology: Anderson [1970]

*See Table 2 of Feves et al. [1977] for modal analyses of these samples.

sections, cleavage cracks and other introduced cracks are not formed during sample preparation. Microcracks in crack sections prepared in this way can be examined with both the scanning electron microscope (SEM) and the petrographic microscope.

Microcrack Textures

Each sample contains very few open microcracks. We infer that each sample previously contained many more open microcracks that have since closed to form the abundant healed and sealed microcracks present in the samples. Where present, crosscutting relationships among former microcracks can be used to interpret successive fracturing and sealing events in the samples.

Granitic Rocks

The Wausau granite, Red River quartz monzonite, and Graniteville granite are characterized by planes of secondary fluid inclusions in quartz, low open crack porosity, and a variety of sealed cracks.

Several episodes of fracturing deformed the Wausau granite. Transgranular healed cracks such as the one shown in Figure 1 are relatively young crack features, since they crosscut other healed cracks and recrystallized grain boundaries. The crack is marked by a plane of secondary fluid inclusions in the quartz and by recrystallized plagioclase in the perthite. Where a similar healed crack intersects an individual albite lamella in a perthitic feldspar that lamella is generally much more altered than the rest of the albite lamellae in the same grain.

Figures 2a and 2b are a pair of optical and SEM micrographs of the same set of cracks in a quartz crystal in the Wausau granite. Three distinct crack-producing events can be distinguished: (1) an older plane of secondary fluid inclusions (labeled A) which stops at the grain boundary; (2) a younger, smaller plane of secondary inclusions (labeled B) which terminates on the older one; and (3) an open microcrack (labeled C) which crosscuts both secondary fluid inclusion planes. The open crack is obviously the youngest feature. The relative timing of the cracking and healing events of the two planes of inclusions is not so clear. The density and the size of the individual inclusions are very different in each. If the diameter of the fluid inclusions is only related to the original crack width, then both A and B may have healed at the same time if they originally had different widths. However, A may have been healed before B had formed.

Figure 3 is a photomicrograph of a pair of quartz crystals. They contain several sets of parallel planes of secondary fluid inclusions, all of which terminate at the grain boundary, a common feature in this and other granites. Dale [1923] found that such sets of planes were parallel to the rift direction of easiest splitting in several New England granites. However, the planes of secondary inclusions in the Wausau granite are perpendicular to the apparent rift.

Cracks in perthite crystals in the Wausau granite are very commonly sealed by quartz. An example is shown in Figure 4, a wedge-shaped crack in perthite sealed by quartz. The quartz in the crack is optically continuous with the neighboring quartz grain but, in some

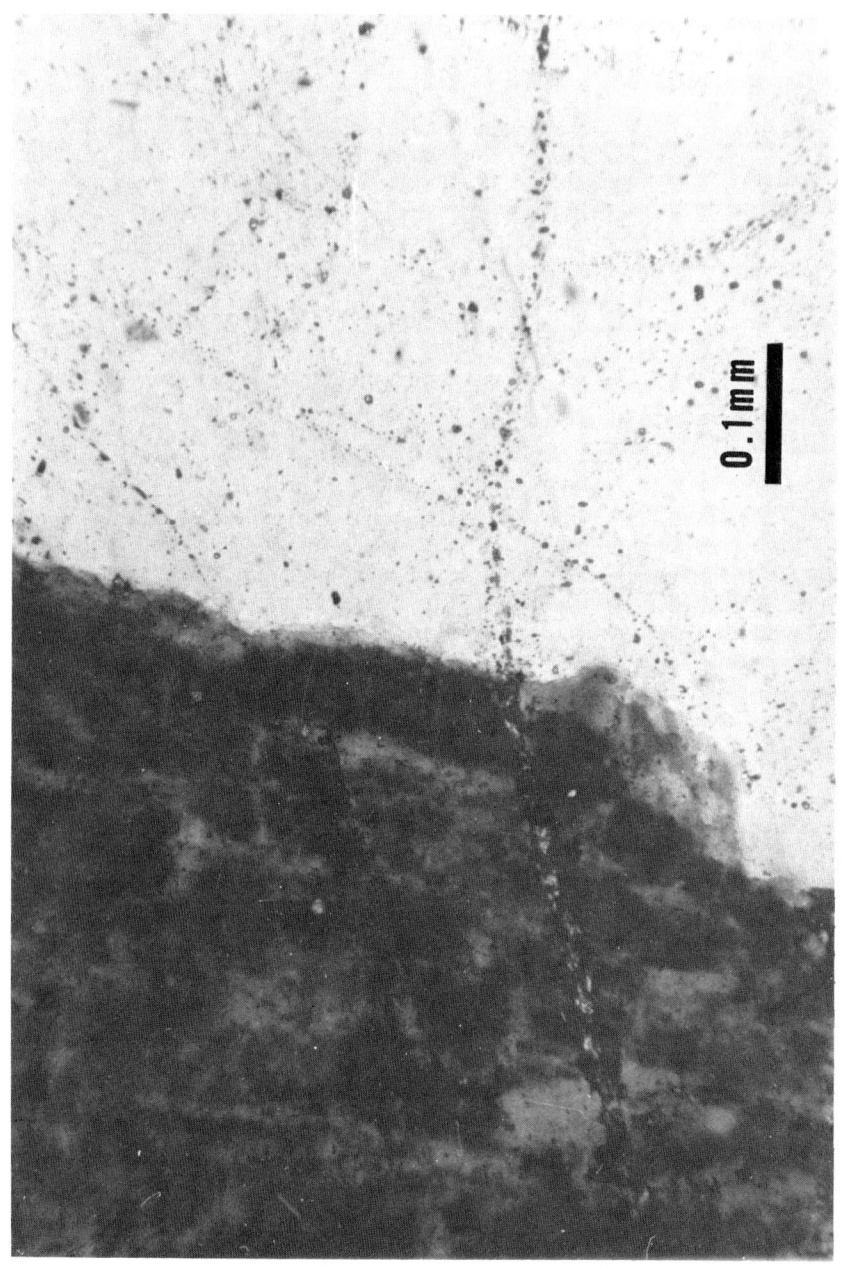

Fig. 1. Healed crack in Wausau granite. Plane-polarized light. The crack is marked by a bubble plane in quartz (right) and recrystallized plagioclase in perthite (left).

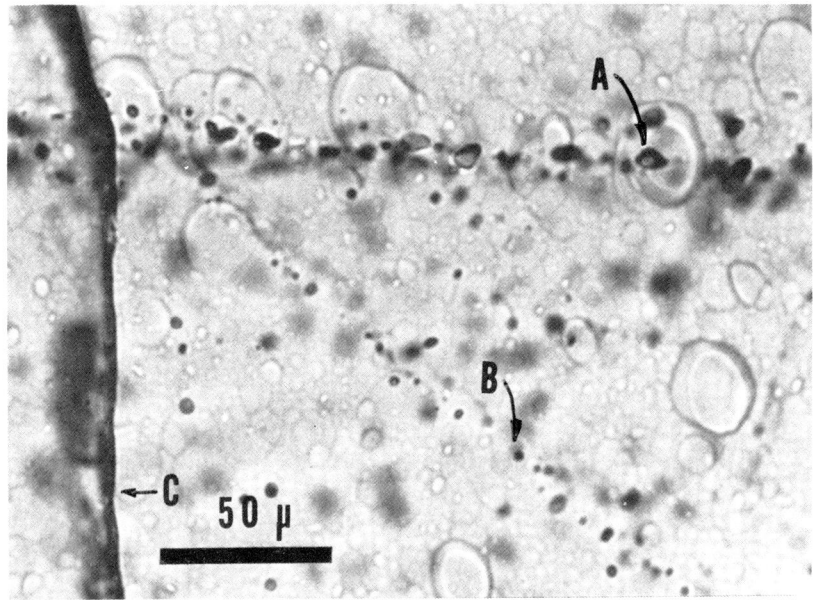

Fig. 2a. Microcracks in quartz, Wausau granite. A and B are healed and C is an open crack. Plane-polarized light.

cases, the quartz is polycrystalline. All of the quartz crystals display undulose extinction, which is considered to be evidence for plastic deformation. It may be, then, that at least some of the wedge-shaped cracks were sealed by plastic flow of adjacent quartz crystals. Longer, thinner polycrystalline quartz-sealed cracks were likely sealed by solution and precipitation of quartz.

Two other crack-sealing materials are less common in the Wausau granite. Chlorite seals some cracks and rutile seals others. Both types of sealed cracks crosscut the other healed cracks in the rock and have no preferred orientation.

The Red River quartz monzonite contains fewer healed and sealed cracks than the Wausau granite. There appear to be more open microcracks in the SEM mosaics of the Red River quartz monzonite, although Feves et al. [1977] found that the Wausau granite and the Red River quartz monzonite have comparable crack porosities. The apparent discrepancy may be due to the difference in the nature of the parameters measured with DSA and explained by considering the difference in the crack with the SEM. From DSA we obtain the distribution of crack porosity with respect to closure pressure of the crack. From the SEM we obtain a visual impression (though it could be measured quantitatively) of the numerical distribution of crack widths.

The hypidiomorphic granular texture of the Red River quartz monzonite is undeformed and unrecrystallized. The weak foliation of the hand specimen was not evident in our thin and crack sections. Planes of secondary fluid inclusions in quartz show no preferred orientation. Figures 5a and 5b are a pair of micrographs of the same field of view

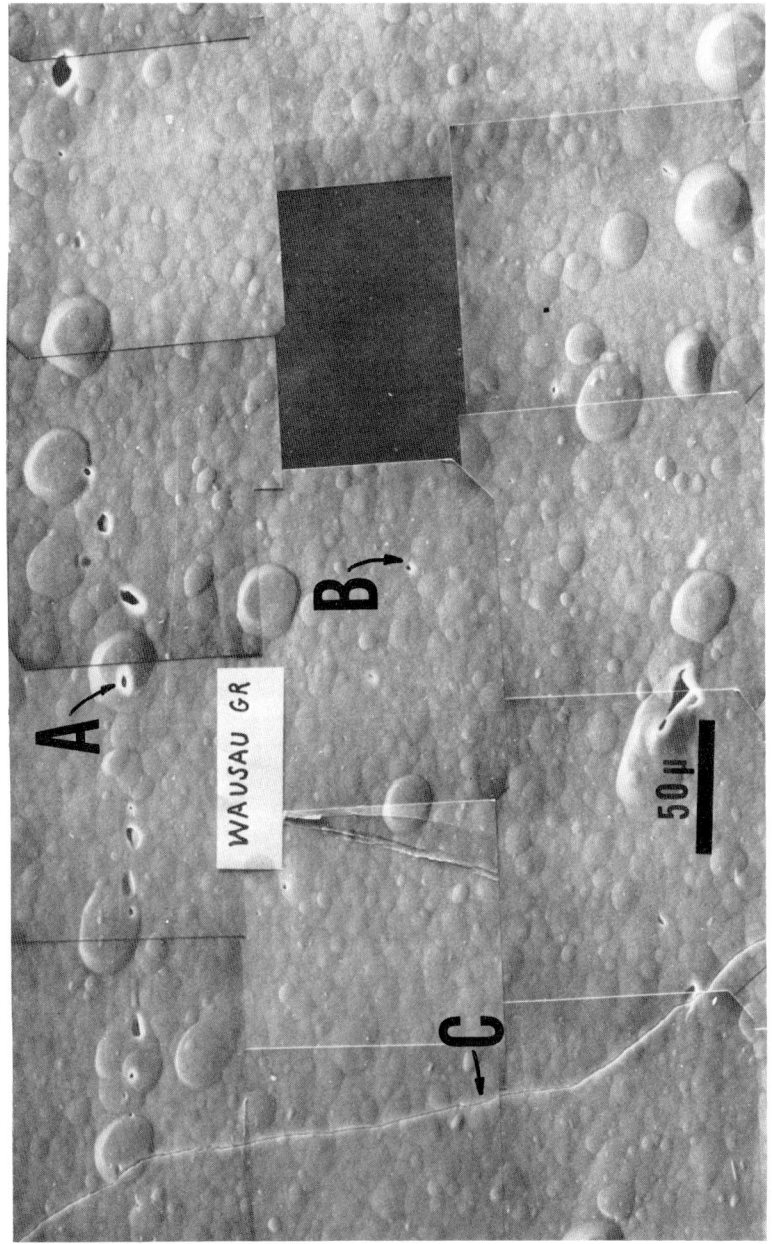

Fig. 2b. SEM micrograph of the same set of cracks in Figure 2a.

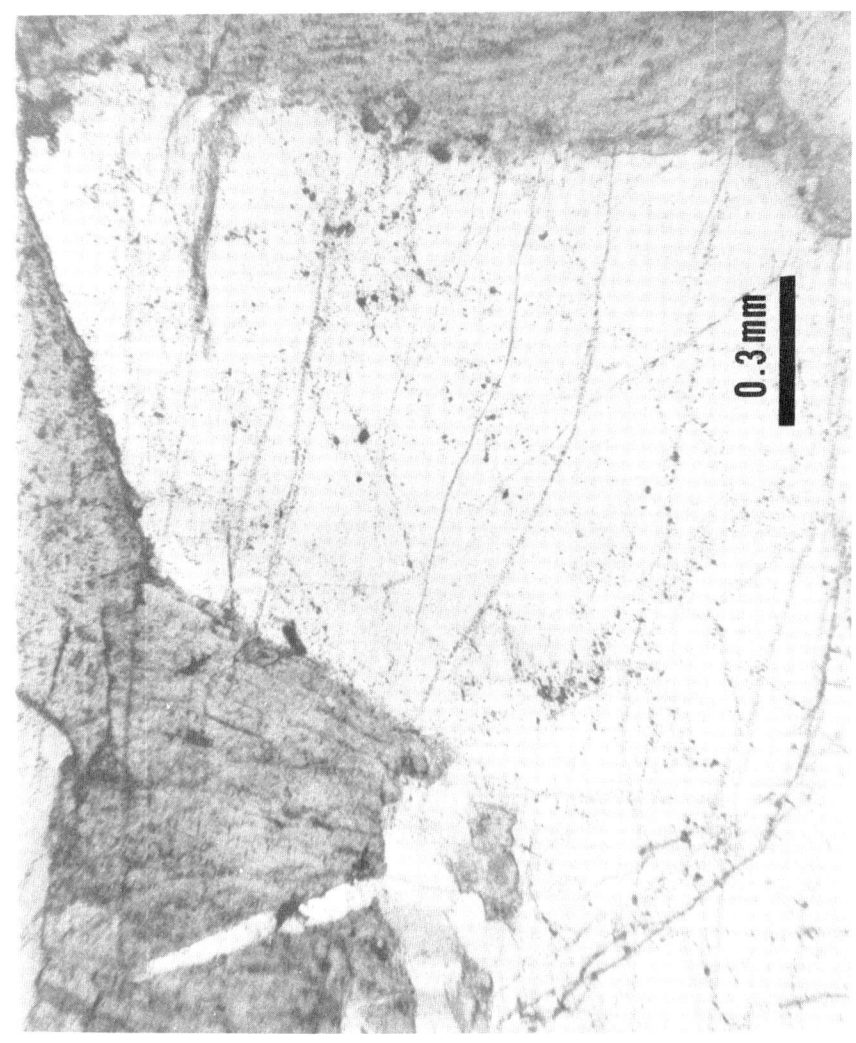

Fig. 3. Parallel healed cracks in quartz, Wausau granite. Plane-polarized light.

Fig. 4. Wedge-shaped crack in perthite which is sealed by quartz, Wausau granite. Plane-polarized light.

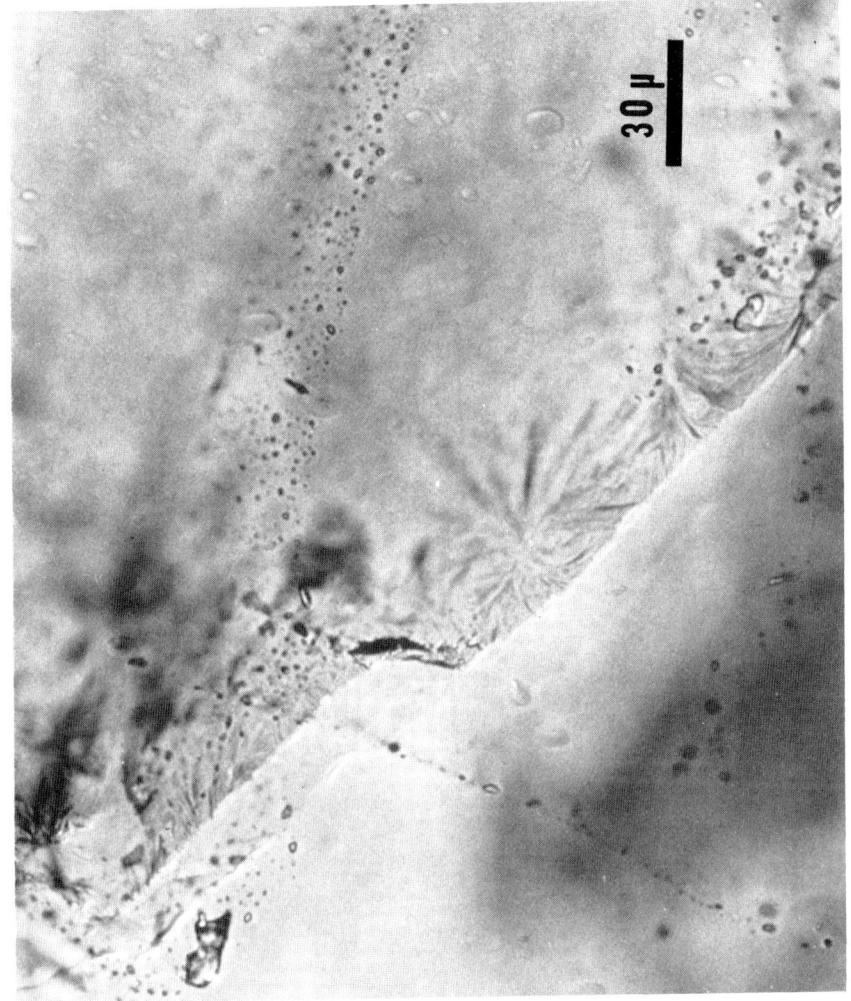

Fig. 5a. Microcrack in quartz sealed with radiating fibrous crystals, Red River quartz monzonite. Plane-polarized light.

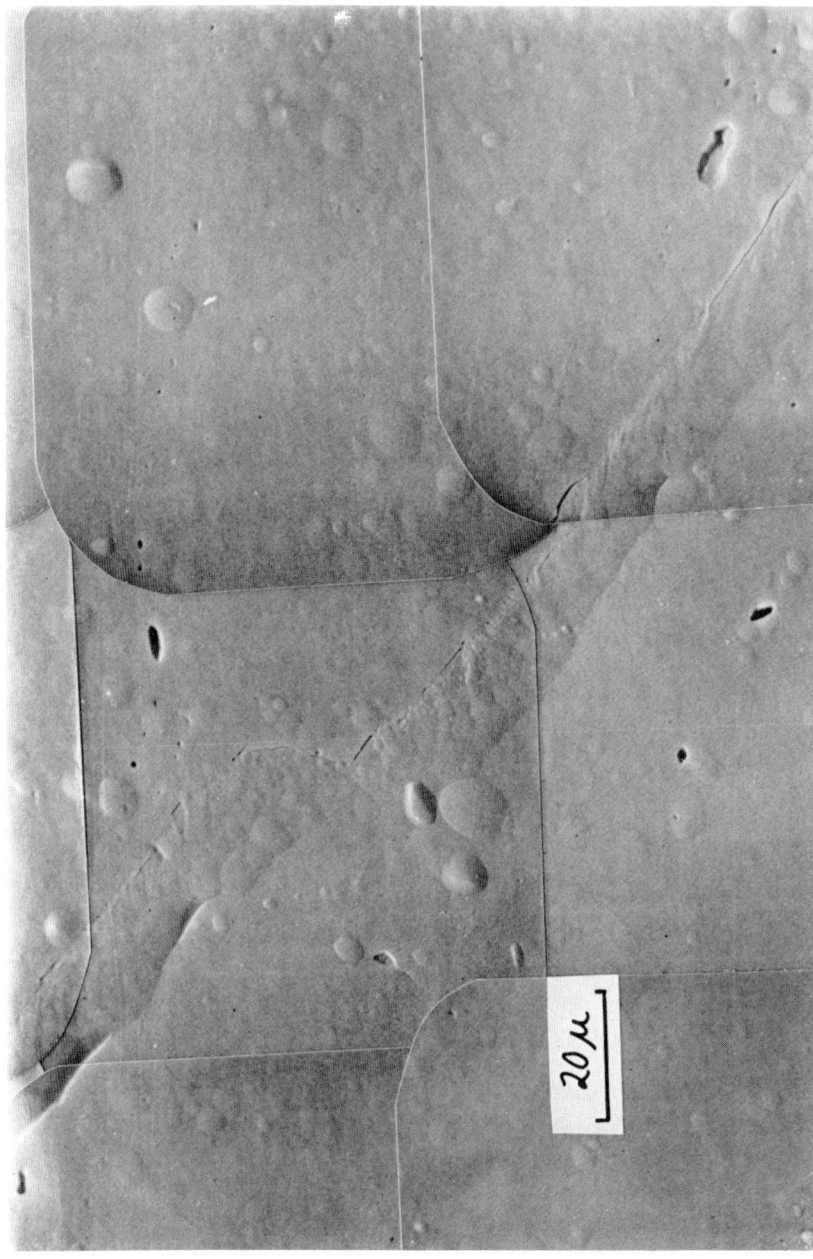

Fig. 5b. SEM micrograph of the same field of view as Figure 5a.

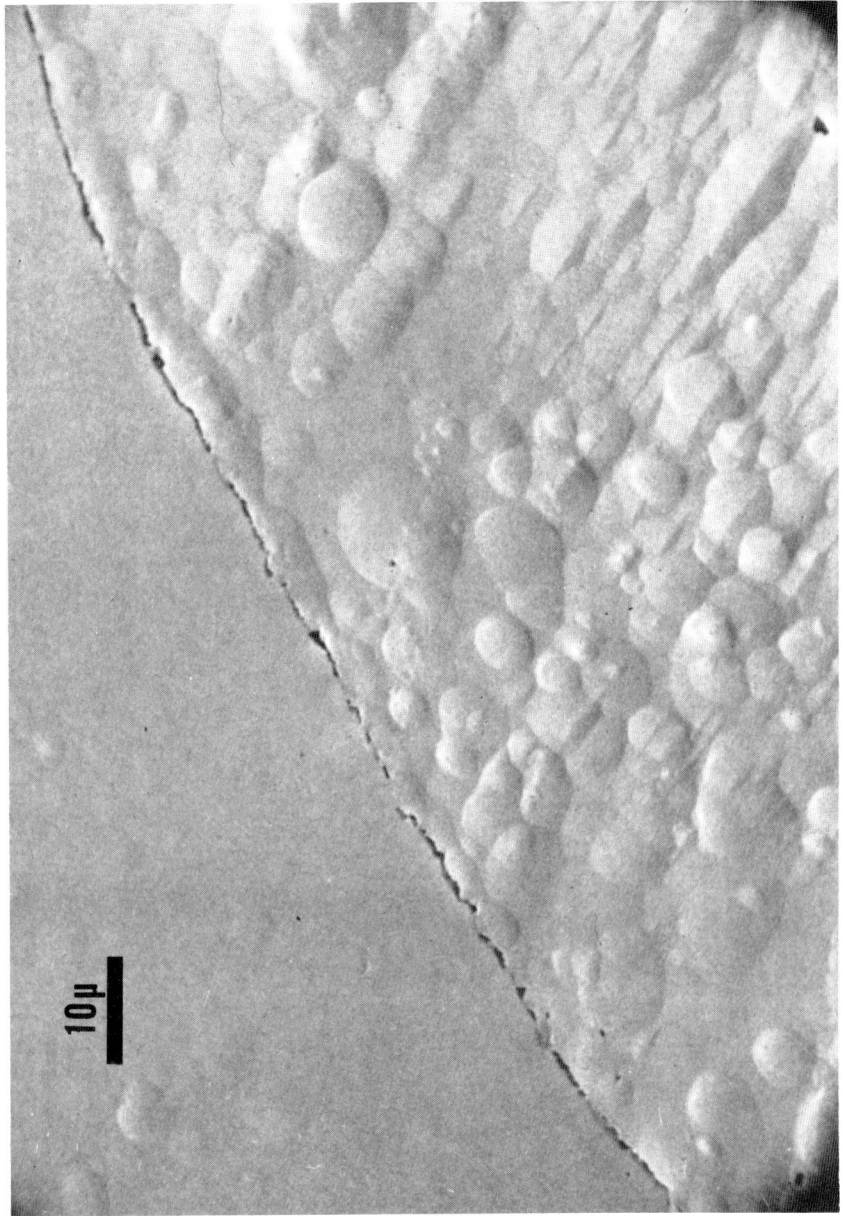

Fig. 6. Partially healed grain boundary crack between quartz and perthite, Red River quartz monzonite. SEM micrograph.

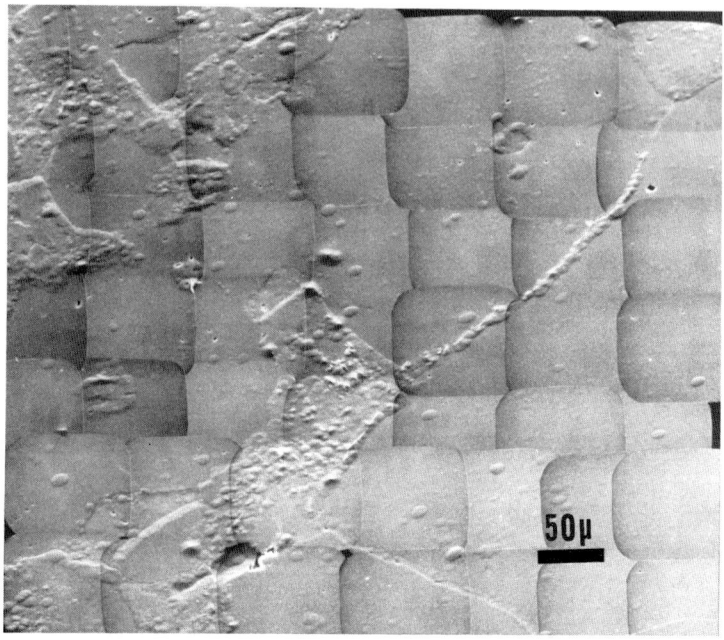

Fig. 7. Typical textures, Graniteville granite. SEM micrograph. The flat sparsely mounded areas are quartz; the highly mounded areas in the upper left and lower center are feldspar. The narrow strip of highly mounded material at the upper right is a feldspar-sealed crack.

taken with the petrographic microscope and the SEM. They show a crack in a quartz grain which dips at about 60° to the surface of the section. The crack is sealed with radial mats of crystals (perhaps chlorite). The crystals do not intersect enough of the surface to be reliably analyzed with the microprobe, but microprobe analyses along the crack show an enrichment of Fe. The crack is completely sealed along most segments. The open segments are 0.1-0.3 µ wide. Several open cracks (0.4 µ wide) and planes of secondary fluid inclusions terminate on the sealed crack, indicating that the sealed crack is the oldest in the sequence of crack-producing events. Quartz containing cracks sealed with the same mineral are common in this rock.

Most of the open microcracks in the rock show evidence of partial healing (Figure 6). Short healed segments are continuous across the cracks. The opposite sides of the grain boundary crack in Figure 6 do not match perfectly. The perthite side of the grain boundary crack is more irregular than the quartz side, and we suggest that fluids circulating in the cracks modified the walls. Such modification of the crack walls is significant to the physical properties of the rock in that the crack would not completely close mechanically under pressure.

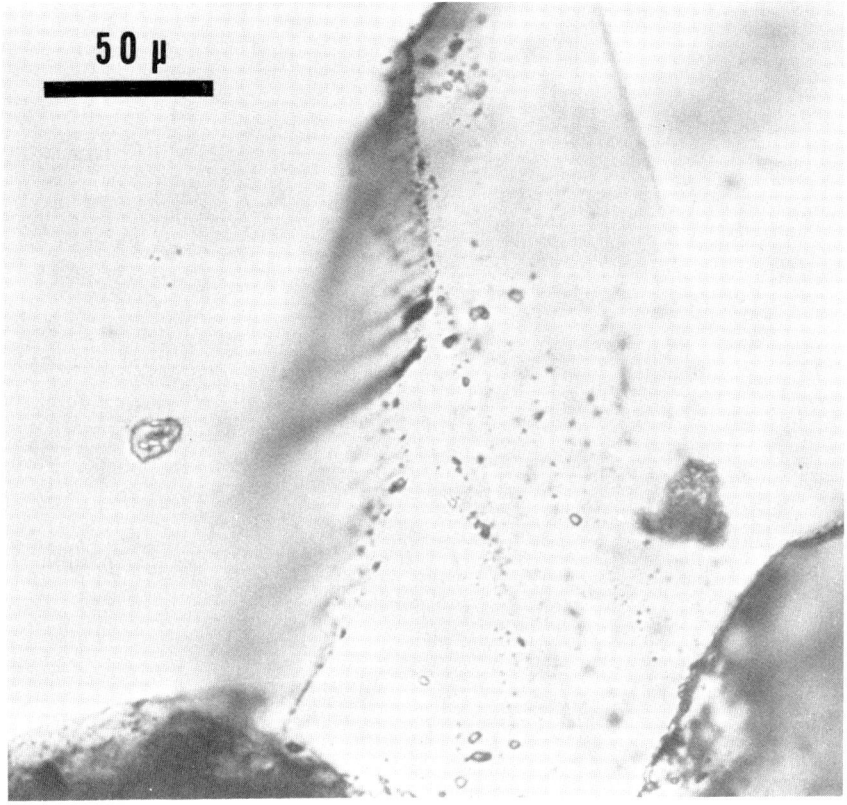

Fig. 8. Microtubes marking a partially healed grain boundary crack between two quartz crystals, Graniteville granite. Plane-polarized light.

The Graniteville granite is coarse grained (average grain size 5 mm) and has sutured grain boundaries. Triple junctions of approximately 120° are common, especially in quartz. The grain boundaries are straight only immediately around the triple junction, however, so the texture is not truly polygonized. Healed and sealed cracks are common but not as abundant as they are in the Wausau granite. Open microcracks coincident with grain boundaries are common in the Graniteville granite but not as common as they are in the Red River quartz monzonite. Feves et al. [1977] measured a lower crack porosity for the Graniteville granite than for either the Wausau granite or the Red River quartz monzonite.

Figure 7 shows textures typical of the Graniteville granite. There is a plane of secondary fluid inclusions in the quartz grain at the upper right. Roughly parallel to the plane of inclusions is a crack which is sealed by K feldspar and which is coincident with a younger open grain boundary crack.

The open microcracks in the Graniteville granite occur most commonly along grain boundaries but show no preferred orientation. They are

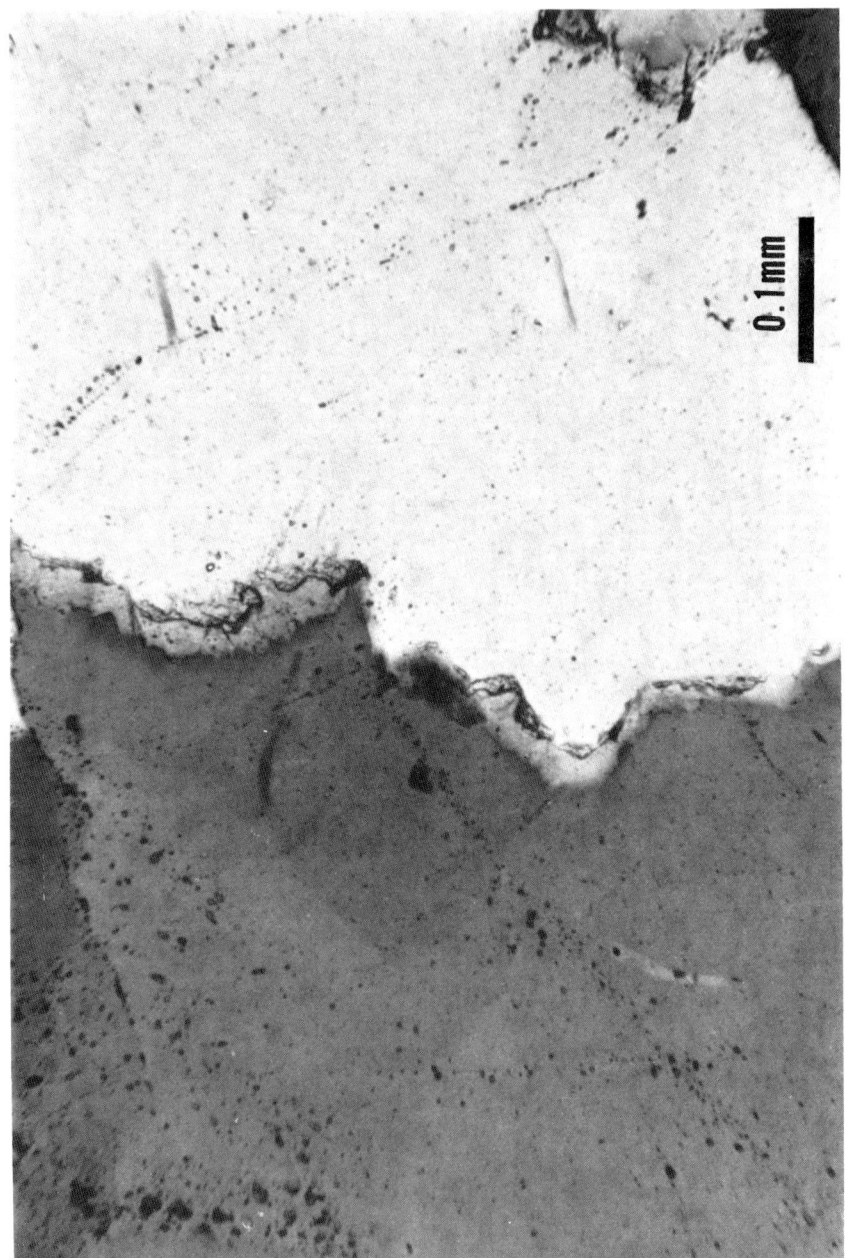

Fig. 9. Fluorite-sealed crack, Graniteville granite. Cross-polarized light.

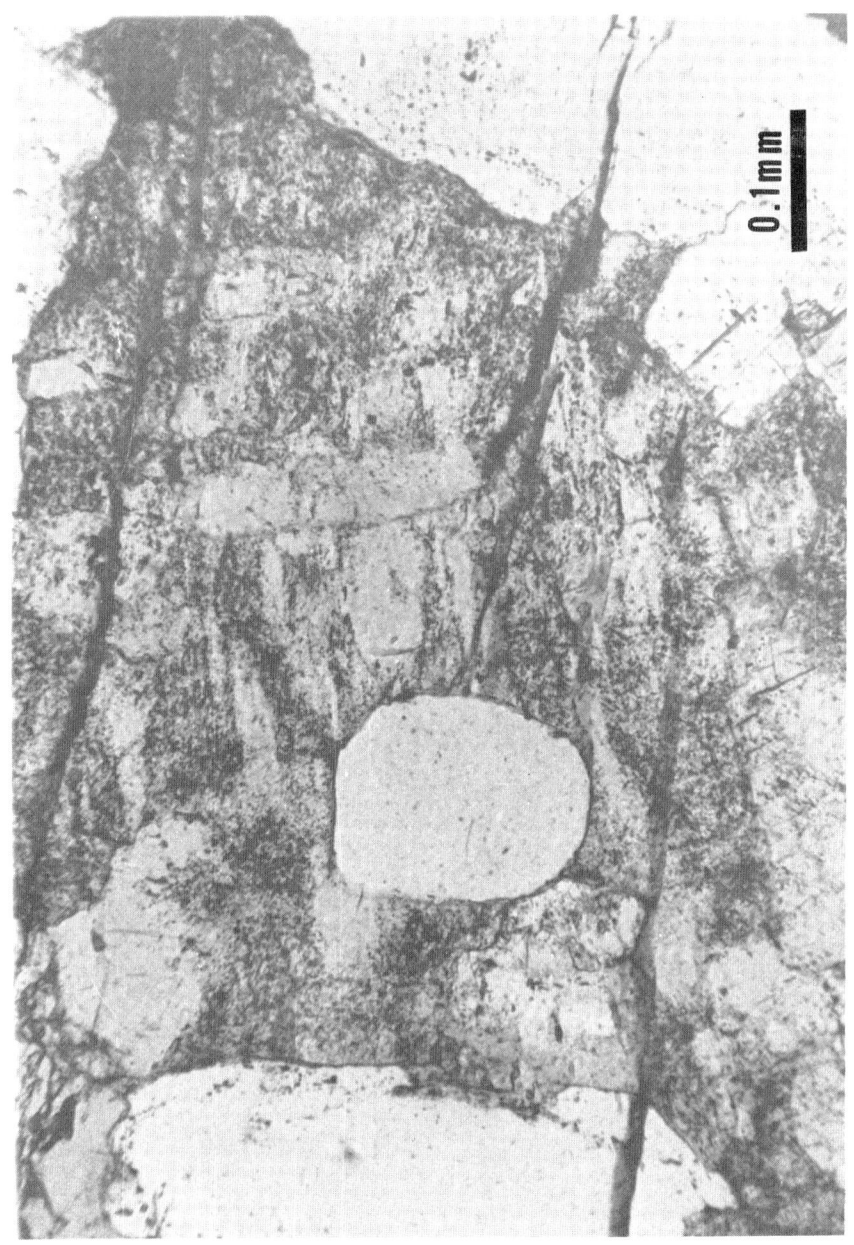

Fig. 10. Parallel rutile-sealed cracks in perthite, Graniteville granite. Plane-polarized light.

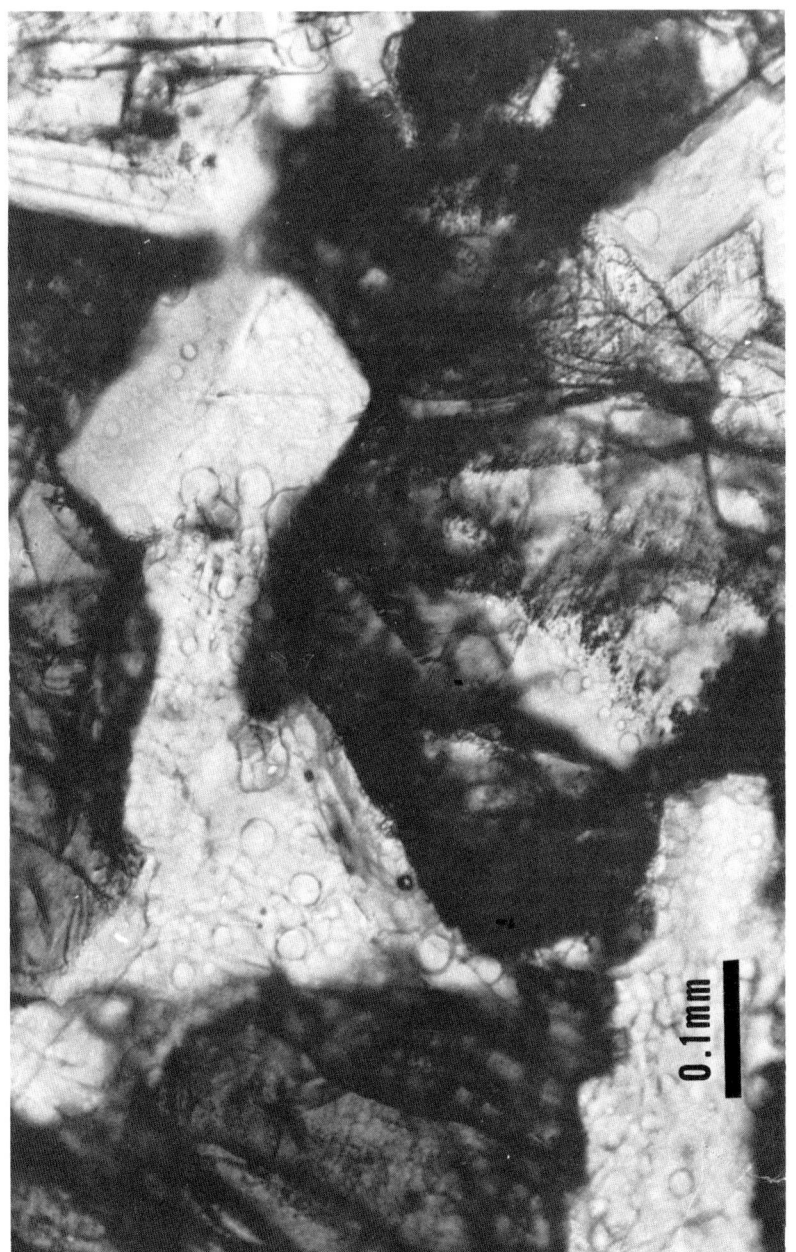

Fig. 11a. Typical textures, Mellen gabbro. The dogbone-shaped cluster of crystals of plagioclase in the upper half is a good landmark for comparison. Plane-polarized light.

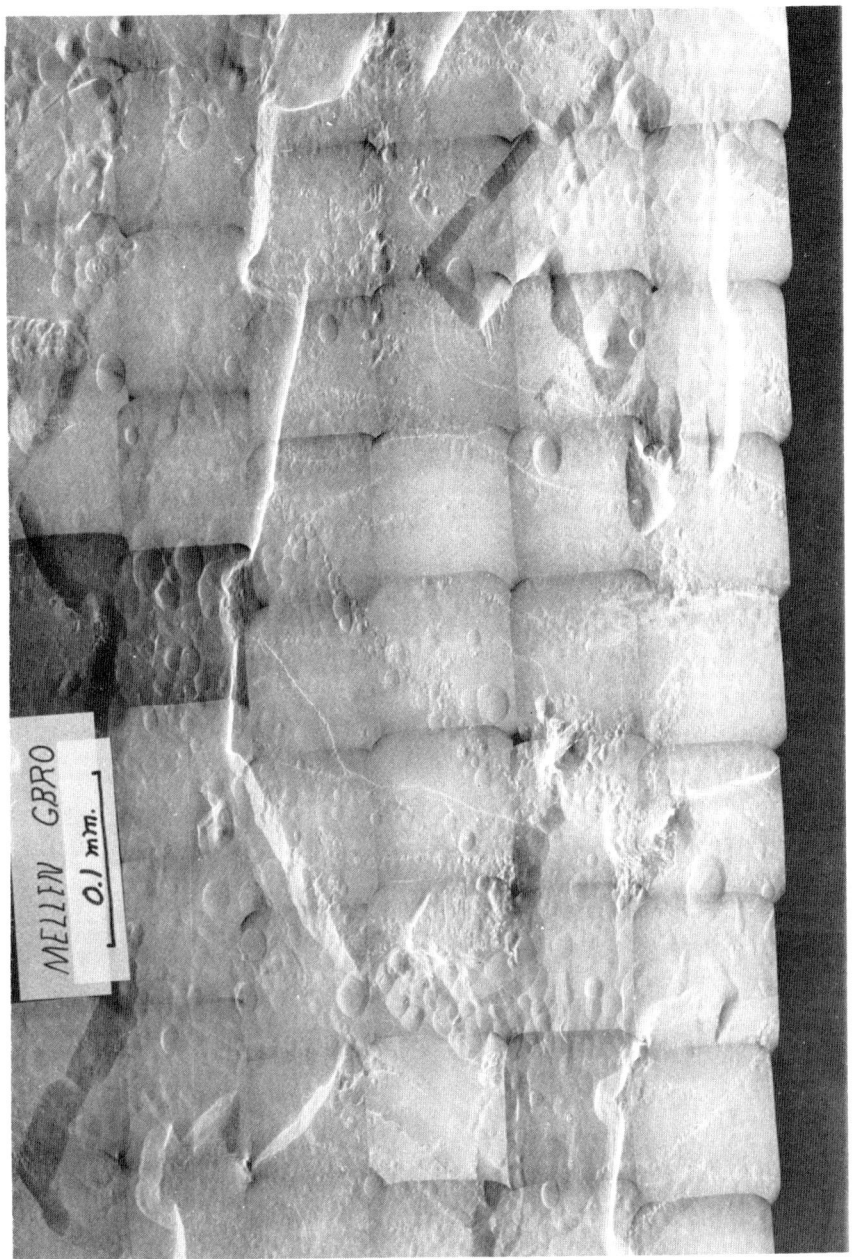

Fig. 11b. SEM micrograph of the same area as Figure 11a.

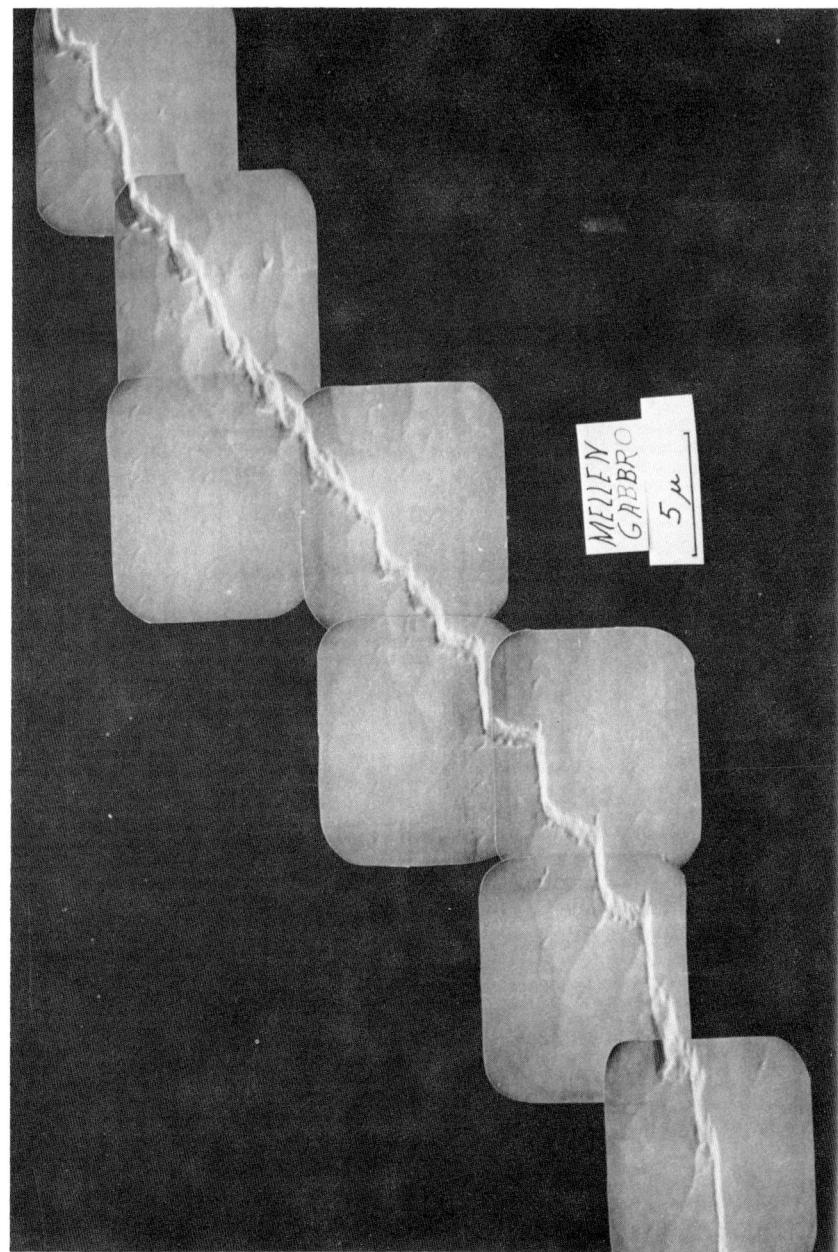

Fig. 12. Healed crack in pyroxene, Mellen gabbro. SEM micrograph.

generally 0.1-0.3 μ wide with short segments that are completely healed and are several grain boundaries long. Healing of some cracks has progressed to the point of forming microtubes (see Figure 8), a step intermediate between an open crack and a bubble plane [Richter and Simmons, 1977].

Fluorite is a late formed accessory mineral which fills interstices between other crystals in the granite. Fluorite also seals thin, short, grain boundary cracks (Figure 9). The fluorite sealed cracks therefore probably opened in the last stages of crystallization of the granite.

Rutile seals later cracks in the granite. The rutile occurs as translucent red plates which give a high Ti peak when analyzed with the microprobe. The rutile-sealed cracks are generally parallel. They generally follow grain boundaries, but a few are parallel to cleavage directions in feldspar (Figure 10). Natural cleavage cracks are not common in igneous rocks [Simmons and Richter, 1976] but clearly were present and later sealed by rutile in the Graniteville granite. There is no primary rutile in the Graniteville granite, so we infer that the rutile in the cracks was introduced by circulating fluids. The rutile-sealed cracks crosscut healed cracks in a few places and are interpreted as relatively young.

The apparent sequence of crack events in the Graniteville granite, then, is as follows. The first grain boundary cracks to be formed were sealed by fluorite. Later grain boundary cracks healed to form planes of secondary fluid inclusions and microtubes. A still later set of parallel cracks opened and were subsequently sealed by rutile. The final fracturing event was the opening of grain boundary cracks, some of which coincide with previously sealed or healed cracks.

Mafic Rocks

The mafic rocks are fresh, unrecrystallized gabbro and diabase. They display an almost complete spectrum of crack features: the Mellen gabbro contains almost no open microcracks and rather few healed cracks, the Tigerton gabbro contains small amounts of both open and healed microcracks, and the Skrainka diabase contains no open microcracks but has extremely abundant sealed and healed cracks.

The Mellen gabbro is one of the most crack-free rocks we have examined. Our sample is very fresh for a terrestrial igneous rock in that the plagioclase shows almost no signs of alteration. The plagioclase does not contain any of the pores abundant in the plagioclase in granites and other rocks [Montgomery and Brace, 1975].

Figures 11a and 11b are a pair of micrographs of the same area. The grain boundaries are straight and uncracked. The pyroxene grains contain several healed cracks that are marked by hairlike arrays of opaque grains. Figure 12 is a high-magnification SEM micrograph of one unusual healed crack that is in the center of Figures 11a and 11b. The former crack is completely healed and is marked by a zigzag-shaped trough. At each step in the trough there is a parallel set of regularly spaced dimples which are too small to be seen with the petrographic microscope. If the steps in the trough are parallel to cleavage directions in the pyroxene crystal, do the dimples represent other, more completely healed cleavage cracks; or are they an expression of submicro-

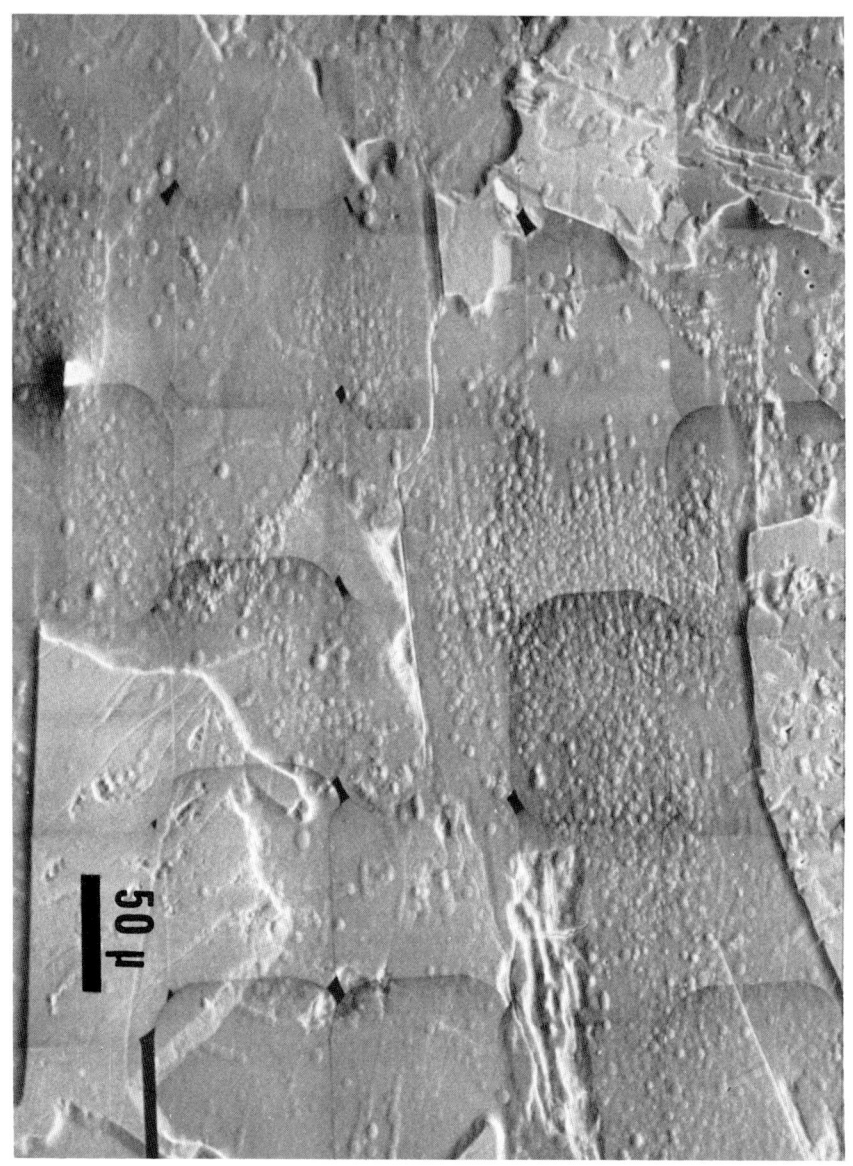

Fig. 13. Typical textures in Tigerton gabbro. SEM micrograph. The sawtooth-shaped grains in the center are chlorite along a healed grain boundary crack between two plagioclase crystals.

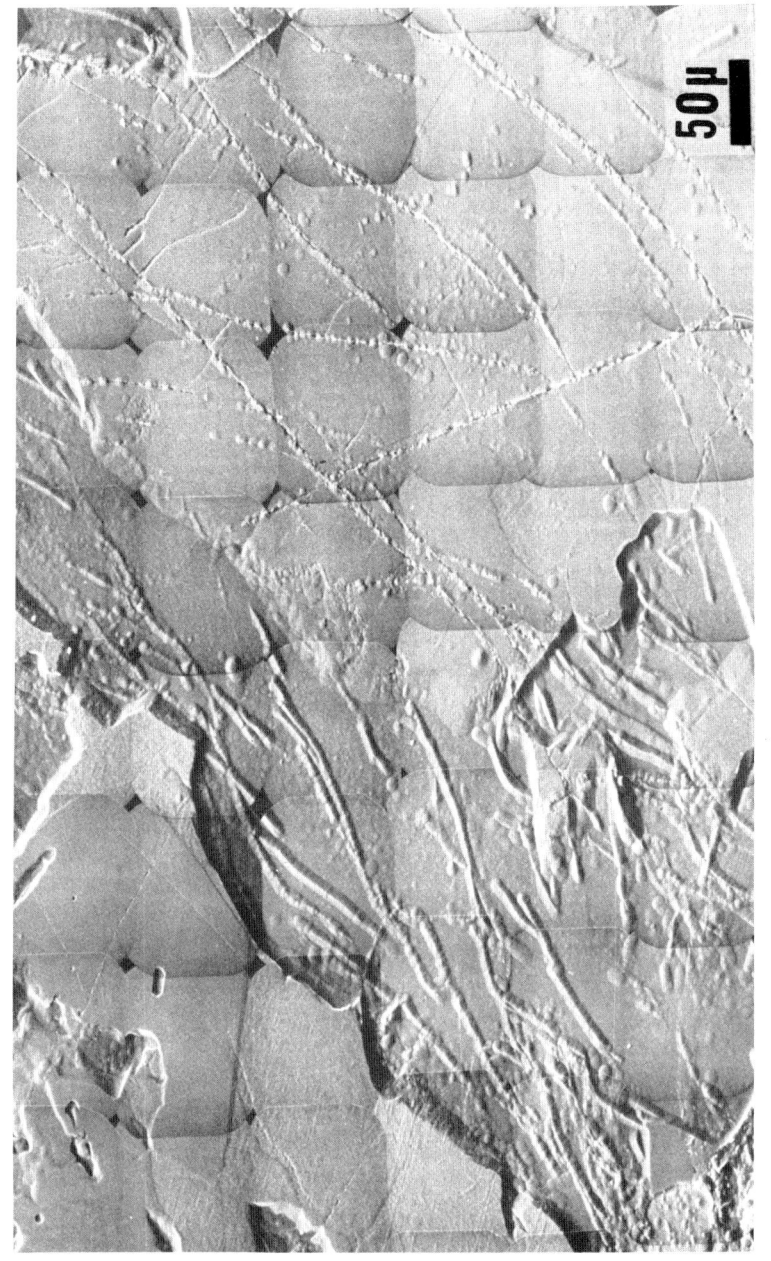

Fig. 14. Sealed cracks in magnetite (left), plagioclase (center), and olivine (right), Skrainka diabase. SEM micrograph. The cracks are sealed with magnetite.

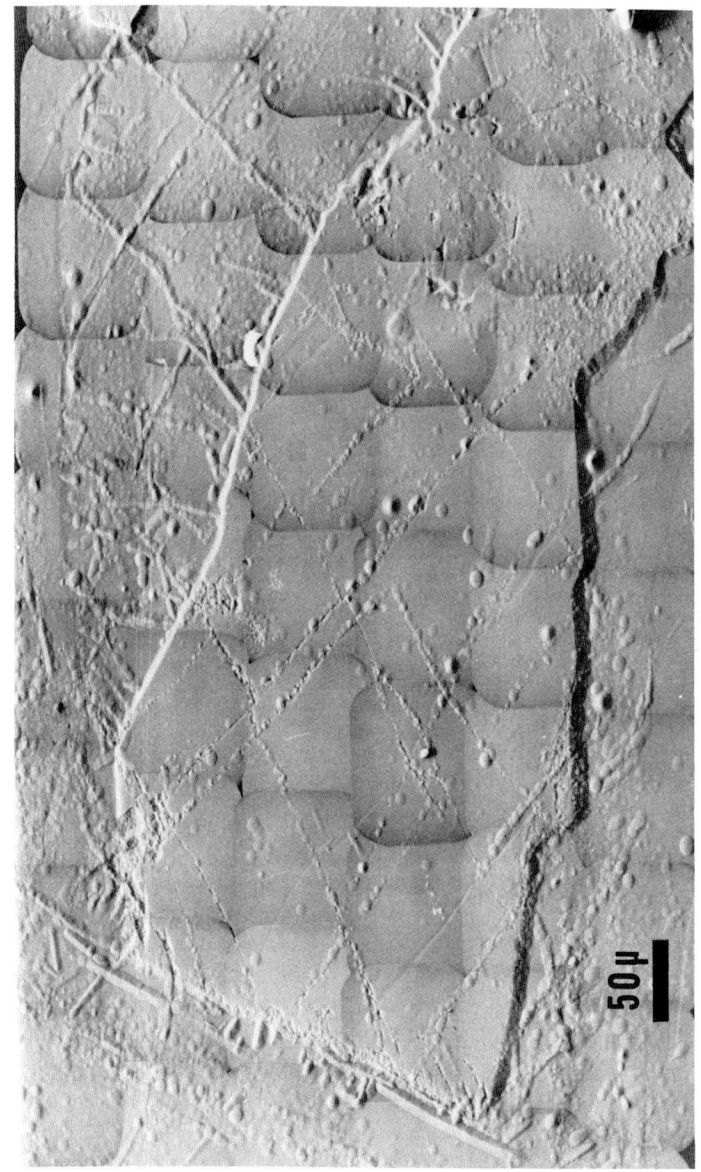

Fig. 15. Transgranular sealed cracks in pyroxene (center) and plagioclase (periphery), Skrainka diabase. SEM micrograph.

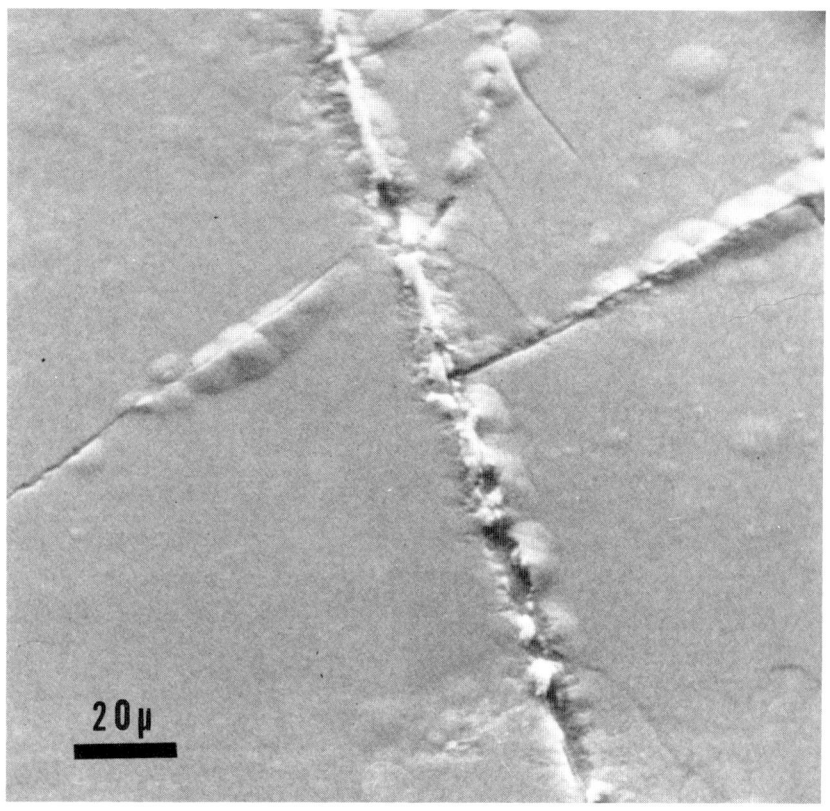

Fig. 16. Sealed crack in olivine, Skrainka diabase. SEM micrograph. The white areas are magnetite, the troughlike low areas are serpentine.

scopic exsolution lamellae of another pyroxene? Similar cracks occur in other pyroxene crystals in the Mellen gabbro, but they are not very common. The paucity of healed cracks and the absence of open microcracks in the Mellen gabbro make it unique among the basement rocks in this study. The evidence suggests that the Mellen gabbro has never contained any significant open crack porosity.

The Tigerton gabbro is from the neighboring Wolf River batholith east of Mellen, Wisconsin. This gabbro contains both open and healed microcracks. Most of the cracks, both healed and open, are intracrystalline and are 0.1-0.2 μ wide. A typical area of this rock is illustrated in Figure 13. Fairly well defined sets of healed and chlorite-sealed cracks are approximately parallel. The pyroxene is commonly partially altered to hornblende. The water necessary for the alteration reactions probably traveled through microcracks. In the center of Figure 13 are two sawtooth-shaped chlorite grains along a healed grain boundary crack between two plagioclase crystals. This form of microcrack is common in the Tigerton gabbro, as was previously

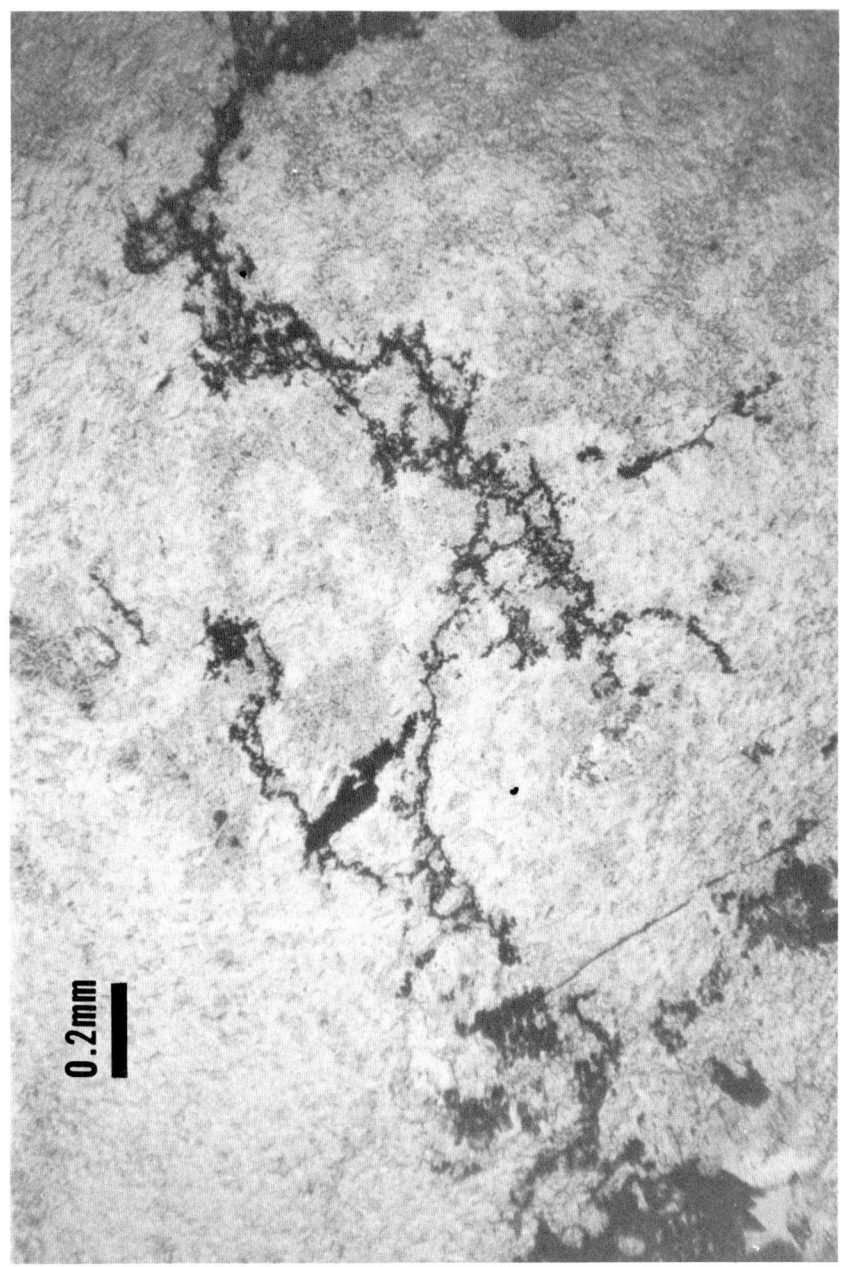

Fig. 17. Rutile-sealed cracks (black), Middlebrook felsite. Plane-polarized light.

noted by Medaris et al. [1973]. We infer that solutions in the crack must have dissolved part of the wall to make it so irregular. The open microcracks in the Tigerton gabbro generally follow grain boundaries and formerly healed cracks. They do not have a pronounced preferred orientation.

The Skrainka diabase was once a highly cracked rock. Figures 14 and 15 are SEM micrographs of typical textures in the Skrainka diabase. Note the large number of subparallel diagonal ridges. Each ridge is a sealed or a healed crack. Figure 16 is a high-magnification SEM micrograph of a portion of one of the sealed cracks in the olivine in Figure 14. The low troughlike segments are probably serpentine. Many of the sealed cracks run through several grains. If we assume that the cracks were 1 μ wide before they were sealed, then the Skrainka diabase would have had a crack porosity of the order of 10^{-2} if the cracks were all open at the same time. The 1-μ estimate for the width is modest, considering the width of the healed and sealed cracks. A crack porosity of 1% is very high, but the cracks are now completely closed. Feves et al. [1977] measured a crack porosity of less than $2-3 \times 10^{-6}$ (the experimental uncertainty) for a block of the same sample. Amos and Desborough [1970] noted that the Skrainka diabase gave a metallic ring when it was hammered, another indication of low crack porosity. The sealed cracks form topographic ridges in plagioclase because the sealing material is more resistant to ion milling. The sealing material appears opaque in transmitted light, is very rich in Fe, and is probably magnetite. In both the olivine and the pyroxene the sealing material is opaque, richer in Fe than the host grain, and probably is magnetite also.

The origin of the cracks in Skrainka diabase is intriguing because of the very high crack porosity. The only rocks with known natural crack porosities as high as 1% are shocked lunar rocks [Todd et al., 1973; Simmons et al., 1975]. We do not suggest that our sample of Skrainka diabase was shocked and then annealed but only wish to signal that the event which formed the cracks in the Skrainka diabase must have been fairly dramatic. Equally remarkable is the fact that the cracks sealed so completely. The Skrainka diabase occurs near Ironton, Missouri, an area of late Precambrian iron mineralization. Perhaps the ore-forming fluids also saturated the diabase and sealed the cracks.

Volcanic Rocks

The two volcanic rocks examined, the Middlebrook felsite and the Stouts Creek rhyolite, contain very few open cracks. Sealed cracks are common in the matrix of the Middlebrook felsite and the phenocrysts of the Stouts Creek rhyolite. Figure 17 is a photomicrograph of cracks in the Middlebrook felsite which are sealed with rutile. The rutile occurs as anastomosing blebs. Figure 18 is a high-magnification SEM micrograph of a portion of a rutile-sealed crack. The white topographic highs are the rutile. The matrix, which is very fine grained (<0.1 mm) quartz and feldspar, is continuous (healed) across the crack in spots, but there are also many irregular voids associated with the rutile. The voids do not occur elsewhere in the matrix.

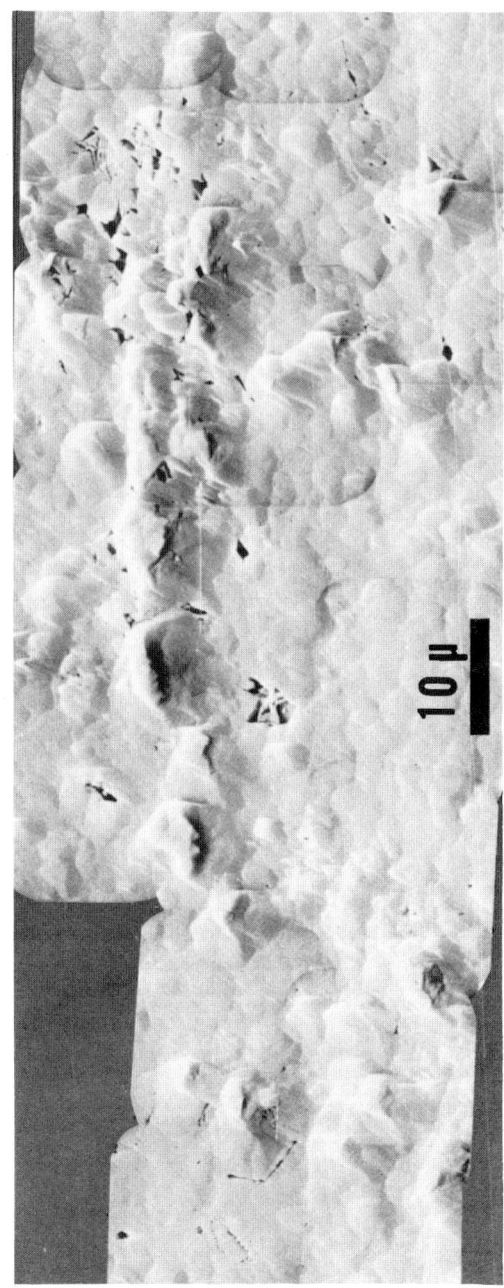

Fig. 18. Rutile-sealed crack (white topographic highs) and associated voids, Middlebrook felsite. SEM micrograph.

The Stouts Creek rhyolite contains healed cracks in the quartz phenocrysts, chlorite-sealed cracks in plagioclase phenocrysts, quartz-sealed cracks in the matrix, and a few open cracks. The bubble planes in the quartz phenocrysts show a preferred orientation. The chlorite-sealed cracks in plagioclase occur along cleavage planes; thus their orientation is controlled by the orientation of the phenocrysts, which is random. The cracks in the matrix of the Stouts Creek rhyolite are partially healed. Figure 19 shows a segment of an open microcrack which is partially healed. The open portions are over 2 µ wide in places. The matrix of the rock contains many irregular small voids in the vicinity of the crack which may have been filled with fluids in situ.

What Happens to Microcracks In Situ

From our data on the microcracks in these Precambrian basement rocks we can infer what happens to microcracks in rocks in situ as a function of time. The DSA spectra of Feves et al. [1977] show that most open cracks close mechanically under a lithostatic load of ≤ 2 kbar.

The mechanically closed cracks anneal with time. Although the details of this process are not understood, the evidence that the process does occur and that it is efficient is both abundant and clear. The fluid and solid inclusion surfaces are common in many minerals. However, we have observed surprisingly few bubble planes in feldspar. We suggest that cracks in feldspars may heal through a process of solution along the crack and precipitation of feldspar or clay minerals at the tip of the crack.

Many cracks fill with secondary minerals: chlorite, magnetite, rutile, quartz, feldspar, and ilmenite. The material for the growth of these minerals may be transported by aqueous fluids or by molecular diffusion over short distances. The annealing process may be interrupted, and some cracks are only partially healed or sealed. We do not know whether insufficient time has elapsed, whether pressures and temperatures were insufficient, or whether compositions of the fluids were unfavorable.

Conclusions

Five main conclusions may be drawn about the microcracks in the Wisconsin and Missouri samples studied.

1. Each sample contains healed and sealed cracks. The evidence for healed cracks is the abundant bubble planes seen most commonly in quartz. The crack-sealing materials include introduced or primary rutile, fluorite, chlorite, magnetite, quartz, and feldspar. The sealed cracks commonly have preferred orientations.

2. The varying abundance of healed and sealed cracks in the different rocks implies that the total crack porosities (with time) were different. The data which would allow us to know the intensity of crack producing events are the crack porosities as a function of time. We suggest that the intensities of the various fracturing events were different in the different rocks. Crosscutting relationships among healed cracks are indicative of multistage fracturing and healing stages for such samples as the Wausau granite. Other

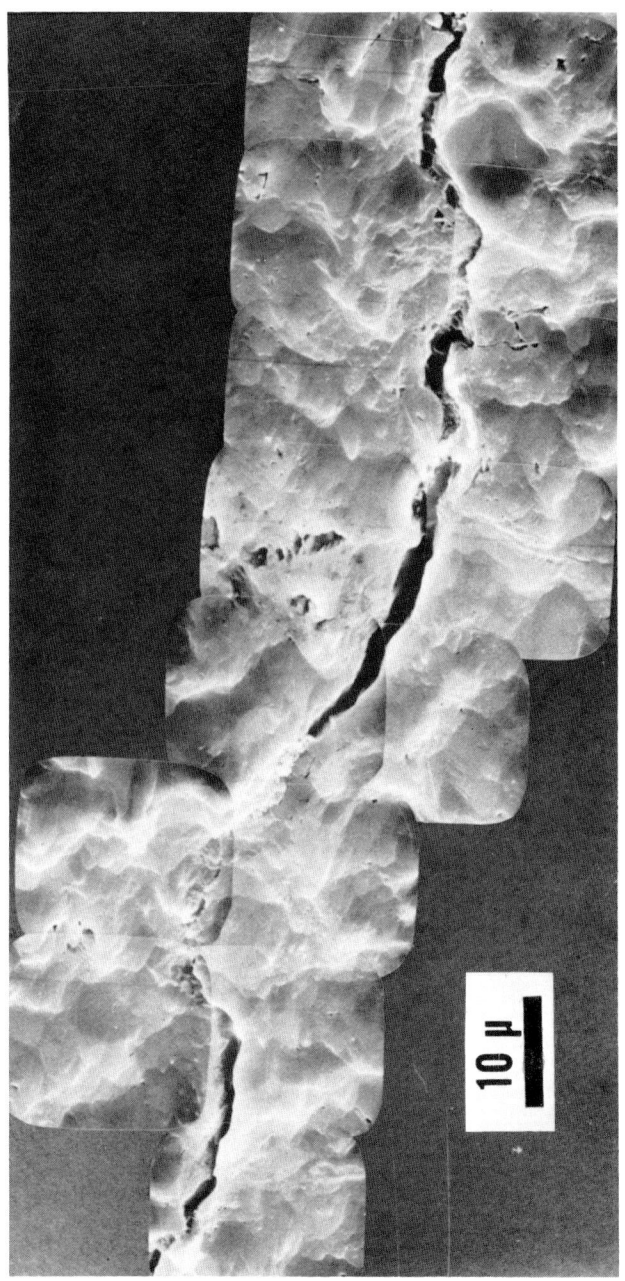

Fig. 19. Partially healed microcrack in the matrix, Stouts Creek rhyolite. SEM micrograph.

samples, such as the Mellen gabbro, apparently were never significantly fractured.

3. Open cracks in each sample are very few, as shown by microscopic examination. The microcrack porosity is very low, as is shown by Feves et al. [1977]. The small density and the low volume of cracks are probably due to the Precambrian basement areas being tectonically stable. The rocks were unloaded extremely slowly, so that cracks formed by lithostatic unloading (common in New England rocks) healed before the rocks finally reached the surface of the earth.

4. Many of the open microcracks appear to be partially healed. If they are present in significant numbers, these cracks should have a significant effect on physical properties as a function of pressure because the pressures needed to close them mechanically are high. Transport properties (electrical conductivity and permeability, for example) would remain high.

5. Finally, the above conclusions indicate that regardless of the intensity and number of stages of Precambrian fracturing of these basement samples, the microcracks are now uniformly healed and sealed. If our suite of samples is representative of basement igneous rocks, then buried Precambrian crustal igneous rocks in the central United States and similar areas elsewhere probably contain almost no 'open' microcracks or cracks which are merely mechanically closed by the lithostatic load. Their physical properties should approach the intrinsic values of the constituent minerals at shallow depths.

Acknowledgements. We benefited from discussions with Michael Feves, Robert Siegfried, Michael Batzle, Herman Cooper, and Michael Fehler. Nancy Stockwell helped process the SEM micrographs. Douglas Danley printed the photographs. Lucille Foley and Ann Harlow typed the manuscript. Financial support was provided by ONR contract N00014-76-C-0478.

References

Amos, D. H., and G. A. Desborough, Mafic intrusive rocks of Precambrian age in southeast Missouri, Rep. Inv. 47, 22 pp., Mo. Geol. Surv. and Water Resour., Rolla, 1970.

Anderson, J.E., M. E. Bickford, A. L. Odom, and A. W. Berry, Some age relations and structural features of the Precambrian volcanic terrane, St. Francois Mountains, southeastern Missouri, Geol. Soc. Amer. Bull., 80, 1815-1818, 1969.

Anderson, R. E., Ash-flow tuffs of Precambrian age in southeast Missouri, Rep. Inv. 46, 50 pp., Mo. Geol. Surv. and Water Resour., Rolla, 1970.

Bickford, M. E., and D. G. Mose, Geochronology of Precambrian rocks in the St. Francois Mountains, southeastern Missouri, Geol. Soc. Amer. Spec. Pap., 165, 1-48, 1975.

Chaudhuri, S., D. G. Brookins, and G. Fanre, Rubidium-strontium ages of Keweenawan intrusions near Mellen and South Range in Wisconsin, paper presented at 15th annual meeting of Inst. on Lake Super. Geol., Oshkosh, Wis., 1969.

Dale, T. N., The commercial granites of New England, U. S. Geol. Surv. Bull., 738, 1-488, 1923.

Emmons, R. C., Guidebook for 17th Annual Tri-State Geological Field Conference, 11 pp., 1953.

Feves, M., G. Simmons, and R. Siegfried, Microcracks in crustal igneous rocks: Physical properties, in The Earth's Crust: Its Nature and Physical Properties, Geophys. Monogr. Ser., vol. 20, edited by J. G. Heacock, AGU, Washington, D. C., this volume, 1977.

Kisvarsanyi, E. B., Petrochemistry of a Precambrian igneous province, St. Francois Mountains, Missouri, Rep. Inv. 51, 96 pp., Mo. Geol. Surv. and Water Resour., Rolla, 1972.

Lockwood, R. P., Petrology of syenites, Wausau, Wisconsin, Wis. Compass, 48, 32-44, 1970.

Medaris, L. G., J. L. Anderson, and J. R. Myles, The Wolfe River batholith--a late Precambrian rapakivi massif in northeastern Wisconsin, in Guidebook to the Precambrian Geology of Northeastern and Northcentral Wisconsin, pp. 9-29, Wisconsin Geol. and Natural History Surv., Madison, 1973.

Montgomery, C. W., and W. F. Brace, Micropores in plagioclase, Contrib. Mineral. Petrol., 52, 17-28, 1975.

Richter, D., and G. Simmons, Microscopic tubes in igneous rocks, Earth Planet. Sci. Lett., 34, 1-12, 1977.

Simmons, G., and D. Richter, Microcracks in rocks, in The Physics and Chemistry of Minerals and Rocks, edited by R. G. J. Strens, pp. 105-137, Interscience, New York, 1976.

Simmons, G., R. Siegfried, and D. Richter, Characteristics of microcracks in lunar rocks, Proc. Lunar Sci. Conf. 6th, 3, 3227-3254, 1975.

Tabet, D. E., Structure and petrology of the Mellen igneous intrusive complex near Mellen, Wisconsin, M.S. Thesis, 81 pp., Univ. of Wis., Madison, 1974.

Todd, T., D. Richter, G. Simmons, and H. Wang, Unique characterization of lunar samples by physical properties, Proc. Lunar Sci. Conf. 4th, 3, 2639-2662, 1973.

Tolman, C. F., and F. Robertson, Exposed Precambrian rocks in southeast Missouri, Rep. Inv. 44, 68 pp., Mo. Geol. Surv. and Water Resour., Rolla, 1969.

Van Schmus, W. R., Chronology of Precambrian rocks in Wisconsin, in Guidebook to the Precambrian Geology and Northeastern and Northcentral Wisconsin, pp. 1-8, Wisconsin Geol. and Natural History Surv., Madison, 1973.

ELECTRICAL CHARACTERISTICS OF IGNEOUS PRECAMBRIAN BASEMENT ROCKS OF CENTRAL NORTH AMERICA

R. M. Housley and J. R. Oliver

Rockwell International/Science Center
Thousand Oaks, California 91360

Abstract. We present ac and dc electrical property data on Precambrian Montello and Wausau, Wisconsin, granites and dc data on Precambrian Mellen, Wisconsin, gabbro. We believe that the dc conductivity values are representative of the crack free rocks at depth in the earth. These rocks naturally have low crack porosities, and care was taken to minimize surface conduction along cracks by cleaning and annealing in appropriate atmospheres. While the granites have low dc conductivities, they show significant ac losses. The Mellen gabbro is much more conducting than the granites or than would be anticipated from its major mineral phases. By comparing our results with other data on crack porosity we find that the crack free rock properties should determine its actual properties at depths below about 2 km in the earth.

Introduction

Richter and Simmons [1977] have argued that Precambrian igneous rocks exposed at several localities in the United States may be representative of a significant portion of the upper continental crust underlying central North America. Feves et al. [1977] have shown that samples of these rocks have very low crack porosities and that most of the cracks present close under very modest hydrostatic pressures. From this they convincingly argue that the cracks currently in the rocks are a result of unloading from depth and could not have been present in the rocks in situ at depths of more than a few kilometers. Although episodes of nonhydrostatic stress must have existed to produce the healed and sealed cracks observed by Richter and Simmons [1977], it seems reasonable to suppose that igneous rocks throughout most of this fairly stable geological region are largely crack free at depths of more than a few kilometers.

It is well known that surface conductivity in insulating materials can, and frequently does, exceed the bulk conductivity by a considerable amount. This is especially likely to be the case if the surfaces are altered or contaminated. Therefore our experimental strategy is to clean cracks of contamination as well as possible and then to anneal them under conditions appropriate to reversing surface alteration and eliminating or minimizing surface damage.

The electrical characteristics of even crack free rocks are expected to depend on the electrical properties of the individual component minerals, the electrical behavior of the interfaces, and the petrographic texture in a complex way. The electrical properties of the minerals and interfaces are in general determined by the presence of trace impurities and defects. The concentration of these impurities and defects is in turn determined by the chemical and physical environment in which the rock last equilibrated. Our annealing conditions are chosen to be too mild to reequilibrate the bulk samples. They are designed only to reverse surface alteration and to allow the surface compositions of grains to approach equilibrium with their interiors. Therefore except in cases where minor phases distributed along partially open cracks contribute significantly, our results are intended to be representative of rocks at the depths at which they last naturally equilibrated in the earth.

The dc conductivity of a rock is generally determined by the major phases, since only these usually constitute interconnected pathways throughout the rock. The ac conductivity, however, can have additional contributions from isolated phases with higher conductivity and from the reorientation of dipoles. These contributions can also lead to increases in the dielectric constant. Thus the ac behavior provides additional information which cannot be inferred from dc conductivity and static dielectric constant values alone.

We present ac and dc results on Montello and Wausau red granites and on Mellen gabbro. We feel that the dc results well represent the values expected from these rocks when they are crack free at the depth in the earth where they last equilibrated.

Samples

We are currently studying a suite of samples supplied by Gene Simmons. They are a subset of the samples described by Richter and Simmons [1977] and Feves et al. [1977] and were collected and processed with great care to avoid the introduction of new cracks. In this paper we report results on a Wausau granite sample bearing the Massachusetts Institute of Technology identification number 1336 and on a Mellen gabbro sample bearing the number 1331. We also present results on a granite sample from Montello, Wisconsin, previously carefully collected for us by Herb Wang.

To prepare specimens for our electrical measurements, we waxed the rocks onto mounting blocks and gently sawed slabs 2-3 mm in thickness from them with an oil-water emulsion lubricated diamond saw. These slabs were carefully washed and repeatedly rinsed in acetone to remove contamination from the wax and oil. They were then washed with distilled water to remove any soluble salts left when the pore fluids previously occupying the cracks evaporated.

Electrical contacts were generally prepared by painting a suspension of colloidal Pt (Hanovia Platinum Bright) on the samples and evaporating away the supporting constituents by heating them to 620°K in dry N_2 for 5 min. Contacts prepared in this way adhered well and did not appear to cause any surface damage. They were not required to have any mechanical strength. The contacts were placed so that all major phases were present in the measured area in roughly the same proportions as they were in the whole slab. Thus we measure the effective conductivity of the major phases in parallel. Since these phases appear to form interconnected pathways through the whole rock, our results should represent the bulk rock within a small numerical factor.

We are making Mössbauer effect measurements on these samples in order to determine the nature of Fe-containing phases and their degree of oxidation. One noteworthy finding from these Mössbauer measurements is that a major fraction of the Fe in the Wausau granite is in the form of hematite, Fe_2O_3. This indicates a highly oxidized state for the rock and is essential information which allowed us to choose a reasonable atmosphere for the equilibration runs and high-temperature electrical measurements.

Atmosphere Control

When mafic rocks or minerals are heated, it is necessary to control the atmosphere carefully so that Fe is neither oxidized nor reduced, since either process causes mineral decomposition and corresponding drastic changes in electrical properties. Our apparatus is designed to allow the atmosphere to be controlled in either of two ways. In one mode, pure gases or gas mixtures flow past the sample. In the other mode, an inert gas, conventionally N_2, flows through a finely powdered solid state buffer maintained at the same temperature as the sample and then flows past the sample. This keeps the partial pressure of oxygen at the sample in equilibrium with the buffer.

The buffer used for the Montello granite results reported in this paper was powdered Rockport, Massachusetts, fayalite, Fe_2SiO_4. This fayalite contains abundant intergrown magnetite, Fe_3O_4, and is expected to reduce the nominal 10 ppm O_2 content of the N_2 stream to a much smaller value close to that determined by the temperature-dependent equilibrium in the reaction $3Fe_2SiO_4 + O_2 \leftrightarrow 2Fe_3O_4 + 3SiO_2$. We refer to this O_2 content as the one determined by the quartz-fayalite-magnetite buffer. Before use, buffers were conditioned by passing N_2 through them for several hours at 1100°K in order to remove any volatile impurities and reverse any changes which could have occurred as a result of low-temperature weathering.

Electrical Measurement

Our experimental apparatus and techniques will be described only very briefly here, since a complete description including drawings is incorporated in another manuscript dealing with the electrical properties of pure forsterite [Morin et al., 1977].

Our sample holder is of the guarded three-terminal configuration recently suggested by Özkan and Moulson [1970] and can accomodate disc-shaped samples up to a few millimeters in thickness. Ohmic contact behavior was verified in all samples by monitoring the dc current-voltage characteristics over 1 or 2 orders of magnitude with both positive and negative polarity.

The dc conductivity measurements were made with a constant applied voltage of the order of 1 V by using a Keithley 610-C electrometer as a virtual ground current amplifier. The output of the electrometer was

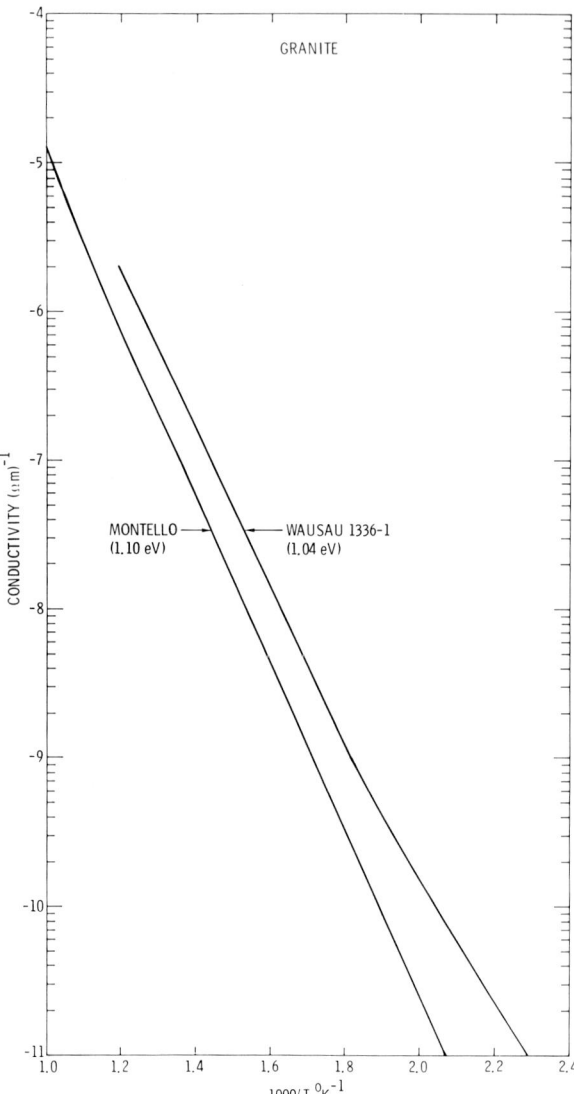

Fig. 1. Arrhenius plot of dc conductivity data for two annealed Precambrian Wisconsin granite samples. Note the low conductivities and similar behavior. The numbers on the ordinate in all figures indicate powers of 10. Activation energies indicated were obtained from the slopes of the lines by expressing kT in electron volts.

fed to a logarithmic voltage converter that permitted us to obtain a continuous plot of log current versus thermocouple voltage on an x-y recorder. The ac measurements were made with a General Radio 1615-A capacitance bridge in a standard three-terminal configuration.

Results

Granites

No Mössbauer data were available when we began the electrical measurements. Therefore the Montello granite sample which we studied first was run in an atmosphere defined by the quartz-fayalite-magnetite buffer. After the first heating cycle to about 830°K the conductivity on cooling had decreased about a factor of 2. On the second heating run to slightly above 1000°K the conductivity on cooling had decreased about another factor of 3, the total decrease at 500°K being brought to about a factor of 7. A third dc conductivity run closely retraced this second cooling curve, however, with some rate-dependent hysteresis above 800°K. The dc data from this third run are presented in Figure 1 as conductivity σ versus reciprocal temperature $1/T$. The activation energy E obtained by fitting the slope to the Arrhenius equation $\sigma = \sigma_o e^{-E/kT}$ is also indicated.

After the third dc run, complete sets of ac data were taken at about 50°K temperature intervals while the sample was being warmed to 1052°K and cooled back down to below 500°K. Generally, the warming and cooling data were in good agreement. However, at certain frequencies the losses were significantly different. The ac conductivity and dielectric constant relative to free space ε' are obtained directly from our ac bridge results by representing the sample as a capacitor with a resistor in parallel and are plotted versus $1/T$ for a representative set of frequencies in Figures 2 and 3, respectively. The loss tangent defined by $\tan \delta = \sigma/\omega\varepsilon'$ calculated from data obtained during the warming cycle is plotted versus frequency in Figure 4.

At the end of this series of runs the sample was examined under a binocular microscope. The most apparent change was a considerable increase in the density of microcracks. As we will argue in the Discussion section, we do not believe that this increase in microcrack density, caused by thermal cycling, has significantly affected our dc conductivity results.

When we made electrical measurements on the Wausau granite, we knew from the Mössbauer results that it had last equilibrated in a quite oxidizing environment. Therefore our initial dc conductivity run to 660°K was in N_2 with about 10 ppm O_2. In this atmosphere the conductivity versus $1/T$ curve increased in slope at the higher temperatures, and the conductivity was about 30% higher on cooling. We then heated the sample to 540°K and rapidly changed the atmosphere to pure O_2 at constant temperature. The conductivity began to decrease with about a 40-s time constant and within a few minutes had stabilized at a value about a factor of 6 lower than that before the O_2 was admitted. We now heated the sample to about 770°K in O_2, and the conductivity dropped about another factor of 2 on cooling. A subsequent dc conductivity run in O_2 gave identical warming and cooling data agreeing perfectly with the previous cooling curve. Complete sets of ac data were then obtained at about 50°K temperature intervals on warming in O_2 to 900°K. Representative results are displayed in Figures 5-7.

Following collection of the ac data the contacts were checked and found to be ohmic. The dc conductivity was then measured as the sample cooled back to room temperature. The results agreed perfectly with the

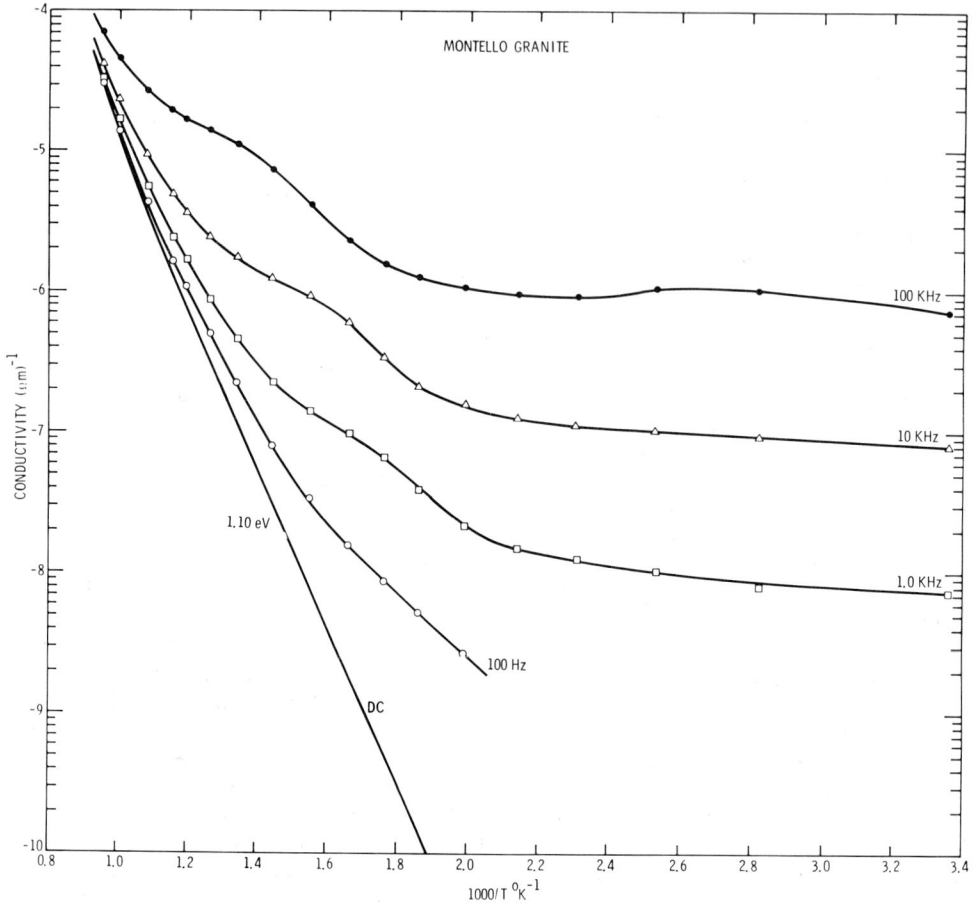

Fig. 2. Comparison of dc and ac conductivities at selected frequencies for Montello granite. Note convergence at high temperatures and large excess ac losses at low temperatures.

previous dc runs and are plotted in Figure 1. The sample was then heated in N_2 containing about 10 ppm O_2 to 870°K with identical warming and cooling curves in perfect agreement with the results obtained in O_2. Similar behavior was found for wet N_2.

Following the above series of runs the Wausau granite sample was warmed in H_2. At about 525°K its conductivity began to increase dramatically. By 670°K it had increased by about 3 orders of magnitude and had a shallow slope corresponding to an activation energy of 0.24 eV on cooling. Following this run a run in O_2 restored the original low conductivity values. The only change evident on examining the sample under the binocular microscope following these runs was a substantial increase in the density of microcracks. As we will argue in the Discussion section, the excellent agreement between dc conductivity results obtained on a cycle to 770°K and on a subsequent cycle to 870°K strongly implies that they are not significantly influenced even by the final microcrack density.

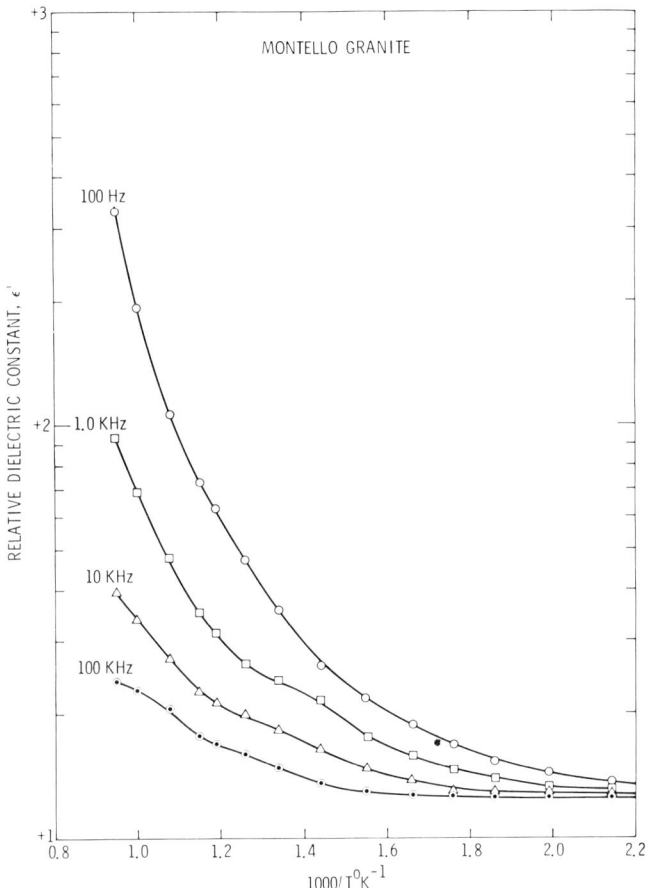

Fig. 3. Relative dielectric constant versus 1/T for Montello granite at selected frequencies. Note convergence at low temperatures and rapid frequency-dependent rise at high temperatures.

These two granites are indistinguishable in appearance and, as can be seen from Figures 1-7, also appear to have very similar electrical properties. At dc they are quite good insulators, but they have significant ac loss mechanisms.

Mellen Gabbro

The Mellen gabbro contains fairly abundant metallic appearing areas, presumably sulfides, visible under the binocular microscope. Measurements on one slab yielded extremely high conductivity values probably due to these sulfides. A small area was found on a second slab which exhibited a much lower but still high conductivity. Figure 8 shows dc conductivity versus 1/T data to 1000°K for that small region. The sample was run in the quartz-fayalite-magnetite buffered atmosphere and after the initial cycle yielded nearly identical warming and cooling curves. For this particular sample, Pt contacts were sputtered on by

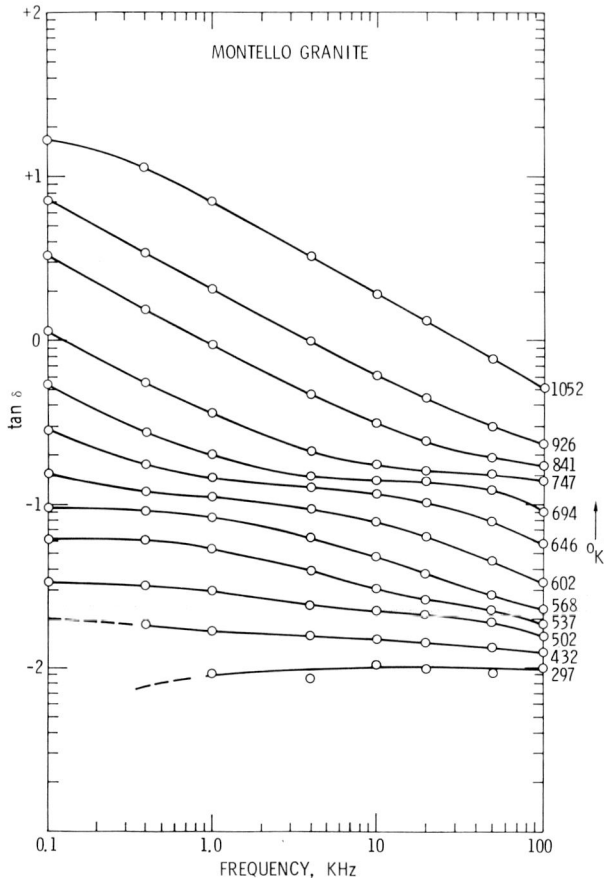

Fig. 4. Log-log plot of the loss tangent defined by $\tan \delta = \sigma/\omega\varepsilon'$ for Montello granite at a selected set of temperatures.

using a dc discharge. This procedure puts a large number of trapped electrons into the sample, so data taken during the first heating cycle would have been useless. Examination of the sample in the binocular microscope following these runs showed no obvious changes. The sample was then heated to 600°K in O_2 for about 1 hour and still showed no change in conductivity.

In view of the difficulties mentioned we were not yet confident that the data shown in Figure 8 were representative of Mellen gabbro. We therefore painted six electrical contacts on our remaining large block of the sample by using colloidal graphite suspended in propanol and heated the sample to 180°C in air in a drying oven. Measurements of the resistance between various pairs of contacts made with a Simpson meter were consistent with each other and with the value predicted from Figure 8. These measurements also indicated that the dominant conduction was through the bulk and not on the surface.

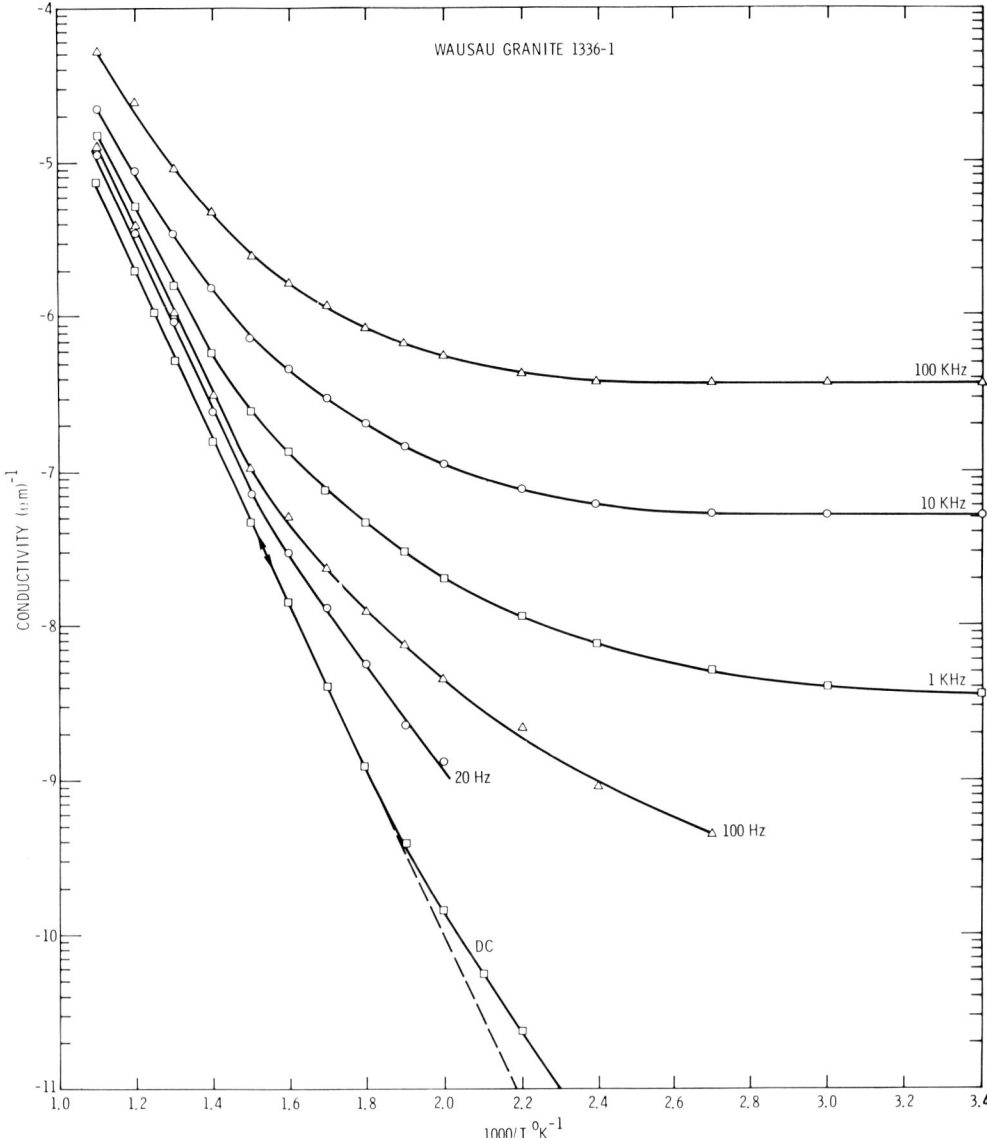

Fig. 5. Comparison of dc and ac conductivities at selected frequencies for Wausau granite. Note the similarity to Figure 2.

Discussion

Granite

The low temperatures at which irreversible conductivity changes were observed during the first heating cycle in O_2 of the Wausau granite indicate that a change taking place along open crack surfaces must have

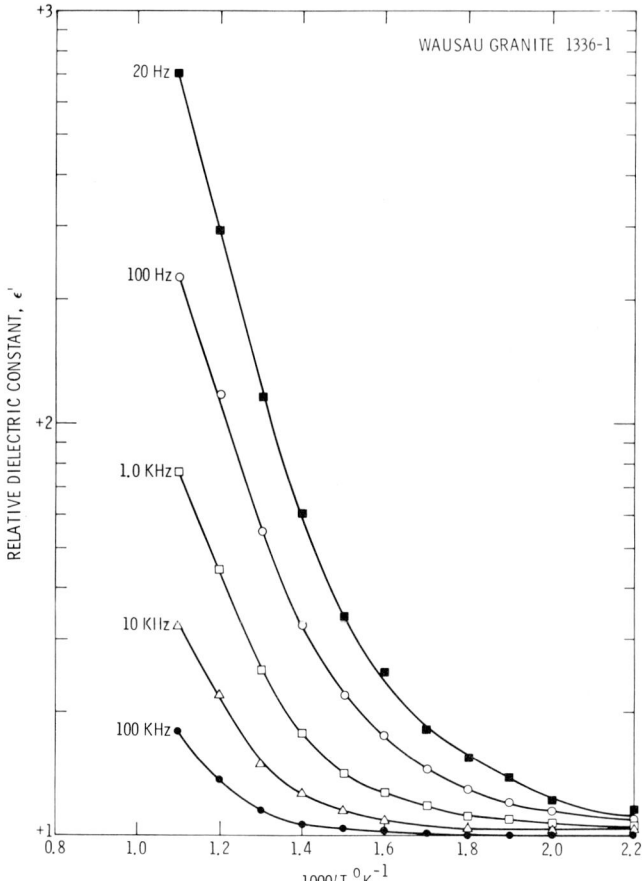

Fig. 6. Relative dielectric constant versus 1/T for Wausau granite at selected frequencies. Note the similarity to Figure 3.

been responsible. Most alteration effects expected from atmospheric or groundwater exposure involve hydration or oxidation and hence could not have been reversed by O_2 while they were stable in dry N_2 at the same temperature. Therefore we feel that the most likely explanation for these results is that carbonaceous material initially in the cracks either naturally or as a result of our sample preparation procedures gave an excess conductivity and that the O_2 treatment removed this carbonaceous material.

We know that a considerable number of excess microcracks were generated by the thermal treatments of this sample. However, after the oxygen anneal at 540°K, a factor of 2 further decrease in conductivity was observed after heating to 770°K in O_2. Subsequent runs to 900°K continued with the same activation energy slope and led to no further changes in the conductivity at lower temperatures. It seems likely that heating in the interval 770°-900°K produced a significant fraction of the excess microcracks observed. Controlled studies by Simmons and Cooper [1977] on Westerly granite suggest that crack porosity would

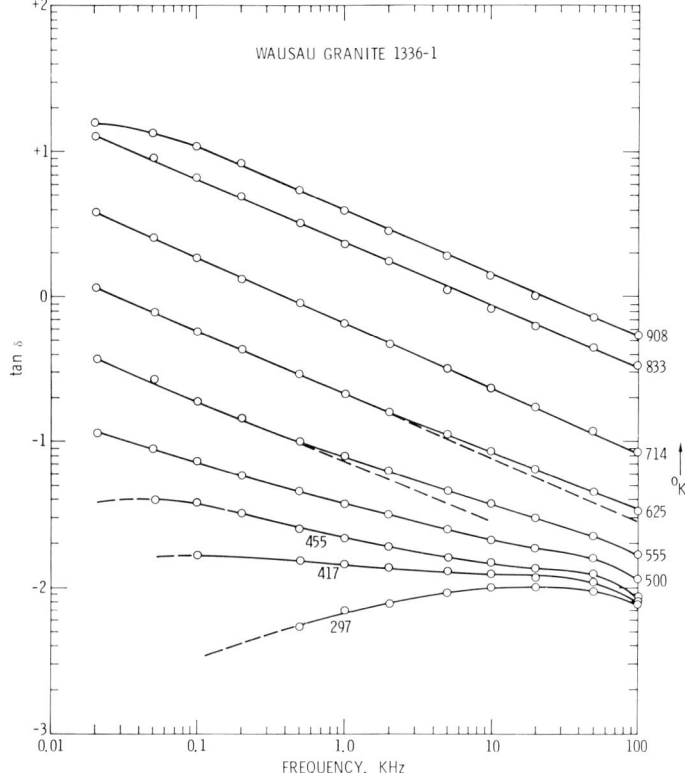

Fig. 7. Log-log plot of the loss tangent defined by $\tan \delta = \sigma/\omega\varepsilon'$ for Wausau granite at a selected set of temperatures. Note the similarity to Figure 4.

increase by almost a factor of 2 in that temperature interval. If this is the case, then these results show that clean cracks even at the density present at the end of our runs have a negligible effect on the dc conductivities measured.

While the results obtained on initial heating cycles of the Montello granite are not so definitive in themselves, they are consistent with the above conclusions. In the absence of sufficient O_2 in the atmosphere, the carbonaceous material was not removed until temperatures of about 1000°K, probably by reduction of hematite in the sample.

It seems very likely that the pure O_2 atmosphere is more oxidizing than the environment in which the granites last equilibrated. However, since the Mössbauer evidence shows that Fe in the rocks was nearly fully oxidized, there is probably little about them that can be changed by exposure to pure O_2. We therefore feel that data obtained in O_2 come closest to representing the rocks as collected.

It is clear that the quartz-fayalite-magnetite buffer is more reducing than the environment in which the granites last equilibrated. Since the major quartz and feldspar phases are nearly Fe-free, it is reasonable to expect that this would have a minor effect on their electrical behavior

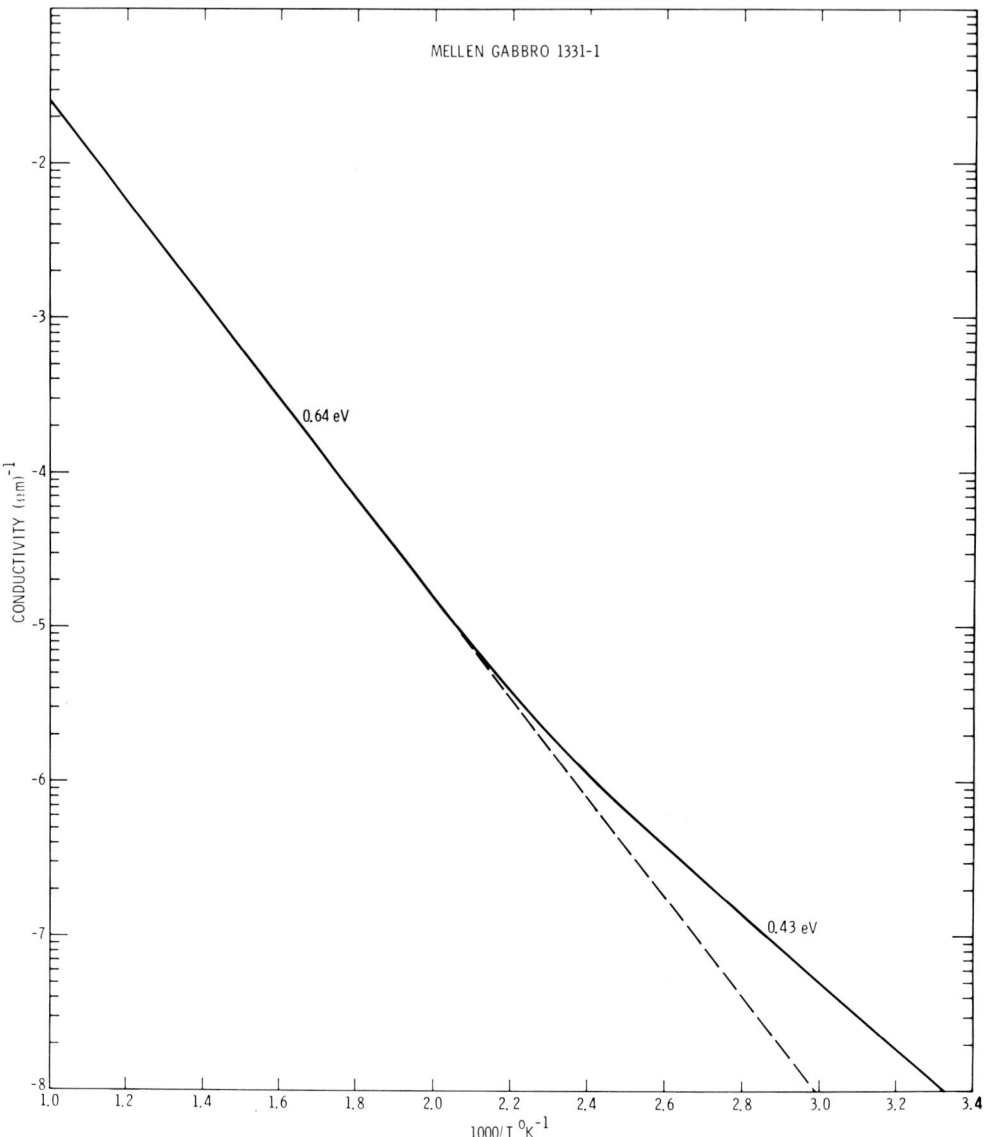

Fig. 8. Arrhenius plot of dc conductivity data for Precambrian Mellen, Wisconsin, gabbro. Note the much higher conductivity values than those shown in Figure 1 for Wisconsin granites.

in the temperature interval studied here. Hematite exposed to this atmosphere will certainly be reduced to magnetite. However, hematite is a very fine grained minor constituent, and both hematite and magnetite are very conducting in relation to the other components of the rocks. Rutile is a more abundant minor phase which fills a number of sealed microcracks [Richter and Simmons, 1977], and its electrical behavior is strongly dependent on its oxidation state.

It seems likely from the Mössbauer data that a magnetite-hematite buffer will come close to simulating the environment in which these rocks last equilibrated, and such a buffer will be used in the future. However, the strong overall similarity of the results obtained in the O_2 and quartz-fayalite-magnetite buffered atmospheres suggests that the major features of the electrical results obtained are insensitive to atmosphere over a wide range of O_2 partial pressures. Weak extra ac loss peaks are evident in the Montello granite data (Figure 2) which are not seen in the Wausau granite data in Figure 5. These extra peaks could possibly be due to partial reduction of rutile by the quartz-fayalite-magnetite buffer.

There are several known processes which may contribute to the large excess ac conductivity which we observe at low temperatures. Dipole reorientation is probably the most familiar one and can occur because of defects and impurities in all the mineral phases and interfaces present. Large dipole effects ascribed to the motion of monovalent ions in channels have been observed in quartz [Snow and Gibbs, 1964]. Rutile exhibits a large dielectric constant and a conductivity strongly dependent on its oxidation state [Grant, 1969].

Isolated inclusions more conducting than the major phases can also lead to ac losses. Among the possible candidates known to be present are hematite, rutile, and trapped pore fluids. A relaxation time defined by $\tau \equiv \epsilon'/\sigma$ can be associated with each inclusion phase. Inclusions of that type only make a significant contribution to ac losses for frequencies such that $\omega\tau \approx 1$. Hematite is so conducting that its contributions to ac losses occur at frequencies much higher than those that we employed. Therefore we can eliminate it or magnetite from further consideration. The same would be true for pore fluids if we assume that only their normal conductivity is involved.

The real part of the dielectric constant is reasonable and nearly independent of temperature and frequency for both samples at low temperatures, as can be seen from Figures 3 and 6. At relatively high temperatures it shows a rapid frequency-dependent rise with increasing temperature. This increase becomes really dramatic at about the temperature where the conduction current equals the displacement current for a given frequency. This suggests that interfacial polarization associated with conduction through the major mineral phases is responsible.

Such interfacial polarization is expected at cracks and along grain boundaries, and because of their shape these structures can have a very large effect even though the volume occupied by cracks is small. If cracks do indeed play a major role in this increase in dielectric constant, then the laboratory values are inappropriate for rocks at depth. Further work should be done to resolve this question, although it is not crucial to any of our discussion, since the high dielectric constant values occur at temperatures higher than those expected in stable regions of the upper crust.

The H_2 run appears to show that the rutile sealing cracks does compose a continuous network through the rock which is capable of high conductivity when the rutile is reduced.

In Figures 4 and 7 our ac results are plotted as loss tangent versus frequency for a range of temperatures. It is hoped that plots of this type will prove convenient for the interpolations or extrapolations required in the planning and interpretation of geophysical experiments.

The ambiguity in the real part of the dielectric constant mentioned above is reflected in these curves but should be of minor importance, since its influence is only large in the region where loss tangents are near 1 and above.

Mellen Gabbro

Although the dc conductivity measured in the Mellen gabbro is much larger than one would expect from the major feldspar and pyroxene phases, the weight of the experimental evidence now strongly indicates that the results are representative of the rock. More work remains to be done to elucidate the conduction paths and mechanisms involved. One possibility might be an interconnected network of magnetite, sulfides, or other high-conductivity minor phase.

Feves et al. [1977] found a very low crack porosity for this rock and hence predict very low conductivities through the pore fluids. Combining the prediction of Feves et al. with our results indicates that the crack-free rock conductivity should dominate the pore fluid conductivity at depths below about 2 km.

Summary

We have obtained and discussed ac and dc electrical property data on two Precambrian Wisconsin granites. We believe that the dc results closely approximate the values which would be found for the crack free rocks. At low temperatures, ac losses considerably exceed those that would be predicted from the dc conductivity values. We have presented the results in a way which we hope may be useful for the planning and interpretation of geophysical experiments.

We have also presented data on the Precambrian Mellen, Wisconsin, gabbro which indicate crack free rock conductivity values much larger than those of the granites or than would be expected from its major mineral phases. These data suggest that conduction through the solid rock should dominate pore fluid conductivity at depths exceeding about 2 km.

Acknowledgments. During the course of this work we have profited greatly from comments and suggestions made by Gene Simmons and John G. Heacock. We are especially grateful for the carefully collected and well-documented samples supplied to us by Gene Simmons. This work was supported by Office of Naval Research Contract N00014-76-C-0653.

References

Feves, M., G. Simmons, and R. W. Siegfried, Microcracks in crustal igneous rocks: Physical properties, in The Earth's Crust, Geophys. Monogr. Ser., vol. 20, edited by J. G. Heacock, AGU, Washington, D.C., this volume, 1977.
Grant, F. A., Properties of rutile (titanium dioxide), Rev. Mod. Phys., 31, 646-674, 1959.
Morin, F. J., J. R. Oliver, and R. M. Housley, Electrical conduction in forsterite Mg_2SiO_4, submitted to Phys. Rev. 1977.

Özkan, O. T., and A. J. Moulson, The electrical conductivity of single crystal and polycrystalline aluminum oxide, J. Phys., $\underline{D3}$, 983-987, 1970.

Richter, D., and G. Simmons, Microcracks in crustal rocks: Microscopy, in The Earth's Crust, Geophys. Monogr. Ser., vol. 20, edited by J. G. Heacock, AGU, Washington, D.C., this volume, 1977.

Simmons, G., and H. W. Cooper, Thermal cycling cracks in three igneous rocks, submitted to J. Geophys. Res., 1977.

Snow, E. H., and P. Gibbs, Dielectric loss due to impurity cation migration in quartz, J. Appl. Phys., $\underline{35}$, 2368-2374, 1964.

INTERNAL FRICTION MEASUREMENTS AND THEIR IMPLICATIONS IN SEISMIC Q STRUCTURE MODELS OF THE CRUST

B. R. Tittmann

Science Center, Rockwell International
Thousand Oaks, California 91360

Abstract. Mitchell (1973) and more recently Herrmann and Mitchell (1975) have presented the first detailed seismic Q profiles (quality factor Q is proportional to reciprocal of internal friction) for a region of the earth's crust. Their profiles show a dramatic increase in seismic Q from values of a few hundred from the surface down to 15 km to as high as 2000 below 15 km. No ordinary rock types measured in laboratory air have shown such high values. Results of laboratory internal friction measurements on strongly outgassed rocks recently made as part of an effort to interpret the high seismic Q values observed in the lunar crust show that moisture absorbed in pores causes the low Q values. This suggests that the sharp increase in Q might mark a boundary below which no appreciable moisture exists in the crustal rocks. Our studies show that the Q of a low-porosity olivine basalt without hydrated mineral phases ranges from about 100 measured dry in normal laboratory air to over 2000 outgassed at moderate temperatures in a high vacuum for about a week. These results suggest that the Q of a dry sample of the same composition and mineralogy would be above 2000. Laboratory data on Q in rocks have characteristics which suggest they are valid for rock in situ.

Introduction

Mitchell (1973) has recently presented the first detailed internal friction profile for a region of the earth's crust. It is based on data from the October 21, 1965, south central Missouri earthquake, recorded at a number of stations located between the Rocky Mountains and the Atlantic coast. The most conspicuous feature of this profile is a sharp increase in Q (decrease in internal friction) from values not exceeding a few hundred from the surface down to about 15 km to a value at least as high as 2000 slightly below 15 km. No ordinary rock types measured in laboratory air have shown such high Q values (Knopoff, 1964). More recently, Herrmann and Mitchell (1975) and Mitchell (this volume) have obtained similar but more detailed results based on data from four earthquakes occurring in the New Madrid seismic region, one earthquake in the northern Hudson Bay region, and three underground nuclear explosions in the western United States. They obtain somewhat lower Q values in the lower crust (Q ~ 1500) but because of relatively large standard deviations do not rule out $Q \approx 2000$ from a depth of 17 km to about 37 km. Lee and Solomon (1975) have combined longer period data with the data of Mitchell (1973) to infer the Q structure of the crust and upper mantle and draw conclusions about lithospheric thickness and properties of the asthenosphere. These examples demonstrate the importance of Q measurements for gaining insights into the natures of the lithosphere and asthenosphere. Complementary studies of material properties in the laboratory are necessary in order to aid in the interpretation of seismic Q values in terms of the nature of material in the upper and lower crust (Housley, et al., 1974).

This paper presents the results of recent laboratory studies of internal friction and velocity made as part of an effort to interpret the high seismic Q values observed near the surface of the moon (Tittmann, et al., 1974). The results of these laboratory measurements carried out in part on strongly outgassed rocks bear on the interpretation of the seismic Q-versus-depth profiles reported by Mitchell (1973) and Herrmann and Mitchell (1975) in terms of the moisture content of crustal rocks.

In the following the technique of laboratory measurements is presented first, followed by the description of three key experiments. The results of these experiments are then discussed and interpreted on the context of the current seismic Q profiles for the crust.

Experimental Approach for Laboratory Internal Friction Measurements

The experimental approach basically embraces the vibrating bar technique in which a uniform bar is excited in one of its natural modes of vibration. The three modes selected by us for the measurements are the flexural, torsional, and longitudinal modes. For excitation in the torsional or longitudinal modes, our specimen is rigidly supported at the mid-point, which becomes the mode for the fundamental free - free mode. For excitation in the flexural mode, our specimen is pivoted at the ends as described in detail in a later section. To eliminate corrections due to dispersion, for example, the bar is given a length-to-radius ratio of at least about 10 to 1 and is further given a circular cross section for torsional resonance or a rectangular cross section for flexural resonance (for longitudinal resonance the cross sectional shape plays a minor role). The sharpness of resonance of a mode of vibration has been observed as a measure of losses in metals and other materials (Quimby, 1928), and this technique has been applied to measuring losses in rock samples by for example, Birch and Bancroft (1938). Many techniques, Norwick and Berry (1972), of exciting the natural mode of the bar are available and Gordon and Davis (1968), for example, used a piezoelectric element bonded to the end of the specimen to form a composite piezoelectric resonator. In the experiments described here, the excitation technique employs the magnetic drive and detection system used by Wegel and Walther (1935) in which the specimen is excited by the use of an alternating magnetic field acting on a ferrous electrode mounted on the specimen end. The The vibrating bar technique was preferred over other techniques for providing information helpful for the interpretation of seismic data for the following reasons:

1) The technique enables measurements over a very large frequency range not accessible by other techniques. In particular with the flexural mode frequencies down to 1 kHz are typical for uniform bars of 3 cm length, but with end loading (Berry, 1955) the effective resonance can be lowered by another order of magnitude to bring the experiment very near to the high end of the seismic range of frequencies. The technique provides therefore, a valuable link between laboratory and seismic field determinations of Q.

2) The vibrating bar technique allows the determination of Q in at least two ways, by measuring the decay of the normal mode vibrations or by measuring the sharpness of the resonance. Since the Q values measured by the two methods should be the same, a valuable self-consistency check is obtained.

3) The vibrating bar technique provides a well-defined mode of vibration with a well-defined nodal point. Positioning of the support at the nodal point assures that essentially no energy is lost in the supports and that the attenuation of vibrational amplitude is accurately reflected in the width of the resonance or the decay of the vibrations.

4) The vibrational mode of excitation provides for a well-defined single frequency ω, which must be known accurately to determine the quality factor Q, i.e., for longitudinal resonance $Q_y = \omega/2\alpha_y c_y$ where c_y is the wave velocity in cm/sec and α_y is the internal friction (attenuation) in nepers/cm. Here, $c_y = (E/\rho)^{1/2}$ where E is Young's modulus in dynes/cm^2 and ρ is the density in gm/cm^3.

5) The method also minimizes errors introduced by the transfer of energy into unobserved vibrational modes which results in an apparent rather than real energy loss.

6) The use of this method allows measurements to be made at very small strain amplitudes to minimize nonlinear processes which are completely absent in terrestrial seismic waves whose strain amplitudes are typically of the order of 10^{-13} to 10^{-10}. Measurements of the strain with a commercial capacity microphone revealed strains in our measurement typically in the 10^{-8} range and as low as 10^{-9}.

Experiment 1: Internal Friction Q = 2400 Achieved in Igneous Rock

Sample Preparation

This experiment was carried out on samples of augite olivine basalt (Quaternary flow from Western Cascade volcanics designated W-8 from Weed, California, near Mt. Shasta). The open pore porosity was measured to be approximately 2.5%.

These rocks were attached to metal blocks with low melting point wax and sawed into bars measuring 8.0 x 0.4 x 0.5 cm with an oil-cooled diamond saw. Preliminary attempts to saw rocks into bars with a gas-cooled diamond saw always resulted in breaking the bars.

After sawing, the bulk of the wax and oil was removed from the samples by washing them in acetone at room temperature. After this procedure the sample typically had a Q of about 20. It was generally followed with a cleaning procedure in which the sample was washed first in boiling trichloroethylene for about an hour, then in boiling ethanol, and finally in boiling distilled water, since this sequence

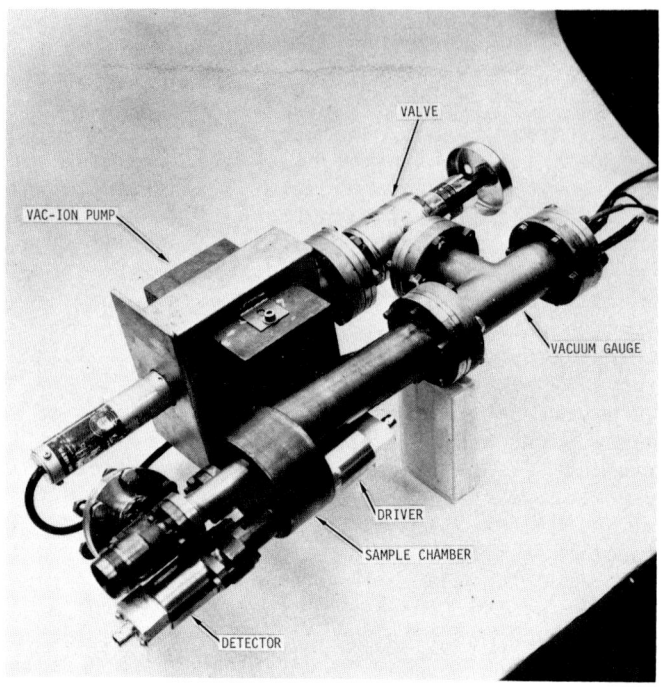

Fig. 1. Ultra-high vacuum apparatus, an all-metal system with 15 liter/sec Varian Vac-Ion pump.

is known from semiconductor technology to leave the least amount of contaminants. Lately we have supplemented or partially bypassed the cleaning procedure by boiling the samples in 30% H_2O_2 for up to several hours, until excess bubbling stops, an indication that oxidizable organics have been consumed. Although no systematic comparison has yet been made on the outgassing behavior of similar samples cleaned by different procedure, our general impression is that the H_2O_2 treatment is the more effective. With this procedure the measured Q after cleaning is about 75 to 100, indicating a sample clean of the oil and wax used in the cutting procedure. (Since the H_2O_2 is an oxidizing agent, it does not further oxidize the rock minerals but affects only the organic and biological material by transforming it to CO_2 and H_2O)

The mode of vibration selected for this experiment was the longitudinal resonance mode and therefore the electrodes consisted of small iron buttons cemented to the ends of the bars. Initially, we used an epoxy to cement the buttons to the bars, but eventually, we learned that at least for small samples, losses in the epoxy dominated the other loss mechanisms above room temperature. We now use either a fast-drying general-purpose household cement, which is satisfactory at room temperature, or a high-temperature silicate cement. The latter is cured by gradually warming the sample to about 150°C in air and holding it there for about a day. This procedure also dries the sample and raises Q at room temperature to as high as ~ 400.

Apparatus

Initially (Tittmann, 1972), the sample, sample holder and associated hardware were placed in a bell jar and a conventional diffusion pump was used to free the rock from any remaining volatiles, such as adsorbed H_2O. More recently (Tittmann, et al., 1975), a second ultrahigh vacuum system was built to provide an even better and cleaner vacuum. This system is shown in Figure 1. Instead of a glass bell jar, diffusion pump, and mechanical forepump, this system is all stainless steel and employs a Vac-Ion pump and a molecular sieve sorption pump so that any backstreaming of oil vapor is eliminated. The system allows for the attachment of a residual gas analyzer to obtain quantitative information on the gases in the sample chamber. Perhaps even more important, the new system allows the sample chamber to be inserted in a liquid helium bath, so that vapor pressures in the chamber can be lowered sufficiently to freeze out all residual gases except He on the chamber walls. This technique should enable us to go to a substantially lower vacuum pressure than we can achieve now with the old system. Figure 1 is a photograph of the apparatus showing a U-shaped tubelike fixture terminated at one end by a 15 ℓ/s Varian Vac-Ion pump and at the other end by the magnetic driver, detector and sample chamber. Connecting these is a 2-in. (5.08 cm) pumping line with a gold seal valve and a Bayard-Alpert ionization gauge. The rock sample is held in the chamber at its mid point by a ring of four spring-loaded ball bearing tipped set screws. The sample is inserted through a port which is sealed with a Cu ring and a flange with a Varian-type stainless steel lip which can be tightened against the Cu seal by means of six bolts. The magnetic driver and detector are mounted on the outside of the chamber and interact with the sample ends through thin walls of nonmagnetic stainless steel. This has the advantage that all the electric circuitry and its associated contaminants (wire coating, solder joints, etc.) do not contribute to the gas load in the vacuum system. The iron buttons which complete the magnetic circuit of the driver and detector are attached to the sample with high-temperature quartz cement (American Fused Silica Corporation) which, after bakeout, does not appear to outgas a substantial gas load.

Results

A typical set of data obtained by the longitudinal resonance technique at room temperature after successive cleaning and outgassing treatments is shown in Table 1 at a frequency of 24 kHz. This frequency was chosen partly to provide a data point to determine the presence of any dispersion. After each outgassing step, Q was observed to increase, the highest value obtained being about 2400. A bandwidth plot at a high Q value is shown in Figure 2. The introduction of dry ($< 10^{-8}$ torr partial pressure of H_2O) reactor grade nitrogen gas into the vacuum chamber after an outgassing treatment did not affect the Q. However, long term exposure to laboratory air at the end of a run brought the Q back down to about 100. These results strongly support the conclusion that igneous rocks display dramatically higher laboratory Q values after they are really freed from volatiles by strong outgassing techniques.

TABLE I.

Internal Friction Quality Factor Q Values in Igneous Rock W-8

Q	Cleaning and Outgassing Conditions
17-20	At room temperature in air at 1 atm after sawing with wet diamond saw, cleaning in acetone bath, and drying.
75-100	At room temperature in air at 1 atm after boiling in 30% H_2O_2 and drying.
400	At room temperature in air after gradual heating to 150°C in air and holding the temperature for about 1 day.
1500	At $\sim 1 \times 10^{-7}$ torr and room temperature after outgassing in 10^{-6} torr for 8 hours at $\sim 300°C$.
2375	At $\sim 1 \times 10^{-7}$ torr and room temperature after outgassing in 4×10^{-7} torr for 20 days at $\sim 400°C$.
100	After long term exposure to laboratory air at end of run.

Experiment 2. Internal Friction Measurements at 50 Hz

In this experiment the mode of vibration selected was the flexural mode for which the resonant frequency is most readily shifted to lower frequencies by end-loading techniques. In particular, by end-loading our specimen 8.5 x 0.6 x 0.4 cm we lowered the resonance to 50 Hz, thus bringing the experiment into the high end of the seismic range of frequencies. (Seismic reflection studies on the upper and lower crust have been conducted at $\sim$ 100 Hz or slightly higher).

Very seldom has this technique been applied to brittle samples which are easily damaged by the conventional sample holder or to the measurement of internal friction on low loss materials because of the generally high "background" damping introduced by the apparatus itself. The present apparatus design was motivated by the desire to measure Q in fragile lunar return samples at zero static stress to shed light on the anomalously low attenuation of seismic waves on the moon. Furthermore, an additional requirement was that the sample be subjected to the low strain amplitudes and frequencies which are characteristic of seismic experiments.

Low Frequency Apparatus

Figure 3 shows a photographic overview of our measurement fixture. Here a rectangular bar-shaped sample of olivine basalt 8.5 x 0.6 x 0.4 cm is shown whose ends are held with a special clamping arrangement

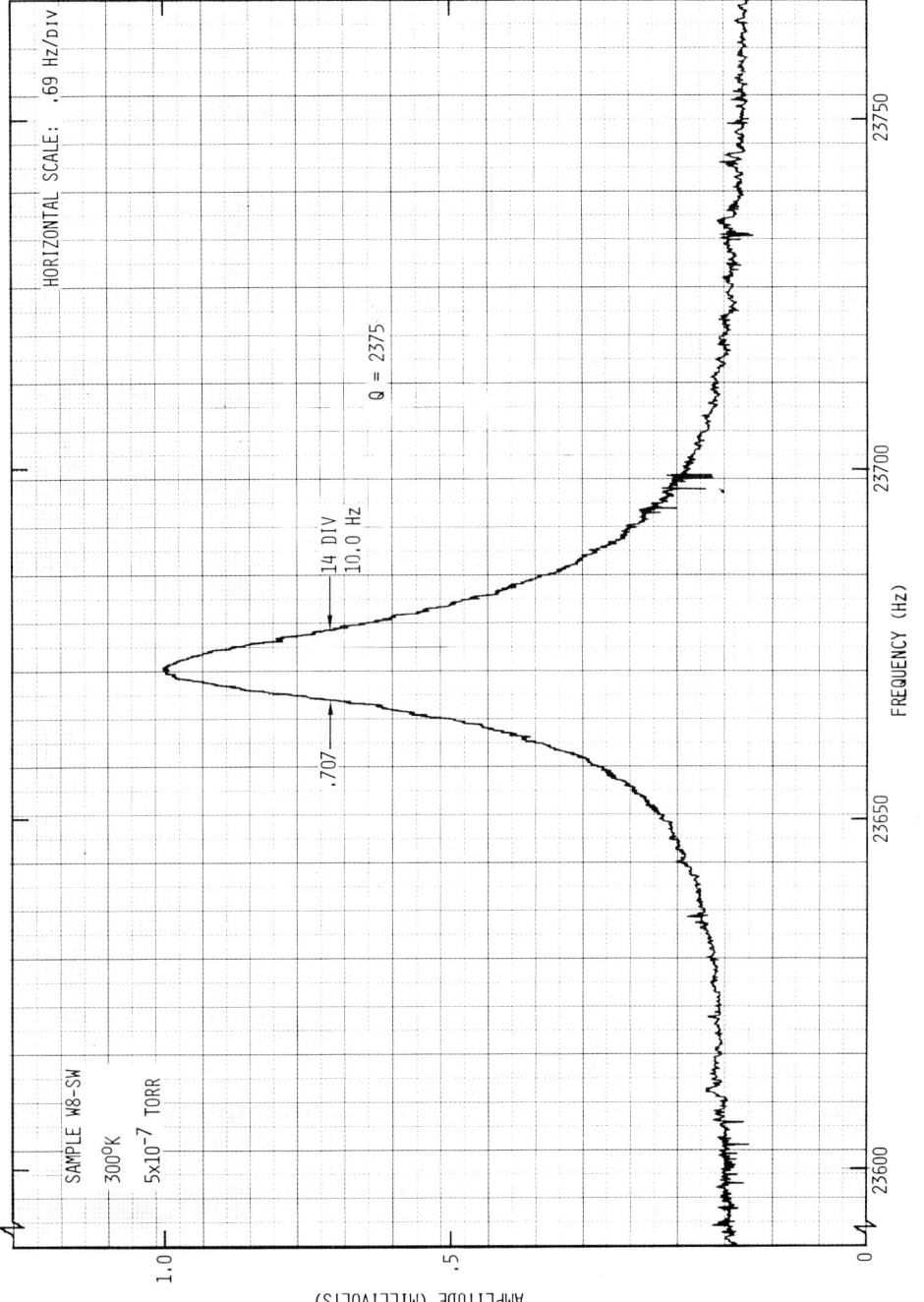

Fig. 2. Bandwidth data on a well-outgassed basalt sample showing a high Q value in the longitudinal mode of vibration.

Fig. 3. Photograph of 50 Hz oscillator with insert showing detail of sample clamp. Sample oscillates around vertical axis formed by supporting fibers at each end.

to a set of adjustable rotors in the shape of a barbell. The sample and the rotors are anchored at the top and bottom by vertical fibers to form vertical axes about which the rotors turn, as shown in Figure 3. A magnetic driver acting on one rotor causes a slight rotation about the vertical axis with a resultant bending of the bar of rock with a deflection of about 10^{-5} cm. The flexure of the bar in turn causes the other rotor to rotate also but out of phase with the first rotor for vibration in the fundamental flexure mode. A magnetic detector senses the slight rotation of the second rotor and transmits an electric signal to the amplifier and oscilloscope. The resonance frequency of the fixture is determined by the stiffness of the bar and the inertia of the rotors. A description of the theoretical and experimental details of operation and construction may be found elsewhere (Tittmann and Curnow, 1976). The fixture is operable in a vacuum and allows for sample heating (with a vacuum furnace) or cooling (with a hollow cylindrical cold finger arranged concentrically around the sample).

In contrast to the design principles described above, most of the previous designs (for example, Sprungmann and Ritchie, 1971) have been based on the end-loaded cantilever. Analysis of the strains during the bending of the sample shows that in the case of the end-loaded cantilever the sample clamp is located at the position of maximum bending strain, a situation inviting difficulties with concentrated stresses. On the other hand, in the present case the clamplike connections between the beam and the rotors are located at positions of minimum bending strain. This result is most easily seen in the limit, where the sample's moment of inertia is very small. Then for the present design the strain distribution is uniform, whereas the cantilever has 50% greater strain at the sample clamp and zero strain at the load end. An additional advantage of the present design over the end-loaded cantilever is the absence of any static stress on the sample.

Fused Silica Sample

The Q value is typically determined by the decay method and Figure 4a shows the signal decay for a bar of fused silica (90 x 6.2 x 4.1 mm). Here the resonance frequency is 70.3 Hz, the time scale is 2 s/cm, and the measured Q = 3434 at 10^{-3} torr and at room temperature is in excellent agreement with Q = 3480 found previously (Sprungmann and Ritchie, 1971), on fused silica with a reed pendulum apparatus in the frequency range 1 to 30 Hz. The smooth monotonic decay of the signal demonstrates the absence of interference from other unwanted modes of vibration.

The maximum typical deflection of the sample at its center was measured by a capacitive microphone which gave an absolute reading for the deflection of $\delta\ell \approx 5 \times 10^{-5}$ cm, which was calculated to be equivalent to a strain of 10^{-7} for a sample with the above dimensions. This strain is considerably lower (factor of 10^2) than that encountered, for example, in the torsion pendulum and places the experiment near the upper end of the seismic range.

Olivine Basalt Sample

Having established the capability of measuring high Q (Q > 3400), we have begun to carry out Q measurements on W-8 samples at 56 Hz. For the rock as received, Q was equal to 51 at room temperature in laboratory air. When the rock was tested at 200 torr-vacuum pressure to remove the air drag on the rotors, the Q increased to 62. Then the rock was exposed to 10^{-6} torr, heated to 200°C for 48 hours, and allowed to cool in vacuum. A room temperature Q value of about 1100 was obtained, as is shown in Figure 4b. This result is analogous to the previous result at kilohertz frequencies in which similar increases in Q were obtained when strong outgassing procedures were carried out. Table II shows high temperature data obtained on the samples after outgassing at 460°C in 10^{-6} torr for about 20 hours. The data reveals high Q values to temperatures as high as 450°C, suggesting that high Q values are also concomitant with hot dry rock.

These results provide valuable evidence for supporting what the kilohertz measurements suggested, i.e., that the removal of volatiles in igneous rocks may be expected to produce dramatic increases in Q even at seismic frequencies. Thus the low-frequency measurements are an important link between the usual laboratory geophysical measurements and the seismic experiments that they are intended to simulate or clarify.

Experiment 3: Q Measurements on Rocks under Confining Pressure

Birch and Bancroft (1939) demonstrated the feasibility of resonating a rock sample under hydrostatic pressure by vibrating the fully encap-

f = 70.3 Hz   Q = 3434

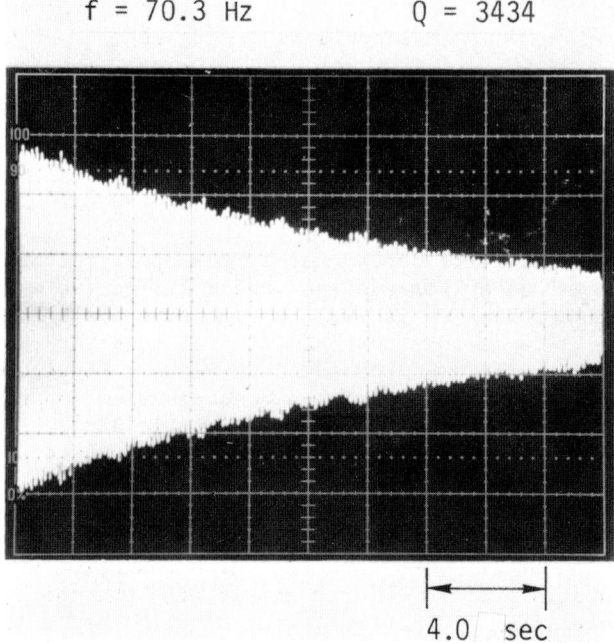

|← 4.0 sec →|

f = 56.0 Hz   Q = 1100   1 sec/cm

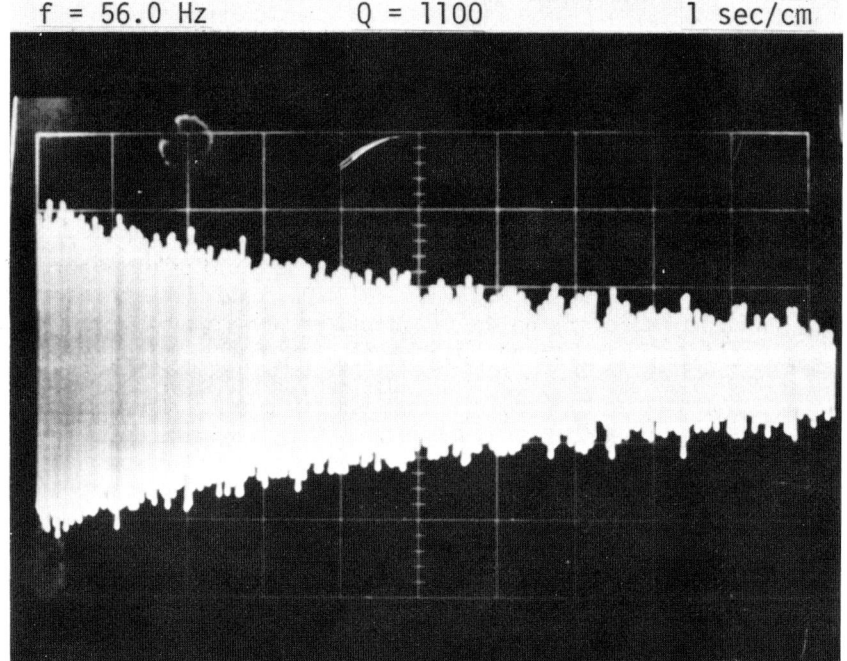

Fig. 4. Oscilloscope trace of signal decay pattern with:
a) quartz sample
b) strongly outgassed olivine basalt sample

TABLE II.

Temperature Dependence of Q at 56 Hz for Olivine Basalt Sample

T (°C)	Q	Conditions
450	900	
350	950	In vacuum of 10^{-7} torr after strong outgassing
250	900	
150	700	
50	1000	

sulated samples in the torsional mode. They used this approach to obtain some Q data and extensive data on the modulus of rigidity as a function of pressure for a wide variety of rock samples. We have begun to explore the feasibility of extending their approach to the measurement of Q as a function of hydrostatic pressure with the additional stipulation that the rock sample environment approach lunar conditions, i.e., a vacuum. Our initial experiments at moderate pressures have been sufficiently successful to make this approach seem promising and some of the results are described below.

Calibration of Background Damping

As was pointed out by Birch and Bancroft (1939), Q measurements are made very difficult by background damping due to such factors as direct pickup from the driving unit, stray induction from other sources, viscous damping due to the high-pressure confining gas, frictional effects of the capsule against the sample, and loss of energy to the supports. Considerable effort was spent in minimizing these factors with the aid of metallic high Q reference samples (polycrystalline Al) until a low background could be established. The progress to the present time is illustrated by high pressure Q values of about 1.2×10^4 which is an approximate measure of the Q of the apparatus (for the unclad Al, $Q \approx 4 \times 10^4$ in the atmosphere). Although this establishes a low loss measurement base, the effort of increasing the apparatus Q is being continued with the goal of making measurements to above 5 kbar, where viscous drag due to the confining gas is expected to be the main source of background damping.

Q of Outgassed Rock

In addition to studying rocks exposed to the atmosphere, our goal is to study rocks under thoroughly outgassed conditions. This is difficult because the rocks are porous and have to be encapsulated under vacuum in a thin wall Cu sheath while they are in a fully outgassed state. We have accomplished this by sealing the capsule in a vacuum of 10^{-5} torr by electron beam welding. Figure 5 shows representative results on a fine grained basalt obtained from an exposed dike in the Santa Monica Mountains in southern California, and having a measured open pore porosity of about 1%. After initial saw cuts the samples were gently ground from square cross-section bars to polygonal bars with n sides (n increasing from 4 to 8) and finally to round cylinders about 1 cm in diameter and 15 cm long. These were then roughly cleaned in trichloroethylene, alcohol, and H_2O_2 and baked out for 3 days in 2×10^{-5} torr at $350°C$. The samples were then inserted into 4 mil wall seamless Cu tubing which was first heated so that when it cooled, it would shrink down tightly against the rock. The sheathed samples and their end caps were baked out again as had been done before and mounted under dry He gas in the welding chamber, where they were held at 1×10^{-5} torr for 12 hours. At this point a 4 kV electron beam was applied to seal the seam between the tube and the end cap. Repeated exposure of the samples, so encapsulated, to 0.5 kbar pressures showed no deleterious effects on the seam due to leaks in the sheath, so that the vacuum inside the capsule appears to have been maintained intact. Correspondingly, the results of Figure 5 show Q ~ 1400, which compares to Q ≈ 200 for samples cleaned as was done before but sheathed while they were exposed to the atmosphere. Also shown in Figure 5 is the pressure dependence of the torsional frequency, which is seen to rise rapidly with pressure, up to 0.1 kbar as the Cu sheath establishes contact with the sample. At higher pressures the resonant frequency is seen to level out with a slope of $v^{-1}(\Delta v/\Delta p) \approx 0.01$ kbar^{-1} at a pressure p ~ 0.25 kbar, where v is the shear velocity. Figure 6 is a photograph showing the details of the sample holder, the Cu clad sample, and the magnetic drive. These results suggest that partial crack closure begins already at low pressures and supports the conclusion of Richter and Simmons (this volume) that most clean high-aspect ratio cracks are already closed at depths corresponding to 2-3 kbars and that these cracks heal over geological time. High pressure runs for the sample in the moist state carried out at 2.5 kbar (Tittmann, 1977) showed that the Q in the moist condition was still much smaller than the dry Q. This implies that the pressure of volatiles keeps the Q low, even when the cracks are partially closed corresponding to some depth below the surface, or alternatively that the presence of dry rock in the crust is concomitant with a high Q.

Discussion of Results

The three experiments discussed above show that the Q of a low-porosity (1-2%) olivine basalt without hydrated mineral phases ranges from about 100 measured dry in normal laboratory air to over 2000 outgassed at moderate temperatures in a high vacuum. Data so far obtained on the frequency dependence down to 50 Hz and limited data on the pressure dependence of the loss associated with small amounts of absorbed water lead us to believe that our laboratory Q values are probably applicable to seismic waves propagating in rocks subjected to high pressures under the earth's surface.

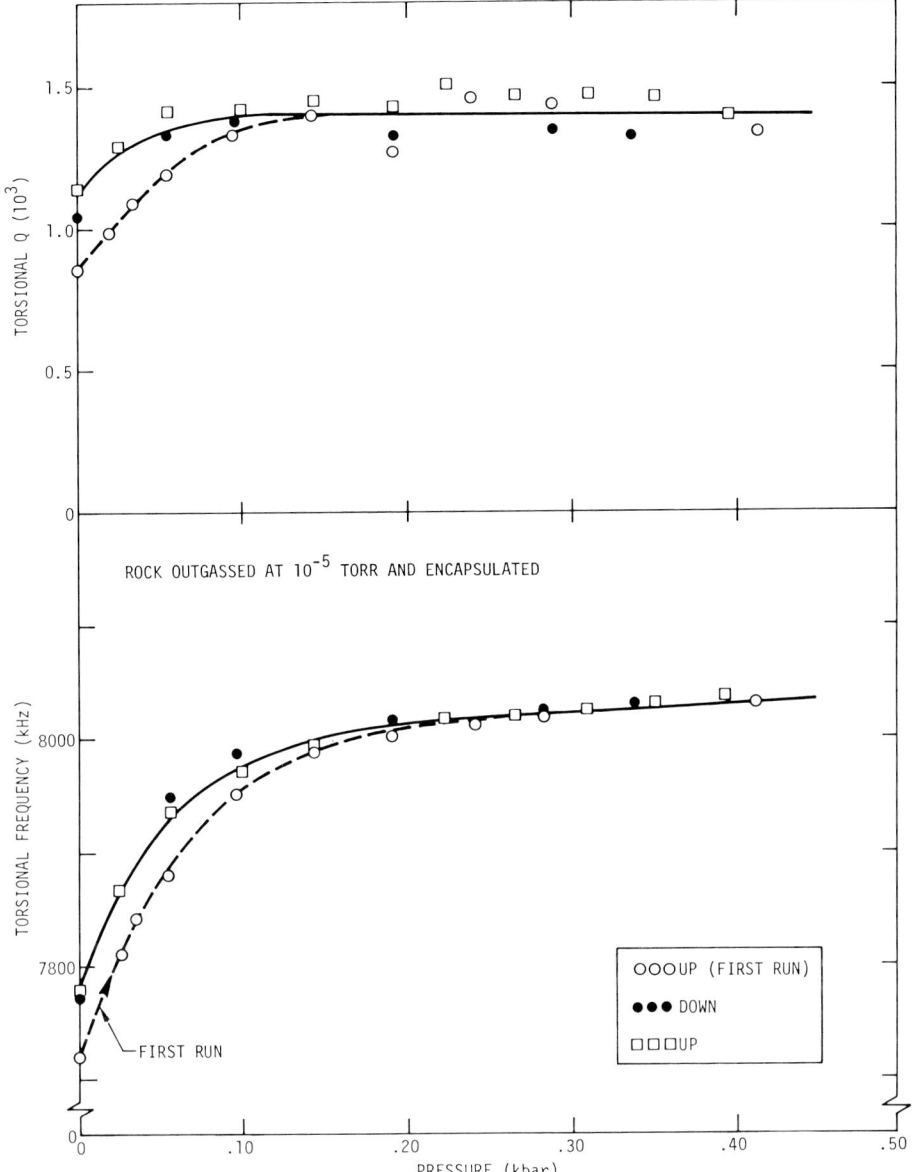

Fig. 5. Q and resonance frequency as a function of confining pressure for strongly outgassed basalt sample sheathed under vacuum of 10^{-5} torr.

Figure 7 shows the results of Hermann and Mitchell's (1975) calculation of Q-depth profile based on seismic field data. They find this profile similar to that of Mitchell (1973) in that it has relatively low values of Q (high Q^{-1}) in the upper crust and a rapid transition, at midcrustal depths, to high Q values (low Q^{-1}) in the lower crust. It differs from the model of Mitchell (1973) mainly by having a less

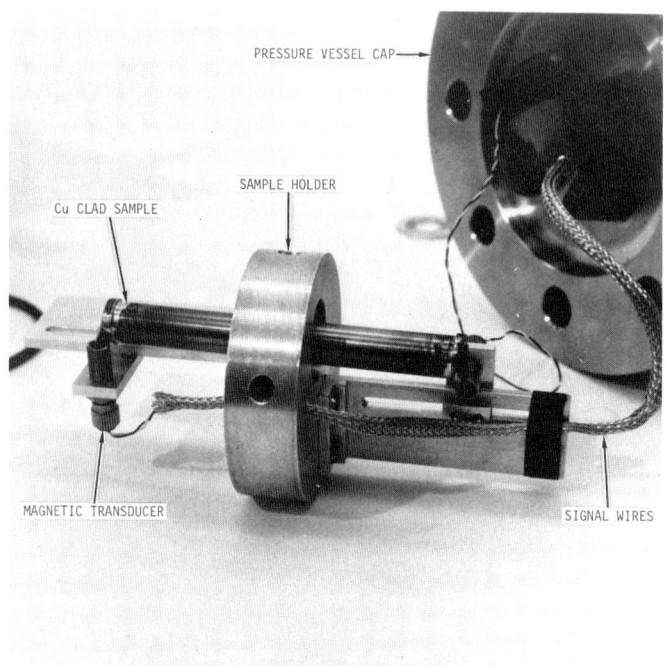

Fig. 6. Photograph of high pressure sample holder.

abrupt transition to high Q values and somewhat lower Q values in the lower crust.

The large standard deviations at greater depths do not permit Hermann and Mitchell (1975) to determine whether Q decreases in value with depth in the lithosphere, as inferred by Mitchell (1973) and Lee and Solomon (1975). However, any model with constant high Q values in the lower crust will predict extremely small values for the internal friction at periods of 40 s and greater.

Hermann and Mitchell (1975) find that the relatively large standard deviations and broad resolving kernels indicate that simpler Q distributions would also satisfy the data. One example is that of the two-layer model shown by the dashed line in Figure 7. It possesses values of 250 from the surface to a depth of 17 km and values of 2000 at all greater depths.

The most conspicuous feature of the profiles of Hermann and Mitchell (1975) and of Mitchell (1973) is the sharp increase in Q (decrease in internal friction) from values not exceeding a few hundred from the surface down to about 15 km to a value as high as 2000 below 15 km. No ordinary rock types measured in the laboratory at room temperature and 1-atm pressure have shown such high Q values (Knopoff, 1964).

As a result of the laboratory measurements on strongly outgassed rocks discussed above we suggest that the sharp increase in Q probably marks a boundary below which no free moisture exists in the crustal

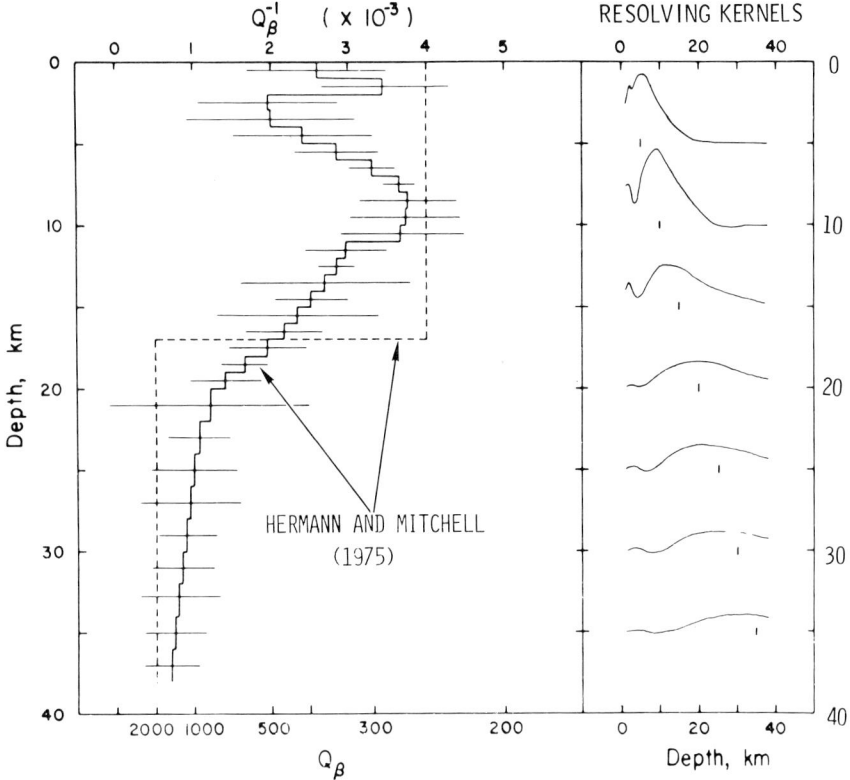

Fig. 7. The derived Q model (solid curve) determined by Hermann and Mitchell (1975). The bars are the standard deviations. The right hand side of the figure gives the resolving kernels at each depth. The simpler Q model (dashed curve) not obtained by the inversion process, provides an adequate, although less optimum fit to the data.

rocks. This boundary might, but need not necessarily, coincide with a transition between chemically distinct rock types. Our laboratory measurements also show that the observation of high Q values are not restricted to dry rocks at room temperature but extend also to temperatures in excess of 400°C, ie. to hot dry rock.

We have shown that Q of a low-porosity olivine basalt changes from about 100 measured dry in normal laboratory air to over 2000 outgassed at moderate temperatures in a high vacuum. The Q returns to roughly its original value after a few minutes reexposure of the sample to laboratory air, and this cycle can be repeated indefinitely. The difficulty which we experience in outgassing the samples well enough to achieve the maximum Q values indicates that just a few monolayers of adsorbed water on pore surfaces are very effective in drastically reducing the Q.

These results imply that the Q of a dry sample of the same composition and mineralogy would be above 2000. They also strongly suggest that the Q values in wet rocks in the earth's crust are drastically reduced from the intrinsic values, because of the presence of even monolayers of H_2O even though the pore volume may be small and the calculated losses due to viscous flow of the water may be negligible. The limited data that we have so far obtained on the frequency dependence of the loss associated with small amounts of adsorbed water lead us to believe that it will be very important at seismic frequencies.

In earlier work along the same direction, Gordon and Davis (1968), Pandit and Tozer (1970), and Pandit (1971) obtained qualitatively similar results on several rock types, although the fractional changes and maximum Q values obtained were considerably lower, most probably because they were not able to outgas the samples as thoroughly.

Some time ago, Birch and Bancroft (1939) showed that Q in the kilohertz frequency range of some, but by no means all, igneous rocks increased from values of a few hundred to values of about 2000 when a pressure of 4000 kg/cm^2 was applied to the jacketed samples, corresponding to a depth of roughly 10 km in the earth. In light of our results we interpret the Birch and Bancroft results as implying that those samples where Q did increase to the vicinity of 2000 were free of moisture.

Based on our interpretation of the seismic data, it appears that at depths below 15 km no free water exists. This conclusion could contribute to understanding the nature of earthquake source mechanisms in the region. This interpretation also implies that in the construction of theoretical electrical conductivity profiles for the region it is appropriate to use values characteristic of dry rocks below depths of about 15 km.

Acknowledgements. The author is grateful to L. Ahlberg, J. Curnow, and H. Nadler for their help in the experiments; and to R. M. Housley, G. A. Alers, and J. M. Richardson for many helpful discussions. The work was partially supported by NASA contract NAS 9-13846.

References

Berry, G.W., Apparatus for the measurement of the internal friction of metals in transverse vibration, Rev. Sci. Instrum., 25, 884-886, 1955.

Birch, F., and D. Bancroft, The effect of pressure on the rigidity of rocks, I, J. Geol., 46, 59-87, 1938.

Gordon, R.B. and L.A. Davis, Velocity and Attenuation of Seismic Waves in Imperfectly Elastic Rock, J. of Geophys. Res., 73, 3917-3935, 1968.

Hermann, R.B., and B.J. Mitchell, Statistical analysis and interpretation of surface-wave anelastic attenuation data for the stable interior of North America, Bull. Seismol. Soc. Amer., 65, 115-1128, 1975.

Housley, R.M., B.R. Tittmann, and E.H. Cirlin, Crustal porosity information from internal friction profile, Bull. Seismol. Soc. Amer., 64, 2003-2004, 1974.

Knopoff, L., Q, Rev. Geophys. Space Phys., 2, 625-660, 1964.

Lee, W.B., and S.C. Solomon, Inversion schemes for surface wave attenuation and Q in the crust and the mantle, Geophys. J., Roy. Astron. Soc., 43, 47-71, 1975.

Mitchell, B.J., Surface-wave attenuation and crustal anelasticity in central North America, Bull. Seismol. Soc. Amer., 63, 1057-1071, 1973.

Pandit, B.I., Experimental studies on the mechanism of internal friction (Q^{-1}) of rocks, PH. D. Thesis, 117 pp., Univ. of Toronto, 1971.

Pandit, B.I., and D.C. Tozer, Anomolous propagation of elastic energy within the moon, Nature, 226, 335, 1970.

Quimby, S.L., An experimental determination of the relation between viscosity and frequency in vibrating solids, Phys. Rev., 31, 1113, 1928.

Tittmann, B.R., J.M. Curnow, and R.M. Housley, Internal friction quality factor $Q \geq 3100$ achieved in lunar rock 70215, 85, Proc. Lunar Sci. Conf. 6th, 3, 3217-3226, 1975.

Tittmann, B.R., and J.M. Curnow, Apparatus for measuring internal friction Q factors in brittle materials, Rev. Sci. Instrum., 47, 1516-1518, 1976.

Sprungmann, K.W., and I.G. Ritchie, An improved reed pendulum apparatus and techniques for the study of internal friction of ceramic single crystals, U.S. At. Energy Comm. AECL-3794, 1971.

Tittmann, B.R., Rayleigh wave studies in lunar rocks, IEEE Trans. Sonics Ultrason. No. 72 CHO 708-8SU, 130-135, 1972.

Tittmann, B.R., R.M. Housley, G.A. Alers, and E.H. Cirlin, Internal friction in rocks and its relationship to volatiles on the moon, Proc. Lunar Sci. Conf. 5th, 3, 2913, 1974.

Tittmann, B. R., Unpublished.

Wegel, R., and H. Walther, Internal dissipation in solids for small cyclic strains, Physics, 6, 141-157, 1935.

IN SITU AND LABORATORY MEASUREMENTS
OF VELOCITY AND PERMEABILITY

H. R. Pratt, H. S. Swolfs,
R. Lingle, and R. R. Nielsen

Terra Tek, University of Utah Research Park, Salt Lake City, Utah 84108

Abstract: In situ velocity and permeability have been measured as a function of stress to 180 bars in a jointed granite; velocity was measured in a porous sandstone. Velocity increased with increasing stress for both rock types, while joint permeability decreased by a factor of 4 over a stress range of 150 bars in the granite. Velocity measurements along paths perpendicular and parallel to joints in granite indicate that jointing significantly attenuates compressional waves up to stress levels of 15 bars. A comparison of in situ and laboratory velocities in granite illustrates the influence of the in situ stress field. Laboratory velocities (15-cm sample) were significantly lower than a small-scale field experiment velocity (3.0-m path length). Velocities obtained from a 100-m long seismic refraction survey were significantly less than velocities obtained in the small in situ experiment but greater than laboratory values. The decrease of a factor of 4 in fluid permeability along joints was the most sensitive parameter to change in stress. The other properties measured included velocity, deformation, and resistivity. The magnitude and orientation of a maximum value for compressional wave velocity in porous sandstone are related to the existing in situ stress field. A velocity decrease of 50% was noted during the stress relief of a 1.6-m block of sandstone. In situ stresses at the site were estimated to be 20 to 60 bars on the basis of strain relief measurements. To obtain the 'average' permeability of a rock mass, both the permeability of the joints and the intact rock must be considered. Laboratory measurements of velocity and permeabilities under confining pore pressures simulating in situ conditions are also given.

Introduction

Measurements of elastic and transport properties of in situ rock masses has important applications to the design of near-surface structures, to the recovery of energy and mineral resources at moderate depths, and to delineating models of the structure of the upper part of the crust. These rock masses contain natural discontinuities such as joints, faults, and bedding planes. The contribution of these discontinuities to the elastic properties of rock such as mechanical strength and seismic velocity and to mass transport properties such as fluid permeability is of great importance. Formulating the in situ contribution of these discontinuities will be an important step in understanding the basic phenomenology of the response of in situ rock masses.

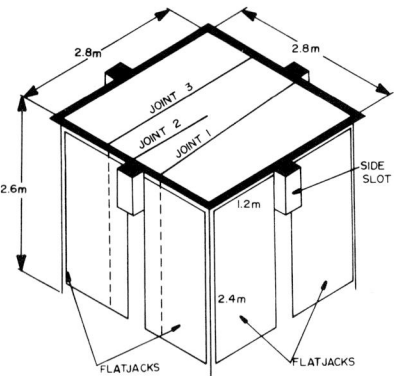

Fig. 1. Schematic diagram of the experimental block and the location of the flatjack loading system.

There are several methods of approaching the problem. Many involve modeling the discontinuities in some way [Jaeger and Cook, 1969; Goodman, 1976]. This may take the form of mathematical analysis of idealized systems [Goodman et al., 1968], tests on model materials [Einstein and Hirschfeld, 1973], or the study of the response of laboratory specimens with artificially induced or natural discontinuities [Byerlee, 1968; Goodman and Ohnishi, 1973]. Another approach is to conduct field tests on actual discontinuities and to compare these results with results of scaled-down tests in the laboratory. The problem of modeling and scaling may therefore be avoided. These data may also add to our interpretation in showing how to scale and model the rock mass response. In situ measurements of velocity and permeability as a function of stress have not been conducted to significant stress levels in the field and, in fact, only a few detailed measurements have been made of velocity and fracture permeabilities in the field under any conditions [Gale, 1975; Jouanna, 1972; Maini, 1971].

Experimental Procedure

In situ tests were conducted in a jointed granite near Laramie, Wyoming [Pratt et al, 1977; Swolfs et al., 1974], and in a porous sandstone near Grand Junction, Colorado [Johnson et al., 1974; Swolfs et al, 1976]. The granite test block is a rectangular prism 2.8-m on a side and 2.6-m deep (Figure 1). The specimen was excavated by using techniques developed previously for the preparation of large in situ samples [Pratt et al. 1974]. Vertical sides of the specimen were formed by line drilling while the entire block remained attached at the bottom. Near the center of each slot a small enlargement was excavated to accommodate the seismic transducers. Oriented samples of rock were removed during this process and saved for laboratory testing. The block contained three parallel vertical joints. Large stainless steel flatjacks were inserted along the sides of the specimen and used as the loading system. A hydraulic pumping system was used to pressurize the flatjacks and to provide a uniform load to the sides of the specimen.

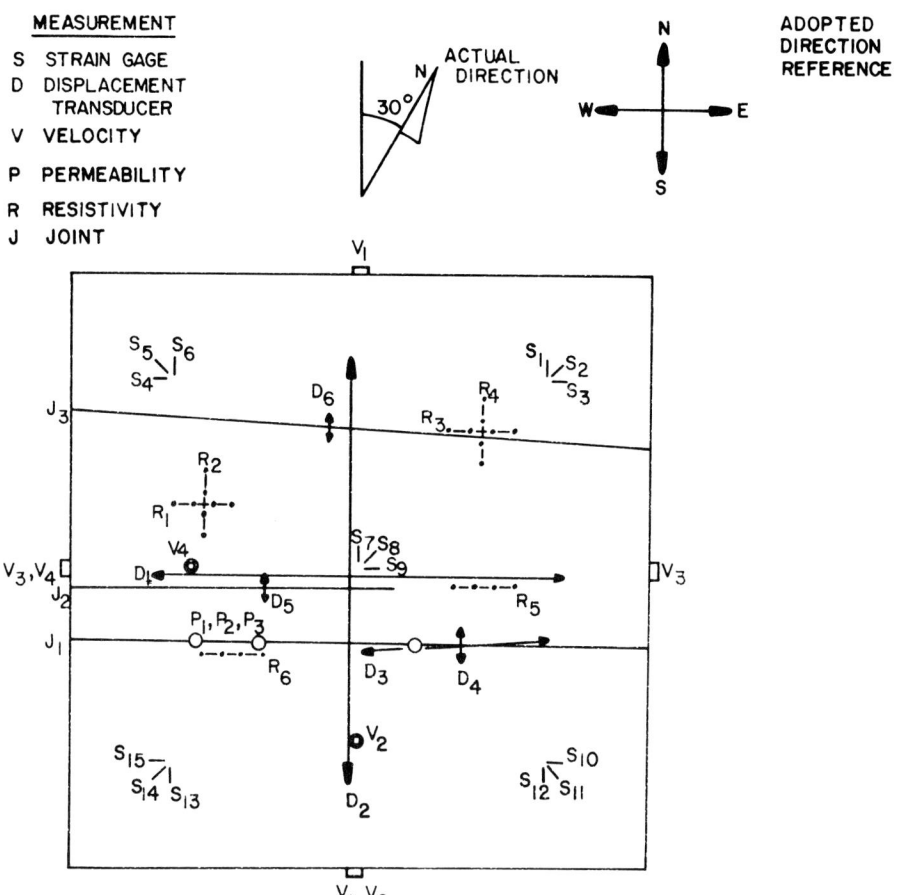

Fig. 2. Sketch of the top surface of the block showing the location of instrumentation relative to the three joints for the granite experiment.

During the series of experiments several loading paths were followed including both biaxial and uniaxial stress normal to the joints and uniaxial stress parallel to the joints. Elastic analysis of the stress distribution during loading showed that the upper two thirds of the block was uniformly loaded. All properties were measured in this part of the block. The same procedure was carried out for the excavation of the porous sandstone block. The sandstone block was smaller, 1.6 x 1.6 x 1.6-m.

The top surface of the granite block was instrumented for the measurement of strain by using Direct Current Differential Transducers (DCDT) and strain gages (Figure 2). Sonic velocity transducers (100 kHz) were mounted on the sides of the block in the four slide slots below the surface. Two shallow holes were drilled into the surface of the jointed granite block 1-m from the sides for additional velocity

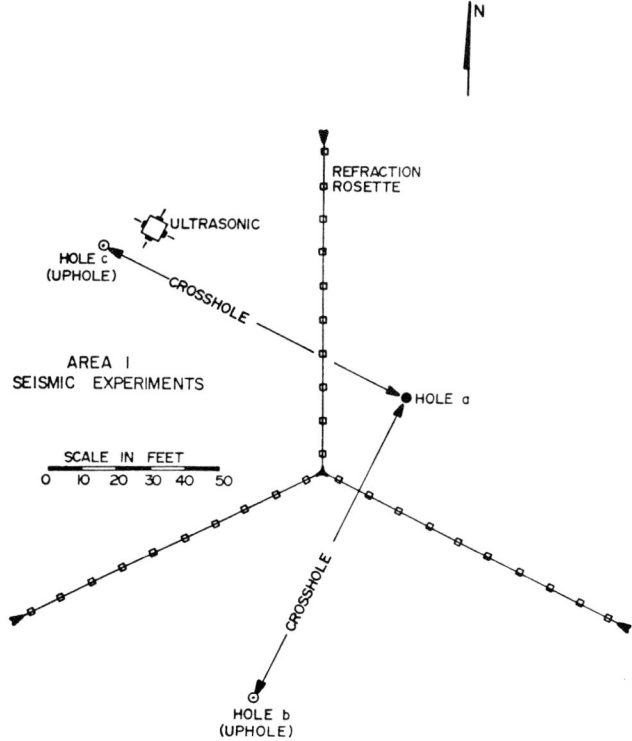

Fig. 3. Location of the field seismic and ultrasonic experiments at the porous sandstone site.

measurements. Compressional velocity was measured in both directions across the entire width of the sample and also across 1-m distances. (Figures 3 and 4). Velocities were measured across the 1.6-m sandstone block.

Permeability along the joint in the granite was measured between pairs of holes 7.6 cm in diameter by 1.8-m in depth drilled along the plane of the joint (Figure 2). After saturating the joints in the granite block by injecting water under pressure for three days, water was injected under 4-bar pressure into one hole and the flow along the joint was measured by several methods: (1) flow rates were measured under steady state conditions, (2) pressure decay was measured between the two holes. Details of the instrumentation and the experimental procedures are given in the work by Pratt et al., [1977].

A laboratory experimental program complemented the field measurements. This program included (1) stress-strain behavior of both the granite and the sandstone (2) ultrasonic compressional wave velocities as a function of pressure for both rock types, and (3) permeability measured perpendicular and parallel to the microfracturing in the granite by using the standard permeability procedures. In addition seismic refraction profiles using 14 Hz geophones were conducted along 100-m lines for comparison with both the in situ blocks and the labora-

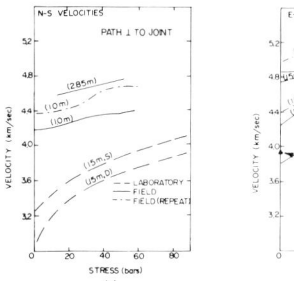

Fig. 4. (a) North-south and (b) east-west velocities as a function of stress for both in situ and laboratory tests. Path lengths (e.g., 2.85 m) are given in meters. Moisture content is shown by symbols D (lab dry) and S (lab saturated).

tory values. The profiles were made along directions parallel and perpendicular to the joints at the granite site, but at an angle to the sides of the test block at the sandstone site (Figure 3). However, crosshole velocity measurements over a 30 foot depth interval were made parallel to the sides of the sandstone block (Figure 3).

Experimental Results

Several experiments were conducted at both field sites along a variety of load paths and at different stress levels. A total of 18 experiments were conducted at both sites. A typical experiment consisted, for example, of an uniaxial stress load perpendicular to the joints while deformation, velocity, and permeability were recorded. The response of velocity and permeability to stress will be discussed in this paper. Deformational and resistivity responses to stress have been discussed previously [Pratt et al., 1977].

Velocity

Compressional velocity measured across the granite block increased with stress (Figure 4). The same was true for the velocities measured in the laboratory under dry and saturated conditions. Velocities measured in the laboratory tests were significantly lower than those measured in the field tests. It was also noted that velocities measured along a 1-m path in the field had a lower velocity than those along the 2.8-m path for measurements both perpendicular and parallel to joints (Figure 4). A velocity of 3.9 km/s was measured during the seismic refraction survey conducted along the path parallel to the east-west velocity measurements and parallel to the dominant joint set. This velocity is lower than either of the measurements on the test block in the field, but is still higher than the laboratory measurements (Figure 4).

In situ compressional wave velocities were also measured across a 1.6-m path in the porous sandstone cube. Prior to excavation, velocities of 2.7 and 3.0 km/s were measured (Figure 5). The figure also shows the orientation of the in situ stress field after overcoring.

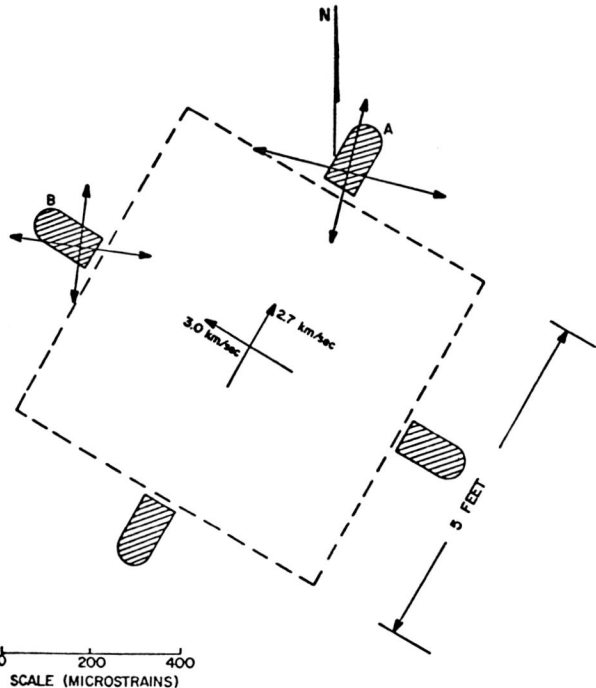

Fig. 5. Porous sandstone experiment showing strain relief of rosettes A and B. Ultrasonic velocities are shown prior to isolation of test block.

After the block was excavated, the velocities were again measured and had decreased significantly by almost 50% to 1.7 and 1.4 km/s for the respective directions. (Figure 6). This was a surprisingly large change, undoubtedly due to the relief of the in situ strains within the block and opening of microfractures. The block was subsequently loaded with a flatjack system. Velocity increased with increasing stress but did not return to the initial velocity measured in the unrelieved block at a strain level equivalent to the strain relief measured in the block (Figure 7). A comparison of the in situ test and the field seismic and laboratory compressional wave velocity measured in the horizontal direction shows that the laboratory ultrasonic velocities are lower for this depth and are only of the order of 2.2 to 2.4 km/s. The ultrasonic velocities are significantly greater on samples from depths greater than 15 ft. (4.57-m). The difference between the cross-hole seismic data and the laboratory ultrasonic data is not yet fully understood but may be due to changes in lithology or changes in jointing which may be present at depths greater than 15 ft (4.57-m) (Figure 8).

Permeability

Fluid flow was measured at the granite along joint 1 in order to obtain the change in fluid permeability with stress (Figure 1). As was expected, the flow rate decreased as a function of increasing stress for loading normal to the joint (Figure 9) and showed a twofold de-

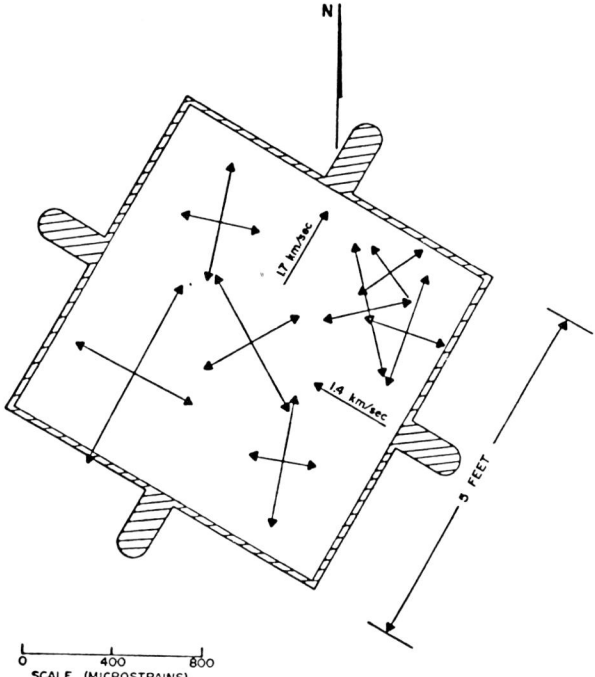

Fig. 6. Porous sandstone experiment showing strain relief of large block. Final ultrasonic velocities are also shown.

crease in biaxial loading. The flow increased during loading parallel to the fractures as the applied stress opened these fractures. A significant decrease of a factor of 4 was noted during the uniaxial stress loading perpendicular to the joints. A residual value of approximately 1.8 cm^3/s was noted up to stresses of the order of 60 bars. A flow rate of the order of 1.8 cm^3/s is approximately 1.2 millidarcies. Permeability was calculated from the flow rate by using both parallel plate theory (1) and Darcy's law (2):

$$Q = \frac{c^3}{12D} \frac{\rho g}{\mu} A \frac{dh}{d\ell} \quad (1)$$

$$Q = k \frac{\rho g}{u} A \frac{dh}{d\ell} \quad (2)$$

where

Q   flow rate, cm^3/s;
k   permeability, (length)^2 darcies;
A   cross-section area, cm^3
dh/dℓ head gradient, bars;
μ viscosity, cP;

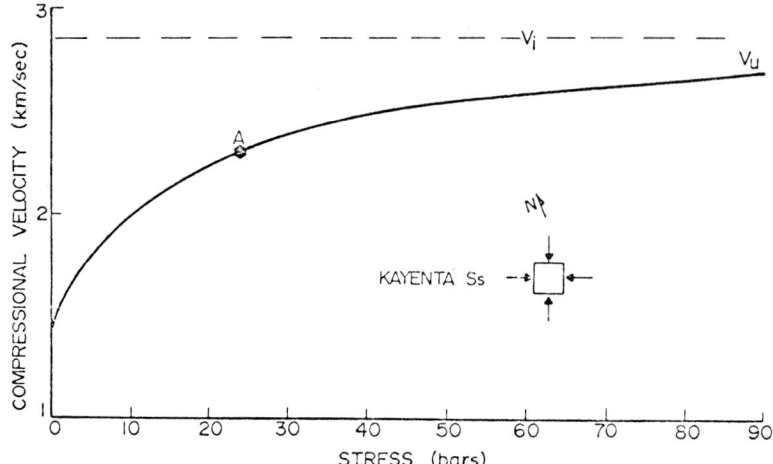

Fig. 7. Compressional wave velocity in the sandstone block as a function of stress. Note that at 90-bar stress the velocity has not reached the original velocity. Point A represents the stress value obtained for the equivalent strain relief of the porous sandstone block.

ρ density of fluid, g/cm^3
g acceleration due to gravity, g/cm
c aperture, cm;
D fracture spacing.

The 1.2-millidarcy value for permeability of jointed granite at stress is significantly greater than the 0.1-millidarcy value obtained for measurement on laboratory samples of granite.

Laboratory studies of porous Navajo sandstone have shown similar decreases in joint and total permeability as a function of stress to confining pressures of 0.75 kbar (Figure 10) [Nelson, 1975]. The total permeability of the rock plus the fracture (k_{fr}) decreases more markedly in comparison to the permeability of the rock (k_r). In the case of the porous fractured rock the permeability of the rock and fracture asymptotically approaches the permeability of the intact rock at high stresses. Figure 11 shows the rapid decrease in permeability along the fracture during the first cycle of loading for three different samples [Nelson, 1975]. Note that during the second cycle of loading the permeability remained constant as a function of confining pressure.

Laboratory testing is currently the only method available for determining permeability and change in permeability at high stresses. Measurements on fused tuff show a dramatic fifteenfold decrease in the permeability with increasing hydrostatic stress to 5 kbar, undoubtedly related to the closing of cracks and the crushing of pores (Figure 12) [Nielsen, et al., 1975]. The role of effective stress on the change in permeability is of interest especially in developing realistic constitutive models for fractured rock. Measurements on limestone show a rapid fivefold decrease in permeability with increasing effective stress from

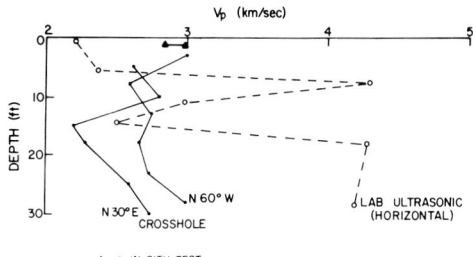

Fig. 8. Comparison of in situ tests, field seismic survey, and laboratory compressional wave velocities in the horizontal direction for the porous sandstone site.

150 to 550 bars (Figure 13). Subsequent reloading after partial relaxation for 7 days is also shown. The effect of decreasing and increasing confining pressure at constant pore pressure (50 bars), and increasing and decreasing pore pressure at constant confining pressure (approximately 600 bars) are similar and show a marked decrease in the permeability (Figure 13).

Recent permeability measurements of laboratory samples of jointed rock taken from geothermal reservoirs indicate the importance of delineating fracture and 'matrix' permeability. Measurements of the matrix permeability of Franciscan graywacke from the Geysers, California, geothermal field at simulated in situ conditions (P_c = 0.34 kbar, P_p = 0.034 kbar) indicate an extremely low matrix permeability of the order of .02 microdarcies, a difference of a factor of 2000 from the permeability measured parallel to a fracture (Table 1) [Pratt and Simonson,

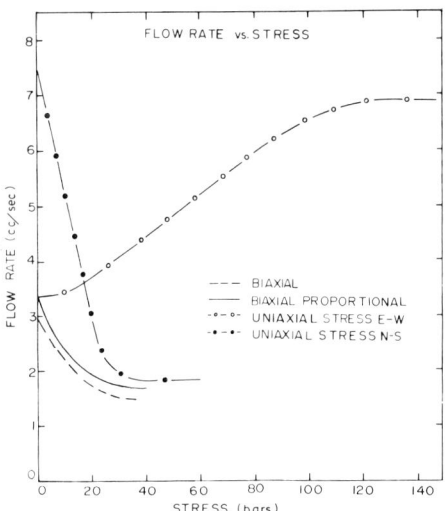

Fig. 9. Flow rate as a function of stress for various loading conditions on the jointed granitic block.

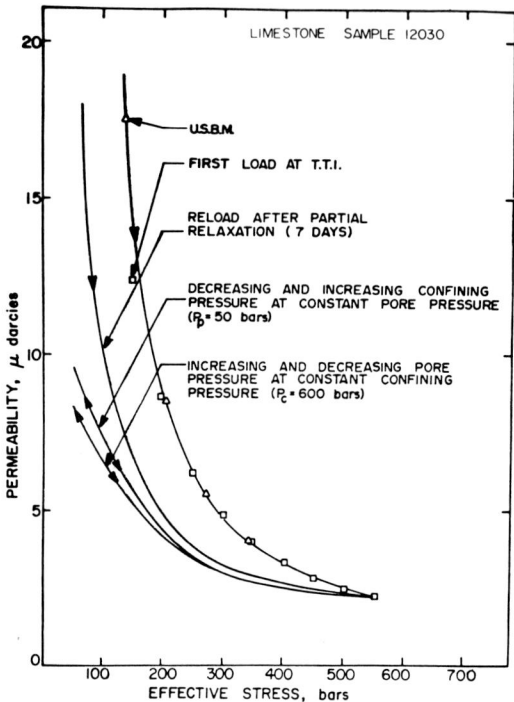

Fig. 10. Permeability versus confining pressure (P_c) showing whole rock permeability and whole rock plus fracture permeability fields for first loading cycle. (After Nelson, 1975).

1976]. Measurements of permeability of jointed rock from the Raft River in Idaho, geothermal site, at estimated in situ conditions give values of 0.03 millidarcy for measurement along tight joints and 0.12 millidarcy for joints filled with sandy material (R. Nielsen, personal communication, 1976). Permeability was also measured in a few samples at ambient temperatures and at 220 degrees F. Only on the specimen with relatively thick joints did permeability decrease significantly from 12 to 0.04 millidarcies as a function of temperature.

Discussion

The relationship between the applied stress field and velocity and permeability is just beginning to be understood through field and laboratory measurements. While field measurements allow analysis of jointed material in its in situ environment, we are not able at the present time to reach the high stresses necessary to simulate conditions at significant depths. In situ measurements, however, will allow us to understand the basic mechanisms and phenomenology of rock mass response of the upper part of the earth's crust. The field tests in jointed granite established the relationship between displacement, velocity, and permeability (Figure 14). The joints shows significant deformation up to a stress level of approximately 15 bars. Above this level the joint stiffened, and the deformation was approximately equal to the in situ modulus of

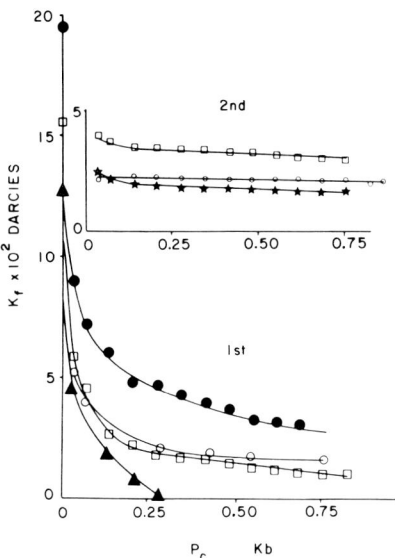

Fig. 11. Fracture permeability versus confining pressure for the first and second loading cycle on laboratory samples of Navajo sandstone. (After Nelson, 1975).

the material. Until that stress level was reached the P wave velocity could not be propagated perpendicular to the joints (curve V_1) because the joints attenuated the P wave. In constrast, a high-amplitude signal was noted in measurements parallel to joints over the same distance. Thus two joints significantly attenuated the compressional wave velocity. This has implications on the attenuation of seismic waves in jointed media over longer paths.

A fourfold change in the flow rate with change in stress normal to the joint (curve P_1) indicates that the flow rate is a most sensitive factor to the stress among the measured parameters, which include deformation, velocity, permeability and resistivity. The residual flow after apparent joint closure remained at a constant value, approximately 1.8 cm^3/s. It is interesting to note that this value of 1.8 cm^3/s equal to 1.2-millidarcies permeability, is an order of magnitude greater than the permeability of the rock matrix which is approximately 0.1-millidarcy. Indications are thus that the significant flow along joints still occurs at reasonably high stress levels in rocks even with stiff jointing systems. This was true even after several loading cycles.

Significant decreases in permeability were noted during closure in laboratory tests on deformable fractures in Navajo sandstone. The permeability of the total rock (jointed and intact) approaches that of the intact rock under pressure (Figure 10) [Nelson, 1975]. It appears that the deformability of the rock has significance in delineating the contributions of both the fracture and the intact permeability. It is also readily apparent from laboratory tests that both the confining pressure and the pore pressure have significant effects on the permeability of fracture in porous materials and that decreases up to factors of 5 are seen under effective stress changes of only a few hundred bars (Figure 13).

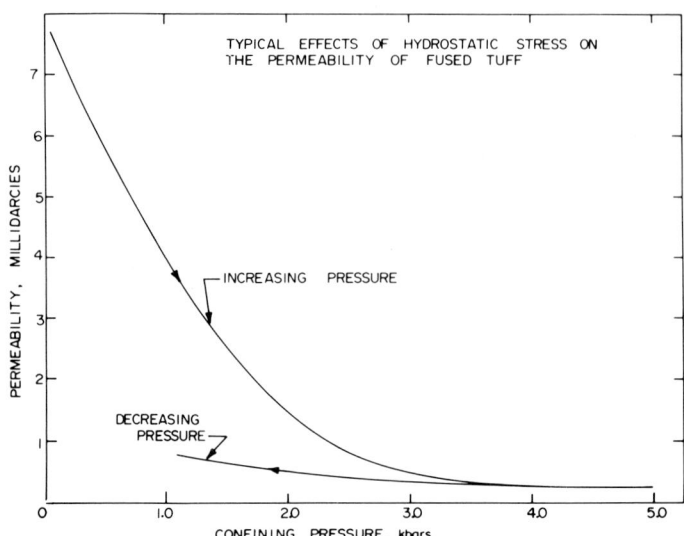

Fig. 12. Permeability as a function of hydrostatic stress for fused tuff.

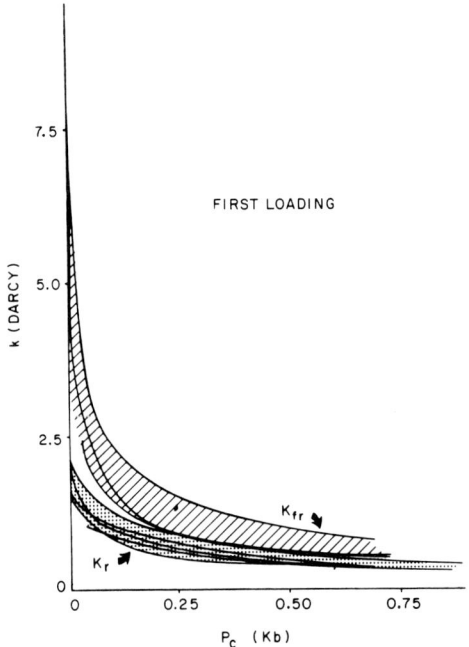

Fig. 13. Permeability as a function of effective stress. Curves show the effect of decreasing and increasing confining pressure at constant pore pressure and increasing and decreasing pore pressure at constant confining pressure. U.S.B.M. refers to United States Bureau of Mines, and T.T.I. refers to Terra Tek, Incorporated.

TABLE 1. Permeabilities of Geothermal Reservoir Rocks

Location	Rock Type	Description	Depth, m	Confining Pressure kbar	Pore Pressure kbar	Permeability microdarcies	
						T = 21°C	T = 104°C
Geysers, Calif.	Franciscan graywacke	Intact rock	1190	0.34	0.034	0.02	
	Franciscan graywacke	Open joint	1190	0.34	0.034	40.0	
Raft River, Idaho RRGE-3	Calcareous sandstone	Healed fracture	1520	0.34	0.149	30.0	6.0
	Silty shale	Healed fracture (intact)	1522	0.34	0.149	1.0	1.0
	Siltstone	Healed fracture	1608	0.34	0.172	30.0	20.0
	Siltstone	Fracture filled with silt	1609	0.34	0.172	120.0	40.0

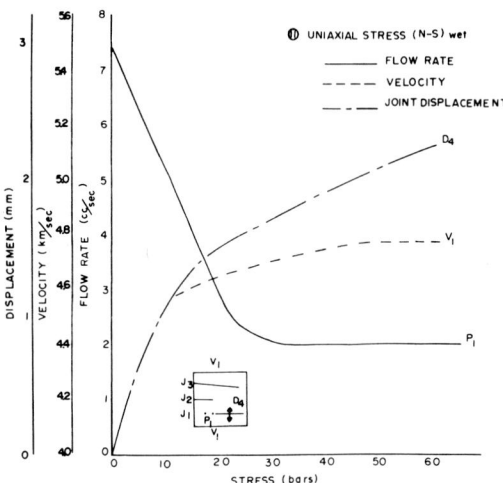

Fig. 14. Displacement, velocity, and flow rate as a function of stress for jointed granite.

The 50% decrease in velocity during the stress relief of the porous sandstone block is significant. The attenuation of seismic waves probably due to the opening of micro-fractures during stress relief indicate that the strain relief is not entirely elastic (Figure 7). The initial measured velocity was not achieved during reloading of the sample to the strain equivalent to the stress level prior to excavation. It is also significant that the velocities obtained during the in situ and laboratory tests on the granite showed a relative size effect which is probably due to two contrasting phenomena. First, the increase in velocity with increasing path length, the 1.0-m velocity path is less than the 3.0-m velocity path in the in situ experiment, is probably due to stress relieving and opening microcracks during excavation of the specimen. The microcracks opened up most significantly in the laboratory sample which was excavated and brought back for testing.

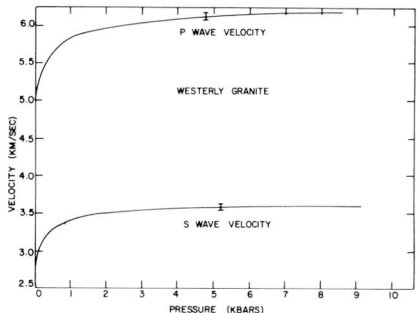

Fig. 15. Longitudinal and shear wave velocities as a function of stress for dry westerly granite.

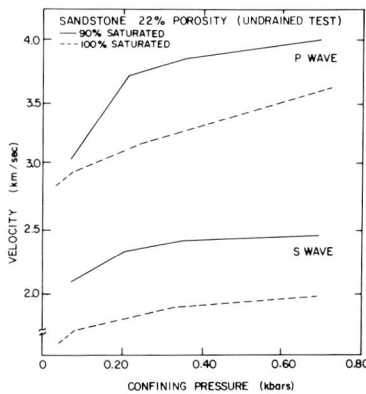

Fig. 16. Compressional and shear wave velocities of porous sandstone as a function of confining pressure for 90 and 100% saturated conditions during an undrained test.

The probable reason that the 1-m path length gave a lower velocity value than the 3-m path is due to the fact that only the outer portion, or rind, of the in situ specimen was stress relieved during excavation, since the block was still attached. It is felt that the center of the specimen over which the longer path traveled was still in a confined state. On the other hand, the intermediate value for the seismic refraction profile was probably due to the presence of discontinuities. A major joint system was delineated perpendicular to the seismic refraction profile.

Other laboratory measurements of longitudinal and shear velocities as a function of stress on nonporous rock such as granite show a marked increase in velocity with pressures up to approximately 1.0 kbar [Birch, 1960; Simmons, 1964] (Figure 15). Beyond 1 kbar the velocity does not appear to increase with increasing pressure. This relationship does not hold for more porous rocks such as sandstone, which shows only a rapid increase to approximately 0.4 kbar (Figure 16). It is also readily apparent that the degree of saturation plays an important role in the velocity obtained during laboratory tests. Note that the absolute velocity and the magnitude of the velocity increase are much greater in the material which is only 90% saturated. This is undoubtedly due to the role of pore pressure in determining the effective stress state in the sandstone (Figure 16).

Permeability values obtained under in situ stress conditions on samples from geothermal reservoir rock demonstrate the significance of fracture permeability under reservoir conditions. The factor of 2000 difference between the fracture permeability and the matrix permeability dominates fluid flow in geothermal reservoirs with low 'matrix' permeability (graywacke, granite, tuff, etc.). The role of fractures is also illustrated by the data from the Raft River area, where fracture permeability was 2 orders of magnitude greater than that of intact rock.

In summary, geologic discontinuities such as fractures, faults, and bedding planes have a significant effect on the velocity and permeability of in situ rock masses. It is felt that the joint spacing and aperture size are critical factors. Whether the role of discontinuities is as significant at higher stress levels remains to be seen, but at least in granitic rock types it is expected that the role of discontinuities persist to moderate depths that must be considered during formulation of models of parts of the earth's crust. It is also apparent that the in situ stress significantly affects the velocities obtained during measurements at least along path lengths of the order of a few meters.

References

Birch, F., The velocity of compressional waves in rocks to 10 kbar, 1, *J. Geophys. Res.*, 65, 1083-1102, 1960.

Byerlee, J. D., Frictional characteristics of granite under high confining pressure, *J. Geophys. Res.*, 73, 6031-6073, 1968.

Einstein, H. H., and R. C. Hirschfeld, Model studies on mechanics of jointed rock, *J. Soil Mech. Found. Div. Amer. Soc. Civil Eng.*, 99, 229-248, 1973.

Gale, J. S., A numerical field and laboratory study of flow in rocks with deformable fractures, *Ph.D. dissertation* Univ. of Calif., Berkeley, 1975.

Goodman, R. E., *Methods of Geological Engineering in Discontinuous Rock*, 277-368, West Publishing, Saint Paul, Minn., 1976.

Goodman, R. E., and N. Y. Ohnishi, Undrained shear tests of jointed rock, *Rock Mech.*, 5, 129-149, 1973.

Goodman, R. E., R. L. Taylor, and T. L. Brekke, A model for the mechanics of jointed rock, J. Soil Mech. Found. Div. Amer. Soc. Civil Eng., SM3, 637-659, 1968.

Jaeger, J. C., and N. G. W. Cook, *Fundamentals of Rock Mechanics*, Meuthen, London, 1969.

Johnson, J. N., et al., Anisotropic mechanical properties of Kayenta sandstone (Mixed Company Site) for ground motion calculations, *Tech. Rep. 74-61*, Terra Tek, Salt Lake City, Utah, 1974.

Jouanna, P., Seepage tests under stress in situ, *Proceedings of the Percolation Through Fissured Rock, Pap. T2-G, International Society of Rock Mechanics*, Stuttgart, Germany, 1972.

Maini, Y. N. T., In situ hydraulic parameters in jointed rock: Their measurement and interpretation, *Ph.D. thesis*, Imperial College, London, 1971.

Nelson, R. A., Fracture permeability in porous reservoirs: An experimental and field approach, *Ph.D. thesis*, Tex. A&M Univ., College Station, 1975.

Nielsen, R., et al., Characterization of rock-glass formed by the LASL subterrene in Bandlier tuff, Tech. Rep. 75-61, Terra Tek, Salt Lake City, Utah 1975.

Pratt, H. R., and E. R. Simonson, Geotechnical studies of geothermal reservoirs, Tech. Rep. 76-2, Terra Tek, Salt Lake City, Utah 1976.

Pratt, H. R., A. D. Black, and W. F. Brace, Friction and deformation of jointed quartz diorite, *Proceedings Third Congress of International Society Rock Mechanics*, 2, part A, 306-310, Nat. Acad. Sci, Washington, D. C., 1974.

Pratt, H. R., and H. S. Swolfs, W. F. Brace, A. D. Black, and J. W. Handin, Elastic and transport properties of in situ jointed granite, *Int. J. Rock Mech. Mining Sci.*, 14, 35-45, 1977.

Simmons, G., Velocity of shear waves in rocks to 10 kbar, *J. Geophys. Res.*, 69, 1123-1130, 1964.

Swolfs, H. S., C. Brechtel, and R. Lingle, Field and laboratory investigations for the Mixed Company and Cedar City sites, Terra Tek Report 76-5, 1976.

Swolfs, H. ., H. R. Pratt, A. D. Black, W. F. Brace, A. S. Orange and K. A. Gronseth, In situ propertoes of a jointed granite (Abs) Trans. *Am. Geophys. Union*, 55, 432, 1974.

GEOTHERMAL SYSTEMS: ROCKS, FLUIDS, FRACTURES

Michael L. Batzle and Gene Simmons

Department of Earth and Planetary Sciences
Massachusetts Institute of Technology
Cambridge, Massachusetts 02139

Abstract. Geothermal systems involve the dynamic interaction among rocks, fluids, and fractures. Some aspects of these systems are typical of the upper crust because they involve water-rock reactions resulting in cementation, diagenesis, or even low-grade metamorphism. Other aspects are atypical of the crust because of their speed, specific reactions, environment, and dependence on structure, particularly fractures. Reactions between fluids and rocks depend on fractures for fluid conduction and reaction surfaces. The morphologies of the fractures are dependent on the cementing, sealing, and alteration reactions. Core samples from several geothermal areas show repeated episodes of fracturing and sealing. Fluid flow and electric currents depend on the conduction paths provided by fractures. Cementation and sealing can lower permeability and raise resistivity by 3-5 orders of magnitude. On the other hand, fluid flow is restricted by sealed fractures. The sealed fractures are occasionally the boundaries between regions of significantly different physical characteristics. Despite the many episodes of fracturing, the sealing processes usually result in low fracture porosity.

Introduction

Core samples from several geothermal areas show an interplay among the rocks, fractures, and interstitial and fracture fluids. Fluid flow through the rock column is dependent on the stratigraphy and structure of the area. Geothermal areas commonly occur in fault zones where open fractures provide flow paths. All the studied areas show repeated episodes of fracturing and sealing. Thus fractures act as both cause and effect in these systems. The physical properties of the rock and fluid circulation are dependent in part on the fracture state of the rock. Fluids react with rocks forming alteration products which modify or seal fractures. The reaction products that remain in the rocks can be used to limit the models of temporal variations in fluid chemistry. Sealing and cementing, in turn, make the rock more susceptible to brittle fracturing.

Figure 1 is an index map of sample locations. Samples from some locations, such as the Dunes and Raft River areas, have been studied in detail. For other areas, such as Marysville, samples have only

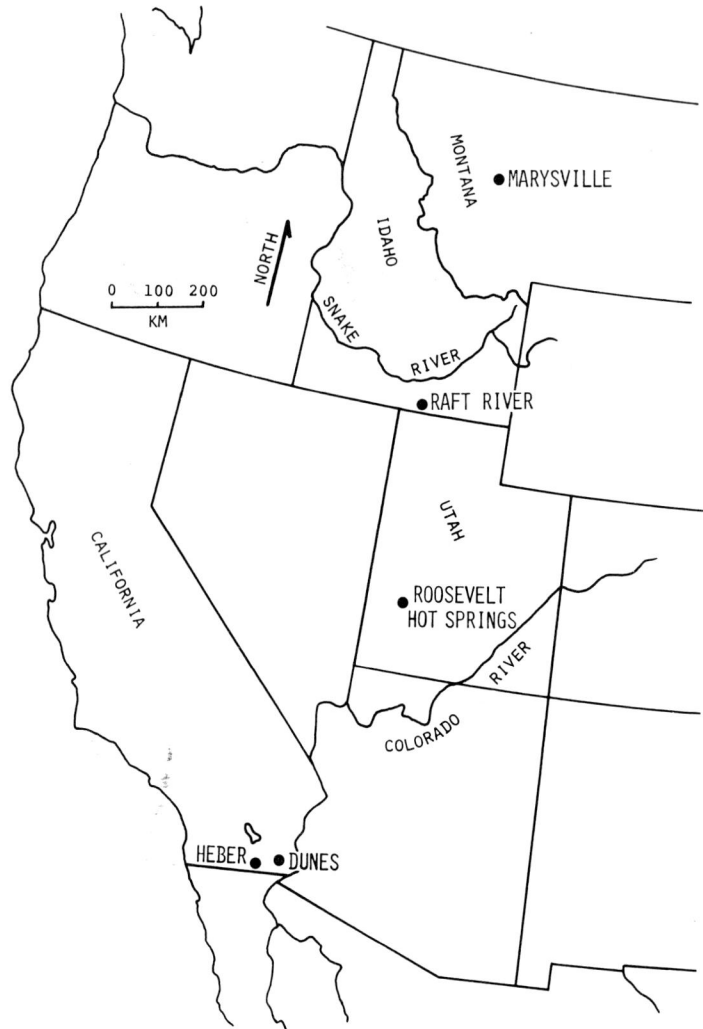

Fig. 1. Index map.

recently been obtained. Samples were generally chosen for high fracture content and indurated nature rather than for stratigraphic completeness. All but the Marysville area are regions of detrital alluvial sedimentation. Usually, the stratigraphy consists of interbedded sands, silts, clays, and conglomerates. In the Raft River and Roosevelt Hot Springs areas, drilling penetrated fractured igneous intrusive rocks. The Marysville well cut metamorphic and granitic plutonic rocks. These igneous rocks may play a significant role in the system. More complete geologic descriptions can be found in Elders and Bird [1974], Bird [1975], Blackwell et al. [1975], and Batzle and Simmons [1976].

In many ways, geothermal systems are not typical of the earth's crust. By definition they possess anomalously high temperatures or heat flow. Because of high temperatures and the availability of reactive fluids, diagenesis, sometimes bordering on low-grade metamorphism, occurs swiftly in some geothermal systems [Muffler and White, 1969]. The reactions and local environments often differ from those that commonly occur in sedimentary basins. These reactions include, for example, the development of adularia and the change from pyrite to hematite stability in the Dunes area. Because of the distribution of fractures and stratified aquifers, successive altered and unaltered zones exist [Elders and Bird, 1974; Bird, 1975]. Typically the first diagenetic reactions involve cementing the sediments, which makes fluid circulation dependent upon open fractures.

In several aspects, geothermal systems have characteristics frequently found in the upper crust. These systems involve several common water-rock interactions. Rocks exhibit authigenesis, replacement, alteration, and breakdown of unstable minerals. Other effects include cementation and compaction.

In this manuscript we will concentrate on the physical aspects of fractures in geothermal systems. This paper is a progress report on our continuing work on numerous core samples from several geothermal areas. A preliminary report on the Dunes and Raft River areas has been published recently by Batzle and Simmons [1976], and the reader is referred to that paper for details on those areas. Material that appears in other publications will only be summarized here.

Fracture Relationships

The five geothermal systems studied all show repeated fracturing superimposed on fluid chemical properties that varied with time. One excellent example of repeated fracturing in a Dunes sample from a depth of 403.9 m is presented in Figure 2. This sample is a sandstone which contains several large clay clasts. At least five episodes of fracturing and resealing have occurred. In this specific case the fracture mineralization remained calcite. The dark wavy bands match almost perfectly with the surface of the clay grain. These dark bands, numbered 1-5, oldest to youngest, are probably due to small particles of clay left behind when the calcite and clay separated during repeated fracturing.

A more complicated history is shown in the Heber sample from 1187 m (Figure 3). This sample is a highly fractured silty sandstone. Probably four or more episodes of fracturing are demonstrated by the figure. Here again calcite remains the sealing material. Crosscutting relationships are commonly used to determine the relative ages of fracturing events (see, for example, Park and MacDiarmid, 1975, p. 71). In this case, the fracture marked '1' is older than both '2' and '4'. The time relationship between '2' and '4' is ambiguous. Some relative age determinations are hindered by the recrystallization of the calcite which has obscured much of the original texture. The area shown in Figure 3 is actually only a small chip which is surrounded by several more sealed fractures. Hence both this sample and the Dunes (403.9 m) sample show repeated fracturing events followed by fracture sealing.

Fractures can also serve as conduits for reactive fluids. This

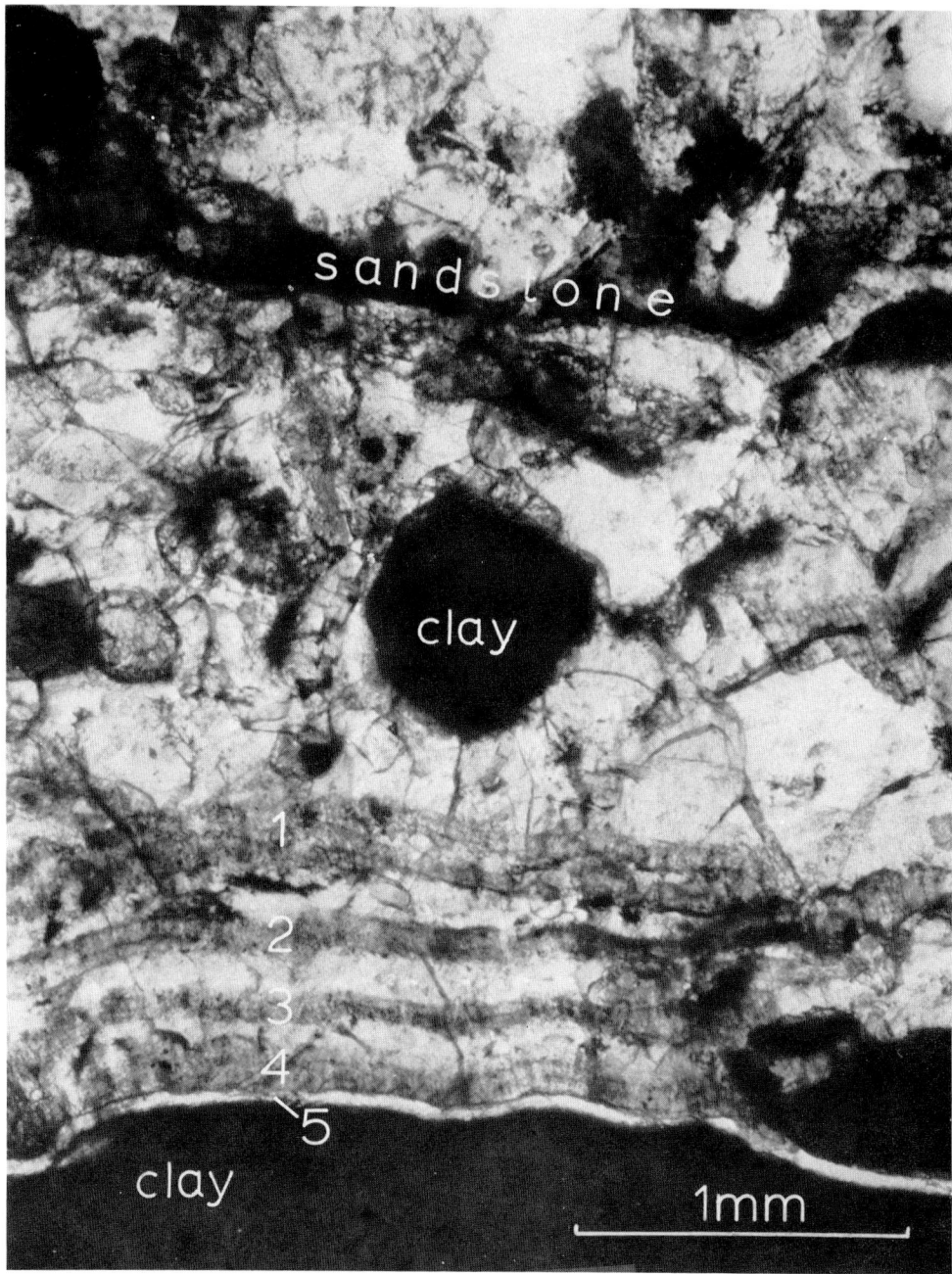

Fig. 2. Dunes sample from a depth of 403.9 m, photographed in transmitted light. The material between the 'sandstone' and the bottom clay fragment is the calcite sealed fracture. Numbers indicate successive fracture boundaries.

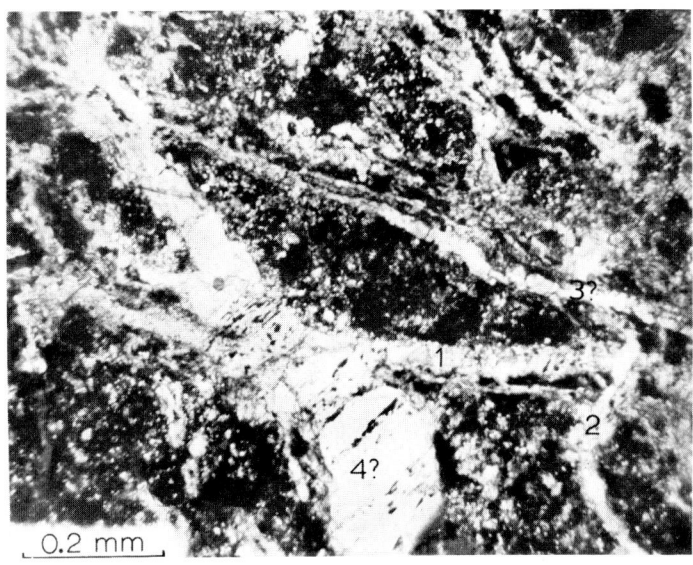

Fig. 3. Heber sample from a depth of 1187 m, photographed with crossed nicols. Dark areas are siltstone. Light areas are calcite sealed fractures.

property is demonstrated by the Roosevelt Hot Springs sample from a depth of 608.7 m. See Figure 4. A zone of alteration surrounds the fracture through this granodiorite. Here again multiple fracturing events may have occurred (numbers '1' and '2' in Figure 4). The relative ages of these 'events' have not yet been determined. Both x-ray diffraction and electron microprobe analysis indicates that hematite is the major fracture sealant. The zone marked '1' is enriched in calcite, feldspar, and quartz. The biotite grains along the fracture (at 'A') are being altered to hematite. Farther from the fracture the mafic grains remain unaltered. At 'B' a replacement rim of potassium feldspar has formed on the plagioclase grain.

Several variations of fluid chemistry with time are indicated in Figure 5. This sample is an argillaceous sandstone from a depth of 345 m in the Raft River area. The first fracture, 'f1,' is now sealed with calcite 'C.' The next two fractures, marked 'f2a' and 'f3a,' are now partially sealed with analcime. A well-developed analcime crystal is shown at 'D.' Another set of fractures is indicated by 'f4a,' 'f4b,' and 'f4c' in the figure. Here the fractures have apparently been widened by etching out the clayey matrix. Analcime remains unetched. The fracture fluids have therefore changed from calcite supersaturation to analcime supersaturation to undersaturation with respect to the clay matrix. This sample is also divided into a well-indurated portion at the top of the figure and a poorly indurated portion at the bottom. The boundary is the calcite sealed fracture f1. This sealed fracture has blocked the circulation of the fluids responsible for the cementation.

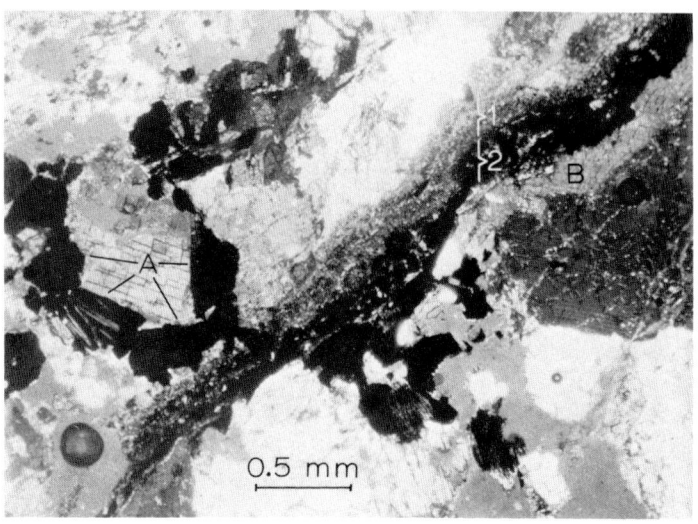

Fig. 4. Roosevelt Hot Springs sample from a depth of 608.7 m, photographed with crossed nicols. The dark band marked '1' and '2' is a sealed fracture. 'A' and 'B' are alteration zones (see text).

Physical Properties

In a self-sealing geothermal system, many physical properties are determined largely by the fracture state of the rock. Such parameters as permeability and electrical resistivity depend on the conduction paths provided by fractures. If a rock is tightly cemented, the only route for significant fluid movement is through fractures. Since the local physical and chemical environment is strongly dependent on the properties of the local fluids and fractures control the distribution of fluids, the environment is indirectly dependent on fractures.

A method modified after Brace et al. [1968] was used to measure permeabilities as small as 1 nanodarcy (10^{-9} darcy). Water in a closed system under pressure is passed through the sample. The decrease of pressure in the system as a function of time allows the permeability to be calculated. Errors in the permeability determinations are probably about 50 to 100% due to the exponential dependence on the several measured factors. Resistivity, which also depends on fracture parameters, was measured on the same saturated samples as permeability. We used a frequency of 100 Hz to minimize polarization effects. These measurements have an error of about 10%.

A sandstone from a depth of 609.2 m from the Dunes area shows the effect of cementation. Figure 6 is a photomicrograph of the boundary between well-cemented and poorly cemented portions. To the right in the poorly cemented region the dark areas between the grains are voids. This region is highly porous and permeable. To the left, in the well-cemented region, the light areas between grains are the calcite cement. This portion has a lower porosity and higher density. The permeability of the poorly cemented region is about 3 milli-

Fig. 5. Raft River sample from a depth of 345 m. This mosaic was made from scanning electron micrographs. The features are as follows: 'B' is a clay rich matrix, 'C' is a calcite sealing fracture, 'D' is a well-formed analcime crystal, 'E' is a void, 'F' is a void filled with epoxy, 'f1'-'f4c' are fractures (see text) [after Batzle and Simmons, 1976].

Fig. 6. Dunes sample from a depth of 609.2 m, photographed with crossed nicols. The poorly cemented portion is to right, the well-cemented portion is to left.

darcies versus 340 nanodarcies for the well-cemented region, a difference of 4 orders of magnitude. The resistivity of the well-cemented region is about double that of the poorly cemented portion. This relatively small change in resistivity is probably due to the presence of clays in the well-cemented region which provide a large surface conduction contribution. New fracturing will be required for any significant fluid movement through the well-cemented region.

The effect of a single fracture is demonstrated by a sample from a depth of 115.8 m in the Dunes area. The permeability of a sample from a tightly cemented unfractured area is about 2.8 nanodarcies. The permeability of a sample of the same size and with a single partially sealed fracture cutting the sample in the direction of fluid flow is about 8.2 millidarcies, a change of 6 orders of magnitude. The resistivity changes by 1 order of magnitude. The difference in the magnitude of change between the permeability and the resistivity is due to the differing dependence of these parameters on the width of the fracture. For a plane slit model, a model well-suited to this single fracture, the permeability depends on the width or aperture to the third power, but the resistivity depends on the width to the first power only. The original unsealed width of the fracture was about 20 μ. Based on the slit model, the 'effective' width of this partially sealed fracture has been reduced to about 5 μ. Hence even a single small fracture is extremely important to the physical properties of the rock.

How the various physical properties depend on the fracture state of a rock is a problem yet to be solved completely. Theoretical models exist, but they need to be refined and tested. Permeability and resistivity will depend, for example, on such factors as the size, shape, distribution, interconnection, and interaction of fractures and pores. Experimental correlations of various other properties with crack parameters are being attempted, and Feves et al. [1977] discuss several.

Fracture State of Rocks

Fractures have a pronounced effect on rock compressibility. From this effect, fracture porosity, effective orientation of cracks in space, and certain aspects of shape can be determined. Differential strain analysis (DSA), a high-precision technique developed by Simmons et al. [1974], was used in this study. This technique reduces error by comparing sample strain with that of a fused silica standard exposed to the same high-pressure environment.

To date, DSA has been used in this particular geothermal study only on Dunes and Raft River samples. For specific details, see the work by Batzle and Simmons [1976]. The samples have a low fracture porosity, and some are strongly anisotropic and inhomogeneous. Although many fractures are apparent to the unaided eye, because of the sealing process the total measured fracture porosities are less than about 0.1% in all samples. For example, the fracture porosity of the 115.8 m sample from the Dunes area is approximately 0.051% (this value was measured on the tightly cemented portion and does not include the open or partially sealed fractures mentioned previously).

The samples from geothermal areas are also often anisotropic and inhomogeneous. Strains due to fractures may differ by values of 30-40% between the axial and radial core directions. Two portions of a single sample from 337.4 m in the Raft River area differ in compressibility by a factor of 6. These two portions are separated by a sealed fracture and thus provide further evidence that sealed fractures can act as barriers to the flow of the cementing fluids. This particular sample is clay rich, and rock compaction obscures the effects of fractures.

Summary

Geothermal systems are a small but interesting part of the earth's crust. These systems involve a dynamic interaction among rocks, fluids, and fractures. The areas studied all showed repeated cycles of fracturing and sealing. Open fractures are important because they provide fluid and electric conduction paths. Sealed fractures however can block fluid flow. Fluid-rock reactions commonly involve mineral precipitation which tends to seal fractures and results in low fracture porosities. Fluid properties commonly change with time and these properties can be investigated through their effects on host rocks and specific fracture events.

Acknowledgements. We gratefully acknowledge the valuable assistance given by Robert Siegfried, Michael Feves, and Dorothy Richter of MIT. The Dunes sample, and much valuable advice, were provided by Wilfred Elders of the University of California at Riverside. Harry Covington of the U. S. Geological Survey and Rodger Stokes and Dennis Goldman of Aerojet Nuclear Company helped to obtain Raft River cores. Phillips

Petroleum Company, through G. Crosby, provided the Roosevelt Hot Springs cores. David Blackwell of Southern Methodist University made the Marysville core samples available. Ann Harlow typed the manuscript. Financial support was provided by NSF-RANN grant AER75-09588.

References

Batzle, M. L., and G. Simmons, Microfractures in rocks from two geothermal areas, Earth Planet. Sci. Lett., 30, 71-93, 1976.

Bird, D. K., Geology and geochemistry of the Dunes hydrothermal system, Imperial Valley of California, M.S. thesis, 123 pp., Univ. of Calif. at Riverside, Riverside, 1975.

Blackwell, D. D., M. J. Holdaway, P. Morgan, D. Petefish, T. Rape, J. L. Steele, D. Thorstenson, and A. F. Waibel, Results and analysis of exploration and deep drilling at Marysville geothermal area, in The Marysville, Montana Geothermal Project, Final Report, edited by W. R. McSpadden (Battelle Pacific Northwest Laboratories - ERDA report #23111-01410), E1-E16, 1975.

Brace, W. F., J. B. Walsh, and W. T. Frangos, Permeability of granite under high pressure, J. Geophys. Res., 73, 2225-2236, 1968.

Elders, W. A., and D. K. Bird, Investigations of the Dunes geothermal anomaly, Imperial Valley, California, 2, Petrological studies, paper presented at meeting of the International Union of Geochemistry and Cosmochemistry, Prague, 1974.

Feves, M., G. Simmons, and R. Siegfried, Microcracks in crustal igneous rocks: Physical properties, in The Earth's Crust, Geophys. Monogr. Ser., vol. 20, edited by J. G. Heacock, AGU, Washington, D. C., 1977.

Muffler, L. J., and D. E. White, Active metamorphism of upper cenozoic sediments in the Salton Sea geothermal field and the Salton Trough, southeastern California, Geol. Soc. Amer. Bull., 80, 157-182, 1969.

Park, C. F., and R. A. MacDiarmid, Ore Deposits, p. 71, W. H. Freeman, San Francisco, 1975.

Simmons, G., R. W. Siegfried II, and M. Feves, Differential strain analysis: A new method for examining cracks in rocks, J. Geophys. Res., 79, 4383-4385, 1974.

COMPLEXITIES OF THE DEEP BASEMENT FROM SEISMIC REFLECTION PROFILING

Jack Oliver and Sidney Kaufman

Department of Geological Sciences, Cornell University
Ithaca, New York   14853

Abstract. Observations of continental basement rocks fall into two categories. For rocks at the surface or within drillable depths, observations are primarily geological in nature, and, in general, they indicate great complexity in structure and great spatial variation in composition. Below drillable depths, information comes largely from geophysical observations that, for lack of resolving power, are commonly interpreted in terms of what must be unrealistically simple models of the earth. An attempt to understand the history, or structure, of the crust by relating one kind of information to the other commonly results in an impasse. Seismic reflection profiling offers hope for bridging this gap. Experiments conducted to date in several foreign countries and in the United States under the Consortium for Continental Reflection Profiling (COCORP) program which involves seismic reflection profiling by using VIBROSEIS sources, indicate basement complexities of much smaller scale than is commonly detected by other geophysical methods. The delineation and mapping of such features should greatly enhance our understanding of the basement and, in turn, of the continents. The data from a COCORP test in Hardeman County, Texas, illustrate a variety of features seismically that appear to be geological in nature and that demonstrate the heterogeneity of the basement. Special processing of the data to extend the duration of the time section from 15 s to 18.5 s resulted in the detection of a deep reflector that may correspond to the crust-mantle boundary and the probable detection of still deeper reflectors within the mantle.

Introduction

The deep basement of the continents is one of the major frontiers of modern earth science. The rocks there are known only in sketchy fashion, yet it seems that detailed knowledge of them will surely be of great importance in the study of such fundamental scientific topics as the formation and evolution of the continents. Furthermore, the deep basement rocks are closely related to the near-surface rocks from which man derives much of his livelihood and may hold the key to understanding, for example, the processes that concentrate certain minerals, the causes and history of the sedimentary basins that hold coal and petroleum, the causes of earthquakes, the mechanisms of volcanoes, and the sources of geothermal energy.

Even at drillable depths, information on the basement is sparse and

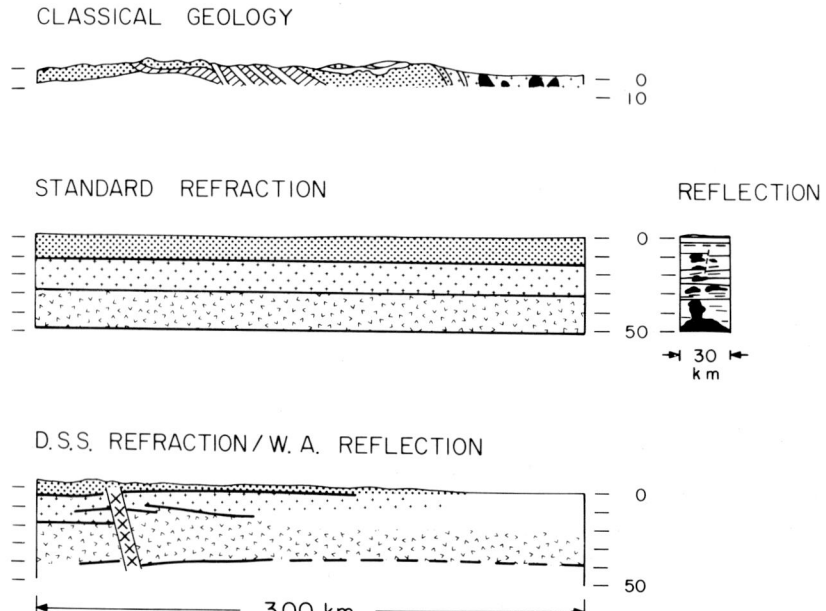

Fig. 1. Comparison of typical crustal sections. The upper drawing is based on geological information; the left center drawing is based on standard refraction procedures; the lower drawing is based on a combination of deep seismic sounding (D.S.S.) and wide angle (W.A.) reflection techniques; the right center drawing is based on near-vertical seismic reflection techniques. All figures are drawn to the same scale but are not intended to represent the same crustal section. Note the contrast in resolution of detail.

irregularly distributed. Below drillable depths, information comes largely from geophysical observations that are mostly of low resolution. Hence our knowledge of deep basement rocks and structure is limited to gross features and to direct measurement of only certain geophysical parameters, such as seismic velocities, density, electrical conductivity, etc. The striking contrast between the necessarily simple geophysical models of the crust and the highly complex geological information on the surface rocks is a major obstacle to better understanding of the basement (Fig. 1).

One way to overcome that obstacle is to study the deep basement by using the geophysical method with by far the highest resolution of structural features and velocity variations. The method is the seismic reflection profiling technique that is the basis for most geophysical prospecting for petroleum and for much of the exploration of the sea floor. The huge operations of the petroleum industry dwarf those of other applications; and as a result of this large effort, seismic reflection profiling has been developed to a high level of sophistication, with both abundant expertise and well-engineered equipment to achieve these goals.

In recent years this method in its modern form has been adapted to the study of the deep basement both abroad and in the United States. Although the number of locations tested is relatively small, the results are sufficient to demonstrate a point that is intuitively obvious but that is sometimes ignored because of the lack of observation of sufficient resolution. The point is that the structure of the deep basement is complex, probably on a scale comparable to that of the shallow basement. This means in turn either that the very common simple layered models of the crust based on most geophysical methods are so grossly oversimplified that they are misleading or that the earth properties measured by such geophysical methods are simple functions of depth and are unrelated to rock type and hence to geologic structures. The latter alternative seems rather unlikely, particularly at shallower depths.

The remainder of this paper describes some of the experiments and some of the results that lead to the conclusion that the basement is complex structurally and that the simple layered models are at best gross approximations of that structure. However, the reader should bear in mind that the application of seismic reflection profiling to the study of the deep basement is in its infancy. Only a few sites have been explored, and there is no comprehensive study of any site. As is true with the application of any new technique to the study of a relatively unknown region, a period of learning and growing skill of interpretation must be anticipated as new data become available. Thus it may well be and in fact it is likely that more will ultimately be gleaned from the raw information that is described in the preliminary discussions given here. However, the point of this note is to emphasize the evidence for great complexity of deep crustal structure, and that point seems to be demonstrated clearly already by the data.

The Method

The basic principle of seismic reflection profiling is simple, deceptively so. A disturbance at or near the surface radiates seismic waves into the earth. Compressional waves reflected from subsurface horizons and traveling along vertical or near-vertical paths are recorded at the surface. The travel times of such waves can be used to determine the depths of the horizons if velocities along the path are known. The source and receiver are moved along the surface to provide a profile of the reflecting horizons, in the manner of an echo sounder aboard ship.

In practice, the application of the method is much more complex. Source elements may be impulsive, as with explosives, air guns, gas exploders, sparks, or weight drops, or they may be vibratory, as in the case of the VIBROSEIS (trademark of Continental Oil Company) method, in which large truck-mounted vibrators excite long near-sinusoidal wave trains of slowly varying frequency. At a particular 'shot point', arrays of source elements are used to enhance the vertically traveling compressional waves and to deemphasize other wave types. Shear waves are considered noise in this application but are the subject of current researches in which they would serve as the signal, providing information on shear wave velocities as well as structure.

The single element of the receiver is the geophone, a seismometer with a natural frequency of several cycles per second. Tens of geophones are distributed in spatial arrays, and their outputs are combined to enhance the reception of reflected waves to form one 'channel'. Typically, 48 or more channels are recorded; hence over 1000 geophones are commonly used in one field operation. The array of sources and the array of geophones move along the profile in increments of perhaps 100 m. Thus there is great redundancy of information. Over a profile a few tens of kilometers in length, information is obtained from waves traveling over millions of different ray paths.

This redundancy of information is used in elaborate data processing to further enhance the signal-to-noise ratio and to provide information on spatial variations in velocity along the profiles. An important technique is common depth point stacking, in which rays incident at various angles on a common reflection point are combined for improved detection. Various complications such as wave diffraction, focusing, side echoes, multiples, and apparent displacement of dipping reflectors arise, and the interpreter must be aware of and alert to those complications. However, the method provides by far the best resolution of structural features that is available from any presently known geophysical technique. For a more detailed discussion of the application of this method to the study of the deep basement see Oliver et al [1976]. For a more detailed discussion of reflection profiling as it is carried out by the petroleum industry see Lindseth [1976] or Dobrin [1976].

Some Results

Although some of the very earliest observations of deep basement reflectors were made in the United States [Junger, 1951] and sporadic activity at a low level followed, most reported studies of this type until recently originated primarily in West Germany, Australia, Canada, and eastern Europe (see Oliver et al. [1976] for a review). In early 1975, field tests involving the application of the modern reflection profiling methods of the petroleum industry, modified to provide information on deep reflectors, began in the United States under the auspices of the COCORP program. In 1975, tests were made at two sites, one in Texas and one in New Mexico, and the early results of the test in Texas form the basis for the discussion of this paper. Presently, more extensive studies at additional sites are being conducted. The data analyzed and presented here have been subjected only to rather routine processing. It is quite possible that more advanced processing, currently underway, will wring additional information from the data on which the following sections are based. However, a number of points are evident from the data at the present stage of analysis.

One simple but important point deals with the relation between the horizontal and vertical scales of the crustal sections. Exaggeration of the vertical coordinate is common in such sections. In studies of the ocean floor, vertical exaggerations as high as 50:1 are common, for example. Geologic cross sections commonly employ large vertical-to-horizontal exaggeration. Crustal sections based on seismic refraction data are commonly shown with vertical exaggerations of about 3:1. The reflection profiles shown here in Figs. 2 and 3, although time sections, are approximately at 1:1 scale and are very tall and thin at that scale. In fact, the data are frequently plotted with horizontal exaggerations of 2 or 3, as is done in Fig. 5. This simple point brings home forcefully the exceptional resolution of the reflection profiling method. While it is difficult to quantify the resolution as the term is used in this general sense, it seems fair to say that at least an order of magnitude improvement over the resolution of vertical detail by other geophysical measurements is involved.

Figs. 2 and 3 show sections from the tests in Hardeman County, Texas. Fig. 2 shows the longer north-south line, and Fig. 3 shows a shorter east-west line crossing the line of Fig. 2. Locations of seismic profiles and stations in Hardeman County are shown in Fig. 4. Several important points are apparent from these data.

1. Take a broad view of the results first. There is good evidence for reflected (or backscattered) energy from a large number of horizons or obstacles at depth; hence the crust must be heterogeneous at least at the scale of seismic wavelengths, typically a few tenths of a kilometer. A particular arrival may be the combined effect of returned energy traveling along more than one path and need not correspond to a single reflector, but the existence of reflectors or scatterers cannot be denied.

2. With some exceptions at relatively shallow depths, the reflections are, in general, not continuous for more than a few kilometers. This observation indicates what should be obvious intuitively: the simple layered models of geophysics with widespread discontinuities of velocity are but the crudest approximations to the crust, and one can be misled by using those models indiscriminantly in geological interpretations.

3. Arrivals that are continuous across a substantial length of the profile, that are concave downward, and that generally are at high angles to the horizontal are fairly abundant. These arrivals are diffractions. Thus they are additional indicators of heterogeneity of rather small scale within the crust. Some of the heterogeneities corresponding to the diffractions of Figs. 2 and 3 fall beyond the line of the section and also out of the plane of the section, but the basic conclusion with regard to crustal non-heterogeneity remains.

4. Direct evidence of the depth to the top of the crystalline basement in this area is available from the bore-hole log from an oil company dry hole located about 300 m from the south end of line 1. The top of the basement is at a depth corresponding to a two-way travel time of 1.6 s. Thus the two strong zones of reflectors, at about 3 s and 4 s, that are more or less continuous over the area are in the shallow part of the basement. Such strong laterally continuous near-horizontal reflectors in the basement are relatively unusual in reflection profiling to date, but they provide some interesting information. If they are traced to outcrops, they could be identified positively. A careful inspection will reveal that they are not perfectly continuous and that they have structure. The structures may be secondary and result from faulting or other forms of deformation, or they may be primary. Near the base of both reflecting horizons are angular unconformities that show up best on line 2 (Fig. 3). These unconformities are important because they indicate something of the nature of the boundaries and of the character of the basement rocks; that is, they demonstrate that the feature is depositional rather than geochemical, as is true in the case of a phase change.

5. In addition to the evidence from reflected waves, there is evidence from the absence of reflected waves. Thus, when it is clear that waves

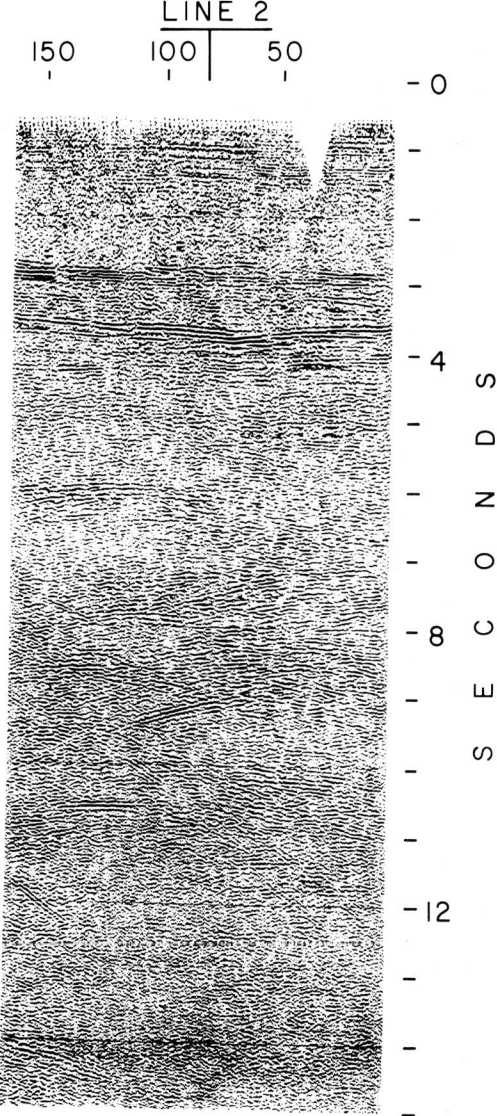

Fig. 2. Seismic section for line 1, Hardeman County, Texas. The location is shown on Fig. 4. The horizontal coordinate is station number. Station interval is 100 m. The vertical coordinate is two-way travel time in seconds. Vertical-to-horizontal scale ratio is approximately 1:1 (a 6 km/s velocity is assumed throughout the section). The location of the intersection of line 2 with line 1 is indicated.

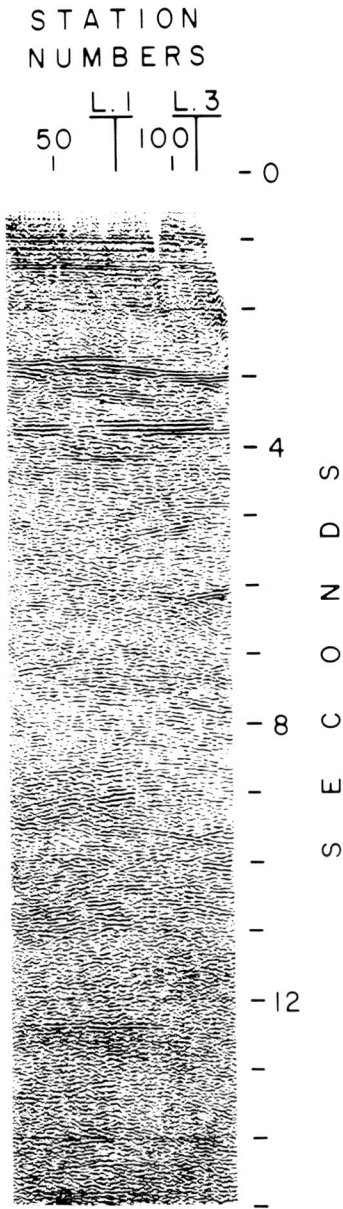

Fig. 3. Seismic section for line 2, Hardeman County, Texas. The coordinates are the same as those in Fig. 2. The locations of intersections of lines 1 and 3 with line 2 are indicated.

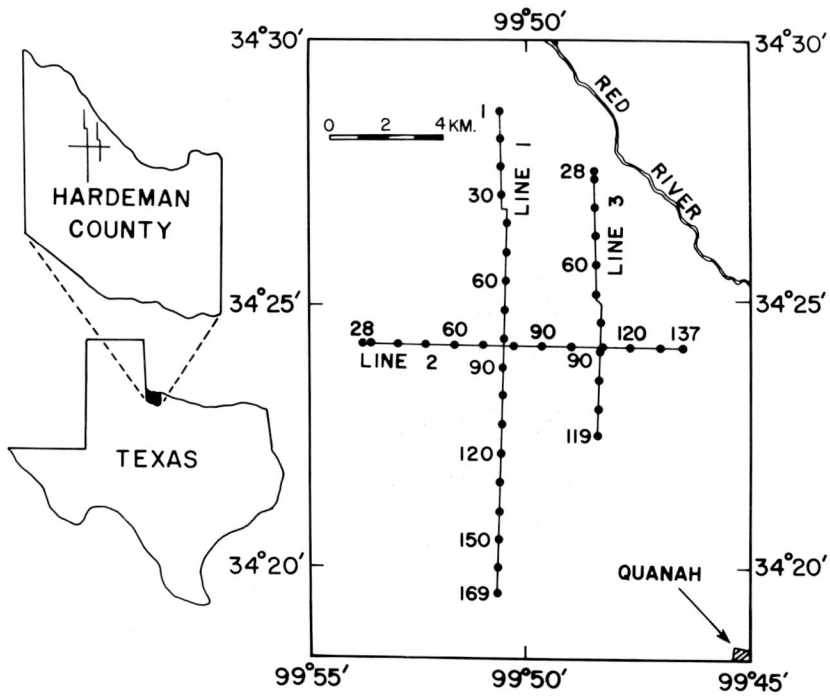

Fig. 4. Location map for seismic profiles in Hardeman County, Texas. Selected station numbers are indicated.

have penetrated a particular zone (that is, it has been traversed by waves traveling to and returning from still deeper reflectors), the absence of reflectors, or the transparency as it is sometimes called, indicates a zone of relative homogeneity seismically and suggests, in turn, relative homogeneity compositionally. The boundaries of such zones are not always clear cut in the data. There are examples of such zones in Fig. 2. One zone is below the gently curved arrival at about 6 s. The dimensions are of the order of several kilometers, and the zones could be indicative of relatively homogeneous intrusions. Of course, delineation of much larger features of this nature would require longer profiles.

6. The fact that the reflection data show information of a geological nature (such as the unconformities) indicates the importance of viewing geophysical observations of the crust in a geological manner. Thus a valid measurement of some geophysical feature, for example, a low-velocity zone, probably indicates something about the geological character of the rocks at that particular site and is not likely to have application more widespread than the occurrence of that particular geological feature. In fact, the information on velocity for this particular site indicates a low-velocity zone beneath the reflector at 3.0 s, and it is probably associated with those rocks that are unconformable below that reflector. If the rocks thicken, thin, or pinch out in other directions, the low-velocity zone will probably do likewise.

7. There are some events, particularly at later times, for which the

curvature (convex upward) is greater than that of curves of maximum convexity calculated for that depth on the basis of simple diffraction from a point discontinuity. Although other explanations are possible and not unlikely, one possible interpretation of this penomenon is that the waves are focused by a strong upward concavity of the deep horizon. In such a case the appropriate curve of maximum convexity is that for the depth of the focal point, not the depth of the horizon. More detailed observations would be required to demonstrate this effect conclusively. However, the point is that reflection data can provide such information and may have in this case. If so, there is additional support for complex structures, perhaps folds, at depths in the crust under Hardeman County.

8. Many of the points made above were presented in slightly different form in an earlier paper [Oliver et al., 1977]. Previously published data for Hardeman County, however, has always terminated after 15 s, and the field observations were designed to provide sections of this duration. By a trick of processing, that is, by correlating with only the low-frequency portion of the vibrator sweep, it is possible to extend the time section by an additional 3.5 s. Thus the section of Fig. 5 shows line 1 with a total time of 18.5 s and also with some horizontal exaggeration. The principal piece of new information is an apparent reflection at a time of about 15.5 s on the part of the section to the left. Velocities at this depth are not known from this study. However, a refraction profile a short distance to the northwest in Oklahoma [Tryggvason and Qualls, 1967] provides some information on the depth to the mantle there. By assuming that the depth and other parameters are about the same in Hardeman County the 15.5 s reflection appears to correspond to the top of the mantle. If so, the data imply that the crust-Moho transition zone is considerably less than 0.3 km in thickness. Whether any significance should be attached to the relative lack of structure in this short segment of the Moho is not clear from this limited piece of data. Indeed, if it were to turn out that the Moho is commonly smooth, continuous, and flat in relation to features of the lower crust, strong evidence in favor of the explanation of the Moho as a phase change or as flow at that depth would be in hand. However, there is insufficient evidence to make this conclusion at present. Furthermore, there is limited evidence for still deeper reflections, suggesting that the upper mantle is also heterogeneous although not necessarily of the same character as the lower crust.

9. Although the Hardeman County site is completely sediment covered and hence provides no data from areas of basement outcrop, a limited quantity of such information from other sites substantiates the claim that reflecting horizons within the basement correlate with known structural or lithologic discontinuities.

Conclusion

The complexity of basement rocks seen at the surface is great. Seismic reflection data suggest that the complexity of deeper basement rocks of the crust is comparable. On the one hand, these observations mean that the use of simple layered models for interpretation of geophysical data can be highly misleading to the unwary earth scientist. On the other hand, the abundant information from the limited seismic observations to date means that there is a wealth of information on the deep basement

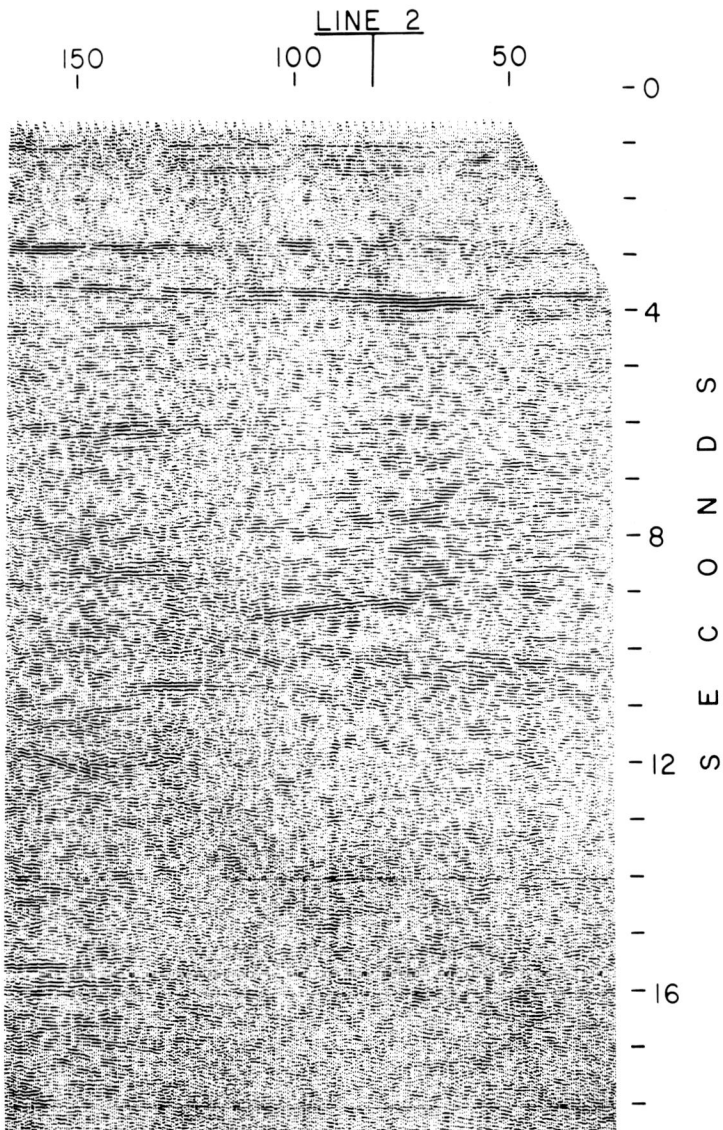

Fig. 5. Seismic section for line 1, Hardeman County, Texas, differing from that of Fig. 2 in that the record two-way travel time has been extended from 15 s to 18.5 s by correlating with only the lower frequency portion of the vibrator sweep. The horizontal distance is about 17 km, and the greatest depth is approximately 55 to 60 km.

that can be obtained through further application of the seismic reflection profiling technique.

Acknowledgments. This paper is based on data acquired under the COCORP program. The program is funded by the National Science Foundation under grant DES 74-22257 to Cornell University. The preparation of this paper was supported by the Office of Naval Research under contract N0D014-75-C-1121 to Cornell University. The members of the COCORP executive committee in addition to the authors are A. W. Bally, J. C. Maxwell, R. P. Meyer, and R. A. Phinney. The assistance of many individuals from industry, government agencies, and universities serving on advisory committees to COCORP is gratefully acknowledged. This is contribution 584 of the Department of Geological Sciences of Cornell University.

References

Dobrin, M., Introduction to Geophysical Prospecting, McGraw-Hill, New York, 1976.

Junger, A., Deep reflections in Big Horn County, Montana, Geophysics, 16, 499-504, 1951.

Lindseth, R., Digital Processing of Geophysical Data--A Review, Society of Exploration Geologists, Calgary, Alberta, Canada, 1976.

Oliver, J., M. Dobrin, S. Kaufman, R. Meyer, and R. Phinney, Continuous seismic reflection profiling of the deep basement, Hardeman County, Texas, Bull, Geol. Soc. Amer., 87, 1537-1546, 1976.

Tryggvason, E., and B. Qualls, Seismic refraction measurement of crustal structure in Oklahoma, J. Geophys. Res., 76, 3738-3740, 1967.

SEISMIC VELOCITY, REFLECTIONS, AND STRUCTURE OF THE CRYSTALLINE CRUST

Scott B. Smithson and Peter N. Shive

Department of Geology, University of Wyoming
Laramie, Wyoming 82071

Stanley K. Brown

Computer Services, University of Wyoming
Laramie, Wyoming 82071

Abstract. After years of speculation about crustal reflections and varying success in finding them, numerous seismic reflections are being mapped within the continental crust. Interpretation of crustal reflections is much more complicated than interpretation of similar data from sedimentary basins, just as exposed highly deformed crystalline rocks are more complicated than flat-lying relatively undeformed sedimentary rocks. Lack of continuity caused by isoclinal folding, disruption of layers, anomalous dips, and intrusions commonly characterizes deformed crystalline rocks, whereas continuity characterizes sedimentary rocks. Crustal velocities range from 6 km/s in granite to over 7 km/s in gabbroic rocks. The maximum velocity contrast is that of granite against gabbro; however, as is well known from synthetic seismograms, larger amplitude reflections may be produced by layers of alternating higher and lower velocity. This is exactly the relationship that is normally found in metamorphic rocks. These reflections are more complex than those from simple interfaces such as would be associated with igneous intrusions. Low-dip reflections showing continuity in the upper crust are most likely caused either by slightly metamorphosed interlayered lava flows and sedimentary rocks or by layered mafic intrusions. Migmatites, which may comprise the major portion of the middle crust, are layered on too small a scale to give reflections except at high frequencies ($\sim$100 Hz). In our seismic studies of crystalline basement rocks, numerous reflections are found at shallow depth where layered metamorphic rocks are exposed, but surprisingly, no reflections are recorded within a large body of amphibolite surrounded by granite gneiss. Igneous intrusions such as an anorthosite complex and a granite batholith are underlain by structure that causes reflections. Complex reflections found beneath a Precambrian batholith suggest layering, and shallower reflections indicate heterogeneity within the intrusion. Granitic intrusions may thus be underlain by metamorphic rocks and give rise to transparent zones devoid of reflections within seismic reflection cross sections. Granitic intrusions, represented by transparent zones, resemble models of gravity intrusions by H. Ramberg. Seismic velocities, crustal refraction studies, and deep crustal reflections all show that the lower crust is highly heterogeneous and that it cannot be gabbroic but must be more felsic, possibly andesitic in mean composition. As is suggested from the Ivrea zone and Jotun nappe, a crude zoning of the crust is present, but the old geophysical picture of a simple layered crust should be abandoned.

Introduction

Geophysicists have long been intrigued by the possibility of obtaining seismic reflections from the crystalline crust, but earlier studies have met with varying success. Because the seismic reflection method is the only high-resolution technique to obtain detailed information from the crust short of deep drilling, interest in crustal reflection studies has been revived [Kanasewich and Cumming, 1965; Dohr and Fuchs, 1967; Oliver et al., 1976; Shive et al., 1975], and numerous reflectors have been mapped within the crust. Now that crustal reflections are commonly being found, the source of these events is a major problem because interpretation of these features is not straightforward. Interpretation of crustal reflections is much more complicated than interpretation of similar data from sedimentary basins because of the much greater structural complexity of crystalline rocks.

We have been conducting seismic reflection experiments in exposed crystalline rocks [Shive et al., 1975; Smithson et al., 1975; Smithson and Shive, 1975] as a type of 'calibration' in order to interpret crustal reflections with less uncertainty. We shall discuss the significance of crustal reflections in terms of known geology, seismic velocity, and reflections that we have recorded in different types of crystalline rocks.

Geology

Exposed crystalline basement rocks provide the major clue for interpretation of crustal reflections. Geologic maps generally show that most crystalline rocks are metamorphic [Smithson and Decker, 1974], and interspersed in this metamorphic framework are scattered igneous intrusions. Major rock types to be expected are metamorphic rocks, migmatites, granite, and mafic intrusions. Metamorphic rocks are typically layered on all scales, and an increasing grade of metamorphism carries a strong depth connotation [Read, 1957]. Low-grade metamorphic rocks associated with shallow depths consist of interlayered recognizable metavolcanic and metasedimentary rocks. Their dips may vary from undeformed near horizontal to vertical, but they are commonly less deformed internally than higher-grade rocks. With increasing depth these rocks pass into higher-grade (amphibolite facies) more deformed metamorphic rocks that probably make up the bulk of the upper crust and commonly show low dips. Amphibolite facies gneisses grade into migmatites [Wegmann, 1935; Mehnert, 1968], mixed rocks, consisting of dark layers interspersed with a mobilized granitic fraction on all scales.

Migmatites and associated rocks (augen gneiss and granitic gneiss) may comprise most of the middle crust [Smithson, 1965; Smithson and Decker, 1974]. Although migmatites may contain some thick (∼1 km) bodies of one rock type [Worl, 1968], granitic and mafic zones are usually interlayered on a scale of centimeters to meters. On a large scale, migmatites may behave as intrusions into their overlying cover of metamorphic rocks [Haller, 1958].

Numerous workers have suggested that the lower crust is predominantly composed of granulite facies rocks [Heier and Adams, 1965; den Tex, 1965; Ringwood and Green, 1966; Ito and Kennedy, 1971; Smithson and Decker, 1974]. Granulites represent the highest-grade minerals, and are commonly associated with similar igneous rocks so that igneous and metamorphic physical conditions overlap here. Granulite facies rocks have low radioactive heat production [Heier and Adams, 1965] and thus agree in this respect with most models of the lower crust [Blackwell, 1971; Heier, 1973; Smithson and Decker, 1974].

Metamorphic rocks are characterized by extreme deformation as plastic flow. Isoclinal folding, attenuation, and disruption of layers give metamorphic rocks a characteristic lack of continuity that contrasts with the lateral extent of sedimentary rocks. Interfaces between layers may be marked by small-scale folding.

Igneous bodies ranging from granite to gabbro are interspersed in a framework of metamorphic rocks. Granites occur as intrusions that may cover thousands of square kilometers and range from 2 to 10 km thick [Bott and Smithson, 1967]. They commonly form sharp contacts with the surrounding metamorphic rocks at higher crustal levels; deeper granites may develop gradational contacts. Granites contain inclusions with dimensions up to kilometers that may have either sharp or gradational borders. Inclusions are more common at the borders of granitic bodies.

Mafic intrusions generally cover much smaller areas than granitic intrusions. They are seldom larger than 10 km wide but may be as large as 200 km across like the Bushveld intrusion [Wager and Brown, 1967]. Mafic intrusions are commonly layered on a scale ranging from centimeters to a kilometer in thickness. Composition of the mafic layers ranges from anorthosite to dunite [Wager and Brown, 1967], although minor amounts of granite may also be present, especially near the top of the mafic intrusion. Form of the intrusions varies from horizontal sills up to 1 km thick to thicker lopoliths and funnels; dips of layering may be high due to later deformation.

Further evidence on the nature of the crust comes from combined geological-geophysical studies of particular areas where deep crustal rocks are exposed. The most important of these is the Ivrea zone in the southern Alps, where a vertical section through the crust is believed exposed [Berckhemer, 1969]. A crustal section here consists of schist, gneiss, and granite near the top, migmatites (mobilized rocks) at greater depths, and finally granulitic rocks. The lower 5-10 km of the section consist of lenticular dioritic and gabbroic bodies interlayered with high-grade gneisses. Transition to mantle material is marked by lenticular peridotites [Schmid, 1967; Berckhemer, 1969].

In the Jotun nappe of the central Norwegian Caledonides, layered anorthosites and related granulites are exposed over an area of 4500 km. Although other interpretations are possible, the granulitic rocks of the nappe probably represent a thick (about 10 km) slice of lower crust that has been emplaced in the upper crust by Caledonian deformation. This occurrence indicates that vast amounts of the deep crust consist of granulite facies rocks.

Seismic Velocity

The most used information available for interpretation of the nature of unexposed rocks comes from measured seismic velocities. Velocities commonly found in the crust range from 6 km/s in granitic rocks to 7 km/s in gabbroic; extreme values are 5.8 km/s in light granite to 7.4 km/s in dense gabbroic rocks. Rocks of intermediate composition have velocities around 6.5 km/s.

Crustal refraction surveys have measured velocities from 6.0 to 6.4 km/s in the upper crystalline crust and from 6.4 to 7.3 km/s in the lower crust [Press, 1966], and these velocities have been associated with a 'granitic layer' and a 'basaltic layer.' Lower crustal velocities are, however, most commonly in the range from 6.5 to 7.0 km/s. A low-velocity zone may be present in the middle crust [Landisman et al., 1971].

Most data on seismic velocity in rocks come from laboratory measurements [Birch, 1960, 1961; Simmons, 1964; Christensen, 1965]. Laboratory measurements in water-saturated samples [Nur and Simmons, 1969] indicate that earlier measurements are systematically low, especially at low confining pressures. Field measurements of P wave velocity [Smithson and Shive, 1975] agree well with laboratory measurements, especially those on water-saturated samples. Relatively high velocities are found at shallow depth so that if these velocities are extrapolated to higher pressures, laboratory measurements of velocity might be slightly low. Velocity generally varies directly as density; therefore the lowest velocities expected in the upper crust under normal physical conditions are 5.8-6.0 km/s in leucocratic granitic rocks [Simmons and Nur, 1968; Smithson and Ebens, 1971; Smithson and Shive, 1975]. Velocities in migmatitic rocks will be slightly higher, 6.0-6.3km/s [Smithson and Ebens, 1971]. Of particular interest are the velocities of intermediate to mafic granulitic rocks, especially granulitic intermediate rocks (Table 1). The data show that syenitic rocks with densities as low as 2.73 g/cm^3 have velocities around 6.5 km/s and that true gabbroic rocks have velocities greater than 7.0 km/s. Granulitic rocks with velocities of 6.4-6.7 km/s have densities of 2.8 g/cm^3 or less; these rocks are closer to being granitic than basaltic. Lower crustal velocities are commonly too low for gabbroic rock.

Reflections

Since an early paper on crustal reflections by Junger {1951}, numerous recent studies have found deep reflections [Fuchs, 1969; Clowes and Kanasewich, 1970; Oliver et al., 1976; Oliver and Kaufman, 1977]. The first significance of these reflections is obviously that numerous discontinuities exist within the continental crust. This is

TABLE 1a. Field Measurements of Velocity [Smithson and Shive, 1975].

Rock Type	Vmin km/s	Vmax km/s	Zmax km	Density gm/cm^3	Mean Atomic Weight	Vlab* km/s
Quartz monzonite	5.1	6.1	1.3	2.68	21.1	6.3
Granite gneiss	5.5	5.8	0.3	2.63	21.4	5.9
Syenite	6.0	6.4	0.6	2.73	21.7	6.6
Anorthosite	6.2	6.4	0.3	2.73	20.6	6.9
Amphibolite	6.2	7.0	0.7	3.01	21.2	6.6-7.2

*At 1 kbar.

TABLE 1b. Laboratory Measurements of Velocity in Granulite Facies Metamorphic Rocks [Christensen and Fountain, 1975]

Rock Type	V_p(1 kbar), km/s	V_p(6 kbar), km/s	Density g/cm^3
Granodioritic gneiss	6.36	6.57	2.68
Granodioritic gneiss	6.06	6.41	2.71
Plagioclase-pyroxene-hornblende gneiss	6.49	6.69	2.73
Syenite gneiss	6.40	6.71	2.83
Syenite gneiss	6.40	6.70	2.84
Plagioclase-pyroxene-hornblende gneiss	6.75	6.95	2.90
Pyroxene syenite gneiss	6.82	7.01	2.93
Hornblende granulite	6.51	6.99	2.98
Pyroxene-hornblende granulite	7.11	7.37	3.03
Pyroxene granulite	6.68	7.40	3.08

V_p in lower crust 6.4-7.3 km/s.

already an important change from typical crustal models. The maximum contrast in acoustic impedance expected within the crust is for granite (gneiss) against gabbro (gneiss). Typical velocities of 6.0 and 7.1 km/s give a reflection coefficient of 0.13, which is considerably less than reflection coefficients for sedimentary rocks; however, as is well known from synthetic seismograms, complex layering may cause much stronger reflections than a single interface (see Fuchs [1969]; Clowes and Kanasewich [1970]; and Figure 1). Complex layering is exactly the situation found in metamorphic rocks described from deep sections of the crust [Schmid, 1967] and in some mafic intrusions [Wager and Brown, 1967]. On the basis of the composite nature of deep reflections, Fuchs [1969] and Clowes and Kanasewich [1970] have suggested an interlayered or transitional nature for major reflectors in the deep crust.

As opposed to other crustal reflection studies, which have been conducted in sedimentary basins, we have conducted seismic reflection studies in areas of exposed Precambrian rocks [Shive et al., 1975] as a means of 'calibrating' seismic crustal studies. Exposed crystalline rocks provide the most obvious possible models for interpretation of crustal reflection; we will discuss sources of reflections in terms of exposed crystalline rocks and our seismic reflection results.

Numerous seismic reflections within a restricted time interval, particularly composite reflections, are probably caused by layers in metamorphic rocks or possibly by a layered mafic intrusion (Figure 1). Reflections from metamorphic layering may have a greater amplitude than reflections from a single interface (Figure 1). Reflections from layers in a mafic intrusion would generally be of restricted lateral extent (less than 10 km) and be weaker than those from metamorphic layering. Records from reflection profiling in exposed layered metamorphic rocks show numerous events within the first 2.8 s (Figure 2).

Near-horizontal reflections showing a number of events and good continuity are most likely caused by interlayered slightly metamorphosed lava flows and/or sedimentary rocks such as those which form the Keweenawan [Thiel, 1956] or by layered mafic intrusions as in the Wichita Mountains of Oklahoma [Widess and Taylor, 1969]. If such deep-seated rocks as migmatites or granulites are obtained from drill cores from the top of the basement beneath sedimentary cover, it is unlikely that we would find slightly metamorphosed rocks at a lower level unless they were emplaced by faulting.

Low-grade (greenschist facies) metamorphic rocks are usually either horizontal if undeformed or dip steeply if deformed. With increasing deformation and metamorphism, recumbent folding is common, and many deep-seated metamorphic rocks paradoxically show low dips. For example, deep-seated granitic gneisses and migmatites show near-horizontal dips through a 4-km vertical section in the Wind River Mountains, Wyoming [Smithson and Ebens, 1971]. Low dips may be locally steepened by cross-folding, doming, or intrusion. Dips in crystalline rocks are often high, so that anomalous move out and reverse move out events may be expected (Figure 3), as has been shown in Soviet deep-seismic-sounding studies of Precambrian shields [Sollogub et al., 1973, p. 72].

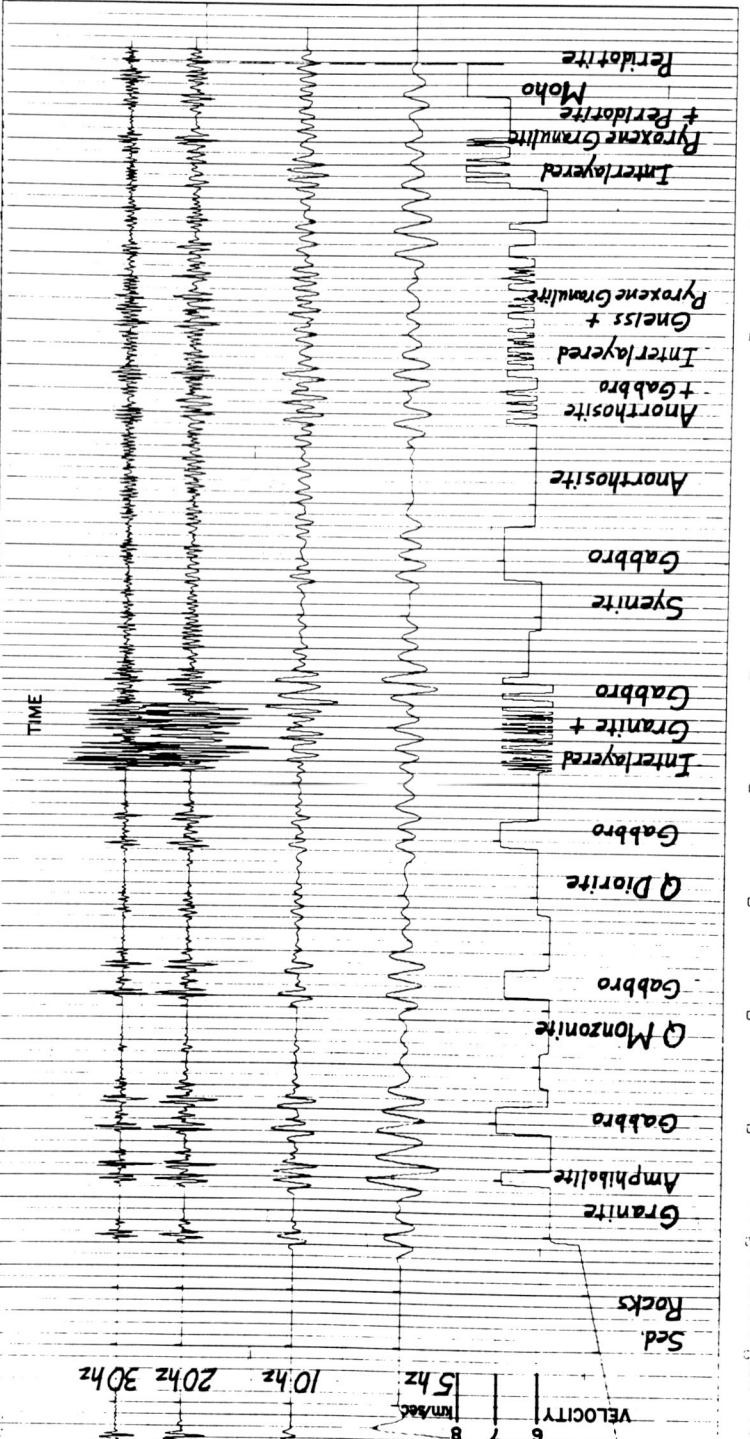

Fig. 1. Synthetic seismograms showing seismic response for different kinds of crustal structure. Time increases to the right. Reflection response for input wavelets of 5, 10, 20, and 30 Hz is shown. Seismograms for 10 and 20 Hz are most realistic because these frequencies are the most usable ones. Granite against amphibolite or gabbro between 2 and 3 s represents the maximum reflection amplitude expected in the crust from a simple interface. Larger reflections from multiple interfaces (layering) are generated by constructive interference at 6 s. At 10 s, interlayered gneiss and pyroxene granulite, which could be an important constituent of the lower crust and only has a moderate contrast in acoustic impedance, may generate a relatively strong reflection through constructive interference. The velocity function is not intended as a specific crustal model but rather to illustrate the seismic response of rock sequences.

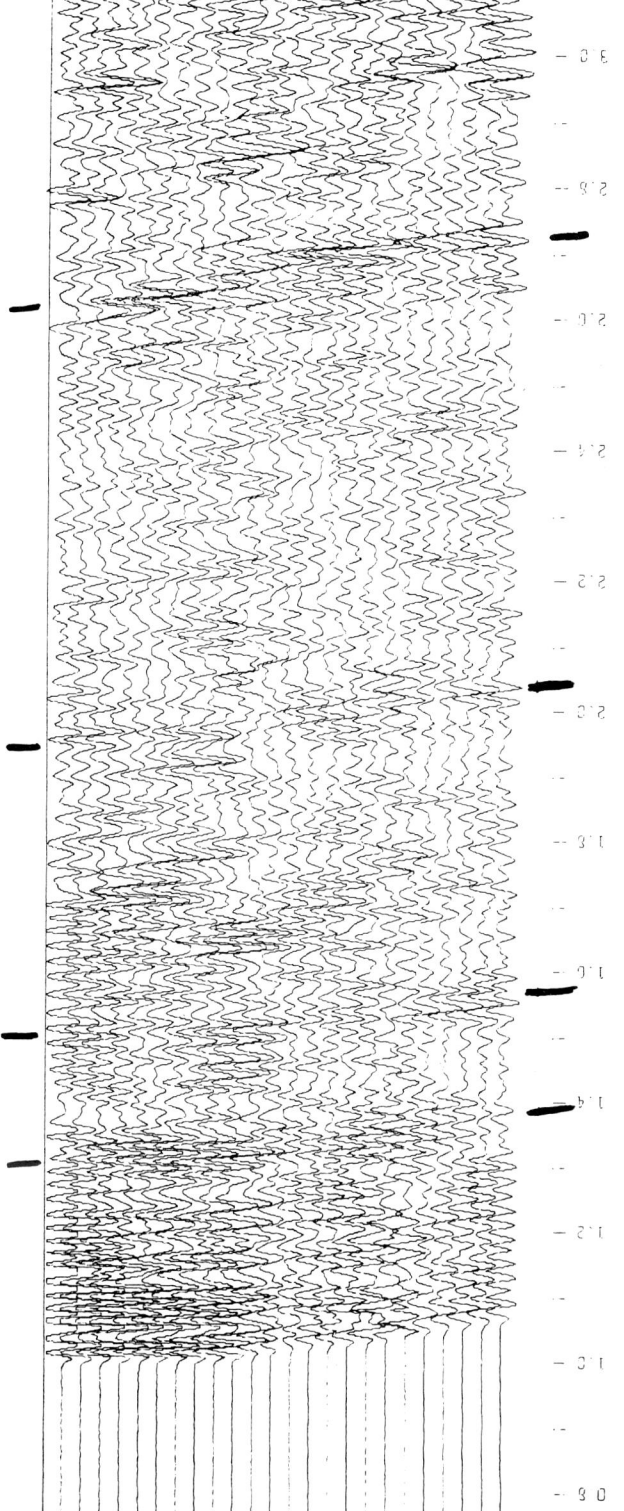

Fig. 2. Seismic record with 1150 m of surface coverage showing numerous reflections in the first 2.8 s. The shot point is offset 3 km. The seismic record is from an area of layered metamorphic rocks.

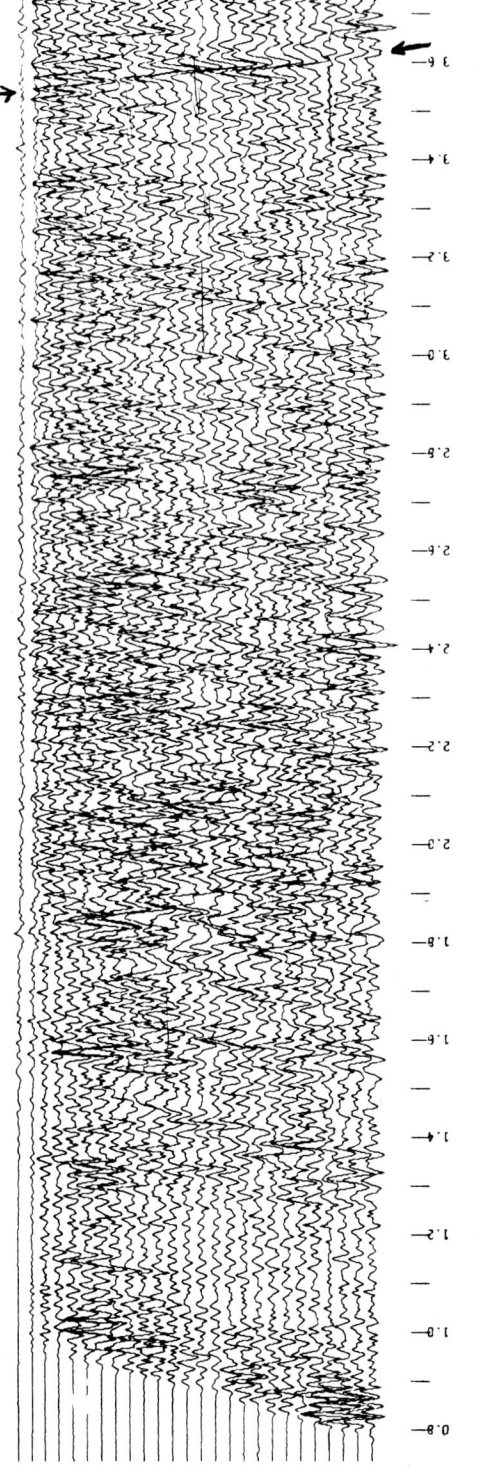

Fig. 3. Seismic record with 1150 m of surface coverage showing a reverse move-out event at 3.6 s. The shot point is offset 2 km. The reverse move-out event indicates high dip, a feature we would commonly expect in the crust.

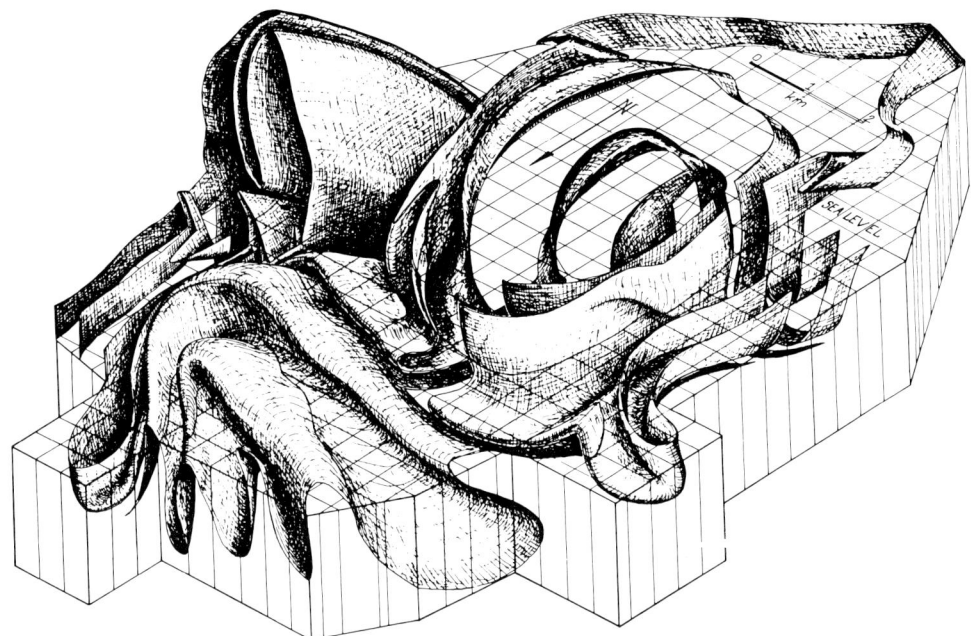

Fig. 4. Diagram of a folded structure from the deeply eroded Precambrian of Greenland [after Berthelsen, 1960]. The folded horizons are pyroxene granulite layers surrounded by granitic gneiss. Although the pyroxene granulite layers would be good reflectors, the possibility of resolving such complex structures with the seismic relection method is slight.

German workers [Liebscher, 1964; Dohr and Fuchs, 1967; Dohr and Meissner, 1975] emphasize the lack of lateral continuity and statistical nature of relections that they have mapped in the deep crust. Lack of continuity is typical of metamorphic rocks, particularly compared with sedimentary rocks. Although metamorphic units may continue for tens and in some cases hundreds of kilometers, discontinuity caused by isoclinal folding, disruption of layers, change in layer thickness, and intrusion is characteristic. Abrupt change in dip and irregularity of interfaces, caused by minor folds, may also interrupt the continuity of reflections. For example, well-exposed extremely complex structure from deep-seated granulites, mapped by Berthelsen [1960], illustrates the difficulty of interpreting such structures from seismic reflections. Complicated structures such as those in Figure 4 will never be resolved by seismic reflection interpretation, much less any other geophysical technique. Fold structures may cause complex interfering discontinuous seismic reflections [Behrens et al., 1972]. Gradational contacts between rock types due to igneous processes may eliminate reflections even though contrasting rock types are present.

Sharp planar igneous contacts should be characterized by simple reflections (Figure 1). Figure 5 shows such a reflections that we believe comes from near the base of an anorthosite complex. Of

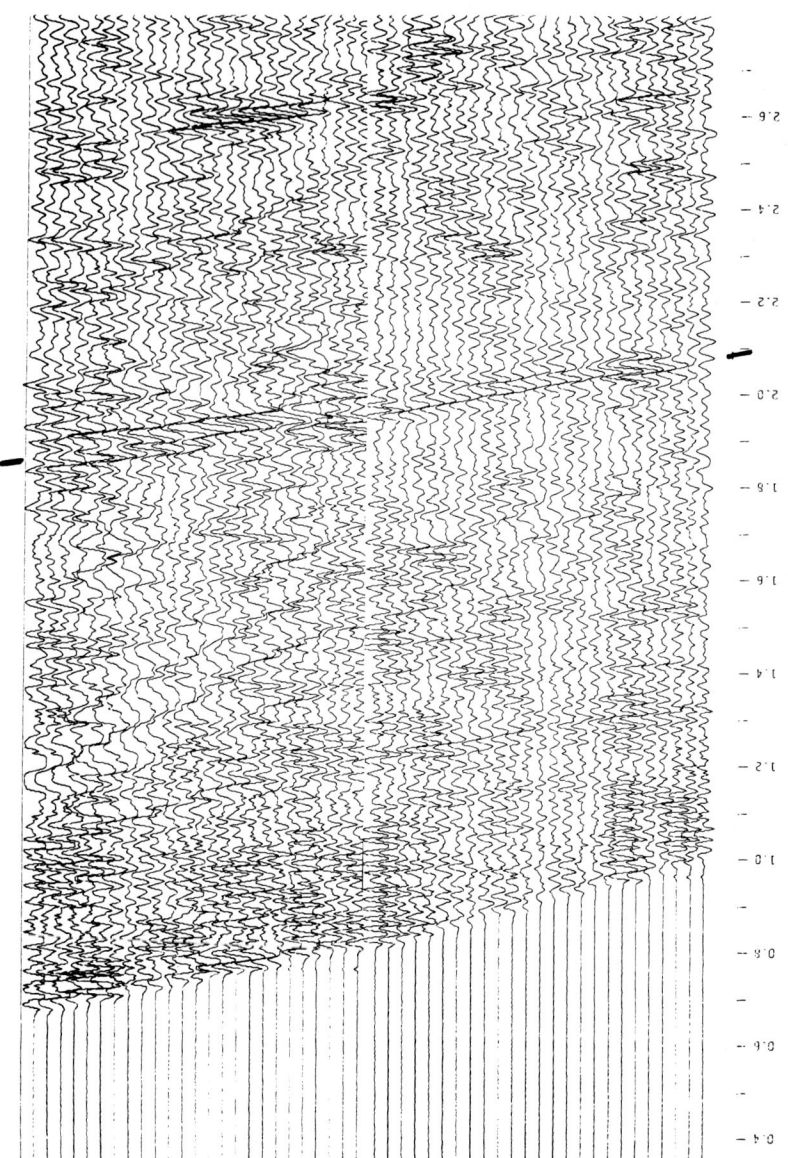

Fig. 5. Seismic record with 2300 m of surface coverage showing a clear reflection at 2.0 s. The shot point is offset 1.5 km. This simple event probably comes from a single interface at a depth of about 5 km near the base of the Laramie anorthosite complex.

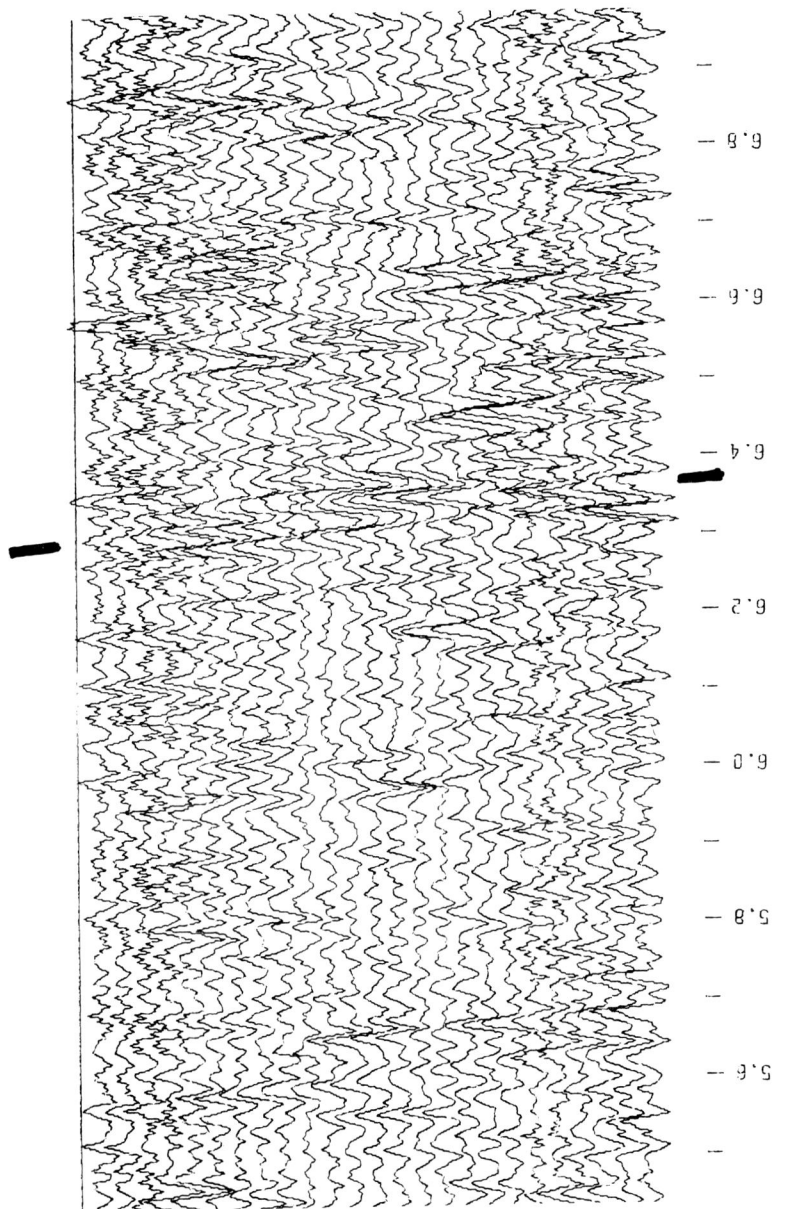

Fig. 6. Seismic record with 1150 m of surface coverage. The shot point is offset 10 km. This seismic line is located in the Precambrian Sherman granite batholith. The reflection which is marked at 6.3 s comes from a depth of 18-20 km, which is below the presumed base of the batholith. The reflection indicates the presence of an interface beneath the batholith, and the composite nature of the reflection (multiple cycles) suggests small-scale layering.

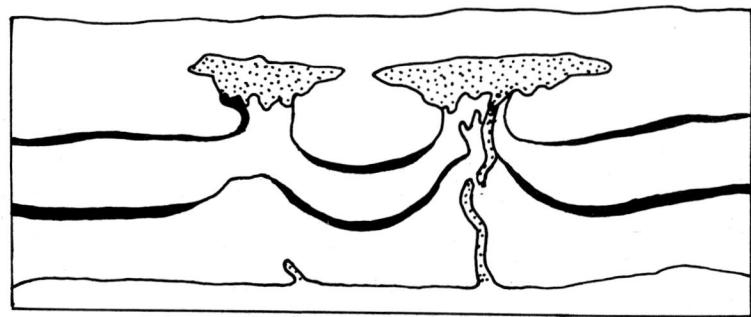

Fig. 7. Model for gravity intrusion of granitic magma. Intrusion is rootless with distrubed layering preserved beneath it [after Ramberg, 1970]. Dots mark the granite.

four different exposed igneous bodies that we have studied, all show reflections from depths of 15-20 km. The most interesting of these is a Precambrian batholith with events at about 9 km and composite reflections from about 20 km (Figure 6). The composite events at 6.3 s (Figure 6) suggest that the batholith is underlain by interlayered higher-velocity, more mafic rocks that are thus probably metamorphic rocks. The gravity intrusion experiments modeled by Ramberg [1970] indicate that granitic intrusions rise through a sequence of layered rocks and are emplaced essentially as rootless bodies (Figure 7). Oliver et al., [1976] and Oliver and Kaufman [1977] have found numerous transparent zones, areas of no reflections, and have interpreted these features as intrusions. Our observations support this interpretation.

Migmatites that may comprise a major portion of the middle crust [Smithson and Decker, 1974] are layered on such a small scale that layering would not be 'seen' except by high-frequency ($\sim$100 Hz) seismic waves. Migmatites may rise in irregular domes to penetrate the overlying metamorphic rocks [Haller, 1958] and would also appear as relatively transparent zones or intrusions in seismic reflection profiles unless high frequencies can be used.

Discussion

Without any speculative interpretation at all, the presence of relections throughout the crystalline crust shows that the crust must be much more heterogeneous and complicated than is shown by simple crustal refraction models. Seismic reflections in the crust may, however, be profitably interpreted on the basis of structure and velocity of exposed crystalline rocks. Also, layering must be present on a relatively small scale throughout the crust. Such layering is suggestive of metamorphic rocks, and transparent zones indicate igneous intrusions.

Crustal refraction studies have commonly found velocities from 6.4 to 7.0 km/s in the lower crust and have found that this velocity varies from place to place. Measured velocities less than 7 km/s cannot be attributed to gabbro, amphibolite, or pyroxene granulite

Fig. 8. A crustal model that shows lateral and vertical heterogeneity. The upper crust consists of metasedimentary and metavolcanic rocks penetrated by a granite batholith (left) and a migmatite dome (right). The upper crust is of intermediate composition [Smithson and Decker, 1974]. The middle crust is a migmatite zone consisting of migmatites, granitic gneiss, and augen gneiss. It is more felsic than overlying metamorphic rocks, has a slightly lower seismic velocity, and contains fewer good reflecting interfaces. The lower crust consists of a framework of metamorphic rocks with scattered igneous bodies. The rocks are heterogeneous and deformed; some contacts may be gradational and others are sharp. Parts of the lower crust are truly gabbroic; parts may even be granite. The lower crust consists predominantly of heterogeneous granulite facies rocks whose mean composition is more felsic than gabbroic.

at high pressures; in areas of normal geothermal gradient, temperatures in the lower crust are probably not high enough [Blackwell, 1971; Smithson and Decker, 1974] to cause a marked decrease in velocity. Velocities in the lower crust correspond to intermediate rocks and even to granitic rocks in the granulite facies. Reflections from the lower crust indicate that the lower crust is heterogeneous with zones of contrasting acoustic impedance. If we eliminate peridotite as an important constituent except in the lowermost crust [Berckhemer, 1969], then granitic, gabbroic, and/or intermediate rocks must be present to cause reflections. Although the lower crust must be heterogeneous on a fairly small scale, the mean composition may be andesitic but must be more felsic than gabbro.

Our model for continental crust (Figure 8) shows a crust that is heterogeneous both vertically and laterally. A crude vertical zonation (Figure 8) probably exists consisting of metamorphic rocks of intermediate composition at the surface, a more felsic migmatite zone in the middle crust, and more mafic rocks in the lower crust. In places the migmatite zone, sandwiched between more mafic rocks, may form a seismic low-velocity zone ($V = 6.0$–6.3 km/s) in the middle crust. The migmatite zone may be arched upward to form domes surrounded by more mafic metamorphic rocks (Figure 8). Where the migmatite zone is broadly uplifted, overlying metamorphic rocks have been eroded away to form vast granite-gneiss terrains so common in shields. The crust consists of a framework of layered metamorphic rocks with structure so complex that its details are probably unresolvable with geophysics. Within this framework are emplaced granitic and mafic intrusions. Mafic intrusions may be layered.

Layering is generally present in the crust on a small scale, and a gross zoning may be present; however, a simple large-scale horizontally layered crustal model as derived from crustal refraction interpretations must be abandoned. The continental crust might more realistically be thought of as a series of blocks of different composition and structure as has been pictured by Beloussov [1965], Kosminskaya and Zverev [1968], and Davydova et al. [1972].

Seismic reflection studies offer the best possibilities for new detailed knowledge of the continental crust. Frequencies up to 30 Hz may readily be used; this would allow resolution of layers 10-20 m thick. Problems such as the nature of the deep crust, suture zones, and continental growth and development may be resolved by reflection profiling. In order to obtain a maximum from interpretation of reflection profiling the sign of the reflection coefficient, its amplitude, and interval velocities should ideally be known.

Acknowledgements. This research was carried out with financial support from U.S. National Science Foundation grants GA-12871 and DES 74-22264.

References

Behrens, J., R. Bortfeld, G. Fommlich, and K. Köhler, Interpretation of discontinuities by seismic imaging, Z. Geophys., 38, 481, 1972.

Beloussov, V. V., The Crust and Upper Mantle of the Continents, 123 pp., Academy of Sciences of the USSR, Moscow, 1965.

Berckhemer, H., Direct evidence for the composition of the lower crust and the Moho, Tectonophysics, 8, 97, 1969.

Berthelsen, A., Structural studies in the Precambrian of western Greenland, Medd. Groenland, 123 (1), 222 pp., 1960.

Birch, F., The velocity of compressional waves in rocks to 10 kilobars, 1, J. Geophys. Res., 65, 1083, 1960.

Birch, F., The velocity of compressional waves in rocks to 10 kilobars, 2, J. Geophys. Res., 66, 2199, 1961.

Blackwell, D. D., The thermal structure of the continental crust, in The Structure and Physical Properties of the Earth's Crust, Geophys. Monogr. Ser. vol. 14, edited by J. G. Heacock, p. 169, AGU, Washington, D. C., 1971.

Bott, M. H. P., and S. B. Smithson, Gravity investigations of subsurface shape and mass distribution of granite batholiths, Geol. Soc. Amer. Bull., 78, 859, 1967.

Christensen, N., Compressional wave velocities in metamorphic rocks at pressures to 10 kilobars, J. Geophys. Res., 70, 6147, 1965.

Christensen, N., and D. M. Fountain, Constitution of the lower continental crust based on experimental studies of seismic velocities in granulite, Geol. Soc. Amer. Bull., 86, 227, 1975.

Clowes, R. M., and E. R. Kanasewich, Seismic attenuation and the nature of reflecting horizons within the crust, J. Geophys. Res., 75, 6693, 1970.

Davydova, N. I., I. P. Kosminskaya, N. K. Kapustian, and G. G. Michota, Models of the earth's crust and M-boundary, Z. Geophys., 38, 369, 1972.

den Tex, E., Metamorphic lineages of orogenic plutonism, Geol. Mijnbouw. 44, 105, 1965.
Dohr, G., and K. Fuchs, Statistical evaluation of deep crustal reflections in Germany, Geophysics, 32, 951, 1967.
Dohr, G., and R. Meissner, Deep crustal reflections in Europe, Geophysics, 40, 25, 1975.
Fuchs, K., On the properties of deep crustal reflectors, Z. Geophy., 35, 133, 1969.
Haller, J., Probleme der Tiefentektonik Bauformen in Migmatitt-Stockwerk der ostgrönlandischen Kalidoniden, Geol. Rundsch., 45, 159, 1958.
Heier, K. S., A model for the composition of deep continental crust, Fortschr. Mineral., 50, 174, 1973.
Heier, K. S., and J. A. S. Adams, Concentrations of radioactive elements in deep crustal material, Geochim. Cosmochim. Acta, 29, 53, 1965.
Ito, K., and G. C. Kennedy, An experimental study of the basalt-garnet granulite transition, in The Structure and Physical Properties of the Earth's Crust, Geophys. Monogr. Ser. vol. 14, edited by J. G. Heacock, p. 303, AGU, Washington, D. C., 1971.
Junger, A., Deep basement reflections in Big Horn County, Montana, Geophysics, 16, 499, 1951.
Kanasewich, E. R., and G. L. Cumming, Near-vertical incidence seismic reflections from the 'Conrad' discontinuity, J. Geophys. Res., 70, 3441, 1965.
Kosminskaya, I. P., and S. M. Zverev, Abilities of explosion seismology in oceanic and continental crust and mantle studies, Can. J. Earth Sci., 5, 1091, 1968.
Landisman, M., S. Mueller, and B. J. Mitchell, Review of evidence for velocity inversions in the earth's crust, in The Structure and Physical Properties of the Earth's Crust, Geophys. Monogr. Ser., vol. 14, edited by J. G. Heacock, p. 11, AGU, Washington, D. C., 1971.
Liebscher, H. J., Deutungsversuche fur die Struktur der tieferen Erdkruste nach reflexionsseismischen und gravimetrischen Messungen im deutschen Alpenvorland, Z. Geophy., 30, 115, 1964.
Mehnert, K. R., Migmatites and the Origin of Granitic Rocks, 393 pp., Elsevier, New York, 1968.
Nur, A., and G. Simmons, The effect of saturation on velocity in low-porosity rocks, Earth Planet. Sci. Lett., 7, 189, 1969.
Oliver, J., and S. Kaufman, Complexities of the deep basement from seismic reflection profiling, in The Earth's Crust, Geophys. Monogr. Ser., vol. 20, edited by J. G. Heacock, AGU, Washington, D. C. this volume, 1977.
Oliver, J., M. Dobrin, S. Kaufman, R. Meyer, and R. Phinney, Continuous seismic reflection profiling of the deep basement, Hardeman County, Texas, Geol. Soc. Amer. Bull., 87, 1573, 1976.
Press, F., Seismic velocities, Handbook of Physical Constants, Geol. Soc. Amer. Mem., 97, 195, 1966.
Ramberg, H., Model studies in relation to intrusion of plutonic bodies, in Mechanism of Igneous Intrusion, edited by G. Newall and N. Rast, p. 261, Gallery Press, Liverpool, 1970.

Read, H. H., The Granite Controversy, 430 pp., Thomas Murby, London, 1957.

Ringwood, A. E., and D. H. Green, Petrological nature of the stable continental crust, in The Earth Beneath the Continents, Geophys. Monogr. Ser., vol. 10, edited by J. S. Steinhart and T. J. Smith, p. 611, AGU, Washington, D. C., 1966.

Schmid, R., Zur Petrographie und Struktur der Zone Ivrea-Verbano zwischen Valle d'Orsola und Val Grande, Schweiz. Mineral. Petrogr. Mitt., 47, 935, 1967.

Shive, P. N., S. B. Smithson, and S. K. Brown, Seismic reflection studies in crystalline crust (abstract), Eos Trans. AGU, 56, 905, 1975.

Simmons, G., Velocities of compressional waves in various minerals at pressures to 10 kilobars, J. Geophys. Res., 69, 1117, 1964.

Simmons, G., and A. Nur, Properties of granites in situ and their relation to laboratory measurements, Science, 162, 789, 1968.

Smithson, S. B., The nature of the 'granitic' layer of the crust in the southern Norwegian Precambrian, Nor. Geol. Tidsskr., 45, 113, 1965.

Smithson, S. B., and E. R. Decker, A continental crustal model and its geothermal implications, Earth Planet. Sci. Lett., 22, 215, 1974.

Smithson, S. B., and R. J. Ebens, Interpretation of data from a 3.05-kilometer borehole in Precambrian crystalline rocks, Wind River Mountains, Wyoming, J. Geophys. Res., 76, 7079, 1971.

Smithson, S. B., and P. N. Shive, Field measurements of compressional wave velocities in common crystalline rocks, Earth Planet. Sci Lett., 27, 170, 1975.

Smithson, S. B., P. N. Shive, and R. Wyckoff, Seismic studies in exposed crystalline basement rocks (abstract), paper presented at 45th Annual International Soc. of Explor. Geophys. Meeting, Denver, Colo., Oct. 12-16, 1975.

Sollogub, V. B. et al., New DSS-data on crustal structure of the Baltic and Ukranian shields, Tectonophysics, 20, 67, 1975.

Thiel, E., Correlation of gravity anomalies with the Keweenawan geology of Wisconsin and Minnesota, Geol. Soc. Amer. Bull., 67, 1079, 1956.

Wager, L. R., and G. M. Brown, Layered Igneous Rocks, 588 pp., W. H. Freeman, San Francisco, Calif., 1967.

Worl, R. G., Taconite in the Wind River Mountains, Sublette County, Wyoming, Prelim. Rep. 10, Wyo. Geol. Surv., Laramie, 1968.

Wegmann, E., Zur Deutung der Migmatite, Geol. Rundsch. 26, 305, 1935.

Widess, M. B., and G. L. Taylor, Seismic reflections from layering within the Precambrian basement complex, Oklahoma, Geophysics, 24, 417, 1959.

CRUSTAL VELOCITIES FROM MARINE COMMON DEPTH POINT REFLECTION DATA

Joel S. Watkins, Richard T. Buffler, Mark H. Houston,
John W. Ladd, Thomas H. Shipley, F. Jeanne Shaub,
John B. Sinton, and J. Lamar Worzel

Geophysics Laboratory, University of Texas, Marine Science Institute
Galveston, Texas 77550

William P. Dillon

Office of Marine Geology, U.S. Geological Survey
Woods Hole, Massachusetts 02543

Abstract. Interval velocities calculated from marine common depth point (CDP) reflection data obtained with a 2.3-km hydrophone array yield consistent results at depths as great as 10 km (6-km subbottom). Data collected from six marine areas in the western North Atlantic Ocean, the eastern Caribbean Sea, and the western Gulf of Mexico are in good agreement with areally coincident refraction and sonobuoy data. Where deep strong reflectors are present below oceanic sediments, interval velocities within oceanic layers 2 and 3(?) were obtained. A 500-km depth section in the western North Atlantic Ocean seaward of the Blake Escarpment illustrates the use of CDP data for the study of the velocity structure of marine rocks.

Introduction

A large part of our knowledge of the subsurface structure of the earth is derived directly or indirectly from determination of seismic wave velocities, especially compressional wave velocities. For four decades, seismic refraction studies have provided a steady flow of information about the velocity structure of the crust and upper mantle of ocean basins. These data now constitute the basis, or standard, for the bulk of our knowledge about the structure of oceanic rocks.
  The use of seismic reflection for the study of the velocity structure of the oceanic crust and upper mantle effectively began in the middle 1960's [see Le Pichon et al., 1968]. This technique, which uses sonobuoys as disposable receivers, resolves details of layer 1 (sedimentary layer) and detects velocity gradients and reversals, capabilities which are either theoretically or practically beyond the capabilities of refraction. The velocity structure in ocean basins as determined from sonobuoy data has been reported by various investigators [e.g., Houtz et al., 1968].

The present paper reports velocities determined from multifold common depth point (CDP) reflection data. This technique differs from the sonobuoy technique in that it uses (1) a fixed length receiver array, usually 2.3-2.4 km instead of a free-floating buoy, constituting an effective array of several tens of kilometers, and (2) multiple receiver arrays which make it possible to operate on data reflecting from approximately the same subsurface point, whereas reflection points in sonobuoy data are linearly arrayed (Fig. 1). Each method has advantages. Common depth point reflections tend to be more coherent than sonobuoy data because of greater uniformity in the vicinity of the reflection points [Mayne, 1962]. But sonobuoy data have the advantage of greater array lengths, which permit recording of wide-angle reflections and refracted arrivals as well as narrow-angle reflections. The costs and the cumbersomeness of the large multireceiver arrays used to collect CDP data limit the array length to a few kilometers; this allows the recording of only narrow-angle reflections in deeper water. On the other hand, the cost of one sonobuoy record equals or exceeds the cost of 200 CDP interval velocity spectra.

The determination of seismic velocities from narrow-angle seismic reflection data is relatively easy when it is done routinely as a part of digitally recorded multifold seismic reflection studies. Determinations are fast and relatively low in cost and can provide detailed velocity data in which lateral changes of velocity can be observed occurring over distances too small for adequate resolution by other techniques. With these velocity estimates computed every 90 m along the ship track, one can see great detail in lateral velocity gradients. With appropriate horizontal averaging, some detail may be lost, but estimates as precise as sonobuoy estimates may be obtained. The amount of averaging necessary for good precision can be judged from the data after collection and can be varied for the purposes of different studies. The geologic potential of closely spaced velocity data appears significant.

Multifold common depth point reflection data have been widely used in shallow water and on land by the petroleum industry to determine the velocity structure but have not been widely used in the deeper waters of ocean basins. The purpose of this paper is to report the preliminary results of an investigation of CDP velocities from six marine areas of different depth and different subsurface geology. These data provide a basis for evaluating the validity of the technique in deepwater investigations.

The points considered in our investigation are the following:

1. Arrival time differences along the array (differences in normal moveout) of reflected rays become progressively smaller as depths of reflectors increase, thereby reducing the accuracy of the velocity determinations. What, then, is the accuracy of the method in the deeper ocean basins? And at what depths can the method be effectively used? To what extent can multifold reflection velocities be used oceanwide? To partially answer this question, we compare multifold velocity results with coincident refraction results obtained from the Gulf of Mexico, the Caribbean Sea, the Southeast Georgia Embayment Shelf, the Blake Plateau, the Blake Basin, and the Hatteras abyssal plain. The accuracy of oceanic refraction results has not been established unequivocably, but

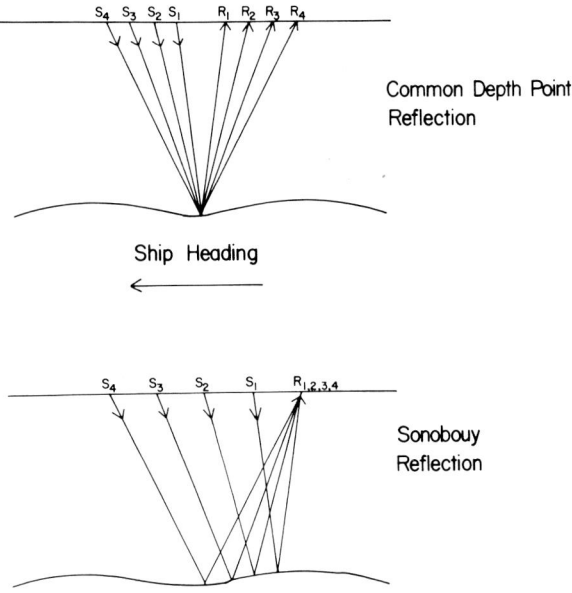

Fig. 1. Idealized common depth point and sonobuoy reflection ray diagrams. Source repetition rate and speed of the ship are synchronized in marine operations, so that successive seismic pulses reflecting from a common subsurface point are detected by receivers at regularly increasing distances from the source. A sonobuoy remains at a fixed location; hence successively received pulses have reflected from different subsurface points. Reflections from common subsurface points tend to exhibit greater coherency [Mayne, 1962].

the wide use of oceanic refraction data makes these data the best available yardstick with which to compare reflection data.

2. As was mentioned previously, narrow-angle reflection data can provide closely spaced velocity determinations, thereby permitting the mapping of lateral velocity variations within individual rock units. This is done by petroleum industry geologists and geophysicists by using land and shallow water data [e.g., Montalbetti, 1971] but has not to our knowledge been attempted with deep water. Is this technique useful in data from the ocean basins? To investigate this problem, we constructed a velocity-depth section extending from the Blake Escarpment northeastward to the Hatteras continental rise.

3. Seismic velocities in oceanic rocks may vary as a function of direction, i.e., rocks may be seismically anisotropic. Indeed, Shor et al. [1973] reported evidence of horizontal anisotropy in rocks of the Pacific crust. If seismic energy traverses rocks of the oceanic crust at one velocity in the vertical direction and at another velocity in the horizontal direction, comparison of reflection and refraction velocities should reveal such anisotropy provided that confidence levels can be established.

Method

Data and results reported herein are from horizontal or nearly horizontal reflectors. Consequently, the discussion of method is limited to the case of horizontal reflectors. Given a multilayered stratiform sequence of rock units, each layer being seismically homogeneous and having a compressional wave velocity v, the arrival time of a reflection from the nth layer can be parametrically expressed as [Slotnick, 1959]

$$t_{x,n} = 2 \sum_{i=1}^{n} \frac{z_i}{v_i \left(1 - p_{x,i}^2 v_i^2\right)^{\frac{1}{2}}} \tag{1}$$

$$x = 2 \sum_{i=1}^{n} \frac{z_i p_{x,i} v_i}{\left(1 - p_{x,i}^2 v_i^2\right)^{\frac{1}{2}}} \tag{2}$$

where t_n is the travel time for a reflection from the nth layer, x is the source to receiver distance, v_i is the velocity of the ith layer, $p_{x,i}$ is equal to $dt/dx|_{x,i}$, and z_i is the thickness of ith layer.

The interval velocity of the nth layer is v_n. Its determination from (1) and (2) is somewhat tedious.

Taner and Koehler [1969] expressed (1) and (2) as a power series:

$$t_{x,n}^2 = C_1 + C_2 x^2 + C_3 x^4 + \ldots \tag{3}$$

They note that for practical use, the first two terms of (3) approximate the results of (1) and (2) within about 2%. Equation (3) can be rewritten as

$$\bar{v}_n^2 = \sum_{i=1}^{n} t_i v_i^2 \Big/ \sum_{i=1}^{n} t_i \quad \text{at } x = 0 \tag{4}$$

where $\bar{v}_n$ is the root mean square velocity and t_i equals $z_i v_i$.

Reflection arrival times in multilayered media can be expressed as

$$t_{x,n}^2 = t_{0,n}^2 + (x^2/\bar{v}_n^2) \tag{5}$$

where $t_{x,n}$ is the arrival time of a reflection from the bottom of the nth layer at a source to receiver distance of x, and $t_{0,n}$ is the arrival time for x = 0. Combining (4) and (5) yields

$$v_n^2 = \frac{\bar{v}_n^2 t_{0,n} - \bar{v}_{n-1}^2 t_{0,n-1}}{t_{0,n} - t_{0,n-1}} \tag{6}$$

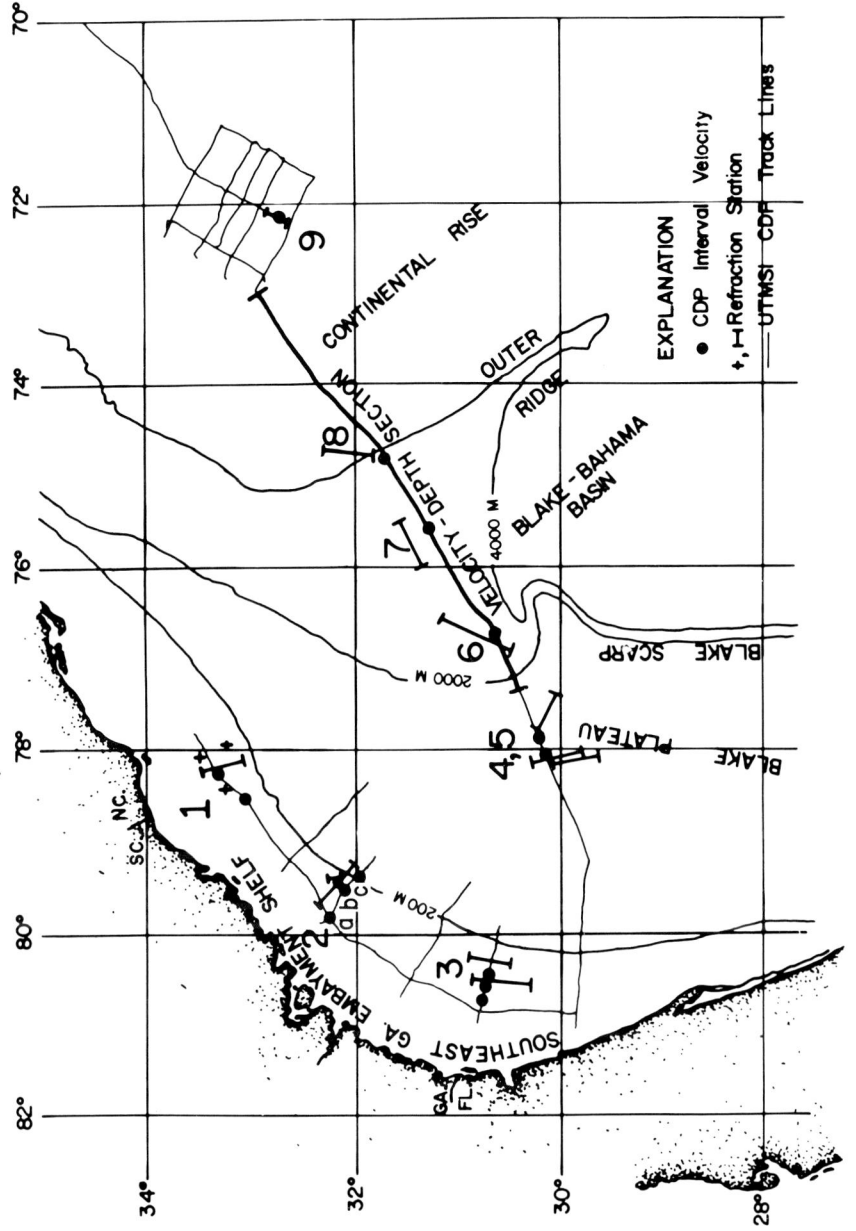

Fig. 2. Locations of velocity data from the western North Atlantic Ocean. The large numbers refer to the sites of the velocity data shown in Figures 3-6 (sites 1-9). The heavy line shows the location of the velocity-depth section (see Figure 11).

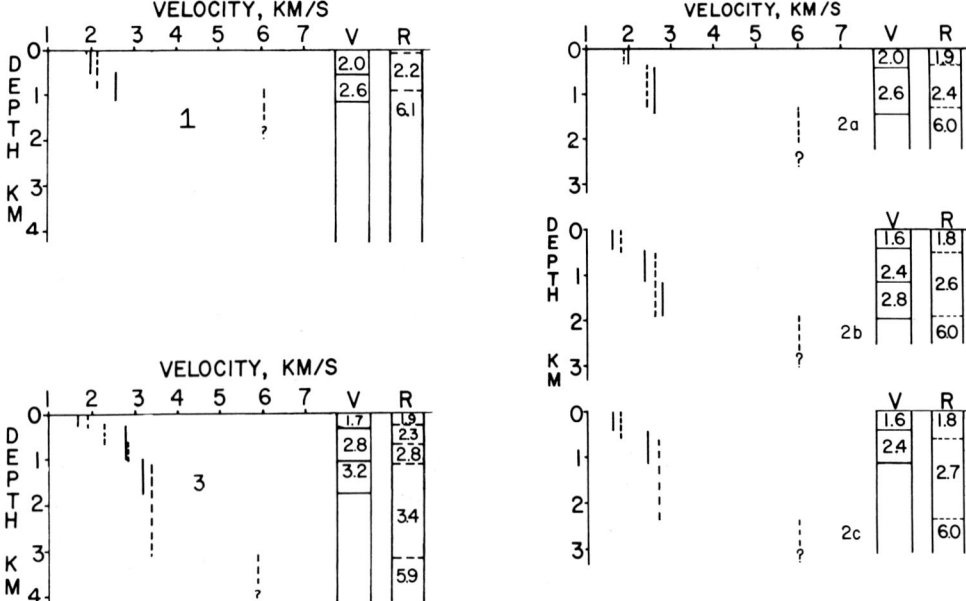

Fig. 3. Velocity data from continental shelf of southeast Georgia Embayment (sites 1-3, see Figure 2 for location). Solid lines and V indicate CDP reflection results, and dashed lines and R indicate refraction results. In site 1, V is the average of two CDP velocity spectra, R is an average of four refraction stations [Hersey et al., 1959, station 11-54; Woollard et al., 1957, stations NC-5, NC-6, NC-7]. In site 2, each V is based on one CDP velocity spectrum and R is based on refraction stations from Hersey et al. [1959] as follows: 2a, station 15-55NW; 2b, stations 15-55SE, 14-55NW and 10-54; and 2c, station 14-55SE. In site 3, V is an average of three CDP velocity spectra, and R is based on refraction stations 5-55N and 11-55 [Hersey et al., 1959].

where v_n is the interval velocity of the nth layer.

Discussion of details of algorithms for computer determination of velocity spectra $\bar{v}_n - f(t_0, z)$ is beyond the scope of this paper. For further information see Taner and Koehler [1969] and Montalbetti [1971]. It is sufficient to note that $\bar{v}_n$ is estimated by one of several correlation methods and then used to calculate v by using the relationship shown in (6). Data in Figures 2-11 were derived from semblance coefficients [Taner and Koehler, 1969].

Results and Discussion

Comparison of Velocities

Comparisons of CDP interval velocities and refraction velocities are shown in the accompanying illustrations (Figures 3-6, 8 and 10). CDP

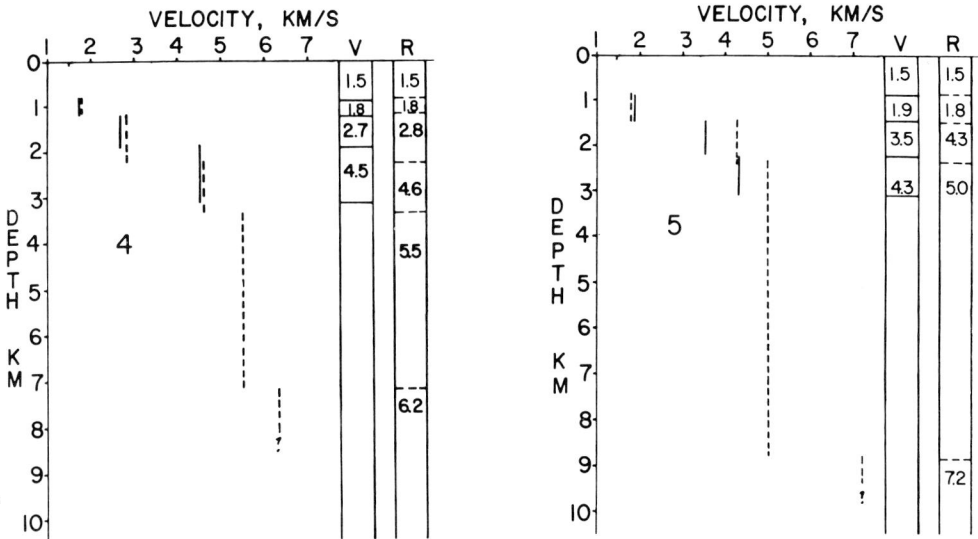

Fig. 4. Velocity data from Blake Plateau (sites 4 and 5, see Figure 2 for location). Solid lines and V indicate CDP reflection results based on one velocity spectrum for each site, and dashed lines and R indicate refraction results from Hersey et al. [1959, stations 1A-56 (site 4) and station 30-55N (site 5)].

interval velocities are plotted as solid lines, and refraction and sonobuoy reflection data are plotted as dashed lines. Velocities are shown in columns on the right of each figure. V indicates multifold reflection velocities, R indicates conventional refraction velocities, R* indicates sonobuoy refraction velocities, and R** indicates sonobuoy reflection velocities. References are given for each figure. CDP data are from the University of Texas Marine Science Institute (UTMSI) cruises IG-12 and IG-15.

It should be noted that refraction velocities represent averages over several tens of kilometers, whereas the CDP velocities reported here are averaged over 0.2 km. Refraction lines are located as much as 20 km from CDP velocity analyses. Areas where location differences may account for different results are noted.

Southeast Georgia Embayment Shelf. Figure 2 shows where data were collected, and Figure 3 gives velocities from the southeast Georgia Embayment Shelf area (sites 1-3). These data reflect a relatively thin section (1-3 km) of low velocity (1.7-3.4 km/s) Cretaceous and Tertiary sedimentary rocks overlying a strong flat-lying 'basement' reflector. This reflector appears to correspond with a high-speed refractor (5.9-6.1 km/s). Overall agreement between reflection and refraction data above basement is good. The apparent increase in velocity to the south (site 3) probably indicates a regional facies change from clastics to carbonates. The high-speed basement reflector tends to mask any deeper subbasement reflectors, if any are present, thus preventing the calculation of any subbasement interval velocities from CDP data. The nature of the basement is not known, although several possibilities have been

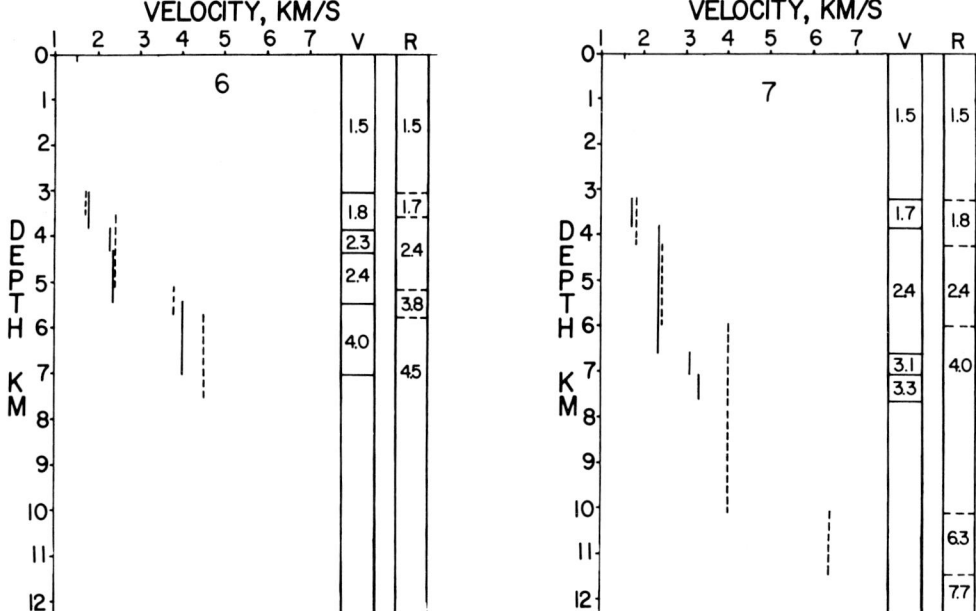

Fig. 5. Velocity data from base of Blake Escarpment and Blake-Bahama Basin (sites 6 and 7, see Figure 2 for location). Solid lines and V indicate CDP reflection velocities based on one velocity spectrum for each site, and dashed lines and R indicate refraction results from Hersey et al. [1959, station 4-56S (site 6) and station 38-55 (site 7)].

proposed, for example, crystalline basement, Triassic or older sedimentary rocks, Jurassic volcanic rocks, or Mesozoic carbonate rocks.

Blake Plateau. The Blake Plateau basically represents a thick section of high-velocity (4-5 km/s) Cretaceous and older shallow water carbonates overlain by lower-velocity (1.8-2.8 km/s) Tertiary sediments that were deposited in water depths similar to that on the plateau today. Two nearly coincident velocity analyses from the Blake Plateau, from site 4 and site 5, are shown in Figure 4. (See Figure 2 for the locations.) The data show a significant lateral velocity change in the second subbottom unit, which may reflect a local carbonate buildup. Agreement with refraction data is excellent at site 4 but less so at site 5. This latter case may indicate differences in the location of the data and thus differences in the geology, or it may be that the refraction data failed to resolve the 3.5-km/s layer. Deep reflectors are locally observed, but interval velocities have not been resolved below about 3-km subbottom in these data.

Blake-Bahama Basin and outer ridge. Data from the base of the Blake Escarpment (site 6, Figure 5) reflect low-velocity Tertiary and Upper Cretaceous rocks (1.8-2.3 km/s and 2.5 km/s, respectively) overlying Lower Cretaceous and older sedimentary rocks (4.0 km/s). Reflection and refraction results agree well to depths of 5-6 km. Below this depth the reflection data fail to distinguish the 3.8- and 4.5-km/s layers.

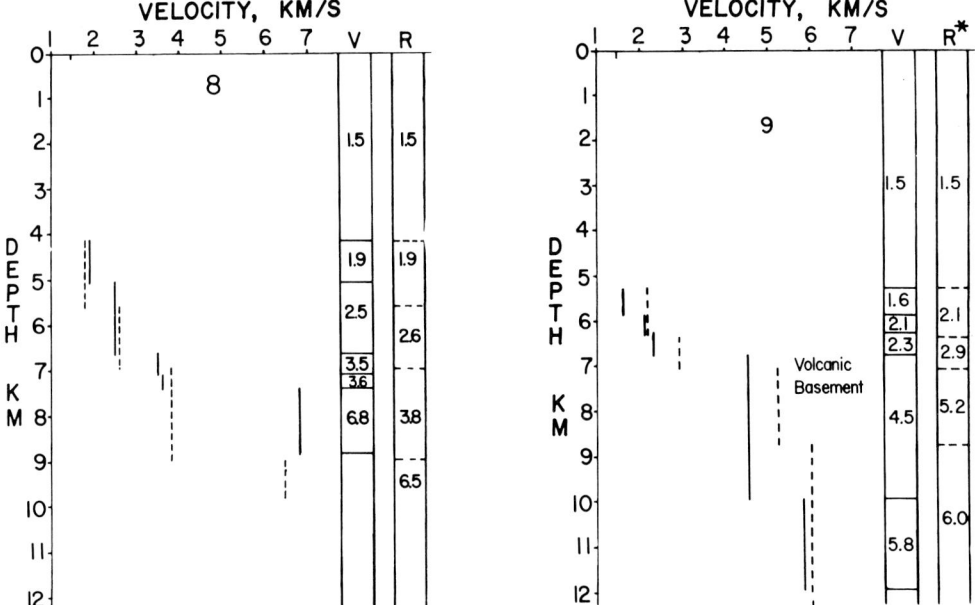

Fig. 6. Velocity data from Blake-Bahama outer ridge and lower continental rise (sites 8 and 9, see Figure 2 for location). Solid lines and V indicate CDP reflection velocities (site 8 based on one velocity spectrum and site 9 based on an average of five velocity spectra). Dashed lines, R, and R* indicate refraction results. Site 8 is from Hersey et al. [1959, station 39-59S], and site 9 is from UTMSI sonobuoy data.

The 4.0-km/s layer may simply represent an average velocity for the two refraction layers or perhaps the differences in the location of the lines.

Reflection results from farther east in the northern Blake-Bahama Basin show a thick Tertiary section (1.7-km/s layer and most of the 2.4-km/s layer), while the upper Cretaceous section has thinned to <1 km and is included within the lower part of the 2.4-km/s layer (site 7, Figure 5). The 3.1- and 3.3-km/s layers probably represent Lower Cretaceous and Jurassic rocks. Again, reflection and refraction data agree well to depths of about 6 km, but below this depth the reflection velocities are significantly lower and show a more complex velocity structure.

Data from beneath the Blake-Bahama outer ridge are in reasonably good agreement (site 8, Figure 6). The 1.9- and 2.5-km/s layers are part of a thick Tertiary sediment accumulation forming the outer ridge, while the 3.5- to 3.6-km/s layers are Cretaceous and Jurassic rocks which thin over a basement high (6.8 km/s). As the refraction station is several kilometers from the seismic line, the discrepancy there probably represents differences in basement topography. The 6.8-km/s layer exemplifies the capability of the reflection technique to resolve subbasement velocities where the basement contains reflectors.

Continental rise. Data collected from relatively deep water (5-6 km) along the continental rise reflect a thin (1-2 km) Jurassic to Holocene sedimentary sequence (layer 1, 1.6 to 2.9 km/s) overlying an irregular reflector thought to be layer 2 (site 9, Figure 6). Layer 1 can be subdivided into at least three units by using CDP data.

Overall agreement between reflection and refraction results in the sediments (layer 1) is good. Here, layer 2 can be resolved into at least two units by using CDP data. Presently, we lack an adequate data base to assess the accuracy and extent to which layer 2 velocities can be investigated by using the multifold reflection method, but the potential implications of these data are great. The agreement between reflection and refraction results for layer 2 is reasonably good considering the depth and possible differences due to location.

Gulf of Mexico. Figure 7 shows locations of the Gulf of Mexico data presented as site 1 through site 4 on Figure 8. These data reveal the velocity structure of the thick sedimentary section (6-7 km) that floors the deep western Gulf of Mexico. In general, the low-velocity units (1.7-2.6 km/s) represent middle to late Tertiary rocks, while the higher velocities represent early Tertiary through Jurassic and possibly older sedimentary rocks [Ladd et al., 1976]. The sedimentary section is too thick to allow resolution of any layers within the oceanic crust.

Site 1 velocities were calculated from relatively noisy reflection data and, consequently, lack the resolution at depth that site 2 and other Gulf of Mexico data display. The agreement with refraction data at site 1 is good to a depth of 5.3 km. Reflection interval velocities between 5.6 and 6.8 km indicate a layer with a velocity intermediate to refraction velocities.

Resolution at site 2 is more typical of the Gulf of Mexico data. Overall agreement is good at site 2. The CDP data reveal a layer between 4.7 and 6.5 km with a velocity intermediate to refraction velocities, but the CDP data do not detect the 4.0-km/s layer identified in the refraction data.

Site 3 velocities are good quality Gulf of Mexico reflection velocities. They show a velocity reversal between 8.6 and 9.1 km and exhibit relatively high velocities at depths of 8-10 km. Site 3 and site 4 reference velocities are from sonobuoy reflection [Houtz et al., 1968] rather than from refraction data. The disagreement in site 4 velocities at depths >3.8 km is probably due to real differences in velocity structure, as this area contains a mobilized salt stratum with a velocity of 4.0-5.0 km/s. The surface of the salt stratum can be seen to undulate as well as form domes in reflection profile data. Hence small differences in location can cause substantial differences in structure. The deep velocity reversals seen in site 3 and site 4 data are not uncommon in western Gulf of Mexico data.

Venezuela Basin. The Venezuela Basin velocity data are in good agreement within the thin sedimentary section above reflector B" (1.8-2.0 km/s) but only in fair agreement within the section beneath B" (site 1, Figures 9 and 10). The nature of the deeper section is complex and not well understood, and it is noteworthy that we are able to resolve velocity structure within this section. Garrison [1972], Hopkins [1973], and Saunders et al. [1973] first reported the existence of reflectors in the Venezuela Basin at depths where refraction data indicated velocities similar to those of oceanic layers 2 and 3. Proprietary multifold data

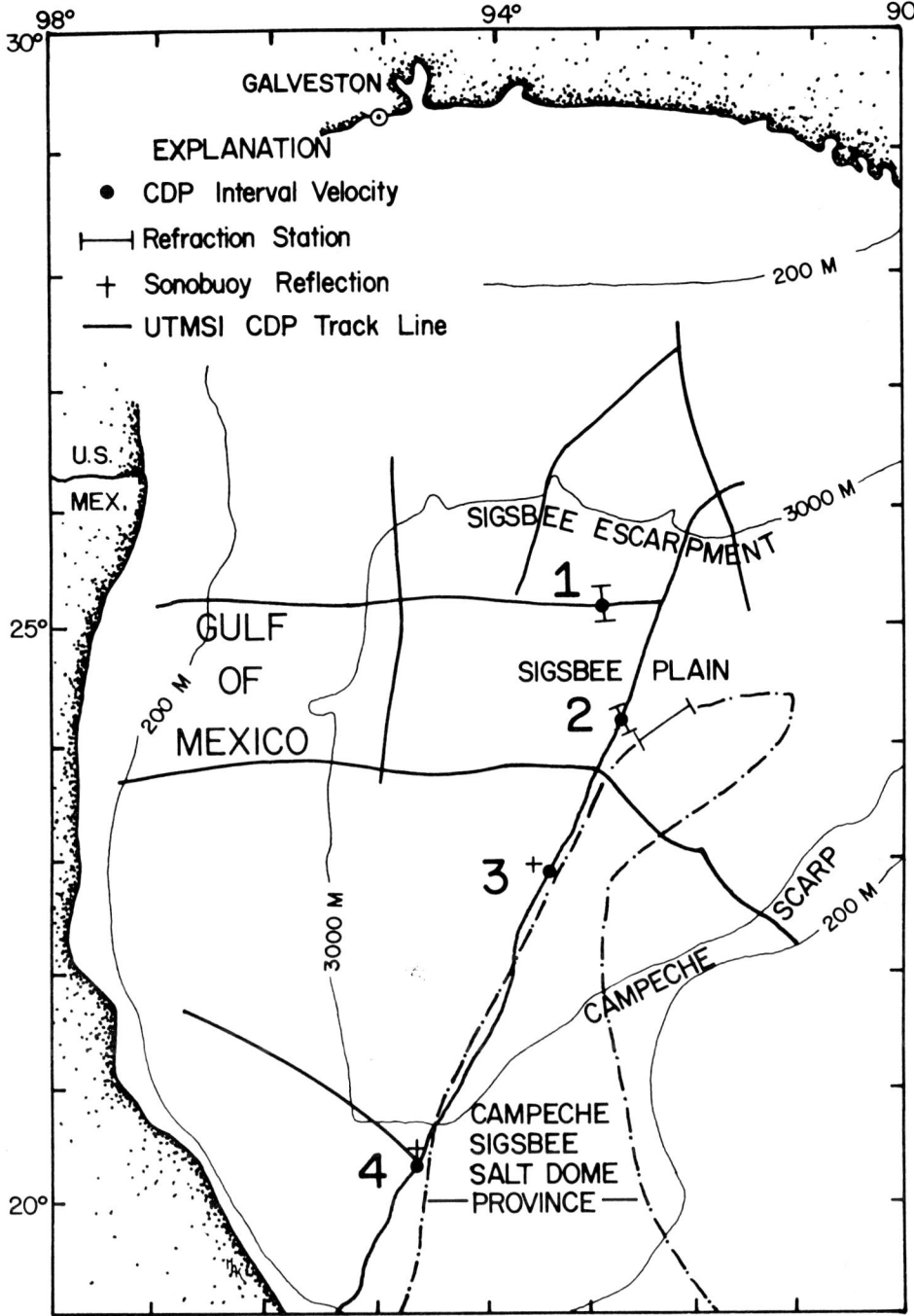

Fig. 7. Locations of velocity data from the western Gulf of Mexico. The large numbers refer to the sites of the velocity data shown in Figure 8 (sites 1-4).

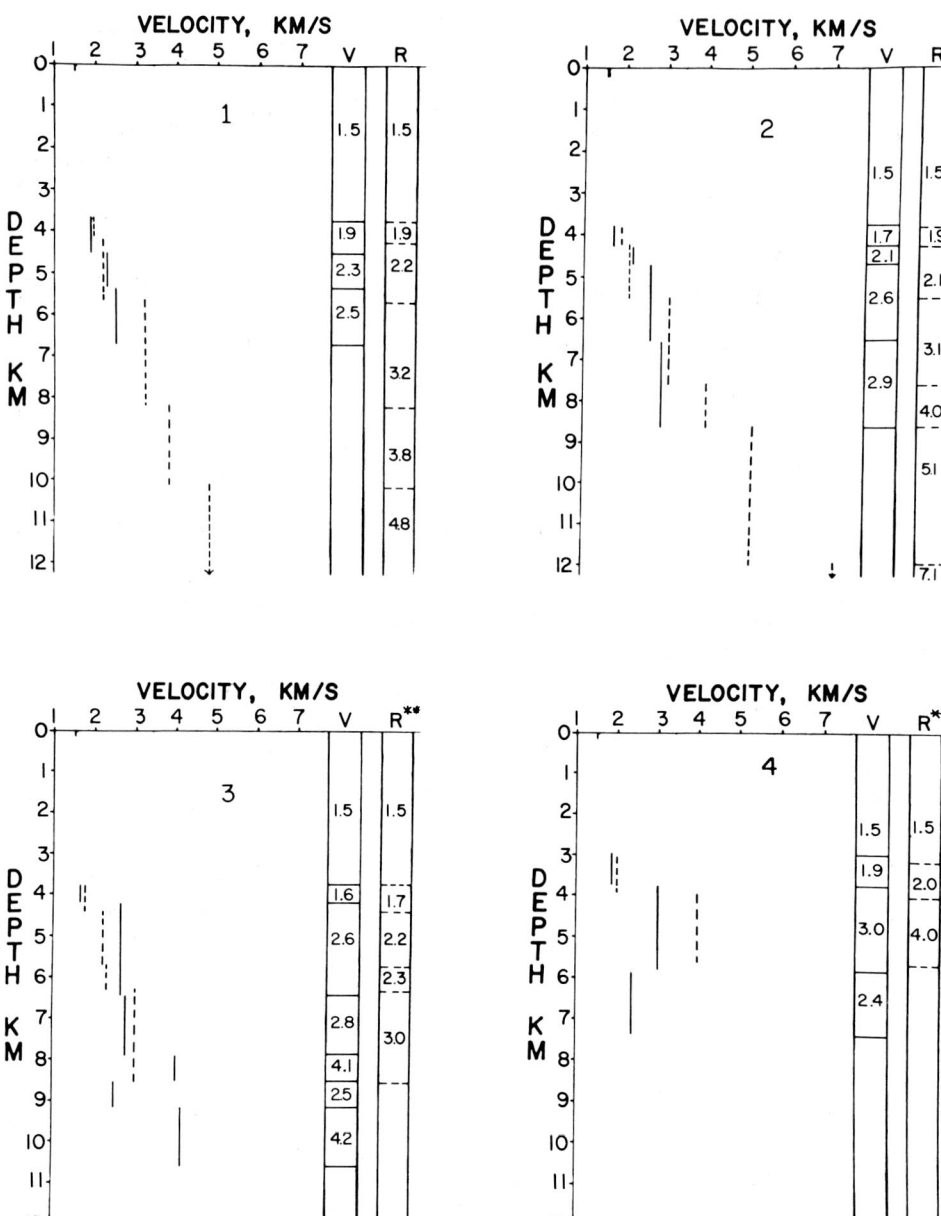

Fig. 8. Velocity data from the western Gulf of Mexico (sites 1-4, see Figure 7 for location). Solid lines and V indicate CDP reflection results based on one velocity spectrum for each site. Dashed lines, R, and R** indicate refraction results. R is from Ewing et al. [1960, station V-24 (site 1) and stations V-23 and V-22S (site 2)]. R** indicates sonobuoy reflection data from Houtz et al. [1968, sonobuoy 12 (site 3) and sonobuoy 16 (site 4)].

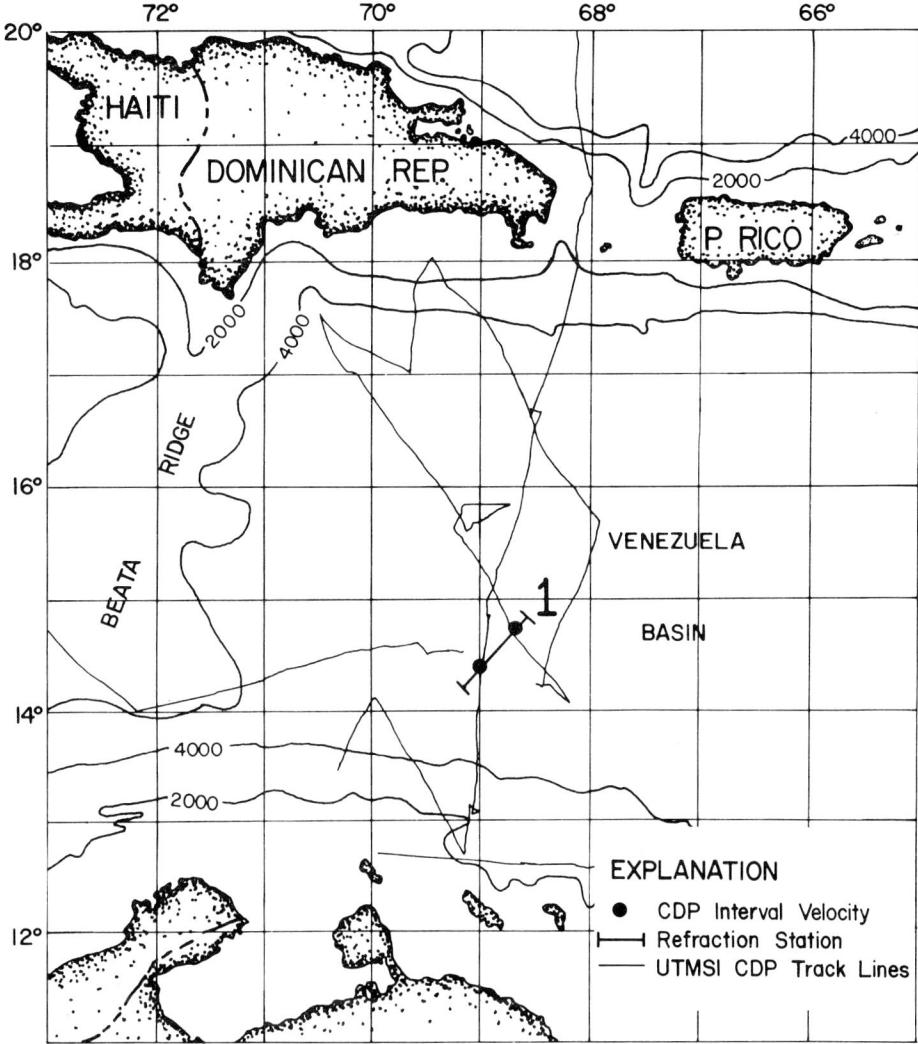

Fig. 9. Location of velocity data from Venezuela Basin, eastern Caribbean Sea. The one refers to the site of velocity data shown in Figure 10 (site 1).

and sonobuoy data further suggested that sub-B" reflectors are widespread in the Venezuela Basin [Ludwig et al., 1975]; this has subsequently been confirmed by our multifold seismic data. This has allowed us to calculate velocities for specific reflector intervals (Figure 10) which can be traced internally for varying distances within the basin.

The discrepancies between velocity analysis results and refraction results may be due to local irregularities in the velocity structure of the Caribbean which were not resolved by the refraction technique, or they may be caused by uncertainties in both methods of velocity deter-

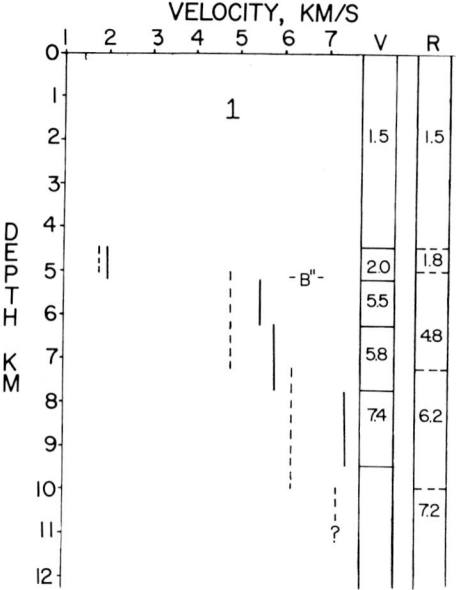

Fig. 10. Velocity data from Venezuela Basin, eastern Caribbean Sea. Solid lines and V indicate CDP reflection results based on two velocity spectra, and R indicates refraction results from Officer et al. [1957, Sutton station 4)].

mination. With velocity analyses at every common depth point, we should be able to determine the source of the discrepancies. However, we have not done this analysis yet.

Blake-Bahama Velocity-Depth Section

A detailed velocity profile is presented here to illustrate how CDP data can be used to study the velocity structure and geology of the deep ocean basin (Figure 11). The section extends for approximately 500 km northeastward across the deep western North Atlantic Ocean from the Blake Escarpment to the Hatteras continental rise (See Figure 2 for the location). It was compiled from 46 velocity analyses determined at approximately 10-km intervals along UTMSI seismic line BB-1 (Figure 2).

The section represents 2-6 km of Jurassic to Holocene deepwater sediments (oceanic layer 1) overlying an irregular basement assumed to be volcanic rocks of oceanic layer 2. The units just above horizon A with velocities of 1.7-2.5 km/s probably represent Eocene and younger age rocks, while the older units with velocities of 2.8-4.9 km/s are largely Cretaceous and older (T. H. Shipley et al., unpublished manuscript, 1977). Intervals represent significant geologic units bounded by major reflectors. Up to eight units can be defined in this manner. Within the sedimentary sequence, interval velocity data from three geologic areas were averaged separately (Blake Basin, Blake outer ridge, and Blake rise) and are presented with their standard deviations on the

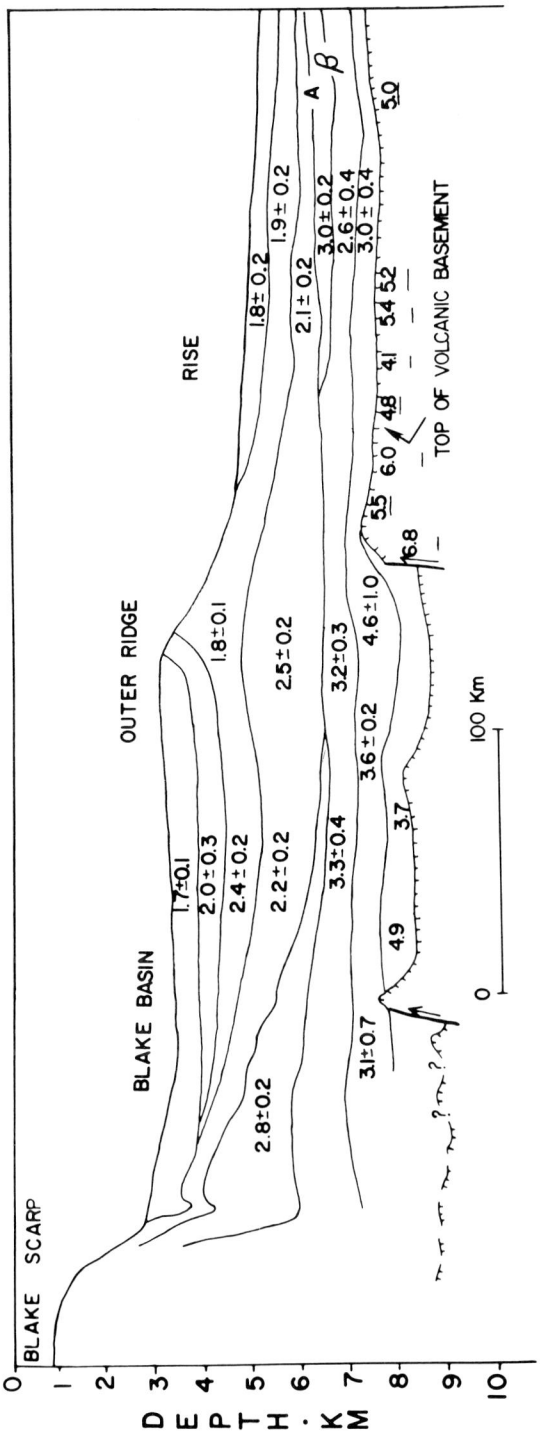

Fig. 11. Velocity-depth section from Blake Escarpment northeast to continental rise that summarizes results of 46 velocity spectra. Average interval velocities and standard deviations are shown for selected reflector intervals. Data are from T. H. Shipley et al. (unpublished manuscript, 1977).

section. In the upper part of the sequence the standard deviations are generally quite small (about 10%) suggesting that the method is quite reliable and internally consistent to depths of about 6 km. It also suggests that variations in velocity within units probably are valid and reflect real changes in the geologic character of the unit. Below this depth the data are somewhat less reliable, although the overall velocity structure and interval velocity changes are evident. The three velocity reversals beneath the Blake Basin and below the continental rise are noteworthy; they apparently reflect geologically significant events. Velocities at the base of the sedimentary section without standard deviations are based on single-velocity spectra.

Beneath the continental rise and within a basement high the volcanic basement (layer 2) contains significant reflectors that dip gently to the east. Velocity values represent individual velocity determinations for intervals between the top of layer 2 and various layers within the unit. Although there is considerable scatter in the values, they represent velocities consistent with velocities previously reported for layer 2 [Houtz and Ewing, 1976]. The scatter may be due to the lack of accuracy of the method at these depths, dipping reflectors, diffractions, and actual differences in velocity within layer 2 or a combination of these. Regardless, these data show that apparently meaningful velocities can be obtained down to depths of 8-9 km in the deep oceans.

Conclusions

1. Velocities calculated from multifold seismic reflection data are in overall good agreement with velocities calculated from refraction and sonobuoy reflection in 15 localities in the western North Atlantic Ocean, the western Gulf of Mexico, and the Caribbean Sea.

2. Multifold reflection velocities resolve significant detail in the marine sedimentary sequences of layer 1 reported here, including velocity reversals. Lateral velocity variations and the relation of velocities to specific reflector intervals have significant potential as tools for the investigation of geologic problems.

3. Multifold reflection data have provided some velocity data from layers 2 and 3(?). The ability to resolve layer 2 and layer 3(?) velocities from multifold reflection data mainly depends on the quality of reflectors within layers 2 and 3. Where reflectors are relatively strong, meaningful velocities can be obtained. Where reflectors are weak or absent, velocities are generally not resolved. If reflectors are sufficiently widespread in layers 2 and 3, the potential of multifold seismic reflection for the study of the velocity structure of these layers is great.

4. The question of anistropy is unanswered in these data. Several reasons, including an inadequate data base, the data scatter, or the lack of anistropy, could account for this result.

Acknowledgments. Financial support for this project was provided by National Science Foundation grants GX 42351, DES 75-06249, and OCE76-14620 and U.S. Geological Survey Office of Marine Geology contract 14-08-0001-14942. Additional financial and material support was provided by Cecil and Ida Green, Exxon Production Research, Chevron Oil Company, Continental Oil Company, Mobil Oil Corporation, Shell Oil

Company, Texaco Inc., the University of Texas Medical Branch Service Computation Center, Western Geophysical Company, and Texas Instruments, Inc. We also would like to thank Otis Murray and the crew of the R/V Ida Green. This is contribution 203 of the Geophysics Laboratory, University of Texas Marine Science Institute.

References

Ewing, J., J. Antoine, and M. Ewing, Geophysical measurements in the western Caribbean Sea and in the Gulf of Mexico, J. Geophys. Res., 65, 4087-4126, 1960.

Garrison, L. E., Acoustic reflection profiles, eastern Greater Antilles, U.S. Geol. Surv. Doc. GD-72-004, Natl. Tech. Inform. Serv., Dep. Commer., Springfield, Va., 1972.

Hersey, J. B., E. T. Bunce, R. F. Wyrick, and F. T. Dietz, Geophysical investigation of the continental margin between Cape Henry, Virginia, and Jacksonville, Florida, Geol. Soc. Amer. Bull., 70, p. 437-466, 1959.

Hopkins, H. R., Geology of the Aruba Gap Abyssal Plain near DSDP site 153, in Initial Reports of the Deep-Sea Drilling Project, vol. 15, edited by N. T. Edgar, J. B. Saunders, et al., pp. 1039-1050, U.S. Government Printing Office, Washington, D. C., 1973.

Houtz, R., and J. Ewing, Upper crustal structure as a function of plate age, J. Geophys. Res., 81, 2490-2498, 1976.

Houtz, R., J. Ewing, and X. Le Pichon, Velocity of deep-sea sediments from sonobuoy data, J. Geophys. Res., 73, 2615-2641, 1968.

Ladd, J. W., R. T. Buffler, J. S. Watkins, and J. L. Worzel, Deep seismic reflection results from the Gulf of Mexico, Geology, 4, 365-368, 1976.

Le Pichon, X., J. Ewing, and R. E. Houtz, Deep-sea sediment velocity determination made while reflection profiling, J. Geophys. Res., 73, 2597-2614, 1968.

Ludwig, W. J., R. E. Houtz, and J. I. Ewing, Profiler-sonobuoy measurements in Colombia and Venezuela basins, Caribbean Sea, Amer. Ass. Petrol. Geol. Bull., 59, 115-123, 1975.

Mayne, W. H., Common reflection point horizontal data stacking techniques, Geophysics, 28, 927-938, 1962.

Montalbetti, J. F., Computer determination of seismic velocities - A Review, J. Can. Soc. Explor. Geophys., 7, 32-45, 1971.

Officer, C. B., J. I. Ewing, R. S. Edwards, and H. R. Johnson, Geophysical investigations in the eastern Caribbean, Venezuelan Basin, Antilles Island Arc, and Puerto Rico Trench, Geol. Soc. Amer. Bull., 68, 359-378, 1957.

Saunders, J. B., N. T. Edgar, T. W. Donnelly, and W. W. Hay, Cruise synthesis, in Initial Reports of the Deep-Sea Drilling Project, vol. 15, edited by N. T. Edgar, J. B. Saunders, et al., pp. 1077-1112, U.S. Government Printing Office, Washington, D. C., 1973.

Shor, G. G., R. W. Raitt, M. Henry, L. R. Bentley, and G. H. Sutton, Anisotropy and crustal structure in the Cocos Plate, Cons. Nac. Cienc. Tech., 13, 337-362, 1973.

Slotnick, M. M., Lessons in Seismic Computing, edited by Richard A. Geyer, p. 194, Soc. Explor. Geophysicists, Tulsa, Okla., 1959.

Taner, M. T., and F. Koehler, Velocity spectra - digital computer derivation and applications of velocity functions, Geophysics, 34, 859-881, 1969.

Woollard, G. P., W. E. Bonini, and R. P. Meyer, A seismic refraction study of the sub-surface geology of the Atlantic Coastal Plain and continental shelf between Virginia and Florida, technical report, contract N7ONR-28512, 128 pp., Dep. of Geol., Univ. of Wis., Madison, 1957.

A NEW MODEL OF THE CONTINENTAL CRUST

Stephan Mueller

Institut für Geophysik, Eidgenössische Technische Hochschule
(Swiss Federal Institute of Technology)
CH-8093 Zurich, Switzerland

Abstract. Overwhelming evidence suggests that a simple two-layered crust is no longer sufficient to explain available geophysical observations. New travel time and amplitude data obtained in seismic refraction experiments, supplemented by subbasement echoes at near-normal incidence, now permit elucidation of the fine structure in a more comprehensive and unified manner. A new model of the continental crust is advanced. The significant features of the new velocity-depth function are as follows. (1) The velocity gradient in the top part of the crystalline basement beneath the sedimentary cover is less steep than has been indicated by laboratory measurements. Increasing pore pressure in the interstices of low-porosity basement rocks is considered to be the most likely explanation for this phenomenon. (2) The sialic low-velocity zone in the upper part (5- to 15-km depth range) of the crust has been associated with a semicontinuous laccolithic zone of granitic intrusions having lower velocities and densities but a higher attenuation than the surrounding basement rocks. Again it must be assumed that it is the water content and the pore pressure in these granites, not the temperature, which play the central role in the observed processes. (3) The middle (10- to 25-km depth range) and lower (20- to 35-km depth range) crustal layers are characterized by surprisingly low average velocities, in contrast to classical models of the earth's crust. In the new model the 'Conrad discontinuity' has degenerated into a thin possibly laminated high-velocity layer of only a few kilometers in thickness which will permit the 'tunneling' of low-frequency waves as observed in near-earthquake studies. (4) The crust-mantle transition zone must be relatively sharp in order to give rise to the observed multiply reflected wide-angle reflections from the Mohorovičić discontinuity. Thin alternating high- and low-velocity lamellas probably make up the uppermost part of the mantle; thus a reasonable explanation for the observed normal incidence reflections and the P_n velocity anisotropy is provided. Although not all block structures within each of the continents of the world show the same details in their velocity-depth functions, the lateral consistency in structure is quite astonishing. The seismic model of the crust presented here results from changes in chemical composition and physical state that occur in a highly varying regime of pressure, temperature, and water content.

An apparent 'layering' of the continental crust can therefore be understood only if quasi-horizontal 'metamorphic fronts' are the dominant processes in the formation of the continental crust.

1. Introduction

The first data on seismic wave velocities in the continental crust were obtained during the first few years of this century by Mohorovičić [1910]. He used records of earthquakes obtained at continental stations within a few hundred kilometers of the epicenter. The velocity-depth function derived from these observations indicated that below a sedimentary cover of varying thickness a layer exists whose compressional (P) wave velocity is about 5.6 km/s and whose corresponding shear (S) wave velocity amounts to approximately 3.3 km/s. The term 'granitic' layer was subsequently introduced by many geophysicists who tacitly assumed that this crustal layer consists primarily of granitic material. On the basis of the same set of observations an abrupt velocity increase was considered by Mohorovičić to occur at a depth of about 60 km.

The 'Mohorovičić (or M) discontinuity' is now generally defined as the more or less sharp interface which separates the crust from the underlying mantle. P waves which have penetrated below this boundary are usually associated with velocities of 8.0 km/s and higher, and the corresponding S waves travel with velocities of 4.4-4.7 km/s through the uppermost mantle. Additional crustal phases were identified by Conrad [1925]. The signals, whose velocities were about 6.5 km/s for the P waves and 3.6 km/s for the S waves, have been associated with a lower crustal layer of 'basaltic' or 'gabbroic' composition. The existence of a so-called 'Conrad (or C) discontinuity' at a depth of about two thirds of the crustal thickness and the continuity of any systematic intermediate crustal layering have been seriously doubted by seismologists for quite some time (see, for example, Båth [1961]).

Two sets of P wave arrivals through the uppermost crust of southwestern Germany were first analyzed by Rothé and Peterschmitt [1950] and Förtsch [1951] in their studies of the 1948 Haslach explosion. In an attempt to explain the two groups of signal onsets these authors deduced the presence of two different materials. The topmost layer with a P wave velocity of 5.6-5.9 km/s classified as granitic by Förtsch [1951], was considered to be separated from a layer of 'diorite' with a P velocity of approximately 6.0 km/s. The interface between these two hypothetical layers (at depths between 2.4 and 6.0 km) was subsequently called the 'Förtsch (or F) discontinuity' by Reich [1957], Schulz [1957], and others. This rather arbitrary definition has created much confusion, because it places the F discontinuity at different depths depending on the method of crustal exploration used.

Over the past 25 years, reliable travel time and amplitude data of seismic waves produced by artificial explosions have accumulated, and velocity-depth functions for the earth's crust can now be determined in much more detail. With the aid of synthetic seismogram sections it is now possible to discriminate between different models which, for instance, on the basis of travel time observations alone, were considered to be equivalent by Tuve et al. [1954]. They stated unmistakably that P velocity-depth functions of types E, F, G, and H in Figure 1 did not fit their seismic refraction data. A one-layer crust of type A left something to be desired, but the simple gradient models of types B, C, and D fit their experimental data equally well, and model C was finally designated as the preferred solution. They concluded that the P wave velocity increased gradually with depth in the crust and that there was no need to introduce layering. At a depth of about 30 km in the Maryland-Virginia region an abrupt velocity change seems to occur within a transition zone of less than 250 m in thickness. This sharp zone of transition is identical to the M discontinuity defined above.

It is the purpose of this paper to demonstrate that there is a considerable variation of the seismic velocities with depth in the crust which can be delineated if travel time and amplitude information obtained by seismic refraction and reflection surveys is utilized in a unified manner. The results presented lead to a consistent new picture of crustal structure, at least in those regions where extensive seismic studies have been carried out. Heacock [1971] has summarized the areas where more detailed velocity-depth functions were obtained, which included velocity inversions of type H in Figure 1. Since then, two regions in particular have been investigated in much more detail: central Europe from the northern margin of the Alps to the northern apex of the South German Triangle (near 51°N latitude) and the Basin and Range province in western North America. The examples discussed in this paper will primarily be taken from recent seismic studies of these two major structural units and their marginal transition to adjacent structures. Striking similarities in the crustal structure will become apparent when the essential features are compared.

In the eastern part of the Basin and Range province, detailed seismic refraction studies have been carried out more recently by Braile et al. [1974] and Keller et al. [1975] which supplemented earlier measurements in that area by Berg et al. [1960] and Eaton et al. [1964]. Because of the excellent quality of the U.S. Geological Survey (USGS) profile Delta-W [Eaton et al., 1964], it has been reinterpreted by Mueller and Landisman [1971] and Landisman et al. [1971] and again by Müller and Mueller [1972]. In these interpretations, first and later refracted and reflected seismic arrivals, critical points, and relative amplitude information are used to determine the fine structure of the crust.

Simultaneously, in central Europe a systematic approach to the unified interpretation of crustal seismic data was attempted by several authors. Fuchs and Landisman [1966] made a first combined travel time interpretation of several refraction profiles in the South German Triangle. This crustal block is tectonically relatively undisturbed, Triassic rocks covering most if its surface and Jurassic sediments outcropping along its eastern and southern margins. It is bounded on the east by the Bohemian Paleozoic massif, on the south by the Molasse basin of the northern foreland of the Alps, and on the west by the Rhinegraben. Five years later the key profile Hilders-S was reinterpreted by Fuchs and Müller [1971], and recently, it has been discussed in more detail by Aichele [1976] and Müller and Fuchs [1976]. The

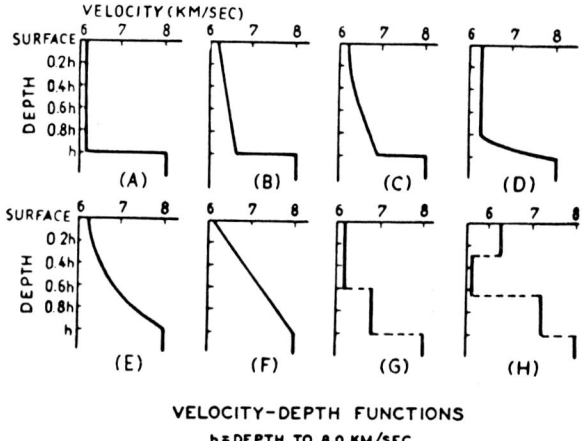

Fig. 1. Various P velocity-depth functions for the continental crust considered in the past [after Tuve et al., 1954]. For discussion of the different models, see text.

western half of the Molasse basin has been studied extensively by Emter [1971, 1976] and Harcke [1972]; this work ultimately resulted in depth contour maps of the F, C, and M discontinuities for that region. At the same time the systematic survey of the Rhinegraben rift system, which forms the western limit of the area under discussion, was continued. The results to date are summarized in papers by Mueller et al. [1973], Mueller and Rybach [1974], Meissner and Vetter [1974], the Rhinegraben Research Group for Explosion Seismology [1974], Edel et al. [1975], Meissner et al. [1976], and Prodehl et al. [1976].

2. Evidence for Fine Structure Within the Crust

In Figure 2c the intermediate part (70 < Δ < 200 km) of the record section is shown for the refraction profile Böhmischbruck-SW, which starts at a quarry near the western margin of the Bohemian massif, runs southwestward along the Franconian-Swabian threshold with outcrops of Jurassic sediments, and passes through the meteorite crater of the 'Nördlinger Ries.' Several attempts have been made in the past to deduce velocity-depth functions for that region on the basis of seismic refraction and wide-angle reflection data [Fuchs and Landisman, 1966; Meissner, 1967a; Giese, 1968; Bram and Giese, 1968].

The number of solutions offered can be narrowed down considerably if subbasement echoes obtained in a special digital reflection survey [Angenheister and Pohl, 1969, 1971, 1976] are taken into account. In Figure 3 (right) the original data of the 9-km-long continuous section are displayed [Dohr, 1972; Dohr and Meissner, 1975]. This profile was recorded with threefold coverage, a mean shot point distance of 240 m, a recording spread length of 1380 m, and a distance of 60 m between the seismometer groups. In general, charges of 12 kg were used. The

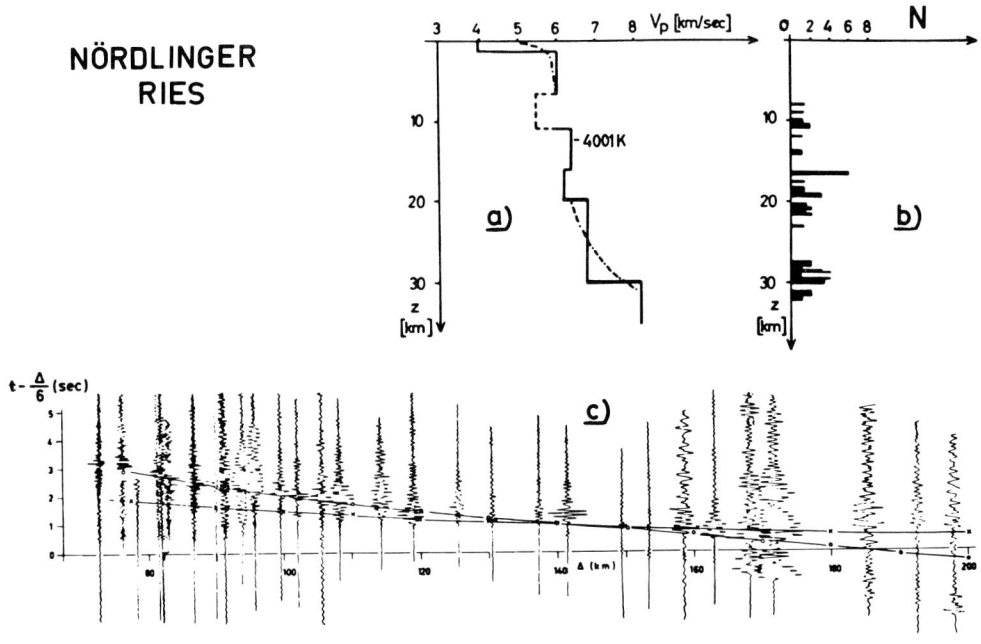

Fig. 2. (a) P velocity-depth functions for the refraction profile Böhmischbruck-SW in the vicinity of the Nördlinger Ries in southern Germany. The solid line represents model 4001K of Emter [1971]; the dash-dot line is after Giese [1968]. (b) Depth distribution of number N of reflecting elements per 0.1-s time interval to the west of the Nördlinger Ries outside the crater [after Angenheister and Pohl, 1971, 1976]. (c) Seismogram section (70 < Δ < 200 km) for the Böhmischbruck-SW profile with calculated reflection hyperbolae for the discontinuities at depths of 20 km (crosses) and 30 km (circles) based on the P velocity-depth model 4001K of Emter [1971] shown in Figure 2a.

dominant frequency of the reflected signals was about 18-20 Hz. In order to emphasize the significant events in the record section the reflecting elements were redrawn in Figure 3 (left) by Angenheister and Pohl [1971, 1976]. Roughly three main reflection bands can be discerned with echo times of 3-5, 6.5-7.5, and 9-10 s. Almost no reflections are found for echo times between 11 and 14 s, i.e., up to the longest recording time.

A histogram of the number N of reflections recorded per 0.1-s time interval as a function of depth is plotted in Figure 2b. The dominant echoes are quite comparable to those obtained in the Molasse basin by Liebscher [1962, 1964] and further to the north by Hehn [1964] (see also Fuchs and Landisman [1966]). In any unified inversion process these dominant echoes have to be brought into agreement with the corresponding refraction results.

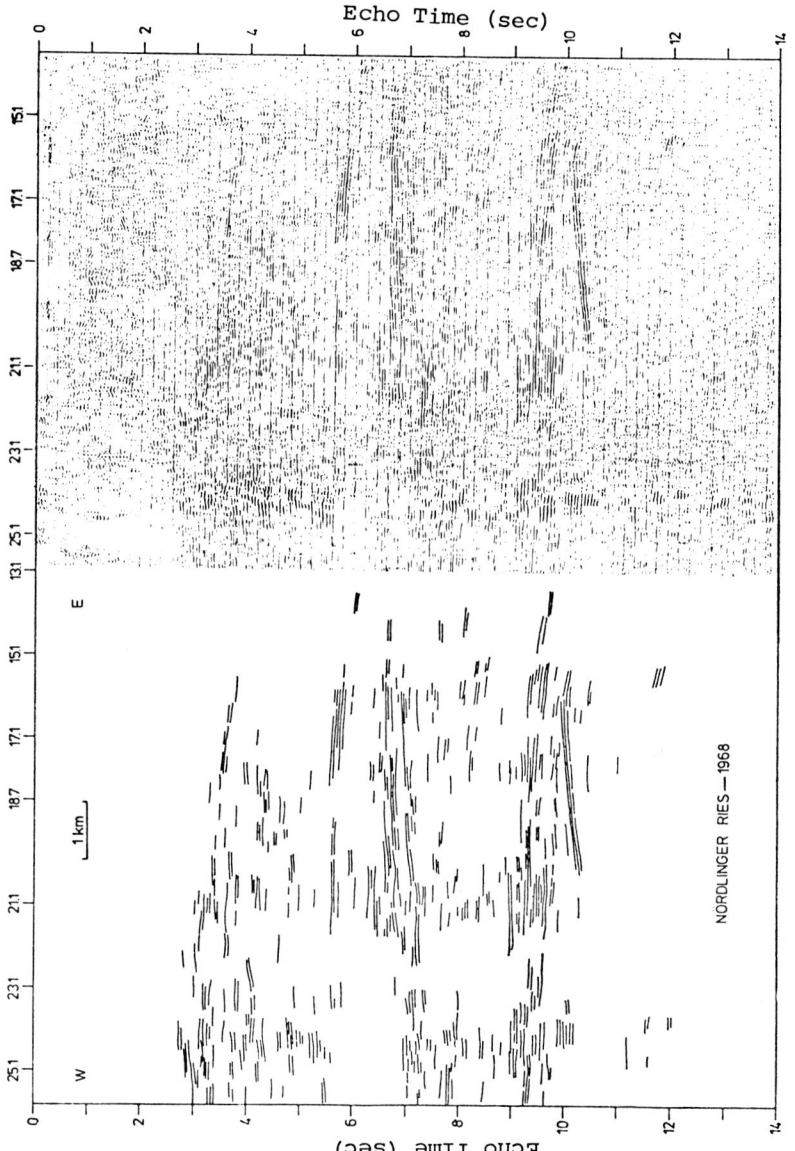

Fig. 3. Seismic reflection section for a 9-km-long continuous profile west of the Nördlinger Ries. For technical details, see text. (Right) Digital play-out with static and dynamic corrections as well as stacking and optimum filtering applied. Constant band-pass filtering is in the range 15-70 Hz. Sampling interval 4 ms [after Dohr, 1972]. (Left) Schematic diagram emphasizing the significant reflection events in the record section. The echo time of 0 s corresponds to an altitude of 400 m above sea level [after Angenheister and Pohl, 1971, 1976].

Except for the velocity-depth function (model 4001K in Figure 2a) deduced by Emter [1971], none of the published models can explain all three groups of deep crustal reflections in Figure 2b. Emter's model also provides a good fit to the wide-angle reflections from the discontinuities at depths of 20 km (C) and 30 km (M) as shown in Figure 2c. Figure 3 clearly indicates that the C and M discontinuities are not just first-order velocity jumps but rather complex transition layers, whose tops lie at depths of about 19 and 27 km, respectively, and whose thicknesses are about 2-3 km. The P wave velocity in the M transition zone increases from roughly 6.8 km/s to slightly over 8.0 km/s, in good agreement with the conclusions reached by Müller and Fuchs [1976] for the Hilders-S profile. This line intersects the Nördlinger Ries between about 180- and 200-km distance from the shot point and therefore permits an independent check on the structure of this M transition layer.

The foregoing discussion has sufficiently justified the need for a more thorough analysis of the various intracrustal features which have evolved by combining seismic refraction and reflection results. In the next few sections the significant features of the velocity-depth function in Figure 2a will be described one by one in more detail.

3. Velocity Gradient at Top of Basement

In areas where there is a section of sedimentary rocks near the surface, as is generally the case, the effects of the velocity gradient in the uppermost part of the crystalline basement are superimposed on the velocity changes because of the change in lithology. The in situ velocity-depth function for the upper part of the basement should therefore be determined in areas where the crystalline rocks are exposed at the surface. Seismic refraction profiles which stay entirely on the same type of basement rocks lend themselves to this kind of investigation.

Giese [1963, 1968] used this type of approach by accurately recording at closely spaced stations the travel times of the direct wave P_g which had traversed the top region of the crystalline basement. He then applied the Herglotz-Wiechert inversion method. The results obtained by Giese in 1963 for the first 24 km of the Böhmischbruck-Eschenlohe profile are depicted by crosses in Figure 4b (see also Landisman et al. [1971, Figure 2]). These data were supplemented in 1968 by scanty observations along the Voggendorf-SE profile. Their inversion led to velocities increasing with depth (circles in Figure 4b), reaching a maximum of 6.25 km/s at a depth of 6.5 km. For comparison, the change in P velocity due to increasing pressure and temperature comparable to conditions in the continental crust is shown for a typical granite (solid curve in Figure 4b) as obtained by Birch [1960, 1961] in laboratory experiments. According to Nur and Simmons [1969] the difference between Giese's field determinations and Birch's laboratory measurements could easily be explained by a different degree of water saturation of the cracks in the crystalline rocks. Smithson and Shive [1975] concluded that field measurements of compressional wave

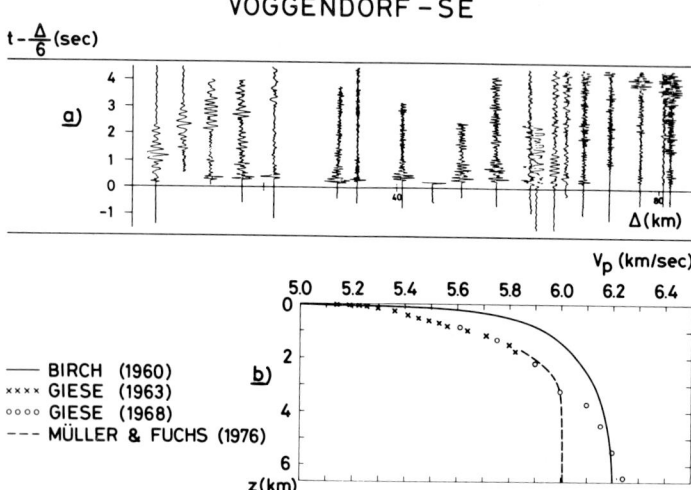

Fig. 4. Velocity gradient in the uppermost part of the crystalline basement. (a) Seismogram section ($0 < \Delta < 85$ km) for the Voggendorf-SE profile [after Wolber, 1968]. (b) P velocity-depth functions for the upper part of the basement. For a detailed explanation, see text.

velocities in common crystalline rocks correspond well with laboratory measurements but that shallow P velocities in the field are much higher than older laboratory measurements.

In contrast to these findings, practically all interpretations of refraction profiles in central Europe and the eastern Basin and Range province of North America gave P_g velocities between 5.8 and 6.0 km/s at depths of a few kilometers beneath the sediments. Müller and Fuchs [1976] have reopened this question for the Voggendorf-SE profile (Figure 4a, as prepared by Wolber [1968]) by searching in a trial and error procedure for a velocity-depth function which both satisfies the admissible travel time correlations and matches the Pg amplitude-distance curve derived by Giese [1968] as closely as possible. The only model (M3) fulfilling both criteria leads to an asymptotic Pg velocity of 6.0 km/s which is reached at a depth of about 3 km (dashed curve in Figure 4b). Müller and Mueller [1972] arrived at a similar result for the Delta-W profile on the basis of travel time and amplitude considerations (see Figure 6 and V_p model N7 in Figure 7). Thermal lowering of the Pg velocity caused by increased heat flow (> 1.5 HFU (1 HFU = 1 μcal cm^{-2} s^{-1})) may be the explanation for the observed relatively low seismic velocities, as was suggested by Smith et al. [1975]. For the Bohemian massif this explanation must, however, be ruled out, since Čermák [1975] has shown that the average heat flow in the Moldanubicum does not exceed 1.2-1.3 HFU. It is more likely that an increase in pore pressure causes the decrease in the P wave velocity of crystalline rocks in the upper part of the basement [see Nur and Simmons, 1969].

4. Sialic Low-Velocity Zone

An increasing number of studies dealing with detailed interpretations of reflected and refracted phases in seismic refraction surveys have reported evidence for a low-velocity layer in the sialic part of the crust at depths ranging from 5 to 15 km whose existence had originally been proposed by Mueller and Landisman [1966]. There are several criteria which will allow the identification of a velocity inversion in the upper crust: (1) the disappearance of the P_g branch, a phenomenon that is greatly influenced by the velocity gradient and the attenuation in the uppermost portion of the crust, i.e., in the lid above the low-velocity layer (see the previous section), (2) the near-vertical and wide-angle reflections from the top of the low-velocity zone provided that the velocity decrease is abrupt enough, (3) the near-vertical and wide-angle reflections from the bottom of the low-velocity zone, and (4) the presence of the P_c phase, a head wave refracted from beneath the bottom interface of the low-velocity layer which emerges from the corresponding wide-angle reflection with increasing distance. It is typically observed with a nearly constant delay of about 1 s after the P_g onset. If the velocity contrast between the bottom and the top of the inversion zone is of the order of 0.2 km/s, P_c will overlap P_g for several tens of kilometers distance. A negligible contrast will lead to the appearance of a 'shadow zone' in distance between P_g and P_c, which has been observed in earthquake intensity studies.

A very valuable guide to the interpretation of crustal refraction profiles has been compiled by Braile and Smith [1975]. It consists of a collection of synthetic seismogram sections for different features in the velocity-depth function. Specifically, record sections were computed for models with a low-velocity layer in the upper crust in order to find possible means of identification of this inversion zone from observed reflected and refracted arrivals.

If for instance, reflections from the top boundary of the low-velocity layer are observed, it can be safely stated that this boundary must be relatively sharp. In most of the refraction profiles shot in central Europe and western North America a reflection of this type is found. As has been pointed out by Mueller [1970] and Landisman et al. [1971], a semicontinuous laccolithic zone of granitic intrusions could well be the most probable explanation for the sialic low-velocity zone in the upper crust.

There is some direct evidence for this suggestion in places where substantial vertical uplift has occurred. In Figure 5 a photograph by Labhart [1975] is shown of the summit region of the Aletschhorn (4195 m above sea level) in the Aar massif of Switzerland. The dark rocks near the top of the mountain are gneisses, and the lighter material underneath is granite which must have intruded into the gneisses. During the subsequent uplift and erosion the sharp contact has been exposed. This example demonstrates one of the few cases where the roof on top of granitic intrusions has been preserved. Traces of probably the same phenomenon can be seen in Figure 3 at echo times of around 3 s, or in Figures 4a and 4b in the paper by Mueller et al.

Fig. 5. Summit region of the Aletschhorn (4195 m above sea level) in the Aar massif of Switzerland (photograph by Labhart [1975]). The dark rocks near the mountain top are gneisses, and the lighter material underneath is granite which must have intruded into the gneisses. This example is one of the rare cases where the sharp upper boundary of the sialic low-velocity zone can be seen at the surface.

[1969], where the top of the low-velocity zone under the western flank of the Rhinegraben appears as a sequence of sharp continuous reflections with echo times of 2.5 and 3.9 s, respectively.

Smith et al. [1975] have proposed that a plausible explanation of the sialic low-velocity zone is the effect of a high temperature gradient on the velocities of upper crustal rocks. They base their argument mainly on the laboratory measurements by Hughes and Maurette [1956, 1957] and the high heat flow observed at the transition from the Basin and Range province to the Colorado plateau. It is quite true that the temperature field in the upper crust plays an important role in shaping the velocity gradients.

Mueller and Landisman [1966] found that the time delay between P_g and P_c measured in widely separated continental areas is approximately proportional to the heat flow; thus the seismic 'prediction' of heat

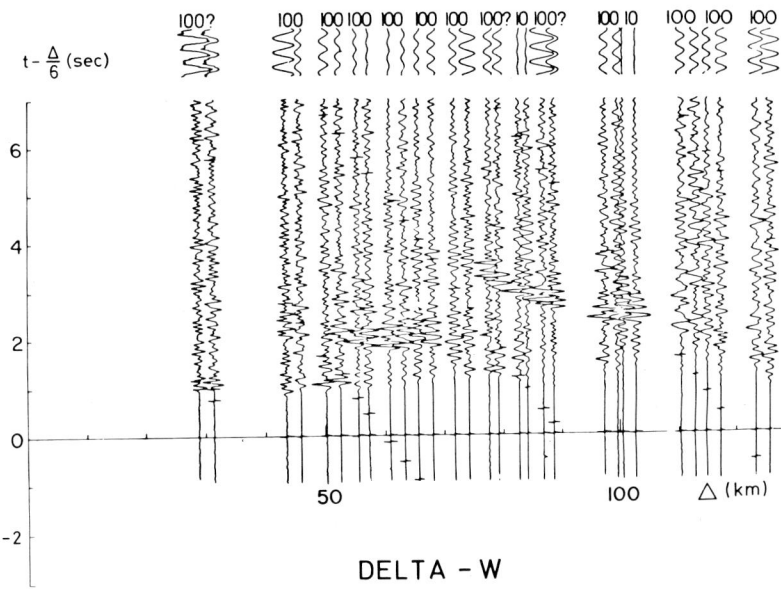

Fig. 6. Seismogram section ($0 < \Delta < 130$ km) for the Delta-W profile crossing the Utah-Nevada state line. On top of each trace, relative calibration signals are given (courtesy of C. Prodehl). Note conspicuous P_c wave group in the distance range between 50 and 70 km from the shot point.

flow values is possible. It is therefore reasonable to suppose that the observed P_c-P_g delay is intimately related to the thermal regime which prevails in the low-velocity region, but it must be stated that the negative velocity gradient determined in the laboratory for granite samples is much too small for realistic crustal temperatures and therefore cannot be taken as the only explanation of the observed P_c-P_g time delay.

A compositional change associated with a more or less abrupt velocity change, which in turn will produce near-vertical and wide-angle reflections, is the more likely explanation for the nature of the sialic low-velocity zone. An additional indication of the relative sharpness of transition is provided by the observations of earthquake-generated P waves converted to S waves [Polshkov et al., 1973]. In these investigations, signal phases are also identified which have been generated by a conspicuous horizon (termed G in the Russian literature) in the central part of the upper crust at a depth between 6 and 12 km.

In the preceding section it has been shown that amplitude information can be used to determine the gradients of the various segments of the velocity-depth function. Müller and Mueller [1972] have reinterpreted the Delta-W profile in the Utah-Nevada region by including in their study not only the travel times but also the amplitudes of the P_g and P_c wave groups. In Figure 6 the original record section of the USGS is displayed with calibration signals for each trace at the top of the section. These data permit the reconstruction of the true signal amplitudes.

With a trial and error procedure using synthetic seismogram sections the gradient in the lid above the low-velocity channel was deduced from the amplitude distance curve for P_g, and similarly, the gradient at the bottom of the channel was found by matching the amplitudes of the synthetic seismograms with the observed amplitudes of the P_c wide-angle reflections, i.e., the conspicuous wave group in Figure 6 between 50- and 70-km distance at a reduced travel time of +2 s. The velocity variation at the upper boundary of the channel was determined by the fact that it apparently produces no clear separate reflection arrivals in the record section of Figure 6. The final preferred model (N7) is shown in Figure 7 (left). It is characterized by a minimum velocity of 5.2 km/s and transition zones at both the top and the bottom whose thicknesses are 2 and 3 km, respectively. The P velocity model in Figure 7 also reproduces the P_g and P_c travel times with an accuracy which is well within observational errors.

At this point it is interesting to compare the velocity-depth function (model N7) in Figure 7 with recent statistical results of surface wave anelastic attenuation data for the stable interior of North America published by Herrmann and Mitchell [1975]. The similarity between the depth profile CUS for the shear wave quality factor Q_S and the P wave velocity-depth function N7 is striking. Although the two sets of data were obtained by completely different procedures, the coincidence of the minimum does not seem to be purely fortuitous.

In fact, even if the Q_S data are separated into two superprovinces, as Mitchell [1975] has done, this feature is retained for western North America. There the Q_S values will be even lower (~100) for the upper crust, and a rapid transition to much higher values for the lower crust occurs at a depth of 17.5 km. The Q_S model of Mitchell [1975] for eastern North America does not show the same detailed features; in contrast, the Q_S value is constant (~250) for the entire upper crust and is about 2000 for the lower crust. All these new Q_S-depth functions differ considerably from the standard anelasticity model MM8 of Anderson et al. [1965], which is also shown for reference in Figure 7. As Mitchell [1975] points out, regional variations in the shear wave quality factor Q_S do not seem to be directly associated with regional temperature variations. Gordon and Davis [1968] have related low-velocity attenuating depth ranges in the earth to interstitial fluid at high pore pressures (see also Nur and Simmons [1969] and Berry [1972]). It is therefore reasonable to assume that lateral variations of the volume of water in pore spaces within the upper crust can explain the changes in the 'intensity' of the reversals in seismic velocities and quality factors.

Finally, Smith et al. [1975] have pointed out that there seems to exist a correlation of the sialic low-velocity zone with a decrease in the frequency of occurrence of earthquakes in the corresponding depth range. This favors the hypothesis [Braile et al., 1974] that vertical motion on normal faults may be taken up in a tectonically 'soft' layer associated with the low-velocity layer shown to have decreased rigidity

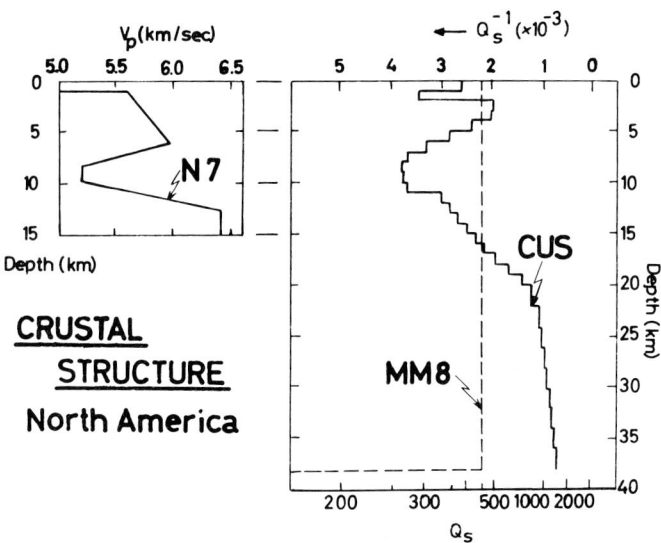

Fig. 7. (Left) Preferred P velocity-depth function (model N7) for the Delta-W profile [after Müller and Mueller, 1972]. (Right) Shear wave quality factor Q_S as a function of depth. Model CUS of Herrmann and Mitchell [1975] is compared to the standard model MM8 of Anderson et al. [1965].

as evidenced by a high Poisson's ratio ($\geqslant 0.3$). More recent investigations by Mueller et al. [1973] as well as Mueller and Rybach [1974] have shown that in the Rhinegraben area and its neighborhood the majority of hypocenters are concentrated close to the top of the sialic low-velocity zone, in accordance with earlier findings.

5. Middle and Lower Crust

In order to reconcile the seismic refraction and reflection results, Emter [1971] was forced to introduce a slight velocity inversion immediately above the C discontinuity at a depth of about 20 km. In this way he could achieve a reasonable fit for the observations in the Molasse basin and the Jura Mountains in southern Germany (see, e.g., Figures 2a and 2b). It is interesting to note that from the interior of this second low-velocity zone, practically no echoes are returned. In Figure 3 (between echo times of about 6 and 7 s) and in Liebscher's [1962, 1964] results this phenomenon is clearly visible. The display of deep reflections in Figure 3 also shows that there is significant fine structure associated with the C discontinuity at echo times between 6.5 and 7.5 s. A rough estimate indicates a thickness for the C transition zone of about 1-3 km. According to Emter [1971], P_b (or P^* in Conrad's notation) velocities ranging from 6.7 to 6.9 km/s have been observed consistently for southwestern Germany. He has taken an average value of 6.8 km/s as the most likely velocity between the C and M

ESCHENLOHE-NW

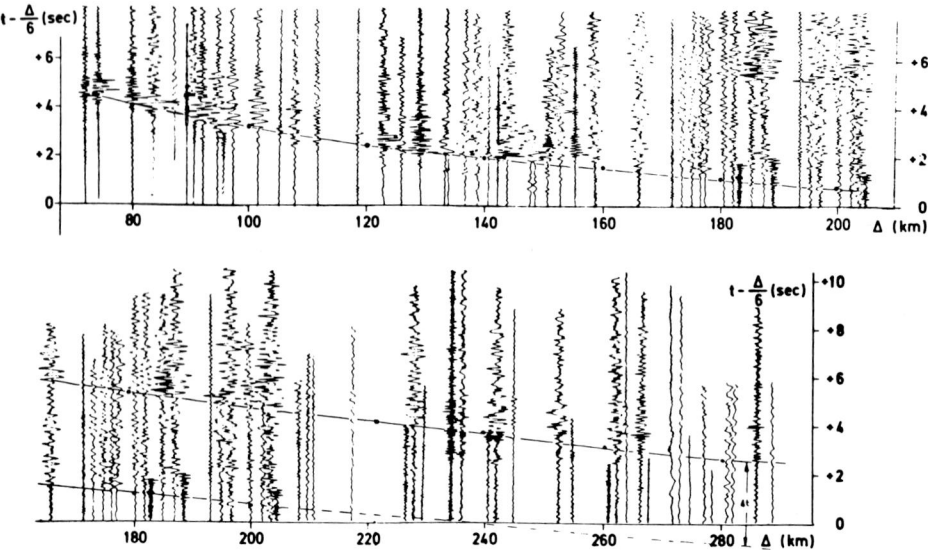

Fig. 8. Two overlapping segments of the seismogram section (70 < Δ < 290 km) for the Eschenlohe-NW profile in southern Germany. The wide-angle reflection hyperbolae corresponding to the regular $(P_MP)_1$ and the first multiple $(P_MP)_2$ phase, delayed by δt, calculated for the crustal model of Figure 9, are indicated [after Emter, 1971].

discontinuities, as was already mentioned in conjunction with the discussion of Figure 2c.

Peterschmitt [1975] has tried for some time to resolve the structural discrepancies between the results obtained from near-earthquake observations and those obtained from deep seismic sounding in Europe. He succeeded in explaining the classical phase $\bar{P}$ at greater distances as a superposition of multiply reflected P_MP waves. In fact, the occurrence of multiple P_MP phases in record sections had been noticed before, particularly for the refraction profiles radiating out of the shot point Eschenlohe at the northern margin of the Alps. An example taken from the Eschenlohe-NW profile [Emter, 1971] is shown in Figure 8. In the upper part the regular $(P_MP)_1$ phase with its critical point at a distance of about 80 km is seen which is duplicated beyond a distance of slightly more than 160 km. The first multiple reflection $(P_MP)_2$ can be unmistakably traced in the lower half of Figure 8 well beyond 280 km. There the profile ends at the rim of the Rhinegraben just north of Heidelberg.

Since the crustal structure in southwestern Germany is sufficiently well known from seismic refraction and reflection measurements and could be checked by gravity calculations [Emter, 1971, 1976; Harcke, 1972], it is relatively easy to trace the P_MP rays along the profile by taking

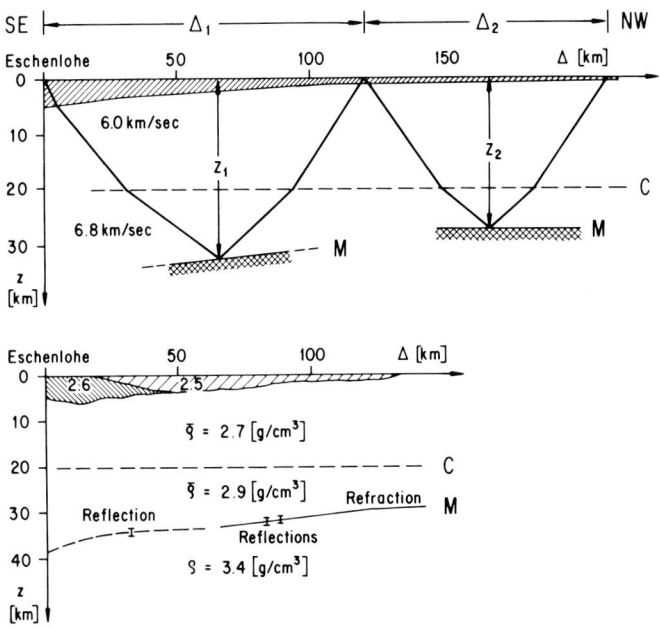

Fig. 9. Simple three-layered crustal model for the Eschenlohe-NW profile traversing southwestern Germany [after Emter, 1971]. (Top) Calculated ray path for the first multiple wide-angle reflection $(P_MP)_2$. (Bottom) Corresponding density model which satisfies the observed Bouguer anomalies along the profile.

the upward slope of the M discontinuity into account (Figure 9). The calculated reflection hyperbolae for $(P_MP)_1$ and $(P_MP)_2$ based on this simple model agree with the observed reflection times extremely well, as can be seen in Figure 8. Peterschmitt [1975] has demonstrated convincingly that multiply reflected P_MP phases will be recorded only if the upper reflection point at the surface lies in sedimentary terrain. If the ray hits the surface in a region of outcropping basement, the effective reflection coefficient will be sizably smaller, and therefore no multiple P_MP phase will be observed.

According to Peterschmitt [1975], the P velocity immediately above the M discontinuity may not exceed 6.4 km/s if the interpretation of $\overline{P}$ in terms of multiply reflected P_MP phases is supposed to function. This condition, of course, leads to a relatively low average P velocity for the crust, which is also indicated by results of t^2, Δ^2 determinations from P_MP observations. A velocity-depth function for the lower crust of the type shown in the inset of Figure 10 and in Figure 13 can resolve the problems outlined above.

In this new picture the C discontinuity would now be a thin (most likely laminated) high-velocity layer of a few kilometers in thickness with a P velocity of about 6.8 km/s. A tunneling effect such as that described by Fuchs and Schulz [1976] might then occur similar to the situation in Figure 12. Low-frequency waves could, on critical

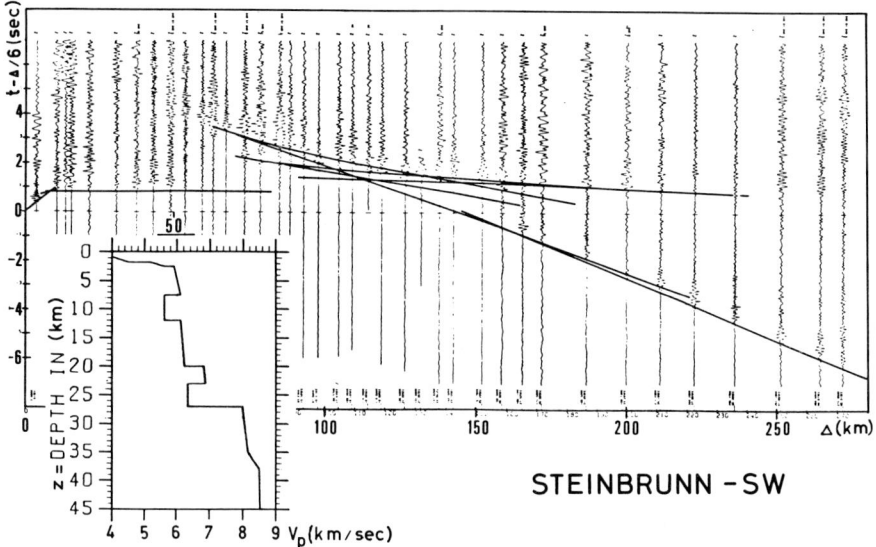

Fig. 10. Seismogram section ($0 < \Delta < 275$ km) for the Steinbrunn-SW profile passing through the folded Jura Mountains between France and Switzerland [after Egloff and Ansorge, 1976]. The signal correlations were calculated on the basis of the P velocity-depth function shown in the inset. Not drawn is the reflection hyperbola associated with the sharp upper boundary of the sialic low-velocity zone, but it can be clearly traced through the strong second arrivals following the weak P_g onsets.

incidence, penetrate this relatively thin high-velocity layer at a depth of about 22 km (see, for example, Figures 10 and 13), while high-frequency signals would effectively be reflected from this presumably laminated transition zone. The depth range of lower velocity underneath this high-velocity 'tooth' would constitute a second zone of velocity inversion in the crust, probably also associated with a higher Poisson's ratio, which will facilitate shearing motions as they occur in overthrusts and subduction zones. A pronounced low-velocity zone for shear waves at the base of the continental crust has recently been postulated by Jordan and Frazer [1975]. Significantly, this channel is characterized by high values of Poisson's ratio (≥ 0.33). The increased velocity contrast immediately above the M discontinuity would also explain the strong near-vertical reflections from that interface better (cf. Figure 3) and at the same time would move the critical point of P_MP and P_n to distances closer to the shot point. In many profiles (as, for example, in the Alps and the Cordillera Betica in Spain) this shift has resulted in a significant improvement of the interpretation (see also Mueller and Landisman [1971]).

The record section of the Steinbrunn-SW profile in Figure 10 is taken from a recent paper by Egloff and Ansorge [1976] on the crustal structure under the folded Jura Mountains between Switzerland and

France. It should serve as an illustration of the scheme of interpretation currently used. The lines of phase correlations drawn correspond to the P velocity-depth function shown in the inset of Figure 10. This crustal model is not necessarily typical. In other cases the lower corner of the sialic low-velocity zone has a slightly higher velocity, which will then cause the shadow zone at about 90 km from the shot point to disappear, possibly leading to an overlap of P_c over P_g for some range of distance. The reflection from the top of the uppermost low-velocity zone is not drawn, although the corresponding second arrivals behind P_g can clearly be seen. As was discussed at the very beginning of this section, a second low-velocity zone is usually identified in southern Germany at a depth somewhat above 20 km (cf. Figure 2a). In the present case, on the basis of seismic refraction observations alone this inversion does not seem to be required. As soon as deep-reflection data become available, it may be necessary to alter the interpretation accordingly. A sub-M gradient zone has been put in at a depth of about 35 km in order to explain the duplication in the P_n arrivals. Fuchs and Müller [1971] as well as Müller and Fuchs [1976] were forced to introduce a similar feature at precisely the same depth.

6. Crust-Mantle Transition

Until rather recently, the M discontinuity was generally considered to be a first-order discontinuity separating layers of the lower crust and the upper mantle. This model of an abrupt velocity change, however, cannot explain the relatively weak coherence of near-vertical signals reflected from the M discontinuity if they are traced over any sizable distance (cf., for example, Figure 3). Another argument shedding doubts on the simplicity of the M discontinuity was the comparatively high amplitude of the P_n phase if it is interpreted as a pure head wave phenomenon. A gradient zone beneath the M interface had to be used to resolve this discrepancy, at least with regard to the major amplitude features.

Davydova [1972] has summarized the different concepts still being discussed about the detailed structure of the crust-mantle transition (Figure 11). She distinguishes three different types of models.

Type 1 is characterized by a sharp interface, i.e., a first-order velocity discontinuity, between rather thick layers of constant or slightly increasing velocity as a function of depth. This type of simple model accounts fairly well for the kinematic features of the waves returned from the M discontinuity. It is therefore still used in the quantitative interpretation of seismic observations, in much the same manner as it has been during the past 67 years since Mohorovičić [1910] first postulated the existence of this important boundary in the earth.

Type 2 corresponds to models of continuous or discontinuous transition zones from the crust to the mantle. The thickness h of these transition zones, regardless of the fine velocity structure, is of the order of less than two wavelengths on the basis of the velocities in the lower

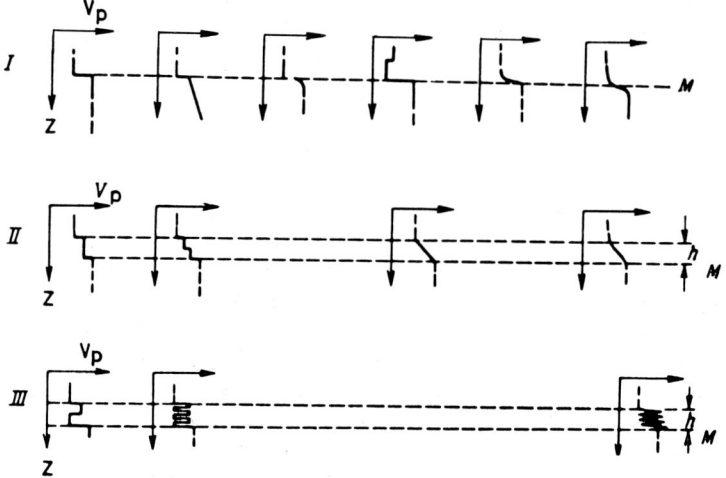

Fig. 11. Types of P velocity-depth functions for the crust-mantle transition zone M: Type 1, first-order discontinuities; type 2, continuous or discontinuous transition zones with thickness h less than two wavelengths; and type 3, laminated transition zones consisting of thin alternating high- and low-velocity layers [after Davydova, 1972].

crust. Models of this type were first discussed by Nakamura and Howell [1964] and were later strongly promoted by Giese [1968]. They are not able to explain the observed near-vertical reflections observed from the M discontinuity, as has been demonstrated by Fuchs [1969].

Type 3 represents models of laminated transition zones consisting of thin alternating high- and low-velocity layers. The significance of this type of transition zone seems to have been pointed out first by Berzon et al. [1962]. The feasibility of applying such models to deep-seated reflectors within the crust and mantle has been discussed by Meissner [1967b] and Fuchs [1968]. Synthetic reflection seismograms calculated by Fuchs [1969] for laminated transition zones show a remarkable similarity to deep crustal reflections identified on field records.

Depending on the dominant frequencies in the signal spectrum the M discontinuity appears as either type 1 or type 3. Peterschmitt [1975] states that for frequencies below 5 Hz, i.e., for wavelengths greater than about 1.2-1.5 km, since they dominate in earthquake-generated waves, the M interface appears to be a first-order discontinuity. The evidence from deep crustal reflections, however, where dominant frequencies well above a 'natural' lower cutoff of around 10 Hz are observed (cf. Figure 3), would heavily favor a type 3 transition from the lower crust to the uppermost mantle, since only a laminated transition zone can account for all known characteristic features of near-vertical reflections.

Recently, Fuchs and Schulz [1976] have postulated that the crust-

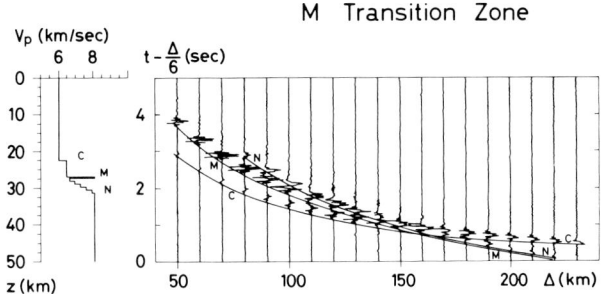

Fig. 12. Synthetic seismogram section for a P velocity-depth model of the crust-mantle transition zone which simulates the observations of low-frequency tunnel waves (adapted from Fuchs and Schulz [1976]). The signal phases C, M, and N correspond to the features with the same symbols in the velocity-depth function (left).

mantle transition zone is even more complex than has been indicated so far. On the basis of observations of low-frequency 'tunnel waves', explosion- and earthquake-generated signals which originate in either the crust or the upper mantle, they deduce a new model which contains one or more thin high-velocity layers, as, for instance, that shown in Figure 12 (left). A similar feature in the crust has been discussed above in conjunction with the nature of the C discontinuity (see Figure 10). Fuchs and Schulz [1976] demonstrate with the aid of synthetic seismograms how tunnel waves, i.e., low-frequency waves beyond critical incidence, penetrate through these thin high-velocity layers. The synthetic record section in Figure 12 was calculated to simulate the detailed P_MP reflections along the crustal profile Florac 03 in France [Sapin and Prodehl, 1973]. As can be seen at a distance of 80 km from the shot point, the 'regular' high-frequency P_MP phase M is followed by a low-frequency arrival N, which obviously was not generated by some anomalous behavior of the near-surface structure beneath the recording stations. The time difference between the regular P_MP phase M and the tunnel phase N decreases with increasing distance, since no significant penetration occurs for large angles of incidence. This effect is clearly visible in Figure 12 (right). According to Fuchs and Schulz [1976] the most reasonable fit to the observations is achieved by a velocity-depth function as is shown in Figure 12 (left). This type of transition structure, if it can be verified in other locations, is a complicated mixture of the three types of transition zones according to the classification of Davydova [1972].

As Fuchs and Schulz [1976] have pointed out, the proposed mechanism of seismic energy tunneling through high-velocity strata does not require that these layers be of unlimited extent. A system of more or less randomly distributed thin high-velocity lamellas embedded in normal crustal or mantle material would provide an effective wave guide with observed phase velocities of P waves reaching nearly 9 km/s in the upper mantle. Such high velocities may be caused by strongly anisotropic minerals with a preferred horizontal orientation. The observed azimuthal

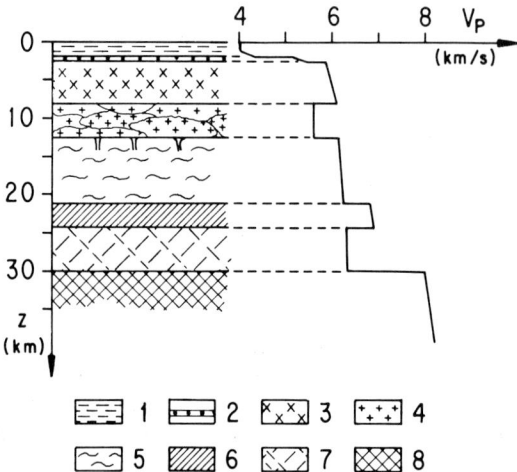

Fig. 13. Schematic model of the continental crust summarizing the various features discussed in this paper: 1, Cenozoic sediments (near-surface low-velocity layer); 2, Mesozoic (and Paleozoic) sediments; 3, upper crystalline basement consisting of metamorphic rocks, such as gneisses and schists (zone of positive velocity gradient); 4, laccolithic zone of granitic intrusions (sialic low-velocity zone); 5, migmatites (middle crustal layer); 6, amphibolites (high-velocity tooth); 7, granulites (lower crustal layer); and 8, ultramafics (uppermost mantle).

dependence of P_n velocities in oceanic areas [Raitt et al., 1969] and the even more pronounced P_n velocity anisotropy under the continental crust [Bamford, 1973, 1976] could thus be explained quite reasonably.

7. New Crustal Model

In general, the earth's crust shows great diversity both laterally and vertically. Deep seismic sounding results have demonstrated a block-type structure of the continental crust with continuous layering extending only over limited ranges. Vertical variations in chemical composition are geologically well documented but as such should not be taken as the exclusive evidence for continuous discrete layering of the crust.

Nevertheless, the crustal model presented in this paper, which is based primarily on data from central Europe, shows some consistent general features which can be derived from purely geophysical observations. The new model is more complex than the classical delineation of the crust under continents into a granitic layer, the upper crust, overlying a basaltic layer, the lower crust. In Figure 13 (right) a schematic P velocity-depth function is depicted which may serve to illustrate the various features of the new crustal model.

Typically, a zone of positive velocity gradient is found beneath a near-surface low-velocity sedimentary layer. It is underlain by the sialic low-velocity zone at a depth of about 10 km. The actual P velocity variation with depth is certainly more complicated than the simplified picture indicates, but there are strong indications that the lower boundary is relatively sharp. It can therefore be considered as being identical to the F discontinuity introduced by Liebscher [1962, 1964], which is equivalent to the G discontinuity cited in the Russian literature (see, for example, Polshkov et al. [1973]). Depending on the velocity contrast between the top and the bottom of this inversion zone, refracted signals and clear wide-angle reflections will be observed. The middle and lower parts of the crust are characterized by a relatively low average velocity with possibly a slight velocity inversion at depths of around 20 km, followed by a high-velocity tooth, the C discontinuity, a few kilometers thick and a depth range of lower velocity between the C and M discontinuities. An abrupt velocity jump occurs at a depth of 30 km, reinforced by a suggested lamination as the dominant feature of the crust-mantle transition.

A crustal model of this type was first proposed by Mueller et al. [1969] for the Rhinegraben area in central Europe, and it has since been tested for consistency in adjacent regions with seismic sources both at the surface and at depth [Peterschmitt, 1975]. The three depth ranges of lowered velocities in Figure 13, i.e., the near-surface sedimentary layer, the sialic low-velocity zone in the upper crust, and the zone of velocity inversion in the lower crust, are presumably also zones of 'weakness' within the crust [Mueller et al., 1976]. Field evidence in the Franco-Swiss Jura Mountains, the central Alps, and the Ivrea zone (see, for example, Schmid [1968] and Bertolani [1968]) supports the concept of 'flake tectonics,' which implies that slivers of crustal material can be sheared off at three levels of depth: (1) at the top of the crystalline basement beneath the sedimentary cover (Jura type), (2) near the lower boundary of the sialic low-velocity zone (central massif type), and (3) immediately above the M discontinuity (Ivrea type [German Research Group for Explosion Seismology, 1968]).

It is, of course, tempting at this point to speculate about the petrological implications of the crustal model presented in Figure 13. In a review of the genetic evolution of the outer silicate shell of the earth, Borchert and Böttcher [1967] discussed a scheme which places an abrupt density and velocity jump at the contact between the sediments and the crystalline basement. The sialic low-velocity zone would have to be identified with a semicontinuous zone of granitic intrusions into the upper crust which, under the appropriate pressure and temperature conditions, must have originated at depths near 20 km (see, for example, Borchert and Böttcher [1967]), i.e., just above the presumably basic high-velocity tooth in Figure 13.

For the lower crust it is difficult to come up with a similar straightforward interpretation. Russian seismologists [Kosminskaya and Riznichenko, 1964] have argued that the apparent layering of the continental crust is primarily due to the advancement of metamorphic fronts. This suggestion has been received with much enthusiasm, particularly by Christensen and Fountain [1975], who state that pressures of 6-10 kbar and temperatures of as much as $600°-800°C$ at the crust-mantle transition zone can combine to provide a suitable environment for medium- and high-grade metamorphism.

Field and laboratory studies of the stability of rocks performed in the past 15 years have demonstrated that metamorphic rocks constitute most of the continental crust (for a recent summary, see Smithson and Decker [1974]). Strong geologic and geophysical evidence suggests that the deeper portion of the upper crust is granitic [Mueller, 1970; Landisman et al., 1971], but little evidence can be advanced that it is homogeneous.

Any typical section of the upper crust under continents is usually composed of crystalline schists, banded gneisses, and migmatitic rocks which surround a series of lenticular or mushroom-shaped metamorphic granitic bodies whose position is determined by the tectonics of the region. In a seismic reflection survey utilizing short wavelengths (<600 m) these rocks will appear heterogeneous because of the rapidly changing composition. If these formations are investigated by refraction seismology with dominant wavelengths of several kilometers, they will be interpreted as being 'homogeneous.'

An abrupt transition from denser gneisses and schists (average density of between 2.75 and 2.85 g/cm^3) to relatively low-density granites (mean density of 2.67 g/cm^3) would be associated with a corresponding change in compressional velocity (decrease from about 6.0 to 5.5 km/s). Normal incidence and wide-angle reflected waves will be produced by such a pronounced contrast in acoustic impedance, in agreement with the seismic observations (see section 4). A quasi-continuous laccolithic zone consisting of numerous acidic intrusions at depths of about 10 km seems to be the most plausible explanation for the sialic low-velocity zone in the upper crust.

The middle part of the crust beneath this first zone of velocity inversion probably consists of a heterogeneous zone of migmatitic rocks, such as granitic gneiss and augen gneiss interlayered with biotite-rich gneiss surrounding small lenses of mafic rocks [Smithson and Decker, 1974]. Their origin is attributed to deep-seated metamorphic processes that may have resulted in partial melting. Typically, their compressional velocities range from 6.0 to 6.3 km/s, and their densities have values between 2.7 and 2.8 g/cm^3 [Smithson, 1971].

There is now ample evidence that the lower part of the continental crust is composed of dense rocks with relatively high seismic velocities. Stability studies of mafic rocks in the field and in the laboratory [Ringwood and Green, 1966] have shown that under the pressure-temperature conditions prevailing in the lower crust they would be recrystallized as metamorphic rocks in the amphibolite or granulite facies. Granulites are generally believed to underlie amphibolites at depth in the crust. Direct evidence for the composition of the lower crust and the crust-mantle transition has been postulated by Berckhemer [1969] to be visible in a crustal section based on a geologic profile through the Ivrea zone of the southern Alps [Schmid, 1968]. This section through an upturned crustal block suggests a zoning of the crust strikingly similar to the schematic model in Figure 13. A comparison illustrates that the upper crust seems to consist of metamorphic rocks, such as gneisses and schists, underlain by post-metamorphic granitic intrusions of Hercynian age [Köppel, 1974]. The middle crust by analogy is then composed of granitic gneiss and migmatites, and the major constituents of the lower crust are mafic metamorphic rocks, which, in the deepest parts of the crust, have been dehydrated and differentiated by partial mobilization.

If this model is correct, it may be surmised that the high-velocity tooth in Figure 13, which is relatively rich in reflecting elements (as seen, for example, at echo times of around 6.5-7.5 s in Figure 3), corresponds to an amphibolite 'layer' on top of granulite facies rocks, which comprise the bulk of the lower crust. Compressional and shear wave velocities for amphibolite facies rocks under confining pressures to 10 kbar measured in the laboratory have been reported by Christensen [1965, 1966]. This experimental work has recently been extended by Christensen and Fountain [1975] to granulite facies rock samples, which on the average exhibited lower seismic velocities and densities than amphibolites under conditions corresponding to those of the lower crust. Laboratory measurements on rock samples from the Ivrea-Verbano and Strona-Ceneri zones by Fountain [1976] have confirmed this trend. The agreement between estimated thicknesses of exposed rock units in those zones and thicknesses determined for crustal layers from seismic refraction experiments in the northern and southern foreland of the Alps is quite reasonable. It can therefore be taken as an additional confirmation of the hypothesis that the metamorphic rock sequence in the Ceneri and Ivrea zones of the southern Alps represents a typical cross section through the continental crust outside the Alpine area.

Variations of seismic velocities with depth in the continental crust can thus be related to changes in the gross chemical composition and/or progressive metamorphism. Water obviously plays an important role during the metamorphism of sedimentary and igneous rocks. Dehydrated granulitic rocks constitute the lower part of the crustal section exposed in the Ivrea-Verbano zone [Berckhemer, 1969]. Estimates of the distribution of water in the middle crust are highly subjective, but there are indications that water will be concentrated in the first-formed granitic melt. It is quite possible that this 'wet zone' may have a velocity lower than the material above and even lower than the dehydrated material below [Berry, 1972]. The release of free water into surrounding pore spaces would provide a consistent explanation for the lowered seismic velocities, the increase in attenuation, the reduction in cohesive strength, and also the higher electrical conductivity discovered in the middle crust of tectonically active areas [Landisman et al., 1971]. Zones of enhanced fluid content in the crust at pore pressures close to the lithostatic pressure may ultimately hold the key to an understanding of the physical mechanisms which are responsible for the major tectonic processes in the continental crust.

Acknowledgments. The author is greatly indebted to D. Emter (Geowissenschaftliches Gemeinschaftsobservatorium, Schiltach, Germany) for giving permission to reproduce several of the figures in this paper.

Particular thanks go to G. Dohr (Preussag AG, Hannover, Germany), T. Labhart (Mineralogisch-Petrographisches Institut der Universität Bern, Switzerland), K. Fuchs and C. Prodehl (Geophysikalisches Institut der Universität Karlsruhe, Germany), as well as J. Ansorge and R. Egloff (Institut für Geophysik, ETH Zürich, Switzerland), who furnished published or unpublished material to be included in this study. Contribution 179, ETH-Geophysics, Zürich, Switzerland.

References

Aichele, H., Interpretation refraktionsseismischer Messungen im Gebiet des fränkisch-schwäbischen Jura, Ph. D. thesis, 105 pp., Univ. of Stuttgart, Stuttgart, Germany, 1976.

Anderson, D. L., A. Ben-Menahem, and C. B. Archambeau, Attenuation of seismic energy in the upper mantle, J. Geophys. Res., 70, 1441-1448, 1965.

Angenheister, G., and J. Pohl, Die seismischen Messungen im Ries von 1948-1969, Geol. Bavarica, 61, 304-326, 1969.

Angenheister, G., and J. Pohl, Deep crustal reflections on a 17 km digital reflection profile in south Germany (Nördlinger Ries), Commun. Observ. Roy. Belg., A13, Sér. Géophys. no. 101, 173-176, 1971.

Angenheister, G., and J. Pohl, Results of seismic investigations in the Ries crater area (southern Germany), in Explosion Seismology in Central Europe-Data and Results, edited by P. Giese, C. Prodehl, and A. Stein, pp. 290-302, Springer, New York, 1976.

Bamford, D., Refraction data in western Germany-A time-term interpretation, Z. Geophys., 39, 907-927, 1973.

Bamford, D., An updated time-term interpretation of P_n-data from quarry blasts and explosions in western Germany, in Explosion Seismology in Central Europe-Data and Results, edited by P. Giese, C. Prodehl, and A. Stein, pp. 215-220, Springer, New York, 1976.

Båth, M., Die Conrad-Diskontinuität, Freiberg. Forschungsh., Ser. C, 101, 5-34, 1961.

Berckhemer, H., Direct evidence for the composition of the lower crust and the Moho, Tectonophysics, 8, 97-105, 1969.

Berg, J. W., Jr., K. L. Cook, H. D. Narans, Jr., and W. M. Dolan, Seismic investigation of crustal structure in the eastern part of the Basin and Range province, Bull. Seismol. Soc. Amer., 50, 511-535, 1960.

Berry, M. J., Low-velocity channels in the earth's crust, Comments Earth Sci., Geophysics, 3, 59-68, 1972.

Bertolani, M., Squardo generale alla petrografia della Valle Strona (Novara)-Guida all'escursione, Schweiz. Mineral. Petrogr. Mitt., 48, 314-328, 1968.

Berzon, I. S., A. M. Epinat'eva, G. N. Pariiskaya, and S. P. Starodubrovskaya, Dynamic features of seismic waves in real media, Izd. AN SSSR, Moscow, 1962.

Birch, F., The velocity of compressional waves in rocks to 10 kbar, 1, J. Geophys. Res., 65, 1083-1102, 1960.

Birch, F., The velocity of compressional waves in rocks to 10 kbar, 2, J. Geophys. Res., 66, 2199-2224, 1961.

Borchert, H., and W. Böttcher, Zur Petrologie der Lithosphäre in ihrer Beziehung zu geophysikalischen Diskontinuitäten, auch der Gesamterde, Gerlands Beitr. Geophys., 76, 257-277, 1967.

Braile, L. W., R. B. Smith, G. R. Keller, R. M. Welch, and R. P. Meyer, Crustal structure across the Wasatch front from detailed seismic refraction studies, J. Geophys. Res., 79, 2669-2677, 1974.

Braile, L. W., and R. B. Smith, Guide to the interpretation of crustal refraction profiles, Geophys. J. Roy. Astron. Soc., 40, 145-176, 1975.

Bram, K., and P. Giese, Die Geschwindigkeitsverteilung der P-Welle in der Erdkruste im Raum Augsburg (Süd-Deutschland)-Ergebnisse und Vergleich zweier seismischer Messungen, Z. Geophys., 34, 611-626, 1968.

Čermák, V., Temperature-depth profiles in Czechoslovakia and some adjacent areas derived from heat-flow measurements, deep seismic sounding and other geophysical data, Tectonophysics, 26, 103-119, 1975.

Christensen, N. I., Compressional wave velocities in metamorphic rocks at pressures to 10 kbar, J. Geophys. Res., 70, 6147-6164, 1965.

Christensen, N. I., Shear wave velocities in metamorphic rocks at pressures to 10 kbar, J. Geophys. Res., 71, 3549-3556, 1966.

Christensen, N. I., and D. M. Fountain, Constitution of the lower continental crust based on experimental studies of seismic velocities in granulite, Geol. Soc. Amer. Bull., 86, 227-236, 1975.

Conrad, V., Laufzeitkurven des Tauernbebens vom 28. November 1923, Mitt. Erdb. Komm. Wiener Akad. Wiss., 59, 1-23, 1925.

Davydova, N. I., Possibilities of the DSS technique in studying properties of deep-seated seismic interfaces, in Seismic Properties of the Mohorovičić Discontinuity, (in Russian), edited by N. I. Davydova, Izdatel'stvo Nauka, Moscow, 1972. (English translation, Israel Program for Scientific Translations, Jerusalem, 1975).

Dohr, G., Reflexionsseismische Tiefensondierung (in German with English summary), Z. Geophys., 38, 193-220, 1972.

Dohr, G., and R. Meissner, Deep crustal reflections in Europe, Geophysics, 40, 25-39, 1975.

Eaton, J., J. Healy, W. Jackson, and L. Pakiser, Upper mantle velocity and crustal structure in the eastern Basin and Range province, determined from Shoal and chemical explosions near Delta, Utah, Bull. Seismol. Soc. Amer., 54, 1567, 1964.

Edel, J. B., K. Fuchs, C. Gelbke, and C. Prodehl, Deep structure of the southern Rhinegraben area from seismic refraction investigations, J. Geophys., 41, 333-356, 1975.

Egloff, R., and J. Ansorge, Die Krustenstruktur unter dem Faltenjura, paper presented at the 36th Annual Meeting, Deutsche Geophysikalische Gesellschaft, Bochum, Germany, April 5-9, 1976.

Emter, D., Ergebnisse seismischer Untersuchungen der Erdkruste und des obersten Erdmantels in Südwestdeutschland, Ph. D. thesis, 108 pp., Univ. of Stuttgart, Stuttgart, Germany, 1971.

Emter, D., Seismic results from southwestern Germany, in Explosion

Seismology in Central Europe-Data and Results, edited by P. Giese, C. Prodehl, and A. Stein, pp. 283-289, Springer, New York, 1976.

Förtsch, O., Analyse der seismischen Registrierungen der Gross-Sprengung bei Haslach im Schwarzwald am 28. April 1948, Geol. Jahrb., 66, 65-80, 1951.

Fountain, D. M., The Ivrea-Verbano and Strona-Ceneri zones, northern Italy: A cross-section of the continental crust.-New evidence from seismic velocities of rock samples, Tectonophysics, 33, 145-165, 1976.

Fuchs, K., The reflection of spherical waves from transition zones with arbitrary depth-dependent elastic moduli and density, J. Phys. Earth, 16, 27-41, 1968.

Fuchs, K., On the properties of deep crustal reflectors, Z. Geophys., 35, 133-149, 1969.

Fuchs, K., and M. Landisman, Detailed crustal investigation along a north-south section through the central part of western Germany, in The Earth Beneath the Continents, Geophys. Monogr. Ser., vol. 10, edited by J. Steinhart and T. Smith, pp. 433-452, AGU, Washington, D. C., 1966.

Fuchs, K., and G. Müller, Computation of synthetic seismograms with the reflectivity method and comparison with observations, Geophys. J. Roy. Astron. Soc., 23, 417-433, 1971.

Fuchs, K., and K. Schulz, Tunneling of low-frequency waves through the subcrustal lithosphere, J. Geophys., 42, 175-190, 1976.

German Research Group for Explosion Seismology, Topographie des 'Ivrea-Körpers' abgeleitet aus seismischen und gravimetrischen Daten, Schweiz. Mineral. Petrogr. Mitt., 48, 235-246, 1968.

Giese, P., Die Geschwindigkeitsverteilung im obersten Bereich des Kristallins, abgeleitet aus Refraktionsbeobachtungen auf dem Profil Böhmischbruck-Eschenlohe, Z. Geophys., 29, 197-214, 1963.

Giese, P., Versuch einer Gliederung der Erdkruste im nördlichen Alpenvorland, in den Ostalpen und in Teilen der Westalpen mit Hilfe charakteristischer Refraktions-Laufzeit-Kurven sowie eine geologische Deutung, Geophys. Abh. Inst. Met. Geophys. Freie Univ. Berlin, 1 (2), 202 pp., 1968.

Gordon, R., and L. Davis, Velocity and attenuation of seismic waves in imperfectly elastic rock, J. Geophys. Res., 73, 3917-3935, 1968.

Harcke, H., Die Struktur der Erdkruste im nördlichen Alpenvorland-Eine Synthese aus seismischen und gravimetrischen Daten, Ph. D. thesis, 68 pp., Univ. of Karlsruhe, Karlsruhe, Germany, 1972.

Heacock, J. G., Intermediate and deep properties of the earth's crust, A possible electromagnetic wave guide, in The Structure and Physical Properties of the Earth's Crust, Geophys. Monogr. Ser., vol. 14, edited by J. G. Heacock, pp. 1-9, AGU, Washington, D. C., 1971.

Hehn, K., Die statistische Auswertung von Reflexionen mit langen Laufzeiten aus dem nordwestdeutschen Raum und ihre Zuordnung zu den bekannten Unstetigkeitsflächen in der Erdkruste, diploma thesis, 54 pp., Tech. Univ. of Clausthal-Zellerfeld, Clausthal-Zellerfeld, Germany, 1964.

Herrmann, R. B., and B. J. Mitchell, Statistical analysis and interpretation of surface-wave anelastic attenuation data for the stable interior of North America, Bull. Seismol. Soc. Amer., 65, 1115-1128, 1975.

Hughes, D. S., and C. Maurette, Variation of elastic wave velocities in granites with pressure and temperature, Geophysics, 21, 277-284, 1956.

Hughes, D. S., and C. Maurette, Variation of elastic wave velocities in basic igneous rocks with pressure and temperature, Geophysics, 22, 23-31, 1957.

Jordan, T. H., and L. N. Frazer, Crustal and upper mantle structure from Sp phases, J. Geophys. Res., 80, 1504-1518, 1975.

Keller, G. R., R. B. Smith, and L. W. Braile, Crustal structure along the Great Basin-Colorado plateau transition from seismic refraction studies, J. Geophys. Res., 80, 1093-1098, 1975.

Köppel, V., Isotopic U-Pb ages of monazites and zircons from the crust-mantle transition and adjacent units of the Ivrea and Ceneri zones (southern Alps, Italy), Contrib. Mineral. Petrol., 43, 55-70, 1974.

Kosminskaya, I. P., and Y. V. Riznichenko, Seismic studies of the earth's crust in Eurasia, Res. Geophys., 2, 81-122, 1964.

Labhart, T. P., Geologie-Einführung in die Erdwissenschaften, p. 108, Hallwag, Bern, 1975.

Landisman, M., S. Mueller, and B. J. Mitchell, Review of evidence for velocity inversions in the continental crust, in The Structure and Physical Properties of the Earth's Crust, Geophys. Monogr. Ser., vol. 14, edited by J. G. Heacock, pp. 11-34, AGU, Washington, D. C., 1971.

Liebscher, H., Reflexionshorizonte der tieferen Erdkruste im Bayrischen Alpenvorland, abgeleitet aus Ergebnissen der Reflexionsseismik, Z. Geophys., 28, 162-184, 1962.

Liebscher, H., Deutungsversuche für die Struktur der tieferen Erdkruste nach reflexionsseismischen und gravimetrischen Messungen im deutschen Alpenvorland, I, II, Z. Geophys., 30, 51-96, 115-126, 1964.

Meissner, R., Zum Aufbau der Erdkruste: Ergebnisse der Weitwinkelmessungen im bayrischen Molassebecken, I, II, Gerlands Beitr. Geophys., 76, 211-254, 295-314, 1967a.

Meissner, R., Exploring deep interfaces by seismic wide-angle measurements, Geophys. Prospect., 15, 598-617, 1967b.

Meissner, R., and U. Vetter, The northern end of the Rhinegraben due to some geophysical measurements, in Approaches to Taphrogenesis, edited by H. Illies and K. Fuchs, pp. 236-243, Schweizerbart, Stuttgart, Germany, 1974.

Meissner, R., H. Berckhemer, and A. Glocke, Results from deep-seismic sounding in the Rhine-Main area, in Explosion Seismology in Central Europe-Data and Results, edited by P. Giese, C. Prodehl, and A. Stein, pp. 303-312, Springer, New York, 1976.

Mitchell, B. J., Regional Rayleigh wave attenuation in North America, J. Geophys. Res., 80, 4904-4916, 1975.

Mohorovičić, A., Das Beben vom 8.X.1909, Jahrb. Meteorol. Observ. Zagreb, 9 (4), 63, 1910.

Mueller, S., Geophysical aspects of graben formation in continental rift systems, in Graben Problems, edited by H. Illies and S. Mueller, pp. 27-37, Schweizerbart, Stuttgart, Germany, 1970.

Mueller, S., and M. Landisman, Seismic studies of the earth's crust in continents, 1, Evidence for a low-velocity zone in the upper part of the lithosphere, Geophys. J. Roy. Astron. Soc., 10, 525-538, 1966.

Mueller, S., and M. Landisman, An example of the unified method of interpretation for crustal seismic data, Geophys. J. Roy. Astron. Soc., 23, 365-371, 1971.

Mueller, S., and L. Rybach, Crustal dynamics in the central part of the Rhinegraben, in Approaches to Taphrogenesis, edited by H. Illies and K. Fuchs, pp. 379-388, Schweizerbart, Stuttgart, Germany, 1974.

Mueller, S., E. Peterschmitt, K. Fuchs, and J. Ansorge, Crustal structure beneath the Rhinegraben from seismic refraction and reflection measurements, Tectonophysics, 8, 529-542, 1969.

Mueller, S., E. Peterschmitt, K. Fuchs, D. Emter, and J. Ansorge, Crustal structure of the Rhinegraben area, Tectonophysics, 20, 381-392, 1973.

Mueller, S., R. Egloff, and J. Ansorge, Struktur des tieferen Untergrundes entlang der Schweizer Geotraverse, Schweiz. Mineral. Petrogr. Mitt., 56, 685-692, 1976.

Müller, G., and S. Mueller, A crustal low-velocity zone in Utah, Geol. Soc. Amer. Abstr. Programs, 4, 204, 1972.

Müller, G., and K. Fuchs, Inversion of seismic records with the aid of synthetic seismograms, in Explosion Seismology in Central Europe-Data and Results, edited by P. Giese, C. Prodehl, and A. Stein, pp. 178-188, Springer, New York, 1976.

Nakamura, Y., and B. F. Howell, Jr., Maine seismic experiment: Frequency spectra of refraction arrivals and the nature of the Mohorovičić discontinuity, Bull. Seismol. Soc. Amer., 54, 9-18, 1964.

Nur, A., and G. Simmons, The effect of saturation on velocity in low porosity rocks, Earth Planet. Sci. Lett., 7, 183-193, 1969.

Peterschmitt, E., Séismes proches et sondages séismiques profonds, paper presented at the 1975 meeting, IASPEI Commission on Controlled Source Seismology, Paris, Aug. 18-23, 1975.

Polshkov, M. K., N. K. Bulin, and B. E. Sherbakova, Crustal investigations of the USSR by means of earthquake-generated converted waves, Tectonophysics, 20, 57-66, 1973.

Prodehl, C., J. Ansorge, J. B. Edel, D. Emter, K. Fuchs, S. Mueller, and E. Peterschmitt, Explosion-seismology research in the central and southern Rhinegraben-A case history, in Explosion Seismology in Central Europe-Data and Results, edited by P. Giese, C. Prodehl, and A. Stein, pp. 313-328, Springer, New York, 1976.

Raitt, R. W., G. G. Shor, Jr., T. J. G. Francis, and G. B. Morris, Anisotropy of the Pacific upper mantle, J. Geophys. Res., 74, 3095-3109, 1969.

Reich, H., In Süddeutschland seismich ermittelte tiefe Grenzflächen und ihre geologische Bedeutung, Geol. Rundsch., 46, 1-16, 1957.

Rhinegraben Research Group for Explosion Seismology, The 1972 seismic refraction experiment in the Rhinegraben-First results, in Approaches to Taphrogenesis, edited by H. Illies and K. Fuchs, pp. 122-137, Schweizerbart, Stuttgart, Germany, 1974.

Ringwood, A. E., and D. H. Green, Petrological nature of the stable continental crust, in The Earth Beneath the Continents, Geophys. Monogr. Ser., vol. 10, edited by J. Steinhart and T. Smith, pp. 611-619, AGU, Washington, D. C., 1966.

Rothé, J.-P., and E. Peterschmitt, Etude séismique des explosions d'Haslach, Ann. Inst. Phys. Globe Univ. Strasbourg, Géophys., 5, 3-28, 1950.

Sapin, M., and C. Prodehl, Long range profiles in western Europe, I, Crustal structure between the Bretagne and the Central Massif of France, Ann. Géophys., 29, 127-145, 1973.

Schmid, R., Excursion guide for the Valle d'Ossola section of the Ivrea-Verbano zone (Novara), Schweiz. Mineral. Petrogr. Mitt., 48, 305-314, 1968.

Schulz, G., Reflexionen aus dem kristallinen Untergrund im Gebiet des Pfälzer Berglandes, Z. Geophys., 23, 225-235, 1957.

Smith, R. B., L. W. Braile, and G. R. Keller, Upper crustal low-velocity layers: Possible effect of high temperatures over a mantle upwarp at the Basin Range-Colorado Plateau transition, Earth Planet. Sci. Lett., 28, 197-204, 1975.

Smithson, S. B., Densities of metamorphic rocks, Geophysics, 36, 690-694, 1971.

Smithson, S. B., and E. Decker, A continental crustal model and its geothermal implications, Earth Planet. Sci. Lett., 22, 215-225, 1974.

Smithson, S. B., and P. N. Shive, Field measurements of compressional wave velocities in common crystalline rocks, Earth Planet. Sci. Lett., 27, 170-176, 1975.

Tuve, M. A., H. E. Tatel, and P. J. Hart, Crustal structure from seismic exploration, J. Geophys. Res., 59, 415-422, 1954.

Wolber, G., Energieverluste elastischer Wellen in der Erdkruste, p. 56, Zulassungsarbeit, Univ. of Karlsruhe, Karlsruhe, Germany, 1968.

THE NATURE OF THE EARTH'S CRUST IN CANADA

M. J. Berry and J. A. Mair

Earth Physics Branch, Department of Energy, Mines and Resources
Ottawa, Ontario, Canada K1A 0Y3

Abstract. A review of seismic crustal studies in three regions of Canada (the northeastern Canadian Shield; the northwestern Canadian Shield near Yellowknife, Northwest Territories; and southern portions of the Canadian Cordillera), reveals some similarities in the derived models but also some important differences. The wide-ranging seismic refraction reconnaissance survey in the Grenville and Superior provinces of the shield shows a surprisingly similar velocity-depth structure in both provinces, which must indicate that the general processes controlling the average velocity structure of the crust have acted uniformly over a wide area. Distinctive features of the general model include a low-velocity layer in the upper crust, a minor discontinuity in the middle crust, and a gradual velocity transition in the upper mantle. A reconnaissance refraction experiment near Yellowknife in the Bear, Slave, and Churchill provinces suggests an apparently simple, flat-lying, and uniform structure; a second more limited but detailed experiment reveals a complex structure in the upper crust bearing a strong correlation to the local geology and exhibiting some of the features seen in the Grenville and Superior provinces. The crust-mantle transition zone is very complex. Surface wave studies and reconnaissance refraction, detailed reflection, and Vibroseis (registered trademark of the Continental Oil Company) surveys in the Canadian Cordillera suggest a highly variable crustal structure with little or no systematic layering that extends over more than tens of kilometers. The detailed structure detected by a Vibroseis experiment suggests significant thrusting and complex laminations at the base of the crust, similar perhaps to the structures seen to the east of the Rocky Mountain Trench in the sedimentary rocks of the Rocky Mountains. A comparison of the results of the surveys suggests that the rocks of the Cordillera still retain the highly complex print written by the tectonic processes of the Phanerozoic. It is suggested, however, that the large-scale lateral structures seen in the shield are largely the result of the state of stress in the crust acting on the crustal material and the presence of water either in chemical combination or as pore fluid.

Introduction

Crustal studies of various types have been undertaken in Canada for nearly 40 years and have included most of the commonly used techniques.

The results of the long- and medium-range refraction experiments have been summarized by Berry [1973], who describes the experiments that have been undertaken and the seismic models that have been derived. Figure 1 shows the location of a seismic refraction experiment in the Grenville and Superior Precambrian provinces, another in the Bear, Slave, and Churchill provinces (including a detailed reflection survey near Yellowknife, Northwest Territories), and the location of the detailed reflection-refraction and Vibroseis surveys that have been recorded more recently in the Canadian Cordillera.

In this paper we shall first discuss the validity of the various models of the crust and then compare the results from the three regions. Specifically, we shall conclude that on the basis of limited data, there do appear to be both similarities and differences between the structure of the crust beneath the Canadian Shield and that found in the geologically younger Cordillera. We shall finish by discussing the nature of the crust as revealed by the seismic experiments.

Grenville Experiment

In 1968 the Earth Physics Branch of the Canadian Department of Energy, Mines and Resources undertook a large-scale seismic refraction survey centered on the low-gravity anomaly along the Grenville Front, or the boundary between the Grenville and Superior provinces, of the northeastern Canadian Shield [Berry and Fuchs, 1973]. The experiment attempted to determine the velocity structure of the crust on each side of the boundary in the two provinces (Figure 1) and to determine the structure along the Front itself. To achieve this, three linear profiles of at least 400 km were recorded with shot points at each of their ends. Three-component tape-recording seismographs were used, stations being spaced approximately 20 km apart. In addition, recordings were made at each shot point as explosions were detonated (individually) at all the other shot points. Thus the field experiment yielded three pairs of record sections and a number of P_n travel time observations linking shot points and observation points arrayed over an area of approximately 300 x 400 km.

In the interpretation of the P_n data, Berry and Fuchs [1973] used the time-term approach of Raitt et al. [1969] to determine variations in crustal thickness and possible anisotropy in the material of the upper mantle. They showed that the crustal thickness was greatest along the Grenville Front, somewhat thinner in the Grenville Province, and thinnest in the Superior Province to the north. Absolute values of crustal thickness of course depend upon an accurate knowledge of the velocity structure within the crust and a knowledge of the upper mantle velocity. The study of P_n data suggested that the upper mantle is weakly anisotropic, its maximum velocity being parallel to the front and its minimum velocity transverse to it. The variation was ± 0.17 km/s about a mean velocity of 8.06 ± 0.06 km/s.

The profile data were arranged into record sections, each trace being normalized to remove the effects of variation of charge weight and seismograph sensitivity. Energy arriving subsequent to the first events was carefully studied for possible correlations among adjacent traces, and velocity-depth models were constructed to fit these. Synthetic record sections were then calculated for the model determined from the

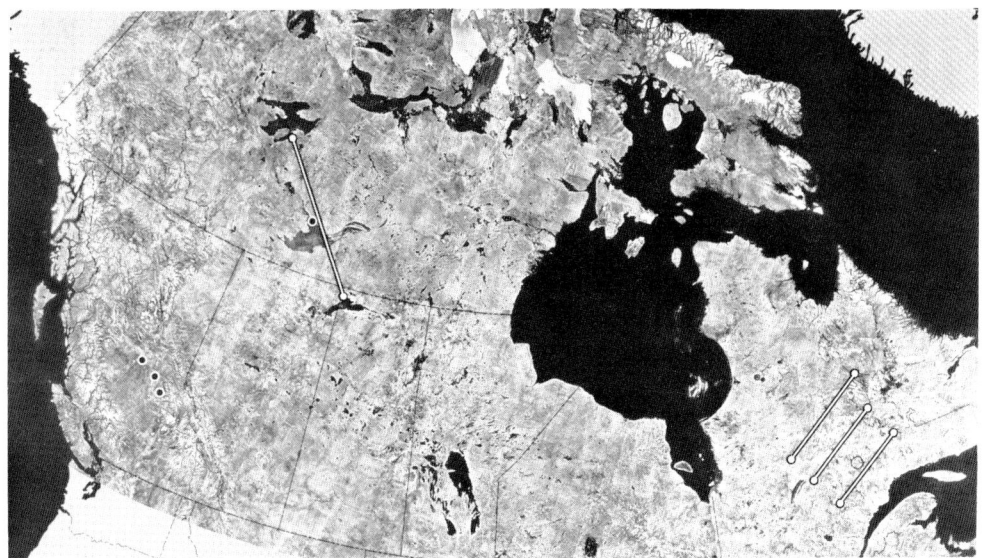

Fig. 1. Erts satellite mosaic of part of Canada. The three lines on the right indicate reversed refraction profiles. The lower line lies in the Grenville Province, the middle line along the Grenville-Superior boundary (Grenville Front), and the upper line in the Superior Province. The line in the upper middle of this figure indicates an areal refraction survey throughout three Precambrian provinces: Churchill on the lower right, Slave in the middle, and Bear at the top. The dot indicates the location of a detailed reflection-refraction survey in the Slave Province. The three dots on the left indicate the locations of reflection profiles in the British Columbia Cordillera.

travel time information. This travel time model was then perturbed to provide a closer fit between the observed and calculated amplitudes. The process was iterated until further perturbation did not appear to provide any significant improvement.

Figure 2 shows a record section recorded along the Grenville Front by using the shot point at the northeastern end of the profile. Superimposed upon the section is the theoretical travel time curve corresponding to the model on the left side of this figure. The bottom section is the theoretical response of the model to the input pulse shown at the side. It is our belief that the model provides a fair, if much simplified, approximation to the observed record section. Considerable effort was spent in attempting to find other models that could explain the main features of the observed seismograms. All such efforts were unsuccessful. Of course, some of the finer details of the model, such as the details of the transition layers, may be incorrect, but in general the model is a reasonable representation of the crust as illuminated by one seismic source and viewed along a line in one direction from that source. Whether or not this complex model is representative remains problematic.

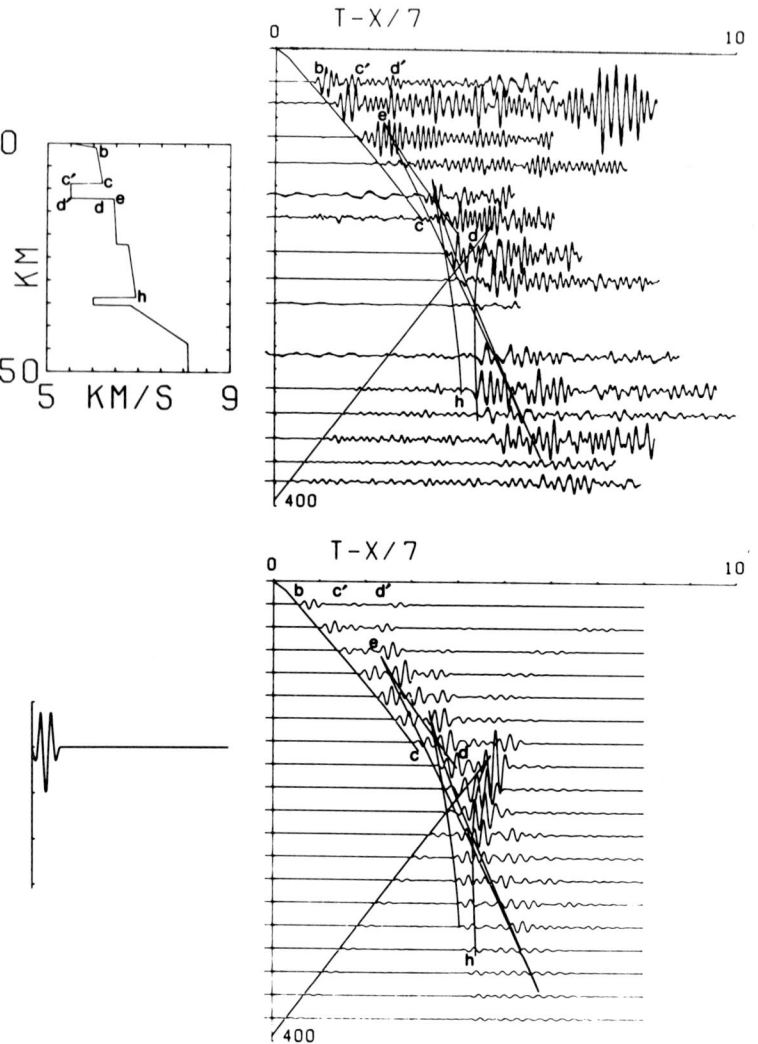

Fig. 2. Refraction data recorded along the Grenville Front (top right), P wave velocity-depth interpretation (top left), and theoretical response of the model when the indicated wave form is assumed (bottom). The observed section, the synthetic section, and the assumed wave form are drawn to the same horizontal scale in seconds. The vertical scale of the sections is in kilometers (from Berry [1972] with permission).

Figure 3 shows the six models that have been developed for the six record sections recorded along the three profiles. The family resemblance of the six models suggests that the main features are generally valid representations of the real crust in the region. Of course, as Berry and Fuchs [1973] point out, the derived models represent the

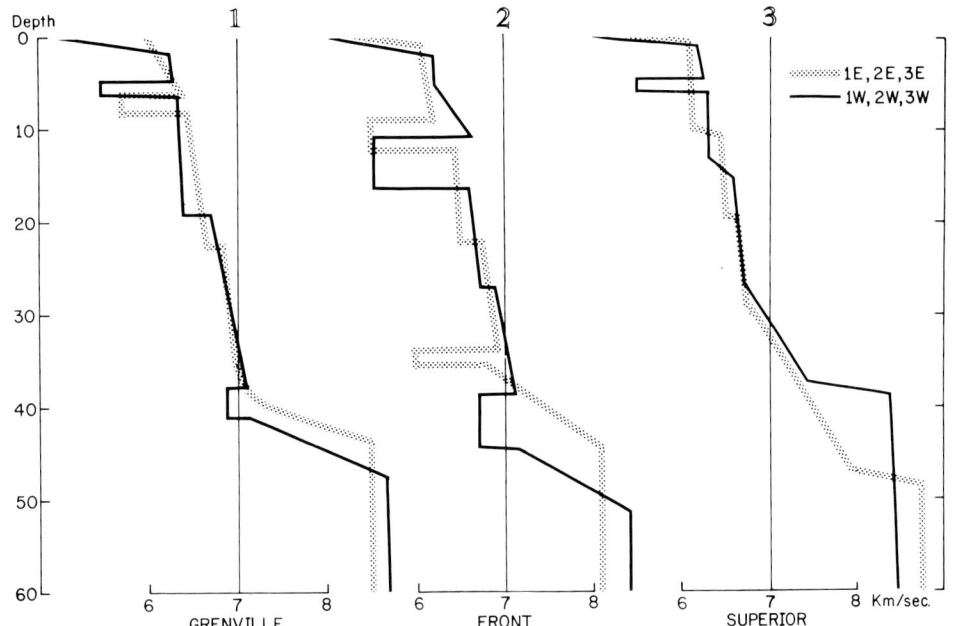

Fig. 3. The six P wave velocity models developed for the three reversed profiles of the Grenville-Superior refraction experiment. The solid lines represent profiles recorded by using eastern shot points, while the stippled lines represent profiles recorded by using western shot points (from Berry and Fuchs, [1973] with permission).

structure of the earth along a cone, starting at the shot point and extending down and along the profile, because the rays that bottom at greater depths are generally observed at greater distances. The effect of this feature of the seismic refraction method is that the derived models do not represent the velocity-depth function at any particular place in the survey area. It is primarily because the models have many features in common that we have confidence that they are relevant to the real earth.

Yellowknife Experiment

The region near Great Slave Lake contains three different Precambrian provinces: Bear, Slave, and Churchill. This region was the locale for a large-scale reconnaissance survey in 1966. The experiment revealed a crustal thickness of 32 km and an average upper mantle velocity of 8.13 km/s. No intermediate refractor was shown by the data, and despite careful analysis, no variation of upper mantle structure between the three different provinces was found. However, there appeared to be a significant difference in upper mantle velocity beneath the Phanerozoic cover rocks overlapping the shield from the west. P_n velocities change from 8.23 ± 0.04 km/s on the west side of this boundary to 8.10 ± 0.03 km/s on the east side [Barr, 1971]. The only other variation from a

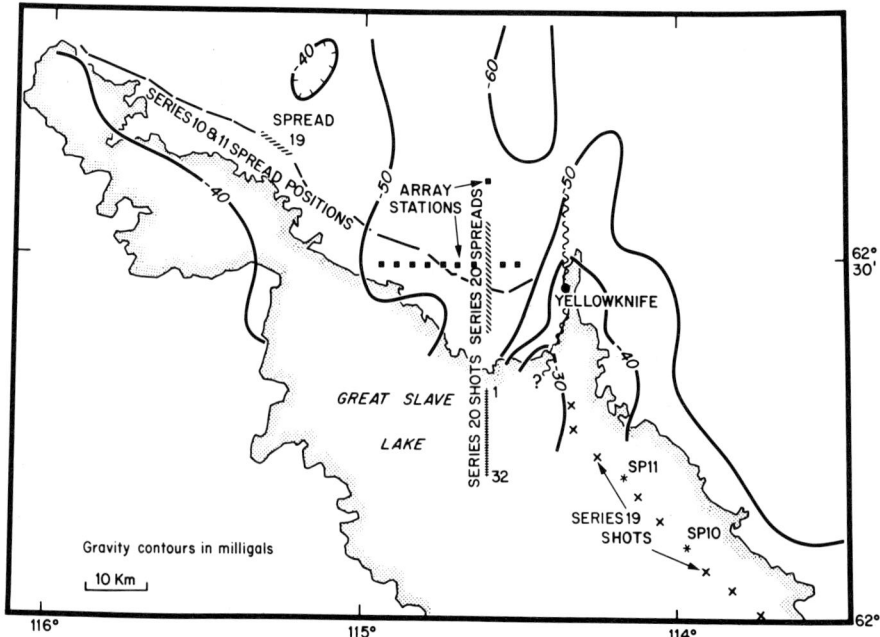

Fig. 4. Location of reflection-refraction experiment near Yellowknife, Northwest Territories, showing shot points (SP10 and SP11), recording lines, and Bouguer gravity contours (from Clee et al. [1974] with permission). The Precambrian Slave Province rocks are exposed to the NE, beginning along the NE edge of the lake. They are assumed to be covered by Phanerozoic rocks to the SW.

uniform model is a depression of 4 km on the Moho boundary, which was detected beneath the east arm of Great Slave Lake, postulated by Hoffman et al. [1974] to be an aulacogen, or failed arm of a triple junction.

In 1969 an extremely detailed set of observations was made in one region of the Slave Province near Yellowknife. This experiment was similar in detail to the deep seismic sounding experiments conducted in the USSR. The data reveal that while the interpretation of the earlier work is a valid simplification, the real structure is more complex. The higher-frequency and shorter-wavelength seismic energy of this second experiment has revealed some unexpected and unusual features.

The locations of shot points, recording lines, and Bouguer gravity contours for the area are shown in Figure 4. The composite recording spread was made up of a total of 432 seismometers at a 250-m spacing. Shots of approximately 200 kg were detonated in up to 30 m of water at shot points 10 and 11 (Figure 4). Although the shot points are only 15 km apart, they produce remarkably different seismic sections, as is shown in Figure 5. A reduced travel time for each trace, calculated from its shot distance in kilometers divided by a velocity of 6 km/s, has been applied to both sections.

The first arrivals, having a velocity of about 6 km/s and the shear

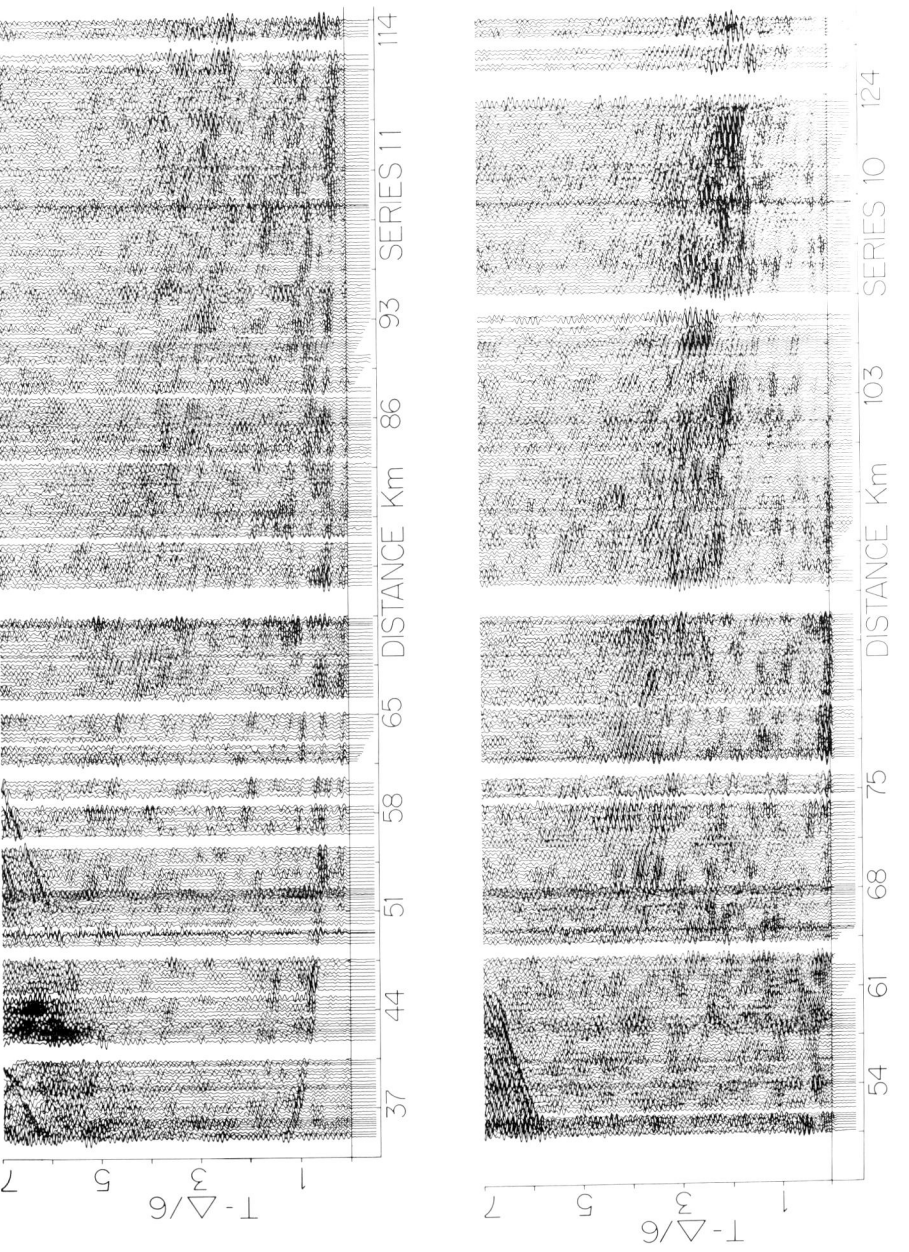

Fig. 5. Reflection-refraction data recorded from shot point 11 (top) and shot point 10 (see Figure 4) by the same array (after Clee et al. [1974] with permission).

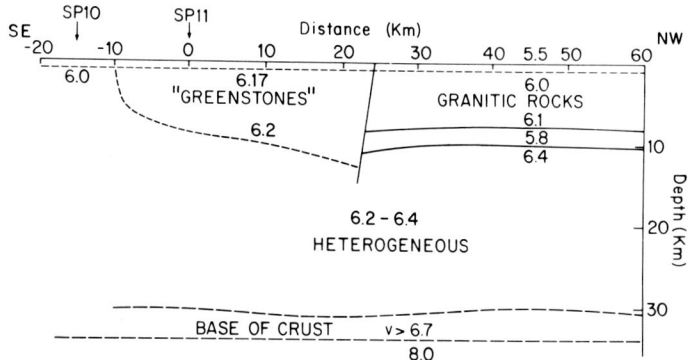

Fig. 6. Interpreted section near Yellowknife, Northwest Territories (from Clee et al. [1974] with permission).

and Rayleigh modes at 5 s and greater, are common to both sections, but there the similarity ends. The strong composite reflection beginning at about 1.2 s and 35 km seen, on the shot point 11 (SP11) series does not appear on the SP10 series. If a continuous horizontal reflector is assumed, this event should, with the time correction that has been applied, occur on the SP10 series, beginning at about 0.57 s and 65 km, for common reflection points. Similarly, the prominent arrivals beginning at about 3.5 s and 70 km on the SP 10 series should be evident from about 5 s and 40 km on the SP11 series, but they are weak or nonexistent. One would expect common reflection point energy from the Moho at these latter approximate times and distances.

The interpretation of Clee et al. [1974], shown in Figure 6, provides an explanation of these differences between the two seismic sections. SP11, interpreted to lie above a complex deposit called the Yellowknife Greenstone Belt, produces prominent early reflections from the top and bottom of a low-velocity layer (5.8 km/s) at 8 and 10-km depths. SP10, however, lies outside the greenstone formation, and if, as is suggested, the base of this formation is irregular and heterogeneous, wave fronts passing obliquely through it will be scattered before they can reach the same reflecting layer. Similarly, to obtain a Moho reflection, seismic energy generated at SP11 must traverse this scattering layer, whereas energy from SP10 does not.

The wave coda from the crust-mantle transition zone, indicated by the SP10 section in Figure 5, is very complex. It shows short time-distance correlations having strongly varying apparent velocities. Character changes are abrupt. The coda has high amplitudes and is of long duration, lasting for intervals of up to 2 s. This suggests that a complex transition zone produces both the high amplitudes and the time duration of the coda. It suggests abrupt changes in the vertical structure of this boundary every few kilometers. For example, migration of the events at 103-110 km in the 2- to 3-s interval of this section, a reflection mode being assumed, favors the interpretation of a near-vertical fault in the transition zone with uplift to the southeast of 2 or 3 km.

Analysis of these data does not indicate a requirement for a refractor/reflector at the intermediate depths typical of the Conrad discontinuity.

Velocities remain typical of granitelike rocks down to the crust-mantle transition zone.

Cordillera Experiment

A mosaic of the Erts imagery of the Canadian Cordillera is shown in Figure 7. Superimposed on this figure are the outlines of the main geological provinces. From west to east these are the Insular Belt (I), the Coast Plutonic Complex (II), the Intermontane Thrust and Fold Belt (III), the Hinterland Belt (IV), the Omineca Crystalline Belt (V), and the Rocky Mountain Belt. Although the units are defined by a detailed study of geology [Wheeler and Gabrielse, 1972], it is clear from the mosaic that they are also physiographically distinct.

According to Monger et al. [1972] the present Canadian Cordillera is the result of a chronological westward stepping of subduction activity. The evidence is recorded in volcanic arc sequences, from which westward shifts of the oceanic basin are inferred. The first good evidence for such an assemblage is the sequence in the area between the Omineca Crystalline Belt (V) and the Intermontane Thrust and Fold Belt (III). This is interpreted to be the result of an island arc-subduction zone complex which evolved during the Late Triassic.

Over the past decade the Earth Physics Branch has undertaken a series of reconnaissance refraction profiles similar in detail to those described in the Superior and Grenville provinces. They have been interpreted by Berry and Forsyth [1975] and Forsyth et al. [1974] to show major features which correlate generally with the regional geological structure. Topography along the Mohorovicic discontinuity appears to have wavelengths corresponding to those of the topography of the major geological provinces and tends to support the thesis of westward stepping subduction. The variation of crustal thickness as deduced by a time-term analysis Berry and Forsyth, [1975] is shown in Figure 8.

The crust under Vancouver Island is anomalously thick, and the role of the island in the formational history of the area is still unclear. The crust thickens eastward under the Coast Plutonic Complex (II) to about 33 km; this depth contour rather faithfully outlines the geologically deduced western boundary of the Intermontane Thrust and Fold Belt (III). It continues to the southeast, crossing into the Hinterland (IV) in the south rather than swinging southward with the geological boundary. The Erts mosaic, however, suggests that the physiographic boundary follows the depth contour rather than the interpreted geological boundary. There is a crustal thickening to the SE and to the NE beneath the Omineca Geanticline (V).

Analysis of the P_n data suggests that the material of the uppermost mantle has a velocity of 7.83 km/s in the southern part of the region and 8.06 km/s in the northern part. Wickens [1977], on the basis of the dispersion of Rayleigh and Love waves, requires a pronounced upper mantle low-velocity channel very close to the base of the crust in the southern part of the Hinterland Belt (Figure 7, IV). It appears quite possible from the refraction results that this low-velocity channel is at the base of the crust in the south and is only slightly deeper in the north.

To reconcile the Bouguer gravity data [Stacey, 1973] with the seismic structural results requires lateral variations in the average crustal

Fig. 7. Erts mosaic of the Canadian Cordillera. The superimposed outlines indicate geological provinces, which are, from left to right, Insular Belt (I), the Coast Plutonic Complex (II), the Intermontane Thrust and Fold Belt (III), the Hinterland Belt (IV), the Omineca Crystalline Belt (V), and the Rocky Mountain Belt. To the right of the dashed line is the Purcell Arch.

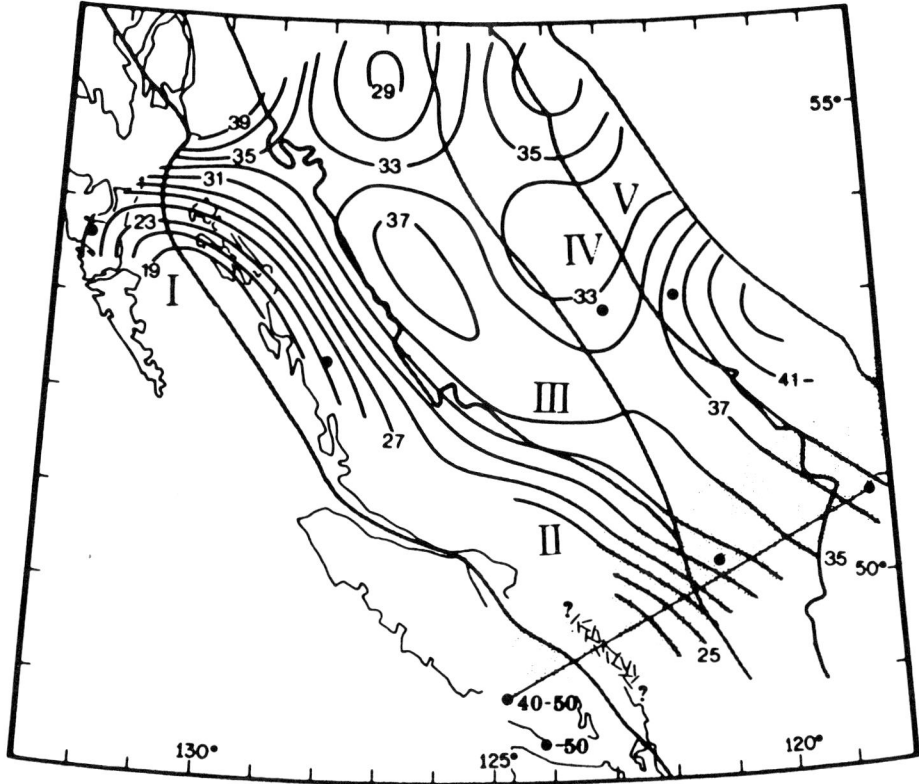

Fig. 8. The contoured time-term surface and interpreted Moho depths in kilometers superimposed on the outlines of the geological provinces of the cordillera (from Berry and Forsyth [1975] with permission).

density. The reconciliation requires that the crust beneath the central Intermontane region be characterized by a mass deficiency, whereas the density of the crust beneath Vancouver Island may be greater than average.

Resolution of the very fine structure of the crust using long-range refraction data is limited by the lower-frequency energy transmitted and the usual unfortunate economic-logistic necessity of limiting recording site spacings to approximately 20 km. Much greater resolving power is provided by near-vertical reflection techniques at the expense, of course, of obtaining a regional interpretation. In 1972 we recorded three reflection profiles using three lakes as shot points, located as is shown in Figure 9. Ahbau and Cariboo lakes are in the Omineca Crystalline Belt, and Naltesby Lake lies in the Hinterland Belt. Refraction results [White and Savage, 1965] had previously indicated the possibility of the existence of a major earth suture between these two provinces.

We first wished to demonstrate that near-vertical reflection techniques could record deep reflections in a cordilleran province. If we were successful, we wished to interpret the data to give a

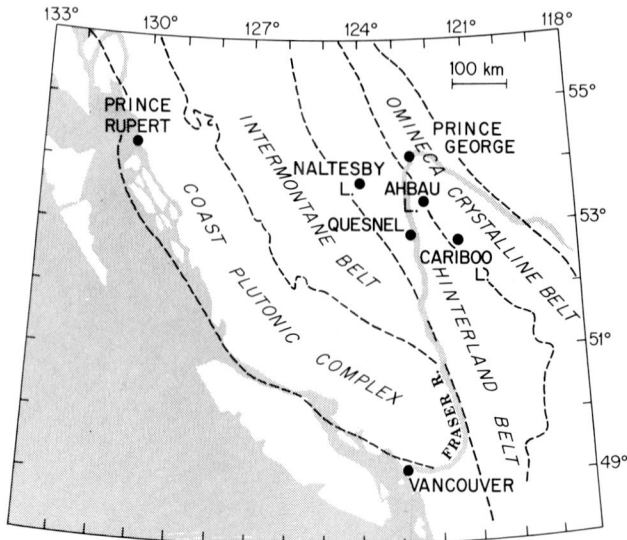

Fig. 9. Location of Naltesby, Ahbau, and Cariboo lakes, the shot point sites for the reflection profiles (from Mair and Lyons [1976] with permission).

velocity-depth profile for each area and to relate these profiles to the crustal structure. It has turned out that we can indeed record reflected energy from depths typical of a crust-mantle transition zone but must use other techniques to estimate velocities. The deep reflection is not sufficiently coherent over the profile length, nor are there any coherent shallow reflections with which to attempt a velocity-depth interpretation.

The field technique and the rather elaborate data processing sequence attempted have been described by Mair and Lyons [1976]. Briefly, the scheme was to employ up to six shot points spaced along each lake, shots from each point to be recorded by closely spaced geophone arrays extended in line along a road. Figure 10 shows the survey at Ahbau Lake: the five shot points in the lake, spaced about 1.5 km apart; and 42 recording sites, spaced about 250 m apart along a road that was more or less in line with the shot points. This geometry allows up to five fold common reflection point (CRP) stacking. Each recording site was composed of 16 geophones, spaced in a quasi-tapered linear array over 130 m. Shot size varied but was generally about 50 kg in various multiple configurations depending upon the available water depth. While the individual seismic sections, generated by shots from each shot point, indicated a reflected event at about 11 s two-way time, our attempts to improve the signal-to-noise ratio by CRP stacking have not been successful. This lack of success results from poor coherence of the traces from the different shot points which make up the CRP groups. It is assumed to be caused primarily by lateral variations in structure and velocity of the overburden encountered by these differing ray paths.

The data from Ahbau Lake are shown in Figure 11. All of the recorded

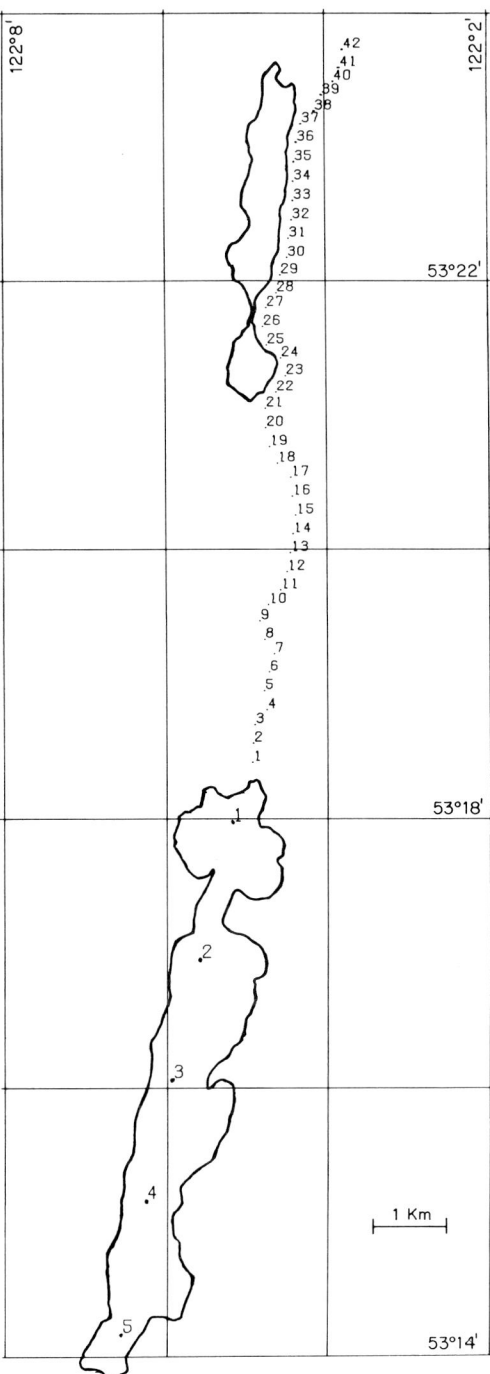

Fig. 10. Survey layout of Ahbau Lake, British Columbia (from Mair and Lyons [1976] with permission).

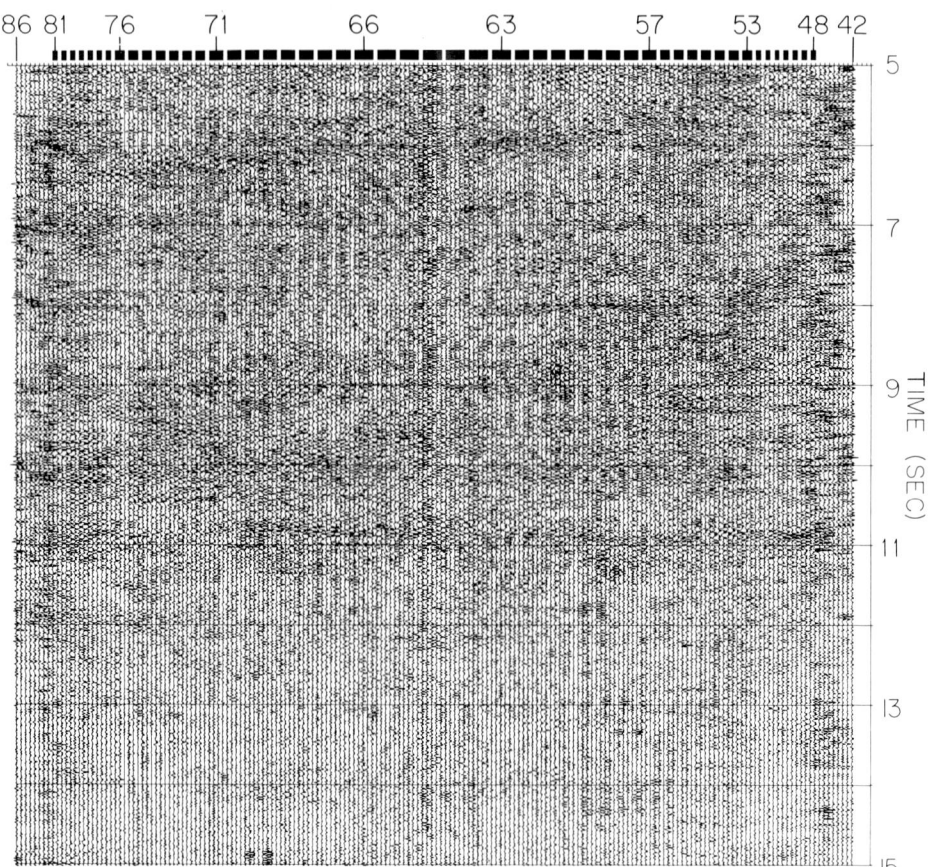

Fig. 11. Ahbau Lake dynamite-produced reflection section. Traces, derived from all five shot points, placed in CRP order but not 'stacked.' A stacking velocity of 5.6 km/s has been applied. The distance axis scale varies with the number of traces which would be involved in a CRP stack. The reflecting point numbers shown relate this section to Figure 12 (from Mair and Lyons [1976] with permission).

traces are shown in CRP stacking order but have not been stacked. While the poor coherence of these CRP groups is apparent, this display does give an indication of crustal structure through an apparent coherence of trace amplitude levels.

Considering only the relatively continuous event at 11 s, we can say that it is not the result of multiple reflected energy developed by shallower formations, since no continuous shallow reflections exist. The event is consistent from all five shot points, which provide geophone to shot point distances ranging from a few hundred meters to over 18 km. The structure producing this event would therefore have to be of the order of 10 km in length, running nearly parallel

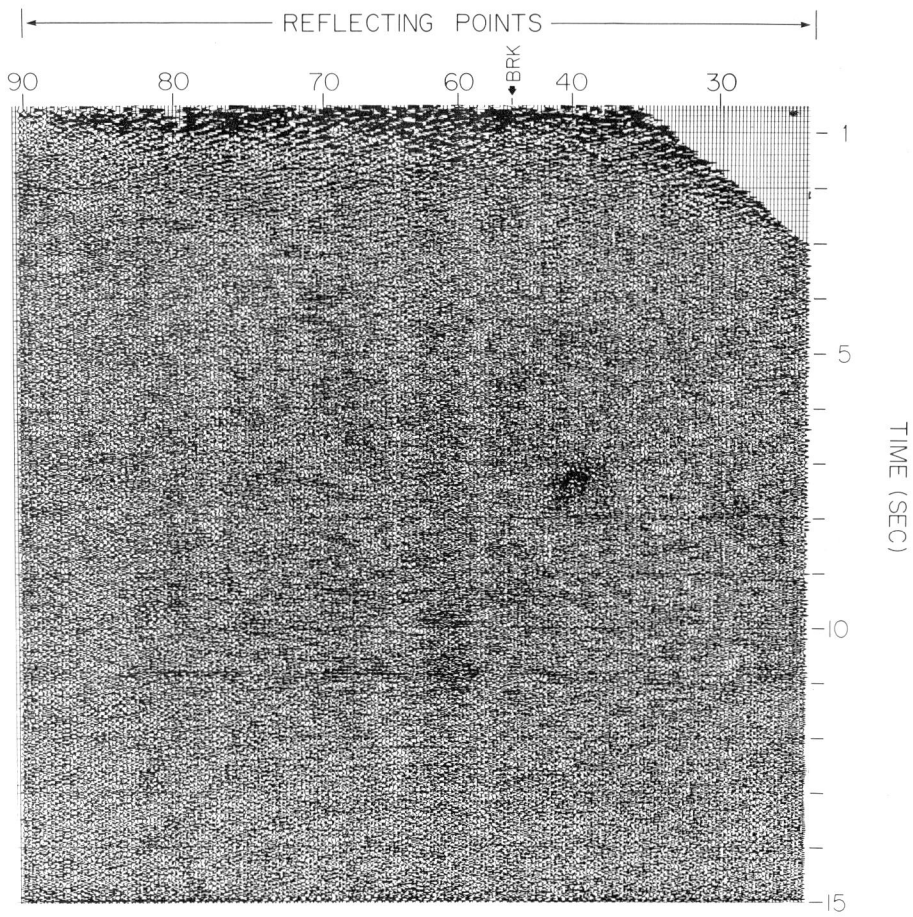

Fig. 12. Ahbau Lake Vibroseis reflection section. Total length of section is approximately 20 km of surface coverage (from Mair and Lyons [1976] with permission).

to the surface profile. Since the profile is actually in a direction that is normal to the major surface features of the area, the most reasonable explanation for the 11-s event is that it results from a near-horizontal reflector lying beneath the profile. On the assumption of an average crustal velocity of 6.3 km/s, the reflection at 11 s is from a depth of 34 km. This velocity and this depth to the Moho are consistent with the model proposed for an area 35 km to the south, the Puntchesakut Lake-Jack of Clubs Lake profile of Berry and Forsyth [1975].

Reflecting angles, a horizontal reflector at 34-km depth being assumed, vary over a range of about 15° within each CRP group for the Ahbau profile. It is possible, for this range of angles, that the nature of the reflecting zone itself can contribute to the poor coherence, and this possibility will be discussed later.

A similar arrival time of about 11 s for a reflected event was obtained for the Cariboo Lake survey, also in the Omineca Crystalline Belt. The deep reflected event obtained at Naltesby Lake, in the Hinterland Belt, appeared at about 10 s two-way time, indicating a reflector about 3 km shallower. This again is consistent with the regional refraction interpretation. Of course, lacking a continuous profile between the two areas, we cannot be sure that these reflected events are from the same horizon.

The distance axis of Figure 11 relates it to the Vibroseis section that follows. Note that the 11-s event disappears to the left of reflecting point 71, possibly reappearing at point 76. Figure 12 is the Vibroseis section obtained at Ahbau Lake in 1973 and shows roughly the same subsurface coverage as Figure 11. The field parameters employed were suggested by the dynamite survey of Figure 11. Each trace here is a twelvefold common reflection point stack of traces resulting from 32 stacked Vibroseis input pulses. Each pulse was a swept sinusoid, varying from 6 to 30 Hz over 15 s. The pulses were provided by three 15-t Vibroseis units working in unison.

An event at 11 s is apparent, and, as is indicated by the dynamite section, the event disappears to the left of reflecting point 70, reappearing at point 75.

The field geometry employed in the two surveys is quite different. A much narrower range of incident angles has been employed in the twelvefold stack producing the Vibroseis section, which may explain its greater apparent success with the CRP technique. However, one can argue that this success is due to a better signal-to-noise ratio of the traces within the stack, to a greater stack complexity, or to both.

Figure 13 illustrates the effect of filtering a broadband section of the Vibroseis data, shown on the left, over narrow passband ranges. We see little useful energy below 10 Hz. Between 9 and 11 s at higher frequencies we see, within the overall cyclic behavior caused by the filtering, apparent reflections which come and go, particularly at 10 s. This technique is a very blunt tool for studies of the nature of reflecting horizons, but the result is suggestive of a complex sequence of layers of varying individual thicknesses, the overall thickness being 3-4 km.

Figure 14 shows an expanded view of a section of the Vibroseis profile. The signal-to-noise ratio is quite good. The event at 10.8 s correlates well, except between reflecting points 71 and 76. Less noticeable but significant, we believe, are several short sequences wich propagate along lines angling upwards to the right at roughly 45° and against which the near-horizontal events terminate or are displaced, for example, the line from beneath reflecting point 74 at 11.8 s through a point beneath reflecting point 70 at 10.8 s, continuing to about reflecting point 68 at 10.6 s. The nature of these lineaments is debatable but suggests interference arising from local fault dislocations.

If one emphasizes all the reasonably good phase correlations apparent on the Vibroseis section, the result is as shown in Figure 15. A general pattern of overthrusting to the right, which is northeast for this profile, becomes evident. Some of these apparent events may, of course, be due to local and to off-line diffraction sources and may not be evidence of real structure along this section. The abrupt changes in reflection character and in the apparent dip and the dis-

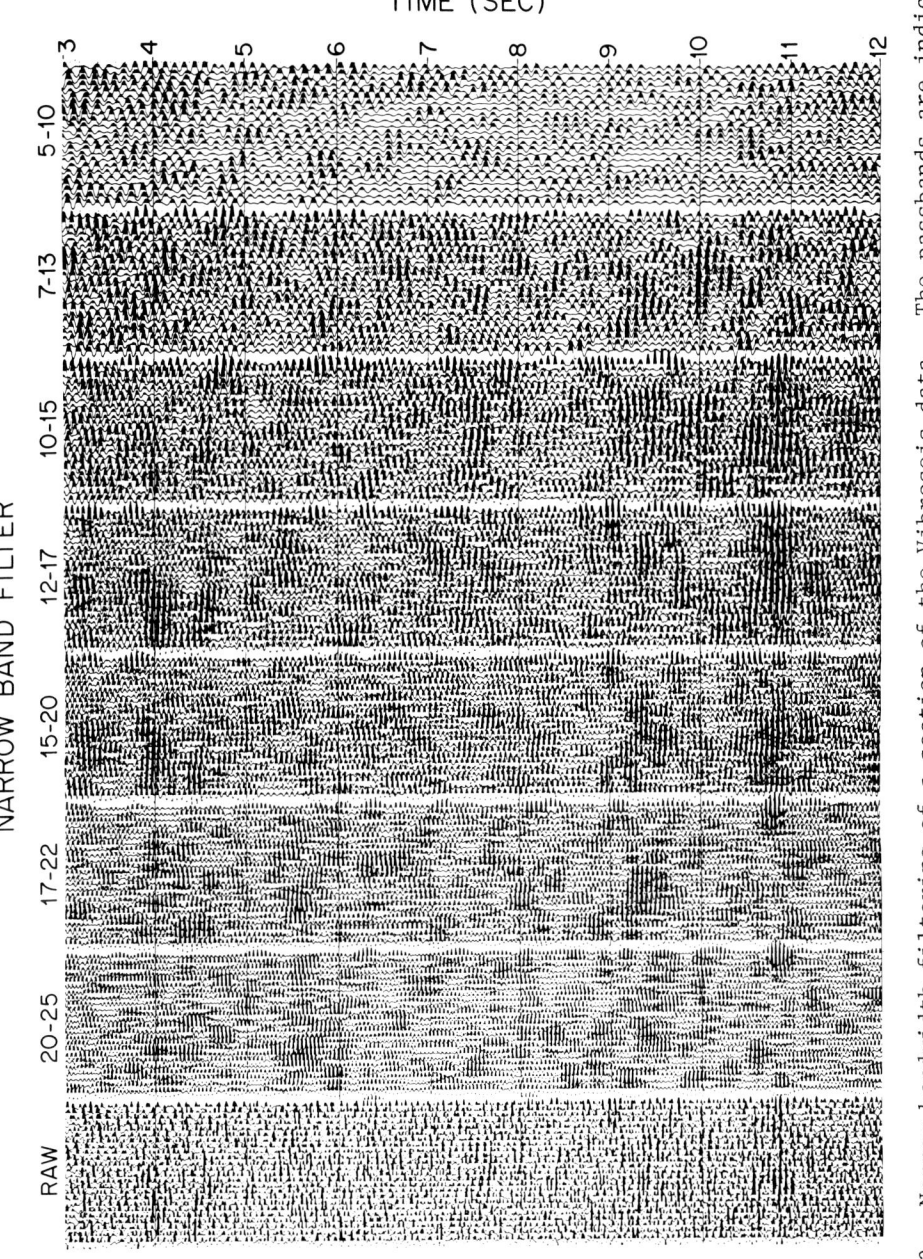

Fig. 13. Narrow bandwidth filtering of a section of the Vibroseis data. The passbands are indicated in hertz (from Mair and Lyons [1976] with permission).

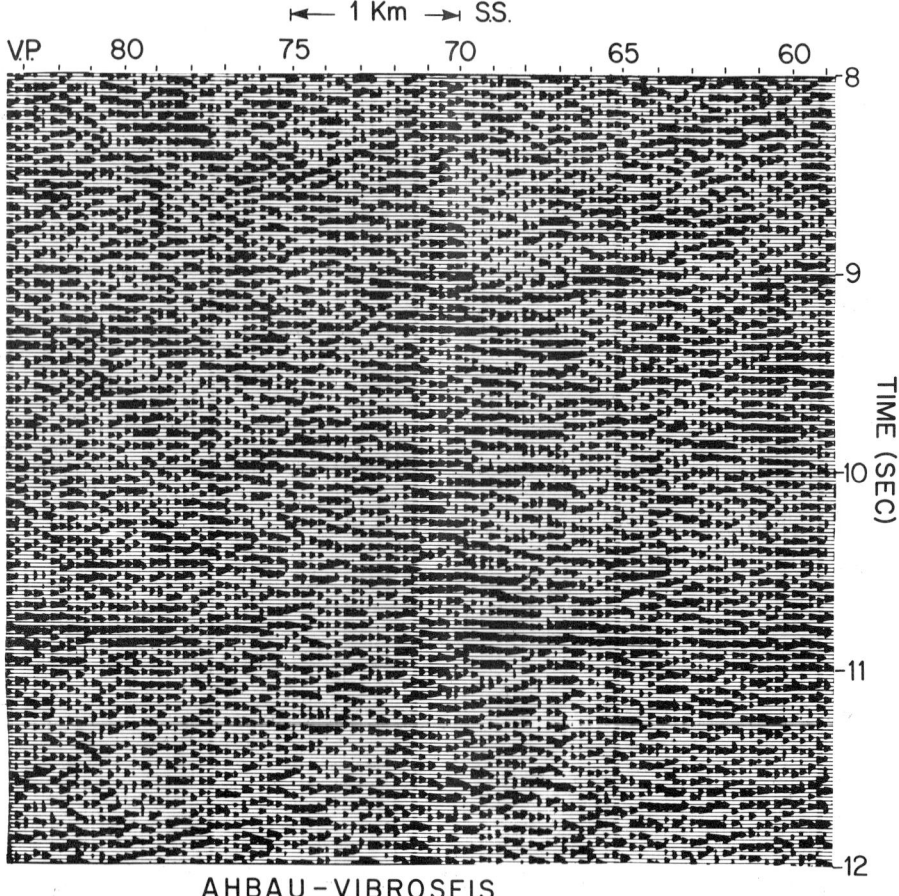

Fig. 14. Expanded view of a section of the Vibroseis data. The subsurface (SS) coverage scale is indicated (from Mair and Lyons [1976] with permission).

continuous nature of the majority of these events suggests, however, that they are indicative of a highly variable local structure. A subparallel near-horizontal system of events is evident in the 9- to 11-s time interval. Continuous segments are generally less than 1 km in lateral extent.

Discussion

The seismic velocity structure of the crust is only of limited interest in itself, however detailed it may be. Its real value lies in what it can tell us of the tectonic history of the region and of the lithology at various depths. In addition, the velocity structure may reveal something of the variation of physical parameters, such as

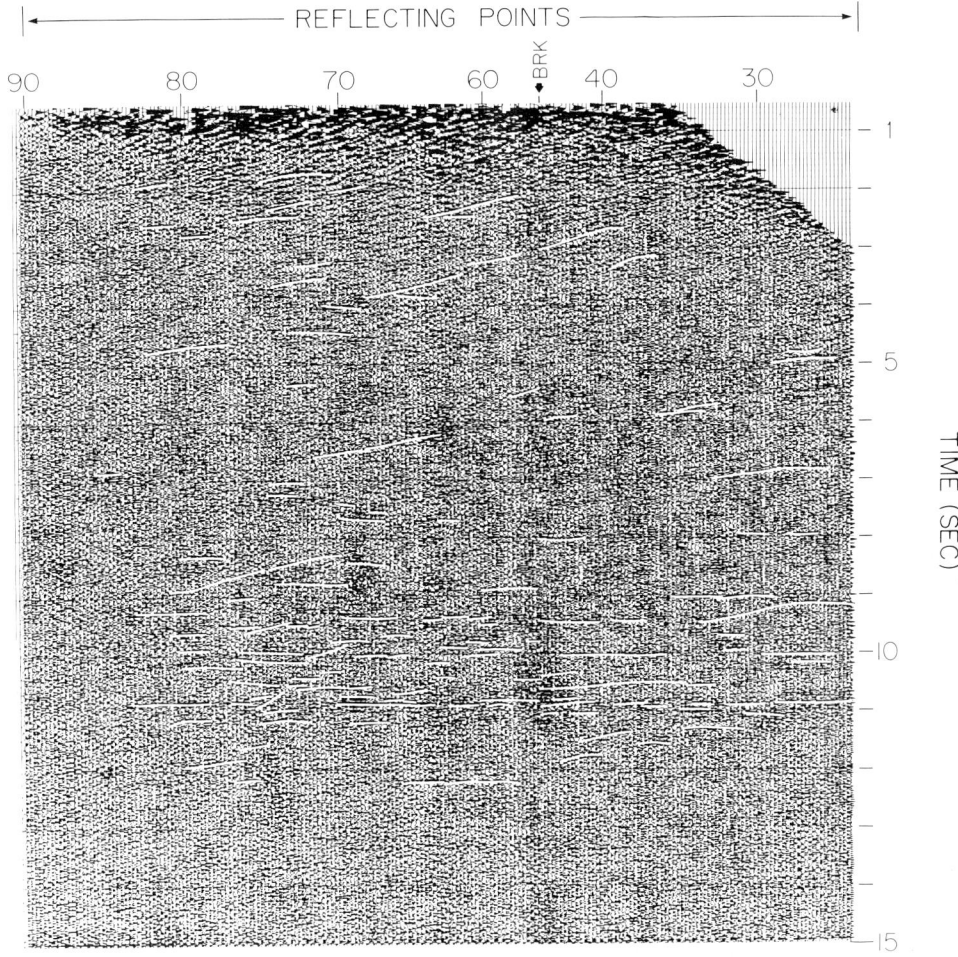

Fig. 15. Vibroseis section with emphasized phase correlations.

stress, porosity, permeability, or temperature, with depth. In the following we shall discuss the nature of the crust in the light of the data that we have described and the results from other fields and laboratory experiments.

Upper Crust

The rapid velocity increase in the top 1-2 km of the bedrock from approximately 5.5 to 6.1 km/s is a commonly observed phenomenon in Canada wherever there have been detailed recordings close to the shot point and wherever the bedrock is close to the surface. It has been observed in the Grenville, Superior, and Slave provinces of the shield and in some parts of the Cordillera. For regions having granitic rocks at the surface it corresponds remarkably well with the velocity-depth curve of Birch [1958] for average granite. The velocity increase is almost certainly caused by the closing of the microfractures in the rock as a result of increasing confining pressure with depth. The same effect is seen in Figure 16, adapted from Nur and Simmons [1969]. The increase in velocity as the stress rises from 0 to 1 kbar is very pronounced. The velocity-depth models of Figures 3 and 6 suggest that beneath the zone of high gradient the velocity becomes nearly constant at 6.0-6.3 km/s showing only a very slight positive gradient with depth. This behavior is certainly consistent with the behavior of either dry or saturated rocks subjected to increasing confining pressure, pore pressure being low.

Each of the two surveys described in the shield has revealed the presence of a well-developed low-velocity zone below the region of nearly constant velocity described above. The zone apparently lies at about 5 km in the Grenville and Superior provinces, at 10 km in the Slave Province, and somewhat deeper, at about 12 km, along the Grenville Front. This variation does not correlate with the age of the province nor with the depth of erosion of the crust as deduced from the metamorphic grade of the rocks at the surface.

Low-velocity zones have been postulated by Landisman and Mueller [1966], and Mueller and Landisman [1966] in the upper crust in such widely separated places as north Germany, west Transvaal, Pennsylvania, southeast Australia, central Japan, east Montana, the Baltic Shield, and the Black Forest of West Germany. Those authors base their interpretation partly upon reflections from the top and bottom of the layer which they typically find at approximately 10 km, having a thickness of 3-4 km. The record section in the upper part of Figure 5 shows such reflections very clearly. Landisman et al. [1971] and Berry [1972] have discussed some of the possible explanations for such a low-velocity layer in the upper crust. The feature appears to be fairly common and is found in a wide variety of geological settings, varying as we have described from regions of the Canadian Shield to the eastern Basin and Range Province [Braile, 1977]. By the use of synthetic seismograms that allow for variation of Q, Braile has interpreted his data to indicate that the low-velocity layer in the Basin and Range Province is also a layer of low Q (approximately 50) and high Poisson ratio (approximately 0.31). As Braile's study is pioneering, we cannot say whether these additional features of the zone are commonplace elsewhere or unique to this region.

The apparently common nature of the zone, coupled with its presence in the upper part of geologic sections that have clearly undergone vastly different histories, suggests to us that its explanation lies in the variation of the physical parameters that affect seismic velocities rather than in the variation of the geochemistry or geologic structure of the crustal section.

If this is correct, then the variation of compressional velocity with depth depicted in Figure 16 offers a likely explanation. The work of Nur and Simmons [1969] shows that in a rock subjected to increasing confining pressure the velocity increases monotonically, provided that the pore pressure is negligible. However, if the pore pressure is allowed to rise and approaches the confining pressure, then the velocity will fall to values close to those experienced at very low confining pressures and hence depths.

Todd and Simmons [1972] have demonstrated that for a number of

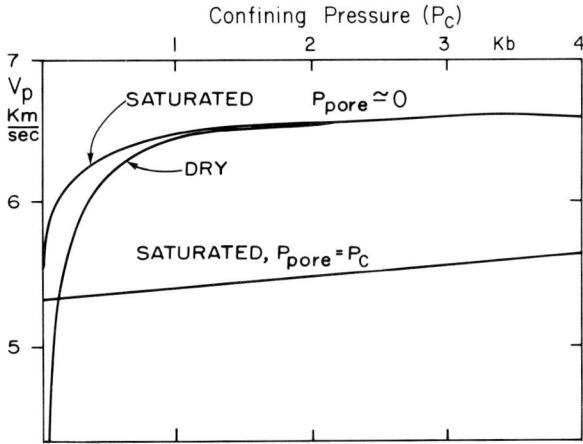

Fig. 16. The effect of pore pressure on compressional velocity as a function of confining pressure in crystalline rocks. The upper curves are for zero pore pressure, while the lower line is for the case in which the pore pressure equals the confining pressure [after Todd and Simmons, 1972].

crystalline rocks, V_p is an increasing function of an effective pressure P_e, where $P_e = P_c - nP_p$. P_c is the confining pressure on the rock due to the weight of the overburden, and P_p is the pore pressure exerted by pore fluid in the rock. The value of n for high pore pressures was found to be about 0.9. For crystalline rocks containing microscopic pore cavities interconnected by microfractures mostly along grain boundaries, one can visualize a situation in which increasing confining pressure with depth squeezes the rock, closes the microfractures, and thereby decreases its permeability. At some depth the rock may become so impermeable as to trap fluid in layers below. The pore pressure would increase at such depths, the result being that the effective pressure would decrease. As is indicated in Figure 16, this effect could give a decrease in velocity of as much as 15%.

The difficulty with such an explanation is that it requires that the rocks maintain some porosity at depths of the order of 10 km but lose all their permeability at a shallower depth in order to provide a cap rock for the layer. Laboratory measurements suggest that at such depths the confining pressure and temperature would tend to close up all microcracks and the pores as well (G. Simmons, personal communication, 1976). This may indeed be the normal case in a hydrostatic earth; however, we know that more frequently than not the state of stress in the earth's crust is not hydrostatic. The existence of scattered seismicity throughout the continents is certain proof of this. Confined fluids may, of course, have pressures that are a significant fraction of the overburden pressure exerted on the host rock. One need look no further than the production of hydrocarbons, in which, even to depths of 8 km, one finds examples of migration into the host rock by ancient fluids having pressures equivalent to the confining pressured exerted on the host rock. In these cases, however, migration has been halted by a local trapping mechanism usually involving a cap rock of composition different from that of the reservoir rock beneath. Without appealing to a change in rock type to form the low-velocity zone we must hypothesize a rapid change in the stress tensor at the top and at the bottom of the layer. Is it possible that the combined effects of temperature, overburden pressure, fluid pressure, and tectonic stress can produce these variations in a total stress tensor otherwise smoothly varying as a function of depth?

The suggestion that significant pore fluids may exist at depths of 10 km or more and that their pore pressure may explain the low-velocity layer is an intriguing one. As Landisman et al. [1971] point out in connection with a postulated deeper low-velocity layer, the presence of such pore fluids would likely produce a zone of low Q and low electric resistivity [Chaipayungpun and Landisman, 1977].

The observation that the boundaries of the low-velocity zone often produce clearly defined reflections suggests strongly that the velocity transitions at the top and bottom are sharp and probably only 100-200 m thick. It remains to be explained how these observations can be made compatible with theory and laboratory measurements of crustal parameters.

Middle Crust

Beneath a low-velocity zone, when it is present, the velocity may be slightly higher than that immediately above the layer. However, whether the velocity increase is present or not, the velocity gradient for the next several kilometers appears to be very similar to that detected in the upper crust. This seems to argue for a relatively uniform material at these depths, at least in a gross sense, the velocity being controlled largely by the pressure and temperature gradients.

In the Grenville and Superior provinces we find a relatively sharp jump in velocity of about 0.3 km/s at depths of about 20 km (Figure 3). A similar feature has been detected by Clowes and Kanasewich [1970] beneath the Alberta Plains and by Hall and Brisbin [1965] beneath the Superior Province west of Lake Superior. We have not detected such a feature in the Yellowknife section (Figure 6) of the Slave province, nor do we find any evidence of it in the Cordillera, as is demonstrated by the Vibroseis section in Figure 12.

This midcrustal refractor and sometime reflector may be related to the Conrad discontinuity in parts of Europe. In Canada it has been named the Riel discontinuity by some authors [Hall and Brisbin, 1965] partly at least to suggest that it may not represent the same feature in Canada as in Europe. Certainly, the characteristics of the discontinuity appear to change significantly from region to region, ranging from a major first-order discontinuity to a minor second-order transition in velocity. When it is present, it is usually interpreted as representing a change of rock type or bulk chemical composition.

In modeling the Q structure of the continental crust beneath Alberta, Clowes and Kanasewich [1970] conclude that the Conrad (Riel) reflector is in fact a complex sandwich of thin (about 200 m) sills of alternating high- and low-velocity material extending over a depth interval of less than 2 km. They require a similar structure to explain the nature of the observed Moho reflector.

The Conrad discontinuity does not appear to be a fundamental property of the earth's crust. It is not reported in many areas of the world and certainly not everywhere in Canada. Even in ancient crustal blocks presumably formed at the same time and therefore presumably under similar conditions it may or may not be present. Perhaps its seismically detectable properties once formed can be erased by subsequent tectonics. Conceivably, an originally stratified chemical boundary can become gradational through prolonged metamorphism. Conceivably, a velocity change can be produced or erased within a generally gradational sequence of crystalline rocks by an anisotropic stress field acting under appropriate temperatures and over a suitable length of time.

Lower Crust

The velocity in the lower crust appears to increase with depth, not necessarily following a smooth function. Refraction data from the Cordillera and from Grenville and Superior provinces require this form of the velocity-depth curve, but the shorter-range profile from near Yellowknife is ambiguous on this point.

Some of the Grenville and Superior profiles suggest that this velocity gradient may be broken by a low-velocity layer (Figure 3). Such a feature has also been discussed by Landisman et al. [1971], who review data from a wide variety of geologic regimes. They suggest that it may be connected with the presence of electrically conductive layers, as was reported,, for example, by Hyndman and Hyndman [1968] in the lower crust of the eastern portion of the Canadian Cordillera; van Zijl [1977] also reports a marked zone of low resistivity (less than 50 Ω m) beneath the Limpopo Mobile Belt of southern Africa.

Landisman et al. [1971] suggest that the zone can be explained in terms of certain dehydration reactions resulting in the release of free water into the surrounding pore spaces. Such a phenomenon, rather similar in concept to that favored earlier for the presence of a low-velocity zone in the upper crust, may well be appropriate for relatively young orogenic regions. However, whether it is relevant to old shield regions remains a moot point. A variation of the concept that water may play a major role in determining the nature of the lower crust has been proposed by van Zijl [1977], who suggests that rather than the occurrence of dehydration reactions, there may be sufficient free water available for the lower crust to contain significant amounts of hydrated minerals such as serpentinite. This may well explain the high-conductivity zone and provides a material having the low velocities that are sometimes observed.

Crust-Mantle Transition

The significant velocity increase at the base of the crust from velocities of about 7 km/s to values greater than 7.8 km/s is a near-universal phenomenon first observed by Mohorovicic. It has not been observed near the axis of spreading ridges, but elsewhere it is almost always an excellent refracting horizon and sometimes an excellent reflector (e.g., Figures 11 and 12).

The velocity transition is considered to represent the transition from crustal to mantle rocks. However, the exact nature of this is still a debatable point. Is the change in velocity caused by a change in bulk chemical composition and hence rock type, by a change in velocity anisotropy, or by some combination of these? Some light may be shed on this question by considering in some detail the nature of the seismic signature within this transition zone.

It has been proposed by many authors, including Fuchs [1969], Clowes and Kanasewich [1970], Meissner [1973], Clee et al. [1974], Dohr and Meissner [1975], and Cumming and Chandra [1975], that the crust-mantle transition zone is a many-layered complex of alternating high- and low-velocity lamina having a total thickness of a few kilometers. Individual layer thicknesses are less than the seismic wavelengths with which the earth allows us to probe. This model is thought by many authors to explain best the observations (1) that near-vertical reflection energy returning from deep within the crust is of much higher amplitude than can be explained by a reasonable first-order discontinuity or by ramp functions of increasing velocity with depth and (2) that the reflected energy is limited in bandwidth, little of the detectable energy having frequencies below 10 Hz. It appears that the model may apply equally well in shield and younger regions.

We add to these requirements that the energy returned from the zone persist at relatively high amplitudes for up to 2-s duration in some areas. The Yellowknife section (Figure 5), for example, showed a short-duration shallow reflection, indicating a short down travelling pulse but a long-duration transition zone signature.

One might consider a model such as that shown in Figure 17. The down going pulse, indicated at the top left, reflected from the model at various angles of incidence produces the changing pulse shapes shown. These pulses contain contributions from all the multiples and converted waves generated by the interfaces, an incident plane wave front being assumed. Numerical experiments indicate that the only effective way to increase the apparent time duration of the reflected energy is to increase the total thickness of the transition layer or the time duration of the incident pulse. In this example it is interesting to note that the highest-amplitude wavelet, apparent at $0°$ incidence, does not correlate with any specific interface but is a result of constructive interference. Under typically noisy field recording conditions, which might obscure all events other than this wavelet, we might be inclined to interpret a sharp first-order transition, whereas we have actually involved a section of 3 km in thickness. If we were able, on a field recording, to discern the first event on this figure and this higher-amplitude wavelet, a $T^2 - X^2$ interpretation, assuming a vertical two-way time of 10 s to the first event, would yield an interval velocity of 7.52 km/s over a section of 2.8-km thickness, whereas the interval velocity is actually 7.06 km/s over a section of 3.05 km, i.e., an anisotropy factor of about 7%.

Such a laminated velocity model may have theoretical appeal, but is it geologically plausible? In the following paragraphs we suggest some mechanisms that may provide an explanation.

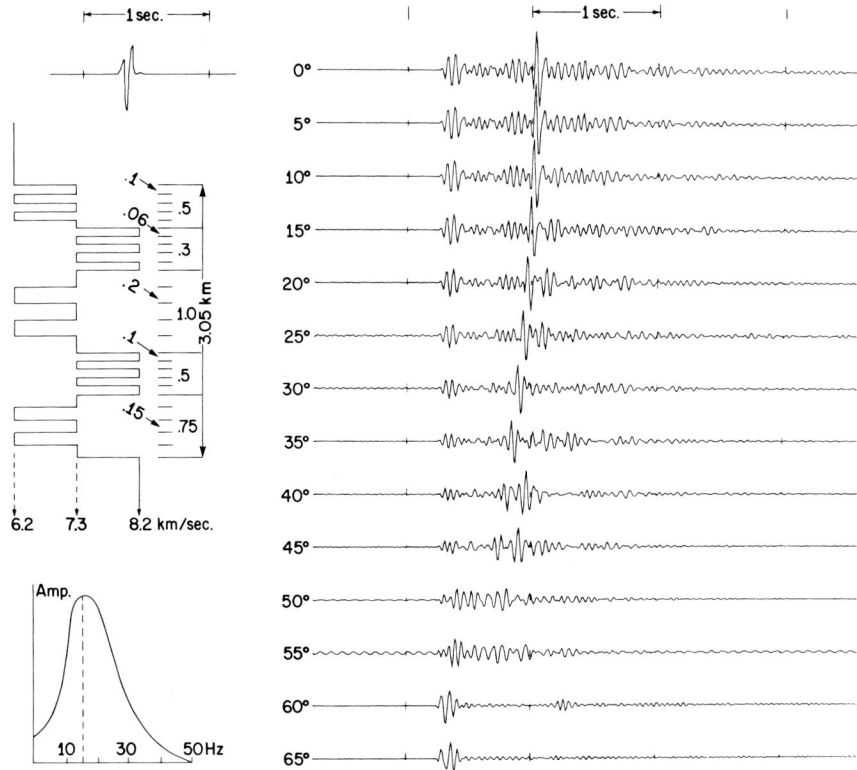

Fig. 17. Theoretical reflection response of the indicated layered sequence to a plane wave pulse (upper left) at various reflecting angles.

It appears that anisotropic stress fields and temperatures suitable to recrystallization with preferred orientation of the crystal axes cannot be considered an uncommon occurrence in the history of the formation of the earth's crust. As a general statement, anisotropy is to be expected in materials formed in nature; isotropy is simply a convenient assumption when we lack the tools and the data to assume otherwise. It appears that refracted velocities of the upper mantle can vary with horizontal direction. Berry and Fuchs [1973] describe a weak dependence of velocity on direction for the eastern Canadian Shield. Bamford [1973] finds this dependence to be true in Germany. Hess [1964] finds this dependence in the uppermost mantle under oceans, as do Raitt et al. [1969]. Birch [1960, 1961] has described the high elastic anisotropy found in some rocks, especially in dunites. Christensen [1972] has indicated that ultramafic rocks frequently display this characteristic. Certainly, it should be a serious consideration in models of crustal structure deduced from horizontal velocity measurements. Unfortunately, it introduces a parameter that can provide velocity variations of 15% or so in the same rock, depending on the direction of measurement, and we are unable to determine accurate velocities in the vertical direction for crustal rocks in situ beyond drillable depths.

J. A. Orcutt et al. (personal communication, 1976) have obtained results on three refraction profiles: (1) along the ridge of the East Pacific Rise, (2) over crust believed to be 2.9 m.y. old, and (3) over crust 5 m.y. old. One of their results indicates that stratification of the crust becomes, seismically, more evident with time. Sleep [1975] indicates some thermal constraints on the formation of oceanic crust. The cooling of an oceanic plate as it moves away from a ridge can result in deviatoric horizontally tensional stresses of several kilobars [Turcotte, 1974]. It seems reasonable that both differentiation and recrystallization with preferred orientation will occur; one suggests the other for such a proposed environment. This differentiation-recrystallization process will tend to orient the cleavage and slip planes of the material horizontally, i.e., normal to the maximum compressive stress. Presumably, over short time intervals or at reduced temperatures these properties are locked into the rock. A change to a deviatoric horizontally compressive stress, for example, as the crust approaches a continental block, could result in slippage and create horizontal laminar structure along these slip planes. To the extent that oceanic crust can be obduced to form part of the continental crust, such a mechanism may explain the presence of laminated boundaries within the continental crust.

To envisage a mechanism that could produce a lamella having the required seismic properties at the crust-mantle transition beneath a continent, we appeal to two further arguments. Haxby and Turcotte [1976] have examined the stresses introduced in the crust by the addition and removal of overburden with associated thermal effects. They find that the deviatoric horizontal stress due to the reduction of overburden is compressional, whereas the stresses produced by uplift on a sphere and the thermally introduced stresses are tensional. The combination leads to a predominantly tensional stress. The record of vertical movements of the earth's crust is well preserved in Phanerozoic sediments. The Cambrian sedimentary record is well documented in the Rocky Mountains of North America. Within the major cycle of submergence beginning about 600 m.y. ago to emergence 100 m.y. later, there is compelling evidence of at least 12 subcycles averaging about 8 m.y. each and further evidence still of many minor vertical movements within each subcycle. The causes of these cycles are much disputed, but the Cambrian record must be accepted as evidence of vertical movements every few million years. During emergence a deviatoric horizontally tensional stress should exist within the crust and perhaps the upper mantle, to be followed by compression during submergence. Given such an anisotropic stress field and a suitable rock material, we can form, by recrystallization within an otherwise uniform stratum, one which, to seismic energy, appears to be laminated. There is no requirement of a 'plating on' of different mantle material to the base of the crust in this model; however, it is not excluded. The same mantle material, being only gradually depleted in lighter low melting point material, is accumulated at the crust-mantle transition in horizontal layers under differing stress conditions and is moved vertically in segments in response to isostatic adjustments of crustal blocks.

Certainly, there is much evidence within recent mountain ranges

of horizontally compressive tectonic forces causing massive thrust
faulting. The fault planes tend to be near horizontal at great
depth, changing as the vertical load decreases to near vertical at
the surface. Old rocks may overlie younger formations, and the
same formation may overlie itself several times. It seems reasonable
to believe that deep structures, including the Moho, have been
involved and that their mode of faulting has been near horizontal
under this stress, perhaps indicated by the Vibroseis section
(Figure 12). Laminar plastic flow may be a better term for the
compensation to this stress at depths of 25-50 km and temperatures
above $300^{\circ}C$ [Turcotte, 1974].

We will not belabor the reader further with conjectural detail
but simply state that the laminated model and velocity anisotropy
concepts provide, in our estimation, the best explanation given to
date for the observed data, recorded from the crust-mantle transition.

Conclusions

We have presented results from three Precambrian provinces and the
Canadian Cordillera. We believe that the derived crustal structures
represent realistic models of the crust of these regions as illuminated
by the seismic waves that we have recorded. In each region our field
technique has been somewhat different, resulting in different types
of seismic models. Nevertheless, the models deduced for the three
Precambrian provinces show a strong resemblance, especially if we
compare the western part of the Yellowknife structure (the 'granitic'
section) to the structures of the Grenville and Superior provinces.
The structure of the Cordillera appears to be somewhat different,
showing no evidence of uniform layering such as is seen in the shield.

In general, one can visualize four factors that will determine the
velocity structure of the crust: (1) the tectonic style of a region,
which will include the geochemistry of the rocks and their orogenic
history, (2) the distribution of water in the crust either chemically
combined to give hydrated minerals or free as pore fluid, (3) the
past and present state of stress, and (4) the past and present
temperature regime.

The tectonic styles of the Slave, Grenville, and Superior provinces
are very different, and hence we must conclude that for Precambrian
rocks that have undergone repeated orogeny the gross crustal structure
that is common to all provinces, as seen by seismic means, is largely
determined by the present and past state of stress and temperature
in the crust and possibly by the distribution of water. The effect
of recent temperature regimes is probably of only minor significance,
except insofar as they may control the hydration of the rocks at the
base of the crust. Small-scale variations from the gross model are
most likely caused by variations of petrology.

In the Phanerozoic rocks of the Cordillera the crust appears to be
largely featureless from the seismic point of view until one detects
a complex signal from the Moho. We suggest that the uniform layering
evident in the shield has not yet established itself in the Cordillera
and that the velocity structure is still dominated by the extremely
complex recent tectonic history, which appears to have produced a
largely heterogeneous crust.

Acknowledgments. The authors would like to thank their colleagues for many stimulating discussions, and Dr. C. Keith for making the calculations shown in Fig. 17. Contribution 674 of the Earth Physics Branch, Canadian Department of Energy, Mines and Resources.

References

Bamford, S. A. D., Refraction data in western Germany - A time-term interpretation, Z. Geophys., 39, 907-927, 1973.

Barr, K. G., Crustal refraction experiment: Yellowknife 1966, J. Geophys. Res., 76, 1929-1947, 1971.

Berry, M. J., Low velocity channels in the earth's crust?, Comments Earth Phys. Geophys., 3, 59-68, 1972.

Berry, M. J., Structure of the crust and upper mantle in Canada, Tectonophysics, 20, 183-201, 1973.

Berry, M. J., and D.A. Forsyth, Structure of the Canadian Cordillera from seismic refraction and other data, Can. J. Earth Sci., 12, 182-308, 1975.

Berry, M. J., and K. Fuchs, Crustal structure of the Superior and Grenville provinces of the northeastern Canadian Shield, Bull. Seismol. Soc. Amer., 63, 1393-1432, 1973.

Birch, F., Interpretation of the seismic structure of the crust in the light of experimental studies of wave velocities in rocks, in Contributions in Geophysics, vol. 6, pp. 158-170, Pergammon, New York, 1958.

Birch, F., Velocity of compressional waves in rocks to 10 kbar, 1, J. Geophys. Res., 65, 1083, 1960.

Birch, F., Velocity of compressional waves in rocks to 10 kbar, 2, J. Geophys. Res., 66, 2199, 1961.

Braile, L. W., Interpretation of crustal velocity gradients and Q structure using amplitude-corrected seismic refraction profiles, in The Earth's Crust: Its Nature and Physical Properties, Geophys. Monogr. Ser., vol. 20, edited by J. G. Heacock, this volume, AGU, Washington, D. C., 1977.

Chaipayungpun, W., and M. Landisman, Crust and upper mantle near the western edge of the Great Plains, in The Earth's Crust: Its Nature and Physical Properties, Geophys. Monogr. Ser., vol. 20, edited by J. G. Heacock, this volume, AGU, Washington, D. C., 1977.

Christensen, N. I., Seismic anisotropy in the lower oceanic crust, Nature, 237, 450, 1972.

Clee, T. E., K. G. Barr, and M. J. Berry, The fine structure of the crust near Yellowknife, Can. J. Earth Sci., 11, 1534-1549, 1974.

Clowes, R. M., and E. R. Kanasewich, Seismic attenuation and the nature of reflecting horizons within the crust, J. Geophys. Res., 75, 6693-6705, 1970.

Cumming, G. L., and N. N. Chandra, Further studies of reflections from the deep crust in southern Alberta, Can. J. Earth Sci., 12, 539-557, 1975.

Dohr, G. P., and R. Meissner, Deep crustal reflections in Europe, Geophysics, 40, 25-39, 1975.

Forsyth, D. A., M. J. Berry, and R. M. Ellis, A 1974 refraction survey across the Canadian Cordillera at 54°N, Can. J. Earth Sci., 11, 533-548, 1974.

Fuchs, K., On the properties of deep crustal reflectors, Z. Geophys., 35, 133-149, 1969.

Hall, D. H., and W. C. Brisbin, Crustal structure from converted head waves in central western Manitoba, Geophysics, 30, 1053-1067, 1965.

Haxby, W. F., and D. L. Turcotte, Stresses induced by the addition or removal of overburden and associated thermal effects: Geology, 4, 181-184, 1976.

Hess, H., Seismic anisotropy of the uppermost mantle under oceans, Nature, 203, 629-631, 1964.

Hoffman, P. F., J. F. Dewey, and K. Burke, Aulacogens and their genetic relation to geosynclines with a Proterozoic example from Great Slave Lake, Canada, Modern and ancient geosynclinal sedimentation, Soc. Econ. Paleontol. Mineral. Spec. Publ., 19, 38-55, 1974.

Hyndman, T. D., and D. W. Hyndman, Water saturation and high electrical conductivity in the lower continental crust, Earth Planet. Sci. Lett., 4, 427-432, 1968.

Landisman, M., and S. Mueller, Seismic studies of the earth's crust in continents, 2, Analysis of wave propagation in continents and adjacent shelf areas, Geophys. J. Roy. Astron. Soc., 10, 539-548, 1966.

Landisman, M., S. Mueller, and B. J. Mitchell, Review of evidence for velocity inversions in the continental crust, in The Structure and Physical Properties of the Earth's Crust, Geophys. Monogr. Ser., vol. 14, edited by J. G. Heacock, pp. 11-34, AGU, Washington, D. C., 1971.

Mair, J. A., and J. A. Lyons, Seismic reflection techniques for crustal structure studies, Geophysics, 41, 1272-1290, 1976.

Meissner, R., The 'Moho' as a transition zone, in Geophysical Surveys, vol. 1, pp. 195-216, D. Reidel, Hingham, Mass., 1973.

Monger, J. W. H., J. G. Souther, and H. Gabrielse, Evolution of the Canadian Cordillera: A plate tectonic model, Amer. J. Sci., 272, 577-602, 1972.

Mueller, S., and M. Landisman, Seismic studies of the earth's crust in continents, 1, Evidence for a low-velocity zone in the upper part of the lithosphere, Geophys. J. Roy. Astron. Soc., 10, 525-538, 1966.

Nur, A., and G. Simmons, The effect of saturation on velocity in low porosity rock, Earth Planet. Sci. Lett., 7, 183-193, 1969.

Raitt, R. W., G. G. Shor, Jr., T. J. G. Francis, and G.B. Morris, Anisotropy of the Pacific upper mantle, J. Geophys. Res., 74, 3095-3109, 1969.

Sleep, N. H., Formation of oceanic crust: Some thermal constraints, J. Geophys. Res., 80, 4037-4042, 1975.

Stacey, R. A., Gravity anomalies, crustal structure and plate tectonics in the Canadian Cordillera, Can. J. Earth Sci., 10, 615-628, 1973.

Todd, T., and G. Simmons, Effect of pore pressure on the velocity of compressional waves in low-porosity rocks, J. Geophys. Res., 77, 3731-3743, 1972.

Turcotte, D. L., Are transform faults thermal contraction cracks?, J. Geophys. Res., 79, 2573-2577, 1974.

van Zijl, J. S. V., Electrical studies of the deep crust in various tectonic provinces of southern Africa, in The Earth's Crust: Its Nature and Physical Properties, Geophys. Monogr. Ser., vol. 20, edited by J. G. Heacock, this volume, AGU, Washington, D. C., 1977.

Wheeler, J. C., and H. Gabrielse, The cordilleran structural province: Variations in tectonic styles in Canada. Geol. Ass. Can. Spec. Pap., 11, 1-72, 1972.

White, W. R. H., and J. C. Savage, A seismic refraction and gravity study of the earth's crust in B.C., Bull. Seismol. Soc. Amer., 55, 463-486, 1965.

Wickens, A. J., The upper mantle of southern British Columbia, Can. J. Earth Sci., 14, 1100-1115, 1977.

THE STRUCTURE OF THE CRUST-MANTLE BOUNDARY BENEATH NORTH AMERICA
AND EUROPE AS DERIVED FROM EXPLOSION SEISMOLOGY

Claus Prodehl

Geophysical Institute, University of Karlsruhe
Karlsruhe, Germany

Abstract. Geophysical studies of the earth's crust and upper mantle during the past 20 years have revealed detailed information on the structure of the crust-mantle boundary in Europe and North America. Mainly, on the basis of explosion seismology the elastic properties of the crust-mantle boundary can be described as follows: the depth to the M discontinuity beneath continents ranges from 20 to 60 km, the velocity of P waves within the uppermost mantle beneath the M discontinuity varies between 7.2 and 8.5 km/s, and the crust-mantle boundary is a discontinuity of the first order only in a few cases. (It is usually a more or less broad transition zone where the thickness varies from 1-2 km to more than 15 km.) The regional investigation of different tectonic units in North America and Europe shows that similar characteristics of the crust-mantle boundary can be related with comparable tectonic units as, for example, graben areas, fold belts, and stable shield areas.

Introduction

Seismic refraction investigations of the past 20 years have shown that the structure of the crust varies broadly between different tectonic areas. This is already evident from some average parameters which can be derived from explosion seismology data. These parameters include the depth to the crust-mantle boundary, the average P wave velocity within the crust, and the velocity of P waves in the upper mantle immediately beneath the Mohorovičić (M) discontinuity. More recent interpretations of seismic refraction data by various authors have shown that the crust cannot be described by a simple two- or three-layer model discontinuously overlying the upper mantle. Rather, the crust shows a more complex structure that includes, in some cases, velocity inversions within the crust. Also, the crust-mantle boundary, instead of being a sharp boundary, is described by a more or less thick transition zone where the velocity increases continuously with increasing depth from velocities typical for the crust to velocities associated with upper-mantle material.

The present article will discuss some average properties of the crust but will be concerned mainly with the appearance of the crust-mantle boundary beneath different tectonic areas of North America and Europe.

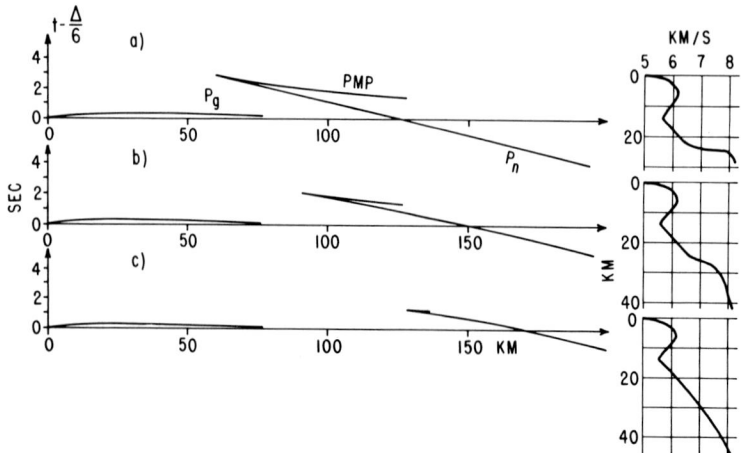

Fig. 1. Travel time diagrams and the corresponding models [from Giese, 1968].

Average Seismic Properties of the Crust

The most simple parameters which can be derived from explosion seismology data are (1) the depth to the crust-mantle boundary, (2) the average P wave velocity within the crust, and (3) the velocity of P waves beneath the M discontinuity.

Crustal cross sections around the world have been published by various authors [e.g., Closs and Behnke, 1961] using these parameters. The most striking feature on such cross sections is the varying thickness of the crust. The depth to the M discontinuity reaches from less than 10 km beneath oceans to more than 50 km beneath orogenic belts.

Also the average velocity changes considerably. Pakiser and Steinhart [1964] have summarized the seismic refraction data for the United States and have shown that, to a certain extent, a mean crustal velocity can be assigned to different tectonic units. For example, the Basin and Range province and the Atlantic Coastal Plains show average velocities generally below 6.2 km/s, while large parts of the Interior Plains are characterized by a mean crustal velocity of greater than 6.5 km/s.

On the basis of the results of various interpretations, Pakiser and Steinhart [1964] also summarized the available P_n velocities, i.e., the P wave velocity immediately below the Mohorovičić discontinuity. Their map clearly indicates that the mobile western United States are characterized by P_n velocities smaller than 8 km/s, while the P_n velocity within the stable part of central and eastern North America is greater than 8 km/s. A similar feature is shown on the map of the P_n velocity for western Germany [Giese and Stein, 1971; Giese, 1976], where areas of young volcanic activity within the Rhenish massif show P_n velocities smaller than 8 km/s.

Giese [1968] has shown that in many cases the P^Mp phase (see, for example, Figures 4 and 5) is not a reflection hyperbola and therefore is not a true wide-angle reflection from a discontinuous crust-mantle

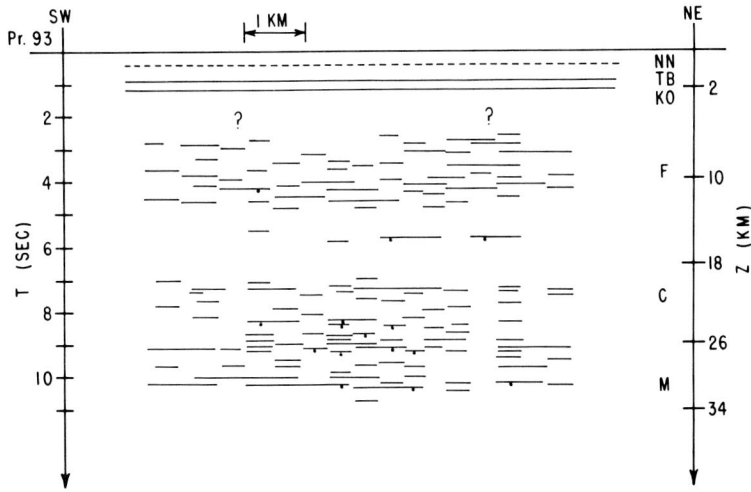

Fig. 2. Travel time profile of near-vertical reflections in the Molasse Basin of southern Germany. The depth scale is not linear. F represents the Fortsch discontinuity; C, the Conrad discontinuity; and M, the Mohovičić discontinuity [from Liebscher, 1964].

boundary where the velocity gradient is infinite. Rather, this P^Mp phase originates at a transition zone of several kilometers in thickness where the velocity increases rapidly but steadily with increasing depth. This causes a retrograde travel time curve which looks very similar to a true reflection hyperbola. Theoretical travel time studies by Giese [1968] demonstrate how a P^Mp phase may be affected for different kinds of transition zones. Figure 1 shows examples of realistic crustal models derived by Giese [1968] for different structures of the crust-mantle boundary. For the first case, model A, a nearly infinite velocity gradient is assumed resulting in a P^Mp curve which can be observed for several tens of kilometers in distance. For model B the corresponding retrograde P^Mp curve is still obtained, but the distance range where it can be observed is smaller. Model C shows an example where the velocity gradient has become so small that almost no retrograde phase is produced.

This conclusion is supported by results obtained from an interpretation of seismic reflection measurements in southern Germany [Liebscher, 1964]. Figure 2 shows a travel time profile published by Liebscher [1964]. This profile shows the observed near-vertical reflections along a 10-km line of continuous observations. Clearly, several bands of reflections can be traced that indicate several reflecting horizons in the earth's crust, including the crust-mantle boundary (M). This profile also demonstrates that the near-vertical reflections cannot be traced continuously but are displaced from each other by up to tenths of a second. If we take into account the lower-frequency content observed by the refraction technique, such a band of reflections as is shown in Figure 2 may well represent a transition zone for waves of greater wavelength.

The computation of synthetic seismograms for different types of a

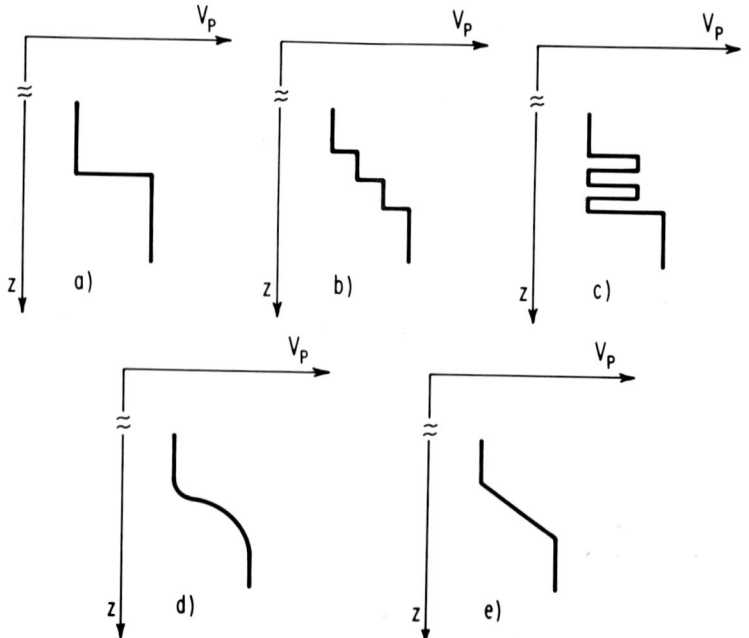

Fig. 3. Different models for the crust-mantle transition zone.

seismic boundary proves that this assumption is correct. Edel et al. [1975] show examples for models A and C of Figure 1, while Hirn et al. [1975] demonstrate that the velocity gradient clearly controls the distance range at which a phase can be observed (model B, Figure 1).

As Figure 3 shows, several possibilities have been proposed by various authors for the nature of the crust-mantle transition zone. In addition to a stepwise transition (model B), a zone consisting of several lamellae of alternating high and low velocities (model C) has been proposed [Giese, 1972; Meissner, 1973]. Giese [1972] has published some examples demonstrating that observed offsets within the P^Mp phase might be explained by velocity inversions within the crust-mantle boundary.

In this paper the emphasis is placed upon total thickness of the crust-mantle transition zone in relation to different tectonic areas in North America and Europe, and therefore only models of types A, D, or E of Figure 3 will be considered.

Main Features of the Structure of the Crust and the Crust-Mantle Boundary in North America

Figures 4 to 6 show examples of seismic refraction observations in North America. The Eureka Fallon profile (Figure 4) is located in the northern part of the Great Basin of the Basin and Range province and shows a well-pronounced P^Mp phase (curve C). The Chinle-Hanksville profile (Figure 5) [Prodehl, 1970b] was recorded through the Colorado plateau. Here a P^Mp phase (curve C) can be correlated, but it is much less clearly developed. The third example (Figure 6)

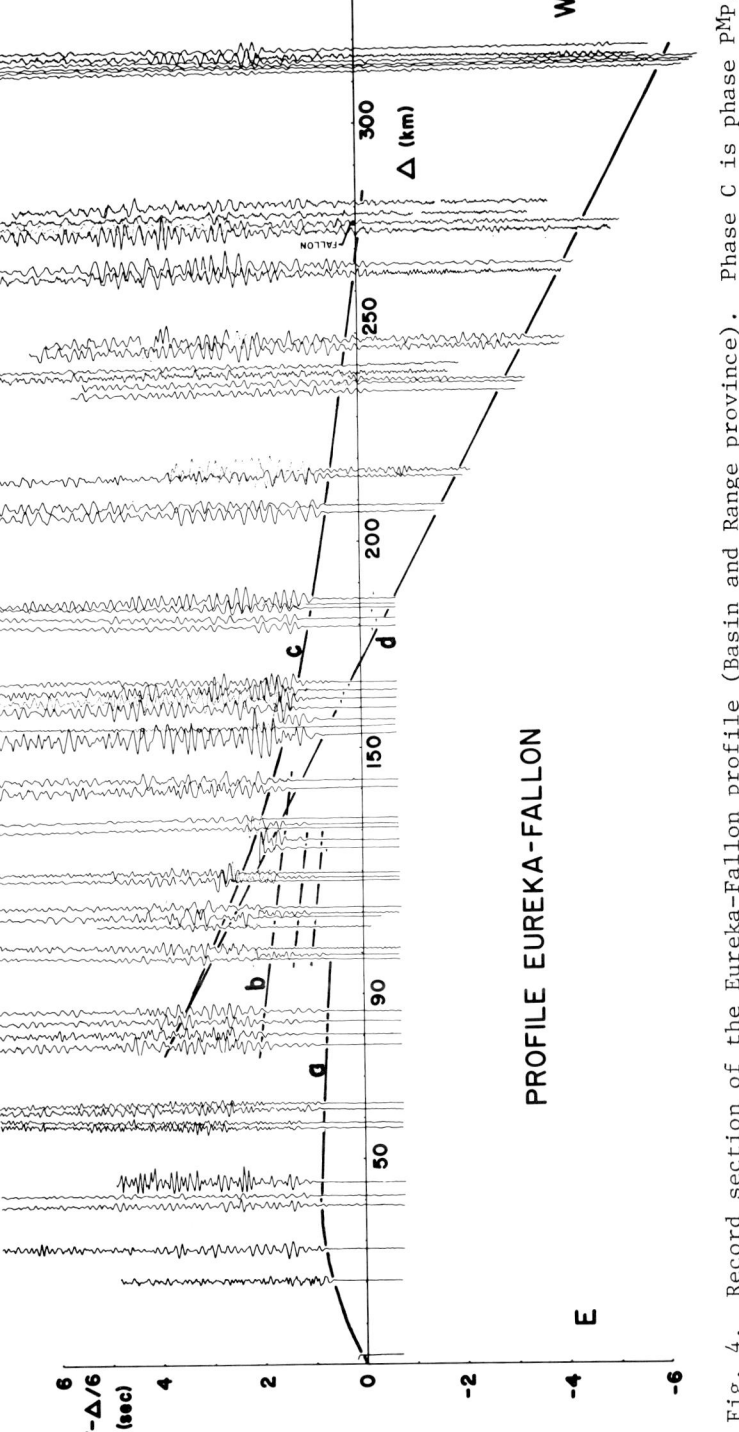

Fig. 4. Record section of the Eureka-Fallon profile (Basin and Range province). Phase C is phase $P^M P$.

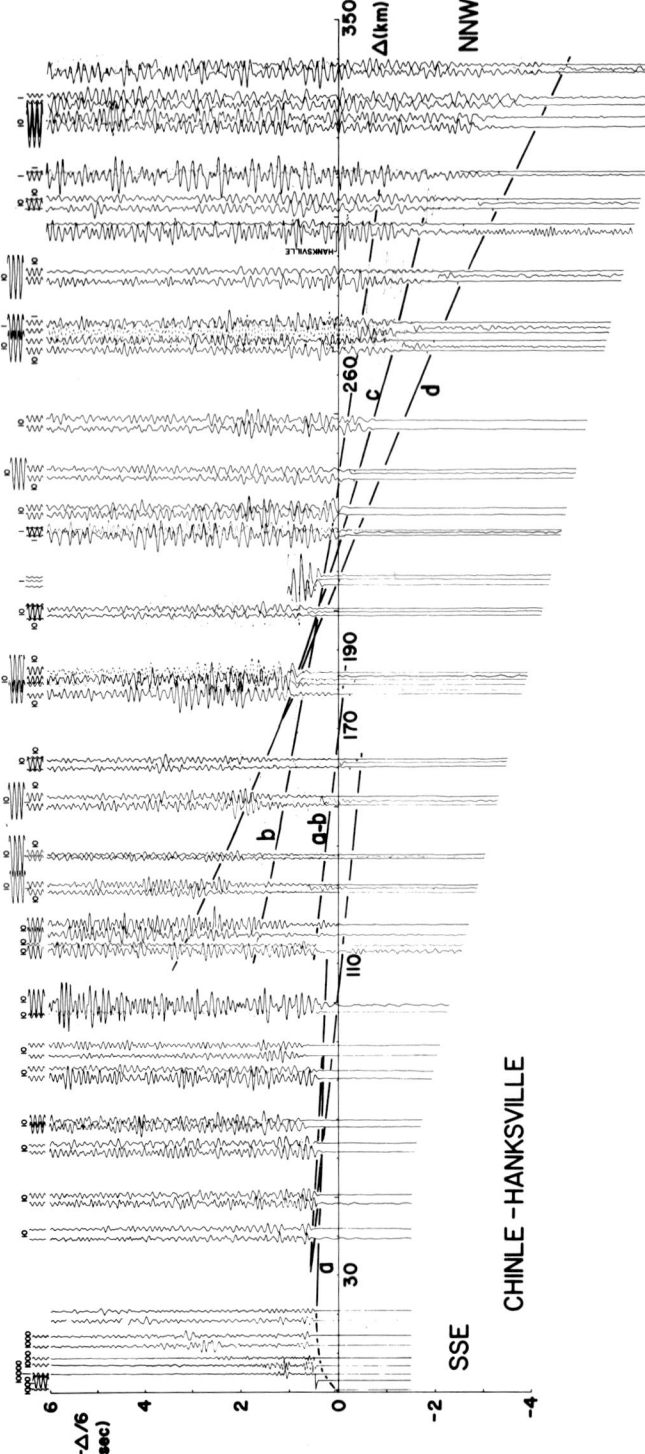

Fig. 5. Record section of the Chinle-Hanksville profile (Colorado plateau). Phase C is phase pMp.

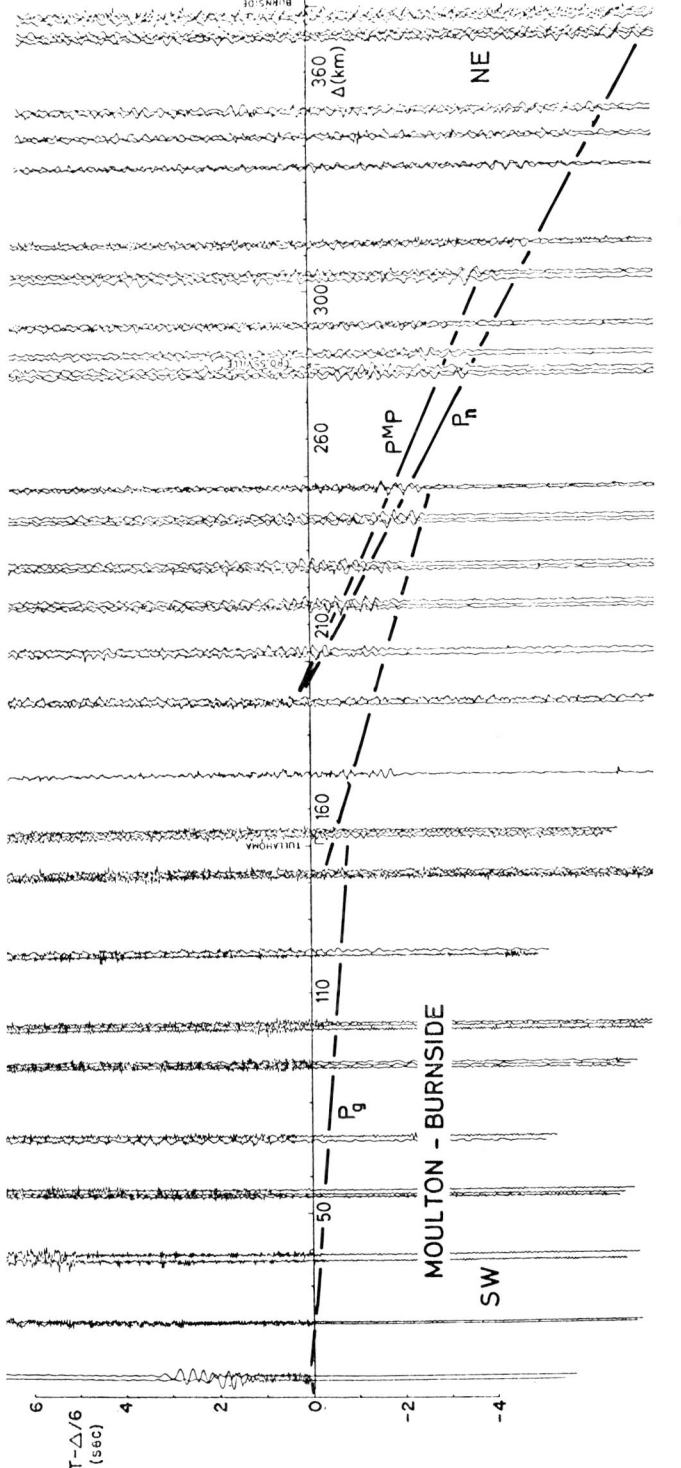

Fig. 6. Record section of the Moulton-Burnside profile (Appalachian Highlands).

is a record section from the profile recorded in the Cumberland plateau in the border area between the Interior Plains and the Appalachian Highlands. This profile shows hardly any secondary arrivals; however, it should be pointed out that in addition to the P_g and the P_n phase, a third, intermediate, phase can be traced in first arrivals which normally cannot be correlated.

The velocity-depth functions calculated for each profile [e.g., Prodehl, 1970a, b, 1976] by using a method proposed by Giese [1968] were used to construct crustal cross sections showing lines of equal velocity. Figure 7 shows some selected crustal cross sections through the United States. The dotted areas in Figure 7 comprise the depth range with velocities less than or equal to 6.2 km/s and so define the main part of the upper crust, while the crosses indicate upper-mantle material with velocities greater than or equal to 7.4 to 7.6 km/s. For each cross section, one or two velocity-depth functions which may be typical for the corresponding province are plotted at the bottom of Figure 7.

The map (Figure 7) seems to indicate a zonal division of the western United States from north to south: in the Lassen Peak area (upper part of velocity-depth function 2 in Figure 7) as well as in the Snake River Plains (function 4 in Figure 7) in the north the upper crustal material is either very thin or may not even exist. In the adjacent northern Basin and Range province (not shown as separate velocity-depth functions), the upper and lower crust are equally well developed, have similar thickness, and can easily be separated from each other. However, in the southern Basin and Range province, (function 5) apparently the lower crust has almost disappeared, the upper crust reaching into a depth of about 25 km with velocities of 6.0 to 6.2 km/s. A similar increase in the sialic crust from north to south can be observed in the Rocky Mountains (from functions 6 to 8) and in the Sierra Nevadas (function 3). In the middle Rocky Mountains (function 6) as well as in the northern Sierra Nevadas (function 3) the velocity contour of 6.4 km/s is found at about 20 to 25 km; also the lower crust with velocities between 6.4 and 7.0 km/s covers a depth range of 20 to 25 km thickness. In the southern part of the Sierra Nevadas as well as in the southern Rocky Mountains (function 8), upper crustal material reaches into depths of 30 to 35 km.

The seismic properties defining the crust-mantle boundary can be correlated clearly with tectonic provinces (Table 1). The Coast Ranges of California west of the San Andreas fault are characterized by a thin crust of 20 to 25 km thickness and a relatively sharp Mohorovičić discontinuity (function 1). The P_n velocity is about 8 km/s. The structure changes considerably for the Sierra Nevadas (functions 2 and 3). There the crust reaches more than 40 km in thickness with a 5 to 10 km thick lower transition zone where the velocity increases continuously with increasing depth from about 6.8 to 7.9 km/s.

East of the Sierra Nevadas in the Basin and Range province (function 5) the depth to the Moho decreases drastically by about 10 km to about 30 to 35 km, the crust-mantle transition zone covering only a narrow depth range of 1 to 2 km. The P_n velocity shows a tendency to decrease toward the east from about 7.8 km/s in western Nevada to 7.4 to 7.5 km/s in western Utah [see Keller et al., 1975].

The thickness of the total crust as well as that of the crust-mantle transition zone beneath areas surrounding the Basin and Range province to the north and the east are very similar to the Sierra Nevadas; beneath the Snake River Plains (function 4) to the north; and beneath the middle Rocky Mountains (function 6) and the Colorado plateau (function 7) to the east a 40 to 45 km thick crust is separated from the upper mantle by a 5 to 10 km thick transition zone.

A similar relationship is true for the southern Rocky Mountains (function 8) and the adjacent Great Plains [see Haggag, 1974]. There the crust reaches a thickness of about 50 km, and the transition zone is about 10 km thick. With the exception of the P_n velocity which reaches 8 km/s in the Great Plains, the crustal structure of the Colorado plateau is very similar to that of the Great Plains. Neither the crustal thickness nor the characteristics of the crust-mantle transition zone change considerably when moving from the Colorado plateau across the southern Rocky Mountains into the Great Plains of eastern Colorado. However, the average velocity within the uppermost 30 km of the crust decreases considerably from about 6.5 km/s in the Colorado plateau and the Great Plains [see Haggag, 1974; Jackson et al., 1963; Mitchell and Landisman, 1970; Stewart, 1968; Warren et al., 1973] to about 6.0 km in the southern Rocky Mountains.

The lack of clear wide-angle reflections on the observations in the westernmost part of the Applachian plateau can only be explained by the existence of very broad transition zones within and at the bottom of the crust, as is shown by the comparison with synthetic seismograms. In comparison with the observations in the central and western United States the observed velocities in the upper (6.3 km/s) as well as in the lower crust (6.9 km/s) are anomalously high, while the P_n velocity is about 8.0 km/s [see Borcherdt and Roller, 1966].

The structure beneath the northeastern Canadian shield, characterized by a thick crust (50km) and a 10 to 15 km thick crust-mantle transition zone [Berry and Fuchs, 1973], is very similar to that of the Interior Plains (function 3 in Figure 7). A significant difference, however, can be seen in the upper-mantle velocity, which reaches 8.5 km/s or more beneath the Canadian shield.

The Crust-Mantle Boundary in Europe

In Europe a great number of seismic refraction experiments have been carried out. In this paper, only a few examples will be discussed in order to describe the general features in more detail (see also Table 1).

The Scandinavian shield was covered by several experiments, the last and most extensive one being the so-called 'Blue Road' experiment. Hirschleber et al., [1975] have published their data in the form of record sections. As was done with the data published by Berry and Fuchs [1973] for the Canadian shield, P^Mp arrivals can be correlated, but they are not as well expressed as they are in profiles through the Basin and Range province for example. C. E. Lund (oral communication, 1976) has reinterpreted the data of the Blue Road experiment, his model for the crust indicating a high average velocity and a smooth transition from crust to mantle over a broad depth range. The total crustal thickness amounts to about 45 km. Also, here the P_n velocity reaches values of 8.4 to 8.5 km/s. Similar results are obtained by

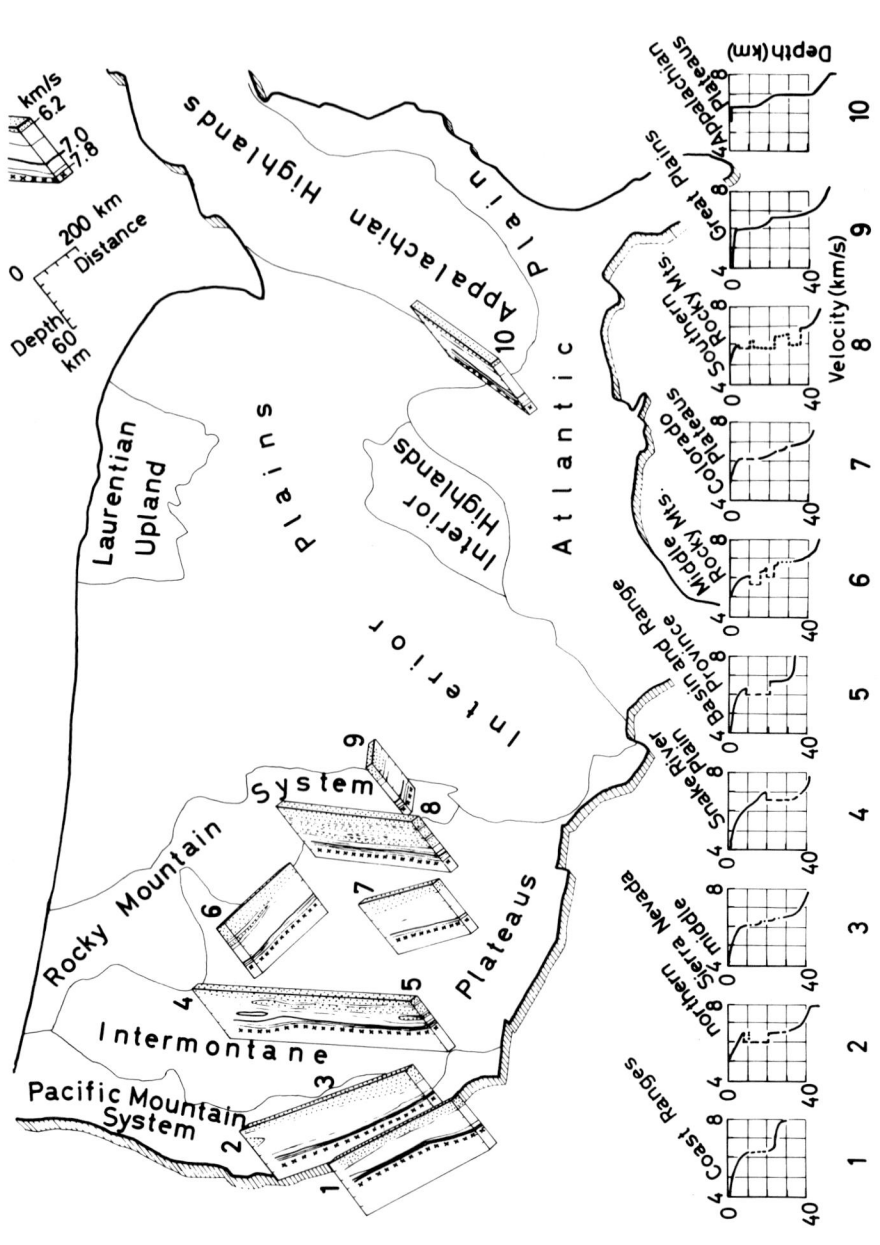

Fig. 7. Map of the United States showing main physical divisions [after Fenneman, 1970], selected crustal cross sections with lines of equal velocity, and typical velocity-depth functions. In the crustal cross sections the dotted areas indicate the depth range with velocities of <6.2 km/s, and the crosses indicate the uppermost mantle with velocities of ≥ 7.4 to 7.6 km/s.

TABLE 1. Characteristics of Crustal Structure

Area	Type of Crust	Depth of Strongest Velocity Gradient	Crust-Mantle Boundary Sharp (<2 km)	Transitional (5-20 km)	$v(P_n)$	Velocity Increases from ~7 to ~8 km/s above	below	Average Crustal Velocity
Canada, Scandinavia Baltics, Ukraine	Shield	40-50		x	8.1-8.4	x		High
Great Plains	Stable continent	40-50		x	8.1-8.4	x		High
Colorado plateau	Part of mobile western U.S.	40-50		x	7.7-7.8	x		High
Appalachia Mts.	Old orogen	50		x	~8.0	x		High
Sierra Nevada Rocky Mountains Alps	Young orogens	40-60		x	7.8-7.9 7.8-8.0 8.0-8.1	x		Medium Low Low
Caledonides, northern England	Old orogene	30-35		x	~8.0	x		Low
Caledonides, northern Scotland	Old orogen	25-30	x		8.0	x		Medium
Hercynian Germany, France, Portugal	Old orogen	30	x		8.0-8.4	x		Low
Basin and Range province	Part of mobile western U.S.	30	x		7.8-7.9	x		Low
Upper Rhine graben Limagne, Auvergne	Graben	20-25	x		8.0-8.1 7.6-8.4		x	Low
Utah graben	Graben	25-30	x		7.3-7.9		x	Low
Snake River Plain†	Graben	20 40		x		x	x	Low High
Appennines	Young orogen	20	x		7.3-8.0		x	Low
Coast Range of CA	Young orogen	20-25	x		8.0	x		Low

*The reference point is the depth of the strongest velocity gradient.
†Two clear boundaries allow two interpretations.

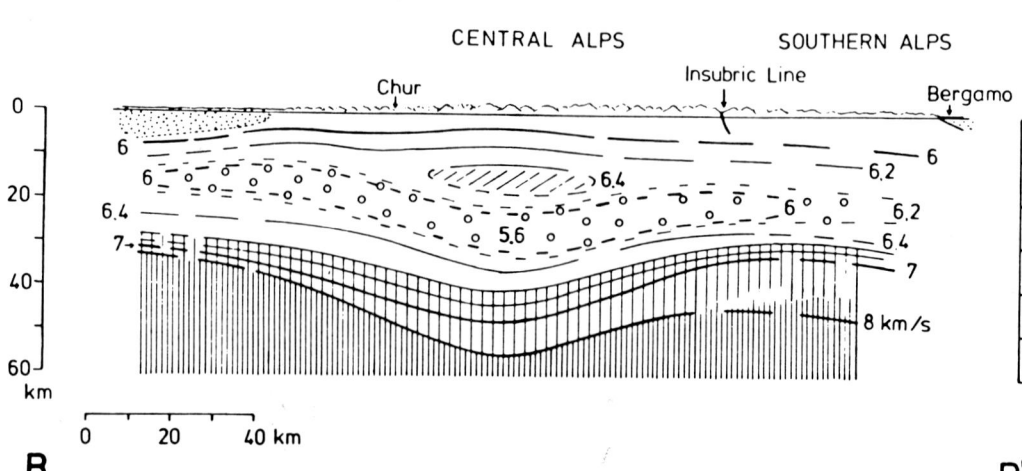

Fig. 8. Simplified north-south cross section through the border between the eastern and western Alps, showing contours of equal velocity. The circles indicate the position of the low velocity zone. The data are from Giese and Prodehl [1976].

Pavlenkova [1973] for the Ukrainian shield; she indicates a crust-mantle transition zone of about 8 to 10 km thickness between 40 and 50 km depth. However, here the reported P_n velocities are not much greater than 8.0 km/s.

The main features of the crustal structure in the Alps are summarized by Choudhury et al. [1971] and Giese and Prodehl [1976]. A typical north-south (N-S) cross section is shown in Figure 8. The main results are the following. The crust reaches a maximum thickness of 50 to 60 km under the axis of the Alps. The lower crust and/or the transition from crust to mantle increases in thickness from the foreland toward the axis of the Alps. The P_n velocity in general is equal to or slightly greater than 8.0 km/s. Within the crust an extensive velocity inversion exists. As is shown in Figure 9, the seismic refraction profiles, crossing the anomalous zone of Ivrea in the western Alps, which is characterized by a strong gravity high, show as particular features a high velocity of 7.4 km/s (phase P_i) and a corresponding retrograde curve P^I that indicates the existence of high-velocity material at only 5 to 10 km depth. The typical P^MP phase (P^M) is recorded with a considerable intercept time delay which can only be explained by assuming an intensive low-velocity zone beneath the Ivrea body. The cross section through the western Alps (Figure 10) shows the corresponding crustal structure. To the west, this low-velocity zone is linked to the general low-velocity zone of the western Alps , while to the east, the high-velocity material (7.2 to 7.4 km/s) of the Ivrea body is linked to the lower crust and mantle under the Po Plain.

As is indicated in Figures 8 and 10, the crust beneath the northern and western foreland of the Alps is thinner than it is beneath the

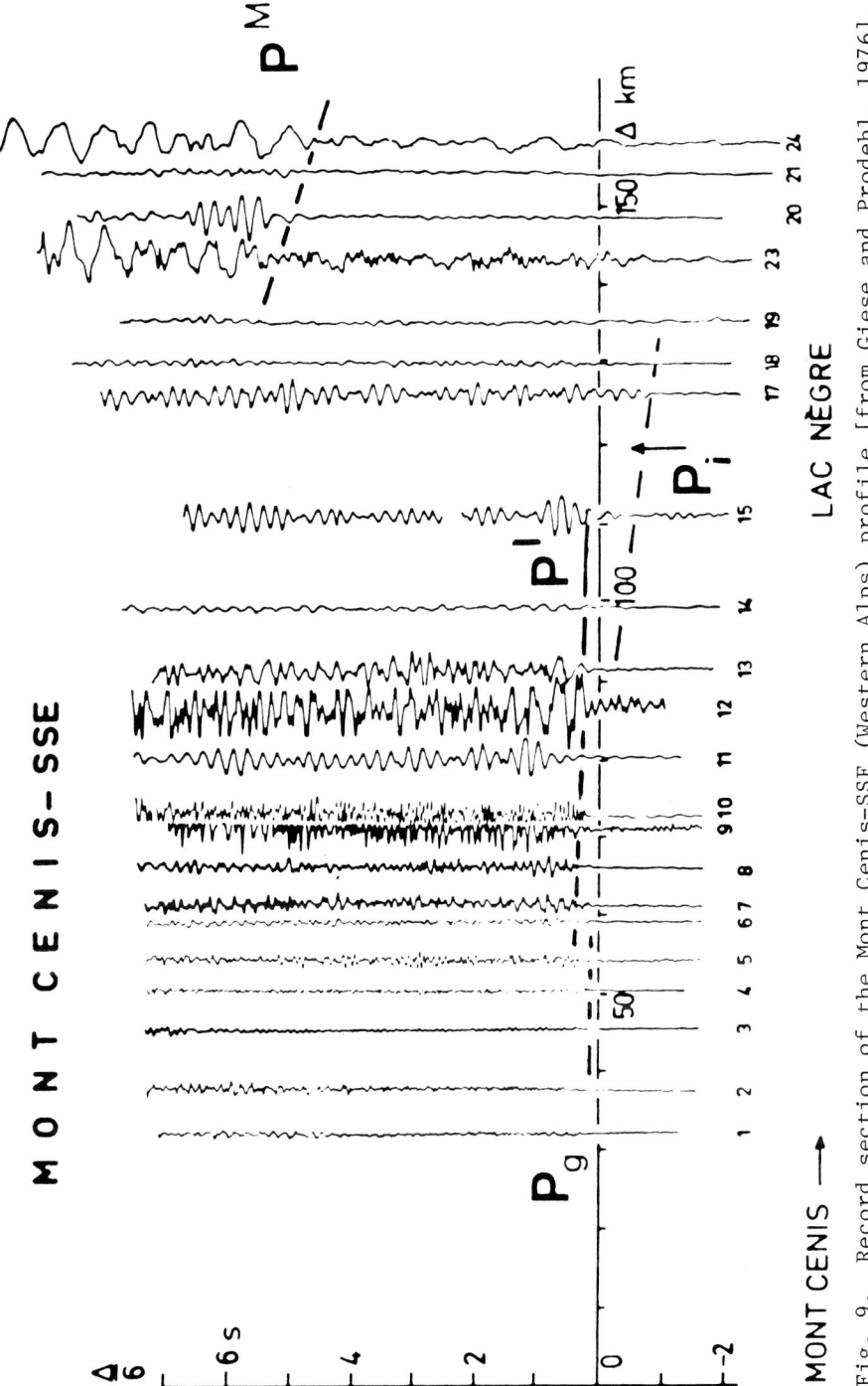

Fig. 9. Record section of the Mont Cenis-SSF (Western Alps) profile [from Giese and Prodehl, 1976].

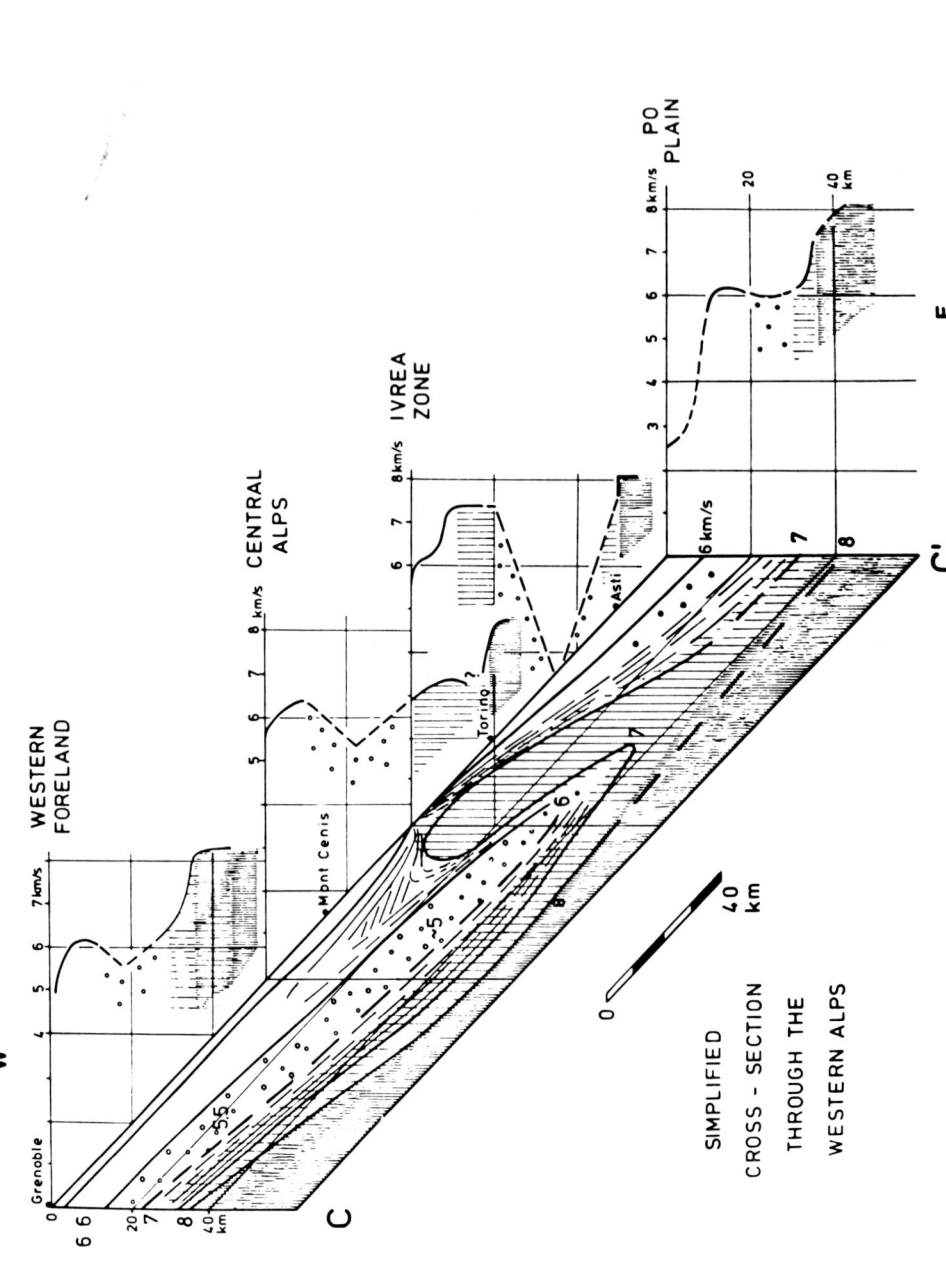

Fig. 10. Simplified east-west cross section through the western part of the Alps, showing contours of equal velocity and typical velocity-depth functions. The circles indicate the position of the low-velocity zone. The data are from Giese and Prodehl [1976].

central Alps reaching a depth of 30 to 34 km. Also the thickness of the crust-mantle transition decreases considerably to only a few kilometers or less. This structure is typical for Hercynian Europe as evaluated for southern Portugal [Mueller et al., 1973b; Prodehl et al., 1974], France [e.g., Sapin and Prodehl, 1973; Perrier and Ruegg, 1973; Sapin and Hirn, 1974], and western Germany [Giese and Stein, 1971; Giese et al., 1976].

The Hercynian system of southwestern Germany and of central France is crossed by several N-S trending graben systems, including the Rhine graben in Germany and the Rhone and Limagne grabens in central France. The structure of the lower crust and the crust-mantle transition zone as derived by Edel et al., [1975] for the Rhine graben area is shown in Figure 11. When approaching the graben area the crust thins from about 30 km in the northwest of Figure 11 (for example, the velocity-depth function of profile 16-195, Figure 11) to 25 to 26 km beneath the Vosges (profiles BA-010, LT-225, and SB-290) and the Black Forest (profiles 24-090 and SB-045, Figure 11). Figure 12 shows the record section of profile SB-045, which is entirely located in the Black Forest beyond 60 km distance. The $P^M P$ phase (phase 1 in Figure 12) is very well defined and results in a more or less sharp Mohorovičić discontinuity. The character of phases changes completely when moving into the graben proper (profiles WI-190-SB and SB-010-WI in Figure 11). As can be seen in Figure 13, phase 1 ($P^M P$) along profile SB-010-WI has disappeared and only phase 2 ($P^I P$) is visible as a strong wide-angle reflection originating at 20 to 21 km depth. The velocity at this depth, where the strongest velocity gradient occurs beneath the graben proper, is only 7.0 to 7.2 km/s. The velocity increases with increasing depth until reaching 8.0 to 8.1 km/s at about 30 km depth without, however, forming a new discontinuity or zone of strong velocity gradient. It is a matter of definition as to which depth range should be defined as the crust-mantle boundary as is discussed in detail by Edel et al. [1975] and Prodehl et al. [1976]. It should be pointed out that for an upper crust the velocity-depth functions in Figure 11 show only an average velocity structure. Only a few of the profiles allow a more detailed resolution of this velocity-depth range, indicating the presence of considerable vertical and horizontal heterogeneities in this zone as found, for example, by Mueller et al. [1973a].

An anomalous crust-mantle transition zone of the Rhine graben proper is also found in other areas. (The depth of strongest velocity gradient is located at a shallow depth of only 20 km, and the velocity only reaches 7.2 to 7.4 km/s between 20 and 25 km depth from the point at which the velocity increases continuously to that of the upper mantle (8.2 km/s).

Giese et al. [1968] report a similar velocity-depth structure for the northern Apennines. The crustal thickness does not exceed 23 km, and the velocity increases from 6.0 to only 7.0 km/s over a depth range of only 2 km. Below 23 km depth the velocity increases only gradually with increasing depth.

It should also be noted that the Ivrea body, whose velocity-depth structure is shown in Figure 10, can be understood as a part of the upper mantle overthrown toward the west over the sialic crust of the western Alps.

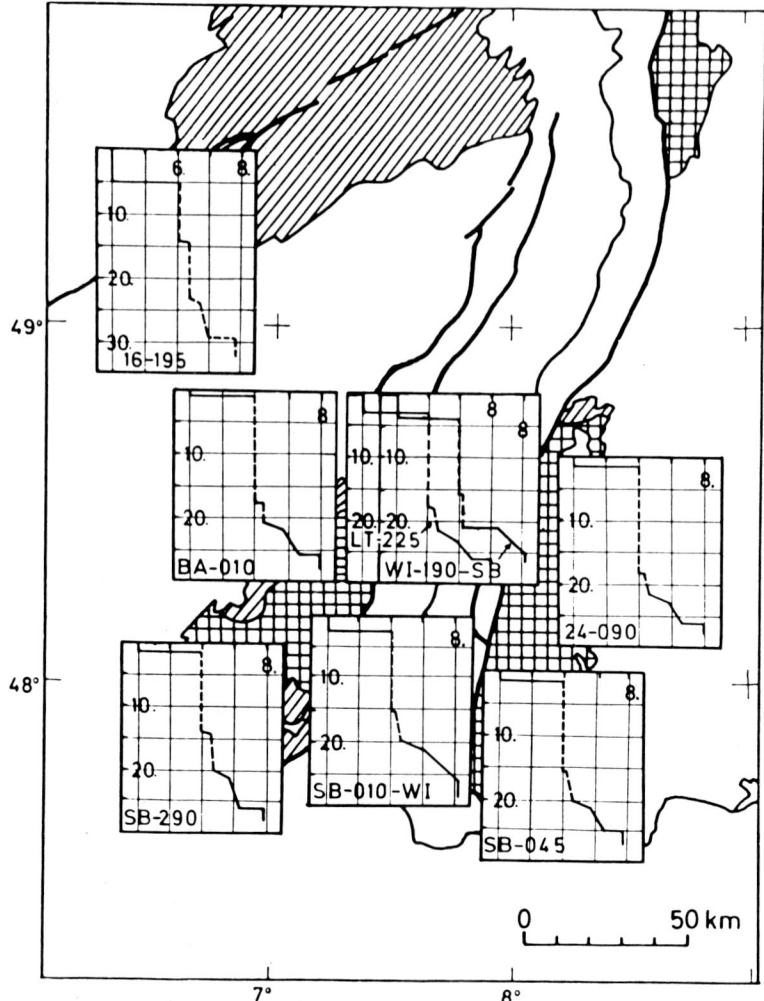

Fig. 11. Map showing velocity-depth distributions for different profiles of the central and southern Rhine graben. The inserts with the models are placed as closely as possible over the regions for which they are typical. The data are from Edel et al., [1975].

Summary and Conclusion

Table 1 summarizes the results of the structure of the crust and the crust-mantle boundary. Generally, there appears to be a relationship between total crustal thickness and thickness of the crust-mantle transition zone. The thinner the crust, the more distinct is the crust-mantle boundary and vice versa. If for grabens and similar anomalous areas the depth of the strongest velocity gradient is defined as the crust-mantle boundary, the above-mentioned relationship is fulfilled here, too. In this case, the P_n velocity is not always the same as the velocity immediately below the depth of the strongest velocity gradient [see Giese and Stein, 1971; Prodehl, 1970a, b].

Another relationship apparently can be generalized: young geosynclinal areas as well as shield areas are characterized by thick crust and generally weak P^MP arrivals. However, the mean crustal velocity is different: it is large (around 6.5 km/s) in shield areas and other areas belonging to the stable continents but it is small (around 6.0 km/s) for young orogens as well as for areas with a thin crust, for example, the Hercynian mountain system and the grabens in Europe and the Basin and Range province in the western United States.

When looking at the crustal thickness of the high mountainous areas of the United States, it seems questionable whether the term 'mountain root' can be used. Beneath the Alps, the crust-mantle boundary as well as the upper-crustal material reach greater depths than in neighboring areas. Beneath North America, however, only one of these features is relevant. The total crust of the Sierra Nevadas is thicker than that of the surrounding provinces; however, a low-velocity zone within the crust does not exist, and the thickness of the upper crust is normal. In the southern Rocky Mountains, total crustal thickness is similar to that of the neighboring areas, but the upper crust with an average velocity of 6 km/s reaches depths of about 35 km in contrast to velocity values found at similar depths beneath the adjacent Colorado plateau to the west and the Great Plains to the east.

The general decrease of mean crustal velocity in the western United States from north to south which appears most clearly in the Great Basin of the Basin and Range province can possibly be correlated with the outward development of western North America during Mesozoic and Cenozoic time. According to Hamilton and Myers [1966], total Cenozoic extension in northern Nevada and Utah may have been 300 km, while further south (northern Arizona) only 50 km are estimated. This strong extension of the crust in the north may have been accompanied by the replacement of light crustal material by dense mantle material, causing the development of a well-established lower crust.

The striking similarity of the crustal structure of the Basin and Range province, an extremely anomalous tectonic area within North America, and the so-called normal structure of Hercynian western Europe leads to the conclusion that the crust of the Hercynian mountain systems is indeed an anomalous one. Similar to the north-south trending grabens and horsts of the Basin and Range province, the Hercynian mountain system has undergone severe block faulting during the Alpine orogeny, including the 'taphrogenesis' of several north-south trending grabens. However, the structure of Hercynian Europe is much more complicated than a simple basin and range pattern. The estimated extension of graben areas, such as the Rhine graben, does not reach values similar to those estimated for the east-west extension of North America. Nevertheless, there seems to exist some probability that the evolution of the crust in Hercynian Europe and in the Basin and Range province might have some similar causes.

Another striking similarity in the structure of the crust-mantle boundary exists for both North American and European graben areas. The structure beneath western Utah is very similar to that of the young graben areas of western Europe. The crustal structure in the Afar depression of Ethiopia also shows a thin crust and velocities of only

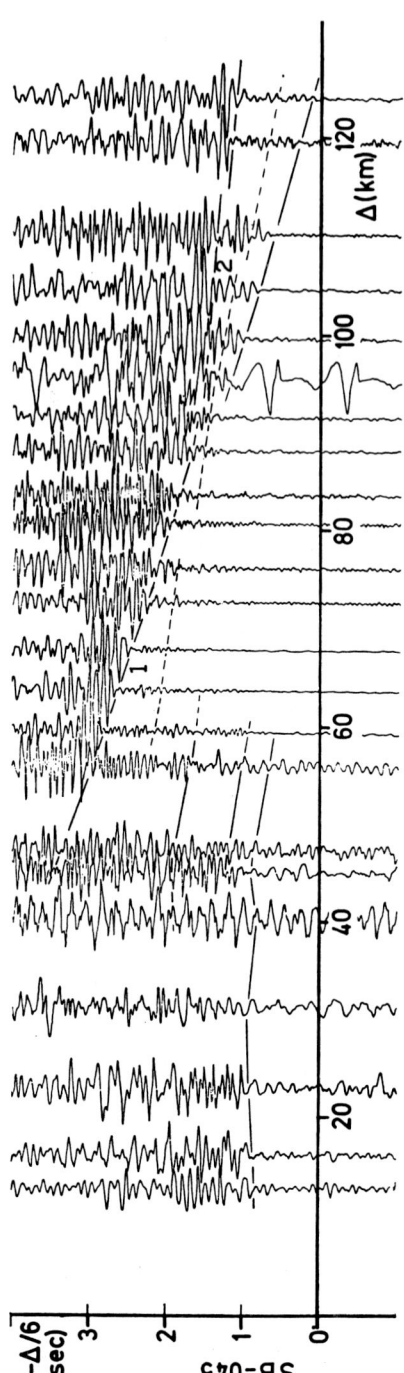

Fig. 12. Record section of the Steinbrunn profile toward northeast (SB-045) (Black Forest). Phase 1 is phase $P^M P$. Data are from Edel et al., [1975].

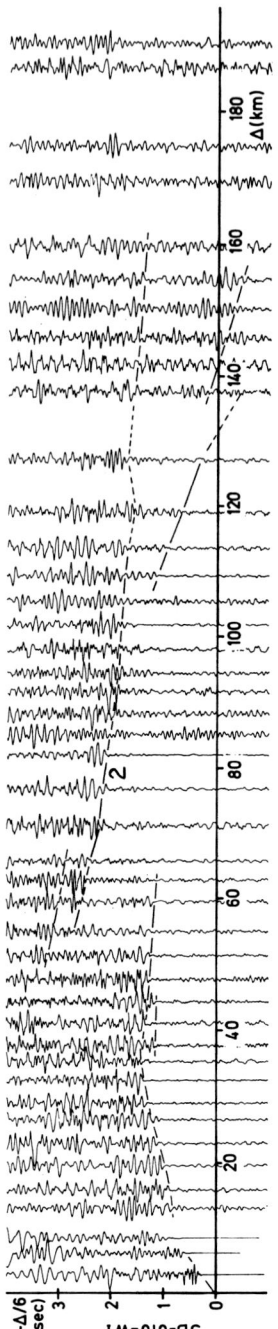

Fig. 13. Record section of the Steinbrunn-Wissembourg (SB-010-WI) (Rhine graben) profile [from Edel et al., 1975].

7.2 to 7.6 km/s at its lower boundary [Berckhemer et al., 1975]. Berckhemer et al. have developed a global systematics of crustal types according to which all three areas, the Basin and Range province, the Rhine graben, and the Afar depression are areas of taphrogenic activity.

Acknowledgments. The study was supported by funds of the Deutsche Forschungsgemeinschaft (German Research Society) and by temporary appointment to the National Center for Earthquake Research of the US Geological Survey in Menlo Park, California, while the author was on leave from Geophysikalisches Institut der Universität, Karlsruhe, Germany. I am indebted to K. Fuchs and R. Kind of the University of Karlsruhe for stimulating advice. I want to thank J. H. Healy, D. P. Hill, L. C. Pakiser, and S. W. Stewart of the US Geological Survey at Denver and Menlo Park for helpful discussions, unpublished reports, and data. I also thank P. Giese of Freie Universität, Berlin, Germany, and S. Mueller, ETH, Zürich, Switzerland, for their interest in this study. Computation facilities were made available by the computer center of Karlsruhe University and by the US Geological Survey.

References

Berckhemer, H., B. Baier, H. Bartelsen, A. Behle, H. Burckhardt, H. Gebrande, J. Makris, H. Menzel, H. Miller, and R. Vees, Deep seismic soundings in the Afar region and on the highland of Ethiopia, in Afar Depression of Ethiopia, edited by A. Pilger and A. Rösler, pp. 89-107, Schweizerbart, Stuttgart, Germany, 1975.

Berry, M. J., and K. Fuchs, Crustal structure of the Superior and Grenville provinces of the northeastern Canadian shield, Bull. Seismol. Soc. Amer., 63, 1393-1432, 1973.

Borcherdt, R. D., and J. C. Roller, A preliminary summary of a seismic refraction survey in the vicinity of the Cumberland Plateau observatory, Tennessee, Techn. Lett. Crustal Stud., 43, US Geol. Surv., Denver, Colo., 1966.

Choudhury, M., P. Giese, and G. de Visintini, Crustal structure of the Alps--Some general features from explosion seismology, Boll. Geofis. Teor. Appl., 13, 211-240, 1971.

Closs, H., and C. Behnke, Fortschritte der Anwendung seismischer Methoden in der Erforschung der Erdkruste, Geol. Rundsch., 51, 315-330, 1961.

Edel, J. B., K. Fuchs, C. Prodehl, and C. Gelbke, Deep structure of the Rhine graben area from seismic refraction investigations, J. Geophys., 41, 333-356, 1975.

Fenneman, N. M., Physical divisions, National Atlas of the United States of America, Sheet No. 59, 1970.

Giese, P., Versuch einer Gliederung der Erdkruste im nördlichen Alpenvorland in den Ostalpen und in Teilen der Westalpen mit Hilfe charakterischer Refraktions-Laufzeitkurven sowie einer geologischen Deutung, Geophys. Abh. Inst. Meteorol. Geophys., 1 (2), Freie Universität, Berlin, Germany, 1968.

Giese, P., The special structure of the P^Mp travel time curve, Z. Geophys., 38., 395-405, 1972.

Giese, P., Results of the generalized interpretation of deep-seismic sounding data, in Explosion Seismology in Central Europe - Data and Results, edited by P. Giese, C. Prodehl, and A. Stein, pp. 201-214, Springer, New York, 1976.

Giese, P., and C. Prodehl, Main features of crustal structure in the Alps, in Explosion Seismology in Central Europe - Data and Results, edited by P. Giese, C. Prodehl, and A. Stein, pp. 347-376, Springer, New York, 1976.

Giese, P., and A. Stein, Versuch einer einheitlichen Auswertung tiefen-seismischer Messunge aus dem Bereich zwischen der Nordsee und den Alpen, Z. Geophys, 37, 237-272, 1971.

Giese, P., K. Günther, and K. J. Reutter, Vergleichende geologische und geophysikalische Bertrachtungen der W-Alpen und des N-Apennin, Z. Deut. Geol. Ges., 120, 151-195, 1968.

Giese, P., C. Prodehl, and A. Stein (Eds.), Explosion Seismology in Central Europe - Data and Results, Springer, New York, 1976.

Haggag, I., Die Geschwindigkeitsverteilung von Kompressionswellen in zwei Gebieten der westlichen USA, abgeleitet aus refraktions-seismischen Messungen. Diplomarbeit, Geophys. Inst. Univ. Karlsruhe, Karlsruhe, Germany, 1974.

Hamilton, A., and W. B. Myers, Cenozoic tectonics of the western United States, Rev. Geophys. Space Phys., 4, 509-549, 1966.

Hirn, A., and G. Perrier, Deep seismic sounding in the Limagne graben, in Approaches to Taphrogenesis, edited by H. Illies and K. Fuchs, pp. 329-340, Schweizerbart, Stuttgart, Germany, 1974.

Hirn, A., C. Prodehl, and L. Steinmetz, An experimental test of models of the lower lithosphere in Bretagne (France), Ann. Geophys., 31, 517-530, 1975.

Hirschleber, H. B., C. E. Lund, R. Meissner, A. Vogel and W. Weinrebe, Seismic investigations along the Scandinavian "Blue Road" traverse, J. Geophys., 41, 135-148, 1975.

Jackson, W. H., S. W. Stewart, and L. C. Pakiser, Crustal structure in eastern Colorado from seismic refraction measurements, J. Geophys., 68, 5767-5776, 1963.

Keller, G. R., R. B. Smith, and L. W. Braile, Crustal structure along the Great Basin-Colorado Plateau transition from seismic refraction studies, J. Geophys. Res., 80, 1093-1098, 1975.

Liebscher, H. J., Deutungsversuche für die Struktur der tieferen Erdkruste nach reflexionsseismischen und gravimetrischen Messungen im deutschen Alpenvorland, Z. Geophys., 30, 51-96, 115-126, 1964.

Meissner, R., The "Moho" as a transition zone, Geophys. Surv., 1, 195-216, 1973

Mitchell, B. J., and M. Landisman, Interpretation of a crustal section across Oklahoma, Geol. Soc. Amer. Bull., 81, 2647-2656, 1970.

Mueller, S., E. Peterschmitt, K. Fuchs, D. Emter, and J. Ansorge, Crustal structure of the Rhine graben area, Tectonophysics, 20, 381-392, 1973a.

Mueller, S., C. Prodehl, A. S. Mendes, and V. S. Moreira, Crustal structure in the southwestern part of the Iberian peninsula, Tectonophysics, 20, 307-318, 1973b.

Pakiser, L. C., and J. S. Steinhart, Explosion seismology in the western hemisphere, in Research in Geophysics, vol. 2, Solid Earth and Interface Phenomena, edited by H. Odishaw, pp. 123-147, MIT Press, Cambridge, Mass., 1964.

Pavlenkova, N. I., Wavefields and Models of the Earth's Crust, 219 pp., Nauka Press, Kiev, Russia, 1973.

Perrier, G., and J. C. Ruegg, Structure profonde du Massif Central français, Ann. Geophys., 29, 435-502, 1973.

Prodehl, C., Seismic refraction study of crustal structure in the western United States, Geol. Soc. Amer. Bull., 81, 2629-2646, 1970a.

Prodehl, C., Crustal structure of the western United States from seismic refraction measurements in comparison with central European results, Z. Geophys., 36, 477-500, 1970b.

Prodehl, C., Comparison of seismic refraction studies in central Europe and the western United States, in Explosion Seismology in Central Europe - Data and Results, edited by P. Giese, C. Prodehl, and A. Stein, pp. 385-395, Springer, New York, 1976.

Prodehl, C., V. S. Moreira, S. Mueller, and A. S. Mendes, Deep-seismic sounding in central and southern Portugal, in Proceedings of the XIVth General Assembly of the European Seismological Commission, edited by H. Stiller and W. Webers, pp. 261-266, Akad, Wiss., Berlin, Germany, 1975.

Prodehl, C., J. Ansorge, J. B. Edel, D. Emter, K. Fuchs, S. Mueller, and E. Peterschmitt, Explosion seismology research in the central and southern Rhine graben - A case history, in Explosion Seismology in Central Europe - Data and Results, edited by P. Giese, C. Prodehl, and A. Stein, pp. 313-328, Springer, New York, 1976.

Sapin, M., and A. Hirn, Results of explosion seismology in the southern Rhône valley, Ann. Geophys., 30, 181-202, 1974.

Sapin, M., and C. Prodehl, Long range profiles in western Europe, 1, Crustal structure between the Bretagne and the Central Massif of France, Ann. Geophys., 29., 127-145, 1973.

Stewart, S. W., Crustal structure in Missouri by seimsic refraction methods, Bull. Seismol. Soc. Amer., 58, 291-323, 1968.

Warren, D. H., J. H. Healy, J. Bohn, and P. A. Marshall, Crustal structure under Lasa from seismic refraction measurements, J. Geophys. Res., 78, 8721-8734, 1973.

INVERSION OF SEISMIC REFRACTION DATA

John A. Orcutt, LeRoy M. Dorman, and Paul K. P. Spudich

Geological Research Division, Scripps Institution of Oceanography
La Jolla, California 92093

Abstract. A variety of methods for analyzing the kinematic data available in a typical seismic refraction profile are investigated. The traditional method of differentiation of a series of sparsely spaced, noisy, travel time data by fitting a series of straight lines is shown to result in an unstable inverse problem. This in turn can lead to unjustified conclusions about the actual velocity variation with depth. Alternative methods, which recognize that travel time data place only integral constraints on velocity models, are stable and permit rigorous calculations of resultant model multiplicity. However, the resolution of the travel time data is shown to be poor. Several proposals for increasing the resolution of available data by incorporating dynamic characteristics of the data and for improving experimental design are discussed.

Introduction

The nonuniqueness of oceanic crustal models derived from travel time data is examined with respect to a recent profile obtained on old oceanic lithosphere by using ocean bottom seismographs. Typically, refraction data are analyzed by assuming that the first arrivals are head waves propagating along the interfaces in an earth consisting of a few thick homogeneous layers. When a layer solution is found, the multiplicity of other velocity models which fit the travel time data is generally not considered. However, the calculation of models given a finite set of data with errors is known not to be unique. That is, an infinite number of models can be found which fit the data. That this nonuniqueness is of more than academic or purely theoretical interest will be clearly shown. The evaluation of the multiplicity of solutions is of paramount importance in understanding the earth's crustal structure.

Data

The data to be examined were obtained with an ocean bottom seismograph array between the E-W trending Clarion and Molokai fracture zones parallel to magnetic anomaly 32 in 5.5 km of water. During the January 1976 Scripps Deepsonde Expedition a long, split refraction line was shot to a range of 600 km by using charges up to 2 tons in weight. Figure 1 illustrates one of the profiles obtained to a range of over 100 km. The ocean bottom seismographs employed

have been described by Prothero [1974], and the use of these
instruments for refraction has been discussed by Orcutt et al. [1976].
The data in Figure 1 have been normalized to a shot size of 100 kg
[Orcutt et al., 1976] and band-pass filtered from 3 to 9 Hz with a
digital zero-phase shift filter, and the amplitudes have been
multiplied by a factor proportional to range to amplify distant
traces. The travel times are reduced by a velocity of 8 km s^{-1}, and
the water layer delay has been removed. A topographic correction has
been applied by assuming that the velocity contours mirror the
topography at the seabed down to a depth corresponding to the highest
apparent velocity [Kennett and Orcutt, 1976].

The travel times on the two sides of the split profile from all
elements of the array matched within the picking errors, so that no
evidence exists in the travel times for significant lateral
heterogeneity. There is, however, some difference in the character
of the seismograms for different elements and directions.

The profile in Figure 1 illustrates the character of most of the
data. The first arrivals within 20 km are strong and impulsive, while
between 20 and 40 km the arrivals become increasingly emergent. The
small emergent P_n wave appears at 40 km followed by a strong second
arrival. The first-arrival times from the various profiles were used
in the inversions of the travel time data.

Data Analysis

The travel time inversion techniques have been discussed previously
by Kennett and Orcutt [1976]. The travel times on all profiles were
picked, and the errors evaluated. All the travel time data were
combined and reparameterized into the $\tau(p)$ curve, where τ is delay
time and p is seismic ray parameter. The advantage of transforming
T(X) data into $\tau(p)$ data is that $\tau(p)$ is a single valued monotonically
decreasing function even in the presence of triplications in the T(X)
curve, and $\tau(p)$ suffers simple jump discontinuities if low velocity
zones are present. By definition, τ = T - pX, so that each travel
time pair (T_0, X_0) plots as a straight line in the τ-p plane with a
τ intercept of T_0 and a slope of $-X_0$, as shown in Figure 2 for
several travel time pairs from the profiles. As Bessonova et al.
[1974] point out, the τ = T_0 - pX_0 lines define the envelope of
the desired curve $\tau(p)$. A graphical construction method described
by Bessonova et al. [1974], as modified by Kennett and Orcutt [1976],
was used to generate the best-fitting $\tau(p)$ curve for the envelope.
Because $\delta\tau$ = δT - $p\delta X$ implies that the bulk of the error in τ
arises from errors in the travel time picks, the upper and lower error
bounds of satisfactory $\tau(p)$ curves can be readily derived. The
bounds on $\tau(p)$ are illustrated in Figure 3 for the data in Figure 2.
When travel time picks for the second arrival beyond 40 km were
reparameterized the branch was shown to be retrograde [Bessonova
et al., 1974]; it shall be referred to as the P_MP branch. The
highest velocity for which the travel time data contain information
is 8.0 km s^{-1} (p = 0.125 s km^{-1}), and the lowest velocity is 5.2 km s^{-1}
(p = 0.192 s km^{-1}). Because the data do not permit the bounds to be
continued to zero delay (the sea floor), 0.1 km of 1.6 km s^{-1}
sediment and 0.3 km of 4.0 km s^{-1} basement were stripped off the data

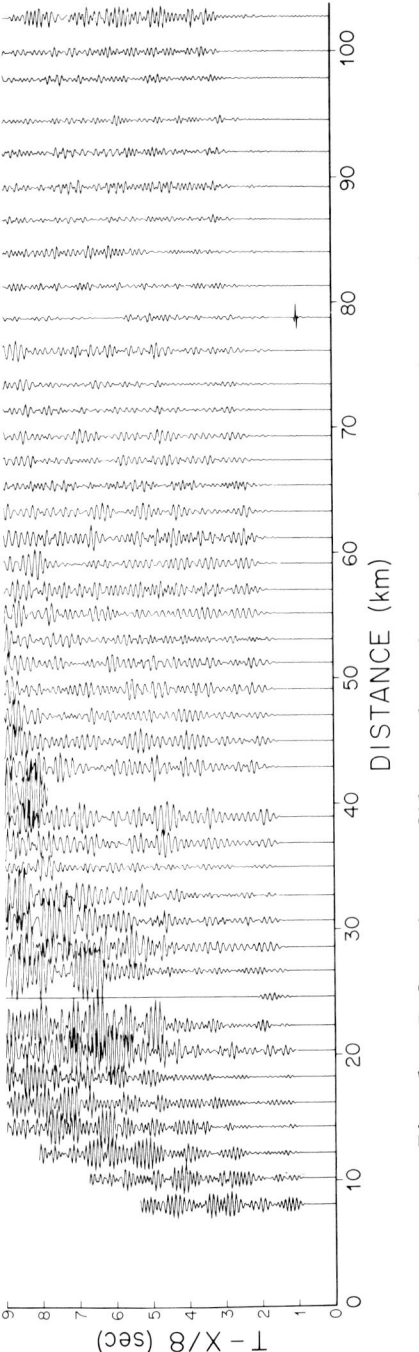

Fig. 1. Refraction profile conducted to ocean bottom seismograph Inez in the Pacific Basin. The data have been band-pass filtered from 3 to 9 Hz and normalized to a single-shot size of 100 kg.

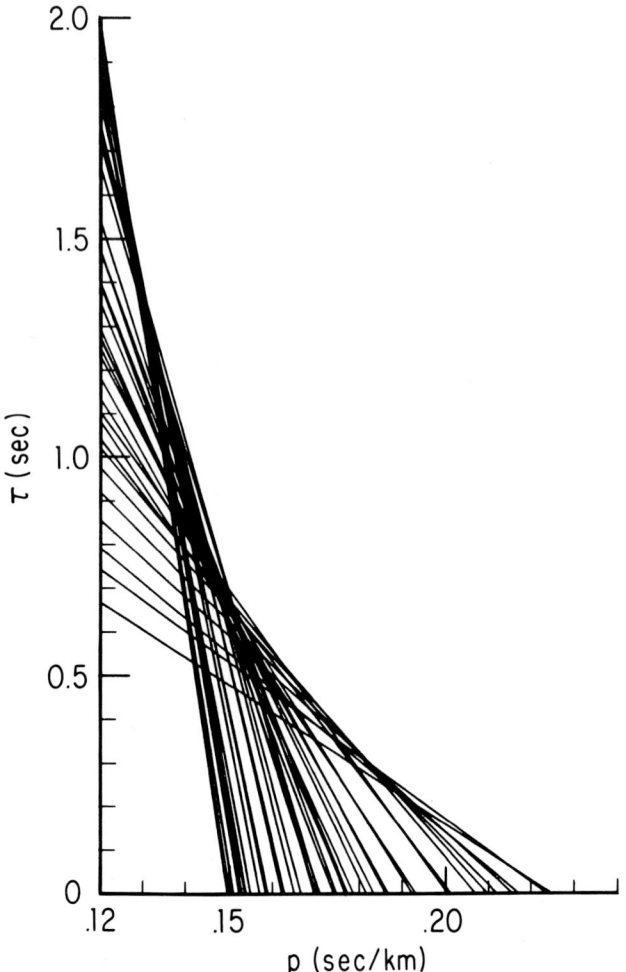

Fig. 2. A plot of several travel time data in the τ-p plane to form the envelope of the $\tau(p)$ function (where $p = v^{-1} \sin i$, i being the angle of incidence and v being the local seismic velocity).

by subtracting their delay time contribution, shown in Figure 3 by the nearly horizontal line near $\tau = 0.25$, from the $\tau(p)$ curve.

With a satisfactory $\tau(p)$ curve and error bounds in hand, several inversion schemes can be used to estimate the resolution and nonuniqueness of velocity models obtained from the travel time data. Such a method is that of Bessonova et al. [1974], which was applied to the $\tau(p)$ bounds in Figure 3 to yield the velocity bounds shown in Figure 4 within which must lie all velocity profiles satisfying the travel time data. Although bounds themselves are not satisfactory velocity models, considerable latitude exists within the bounds for choosing satisfactory models.

The delay time data were also inverted by using the generalized inversion scheme of Backus and Gilbert [1968] as described by Johnson and Gilbert [1972] and Kennett and Orcutt [1976]. Because the

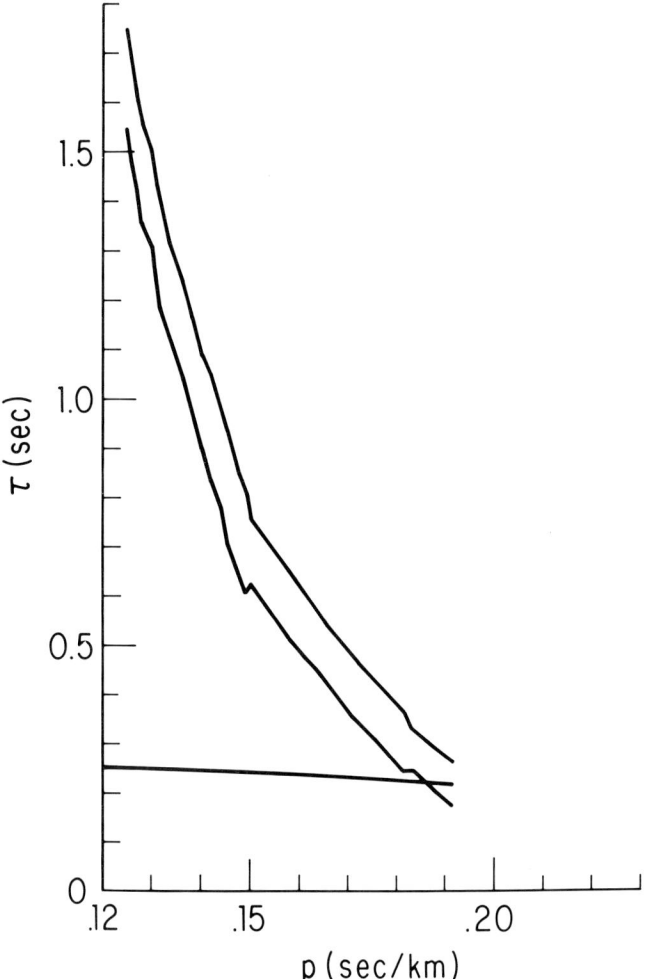

Fig. 3. Bounds on the delay time function for all the data from a three-element ocean bottom seismograph array to a range of 100 km.

problem is nonlinear, a starting model is perturbed until a fit to the data is obtained. Figure 5 illustrates the results of this inversion when the starting model was the mean of the Bessonova bounds in Figure 4. The functions to the right of the figure, called the resolving kernels, describe the averaging of the true earth by the finite, erroneous data. That is, one's observation $\tilde{\alpha}(Z_o)$ of the velocity at depth Z_o, is given by

$$\tilde{\alpha}(Z_o) = \int_0^{r_{earth}} R(Z, Z_o)\alpha(Z) \, dZ \qquad (1)$$

where $\alpha(Z)$ is the true velocity of the earth at depth Z and $R(Z,Z_o)$ is the resolving kernel for target depth Z_o. The resolving kernel

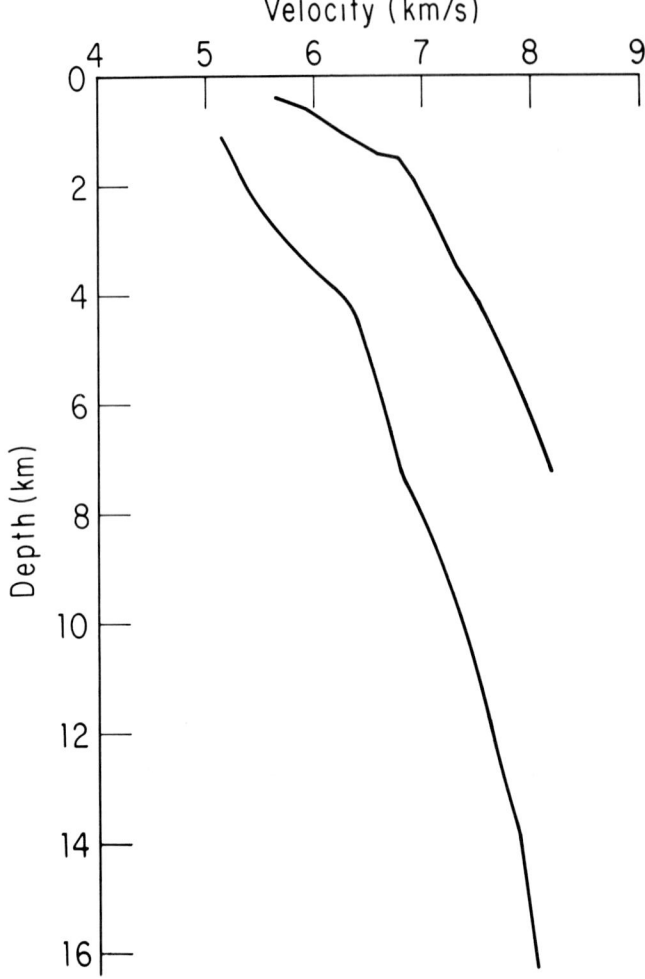

Fig. 4. Bounds on velocity resulting from extremal inversion of the delay time bounds of Figure 3.

is a function determined by the physics of the problem and by the number of data. If one had an infinite number of error free data, he could construct his resolving kernels to be Dirac delta functions, $\delta(Z - Z_o)$, and his estimates $\tilde{\alpha}(Z_o)$ of $\alpha(Z)$ would be perfectly accurate. However, with a finite number of inaccurate data the resolving kernels have a finite width and hence one can associate his estimate of velocity only with a certain depth range. The vertical bars superimposed upon the final model at the left in Figure 5 are measures of the width of these kernels and are called the 'spread'. The horizontal bars represent the error in the estimated velocity at any depth due to errors in the data. Because the problem is nonlinear, the resolving kernels and errors become model dependent as well as data dependent, and such estimates of model ambiguity are generally smaller than the estimates obtained by using a fully nonlinear extremal inversion [Wiggins et al., 1973]. The final model obtained in this case fits the first-arrival time

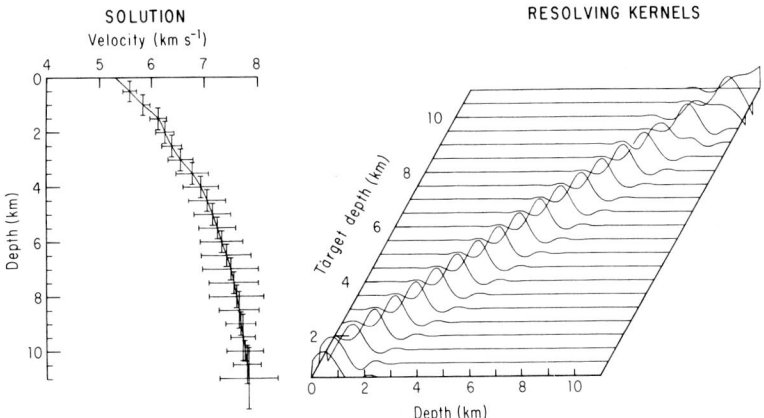

Fig. 5. Generalized inversion of the delay time data by using the mean of the extrema in Figure 4 as the starting model.

data nicely yet contains no discontinuities. The resolving power of the data appears to be quite good, 1.0-1.5 km, until mantle velocities are reached where the lower gradient results in a small geometric spreading factor dp/dX and the resolving power diminishes as indicated by the broadening of the resolving kernels.

If we assume a discontinuity at the Moho as a starting model, the result of the linearized inversion is shown in Figure 6. The resolving kernels in this case have discontinuities at the Moho depth and deteriorate where the gradient is low. The discontinuity is necessary in order to account for the retrograde P_MP branch of the travel time curve between 40 and 80 km as shown in Figure 7, where the travel time curve is superimposed upon the data. Several models resulting from the linearized inversion, all of which fit the travel time data, are shown in Figure 8.

Alternately, even if one wished to constrict himself to thick layered rather than gradational velocity models, a range of nearly equally acceptable layer solutions could be found. The 'best-fitting' one- and two-layer crustal solutions were sought by fitting straight line segments through the T,X data and allowing the crossover distances of line segments to assume any position along the profile which minimized

$$\chi^2 = \sum_{i=1}^{N} (E_i/\sigma_i)^2 \qquad (2)$$

where χ^2 is chi squared [Matthews and Walker, 1970], E_i is the residual time difference between the fit straight line and T_i, σ_i is the standard deviation of the travel time pick, and N, the number of data, is 41. If the residuals from the straight line fit are approximately equal to the quoted travel time errors, $\chi^2 \simeq N$, and the fit is reasonably acceptable. The two solutions are shown

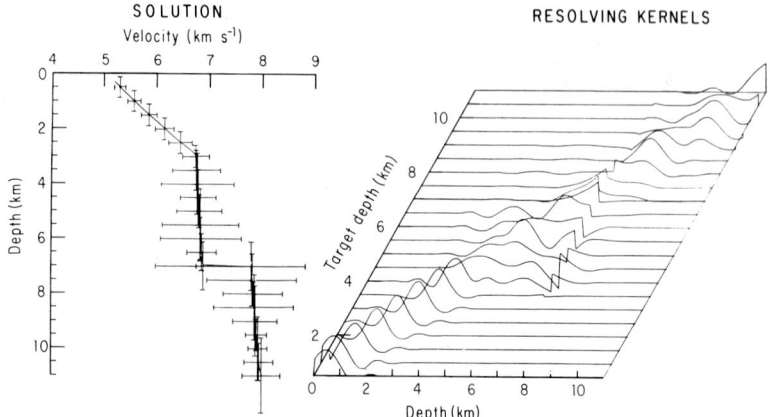

Fig. 6. Generalized inversion of the delay time data by using a starting model with a discontinuity at the Moho.

superimposed upon the extremal bounds in Figure 9. The behavior, as noted by Kennett and Orcutt [1976], is general in that a proper layer solution found by least square fitting a small number of straight lines to the travel time data will move back and forth between extremal bounds. The delay time data from the layer solutions determined by the travel time axis intercepts and line segment slopes are plotted in Figure 10. Note that the points lie along the lower τ bound as would be expected, since a layer solution is a minimum thickness solution for a data set [Kennett and Orcutt, 1976]. The travel time curves for the two-layer crust are shown in Figure 7.

In Figure 11, χ^2 is contoured on a plot of the first versus the second crossover distance (called breakpoint one and breakpoint two in the figure). The extremely high χ^2 to the right of the figure represents a forbidden area where the second-layer velocity is lower than the first-layer velocity. The global minimum is indicated by a plus. It can be seen that fixing one crossover distance at a minimum leaves the second crossover distance free to assume nearly any value without changing χ^2. In fact, χ^2 is only marginally improved by introducing a second layer into the crust.

Conclusions

The traditional practice of reducing seismic refraction data by fitting straight line segments to first arrivals as done earlier in searching for the best χ^2 fit can be shown to be a simple application of the familiar Herglotz-Wiechert-Bateman integral

$$Z(p) = \frac{1}{\pi} \int_p^{U(o)} \frac{X(q)}{(q^2 - p^2)^{1/2}} \, dq \qquad (3)$$

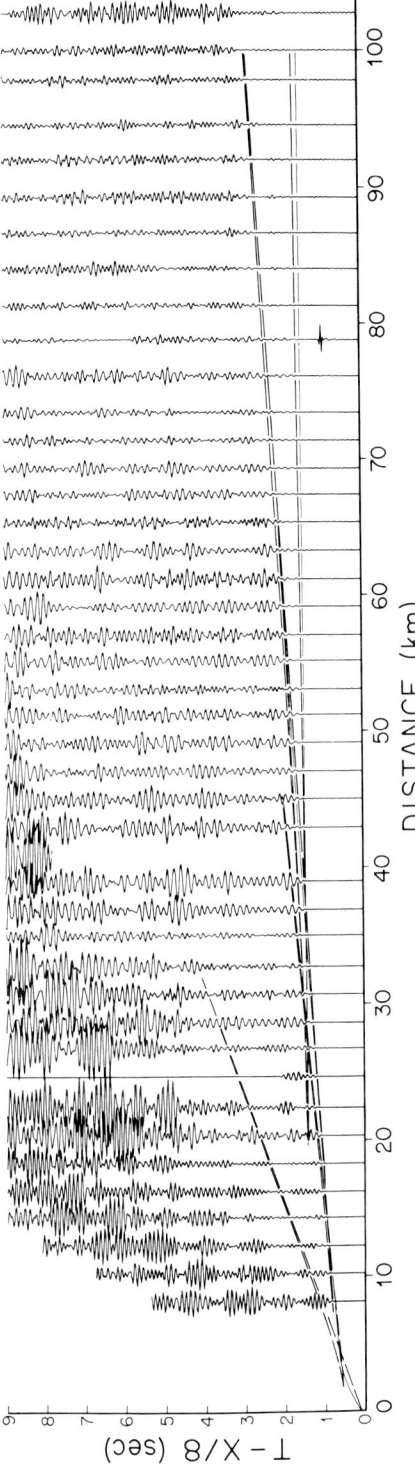

Fig. 7. Data of Figure 1 with the travel times of the linearized solution of Figure 6 (heavy solid line) and the two-layer crustal solution of Figure 9 (light dashed line) superimposed.

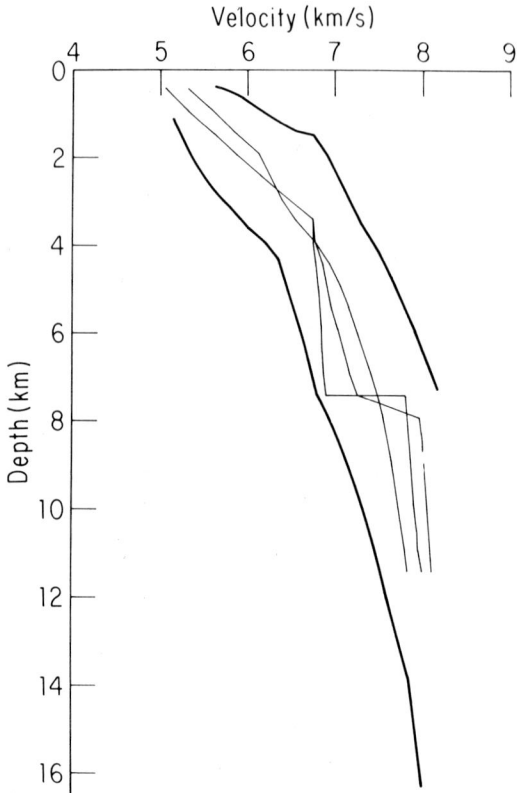

Fig. 8. The extremal bounds of Figure 4 with several models constructed with the linearized inversion scheme. All models fit the first-arrival travel time data.

where $Z(p)$ is the depth at which a ray with ray parameter p bottoms and $U(o)$ is the surface slowness. The travel time data themselves are not used in the inversion; rather the data are differentiated by fitting straight lines to form $X(p)$. The process of differentiation of travel time data is not stable in that two curves passing through the travel time points may be arbitrarily close and yet their first derivatives may be wildly different. The Herglotz-Wiechert-Bateman inversion is a poorly posed inverse problem since a small change in the data can result in a large change in the derived model. On the other hand, because errors in τ depend upon errors in p only to the second order [Brune, 1964], the $\tau(p)$ reparameterization is very stable, and the resultant inversions yield good estimates of the multiplicity of the solutions.

The wide variety of solutions available which fit the first-arrival

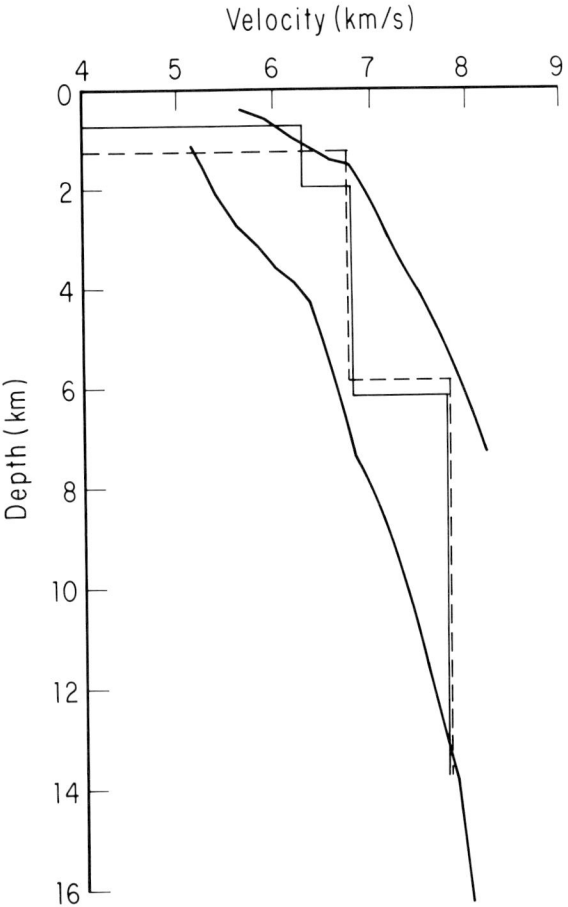

Fig. 9. Optimum one- and two-layer solutions which provide a minimum χ^2. The behavior of the solutions is general in that they move back and forth between the extrema. The velocity of 4 km s^{-1} is assumed. (_____ for two-layer; _ _ _ _ for one-layer.)

travel time data is perhaps disheartening. Nevertheless, the travel time data themselves provide only integral constraints on the models, and in the face of error and sparse, inadequate data the constraint is not tightly restrictive. Constraints on the solution are improved if second arrival travel time data are available, but the small temporal separation between various branches of the travel time curves in marine refraction profile, compared to the long source function duration, make picking of second arrivals extremely difficult. The situation can be improved in two ways. First, small arrays with instrument spacings of less than one seismic wavelength can be used to provide bounds on the $X(p)$ function by directly measuring the incident ray parameter p rather than attempting to obtain p from differenti-

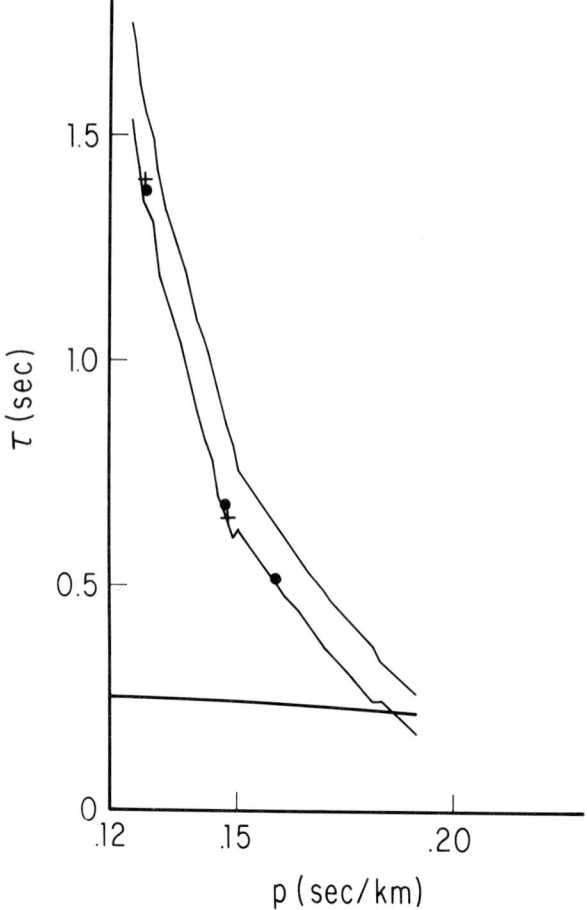

Fig. 10. Delay time data from the optimum layer solutions. The pluses represent the one-layer crust, and the circles the two-layer crust.

ation of widely spaced travel time data. An extremal technique such as that of Wiggins et al. [1973] which jointly inverts delay time and X(p) envelopes can be used to place much tighter bounds on the earth models. Second, although data errors are large, amplitude information can be used to constrain the velocity gradients within the model and can readily discriminate between the smooth and discontinuous models in Figure 8, all of which are indistinguishable on the basis of first arrival travel time data. The conduct of a formal inversion by using amplitude data, however, is not straightforward, since in the crustal regime with high wave numbers and frequencies, geometrical ray theory is very inadequate and complex wave theoretical techniques must be used.

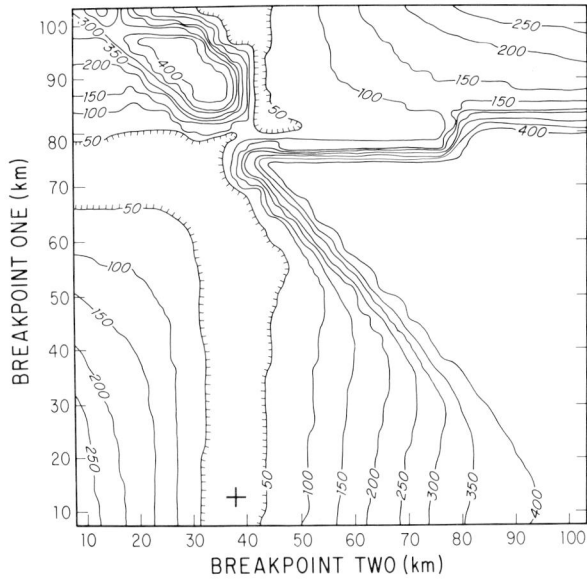

Fig. 11. Contours of χ^2 on a plot of the first versus the second breakpoint (crossover distance) in a two-layer crustal solution.

Acknowledgements. The authors would like to thank Brian Kennett, Tom Jordan and Bob Parker for helpful discussions related to the interpretation of seismic data. This research was sponsored by the National Science Foundation under Grant DES74-11909 and by the Office of Naval Research.

References

Backus, G., and F. Gilbert, The resolving power of gross earth data, Geophys. J. Roy. Astron. Soc., 16, 169-205, 1968.

Bessonova, E. N., V. M. Fishman, V. Z. Ryaboyi, and G. A. Sitnikova, The tau method for the inversion of travel-times, I., Deep seismic sounding data, Geophys. J. Roy. Astron. Soc., 36, 377-398, 1974.

Brune, J. N., Travel-times, body waves and normal modes of the earth, Bull. Seismol. Soc. Amer., 54, 2099-2128, 1964.

Johnson, L. E., and F. Gilbert, Inversion and inference for teleseismic ray data, Methods Comput. Phys., 11, 231-266, 1972.

Kennett, B. L. N., and J. A. Orcutt, A comparison of travel time inversions for marine refraction profiles, J. Geophys. Res., 81, 4061, 1976.

Matthews, J., and R. L. Walker, Mathematical Methods of Physics, W. A. Benjamin, Inc., Menlo Park, 1970.

Orcutt, J. A., B. L. N. Kennett, and L. M. Dorman, Structure of the

East Pacific Rise from an ocean bottom seismometer survey, Geophys. J. Roy. Astron. Soc., 45, 305-320, 1976.

Prothero, W. A., A short period ocean bottom seismograph, Bull. Seismol. Soc. Amer., 64, 1251-1262, 1974.

Wiggins, R. A., G. A. McMechan, and M. N. Toksöz, Range of earth structure nonuniqueness implied by body wave observations, Rev. Geophys. Space Phys., 11, 87-113, 1973.

GEOPHYSICAL EVIDENCE FOR A MAGMA BODY IN THE CRUST IN THE VICINITY OF SOCORRO, NEW MEXICO

A. R. Sanford, R. P. Mott, Jr., P. J. Shuleski, E. J. Rinehart, F. J. Caravella, R. M. Ward, and T. C. Wallace

New Mexico Institute of Mining and Technology
Socorro, New Mexico 87801

Abstract. An analysis of S_xP and S_xS reflections on microearthquake seismograms indicates the existence of magma beneath the Rio Grande rift in the vicinity of Socorro, New Mexico. As presently mapped, the magma occurs beneath 1200 km^2 of the central part of the rift at depths ranging from 18 to 20 km. The general configuration of the upper surface of the magma body near Socorro suggests its shape is that of an elongate laccolith with a north-south trend, a width of 8 km, and a thickness of 2 km. Several observations suggest that magma is being injected into the crust at the present time. The microearthquake activity is diffusely distributed over an area of 2000 km^2 that is roughly centered on the magma as presently mapped. Most earthquakes in the Socorro area have occurred in swarms, and one of these in 1906-1907 appears to have been comparable to the Matsushiro swarm. Level line data indicate surface uplift roughly coincident with the detected magma.

Introduction

Geophysical observations made over a period of years indicate the existence of magma at intermediate depths in the crust beneath the Rio Grande rift in the vicinity of Socorro, New Mexico. Important geophysical characteristics of the region are a diffuse geographical pattern of seismic activity, earthquake swarms, exceptionally strong S to P (S_xP) and S to S (S_xS) reflections from depths of 18-20 km (below mean surface elevation), high heat flow, and historical surface uplift. The rift at Socorro also has some unusual geologic features such as intragraben horsts. Each of the geophysical and geologic observations can be explained in a number of ways, but taken collectively, they make a strong case for the existence of a rather extensive layer (~1200 km^2) of magma in the crust along the axial region of the rift.

Research on this magma body, as well as a search for smaller magma bodies at depths of less than 13 km, is being actively pursued. This paper is a progress report on work completed to mid-1976 on the deeper magma body. Because the position of the magma relative to crustal structure is important, a summary of the known characteristics of the crust beneath the rift has been included in the paper.

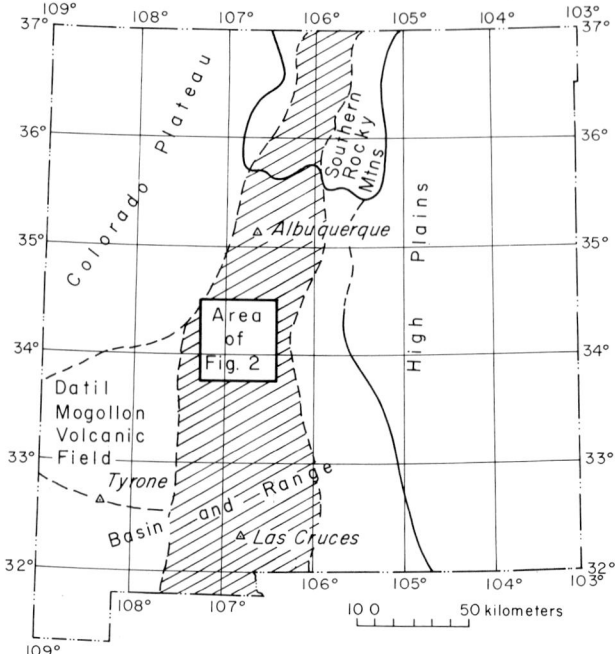

Fig. 1. Physiographic provinces and the Rio Grande rift in New Mexico (after Chapin, 1971).

The Rio Grande Rift

Geological Characteristics

Socorro, New Mexico, is located within the Rio Grande rift (see Figures 1 and 2), a major structure formed by an east-west crustal extension beginning about 25 to 29 m.y. ago and continuing to the present [Chapin and Seager, 1975]. North of Socorro the rift consists of a series of linked north trending structural depressions arranged in echelon in a NNE direction into central Colorado [Kelley, 1952, 1956; Chapin, 1971]. In southern Colorado and northern New Mexico the rift penetrates the Southern Rocky Mountains. In central New Mexico the rift lies between the Colorado Plateau and the High Plains. South of Socorro the rift merges in a complex and unknown way with the Basin and Range province.

The character of the rift changes rather abruptly 40-50 km north of Socorro (see Figure 2). North of Bernardo the rift consists of a single basin (maximum length and width 160 and 60 km, respectively) with raised margins. South of Bernardo the rift consists of a series of basins and ranges. The central ranges, the Socorro-Lemitar and Chupadera mountains, which separate the La Jencia basin from the Socorro basin, are intragraben horsts that formed relatively late in the history of the rift, 9-10 m.y. ago [Denny, 1941; Chapin and Seager, 1975]. Development of horsts through several thousand feet of Cenozoic sedimentary fill is known to occur at only two locations along the rift, Socorro and Las Cruces.

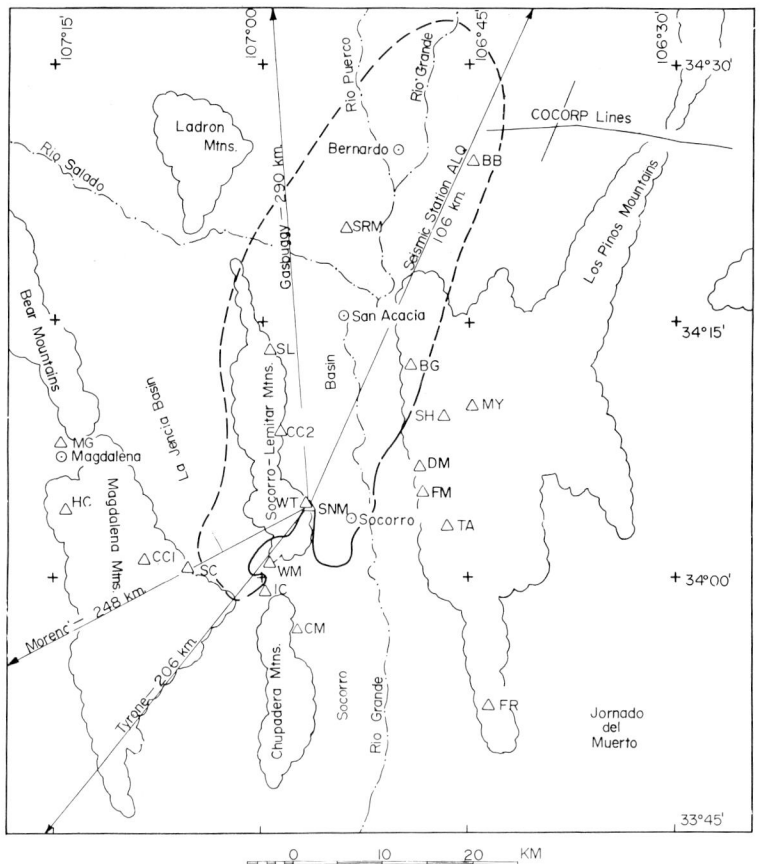

Fig. 2. Map showing the basins and ranges of the Rio Grande rift near Socorro and the location of seismograph stations. Outline of the magma body is shown approximately.

A gravity survey of the rift in the vicinity of Socorro [Sanford, 1968] indicates that the long narrow Socorro basin is composed of three linked structural depressions. The two depressions adjacent to the Socorro-Lemitar Mountains are highly asymmetrical. Narrow fault zones with large displacements, perhaps as great as 3.5 km, border the western margins of these downdropped crustal blocks. From their deepest points west of the Rio Grande these structural basins rise fairly gradually to the east, probably by a combination of step-faulting and tilting. The La Jencia basin is not as narrow as the Socorro basin nor as structurally asymmetrical. However, the total structural relief for the La Jencia basin is comparable to that of the structural depressions comprising the Socorro basin, i.e., of the order of 3 km.

A great deal of volcanic activity has accompanied the formation of the Rio Grande rift. The periods of greatest activity were from 26 to 20 m.y. ago and from 5 m.y. ago to the present [Chapin and Seager, 1975]. Basaltic andesites were most abundant during the first period of volcanism, whereas true basalts dominate the later period. Basalt flows of a very recent age occur in the Socorro area.

Numerous thermal springs occur along the entire western margin of the rift [Summers, 1965]. Coincident with these springs is a ribbon of high heat flow, >2.5 HFU (μcal cm^{-2} s^{-1}) [Reiter et al., 1975]. In the Socorro area the largest and hottest thermal springs are located in the southern end of the Socorro-Lemitar Mountains, which is also an area of very high heat flow.

Crustal Structure

Figure 3 is a north-south crustal profile of the Rio Grande rift through the Socorro region. The 5.8-km/s upper crust velocity is significantly lower than that found beyond the margins of the rift, 6.15 km/s. Both velocities were observed on a refraction profile south from the Gasbuggy underground nuclear explosion [Toppozada and Sanford, 1976], the lower velocity occurring only after the profile entered the rift 100 km north of Socorro. Additional evidence for the low upper crustal velocity was obtained by Toppozada [1974] from measurements of Pg velocities (5.86 km/s (stan. dev. ±0.10)) between two seismic stations within the rift, Socorro (SNM) and Albuquerque (ALQ). Also a Pg velocity of 5.8 km/s was found to fit best the travel times and distances between station ALQ and microearthquakes accurately located in the vicinity of Socorro [Sanford et al., 1973].

In Figure 3 two values of Poisson's ratio are shown in the upper half of the 5.8 km/s layer, 0.22 north of Socorro and 0.26 in the immediate vicinity of Socorro. The lower value was obtained by Sakdejayont [1974] from an analysis of S-P intervals of natural earthquakes recorded at stations ALQ and SNM (see Figure 2). Most of the events used in Sakdejayont's study had hypocenters within or bordering the rift and had average distances from SNM and ALQ of 28 and 82 km, respectively. Because the travel was greatest to ALQ, the Poisson ratio obtained is more representative of the upper crust on the northern end of the profile in Figure 3 than beneath Socorro. Poisson's ratio directly beneath Socorro, 0.26, was calculated from the slope of Wadati diagrams for 50 nearby microearthquakes recorded on local networks of 5 to 6 seismic stations. An explanation for the increase in Poisson's ratio beneath Socorro is the possible existence of several small magma bodies at shallow depths in the upper crust [Sanford et al., 1976].

The starting point for determining the thickness of the crust in Figure 3 was the refraction profile southward from Gasbuggy. Toppozada and Sanford [1976] obtained a total crustal thickness of 39.9 ± 1 km at the crossover distance 100 km north of Socorro. Reversed Pn interval velocities obtained by Toppozada [1974] indicate that the Moho dips about 2° northward from Las Cruces to Albuquerque. Extrapolating southward 100 km along the Gasbuggy profile with this dip yields a depth of 36.4 ± 1 km beneath station SNM at Socorro.

Other geophysical data suggest that the crustal thickness obtained in this matter is a reasonable estimate for Socorro. Toppozada [1974], using reversed Pn velocities between SNM and ALQ, determined that the crust could be 3.6 km thicker beneath ALQ than SNM. Phinney [1964], using spectra of long-period body waves, placed limits on the total crustal thickness beneath ALQ from 36.2 to 41.1 km. If the thickness at SNM is assumed to be near the lower limit, i.e., 35.4 km, then the crustal thickness beneath ALQ is 39.0 km, which is near the average Moho depth obtained by Phinney.

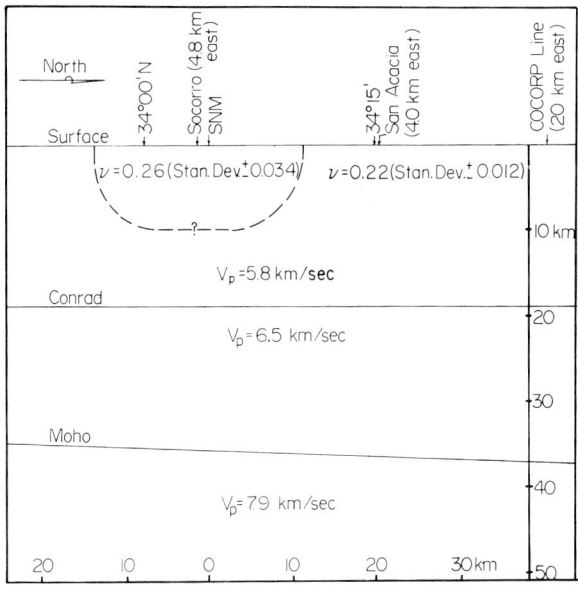

Fig. 3. North-south crustal profile of the Rio Grande rift through the Socorro region. The profile is along a meridian through station SNM which has coordinates 34°04.21'N, 106°56.79'W.

As a further check, travel times of Pn phases to SNM for mining explosions at Tyrone, New Mexico, and Morenci, Arizona [Dee, 1973], and the delay time for the crustal structure shown in Figure 3 were used to calculate crustal depths at these two open pit copper mines. A minimum depth of 25 km to the Moho was obtained at both locations, which is not far from the only published crustal thickness for southwestern New Mexico, 30 km [Tatel and Tuve, 1955].

The poorest known features of the crustal structure near Socorro are the depth and the dip of the Conrad discontinuity and the velocity of the lower crust. The depth of the Conrad, 18.9 km, and the velocity of the lower crust, 6.5 km/s, were taken from the interpretation of the unreversed Gasbuggy profile [Toppozada and Sanford, 1976]. The Conrad was assumed to parallel the surface because most published crustal sections for the southwestern United States show that crustal thinning occurs at the expense of the lower crust [Toppozada, 1974]. The P* arrivals on the record section for the Gasbuggy profile indicate that the Conrad discontinuity is continuous over the length of profile.

Changes in crustal thickness to the east and west of Socorro can be calculated from regional gravity. Figure 4 is a regional Bouguer gravity map from a study by Sanford [1968]. Differences in crustal thickness were calculated by assuming that variations in the observed Bouguer anomalies could be totally attributed to a 0.2-g/cm^3 density contrast across the Moho. (The latter value was found from the velocities above and below the Moho (Figure 3) using the relation between velocity and density obtained by Nafe and Drake [1963].) These calculations indicate very little change in crustal thickness east of the Rio Grande

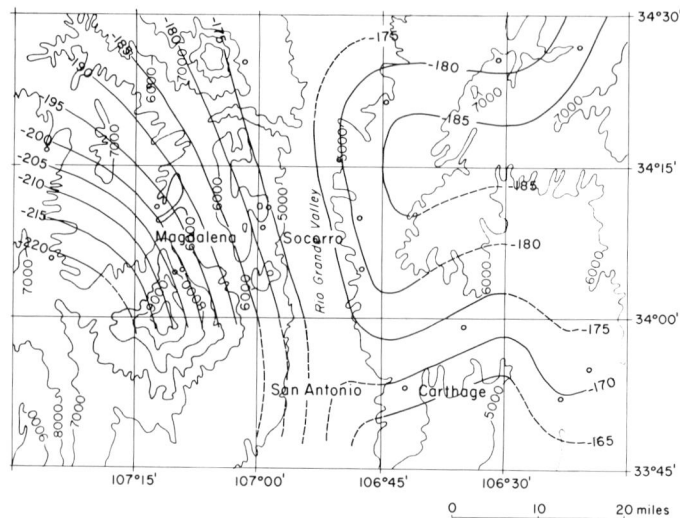

Fig. 4. Regional Bouguer gravity map [from Sanford, 1968].

rift and a maximum increase of 6 km in thickness about 40 km west of Socorro.

The calculated increase in crustal thickness to the west from gravity data is substantially greater than that calculated from seismic data. The apparent velocities of Pn arrivals from Tyrone and Morenci mining explosions [Dee, 1973] can be used with the crustal structure in Figure 3 to calculate the dip on the Moho surface in two directions, S40°W and S63°W. Both dips, 3.4° and 1.1°, respectively, are substantially less than the 8° westward dip from gravity data. Regardless of which dip is used, the geophysical data do not indicate a sharp symmetrical mantle upwarp beneath the rift in the vicinity of Socorro.

The Magma Body

The primary evidence for a magma body beneath the Rio Grande rift near Socorro comes from an analysis of S_xP and S_xS reflections on microearthquake seismograms. The results of studies on these reflections appear in two papers, Sanford and Long [1965] and Sanford et al. [1973]. Presented below is material from these two papers along with new data obtained since the last publication.

Seismograph Stations and Records

The locations of stations that provided data for the study of reflection phases are shown in Figure 2. For most stations the nominal magnification of seismographs at 10 Hz was greater than 1×10^6.

Microearthquake seismograms with clearly defined reflection phases from stations SNM, CC1, and ALQ appear in the two earlier papers. Shown here in Figure 5 are vertical component seismograms for stations at CC2, WM, CM, WT, and IC. When they are well recorded, as in these examples, the reflected phases are impulsive and have the same general

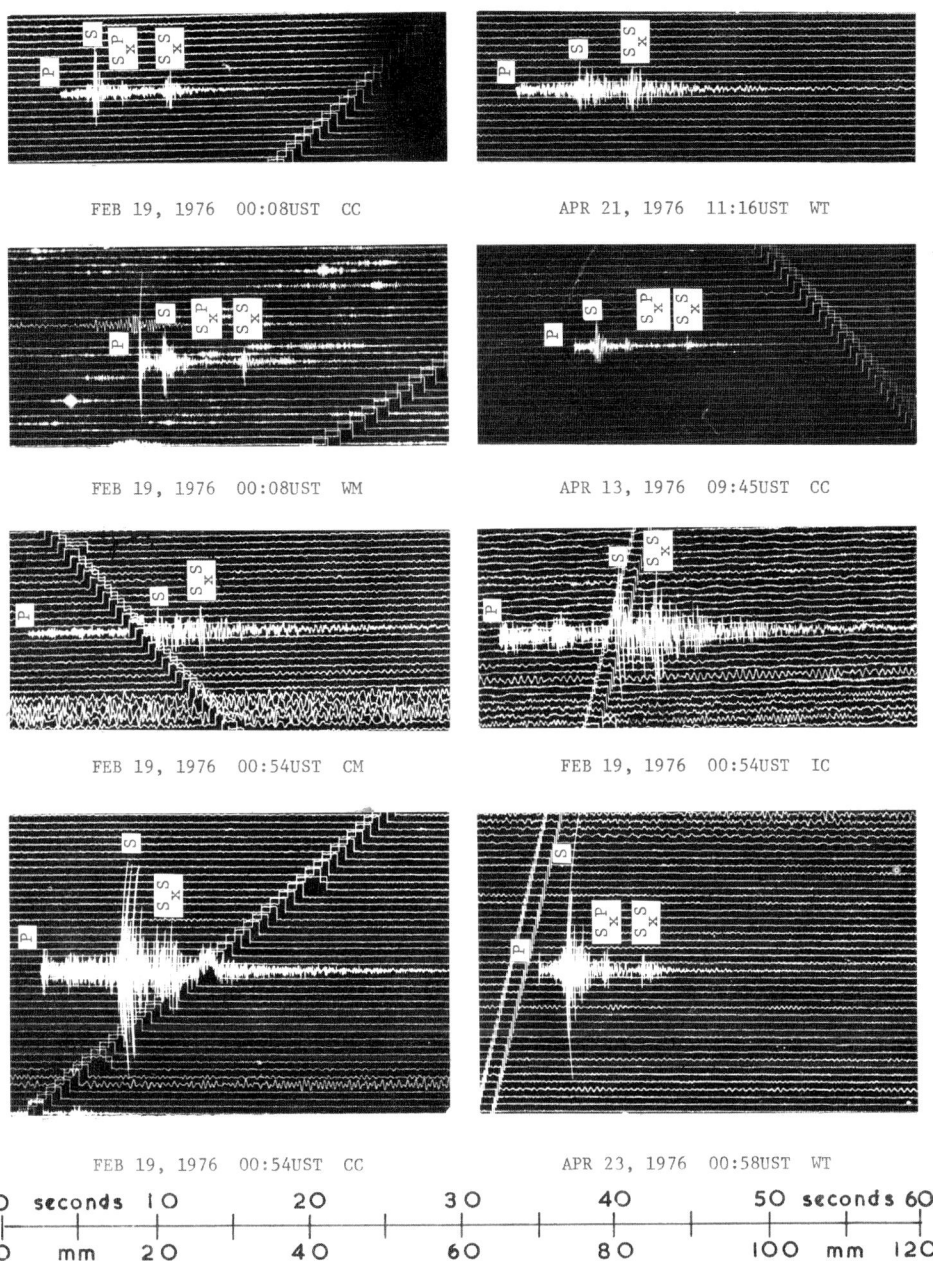

Fig. 5. Vertical component seismograms of microearthquakes with S_xP and S_xS reflection phases. Given beneath each seismogram is the date, time, and station.

frequency content as the S phase. On a statistical basis, S_xP is the weaker of the two phases and it is identified on far fewer seismograms than S_xS because its arrival is obscured by S wave energy which follows a direct path from the focus.

Because the strength of the reflections depends on a number of factors, e.g., the fault mechanism and the location of the station relative to the focus, not all microearthquakes recorded in the Socorro area have identifiable S_xP or S_xS phases. When vertical component instruments are used in an area where the magma body exists, about one half of the recorded microearthquakes have S_xS reflections. If horizontal component instruments are used in the same areas, a much higher percentage of S_xS reflections can be identified, e.g., up to 90% for station CC1.

Identification of the Reflection Phases

The primary data used to identify the late phases as S_xP and S_xS reflections were arrival times as a function of distance. In the work by Sanford and Long [1965] it was shown that the observed arrival times of the late phases could be bracketed by theoretical curves for $S_{18}P$ and $S_{18}S$ reflections constructed for two depths of focus, 4 and 9 km. In the work by Sanford et al. [1973] the correspondence between observed arrival times and theoretical arrival times was improved by introducing a moderate amount of dip, 6°, on the 18 km deep reflecting surface. Also in this later paper the identification of the early phase as an S_xP reflection off the same discontinuity as S_xS was strengthened by showing that observed and theoretical $S_{18}P$ times do not differ by more than 0.4 s in the distance range $1.2 \leq S-P \leq 3.0$ s.

Amplitudes of Reflection Phases

An unusual feature of the reflections is their large amplitudes relative to the direct S phase. For 150 vertical component seismograms at station SNM the S_xS to S amplitude ratios were found to range from 0 to 4.4 and average 1.3 [Sanford et al., 1973]. These large ratios can be obtained if the S wave energy radiated downward is much greater than that radiated outward from the focus and if the reflecting discontinuity has a large coefficient of reflection.

The tectonics of the Rio Grande rift indicate that contemporaneous faulting should be predominantly of the normal type. This has been confirmed by a composite fault plane solution for microearthquakes in the Socorro area. Figure 6 shows the fault plane solution and the position of P, T, and X axes on the upper focal sphere. The orientation of stress axes (P and T) and the direction of movement (X) indicate predominantly normal fault movement. If slip on normal faults propagates unilaterally downward at 0.9 of the shear wave velocity, the SV amplitude in the direction of the slip will be 10 times greater than that normal to the fracture [Savage, 1965]. Because this type of fault-slip should be common in the rift, the S wave energy leaving the focus along the ray path of the reflections will be in many instances much stronger than that along direct ray paths to the stations.

The reflected S_xS phase loses strength relative to the direct S wave because its longer distance of travel leads to greater geometrical spreading and attenuation. These factors diminish the amplitude of S_xS

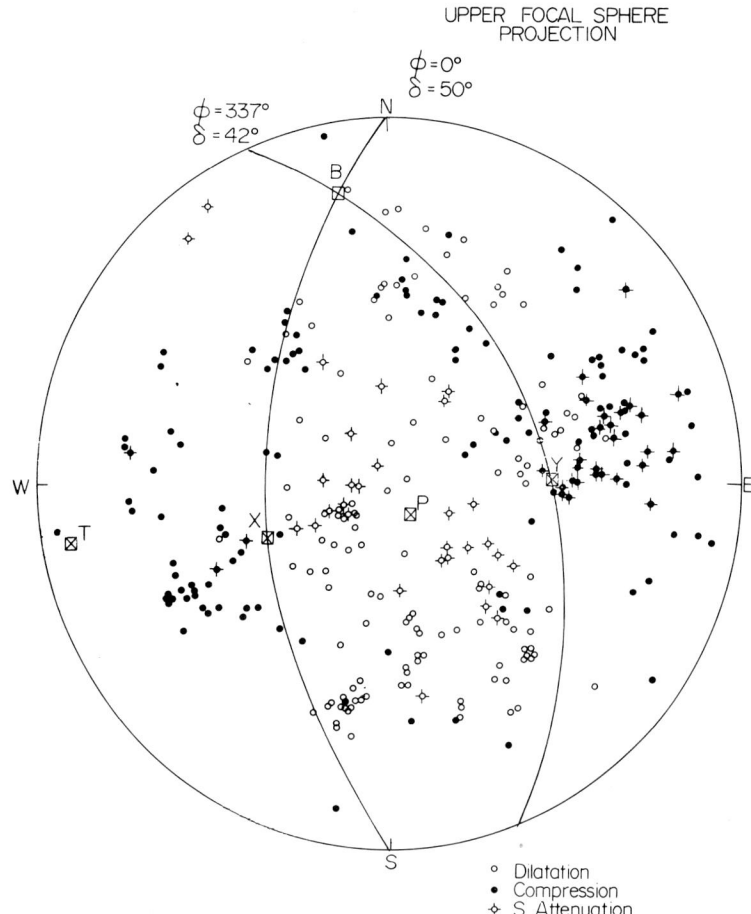

Fig. 6. Composite fault plane solution for microearthquakes in the vicinity of Socorro. Projection is on the upper focal sphere (Wulff stereographic projection). Orientation of minimum T and maximum P compressive stresses and direction of slip X are given.

relative to S by a factor of at least 2.0. The observed S_xS to S amplitude ratio also depends on the coefficient of reflection, which for normal crustal discontinuities would be about 0.01-0.2. However, the observations discussed below indicate much higher coefficients of reflection than would be expected for a normal Conrad or Moho discontinuity. Thus with asymmetrical radiation from the earthquake foci and unusually high coefficients of reflection, large S_xS to S amplitude ratios are possible despite S_xS losses from attenuation and geometrical spreading.

Properties of the Discontinuity

By assuming that plane-wave theory is applicable, the ratio of S_xP to S_xS amplitudes can be used to estimate the general magnitude of the re-

TABLE 1. Theoretical and Observed

		Theoretical		
		$\alpha_1 = 5.80$, $\alpha_2 = 6.50$, $\beta_1 = 3.35$, $\beta_2 = 3.75$, $\rho_1 = 2.81$, $\rho_2 = 2.89$		
Angle of Incidence S_xP	Angle of Incidence S_xS	Reflection Coefficient P	Reflection Coefficient S	S_xP/S_xS
6.0	8.8	0.016	0.060	0.27
11.9	17.3	0.030	0.029	1.03
14.1	21.1	0.034	0.0096	3.54
15.6	23.3	0.036	0.0034	10.59
20.2	30.5	0.038	0.062	0.61

*Average plus or minus the standard deviation.

flection coefficients which are functions of the velocity and density contrasts across the discontinuity and the angle of incidence. S_xP and S_xS are generated by S phase energy traveling slightly separated ray paths from the focus, and thus their amplitude ratio is relatively unaffected by differences in energy radiated from the focus. Table 1 lists five S_xP to S_xS amplitude ratios, each obtained by averaging many observed ratios narrowly grouped about the tabulated angles of incidence. The ratios are based on amplitudes measured on vertical component seismograms from station SNM. In order to compare with theoretical amplitude ratios, the measured ratios were corrected by multiplying by the sine of the angle of incidence for the S_xS wave and dividing by the cosine of the angle of incidence for the S_xP phase.

Theoretical S_xP to S_xS amplitude ratios based on plane-wave theory are presented in Table 1 for two types of discontinuity. In the first the velocity and density contrasts are selected to duplicate the Conrad discontinuity in Figure 3. The theoretical ratios for this type of discontinuity are greatly different from the observed ratios. In the second example the velocity and density contrasts approximate the conditions at the boundary of a magma layer. The theoretical values for this case still exceed the observed values, but the discrepancy is greatly reduced. In addition, the steady increase in the observed ratios with increase in angle of incidence is duplicated with this type of discontinuity.

The broad frequency content of the S_xS reflections indicates that the top of the magma body is a single sharp discontinuity. Most S_xS phases on microearthquake seismograms have a dominant frequency on the order of 15 Hz, but with clear evidence that lower frequencies also

Ratios of S_xP to S_xS Amplitudes

Theoretical			Observed	
$\alpha_1 = 5.80$, $\alpha_2 = 5.80$, $\beta_1 = 3.35$, $\beta_2 = 0.00$, $\rho_1 = 2.81$, $\rho_2 = 2.81$				
Reflection Coefficient P	Reflection Coefficient S	S_xP/S_xS	Measured S_xP/S_xS*	Corrected S_xP/S_xS*
0.12	0.94	0.13	0.11 ± 0.023	0.017 ± 0.0035
0.22	0.82	0.27	0.15 ± 0.057	0.046 ± 0.017
0.25	0.75	0.33	0.24 ± 0.20	0.088 ± 0.074
0.27	0.72	0.38	0.27 ± 0.22	0.111 ± 0.086
0.32	0.67	0.48	0.58 ± 0.10	0.314 ± 0.056

exist. In addition, recordings of strong microearthquakes on instruments with poor high-frequency response have clearly defined reflection phases with dominant frequencies as low as 3 Hz.

Extent of the Magma Body

The method used to determine the extent of the magma layer is illustrated in Figures 7 and 8. Figure 7 shows the approximate geographic position of points of reflection for all identified S_xS phases. The difficulty with using Figure 7 alone to locate the boundary of the magma layer is that one is not certain that the absence of reflections results from the disappearance of the magma body or from a lack of data. This ambiguity can be removed by plotting the hypothetical reflection points for microearthquakes which do not have identifiable reflections. The geographic distribution for these points is shown in Figure 8. Note that the region southeast of Socorro has numerous points in Figure 8 but no true reflection points in Figure 7. Thus for this area the margin of the magma body can be mapped with some confidence and is shown as a solid line in Figure 7. As one might expect, the boundary is quite irregular.

Elsewhere, the number of points on both figures is inadequate to place the boundary accurately. For these areas a smooth boundary enclosing the outermost known reflection points has been drawn, the dashed line in Figure 7. The estimated northern limit of the body approximately 60 km north of station WM, is based on reflections observed on ALQ seismograms and reported in the paper by Sanford et al. [1973]. The minimum areal extent of the magma body is about 1200 km^2.

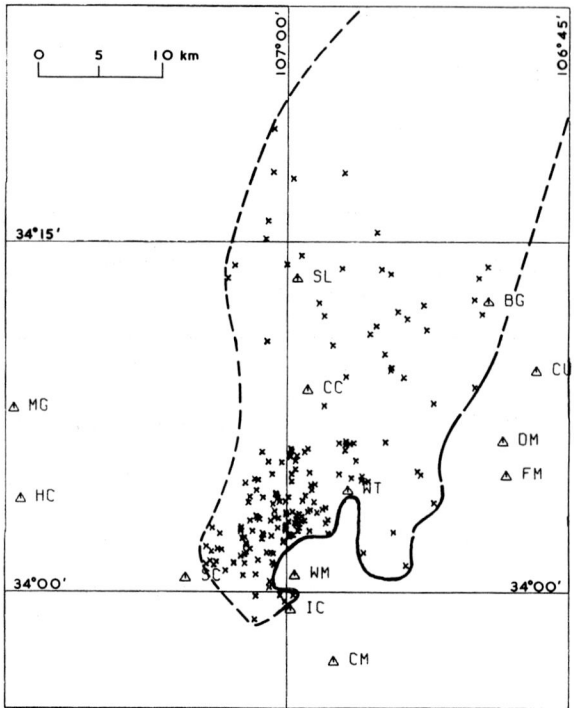

Fig. 7. A map showing the approximate location of points of reflection for all identified S_xS phases.

Depth of the Magma Body

The first depth to the magma layer, 18 km, was obtained by matching theoretical to observed arrival times [Sanford and Long, 1965]. This depth, which now appears close to the minimum for the magma body, was confirmed by a different method in a later paper [Sanford et al., 1973]. From nearly 10 years of records at station SNM, about 150 microearthquakes with exceptionally clear reflection phases were selected. For each of these events, two depths to the magma were calculated, one by assuming that the depth of focus was zero and the other by assuming that the epicentral distance was zero. All depths obtained by assuming that h = 0 were less than 17.8 km; all those obtained by assuming that Δ = 0 were greater than 17.8 km. Other depth information presented in the 1973 paper was time-distance data indicating 6° of northward dip for a distance of 30 km and 13 individual depth points calculated for events with accurately determined hypocenters. The latter depths appeared to show dip in nearly all directions from a minimum depth point of 18 km very near Socorro.

Since the 1973 study, the number of depths calculated from individual reflection arrivals has increased to 142, most of these for the southern end of the magma body (Figure 7). These data initially contained considerable scatter which was attributed to errors in the depths of focus obtained in the location program. The crustal model

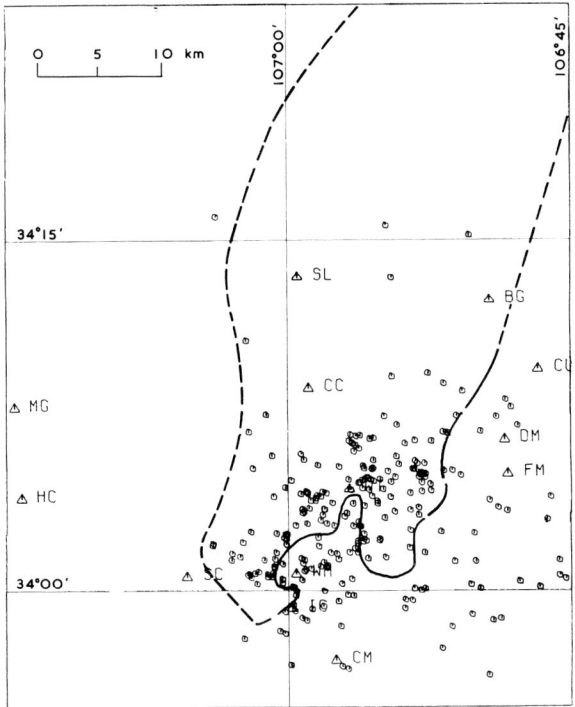

Fig. 8. A map showing the hypothetical reflection points for microearthquakes which do not have S_xS phases.

for this program, a simple half space with a velocity of 5.8 km/s, is only a first approximation to the actual crustal structure of the area.

A decrease in the scatter of depths to the magma body was obtained by limiting the data set to events located by a seismic network having at least one station within 13 km of the hypocenter. The close station minimizes the errors in depth of focus arising from the simple crustal model. The limited data set was used to estimate the shape of the upper surface of the magma body. This surface in combination with reflection arrivals from the nearest station was used to recalculate depths of focus for the entire data set. Depths of focus obtained from reflections are less susceptible to errors arising from changes in crustal velocity than those obtained by the location program. The new depth of focus and the reflection times at more distant stations were used then to recalculate depths to the magma body.

Although scatter remain in the data, the general shape of the upper surface of the southern end of the magma body is a north trending hill with a height of about 2 km and a width of about 8 km. Depths beneath the mean surface elevation range from 18 to 20 km. The configuration of the upper surface suggests the magma body could be an elongate laccolith extending in a north direction beneath the intragraben horsts. The laccolith would have a width of about 8 km and a thickness of about 2 km.

Accurate depths to the magma have not been obtained for the central

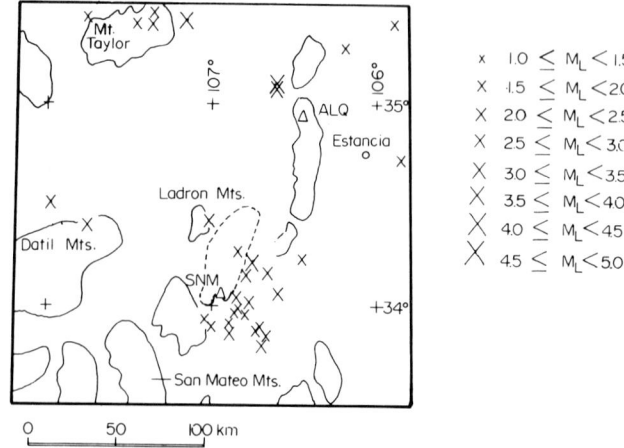

Fig. 9. Earthquake activity in central New Mexico, 1962-1972. Only shocks above the minimum detectable level throughout the area, $M_L \geq 1.8$, are plotted. The outline of the magma body is shown with dashed lines.

and northern segments of the body (Figure 7). In the work by Sanford et al. [1973] the depths for the northern 30 km of the body were calculated from S_xS reflections observed at ALQ. These calculations indicated that the upper surface of the magma body dipped northward and reached a depth of 30 km at a distance of 60 km from Socorro. The depths calculated from these wide-angle reflections depend critically on the choice of the S wave velocity for the upper crust. Until this velocity is very accurately known, these depths should be considered tentative.

Recent Magmatic Intrusion

Two general lines of evidence suggest that magma may have been intruded into the crust in very recent times. The first is the discovery by Reilinger and Oliver [1976] of historical surface uplift in the Socorro region roughly coincident with the spatial extent of the proposed magma body. On the basis of an analysis of level line data for the period 1909-1952 they were able to establish average rates of uplift as great as about 6 mm/yr. (Geodetic measurements over a 12-station network centered on Socorro have not revealed any horizontal movements in the period 1972-1976 (J. C. Savage, personal communication, 1976).)

The other evidence suggesting recent intrusion of magma is the pattern of natural seismic activity in space and time. The latter type of data is discussed in some detail in the following two sections.

Spatial Distribution of Earthquakes

Figure 9 is a map showing the location of epicenters in the central Rio Grande rift of New Mexico for the time period 1962-1972. Only shocks above the minimum detectable level throughout the area, $M_L \geq 1.8$, are plotted in Figure 9. The map shows substantially more shocks in the Socorro region than elsewhere along the rift in central New Mexico.

However, these shocks are weak and their total energy release is somewhat less than it is for the two earthquakes that occurred near Albuquerque during the same time period. Thus tectonic strain in the Socorro area may be no greater than elsewhere along the central part of the rift but it is being relieved in a different manner.

Figure 10 shows locations of microearthquakes in the immediate vicinity of Socorro. The epicenters shown on this map were established by local networks of seismic stations and thus are far more accurate than those shown in Figure 9, which were determined from regional networks. The distribution of earthquakes in Figure 10 suggests that epicenters in Figure 9 are systematically shifted to the southeast.

The earthquake activity in the Socorro area is diffusely distributed over a 2000-km^2 area. Considering the tectonics of the region, this pattern of activity appears somewhat unusual. Geologic mapping [Chapin and Seager, 1975] and gravity observations [Sanford, 1968] indicate that the rift is an extensional structural feature formed by a stress field having a maximum compressive stress (P) in a vertical direction and a minimum compressive stress (T) in an east-west direction. The north-south strike of normal faults offsetting geomorphic surfaces of Quaternary age [Sanford et al., 1972] indicates this stress field had not changed up to a few tens of thousands of years ago. Evidence that the same stress field continues to the present comes from the orientation of P and T axes in the composite fault plane solution obtained for microearthquakes in the Socorro area (see Figure 6).

With simple extensional tectonics, the expected distribution of earthquakes is an alignment of epicenters along some of the major faults bordering the basins and ranges such as was obtained by Stauder and Ryall [1967] in the Fairview Peak area of Nevada. The diffuse pattern of activity in the Socorro area may be the result of a stress perturbation on the preexisting east-west tensile field (lease compressive stress) arising from the intrusion of magma into the crust. Stuart and Johnston [1975] have used a similar crustal stress model to explain many of the geophysical observations during the Matsushiro earthquake swarm in central Honshu, Japan. Near Socorro the effect of a north trending intrusion of laccolithic shape would be to make the east-west regional stress more tensile along the axis of the uplift and less tensile along the flanks of the uplift. Thus the effect of such an intrusive body would be to enhance seismic activity over the body and to inhibit it near the margins. The geographic distribution of epicenters relative to the outline of the magma body follows this pattern in a general way (see Figure 10).

Time Distribution of Earthquakes

A characteristic of historical as well as recent seismic activity in the Socorro area is the occurrence of a majority of earthquakes in swarms. As has been noted for many years, earthquake swarms are observed in the vicinity of active volcanoes and in regions that have had volcanic activity in geologically recent times [Richter, 1958].

The earliest report of earthquakes anywhere in New Mexico is a description by an Army surgeon of an earthquake swarm at Socorro [Hammond, 1966]. The swarm, which contained 22 felt shocks, commenced on December 11, 1849, and lasted until February 8, 1850. The next known swarm near

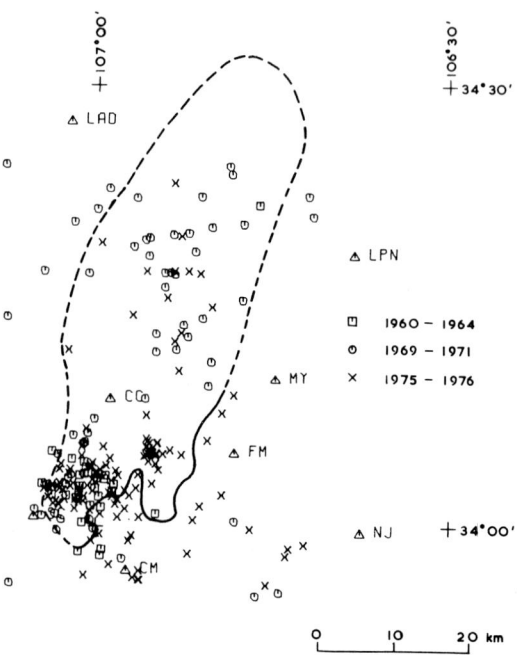

Fig. 10. Epicenters for microearthquakes in the vicinity of Socorro, New Mexico, 1960 through May 1976. The outline of the magma body is shown approximately.

Socorro consisted of 34 felt shocks in the 3-month interval starting January 19, 1904 [Bagg, 1904]. The latter swarm was followed 2 years later by an intense and prolonged sequence of shocks [Reid, 1911], which has not been duplicated in the Socorro region to the present time. This swarm commenced on July 2, 1906, and lasted well into 1907. During this period, shocks were felt almost every day and at times reached a frequency of one perceptible tremor an hour. Three shocks of this swarm (0510 MST, July 12; 1200 MST, July 16; 0520 MST, November 15) reached a Rossi-Forel intensity of VIII at Socorro and were felt over areas of about 275,000 km^2. The Richter magnitude of these quakes is estimated to be near 6.

The 1906-1907 Socorro earthquake swarm is similar in many respects to the Matsushiro swarm which some believe may have been caused by a magmatic intrusion at shallow depth [Stuart and Johnston, 1975]. The duration of the Matsushiro swarm was about 1½ years (from early August 1965 to early January 1967), and magnitudes were low with no shocks appearing to exceed 5 [Båth, 1973]. Total energy release for the two swarms must have been comparable [Nur, 1974].

Approximately 75% of the microearthquake activity in the Socorro area is occurring in swarms (defined as six or more shocks from the same focal region within a 24-hour period). Figure 11 shows the distribution of detected swarms for the intermittent recording from April 1975 through May 1976. Most swarms during this period occurred in the southern ends of the Socorro-Lemitar Mountains and the La Jencia basin.

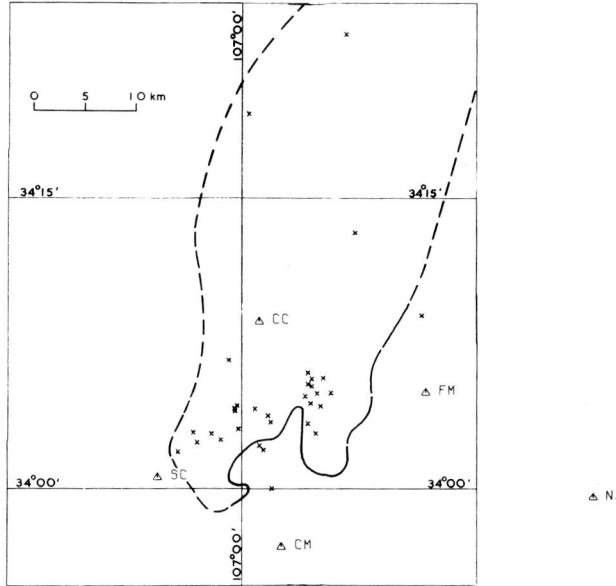

Fig. 11. Distribution of microearthquake swarms, April 1975 to May 1976, and the outline of the magma body. Dashed lines indicate that the boundary of the magma body is uncertain.

Earlier studies [Sanford and Holmes, 1961; Sanford et al., 1972] indicate that swarms have occurred over the magma body as far north as the vicinity of Bernardo (Figure 2).

Discussion

The upper surface of the magma body is located at depths of 18 to 20 km which is close to the Conrad discontinuity in Figure 3. Although the position of the body may be related to the Conrad, several observations indicate that the reflections are not from the Conrad. The reflection coefficients for the Conrad (Table 1) cannot explain the absolute amplitudes of the S_xP and S_xS reflections or their ratios. The P* arrivals on the Gasbuggy profile indicate a continuous Conrad with little relief. On the other hand, the S_xP and S_xS reflections define a discontinuous interface of limited extent with considerable relief.

In addition to the intrusion of magma at depths near the Conrad discontinuity the unusual characteristics of the crust beneath the rift at Socorro (and elsewhere) are the low upper crust and upper mantle velocities, 5.8 and 7.9 km/s, respectively. The low upper crust velocity is likely to be a consequence of the formation of the rift, i.e., the large-scale fragmentation of the upper crust by normal faults [O'Connell and Budiansky, 1974]. On the other hand, the low upper mantle velocity is likely to be related to the mechanism which has produced crustal extension for the past 25 to 29 m.y.

Acknowledgements. The research described in this paper was sponsored jointly by the National Science Foundation (Grant Number DES74-24187) and the energy Resource Board of the State of New Mexico (Grant Number ERB-75-300).

References

Bagg, R. M., Earthquakes in Socorro, New Mexico, Amer. Geol., 34, 102-104, 1904

Båth, M., Introduction to Seismology, Birkhäuser-Verlag, Basel, Switzerland, 1973.

Chapin, C. E., The Rio Grande rift, Part I: Modifications and additions, N. Mex. Geol. Soc. Field Conf. Guideb., 21, 191-201, 1971.

Chapin, C. E., and W. R. Seager, Evolution of the Rio Grande rift in the Socorro and Las Cruces areas, N. Mex. Geol. Soc. Field Conf. Guideb., 26, 297-321, 1975.

Dee, M., A crustal and P-wave velocity study of portions of SW New Mexico and SE Arizona using open pit mining explosions, M. S. Independent Study, Geosci. Dep., N. Mex. Inst. of Mining and Technol., Socorro, 1973.

Denny, C. S., Quaternary geology of the San Acacia area, New Mexico, J. Geol., 49, 225-260, 1941.

Hammond, J. F., A Surgeon's Report on Socorro, N. M., 1852, Stagecoach Press, Santa Fe., New Mexico, 1966.

Kelley, V. C., Tectonics of the Rio Grande depression of central New Mexico, N. Mex. Geol. Soc. Field Conf. Guideb., 3, 93-105, 1952.

Kelley, V. C., The Rio Grande depression from Taos to Santa Fe, N. Mex. Geol. Soc. Field Conf. Guideb., 7, 109-114, 1956.

Nafe, J. E. and C. L. Drake, Physical properties of marine sediments, in The Sea, vol. 3, ed. by M. N. Hill, pp. 794-815, Interscience Publishers, New York, 1963.

Nur, A., Matsushiro, Japan, earthquake swarm: Confirmation of the dilatancy-fluid diffusion model, Geology, 2, 217-221, 1974.

O'Connell, R. J., and B. Budiansky, Seismic velocities in dry and saturated cracked solids, J. Geophys. Res., 79, 5412-5426, 1974.

Phinney, R. A., Structure of the earth's crust from spectral behavior of long-period body waves, J. Geophys. Res., 69, 2997-3017, 1964.

Reid, H. F., Remarkable earthquakes in central New Mexico in 1906 and 1907, Bull. Seismol. Soc. Amer., 1, 10-16, 1911.

Reilinger, R., and J. E. Oliver, Modern uplift associated with a proposed magma body in the vicinity of Socorro, New Mexico, Geology, 4, 583-586, 1976.

Reiter, M. A., C. L. Edwards, H. Hartman, and C. Weidman, Terrestrial heat flow along the Rio Grande rift, New Mexico and southern Colorado, Geol. Soc. Amer. Bull., 86, 811-818, 1975.

Richter, C. F., Elementary Seismology, W. H. Freeman, San Francisco, Calif., 1958.

Sakdejayont, K., A study on Poisson's ratio and V_p/V_s ratio in the Rio Grande rift, M. S. Independent Study, Geosci. Dep., N. Mex. Inst. of Mining and Technol., Socorro, 1974.

Sanford, A. R., Gravity survey in central Socorro County, New Mexico, Circ. 91, N. Mex. Bur. of Mines and Miner. Resour., Socorro, 1968.

Sanford, A. R., and C. R. Holmes, Note on the July 1960 earthquakes in central New Mexico, Bull. Seismol. Soc. Amer., 51, 311-314, 1961.

Sanford, A. R., and L. T. Long, Microearthquake crustal reflections, Socorro, New Mexico, Bull. Seismol. Soc. Amer. 55, 579-586, 1965.

Sanford, A. R., A. J. Budding, J. P. Hoffman, O. S. Alptekin, C. A. Rush, and T. R. Toppozada, Seismicity of the Rio Grande rift in New Mexico, Circ. 120, N. Mex. Bur. of Mines and Miner. Resour., Socorro, 1972.

Sanford, A. R., O. S. Alptekin, and T. R. Toppozada, Use of reflection phases on microearthquake seismograms to map an unusual discontinuity beneath the Rio Grande rift, Bull. Seismol. Soc. Amer., 63, 2021-2034, 1973.

Sanford, A. R., R. P. Mott, Jr., P. M. Shulseki, E. J. Rinehart, F. J. Caravella, and R. M. Ward, Microearthquake investigations of magma bodies in the vicinity of Socorro, New Mexico, Geol. Soc. Amer. Abstr. Programs, 8, 1085-1086, 1976.

Savage, J. C., The effect of rupture velocity upon seismic first motions, Bull. Seismol. Soc. Amer., 55, 263-275, 1965.

Stauder, W., and A. Ryall, Spatial distribution and source mechanism of microearthquakes in central Nevada, Bull. Seismol. Soc. Amer., 57, 1317-1345, 1967.

Stuart, W. D., and M. J. S. Johnston, Intrusive origin of the Matsushiro earthquake swarm, Geology, 3, 63-67, 1975.

Summers, W. K., A preliminary report on New Mexico's geothermal energy resources, Circ. 80, N. Mex. Bur. of Mines and Miner. Resour., Socorro, 1965.

Tatel, H. E., and M. A. Tuve, Seismic exploration of the continental crust, Geol. Soc. Amer. Spec. Pap., 62, 35-50, 1955.

Toppozada, T. R., Seismic investigation of crustal structure and upper mantle velocity in the state of New Mexico and vicinity, Ph.D. dissertation, Geosci. Dep., N. Mex. Inst. of Mining and Technol., Socorro, 1974.

Toppozada, T. R., and A. R. Sanford, Crustal structure in central New Mexico interpreted from the Gasbuggy explosion, Bull. Seismol. Soc. Amer., 66, 877-886, 1976.

A SUMMARY OF SEISMIC SURFACE WAVE ATTENUATION AND ITS REGIONAL VARIATION ACROSS CONTINENTS AND OCEANS

Brian J. Mitchell, Nazieh K. Yacoub, and Antoni M. Correig

Department of Earth and Atmospheric Sciences, Saint Louis University
Saint Louis, Missouri 63103

> Abstract. Presently available surface wave attenuation data obtained for paths entirely within either continents or oceanic regions are summarized and compared. Attenuation coefficients obtained for the stable regions of North America and Eurasia are nearly identical within the limits of observation and are lower than values obtained for tectonic or oceanic regions over much of the common period range.
> Shear wave internal friction values Q_β^{-1} for the uppermost 15-20 km of the crust of North America range between about 4×10^{-3} and 9×10^{-3}, the lower values occurring in eastern North America and the higher values (or lower Q_β) occurring in western North America. Greater depths in the crust have smaller Q_β^{-1} values, about 1×10^{-3} or less, and exhibit much less (perhaps negligible) regional variation.
> Average Pacific crustal values for Q_β^{-1} are higher than those of the lower crust beneath continents. A zone of low Q_β^{-1} (or high Q_β) may exist at depths between 15 and 20 km, and a well-developed zone of high Q_β^{-1} values (or low Q_β) begins at a depth of about 60 km. Although Rayleigh wave attenuation coefficients obtained for the eastern Pacific exhibit large confidence limits, they appear to be slightly greater than average values for the entire Pacific. A narrow zone near the East Pacific Rise seems characterized by even greater attenuating properties.
> Higher-mode Rayleigh and Love wave attenuation coefficients have been computed for a shield model, a Basin and Range model, and a Pacific model. These should be useful input to theoretical multimode seismogram computations for these regions.

Introduction

Seismic surface wave velocities have been used extensively to make inferences concerning the elastic properties of the earth as a function of depth. The decay of surface wave amplitudes with distance has been studied much less frequently because of the difficulty in making meaningful determinations. The difficulty may be produced by a number of factors, including scattering, mode conversion, lateral refraction, multipath propagation, and nonideal source characteristics.

In spite of these difficulties, it is important to try to obtain amplitude attenuation data because they provide a measure of the anelastic properties of the earth, which are strongly related to factors

such as fluid content, temperature, and phase change. Recent work
[e.g., Jeffreys, 1965; Liu et al., 1976] indicates that anelasticity
affects surface wave phase velocities in a significant way. Those
results provide additional stimulus for studying the attenuation of
surface waves and to obtain Q models which are consistent with those
data.

The difficulties of measurement mentioned above suggest that the
experimental uncertainty for any surface wave attenuation coefficient
determination will be much larger than the uncertainty associated with
a velocity determination. For this reason it will be important to
assign confidence limits to any attenuation coefficient data with are
obtained and to consider these in any attempt to invert the data to
obtain a model of crustal or mantle anelasticity. Fortunately,
statistical methods are now available to assist in the analysis of
surface wave attenuation data, and inversion techniques have been
developed which permit us to judge how well any feature of an earth
model can be resolved.

Only those attenuation studies which have been restricted to purely
continental or purely oceanic paths will be considered in the sections
which follow. Our studies, so far, have been restricted to three
broad regions: North America, Eurasia, and the Pacific. Attenuation
coefficients obtained for the stable regions of North America and
Eurasia are similar [Yacoub and Mitchell, 1977]. Similarly, attenuation coefficients obtained for the Pacific are similar to previously
determined values for the Atlantic [Mitchell et al., 1976; Ben-Menahem,
1965], at least at longer periods. These similarities suggest that the
data and models for the specific regions presented in this study may
have applications to other continents and oceanic regions.

Methods of Data Processing and Inversion

Attenuation Coefficient Determinations

The methods described in this section have been discussed in detail
in several studies. Consequently, only a brief summary of these
methods will be given here.

The surface wave attenuation coefficients presented in the following
sections were obtained in two ways. All of the determinations for the
Pacific and most of those for North America were made by using the
method of Tsai and Aki [1969], whereas a few of the Rayleigh wave
determinations for North America and all of those for Eurasia employed
the method of Mitchell [1975]. Both methods (as described below)
require several amplitude observations for a single determination of
an attenuation coefficient value. Consequently, statistical methods
can be employed to assign confidence limits to and to test the reliability of each determination.

Both methods use spectral amplitudes as observed at several seismograph stations located at various distances and azimuths from the
source. The seismograms are digitized (the vertical component for Rayleigh waves and the transverse component as determined from a coordinate transformation of the north-south and east-west components for
Love waves), and spectral amplitudes are determined by a method such as
the multiple-filter technique [Dziewonski et al., 1969].

The method of Tsai and Aki [1969] requires that the focal mechanism and depth of the earthquake source, as well as the average structure of the medium of propagation, be known. Theoretical surface wave radiation patterns for various frequencies can then be calculated by a formulation such as that given by Ben-Menahem and Harkrider [1964]. A linear least squares fit of the observed amplitudes to the theoretical ones for each period yields an estimate of the attenuation coefficient and the source spectrum.

The method of Mitchell [1975] also fits observed to theoretical amplitudes by least squares. It differs from that of Tsai and Aki [1969] by being a nonlinear process, by determining values for four rather than two parameters, and by requiring no previous knowledge of the crustal structure or the earthquake focal mechanism. It makes use of the idea that all Rayleigh wave radiation patterns can be approximated by an equation which expresses those patterns as the sum of the separate patterns due to an azimuthally symmetric source and that due to a horizontal double couple. Details of this method are given by Mitchell [1975] and Yacoub and Mitchell [1977].

Both methods use several observations (between 12 and 4 for all cases in the present study) for each attenuation coefficient determination. Statistical techniques can therefore be applied to the data to obtain confidence limits and to test the data for reliability. Herrmann [1974] and Herrmann and Mitchell [1975] use the inferences of correlation analysis to establish a criterion for accepting or rejecting any attenuation coefficient determination. A correlation coefficient between observed and theoretical amplitude radiation patterns is obtained and compared with a critical correlation coefficient value determined by the number of observations and the confidence level. The attenuation coefficient value is accepted only if the observed correlation coefficient exceeds the critical value.

Yacoub and Mitchell [1977] developed a new method to attempt to observe regional differences in Rayleigh wave attenuation across Eurasia. They developed a linear least squares method for separating the effect of two regions with different attenuating properties. Their results are described in a later section.

Inversion of Data

It is hoped in surface wave amplitude studies that attenuation due to true energy dissipation can be isolated from the adverse effects mentioned earlier. Unfortunately, it is unlikely that this situation can ever be achieved in practice. The best that can be hoped for is that errors produced by those effects will influence data in a random way and can be accounted for in the confidence limits associated with the attenuation coefficient values. We will assume this to be case and apply modern inversion theory [Backus and Gilbert, 1970] to the attenuation data from North America and the Pacific to obtain models of Q_β^{-1}, the shear wave internal friction, as a function of depth beneath those regions. A stochastic form of inversion theory, described by Wiggins [1972] and applied to Q_β^{-1} inversions by Mitchell [1976], has been used to obtain the models of the present study. The equations of Anderson et al. [1965], which assume frequency independence for both Q_β^{-1} and Q_α^{-1}, the intrinsic internal friction values for shear and compressional waves, provide the framework for the inversion calculations. Since surface wave amplitudes are much less sensitive to Q_α^{-1} than to Q_β^{-1}, we assume (in accordance with the results of Anderson et al.) that $Q_\beta^{-1} = 2Q_\alpha^{-1}$.

Resolving kernels and standard deviations are computed for each layer of the models and are presented along with the Q_β^{-1} values. The width of the resolving kernel indicates the depth range over which each determination is averaged, and the standard deviation bars provide a measure of uncertainty of the model averaged over the same depth interval.

Attenuation Across Continental Regions

Although surface wave attenuation measurements are more prevalent for continental regions than they are for oceanic regions, they are still very few in number in comparison to surface wave velocity studies. Attenuation data for purely continental paths are available only for North America and Eurasia at the present time.

North America

The attenuation data for this region are more numerous than they are for any other region of the world. The attenuation coefficients obtained by Herrmann and Mitchell [1975] for eastern and central North America were obtained from data for eight events. Surface waves from those events were recorded by between 24 and 42 seismograph stations in the United States and Canada. Mean attenuation coefficient values and the associated standard deviations for Rayleigh and Love waves which traverse eastern and central North America appear in Figure 1. We have plotted standard deviations rather than 95% confidence limits because standard deviations are used for the inversion for Q_β^{-1} which will be discussed later. The attenuation coefficient values given by Solomon [1972a] for the east central United States and those obtained by Hasegawa [1973] for the Canadian shield also appear in Figure 1.

The only direct determinations of surface wave attenuation coefficients for western North America appear to be those of Solomon [1972a] and are restricted to periods greater than 15 s. Nuttli [1973] noted that surface waves at very short periods must be attenuated more rapidly in western North America than they are in eastern North America. Mitchell [1975] observed that significant regional differences in attenuation occur at periods less than about 16 s and that the differences become more pronounced with decreasing period. All amplitudes were converted to values which would be expected if the recording station were at a distance of 1000 km. Ratios of these equalized Rayleigh amplitudes were determined and converted to differences in attenuation between eastern and western North America. These can be used to deduce average values for Rayleigh wave attenuation coefficients throughout western North America. Those values appear in Table 1 along with values representative of eastern North America [Herrmann and Mitchell, 1975].

The values of Herrmann and Mitchell [1975] are of sufficient quality for inverting to obtain a model of Q_β^{-1}. In addition, the attenuation coefficient differences between eastern and western North America have

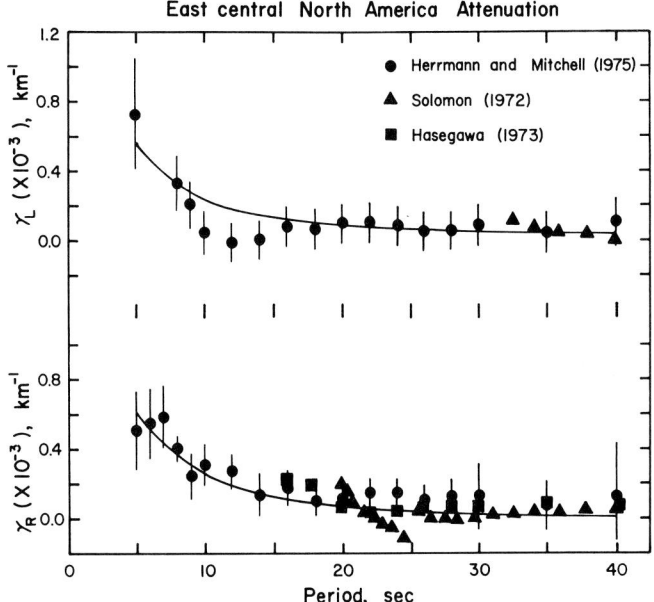

Fig. 1. Fundamental mode Love and Rayleigh wave attenuation coefficients for eastern and central North America. Circles indicate data of Herrmann and Mitchell [1975] along with their associated standard deviations for a broad region of east central North America; triangles, data of Solomon [1972b] for a path between Rapid City, South Dakota, and Atlanta, Georgia; and squares, data of Hasegawa [1973] for a path across canada. The solid lines through the data are theoretical attenuation coefficient values for the Q_β^{-1} model of Herrmann and Mitchell [1975].

been inverted to obtain a model for ΔQ_β^{-1}, the difference in internal friction as a function of depth between eastern and western North America. That difference can be used to infer a Q_β^{-1} model for western North America [Mitchell, 1975]. As is indicated by Figure 2, the two models differ by about a factor of 2 in the upper crust, western North America being more highly attenuating than eastern North America. The lower crust exhibits very low values for both regions. It is difficult to distinguish between the different Q_β^{-1} values, which are both very low, but it is possible that Q_β^{-1} in the lower crust may not be greatly different for the two regions. Mitchell [1973b] and Lee and Solomon [1975] also obtained models for which Q_β^{-1} decreased to low values at midcrustal depths.

Higher-Mode Attenuation

Little is known concerning the rate of attenuation of higher-mode surface waves. Because of observational difficulties, such as overlap of modes or poor higher-mode excitation, the only observed values [Mitchell, 1973a; Herrmann, 1973] have been restricted to a few periods for single earthquakes. A knowledge of the attenuation coefficient

TABLE 1. Amplitude Ratios and Attenuation Coefficient Differences for Rayleigh Waves Recorded in Eastern and Western North America

Period, s	Amplitude Ratio* West/East	$\gamma_W - \gamma_E$, 10^{-4} km^{-1}	γ_E† 10^{-4} km^{-1}	γ_W, 10^{-4} km^{-1}
5	0.19	16.60	5.10	21.70
6	0.22	15.10	5.50	20.60
7	0.31	11.70	5.87	17.60
8	0.53	6.35	4.06	10.40
9	0.60	5.11	2.45	7.56
10	0.77	2.61	3.12	5.73
12	0.76	2.74	2.70	5.44
14	0.79	2.36	1.37	3.73
16	0.85	1.63	1.76	3.39
18	0.86	1.51	1.01	2.52
20	0.87	1.39	1.19	2.58
25	1.00	0.00	1.28	1.28
30	0.97	0.30	1.31	1.61
35	0.96	0.41	0.75	1.16
40	0.91	0.94	1.25	2.19

*Values taken from Mitchell [1975].
†Values taken from Herrmann and Mitchell [1975].

values over an extended period range for several surface wave modes is still lacking. That information will be useful for theoretical seismogram computations of multimode surface waves, which are becoming more and more important in seismology. Since Q_β^{-1} models are now available, theoretical surface wave attenuation coefficient values for all modes can be computed by using a formulation such as that of Anderson et al. [1965]. Figures 3 and 4 present results of these computations for Rayleigh and Love waves for models taken to represent portions of eastern and western North America.

Eurasia

Determinations of attenuation coefficients for paths across Eurasia are more difficult than those for North America because of the smaller number of stations available and because the paths between sources and stations are necessarily longer and often cross several tectonic provinces. These factors produce larger confidence limits for the observed attenuation coefficient values in comparison to those which are observed for the North American data. Yacoub and Mitchell [1977] present Rayleigh wave attenuation data for Eurasia using six earthquakes and two nuclear explosions. In addition to mean attenuation coefficient values, their regionalization method, mentioned earlier, yields somewhat lower values for stable regions than for regions which they broadly classify as being tectonic.

Mean attenuation coefficients obtained for four events located in central Asia appear in Figure 5. The mean Rayleigh wave attenuation coefficient values for the shield regions of Eurasia are nearly the

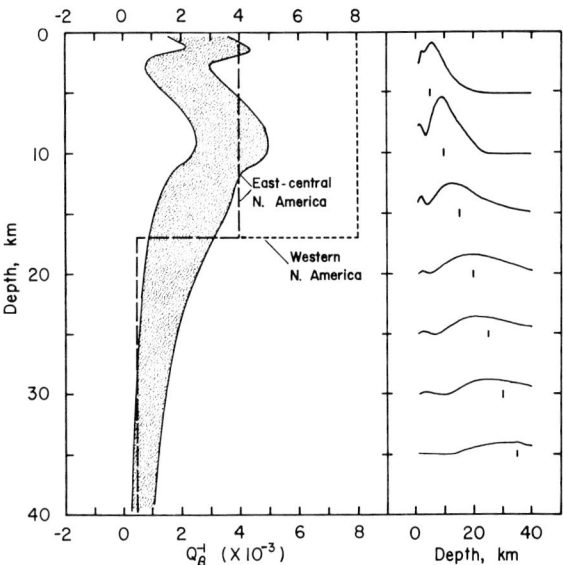

Fig. 2. Q_β^{-1} model for the crust of eastern and central North America. The shaded region indicates standard deviations for a model which results from inversion of the data in Figure 1. Resolving kernels obtained from the inversion process appear on the right. The width of each kernel indicates a depth interval over which the determined value of Q_β^{-1} at that depth is averaged. The depth to which each kernel corresponds is indicated by a short vertical line and by the ordinate on the figure. The long dashes delineate a two-layer Q_β^{-1} model (adapted from Herrmann and Mitchell [1975]) which is compatible with the attenuation data for east central North America, and the short dashes delineate a two-layer model inferred for western North America as adapted from Mitchell [1975].

same as those obtained for eastern North America, although the standard deviations are larger. This result implies that the distributions of Q_β^{-1} with depth in the crust of both regions cannot be greatly dissimilar.

It is interesting to note that regional variations in attenuation not only can increase the uncertainty of the observations but can affect the mean as well. Figures 6a and 6b, for an event in Eurasia, indicate a drastic example of this effect. Here the mean attenuation coefficients at short periods take on apparent negative values before regional variation is considered, whereas the regionalized values for both stable and tectonic regions are positive. Such unrealistic negative values can occur for the unregionalized determinations if the average path length through highly attenuating material is substantially shorter than the average path length through material characterized by lower attenuation. Apparent negative values in laterally varying regions can occur for any method which fits theoretical and observed radiation patterns to obtain attenuation coefficients.

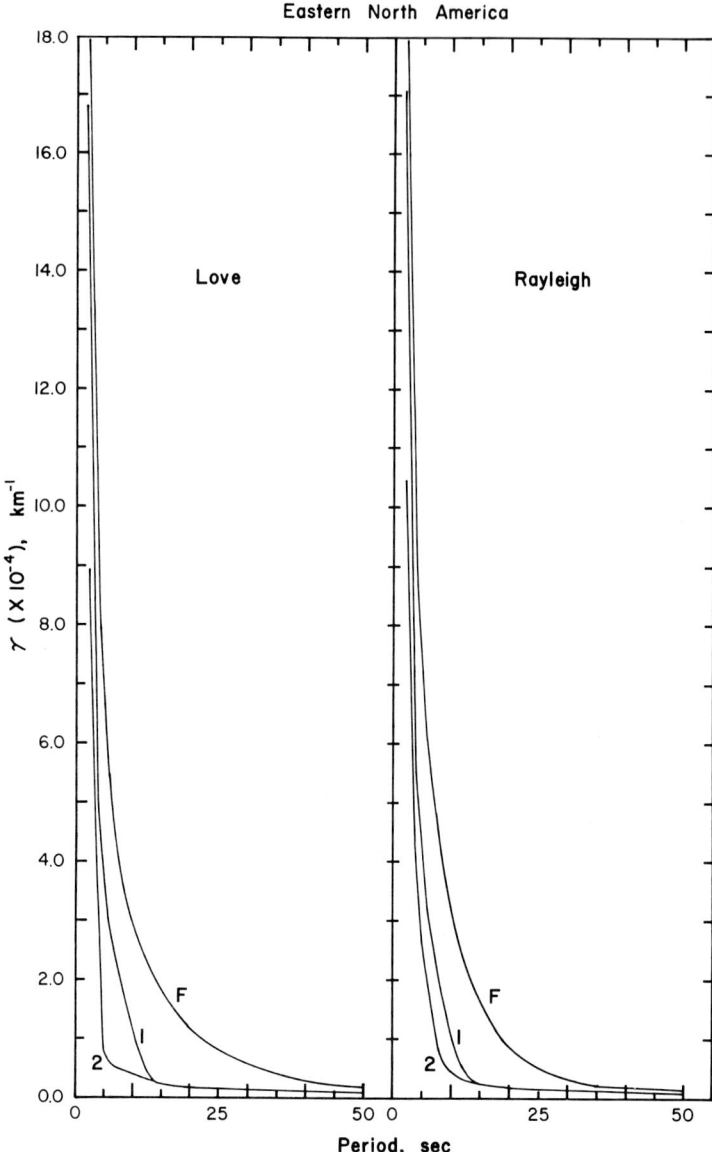

Fig. 3. Theoretical attenuation coefficient values for the fundamental and two higher-mode surface waves propagating through a crustal model for eastern North America. The velocity model is from McEvilly [1964], and the Q model is taken from the long-dash line in Figure 2.

Attenuation Across Oceanic Regions

Nature of Data Available

Surface wave attenuation coefficient determinations for oceanic regions are even less common than determinations for continental

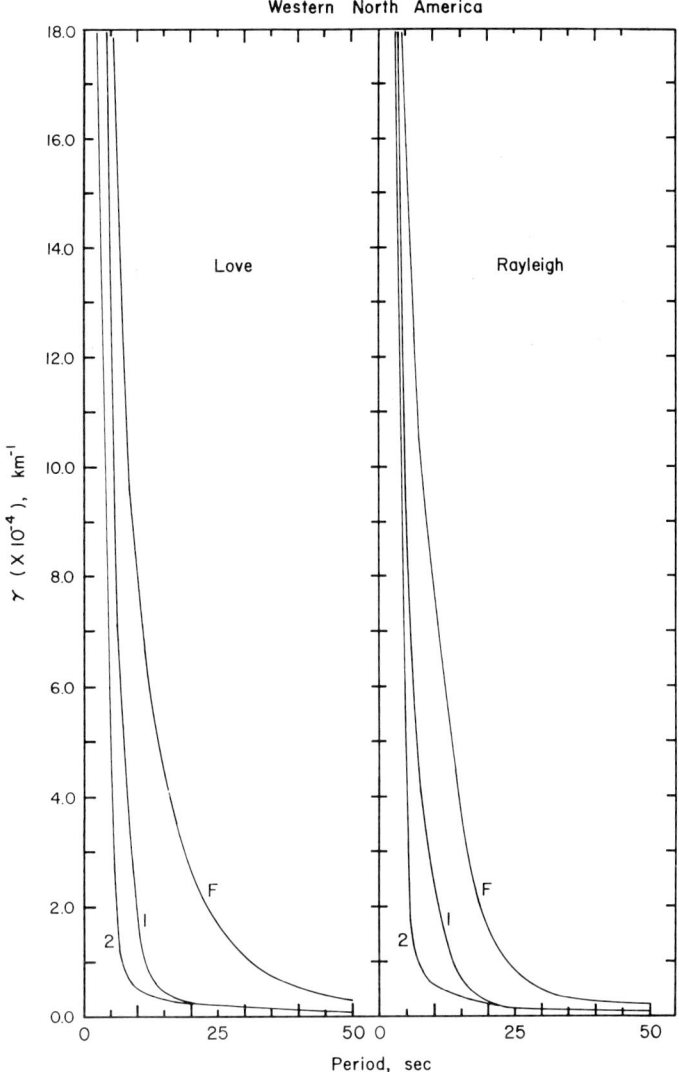

Fig. 4. Theoretical attenuation coefficient values for the fundamental and higher-mode surface waves propagating through a crustal model for western North America. The velocity model is from Braile et al. [1974], and the Q model is taken from the short-dash line in Figure 2.

regions. Strictly speaking, we cannot, at the present time, determine the rate of surface wave attenuation for a purely oceanic path, since the seismograph stations are ordinarily located on the continents. Stations can be selected for which the segment of continental path is small, but the presence of the continental margin leads to several difficulties in amplitude observations. One difficulty results because of the transmission coefficient of surface waves at the ocean-continent

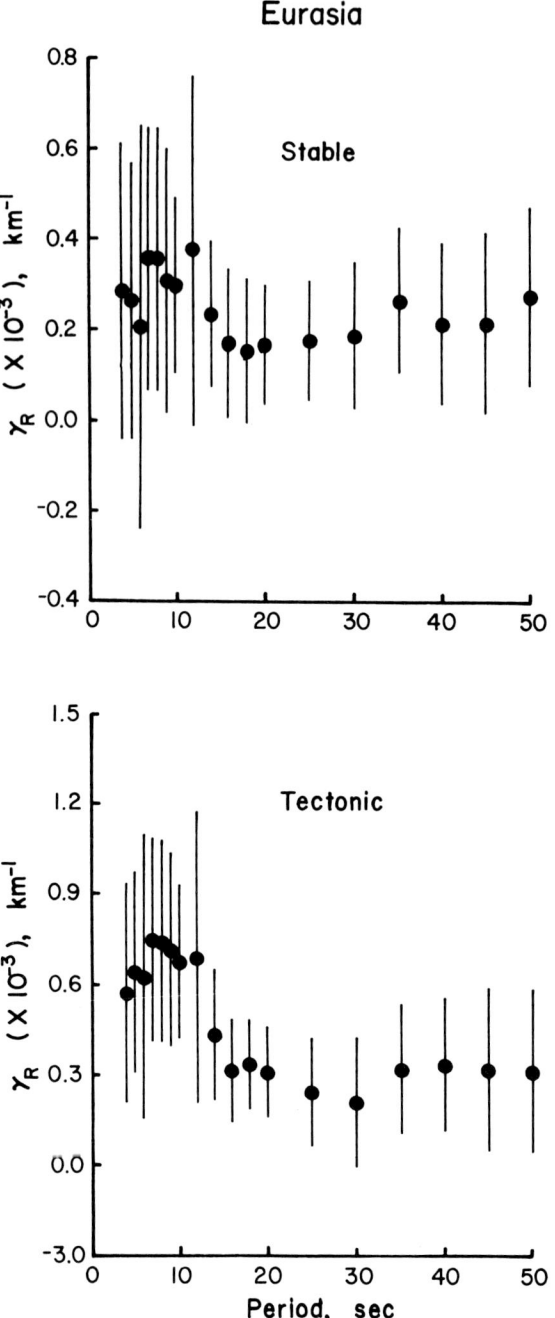

Fig. 5. Mean Rayleigh wave attenuation coefficient values and standard deviations obtained for stable regions and for tectonic regions of Eurasia (adapted from Yacoub and Mitchell [1977]).

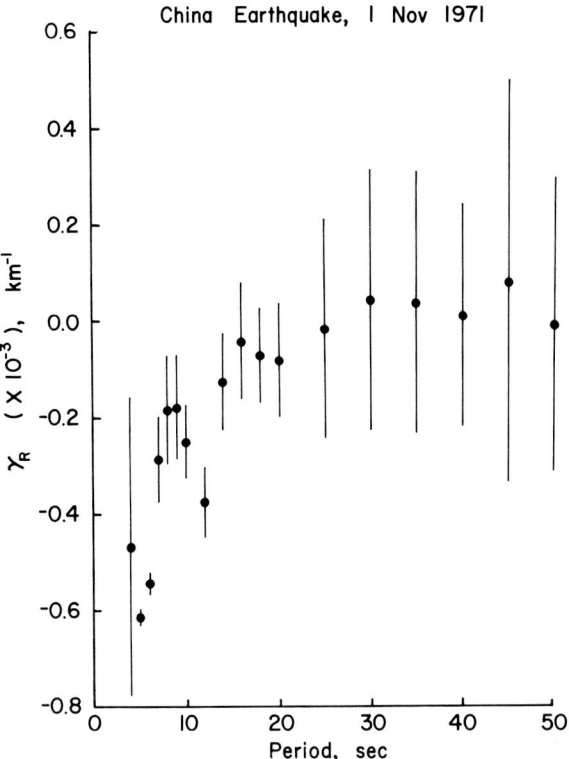

Fig. 6a. Mean Rayleigh wave attenuation coefficient values obtained for Eurasia from the earthquake of November 1, 1971 (adapted from Yacoub and Mitchell [1977]).

transition. Finite element calculations of Love and Rayleigh wave transmission coefficients at various types of lateral transitions [Lysmer and Drake, 1971; Drake, 1972] indicate that the coefficients may vary for different coastal regions. Continental margins may also produce lateral refractions and multipathing [McGarr, 1969; Capon, 1971], which can greatly affect surface wave amplitudes. In addition, mode conversion from fundamental to higher modes, and vice versa, is apt to be important for Love waves (but perhaps not for Rayleigh waves) at continental margins [Drake, 1972].

Mitchell et al. [1976] compiled a body of attenuation coefficient data for the Pacific using records from seismograph stations which are located on islands or near the edge of continents. They assumed that the transmission coefficient values associated with the ocean-continent transition are the same for all stations. If that is the case, the method of Tsai and Aki [1969] might yield source spectrum values which are incorrect, but the attenuation coefficient values will not be affected. Any variation in the value of the transmission coefficient among the various stations is assumed to be reflected in increased confidence limits for the observed data.

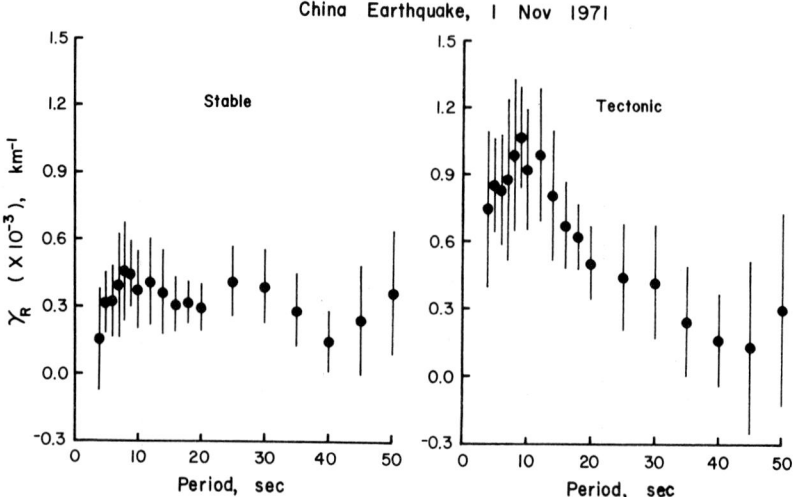

Fig. 6b. Rayleigh wave attenuation coefficient values obtained for stable regions and tectonic regions of Eurasia from the earthquake of November 1, 1971 (adapted from Yacoub and Mitchell [1977]).

Interference between fundamental and higher modes may cause severe problems in Love wave amplitude studies but is probably less important for Rayleigh waves [Mitchell et al., 1976]. The reason for this difficulty is that higher-mode group velocity curves often intersect or overlap with those of the fundamental mode. We are therefore unable to study the two modes individually. For this reason, Mitchell [1976] did not use Love wave data for his favored Q_β^{-1} model for the Pacific.

Average Q_β^{-1} Model for the Pacific

The Rayleigh wave attenuation coefficient versus period data of Mitchell et al. [1976] for the Pacific appear in Figure 7 along with the data of Ben-Menahem [1965] and Tsai and Aki [1969]. The data of Ben-Menahem are for a single path across the Atlantic.

Mitchell [1976] inverted these data to obtain a model of Q_β^{-1} as a function of depth for the crust and upper mantle beneath the Pacific. Figure 8 indicates that Q_β^{-1} values for the oceanic crust are higher than the Q_β^{-1} values for the lower crust of eastern North America. This result is interesting, since compressional wave velocities in those regions are similar for continents and oceans (about 6.7 km/s).

At greater depths (between 15 and 20 km), Q_β^{-1} becomes much smaller. This decrease appears to be only marginally required by the data. If it is real, the high Q values which may be associated with this depth range may provide an explanation for the great distances over which Pn and Sn propagate beneath the Pacific [Sutton and Walker, 1972; Molnar and Oliver, 1969].

An increase in Q_β^{-1} occurs at a depth of about 60 km for this average Pacific model. Shear wave velocities inferred from surface wave velocity studies of the Pacific often show decreases at that same depth. The Q_β^{-1} values increase to a depth of about 150 km and then decrease

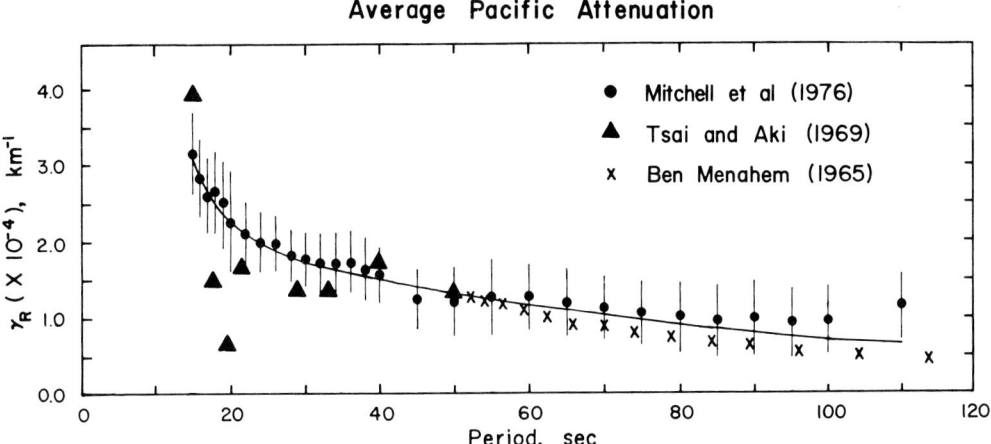

Fig. 7. Fundamental mode Rayleigh wave attenuation coefficients for the Pacific Ocean. The values are from Mitchell et al. [1976], Tsai and Aki [1969], and Ben-Menahem [1965]. The solid line through the data indicates theoretical values for the Q_β^{-1} model of Figure 8.

again. It should be pointed out that the resolving kernels become increasingly broad at greater depths because the long wavelengths which penetrate to those depths sample large depth intervals. Consequently, it is more difficult to resolve features of the model at great depth than it is to resolve features nearer the surface.

Higher-Mode Attenuation

Since a Q_β^{-1} model (Figure 8) is now available for the Pacific, attenuation coefficients for the higher modes can be computed in the same way that they were for the continental models of the previous section. The results of these computations for both Rayleigh and Love waves, when velocity model 8-3-2 of Saito and Takeuchi [1966] is used, appear in Figure 9.

Eastern Pacific

Our first attempts to observe regional variations of surface wave attenuation in the Pacific were influenced by the work of Kausel et al. [1974] and Forsyth [1975]. They observed lower surface wave phase velocities in the young eastern Pacific than they observed in other regions of the Pacific. We have therefore investigated the attenuation of Rayleigh wave amplitudes for the same region. The stations used for this portion of the study were restricted to those on the western coasts of South America, Central America, Mexico, and the United States and to islands in the eastern and central Pacific (Figure 10). The paths to those stations traverse oceanic crustal material which is less than 50 m.y. in age [Sclater and Francheteau, 1970] in all cases except for portions of the paths to stations KIP and RAR in the central Pacific.

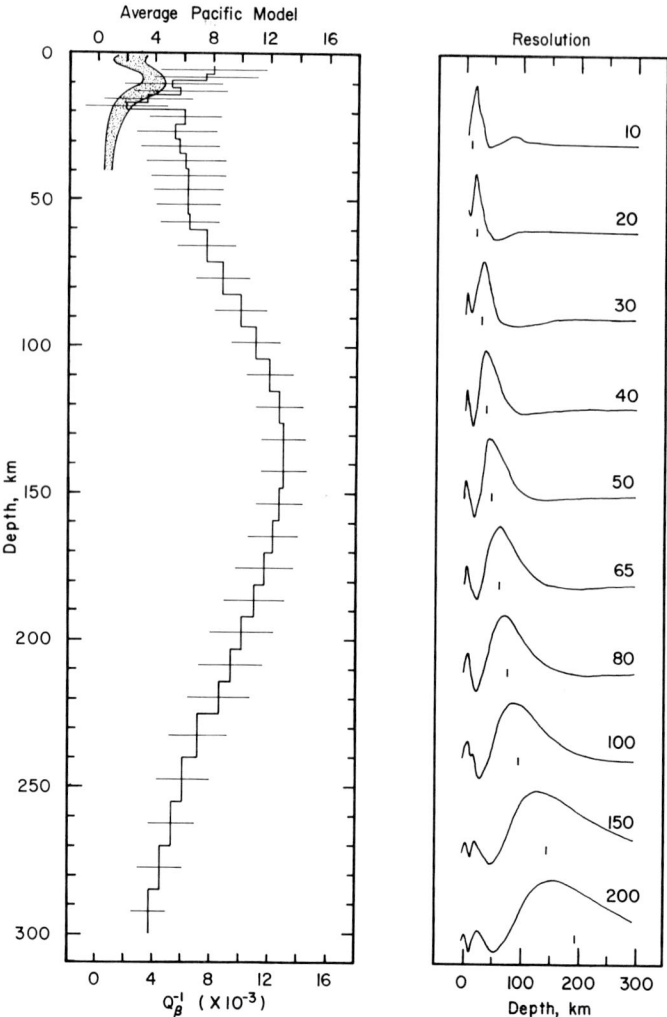

Fig. 8. Q_β^{-1} model of the Pacific crust and upper mantle obtained by Mitchell [1976] from the inversion of Rayleigh wave attenuation coefficient data. Resolving kernels appear on the right. The number by each kernel indicates the depth to which it pertains. Each kernel has been normalized to the same maximum amplitude. The horizontal bars with the model indicate 1 standard deviation and pertain to the depth range indicated by the width of the appropriate resolving kernel. The shaded area indicates the model and standard deviations for the east central North American model of Herrmann and Mitchell [1975].

The method of Tsai and Aki [1969] applied to the events of Table 2 yields the Rayleigh wave attenuation coefficients in Table 3. Their mean values are plotted in Figure 11. The trangles represent a highly attenuating path along the East Pacific Rise. It is apparent that few of the determinations can be considered very reliable according to the

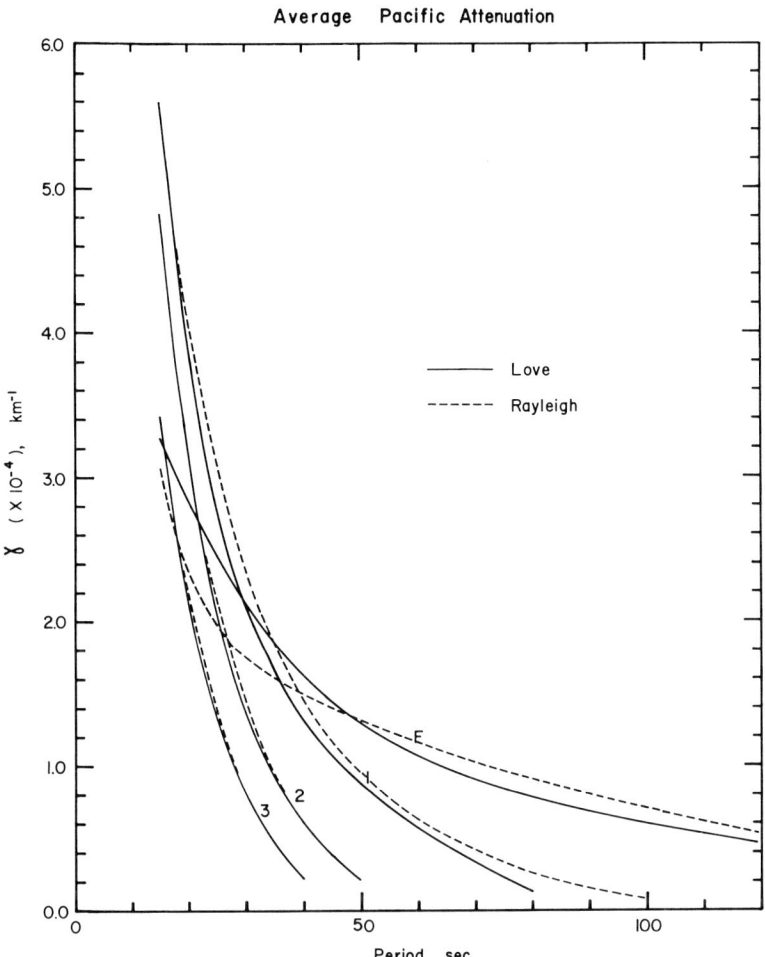

Fig. 9. Theoretical attenuation coefficient values for the fundamental and first three higher-mode surface waves propagating through a model for the Pacific crust and upper mantle. The velocity structure is taken from model 8-3-2 of Saito and Takeuchi [1966], and the Q_β^{-1} model is that of Mitchell [1976].

criteria of Herrmann [1974] and that relatively large standard deviations are associated with the derived values. The results plotted in Figure 11 indicate that Rayleigh wave attenuation coefficient values in the eastern Pacific may be somewhat higher than average values obtained for the entire Pacific [Mitchell et al., 1976]. There is a large overlap of the standard deviation bars, however, except at the shorter periods.

East Pacific Rise

One path which lies between the earthquake of November 6, 1965, and station LPS, La Palma, El Salvador, seems characterized by much higher

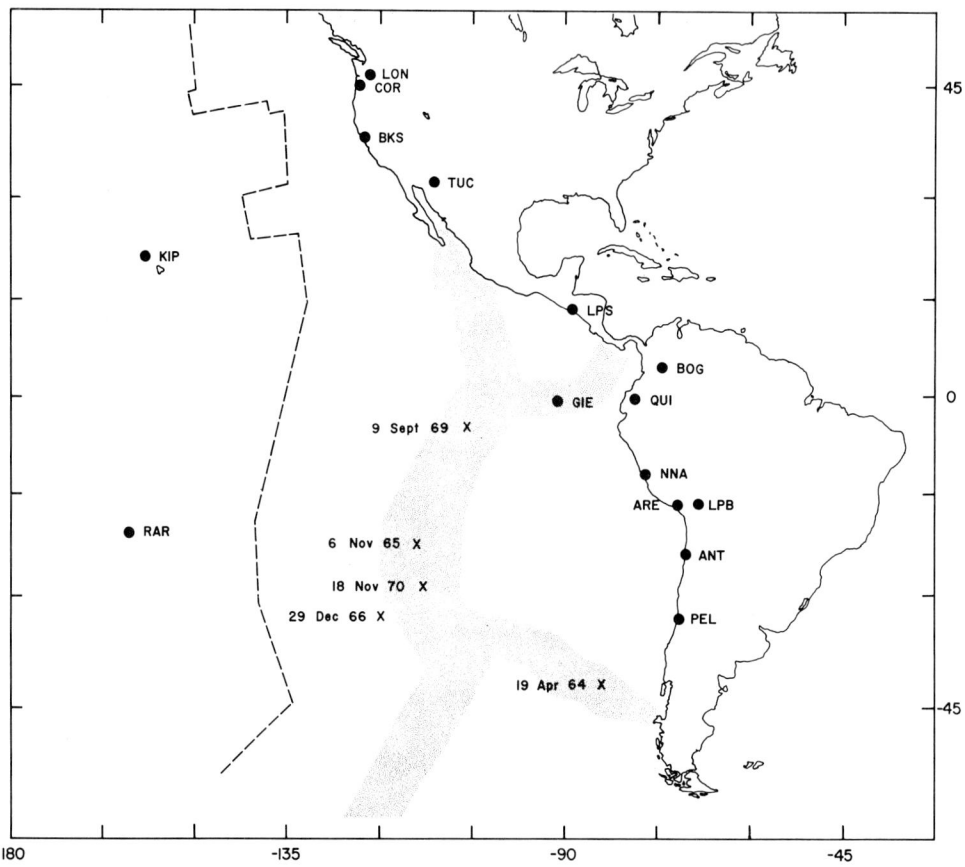

Fig. 10. Map of the eastern Pacific indicating events and stations used for the attenuation study of that region. The shading indicates the approximate extent of crustal material which is less than 10 m.y. in age. The dashed line separates material which is less than 50 m.y. of age from older material.

attenuation than other paths in the eastern Pacific for all periods. Much of that path lies along or near the East Pacific Rise. That source-receiver combination is the only one in the present study that has such a large portion of its path near the East Pacific Rise. The amplitudes observed at that station were used to calculate apparent attenuation coefficients for a path along the ridge. As was noted above, these values are plotted as triangles in Figure 11. The data are too few and uncertain to obtain a detailed model of the Q distribution in this region. However, they suggest that a narrow zone of low Q values such as that inferred by Solomon [1973] for the mid-Atlantic ridge might also exist beneath the East Pacific Rise.

Conclusions

Amplitude studies of surface waves are generally characterized by much uncertainty, and most inferences concerning the properties of the

TABLE 2. Earthquake Parameters [Forsyth, 1972]

| Date | Origin Time | Latitude, deg | Longitude, deg | Fault Parameters | | | Magnitude |
				Strike	Dip	Slip	
Sept. 9, 1969	1523:10.8	-4.43	-105.93	100	80	-165	5.2
Nov. 6, 1965	0921:48.6	-22.13	-113.76	52	60	166	6.2
Nov. 18, 1970	2010:58.2	-28.72	-112.74	119	80	-6	5.6
Dec. 29, 1966	1156:22.9	-32.66	-119.79	50	60	-160	5.1
April 19, 1964	0513:01.6	-41.70	-84.00	271	62	-11	5.5

TABLE 3. The Rayleigh Wave Attenuation Coefficient γ_R (10^{-4} km^{-1}) With Its 95% Confidence Levels for the Eastern Pacific

T	Sept. 9, 1969	Nov. 6, 1965	Nov. 18, 1970	Dec. 29, 1966	April 19, 1964
15	3.27 ± 5.33	0.84 ± 4.48*	1.50 ± 5.22*	5.03 ± 3.44	2.19 ± 3.49*
16	4.02 ± 3.99	0.69 ± 4.60*	1.02 ± 4.67*	7.83 ± 3.02	1.22 ± 7.75*
17	3.73 ± 3.80	1.18 ± 3.79*	0.76 ± 4.29*	8.02 ± 3.41	1.59 ± 7.07*
18	3.51 ± 3.65*	1.63 ± 3.95*	1.08 ± 4.37*	6.55 ± 3.17*	2.13 ± 6.18*
19	2.51 ± 3.57*	1.77 ± 3.60*	0.99 ± 4.43*	5.35 ± 3.32*	2.30 ± 6.68*
20	2.97 ± 3.62*	1.84 ± 2.97*	0.62 ± 4.43*	4.44 ± 3.49*	2.89 ± 6.47*
22	2.43 ± 4.06*	1.68 ± 2.31*	0.11 ± 4.44*	3.35 ± 3.53*	3.80 ± 4.58*
24	1.95 ± 3.93*	1.49 ± 2.42*	0.11 ± 4.07*	2.52 ± 3.66*	3.42 ± 3.76*
26	1.79 ± 3.21*	1.32 ± 2.31*	-0.03 ± 3.88*	1.88 ± 3.83*	3.21 ± 3.27
28	1.63 ± 2.66*	1.14 ± 2.10*	-0.08 ± 3.60*	1.48 ± 3.88*	3.17 ± 3.29
30	1.64 ± 2.23*	0.91 ± 2.19*	0.08 ± 3.10*	1.31 ± 3.81*	3.10 ± 3.47*
35	2.03 ± 1.30	1.09 ± 2.21*	0.45 ± 2.91*	1.48 ± 4.34*	2.91 ± 3.34*
40	1.73 ± 1.57	1.86 ± 1.81	0.54 ± 3.59*	1.26 ± 4.24*	2.60 ± 2.77
45	1.46 ± 1.85	2.02 ± 1.77	0.63 ± 4.18*		2.37 ± 2.30
50	1.39 ± 1.84*	1.85 ± 1.70	0.87 ± 3.57*	0.80 ± 3.95*	2.31 ± 2.06
55	1.27 ± 2.53*	1.80 ± 1.70	1.26 ± 3.11*	0.47 ± 4.04*	2.33 ± 1.98
60	1.36 ± 3.10*	1.67 ± 1.80	1.75 ± 3.03*	0.27 ± 4.70*	2.49 ± 2.08
70	0.28 ± 2.37*	1.30 ± 2.17	1.27 ± 3.06*		2.43 ± 2.30
80	-1.34 ± 2.25*	1.91 ± 2.48		0.08 ± 4.10*	1.86 ± 2.45
90	-2.02 ± 3.09*	3.13 ± 2.87*	0.14 ± 3.19*		1.86 ± 2.54
100	1.29 ± 4.31*	4.62 ± 3.55*	-0.05 ± 2.96*		2.29 ± 3.37
110		5.01 ± 3.76*	-1.03 ± 3.19*	-0.30 ± 4.42*	2.35 ± 3.77

*Here r < r_c, and low confidence is associated with the corresponding attenuation coefficient. Parameter r_c is the critical correlation coefficient between observed and theoretical radiation patterns [Herrmann, 1974].

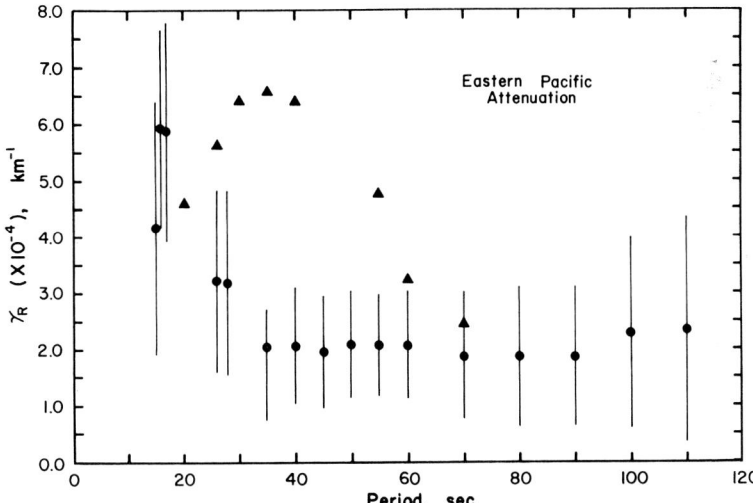

Fig. 11. Rayleigh wave attenuation coefficients and their standard deviations for paths through the eastern Pacific. The triangles indicate apparent attenuation coefficients for a path along the East Pacific Rise (November 6, 1965, earthquake to LPS).

earth are less detailed than those obtained from velocity studies. However, the following conclusions seem to be justified by presently available surface wave attenuation data.

1. Continental Q_β^{-1} values are relatively high (between 4×10^{-3} and 9×10^{-3}) for the upper 15-20 km of the crust and decrease rapidly over a small depth range to much smaller values (1×10^{-3} or less) in the lower crust.

2. Resolvable differences in Q_β^{-1} between eastern and western North America are confined to the upper crust, where Q_β^{-1} values are about twice as great in western North America as they are in eastern North America.

3. Attenuation data for stable regions, at least in North America and Eurasia, can be explained by similar Q_β^{-1} distributions.

4. Q_β^{-1} values for the Pacific crust are higher (averaging about 7×10^{-3}) than values for the lower crust of continents (averaging 1×10^{-3} or less).

5. A zone of low attenuation (or high Q) might exist at depths between 15 and 20 km beneath the Pacific.

6. The top of a low Q zone for an average model of the Pacific begins at a depth of about 60 km.

7. Surface wave attenuation coefficients for the eastern Pacific are somewhat greater than average values for other regions of the Pacific, and surface waves which propagate near the East Pacific Rise indicate even greater attenuation.

These conclusions are necessarily quite broad in nature and are intended to serve as a basis with which to compare future more detailed studies. As additional high-quality data become available, hopefully for smaller regions, we might expect to obtain more detailed models of the distribution of Q_β^{-1} with depth in the earth.

Acknowledgements. The authors thank Guey-kuen Yu for assistance with some of the computations in this study. Robert Herrmann made his surface wave programs available to us. The work was supported in part by the National Science Foundation under grant DES 74-22144 and in part by the Advanced Research Projects Agency of the Department of Defense and was monitored by the Air Force Office of Scientific Research under contract F44620-73-C-0042.

References

Anderson, D.L., A. Ben-Menahem, and C.B. Archambeau, Attenuation of seismic energy in the upper mantle, J. Geophys. Res., 70, 1441-1448, 1965.

Backus, G., and F. Gilbert, Uniqueness in the inversion of gross earth data, Phil. Trans. Roy. Soc. London, Ser. A, 266, 123-192, 1970.

Ben-Menahem, A., Observed attenuation and Q values of seismic surface waves in the upper mantle, J. Geophys. Res., 70, 4641-4651, 1965.

Ben-Menahem, A., and D.G. Harkrider, Radiation patterns of seismic surface waves from buried dipolar point sources in a flat stratified earth, J. Geophys. Res., 69, 2605-2620, 1964.

Braile, L.W., R.B. Smith, G.R. Keller, and R.M. Welch, Crustal structure across the Wasatch front from detailed seismic refraction studies, J. Geophys. Res., 79, 2669-2677, 1974.

Capon, J., Comparison of Love- and Rayleigh-wave multipath propagation at LASA, Bull. Seismol. Soc. Amer., 61, 1327-1344, 1971.

Drake, L.A., Rayleigh waves at a continental boundary by the finite element method, Bull. Seismol. Soc. Amer., 62, 1259-1268, 1972.

Dziewonski, A.M., S. Bloch, and M. Landisman, A technique for the analysis of transient seismic signals, Bull. Seismol. Soc. Amer., 427-444, 1969.

Forsyth, D.W., Mechanisms of earthquakes and plate motions in the east Pacific, Earth Planet. Sci. Letters, 17, 189-193, 1972.

Forsyth, D.W., A new method for the analysis of multi-mode surface wave dispersion: Application to Love wave propagation in the east Pacific, Bull. Seismol. Soc. Amer., 65, 323-342, 1975.

Hasegawa, H.S., Surface and body-wave spectra of Cannikin and shallow Aleutian earthquakes, Bull. Seismol. Soc. Amer., 63, 1201-1225, 1973.

Herrmann, R.B., Surface wave generation by the south central Illinois earthquake of November 9, 1968, Bull. Seismol. Soc. Amer., 63, 2121-2134, 1973.

Herrmann, R.B., Surface wave generation by central United States earthquakes, Ph.D. thesis, 262 pp., St. Louis Univ., St. Louis, Mo., 1974.

Herrmann, R.B., and B.J. Mitchell, Statistical analysis and interpretation of surface wave anelastic attenuation data for the stable interior of North America, Bull. Seismol. Soc. Amer., 65, 1115-1128, 1975.

Jeffreys, H., Damping of S waves, Nature, 208, 675, 1965.

Kausel, E.G., A.R. Leeds, and L. Knopoff, Variations of Rayleigh wave phase velocities across the Pacific Ocean, Science, 186, 139-141, 1974.

Lee, W.B., and S.C. Solomon, Inversion schemes for surface wave attenuation and Q in the crust and the mantle, Geophys. J. Roy. Astron. Soc., 43, 47-71, 1975.

Liu, H., D.L. Anderson, and H. Kanamori, Velocity dispersion due to anelasticity: Implications for seismology and mantle composition, Geophys. J. Roy. Astron. Soc., 47, 41-58, 1976.

Lysmer, J., and L.A. Drake, The propagation of Love waves across nonhorizontally layered structures, Bull. Seismol. Soc. Amer., 61, 1233-1251, 1971.

McEvilly, T.V., Central U.S. crust-upper mantle structure from Love- and Rayleigh-wave phase velocity inversion, Bull. Seismol. Soc. Amer., 54, 1997-2015, 1964.

McGarr, A., Amplitude variations of Rayleigh waves-Horizontal refractions, Bull. Seismol. Soc. Amer., 59, 1307-1334, 1969.

Mitchell, B.J., Radiation and attenuation of Rayleigh waves from the southeastern Missouri earthquake of October 21, 1965, J. Geophys. Res., 78, 886-899, 1973a.

Mitchell, B.J., Surface wave attenuation and crustal anelasticity in central North America, Bull. Seismol. Soc. Amer., 63, 1057-1071, 1973b.

Mitchell, B.J., Regional Rayleigh wave attenuation in North America, J. Geophys. Res., 80, 4904-4916, 1975.

Mitchell, B.J., Anelasticity of the crust and upper mantle beneath the Pacific Ocean from the inversion of observed surface wave attenuation, Geophys. J. Roy. Astron. Soc., 46, 521-533, 1976.

Mitchell, B.J., L.W.B. Leite, Y.K. Yu, and R.B. Herrmann, Attenuation of Love and Rayleigh waves across the Pacific at periods between 15 and 110 seconds, Bull. Seismol. Soc. Amer., 66, 1189-1201, 1976.

Molnar, P., and J. Oliver, Lateral variations of attenuation in the upper mantle and discontinuities in the lithosphere, J. Geophys. Res., 74, 2648-2682, 1969.

Nuttli, O.W., Seismic wave attenuation and magnitude relations for eastern North America, J. Geophys. Res., 78, 876-885, 1973.

Saito, M., and H. Takeuchi, Surface waves across the Pacific, Bull. Seismol. Soc. Amer., 56, 1067-1091, 1966.

Sclater, J.G., and J. Francheteau, The implications of terrestrial heat flow observations on current tectonic and geochemical models of the crust and the upper mantle of the earth, Geophys. J. Roy. Astron. Soc., 20, 509-542, 1970.

Solomon, S.C., Seismic wave attenuation and partial melting in the upper mantle of North America, J. Geophys. Res., 77, 1483-1502, 1972a.

Solomon, S.C., On Q and seismic discrimination, Geophys. J. Roy. Astron. Soc., 31, 163-177, 1972b.

Solomon, S.C., Shear wave attenuation and melting beneath the mid-Atlantic ridge, J. Geophys. Res., 78, 6044-6059, 1973.

Sutton, G.H., and D.A. Walker, Oceanic mantle phases recorded on seismographs in the northwestern Pacific at distances between 7° and 40°, Bull. Seismol. Soc. Amer., 62, 631-655, 1972.

Tsai, Y.B., and K. Aki, Simultaneous determination of seismic moment and attenuation of surface waves, Bull. Seismol. Soc. Amer., 59, 275-287, 1969.

Wiggins, R.A., The general linear inverse problem: Implication of surface waves and free oscillations for earth structure, Rev. Geophys. Space Phys., 10, 251-285, 1972.

Yacoub, N.K., and B.J. Mitchell, Attenuation of Rayleigh wave amplitudes across Eurasia, Bull. Seismol. Soc. Amer., in press, 1977.

INTERPRETATION OF CRUSTAL VELOCITY GRADIENTS AND Q STRUCTURE USING AMPLITUDE-CORRECTED SEISMIC REFRACTION PROFILES

Lawrence W. Braile

Department of Geosciences, Purdue University
West Lafayette, Indiana 47907

Abstract. Amplitudes of refracted and reflected seismic waves are greatly affected by velocity gradients and anelasticity (Q^{-1}). Combined interpretation of the amplitudes of upper crustal phases, the head wave in the upper crust (P_g), and the reflection from the top of the lower crustal layer (P_{cr}) allows inference of both velocity and Q structure from amplitude-corrected seismic refraction data. Synthetic seismograms were computed by using the reflectivity method modified to include effects of anelasticity for several upper crustal models. Modeling of the amplitude-distance variations of the P_g and P_{cr} phases allows simultaneous determination of velocity gradients and Q structure. Interpretation of an amplitude-corrected seismic refraction profile from the eastern Basin and Range Province of the western United States suggests anomalously low compressional wave Q for the upper crust. Rapid attenuation with distance of the P_{cr} phase for this refraction profile suggests that seismic wave energy absorption may occur primarily in the upper crustal low-velocity high Poisson ratio layer, where a Q of ~50 is inferred.

Introduction

While the general velocity structure of the earth's crust can be inferred from studies of the travel times of seismic waves, detailed velocity structure, including identification of velocity gradients within layers and recognition of low-velocity layers, can best be determined by a combined interpretation of the travel times and amplitudes of refracted and reflected waves. Seismic wave amplitudes are sensitive to the presence of velocity gradients and anelasticity (Q^{-1}). Cerveny [1966] has shown that even small velocity gradients (± 0.01 km s^{-1} km^{-1}) are sufficient to affect amplitudes of head waves by a factor of 1 or 2 orders of magnitude. Such velocity gradients, however, result in only slight curvature of the travel time curve. Hill [1971] has shown that the effects of velocity gradients and anelasticity cannot be separated by using the amplitude decay with distance of a head wave phase.

The purpose of this paper is to present a method for determining both velocity and Q structure in the earth by using combined interpretation of amplitudes of refracted and reflected phases and to apply the method to determination of the compressional wave Q in the upper crust in the eastern Basin and Range Province of the western United States. The

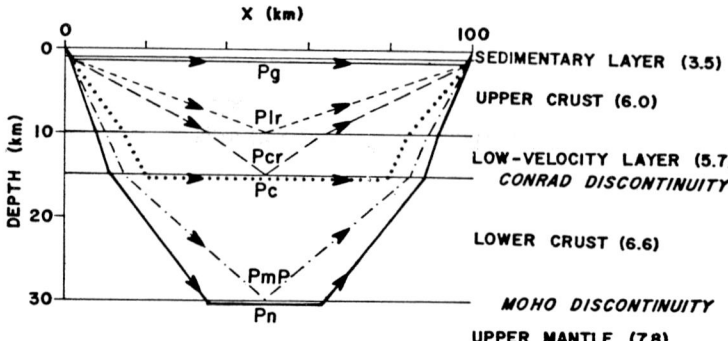

Fig. 1. Schematic diagram of prominent crustal and uppermost mantle compressional wave phases. Characteristic velocities for Basin and Range Province crustal structure are given in kilometers per second for each layer.

method employs modeling of observed amplitudes from seismic refraction profiles with theoretical amplitudes from synthetic seismograms. Synthetic seismograms are computed by using a modification of the reflectivity method [Fuchs and Müller, 1971] in which the effect of attenuation on body waves is included by introducing complex velocities to calculate the plane wave reflection coefficients. By following Schwab and Knopoff [1972], if c and Q are the phase velocity and the quality factor, respectively, for either the compressional or the shear waves, then the complex velocity is

$$V = V_1 + iV_2$$

where

$$V_1 = c[4Q^2/(4Q^2 + 1)]$$

$$V_2 = c[4Q^2/(4Q^2 + 1)]/2Q$$

Kennett [1975] has described a similar modification of the reflectivity method, which he uses to investigate attenuation in the upper mantle. In an alternative approach, Helmberger [1973] has employed an empirical attenuation filter in order to introduce anelasticity into the calculation of synthetic seismograms by the Caignard-deHoop technique.

Model Studies

Synthetic seismograms are calculated for several upper crustal velocity models both with and without the effects of attenuation. A 3-Hz dominant frequency wavelet is used. Shear wave velocities are computed from compressional wave velocities, a Poisson ratio of 0.25 being assumed. Densities are computed from compressional wave velocities by following Birch [1964]. Shear wave Q was specified to be 4/9 of compressional wave Q by following the hypothesis of Anderson and Archambeau [1964] of no attenuation in pure compression.

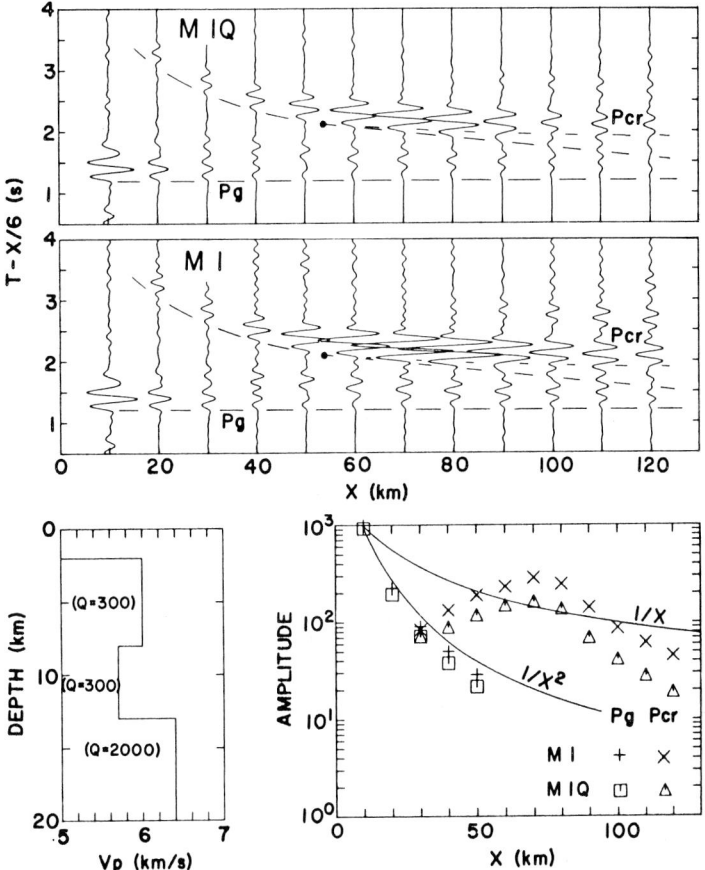

Fig. 2. Synthetic seismograms and amplitude-distance curves for models M1 and M1Q. Velocity model is shown in lower left. Q values are infinite for model M1 and have values shown in the figure for model M1Q. Synthetic seismograms are for the vertical component. Amplitudes are multiplied by distance for convenient plotting. Solid circles indicate the ray-theoretical critical point for the head wave from the lower crustal layer. Phases are identified according to the notation in Figure 1. The solid lines in the relative amplitude versus distance plot are reference lines showing $1/x$ and $1/x^2$ amplitude decay referenced to 10^3 at a distance of 10 km.

The crustal and uppermost mantle phases which are considered in this study are shown schematically in Figure 1. The crustal model shown is similar to models which have been proposed for the Basin and Range Province and includes an upper crustal low-velocity layer (LVL).

Six upper crustal velocity models, synthetic seismograms calculated for both the anelastic and the perfectly elastic case, and amplitude-distance curves are shown in Figures 2-7. It will be seen that modeling of the amplitudes of the head wave phase (P_g) and the reflection from the top of the lower crustal layer (P_{cr}) allows separation of the ef-

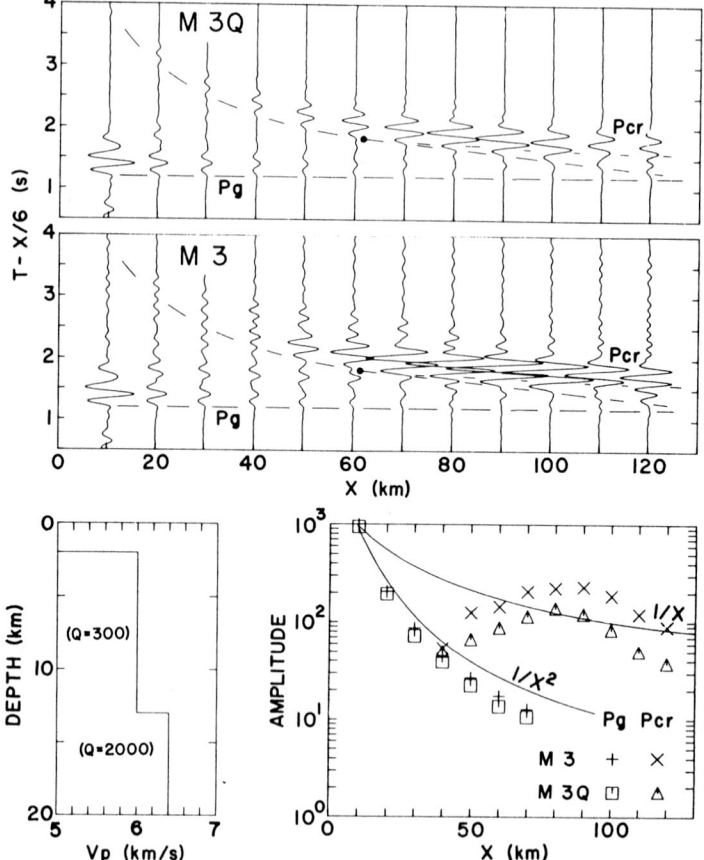

Fig. 3. Same as Figure 2 for models M3 and M3Q.

fects of velocity gradients and Q structure. However, inference of both velocity and Q structure from P_g and P_{cr} amplitude decay requires accurate amplitude data and recognition of distinct phases over a considerable distance range.

The amplitude-distance curves of Figures 2-7 illustrate several features which bear on the problem of recognition of velocity gradients and Q structure of the upper crust.

P_g amplitudes. A positive velocity gradient in the upper crustal layer, as is seen in models M8 and M11 (Figures 4 and 7), produces a much less rapid amplitude decay than is produced in the homogeneous layer case, model M3 (Figure 3). Both a negative velocity gradient, M9 (Figure 5), and the presence of anelasticity (Q models) produce more rapid amplitude decay.

P_{cr} amplitudes. The distance at which the P_{cr} amplitude-distance curve peaks (about 60-100 km for the models shown) is affected by the velocity contrast at the upper crust-lower crust boundary. Thus the P_{cr} amplitude peak shifts from near 90 km for the homogeneous upper crustal model (Figure 3) to about 70 km (model M1, Figure 2) owing to the

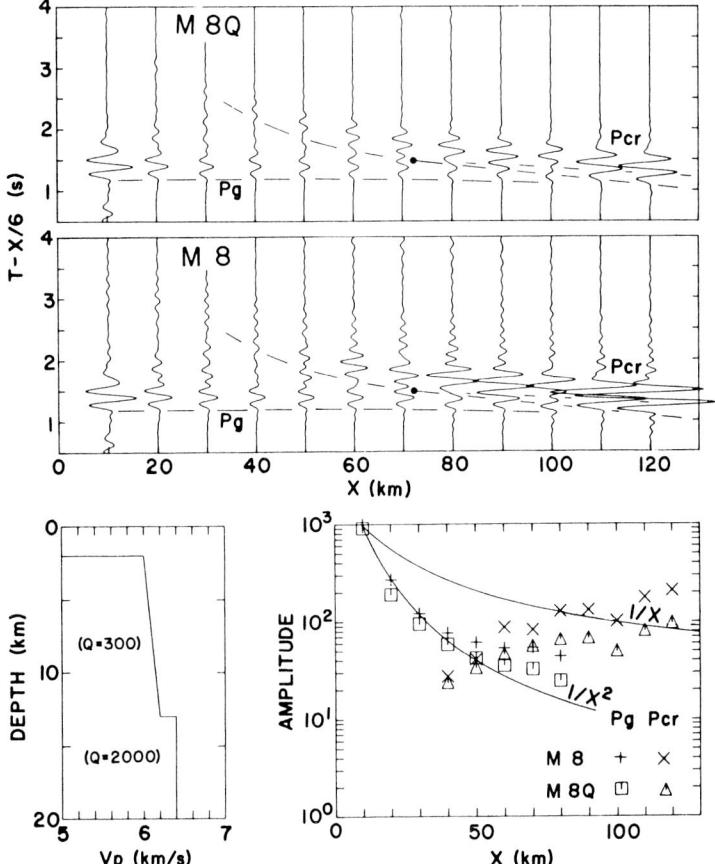

Fig. 4. Same as Figure 2 for models M8 and M8Q.

presence of an LVL. The P_{cr} amplitude curve peaks sharply near 100 km owing to a velocity transition zone at the base of the upper crust (model M10, Figure 6). A positive velocity gradient in the upper crust (Figures 4 and 7) produces large amplitudes at distances greater than 100 km, well beyond the critical distance.

Separation of the effects of negative velocity gradients and anelasticity is illustrated by models M3 and M9 (Figures 3 and 5, respectively). The negative velocity gradient infinite Q model (M9) can be seen to have a P_g amplitude decay which is indistinguishable from the homogeneous layer anelastic model M3Q (Figure 3) However, the P_{cr} amplitudes for M9 (Figure 5) and M3Q (Figure 3) differ by approximately a factor of 2. Similarly, a positive velocity gradient low Q model (similar to M11Q, Figure 7) may have a P_g amplitude decay equivalent to the homogeneous perfectly elastic upper crustal model (M3, Figure 3). However, the two models may be differentiated by their P_{cr} amplitudes.

Application of this modeling method to determination of crustal velocity and Q structure is limited by several factors. Accurate amplitude corrections and close station spacing are necessary so that local site

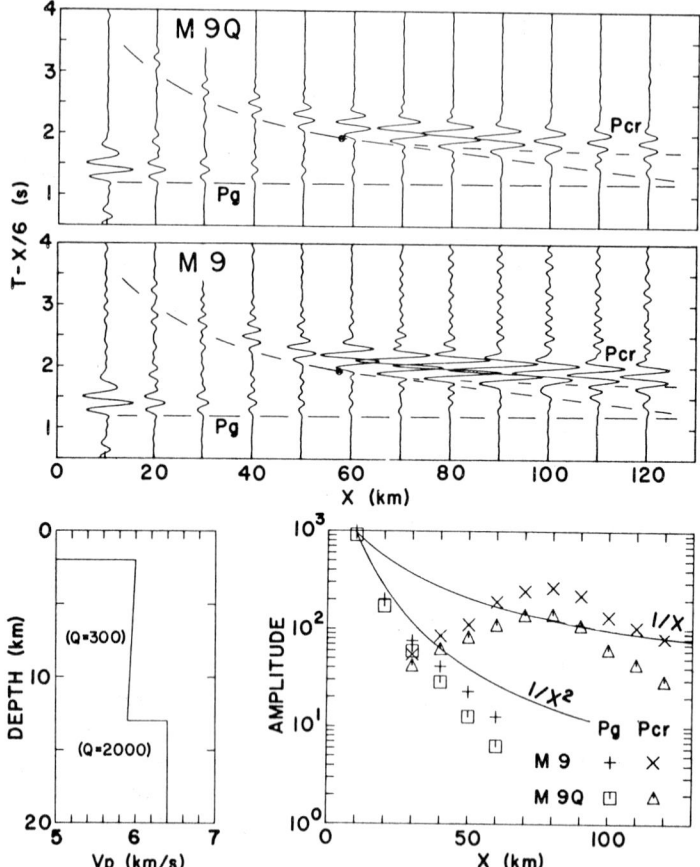

Fig. 5. Same as Figure 2 for models M9 and M9Q.

responses may be 'averaged out.' Lateral homogeneity in both velocity and anelasticity is assumed. Uniqueness of the derived model is not assured by the modeling procedure, and since the interpretation is not based on an inversion method, estimates of resolution are qualitative and may only be determined by a lengthy perturbation analysis of the derived crustal model. Despite these limitations, modeling of amplitude data for crustal phases can yield important information on crustal velocity and Q structure. Significant improvements in interpretation should be possible with high-quality amplitude data.

Application to Eastern Basin and Range Province

A detailed seismic refraction profile from the eastern Basin and Range Province was presented by Keller et al. [1975]. These data, herein called the 1972 Utah refraction profile (Figure 8), were recorded along a line due south of the Bingham copper mine southwest of Salt Lake City, Utah. This record section is conducive to amplitude interpretation because of high signal to noise ratio, the even distribution of stations

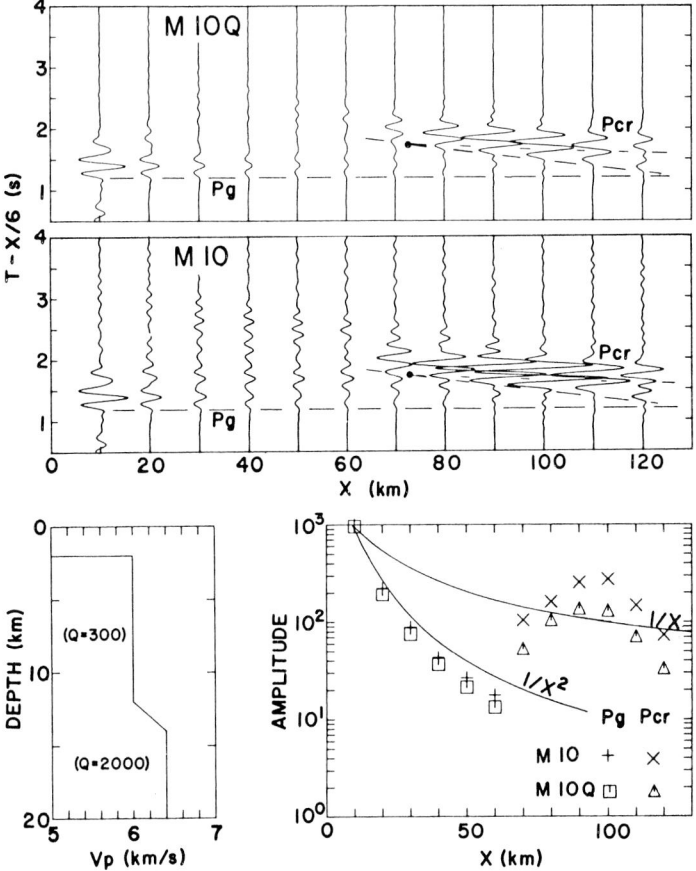

Fig. 6. Same as Figure 2 for models M10 and M10Q.

over a distance range of 240 km, and the existence of distinct phases. The record section was previously interpreted by Keller et al. [1975] primarily on the basis of travel times.

The data in Figure 8 were amplitude-corrected by removing the effect of the instrument and accounting for differences in source sizes. This correction was accomplished by deconvolving the seismograms within the frequency range of 1-15 Hz by using

$$A_c(f) = k \frac{A_f(f) I_f(f)}{S(f) I_b(f)}$$

where

$A_c(f)$ amplitude spectrum of the corrected field seismogram;
$A_f(f)$ amplitude spectrum of vertical ground motion at the field station;

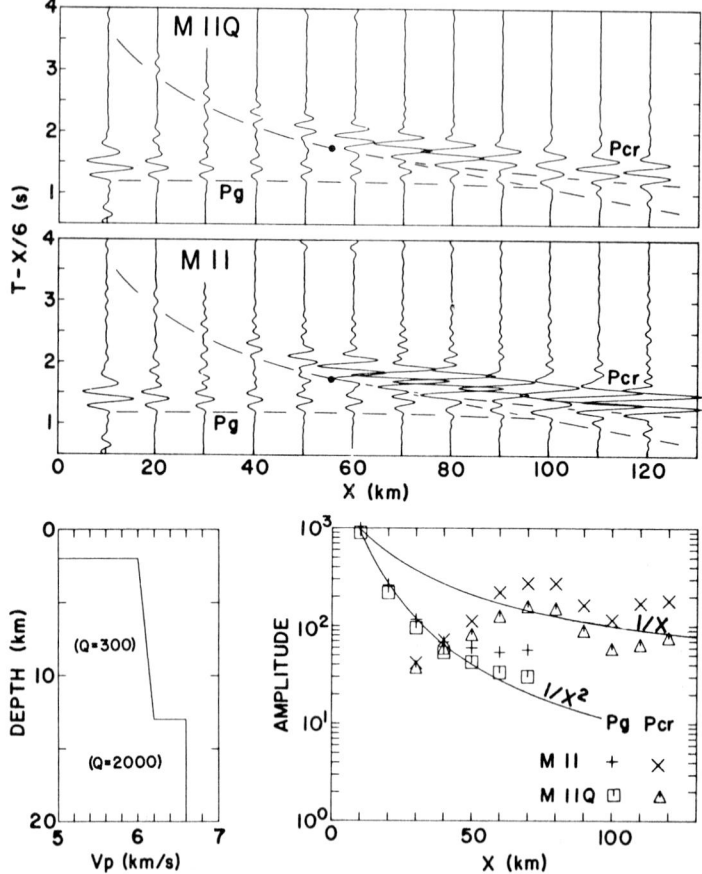

Fig. 7. Same as Figure 2 for models M11 and M11Q.

$I_f(f)$ instrument response of the field seismograph;
$S(f)$ amplitude spectrum of the vertical component of the source function at the base station;
$I_b(f)$ instrument response of the base station seismograph;
k constant relating the relative amplifications of the base and field station seismographs.

The base station for all sources was located approximately 12 km NNE of the sources. The product $A_f(f) \cdot I_f(f)$ is the Fourier transform of the field seismogram, and $S(f) \cdot I_b(f)$ is the Fourier transform of the first 0.3 s of the base station seismogram. Inspection of the base seismograms suggested that the source function was best represented by the first 0.3 s of the seismogram. This length was equal to the first 1½-2 cycles for most sources. Since the seismographs were matched instruments, it is assumed that $I_f(f) = I_b(f)$ and that the effect of the instrument response is given by the constant k. This assumption was substantiated by daily instrument calibrations. The amplitude-corrected seismograms are shown in record section form in Figure 9, and amplitudes

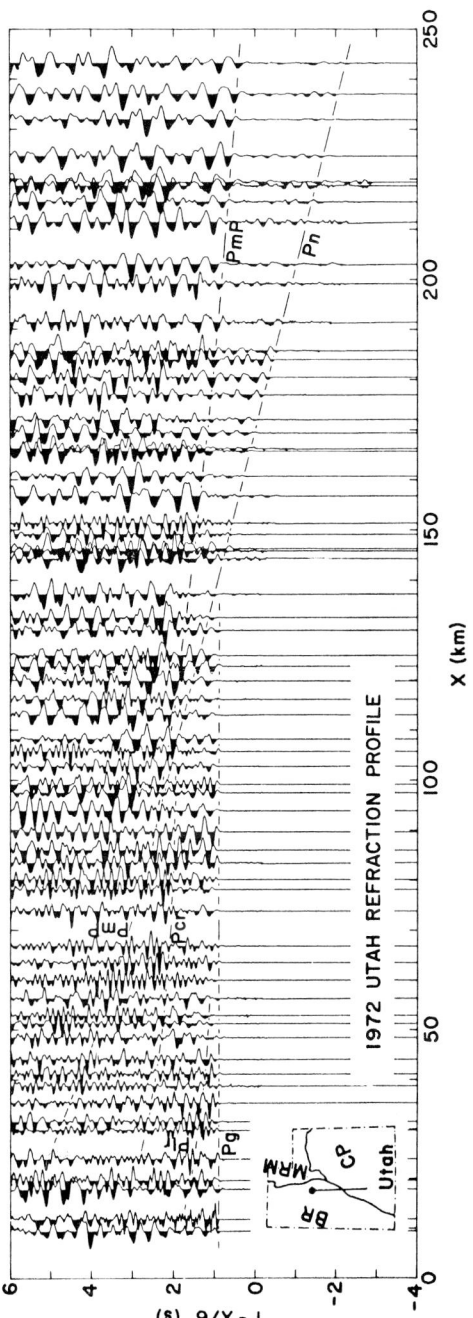

Fig. 8. Record section of vertical component seismograms for the 1972 Utah refraction profile. Phases are identified according to the notation in Figure 1. Inset map shows location of profile: BR, Basin and Range Province; CP, Colorado Plateau Province; MRM, Middle Rocky Mountain Province.

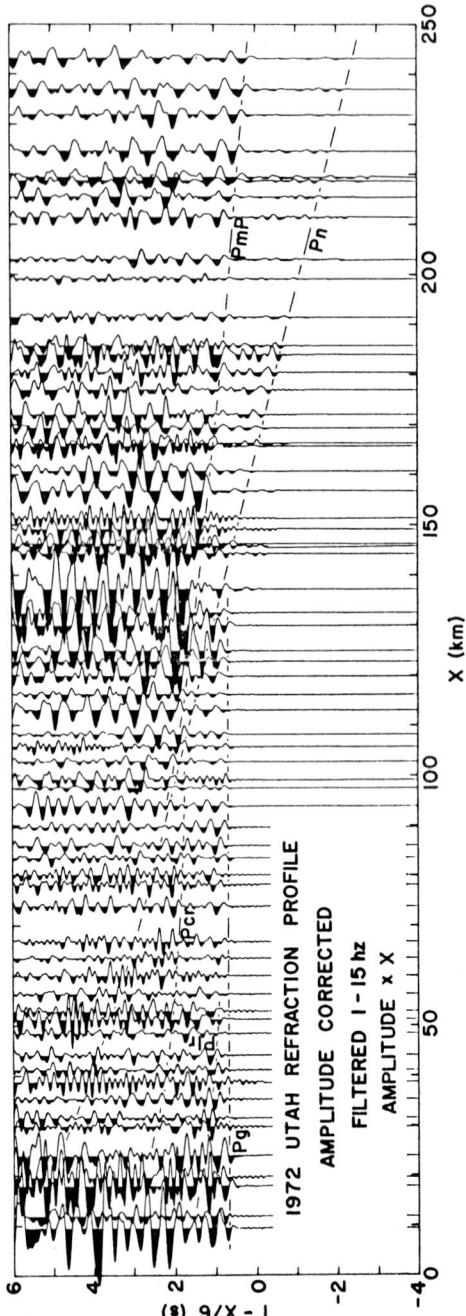

Fig. 9. 1972 Utah refraction profile after amplitude correction. Amplitudes have been multiplied by distance for convenient plotting.

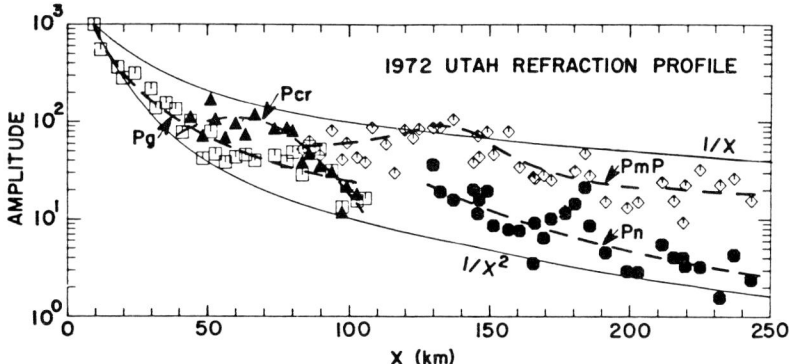

Fig. 10. Relative amplitude versus distance plot for prominent phases in Figure 9. Solid lines show $1/x$ and $1/x^2$ amplitude decay referenced to 10^3 at a distance of 10 km. Heavy dashed lines are inferred amplitude-distance curves fit through the data points. Theoretical amplitudes for the model shown in Figure 11 fall on the amplitude-distance curves for the P_g and P_{cr} phases, except near 10 km.

of prominent phases are plotted versus distance in Figure 10. Some scatter exists in the amplitude-distance curves owing to local site responses and possible inaccuracy in the amplitude correction. However, because of the close station spacing and the large number of amplitude observations the trend of the amplitude decay for each phase is reasonably well determined.

Two important features of the amplitude-distance (Figure 10) plot of upper crustal phases (P_g and P_{cr}) are that the amplitudes decay less rapidly than those of the homogeneous layer case (Figure 3) and that amplitude curve peaks at small distances (~60 km) and decays rapidly with distance at greater distances. The P_g decay suggests the presence of a positive velocity gradient in the upper crustal layer. However, the P_{cr} amplitude peak at short distances and the rapid amplitude decay at greater distances preclude the presence of a significant positive gradient. In fact, in order to explain the P_{cr} amplitudes an LVL, as previously inferred for this area by Keller et al. [1975], and a low compressional wave Q of about 100 are suggested for the upper crust. However, a Q of 100 produces a rapid decay of the P_g phase, which is not observed in Figure 10. A possible explanation of the P_g amplitude curve is that the P_{lr} phase, which is produced by reflection from the top of the LVL and which has arrival times only about 0.2 s after P_g at distances greater than about 40 km, interferes with P_g and causes a smaller apparent decay of P_g with distance.

In order to explain both the P_g and the P_{cr} amplitude-distance curves the crustal model in Figure 11 is proposed. The velocity structure is identical to that given by Keller et al. [1975]. The model produces calculated amplitudes of P_g and P_{cr} phases which match the observed amplitudes except in the short distance range (near 10 km). It is possible that the P_g and P_{cr} amplitudes observed in the 1972 Utah refraction profile could be explained by a model having a constant Q of about 100 in the upper crust instead of the variable Q model shown in Figure 11.

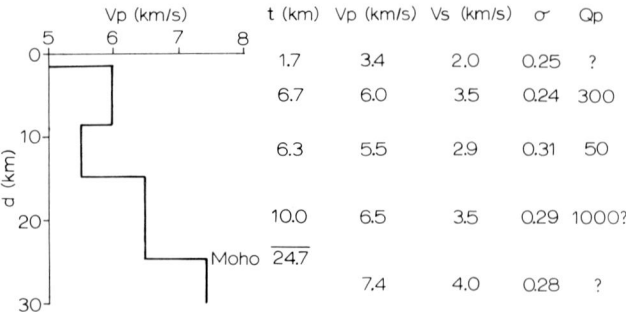

Fig. 11. Proposed crustal velocity and Q model for the eastern Basin and Range Province based on travel time and amplitude interpretation of the 1972 Utah refraction profile.

However, such a model would have to include a velocity structure more complex than the structures in the models shown in Figures 2-7, and the existing data are not sufficient to deduce such a feature. Furthermore, an anomalous Poisson ratio has previously been inferred for the LVL, and thus very low Q would not be unexpected. Two mechanisms which have been shown, under 'crustal' conditions, to produce an increase in Poisson ratio and a decrease in Q are increased temperature and increased pore fluid pressure [Spencer and Nur, 1976; Kissell, 1972; Volarovich and Gurvich, 1957; Grosenbaugh and Nur, 1976].

A complete analysis of the P_mP and P_n amplitude curves for the 1972 Utah refraction profile (Figure 10) will require considerable modeling and will be deferred for a future study. However, preliminary modeling using synthetic seismograms indicates that Q in the lower crust in the eastern Basin and Range Province must be at least 1000 to explain the P_mP amplitude curve.

Discussion

Few determinations of Q have been made in the upper crust. Press [1964] measured average values of Q for the upper crust in the Basin and Range Province from seismic profiles from the Nevada nuclear test site. He reported Q values of 450 and 260, inferred from L_g and P_g phases, respectively. However, the lower value (for P_g) was attributed to partial loss of P wave energy by mode conversion. Regional values of shear wave Q for the upper crust have been determined by Mitchell [1975] and by Lee and Solomon [1975] using Rayleigh wave attenuation. Mitchell reports shear wave Q values of ~125 and ~250 for the upper crust in the western and eastern United States, respectively. Lee and Solomon suggest crustal shear wave Q values of ~200 for a path in the western United States and ~300 for paths in the eastern United States. Compressional wave Q values for the upper crust consistent with averages of Mitchell's and Lee and Solomon's results are about 300 for the western United States and 750 for the eastern United States. Thus the proposed upper crustal Q values from this study (Figure 11) are exceptionally low and provide further evidence for an anomalous crust in the eastern Basin and Range Province [Smith et al., 1975].

Acknowledgements. I am grateful to Karl Fuchs and Gerhard Muller for providing me with a copy of the reflectivity computer program, which was modified for use in this research. I thank Robert B. Smith for his assistance in obtaining the information necessary to make amplitude corrections on the 1972 Utah refraction profile. This research was supported by the Office of Naval Research, Earth Physics Program, contract N00014-75-C-0972, and was presented at the Office of Naval Research symposium, 'The Nature and Physical Properties of the Earth's Crust,' held at Vail, Colorado, August 2-5, 1976.

References

Anderson, D. L., and C. B. Archambeau, The anelasticity of the earth, J. Geophys. Res., 69, 2071-2084, 1964.

Birch, F., Density and composition of the mantle and core, J. Geophys. Res., 69, 4377-4387, 1964.

Cerveny, V., On dynamic properties of reflected and head waves in the n-layered earth's crust, Geophys. J. Roy. Astron. Soc., 11, 139-147, 1966.

Fuchs, K., and G. Müller, Computation of synthetic seismograms with the reflectivity method and comparison with observations, Geophys. J. Roy. Astron. Soc., 23, 417-433, 1971.

Grosenbaugh, M., and A. Nur, Crustal low-velocity models (abstract), Eos Trans. A.G.U., 57, 961, 1976.

Helmberger, D. V., On the structure of the low-velocity zone, Geophys. J. Roy. Astron. Soc., 34, 251-263, 1973.

Hill, D. P., Velocity gradients and anelasticity from crustal body wave amplitudes, J. Geophys. Res., 76, 3309-3325, 1971.

Keller, G. R., R. B. Smith, and L. W. Braile, Crustal structure along the Great Basin-Colorado Plateau transition from seismic refraction studies, J. Geophys. Res., 80, 1093-1098, 1975.

Kennett, B. L. N., The effects of attenuation on seismograms, Bull. Seismol. Soc. Amer., 65, 1643-1651, 1975.

Kissell, F. N., Effect of temperature variation on internal friction in rocks, J. Geophys. Res., 77, 1420-1423, 1972.

Lee, W. B., and S. C. Solomon, Inversion schemes for surface wave attenuation and Q in the crust and the mantle, Geophys. J. Roy. Astron. Soc., 43, 47-71, 1975.

Mitchell, B. J., Regional Rayleigh wave attenuation in North America, J. Geophys. Res., 80, 4904-4916, 1975.

Press, F., Seismic wave attenuation in the crust, J. Geophys. Res., 69, 4417-4418, 1964.

Schwab, F. A., and L. Knopoff, Fast surface wave and free mode computations, Methods Comput. Phys., 11, 87-180, 1972.

Smith, R. B., L. W. Braile, and G. R. Keller, Upper crustal low-velocity layers: Possible effect of high temperatures over a mantle upwarp at the Basin Range-Colorado Plateau transition, Earth Planet. Sci. Lett., 28, 197-204, 1975.

Spencer, J. W., Jr., and A. M. Nur, The effects of pressure, temperature, and pore water on velocities in Westerly granite, J. Geophys. Res., 81, 899-904, 1976.

Volarovich, M. P., and A. S. Gurvich, Investigation of dynamic moduli of elasticity for rocks in relation to pressure, Bull. Acad. Sci. USSR, Geophys. Ser., Engl. Transl., no. 4, 1-9, 1957.

DETERMINING THE RESISTIVITY OF A RESISTANT LAYER IN THE CRUST

George V. Keller and Robert B. Furgerson

Colorado School of Mines
Golden, Colorado 80401

Abstract. Direct current resistivity surveys are capable of determining only the resistivity-thickness product for the resistant portion of the earth's crust. In order to estimate the resistivity in this resistant zone it is necessary to combine the results of direct current soundings with the results of electromagnetic or magnetotelluric soundings, which are capable of determining only the thickness of the resistant zone. Knowing the thickness and the thickness-resistivity product provides an unambiguous value for resistivity, at least in principle. However, distortions of direct current soundings by lateral changes in resistivity usually limit the determination of the resistivity-thickness product to a minimum possible value, and often this minimum constraint is very loose. In an attempt to improve the reliability with which the resistivity-thickness product can be determined a rotating dipole survey procedure has been devised and tried in central Wisconsin. The results, when combined with magnetotelluric interpretations in the same areas provided by Bostick, et al. (1977), indicate the maximum resistivity in the crust to be at least several million ohm meters. These results offer the prospect that further improvements in the joint interpretation of direct current and electromagnetic survey data will permit reasonably accurate determinations of the maximum resistivity in the crust.

Introduction

The problem of determining the maximum resistivity in the earth's crust has proven to be particularly difficult to solve. Attempts to study the distribution of resistivity values on a crustal scale [Keller, 1971; Brace, 1971] have yielded the information that resistivity exhibits a maximum in the upper or middle part of the earth's crust. Near-surface rocks are rendered relatively conductive by the presence of electrolyte in pore spaces in the rock. Below the zone of maximum resistivity it has been recognized that rocks become progressively more conductive with increasing depth. It is generally assumed that this increase in conductivity is caused by increasing temperature, which liberates ions in the solid phase of a rock, but other phenomena, including increased hydration of the rock or development of electronically conducting minerals, may occur also.

Knowledge of the maximum resistivity in the crust is of growing importance in two applications: earthquake prediction and exploration for geothermal resources. Laboratory studies [Brace and Orange, 1968] indicate that diagnostic changes in electrical properties take place when a rock is loaded almost to its breaking point. The strongest rocks, which are usually the ones which determine when and where an earthquake occurs, tend to be the rocks with the highest resistivity. Thus in the use of resistivity surveys for earthquake prediction it would be highly desirable to be able to determine the resistivity in the resistant portion of the crust with reasonable accuracy and reliability.

In geothermal exploration we are concerned with locating areas in which the resistivity of the rock has been lowered from normal values by unusually high temperatures. At very shallow depths, where hot water-laden rocks have extremely low resistivities, or at very great depths, where rocks at or near the melting point have extremely low resistivities, exploration is relatively straightforward; it is far easier to detect and measure the resistivity of a mass of conductive rock than of a resistive rock. The problem which remains unsolved is that of recognizing the effects of elevated temperature at moderate depths in the resistant part of the crust. Here, we are capable of determining the actual resistivity distribution only within an order of magnitude at best, so that the effect of elevated temperature usually cannot be recognized.

The Problem

The nature of the problem can be seen by considering the capabilities of various electrical sounding techniques for resolving a simplified version of the crustal resistivity profile. Consider the simple three-layer profile indicated in Figure 1; here, the crust and upper mantle are represented by three layers, each of which is characterized by a single value of resistivity.

Many investigations of crustal conductivity profiles have been carried out by using the magnetotelluric method. The magnetotelluric sounding method has come into use over the past decade with development dating from a paper by Cagniard [1953]. Cagniard demonstrated mathematically that if natural electromagnetic noise could be assumed to consist of planar electromagnetic waves propagating vertically into the earth, the apparent conductivity of the earth can be computed with a simple formula from measurements of the electric field and magnetic field intensities. Inasmuch as the distance that an electromagnetic wave can penetrate into a conductor depends upon frequency, soundings are made by measuring the electric and magnetic field intensities at a variety of frequencies.

The depth to which the conductivity profile can be

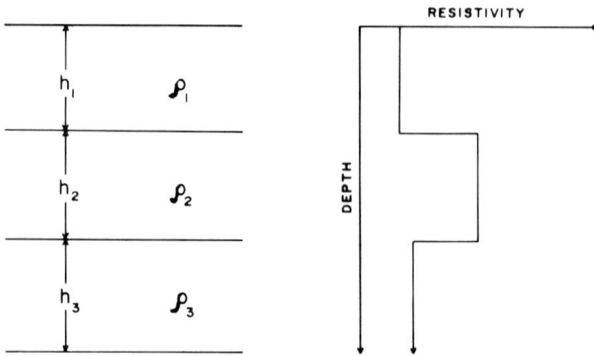

Fig. 1. Simplified crustal resistivity profile.

determined with the magnetotelluric method is determined by the frequencies analyzed. For the earth resistivity profile shown in Figure 1, where the resistivity contrasts between layers are assumed to be very large (greater than 100:1), a magnetotelluric curve exhibits ambiguities in interpretability caused by large contrasts in resistivities. When the middle layer in the sequence of three has a much higher resistivity than the outer two, the magnetotelluric sounding curve is nearly independent of the resistivity of the middle layer and depends only on the thickness of that layer. The maximum value of apparent resistivity that can be measured in such a three-layer sequence is

$$\rho_{a,max} = (h_1 + h_2)/2S_1 \qquad (1)$$

where $h_1 + h_2$ is the combined thickness of the top two layers and S_1 is the ratio h_1/ρ_1 for the first layer. The quantity S is termed the conductance of a layer. In areas where the surface layer conductance is only a few mhos the maximum apparent resistivity in a crustal scale survey may be several tens of thousands of ohm meters. When the conductance is a few thousand mhos, as in a sedimentary basin, the maximum apparent resistivity will be less than 100 ohm meters. The frequency at which the maximum apparent resistivity is observed is given by

$$\omega_{max} = \frac{1}{\mu(h_1 + h_2)S} \qquad (2)$$

The frequencies for maximum apparent resistivity in a three-layer H equivalent section are given graphically in Figure 2.

Thus it appears that the magnetotelluric (MT) method is effective in detecting the presence and properties of the first and third layers in our hypothetical crustal model but is quite ineffective in determining the resistivity in the second layer. This same limitation will hold true for all electromagnetic methods in which the effects of eddy currents induced in the

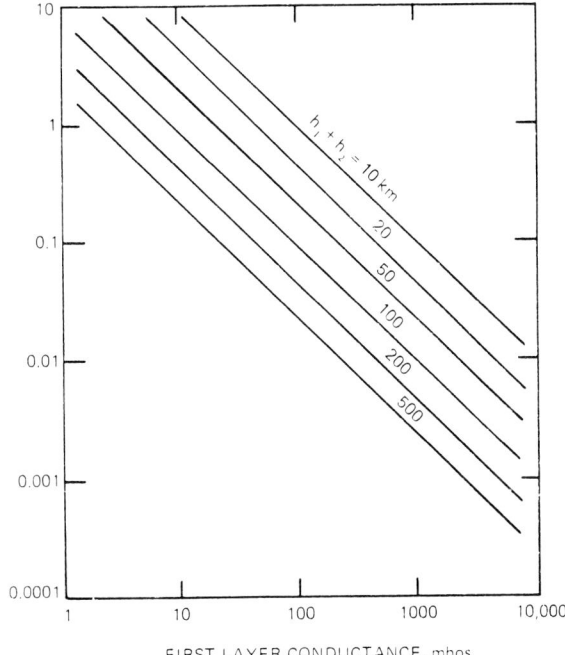

Fig. 2. Chart for determining the frequency at which the maximum apparent resistivity will be observed with the magnetotelluric method for the resistivity profile shown in Figure 1.

earth are measured; in highly resistant zones, only very small induction currents are generated, and these are difficult to detect.

Another approach to determining the crustal resistivity profile, which has been used to a lesser extent than the MT method, is the direct current resistivity sounding. The dc sounding method is the best known and understood of the electrical probing methods, perhaps because it has been in use for many years or because it can be treated mathematically at an easier level than the ac probing methods. The essential feature of a dc probing method is that the current supplied to the ground has a spectrum of frequencies low enough that the flow of current may be described solely with Laplace's equation and with no consideration of magnetic effects from the current flow. It is difficult to predict at what frequency the magnetic effects become important even when the resistivity distribution in the ground is known, and usually, a frequency far below a reasonable limit is used in dc soundings to assure that there is no error contributed by ac coupling effects.

Many specific field techniques have been used in dc soundings, but for crustal scale surveys, only two of these have found favor: the Schlumberger technique and the dipole technique. The Schlumberger electrode array consists of four colinear electrode contacts [Kunetz, 1966]. In making a sounding the outer two electrodes, which are used to supply current to the ground, are moved progressively away from the center of the spread. The inner two electrodes, at the center of the array, are used to measure the voltage drop developed by the current. In discussing the Schlumberger array the assumption is made that the separation between the inner measuring electrodes is small in comparison with the separation between the outer current electrodes. If this assumption is valid, the ratio of voltage drop to electrode separation can be said to be approximately equal to the electric field intensity.

With the dipole array, four electrodes are used also [Alpin, 1966], but they are not arranged geometrically in the same manner as the Schlumberger array. Current is supplied to the ground with one pair of electrodes, usually fixed in location, while a component of the electric field is mapped as a function of distance from this current source with a second pair of electrodes. If the separation between the current electrodes is much less than the distance to the location at which the electric field is being detected, then it is possible to characterize the source solely in terms of its dipole moment: current intensity times electrode separation.

Normally, the electric field is mapped away from the dipole source along one of the principal directions. When the component of electric field parallel to the source axis is mapped outward along the equatorial axis of the source, this sounding is termed an equatorial dipole sounding. For a horizontally layered earth this procedure produces the same apparent resistivity curve as a function of separation as does the Schlumberger array, though this similarity does not hold for any other earth geometry. When the electric field is measured at locations along the axis of the source, the sounding is termed a polar dipole sounding, and the apparent resistivities behave quite differently from those measured with a Schlumberger or an equatorial dipole array. Other locations of the receiving electrodes may be used also, and the electric field may even be mapped in detail in the plane about the source dipole [see Alpin, 1966; Keller, et al., 1975].

Both the Schlumberger and the dipole arrays have advantages and disadvantages relative to one another when they are used for crustal scale resistivity soundings, and so the choice of one or the other is not always clear-cut. The principal disadvantage of the Schlumberger array is that the span between current electrodes must be increased to hundreds of kilometers to provide information about the lower crust and upper mantle, even under favorable conditions. When this can be done, considerable information can be obtained on the resistivity profile through the crust, as may be seen in the paper by van Zijl [1977].

In order to provide a detectable signal at the receiving

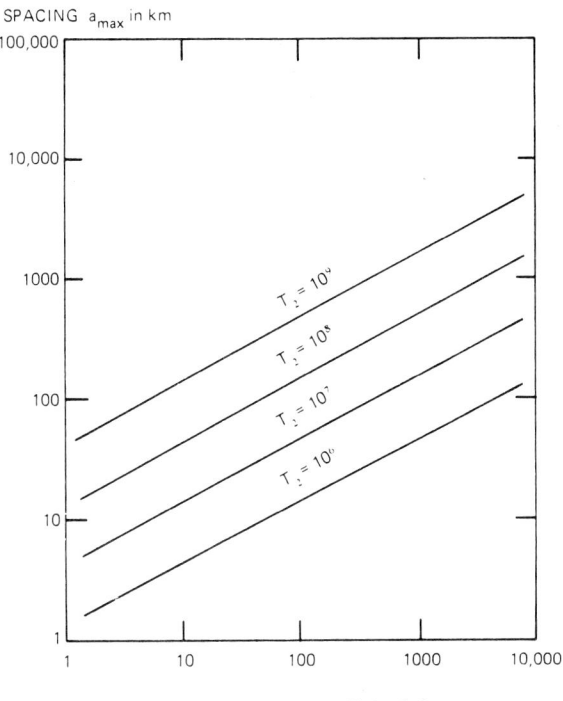

Fig. 3. Chart for determining the electrode spacing at which the maximum apparent resistivity will be observed with a Schlumberger or equatorial dipole array for the resistivity profile shown in Figure 1. S_1 is the ratio h_1/ρ_1 for the surface layer; T_2 is the product $h_2\rho_2$ for the second layer.

electrodes at such large spacings the voltages applied to the current line must be of the order of hundreds to several thousands of volts. Extensive precautions must be taken so that such a length of current-carrying wire does not comprise a hazard to life. One satisfactory solution to this problem is the use of out-of-service power lines, which are already protected. In recent years, considerable work has been done during grounding tests of high-voltage direct current transmission lines. A limitation of such an approach is the fact that soundings must be made at locations where there are available power lines rather than at locations which may be the most interesting from the geological point of view.

Advocates of the use of dipole arrays for crustal scale resistivity soundings feel that the operational ease of the method is the chief advantage over the Schlumberger array. With a dipole source length of 1-10 km, it is usually possible to make soundings to depths of tens of kilometers even in densely inhabited areas with judicious choice of the source location. Very large currents are required to provide measurable signals at the maximum spacings.

In the crustal model (Figure 1), because the surface layer is much more highly conductive than the crystalline basement, apparent resistivity values observed at relatively short spacings begin rising with a slope of nearly +1. At somewhat larger spacings the apparent resistivity reaches a maximum and then decreases rapidly at larger spacings. For the Schlumberger and equatorial dipole arrays the spacing at which the maximum occurs is [Keller, 1968]

$$a_{max} = (S_1 T_2/2)^{1/2} \qquad (3)$$

where S_1 is the longitudinal conductance of the surface layers and T_2 is the product $\rho_2 h_2$, termed the transverse resistance of the second layer.

It is readily apparent that the spacing to which one must go to establish the maximum on the apparent resistivity curve for a crustal scale sounding increases both with the conductance of the surface layers and with the transverse resistance of the crust. The results of such soundings which have previously been reported in the literature provide values for T_2 ranging from 10^6 to 10^9 ohm m^2. The conductance of surface rocks may range from a few mhos where the rocks consist only of weathered crystalline basement to some thousands of mhos in deep sedimentary basins. The spacings required to measure maximum apparent resistivity are shown graphically in Figure 3 as a function of S_1 and T_2.

The value of apparent resistivity observed at the maximum depends on S_1 and T_2 and is essentially independent of the resistivity in the second layer. The equation may be rewritten as

$$\rho_{a,max} = (2T_2/S_1)^{1/2} \qquad (4)$$

The maximum value for apparent resistivity in a crustal scale sounding is shown graphically in Figure 4 as a function of S_1 and T_2.

Ideally, for the simplified three-layer model of the crust it is possible to determine the resistivity of the middle layer by making measurements both with a dc sounding system, which detects the product $T_2 = \rho_2 h_2$, and an ac sounding system, such as the magnetotelluric method, which detects only h_2. It is to be expected that less than perfect results will be obtained in practice because the earth conductivity profile is not as simple as that shown in Figure #1, and because lateral changes in the properties of any of the layers, especially the surface layer, will complicate the patterns of apparent resistivity measured in the field. In order to evaluate the merits of a combined ac-dc sounding program for studying the properties of the resistant part of the crust the Office of Naval Research sponsored a study in north central Wisconsin carried out jointly by investigators from the University of Texas (UTA), and University of Wisconsin (UW), and the Colorado School of Mines (CSM). This paper will describe the dc resistivity studies

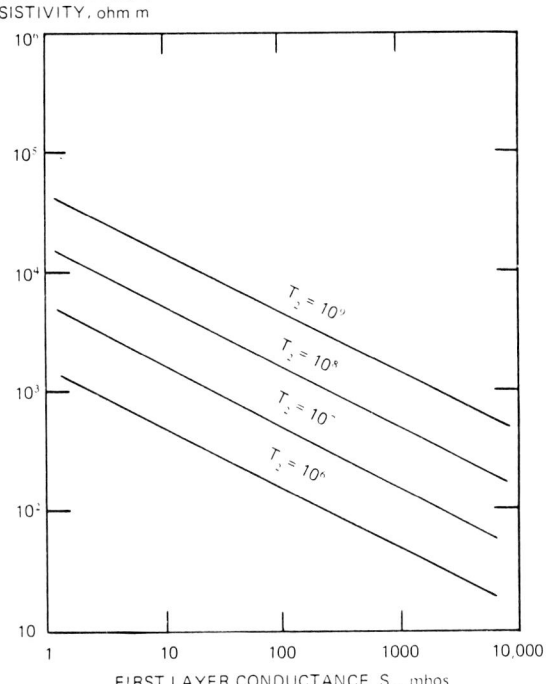

Fig. 4. Chart for determining the maximum apparent resistivity that will be measured with the Schlumberger or equatorial dipole arrays for the resistivity profile shown in Figure 1. S_1 is the ratio h_1/ρ_1 for the surface layer; T_2 is the product $h_2\rho_2$ for the second layer.

carried out by CSM; the results of other studies are described by Bostick, et al. [1977] (UTA) and by Sternberg and Clay [1977] (UW).

Rotating Dipole Method

A serious limitation to the use of dipole surveys for crustal scale resistivity soundings is the high degree of sensitivity of such arrays to errors caused by lateral changes in resistivity of near-surface rocks. Surface inhomogeneities in resistivity cause effects which are often indistinguishable from the effects of layering at depth, so that simple dipole sounding surveys provide highly ambiguous results. These ambiguities can be minimized by carrying out highly redundant dipole surveys with a technique which was called "dipole mapping method" Keller et al., [1975]. A dipole mapping survey is conducted by making measurements of electric field intensity over the entire surface area reached by measurable currents from the source dipole rather than only along specific traverses. In this way, areas of anomalously high or low surface resistivity can be recognized from the patterns that they evoke on a contour map of values of apparent resistivity, and measurements that are relatively little affected by surface effects can be selected for interpretation as resistivity sounding curves. An extensive set of dipole resistivity maps obtained by computer simulation has been prepared to show how dipole resistivity maps reflect surface inhomogeneities of various sorts [Furgerson and Keller, 1974]. Interpretation could be done using two-dimensional fitting of the data. Such an interpretation procedure will help in reducing the nonuniqueness of response discussed earlier.

Examination of these computer-generated simulations indicates not only what the extent of the anomalies associated with surface irregularities in resistivity is but also that the extent of an anomaly is strongly dependent on the mutual geometric relationship of the electrode array and the position of features with anomalous resistivity. In other words, the effect of a surface irregularity in resistivity can be maximized or minimized by an appropriate selection of the orientation of the source and receiver dipoles.

One approach to evaluate the maximum and minimum effects from surface irregularities is to measure the electric field using several sources lying in various directions from the observation point. This type of redundancy is very expensive in terms of field effort and is ineffective unless the approximate locations of the surface irregularities are known ahead of time so that appropriate source locations can be selected.

An approach to obtaining multiple coverage from dipole sources which appears to be more practical is the use of a single rotating dipole source. Measurements are made at a receiver location as a function of the orientation of the source dipole as the source dipole is swung through a 360^0 rotation. As the source rotates, the direction of current flow at the receiver site will rotate through all possible directions, and apparent resistivity values which are maximally and minimally affected by surface irregularities will be measured.

It is not necessary to rotate the dipole source physically; if it were, the complexity of the surveying technique would be too great. However, the source dipole can be rotated in effect by using only two dipoles at the source, oriented in directions which preferably are orthogonal to one another. These two source dipoles are then powered simultaneously with current intensities, to provide a resultant dipole moment in the desired direction. Thus by changing the ratio of currents to the two source dipoles the resultant moment can be rotated through a full circle.

A still more practical field procedure consists of making only two sets of electric field measurements at a receiver site, one for each of the two orientations of the source dipole. Then, the two electric field vectors can be added in the proper proportions to find the electric field that would have been measured with any orientation for the source dipole.

Apparent resistivities for a given source location-receiver location combination may be computed in a variety of ways. For example, one might use only the component of the electric field parallel to the source dipole to compute a so-called parallel

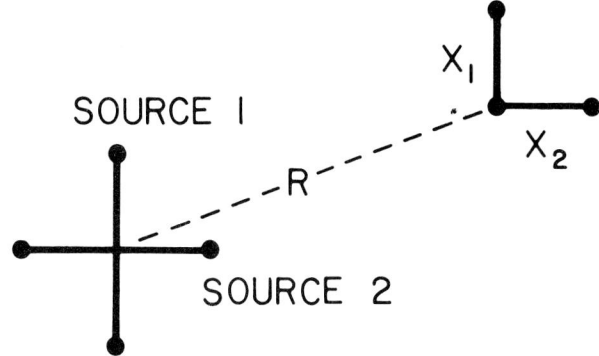

Fig. 5. Layout used for a rotating dipole survey.

field apparent resistivity, or only the component of the electric field in a direction perpendicular to the axis of the source dipole, or only the component directed radially away from the source dipole (all of these ways of computing apparent resistivity have been described by Alpin [1966]). On the basis of our experience with the dipole mapping method up to the present time we have elected to use the 'total field apparent resistivity' method of expressing resistivities measured in Wisconsin with the rotating dipole technique. This definition of apparent resistivity is based on the use of the magnitude of the electric field vector at the receiver site, regardless of the direction in which it lies. The expression for computing apparent resistivity using the magnitude of the electric field is [Keller, et al, 1975]:

$$\rho_T = \frac{2\pi R^3}{[(3\cos\theta\sin\theta)^2 + (3\sin^2\theta - 1)^2]^{1/2}} \frac{E_T}{M} \quad (5)$$

where R is the distance between source and receiver, θ is the angle between the resultant axis of the dipole source and the radius vector from the source to the receiver, M is the moment of the source, defined as the product of current and source dipole length, and E_T is the magnitude of the resultant electric field at the receiver site. The geometry described by this equation is shown in Figure 5.

Equation (5) is strictly applicable only if the separation between current electrodes, as well as the separation between receiver electrodes, is vanishingly small. This requirement is met within reasonable limits of accuracy when the distance R is more than 5 times as great as the length of either the source dipole or the receiver dipole. This condition poses no particular limitation in crustal scale surveys, where dipole lengths are rarely greater than 1-2 km, while dipole separations up to 100 km may be used.

The layout of two dipole arrays for use in a rotating dipole survey and the way in which rotation is accomplished are indicated in Figure 5. In this illustration the two sources are taken to have real moments M_1 and M_2, which are the products of current intensity and length for the two sources. Two total field vectors ET1 and ET2 are measured, since each of the two sources is powered individually. Then, several directions α are selected as normals to the direction of a composite source. The moment of one of the sources, say M_2, is scaled by some factor β so that when the scaled moment βM_2 is added to the actual moment M_1, the components of the two source moments in the normal direction α exactly cancel. The resultant moment is then in the desired direction. The total electric field that would be measured with this combined source is determined by adding ET1 to the scaled electric field βET2. The resistivity is then computed by using (5) as though a single-source dipole measurement had actually been made with the combined source.

Rotating dipole survey data may be presented in many formats. For example, the direction α may be selected so that the measurement made at any receiver station is along a principal axis of the combined source; that is, all points on a dipole map can be computed for an exactly equivalent polar or equatorial dipole array. Alternatively, apparent resistivity values are computed for a series of directions covering an angle of $180°$ and the maximum and minimum values are selected to represent the possible range of values that can be measured at a given receiving point. We have not yet carried out a study extensive enough to determine the best format for presenting rotating dipole resistivity data, but a few calculations which have been done are useful in describing the behavior of such data.

Figures 6-13 show several ways of presenting rotating dipole data for computed values of apparent resistivity for a simple model, that of a vertical plane separating two regions with true resistivities that differ by a factor of 10. In one case the crossed dipole source is located in the more resistive medium (Figures 6-9), and in the other case the crossed dipole source is located in the more conductive medium (Figures 10-13). Apparent resistivity values can be presented as though each of the dipole sources in a crossed pair were used to make an independent dipole mapping survey. The four single-source dipole maps are shown in Figures 6, 7, 10, and 11. The major shortcoming of single-source dipole coverage is apparent on these four maps; the apparent resistivity values observed close to the boundary on the side toward the source show a complicated pattern. Some values of apparent resistivity are greater than any real resistivity in the model, while others are lower than any real value.

In our experience to date, the simplest contour maps seem to be those in which the average of the maximum and minimum values of apparent resistivity obtained on rotation is plotted (Figures 8 and 12). On these maps the resistivity measured on the side of the boundary toward the source varies only slightly from the correct value. The apparent resistivity measured on the far side of the boundary is constant but not equal to the actual resistivity.

Another parameter which can be used to describe the results

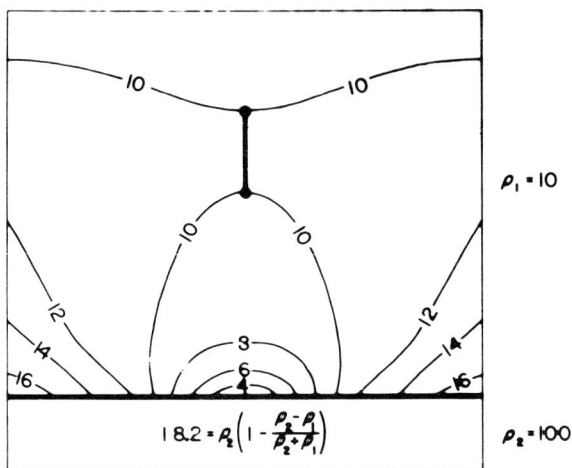

Fig. 6. Single-source bipole-dipole resistivity map for the case of a vertical faultlike boundary separating regions with true resistivities differing by a factor of 10. The source is normal to the boundary and located in the more resistive medium.

of a rotating dipole survey is the ellipticity of the pattern of apparent resistivity values obtained on rotation. This is defined as the square root of the ratio of the maximum and minimum values of apparent resistivity for a single receiver station. Contour maps of the values for ellipticity (Figures 9 and 13) show

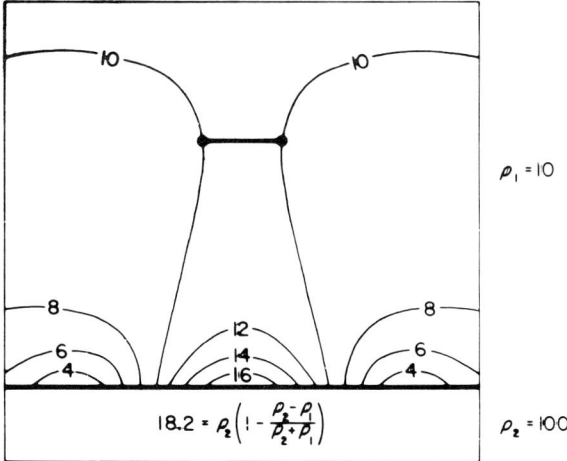

Fig. 7. Single-source bipole-dipole resistivity map for the case shown in Figure 6 except that here the source is oriented parallel to the boundary. Note that rotation of the source has markedly changed the patterns of apparent resistivity along the boundary.

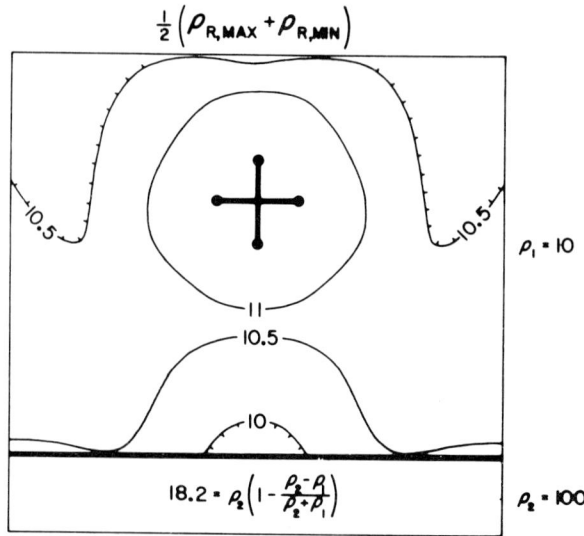

Fig. 8. Average of maximum and minimum values of apparent resistivity obtained by source rotation for the data shown in Figures 6 and 7. Note that the complexity of patterns in apparent resistivity along the fault is much reduced.

that the maximum ellipticity is measured along the boundary between the two regions with different resistivity and closely defines the boundary. It should prove to be useful in locating such boundaries.

Field Surveys

During the summer of 1973 and 1974, rotating dipole surveys were carried out in north central Wisconsin in the areas outlined in Figure 14. Much of the area where the surveys were carried out is covered by glacial till with an average thickness of 15 m, but it ranges in thickness from 0 to 100 m. Because of the glacial cover the geology of the bedrock is poorly known, but the bedrock is believed to be predominantly granitic. According to Weidman [1907] the suboutcrop should be 85% granite and syenite, 5% greenstone, diorite, and gabbro, 2% slate and graywacke, 2% quartzite and conglomerate, and 2% gneiss and schist. On the basis of this information one would expect central Wisconsin to be an area in which high resistivities would be present in the shallow crust.

In the field surveys, source dipoles with a length of approximately 1600 m were used. Metal culverts were used to obtain relatively low ground contact resistances. At locations where three culverts could be found to form an L shaped combination of source dipoles the two sources were set up and energized sequentially. That is, the power source was first connected to

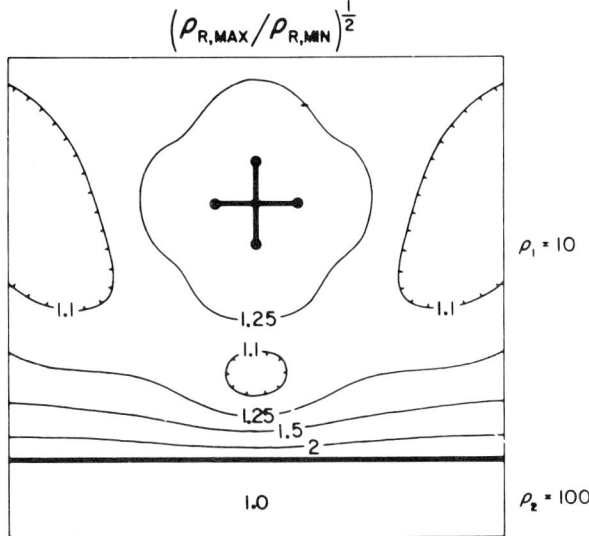

Fig. 9. Ellipticity (square root of the ratio of maximum to minimum resistivity) of rotated dipole resistivities for the case used in Figures 6-8.

one dipole, and electric field observations were made at all the receiver sites where measurements could be made. Then, those stations were reoccupied as the second source was energized.

The primary energy was a gasoline engine motor-generator set

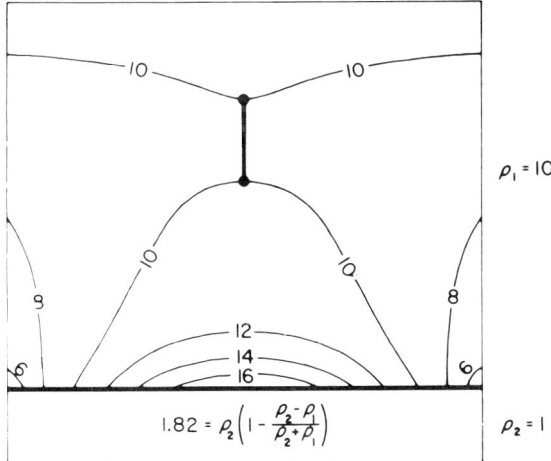

Fig. 10. Single-source bipole-dipole resistivity map for the case of a vertical faultlike boundary separating regions with true resistivities differing by a factor of 10. In contrast to the case shown in Figure 6 the source is located in the more conductive medium.

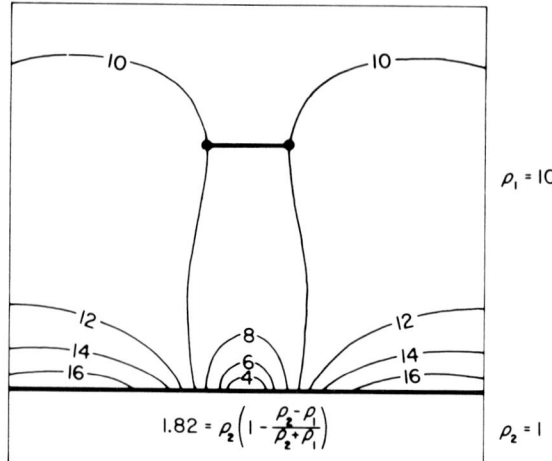

Fig. 11. Single-source bipole-dipole resistivity map for the case shown in Figure 10 except that here the source is oriented parallel to the boundary.

with a capacity of 25 kVA at 235 V 60 Hz. The output of the generator set was stepped up to 1000 V using a variable transformer and then rectified and switched to supply alternating positive and negative dc steps of current to the cables connecting the power supply to the electrodes. The switching was done in such a way as to form an asymmetric wave form; that is, the length

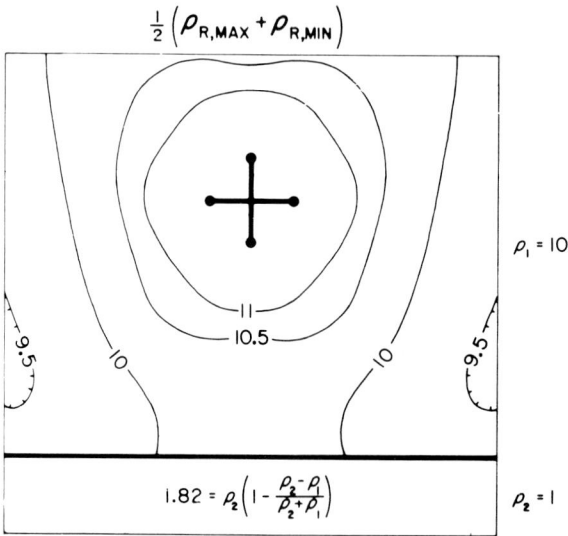

Fig. 12. Average of the maximum and minimum values of apparent resistivity obtained on dipole rotation for the data shown in Figures 10 and 11.

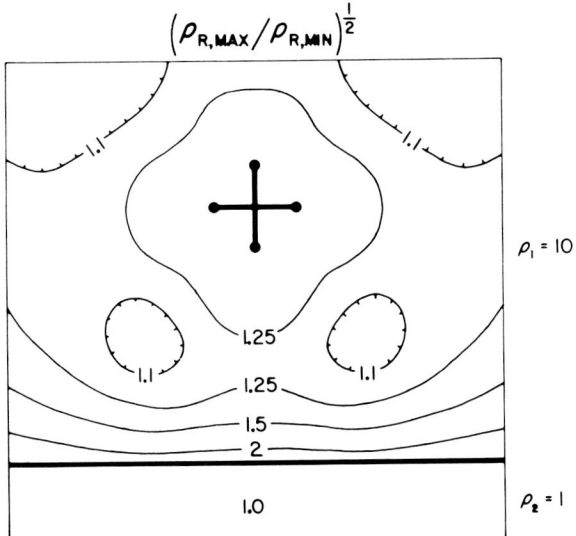

Fig. 13. Ellipticity of rotated dipole resistivities for the case used in Figures 10-12.

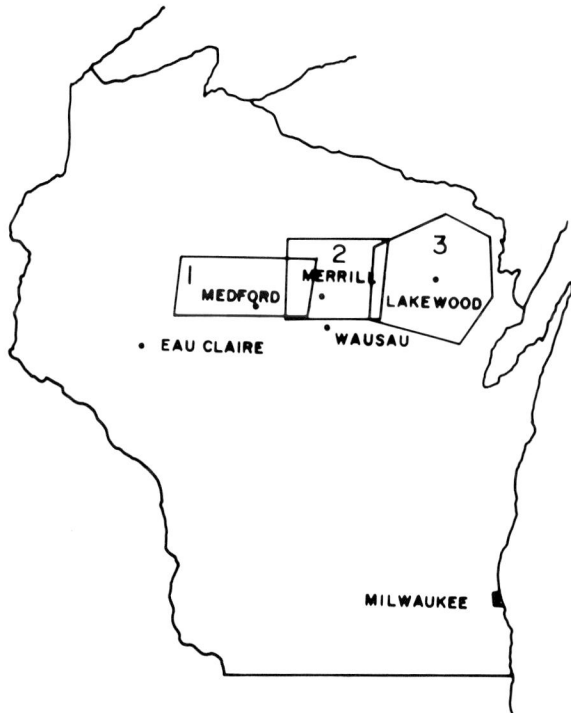

Fig. 14. Locations of three rotating dipole surveys carried out in Wisconsin.

of a current step in one direction was about twice the length in
the opposite direction. This made it possible to specify the
polarity of the signals recorded at a receiver site. The
periodicity of the current transmissions was once per 20 s. The
amplitude of the current steps ranged from a low of about 10 A to
a high of about 60 A, depending on the contact resistance for a
particular source.

At a receiver site, voltages were measured with two orthogonal
electrode pairs in order to determine two components of the total
electric field vector. The separation between receiver electrodes
in each pair was 30 m. The voltage from a pair of receiver
electrodes was filtered, amplified, and recorded on an analog
recorder.

Three pairs of dipole sources were used to provide coverage
over most of the crystalline suboutcrop area of north central
Wisconsin. Some overlapping coverage was provided between
adjacent sources, as shown on the maps locating the position of
the individual source dipoles and receiving stations (Figure 15).

The use of a pair of orthogonal dipole sources combined with
the detection of both magnitude and direction of the electric
field vector at a receiver station provides a highly redundant
measurement scheme; that is, more quantitites are measured than
are needed to compute a single value of apparent resistivity.
As a consequence, apparent values of resistivity can be computed
in many different ways. Unless the earth is completely uniform,
these various computed values are likely to vary widely. The
field data obtained from the three pairs of dipole sources used
in Wisconsin were reduced by using the rotating source concept
described earlier. The two electric field vectors observed at
each station (one for each direction of the source dipole) were
added together in various proportions to provide single resultant
vectors pointing in a variety of directions. Apparent resistivity
values were computed by using the total field expression (equation
(5)), which discards information on the direction of the E field
and requires knowledge only of the magnitude. The resulting group
of apparent resistivity values, one for each direction of a
composite E field vector, when plotted as a function of the
direction of the E field, forms an ellipse. These ellipses
may then be characterized by a relatively small number of
parameters. The ones used here in presenting the data visually
are the average resistivity (formed by averaging the maximum
and minimum radii of the ellipse of resistivity), the direction
of the major diameter of the ellipse, and the ellipticity
(arbitrarily defined as the square root of the ratio of the
lengths of the major and minor axes of the ellipse).

Contour maps of the average resistivity obtained on rotation
are shown in Figure 16 for each of the three surveys. It should
be noted that lateral changes in resistivity dominate the
contour patterns. If the area were laterally uniform with resistivity varying only as a function of depth, apparent resistivities
would contour more or less concentrically about the source. Here,
no strong concentric pattern is easily recognized, and in fact,
measurements made from source 1 (top) indicate a faultlike boundary

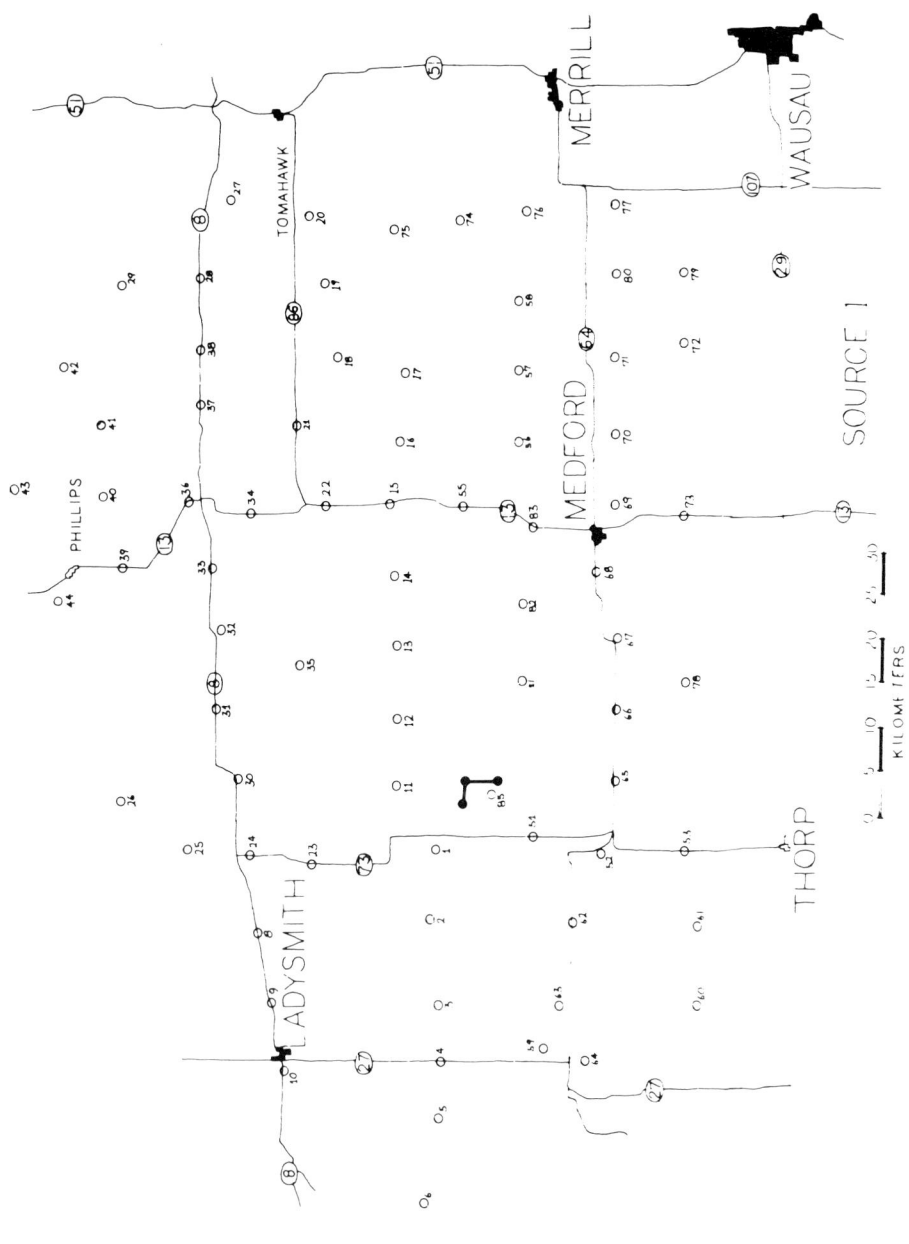

Fig. 15. Locations of dipole sources and receiver stations for three rotating dipole surveys carried out in central Wisconsin.

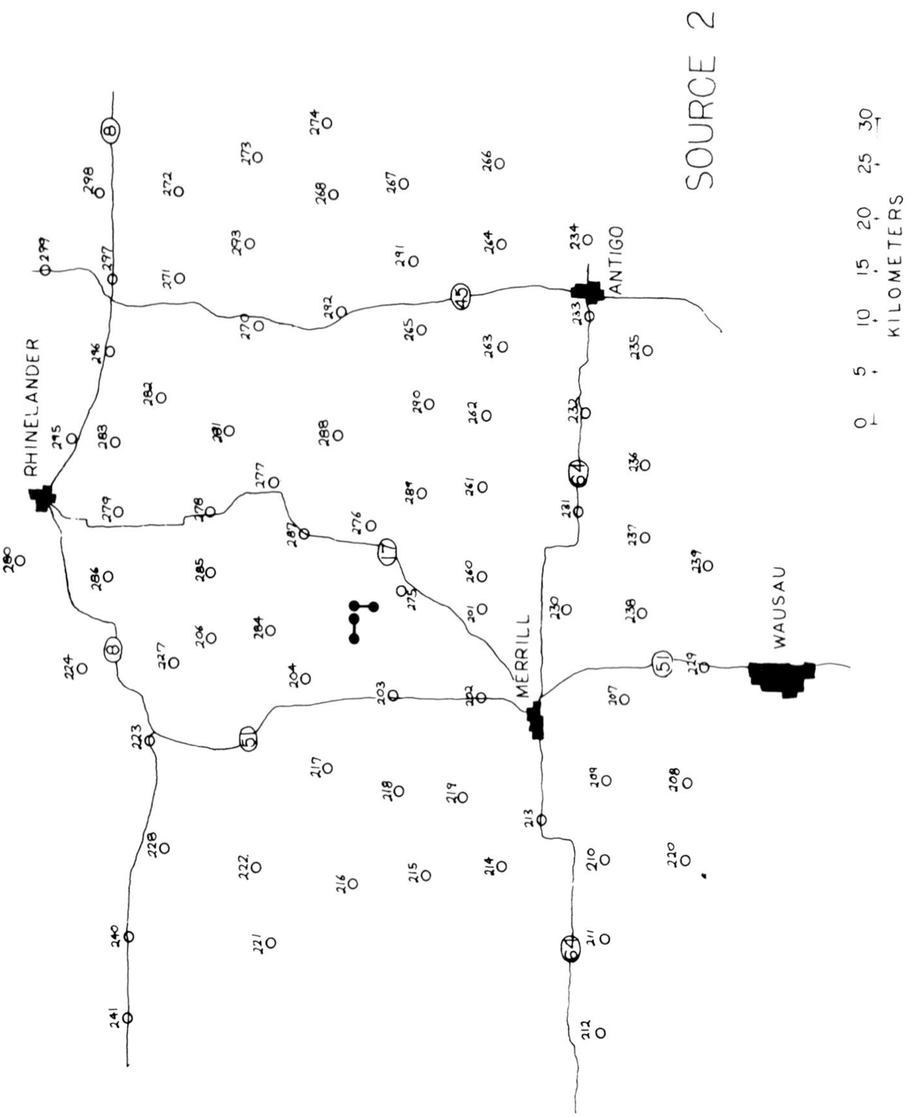

Fig. 15. (continued)

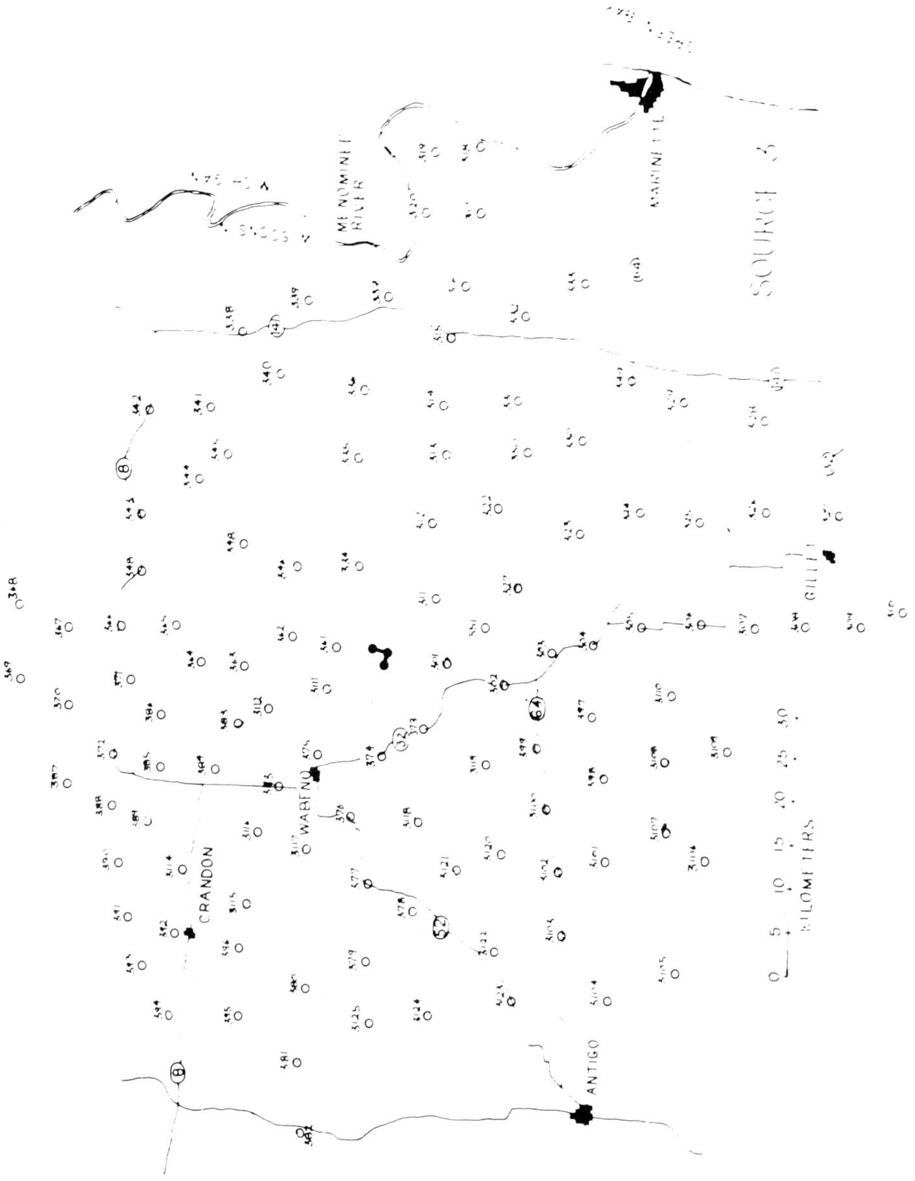

Fig. 15. (continued)

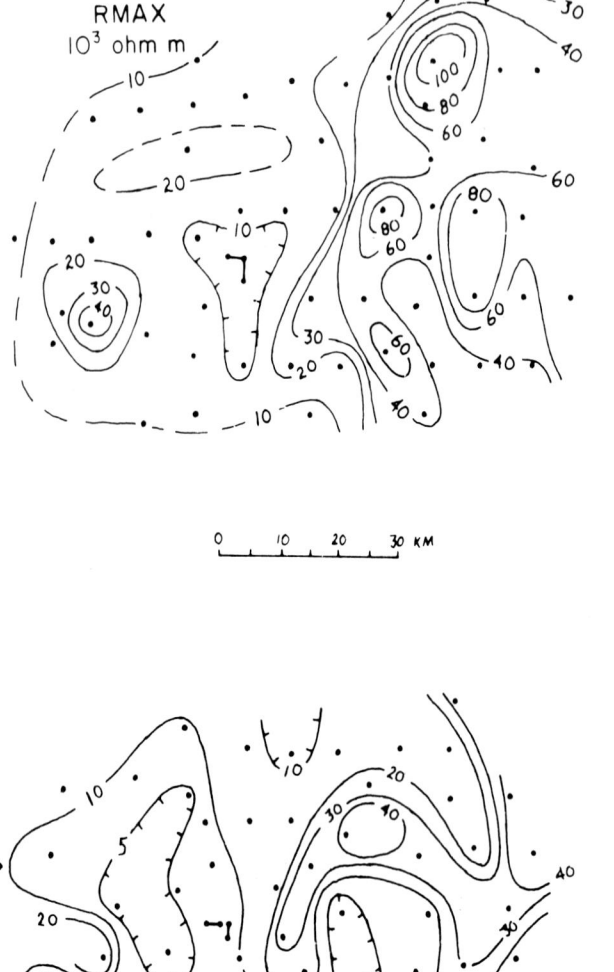

Fig. 16. Contour maps of the maximum resistivity obtained on rotation of each of the three dipole sources.

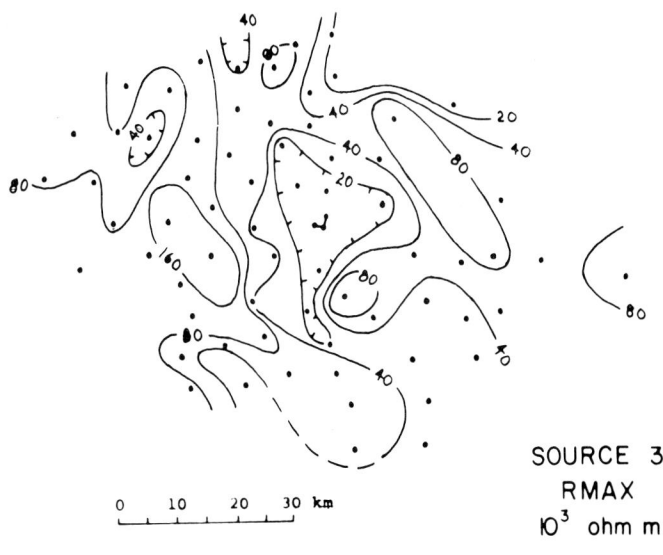

Fig. 16. (continued)

between areas of relatively low and relatively high resistivity about 20 km east of the source location.

Maps giving contours of the value of ellipticity and the directions of the major axes of the resistivity ellipses are shown in Figure 17. Local areas of high ellipticity are associated with lateral changes in apparent resistivity.

Our major interest is in the information that these data can provide about the resistivity at depth. The data in Figures 16 and 17 indicate that considerable caution must be used in attempting to determine resistivities at depth in view of the strong lateral effects that are present. One approach in extracting information about resistivities at depth is to plot apparent resistivity values as a function of distance from the source on the assumption that measurements made at greater distance reflect resistivities at greater depths in the earth. Such plots are shown in Figure 18 for the three dipole surveys carried out in Wisconsin.

Although the data plotted in Figure 18 are widely scattered, there is a tendency for apparent resistivity to increase with distance, as would be the case if actual resistivity were to increase with depth. A similar plot for synthetic data, obtained by computing apparent resistivities for a grid of observation points for a two-layer model in which the substratum is 10 times as resistive as the overburden, is shown in Figure 19 for comparison. Some scatter is present in this synthetic plot, but the scatter is far less than that shown by the real data in Figure 18.

Interpretation

As a step in interpretation it is tempting to average these highly redundant data in an effort to mask the scatter and

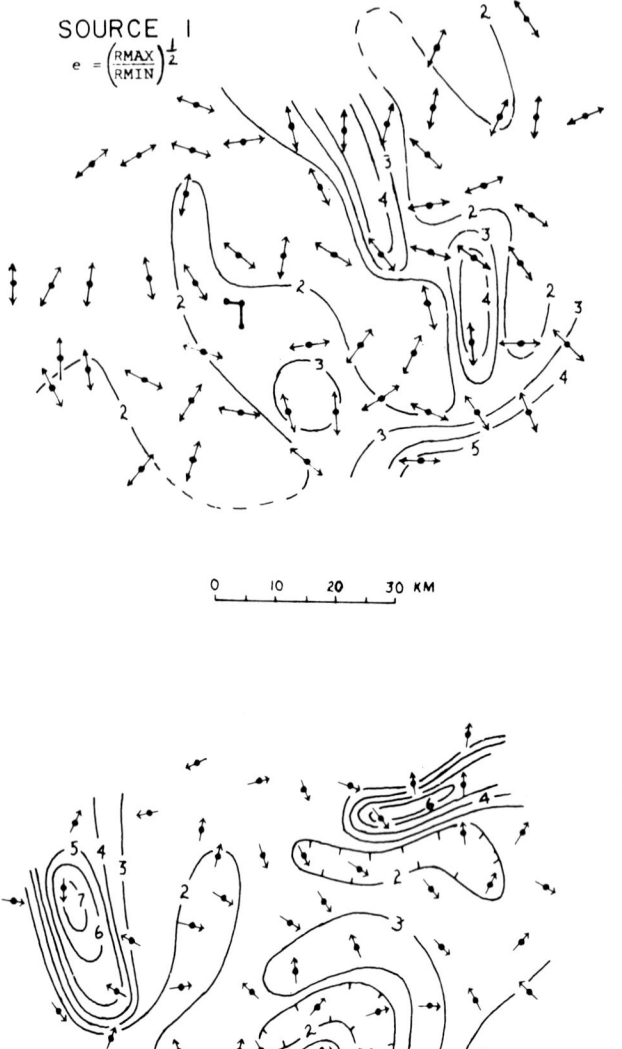

Fig. 17. Directions and magnitudes of the ellipticities in apparent resistivity obtained in rotation of each of the three dipole sources.

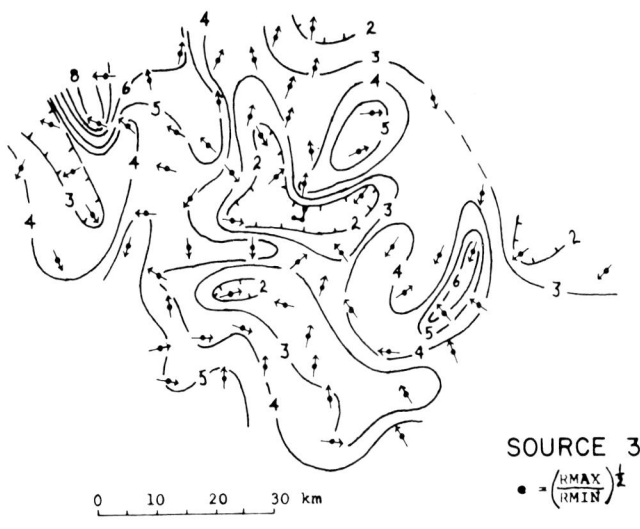

Fig. 17. (continued)

effects of lateral inhomogeneities. Such averaging requires the assumption that effects caused by lateral inhomogeneities will be randomly distributed. In fact, this is not the case; apparent resistivity values are lowered more by a conductive inhomogeneity than they are raised by a resistive inhomogeneity. As a consequence, any average of a large number of observations will be consistently lower than the resistivity which would be observed in the absence of the inhomogeneities. If we recognize that this biasing of averages is present, we can still use the averages to characterize the gross behavior of apparent resistivity with distance from the source, as is done in Figure 20. Here, all values of apparent resistivity that we measured (each value being the average of the major and minor axes of a resistivity ellipse) were grouped in three classes: values determined at distances of 10-20 km, 20-40 km, and greater than 40 km. Within each group the geometric mean distance and the geometric mean resistivity were computed. These values are shown as three circles in the upper right corner of Figure 20. The vertical bars shown for the shorter distance are taken from a report by B. K. Sternberg (private communication, 1975). The bars summarize resistivity surveys done in this area by a number of earlier investigators using the Schlumberger and dipole sounding techniques, the width of each bar indicating the principal spread of data for eight soundings. The curve was computed by Sternberg for an earth model in which the resistivity becomes infinite at a depth of 7 km. This 'basement' is overlain by two layers, one extending to 10-m depth with a resistivity of 250 ohm m and the other extending from there to 7-km depth with a resistivity of 4000 ohm m. The fit is good, but a wide range of models might have been selected which also would have provided a good match to the data.

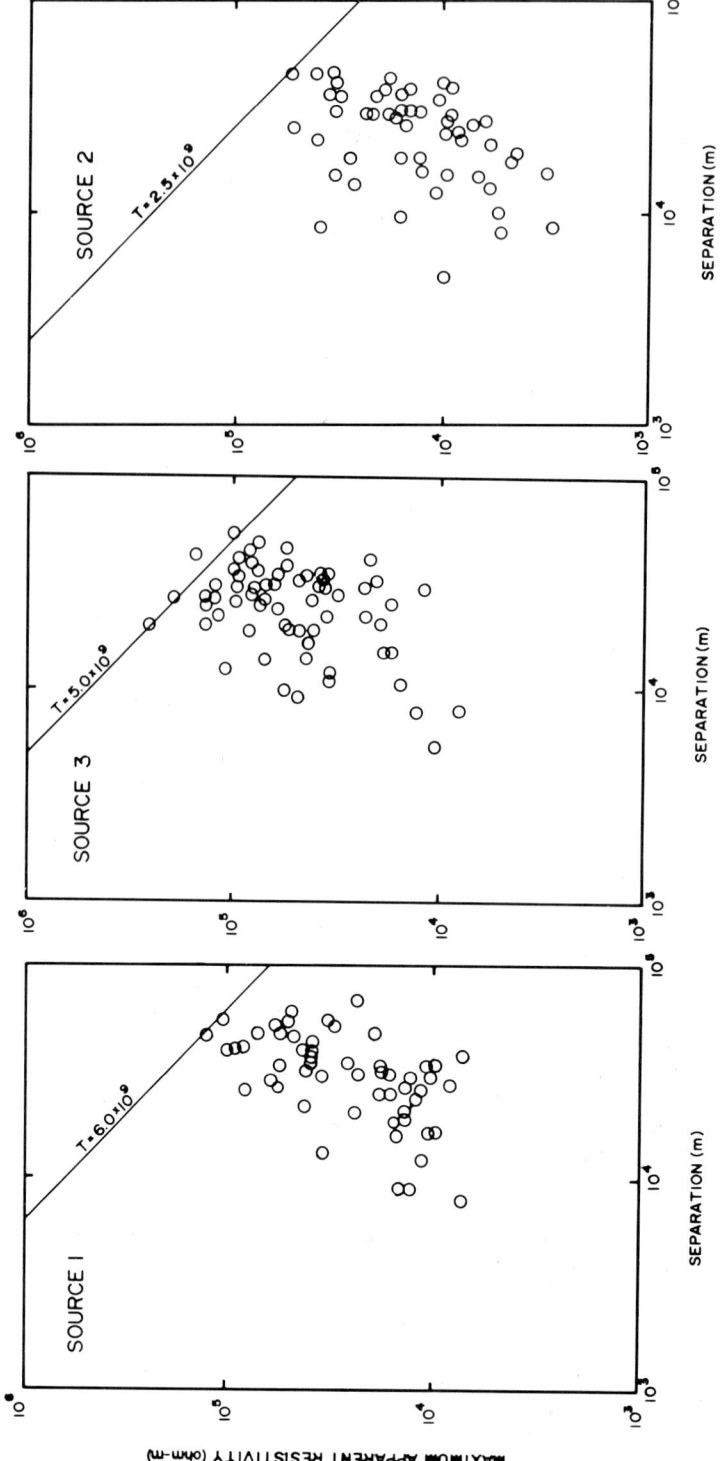

Fig. 18. Plots of apparent resistivity as a function of distance from the source for each of the three dipoles sources.

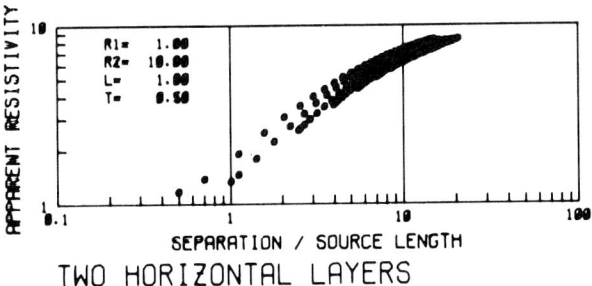

Fig. 19. Plot similar to that in Figure 18 for values of apparent resistivity computed for a simple model of a uniform surface layer resting on a substratum which is 10 times more resistive than the layer.

If observations could have been carried to greater distances, we would expect to see the data shown in Figure 20 pass through a maximum. The product of apparent resistivity and spacing at that maximum would be very nearly equal to the transverse resistance (the product of resistivity and thickness) for the resistive portion of the crust. Because the data do not appear to reach a maximum, we can say only that the last point specifies a minimum value for that maximum, which lies between 5 and 6 x 10^9 ohm m^2.

Magnetotelluric soundings carried out in this same area by Bostick, et al. [1977] provide an estimate of 15-20 km for the thickness of this resistant zone. For this thickness the resistivity would have to average 250,000-400,000 ohm m over the entire interval to provide a transverse resistance of 5-6 x 10^9 ohm m^2.

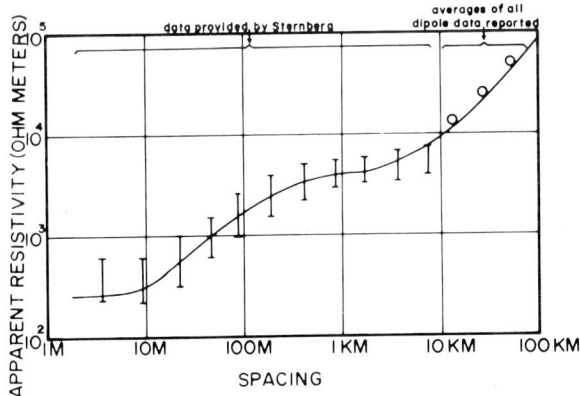

Fig. 20. Averaged resistivities plotted as a function of spacing or distance from the source. Measurements at less than 10 km were made by previous investigators and summarized by B. Sternberg (private communication, 1975).

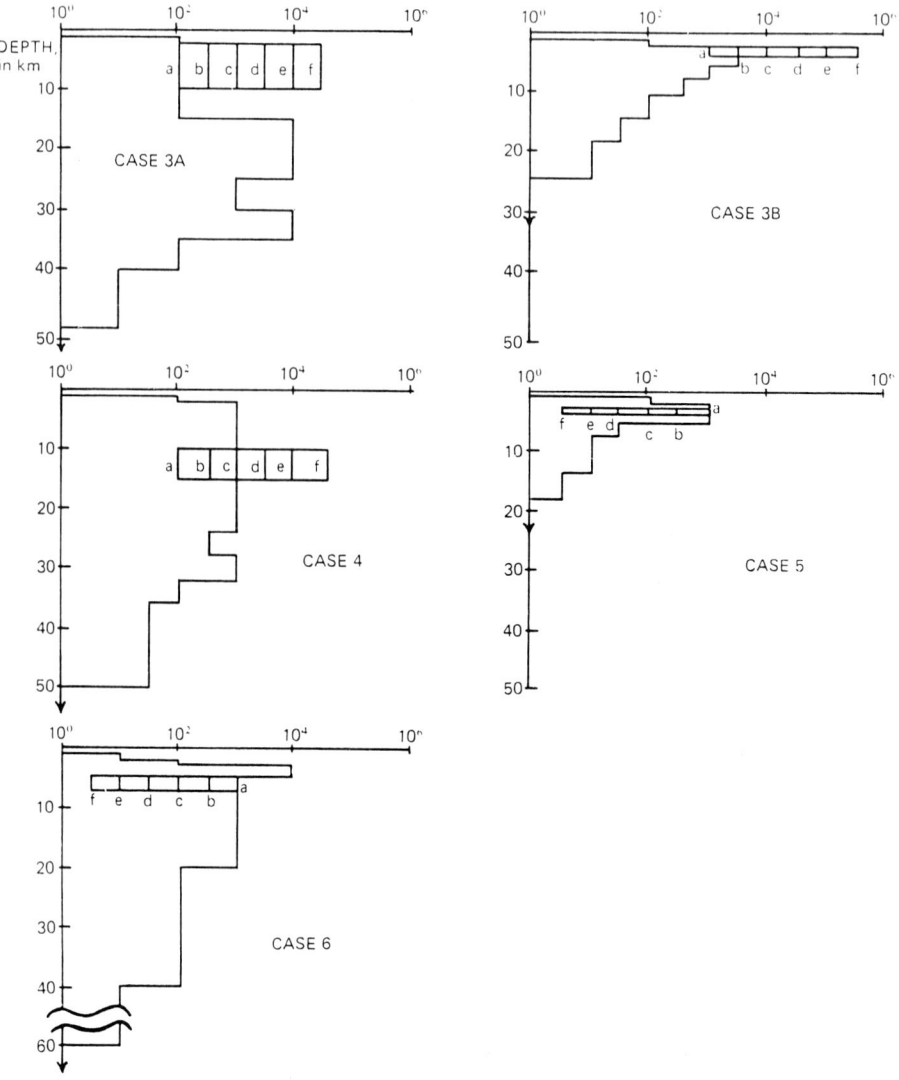

Fig. 21. Crust/mantle resistivity profiles used in computer study.

In fact, it is highly unlikely that the resistivity is uniform over such a depth range in the earth. The effects of elevated temperature will cause a gradational decrease in resistivity from its maximum value in the crust as one goes deeper in the crust or mantle. The upper boundary of the conductive regions will be sensed at shallower depths with a dc sounding method than with a magnetotelluric sounding method. Therefore the straightforward calculation of average resistivity in the resistant part of the crust is not meaningful.

The problem can readily be demonstrated in more analytical terms. First, consider the sample case of an earth model in

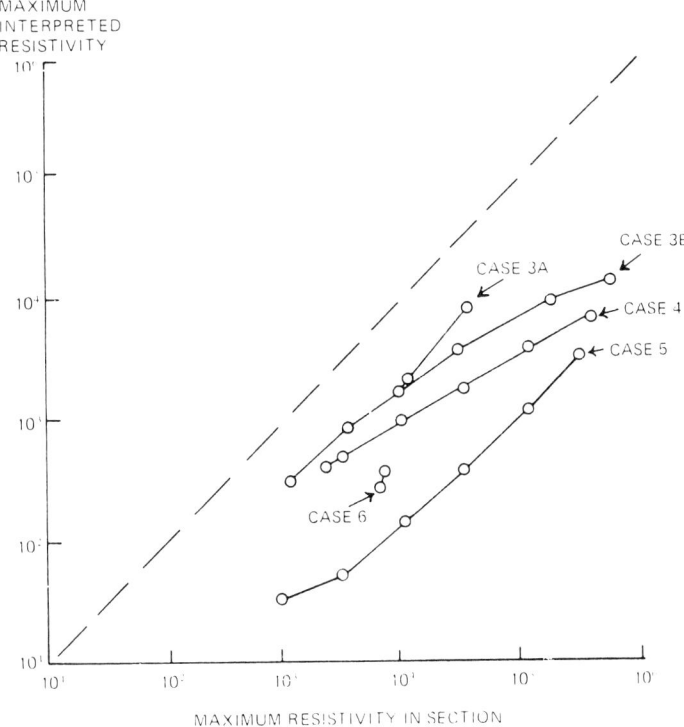

Fig. 22. Apparent resistivity for the resistant portion of the crust and mantle obtained from three- or four-layer interpretations of magnetotelluric and dc resistivity sounding curves computed for the models in Figure 21.

which the actual resistivity rises to a very large value in the upper part of the crust and then decreases gradually over a long interval extending into the mantle. The transverse resistance derived from dc soundings might pertain to only a few kilometers of this zone, while the total thickness derived from magnetotelluric data would be much greater, perhaps several hundred kilometers. To investigate this phenomenon, synthetic sounding curves were computed for a number of hypothetical crustal models (the model profiles are shown in Figure 21). These synthetic soundings, both dc and magnetotelluric, were then interpreted by a conventional least squares inversion technique with the constraint that the synthetic soundings be matched with a three- or four-layer earth model. There was no significant difference between the closeness of fit for the three-layer and four-layer models. This is characteristic of true equivalence between interpretations and indicates that for these models a three-layer interpretation is the best that can be done in practice. The rms error in fit ranged from a low value of about 0.33% (case 4) to a high value of 1.82% (case 5). These errors in fit are less than the error which is normally obtainable in fitting real data, which contain some experimental error. Normal errors in fit obtained after least squares inversion are typically 2-3% for good dc resistivity data and 3-5% for good magnetotelluric data.

As expected, it was found that the average resistivity for the resistant zone in these three-layer interpreted models, using the combination of dc soundings to find T_2 and magnetotelluric soundings to find h_2, was much less than the maximum resistivity assumed in the initial profile. This phenomenon is shown graphically in Figure 22, which demonstrates the relationship between

$$\rho_2 = \frac{T_2 \text{(dc sounding)}}{h_2 \text{(MT sounding)}}$$

and the maximum resistivity in the initial profile. The interpretations for each set of models are connected by lines, so that the effect of the one variable parameter in each model can be seen.

In Figure 22 the maximum resistivities interpreted with a three-layer model fall below the actual maximum resistivity by factors ranging from 5 to 100, the median factor being in the neighborhood of 10. If these results are representative, we can conclude that for the dc and MT surveys carried out in Wisconsin the maximum resistivity in the crust should be 10 times as great as the value derived from a straightforward combination of the interpretations of the dc and MT surveys. The maximum resistivity in the crust over some short interval of several kilometers is probably 2.5-4 million ohm m.

References

Alpin, L. M., The theory of dipole sounding, in Dipole Methods for Measuring Earth Conductivity, 302 pp., Consultants Bureau, New York, 1966.

Bostick, F. X., H. W. Smith, and J. E. Boehl, Magnetotelluric DC Dipole-Dipole Soundings in Northern Wisconsin, technical report, Office of Nav. Res., Nat. Sci. Found., Washington, D. C., 1977

Brace, W. F., Resistivity of saturated crustal rocks to 40 km based on laboratory studies, in The Structure and Physical Properties of the Earth's Crust, Geophys. Monogr. Ser., vol. 14, edited by J. G. Heacock, pp. 243-256, AGU, Washington, D. C., 1971.

Brace, W. F., A. S. Orange, Electrical Resistivity Changes in Saturated Rocks During Fracture and Frictional Sliding, Jour. Geophys. Research, V. 73, No. 4, p. 1433-1445, 1968.

Cagniard, L., Basic theory of the magnetotelluric method of geophysical prospecting, Geophysics, 18, 605-635, 1953.

Furgerson, R. B., and G. V. Keller, Computed dipole resistivity effects for an earth model with vertical and lateral contrasts in resistivity, technical report, 194 pp., Office of Nav. Res., Washington, D. C., March 12, 1974.

Keller, G. V., Electrical prospecting for oil, Quart. Colo. Sch. of Mines, 63 (2), 1968.

Keller, G. V., Electrical properties of the earth's crust - A survey of the literature, ONR contract N00014-70-C-0290, Colo. Sch. of Mines, Golden, 1971.

Keller, G. V., R. B. Furgerson, C. Y. Lee, N. Harthill, and J. J. Jacobson, The dipole mapping method, Geophysics, 1975.

Kunetz, G., Principles of Direct Current Resistivity Prospecting, 103 pp., Gebruder Borntraeger, Berlin, 1966.

Sternberg, B. K., and C. S. Clay, Flambeau anomaly - A high-conductivity anomaly in the southern extension of the Canadian shield, in The Earth's Crust, Geophys. Monogr. Ser., vol. 20, edited by J. G. Heacock, AGU, Washington, D. C., this volume, 1977.

van Zijl, J. S. V., Electrical studies of the deep crust in various tectonic provinces of southern Africa, in The Earth's Crust, Geophys. Monogr. Ser., vol. 20, edited by J. G. Heacock, AGU, Washington, D. C., this volume, 1977.

Weidman, S., The geology of north central Wisconsin, Bull. Wis. Geol. Natur. Hist. Surv., 16, 1907.

ELECTRICAL STUDIES OF THE DEEP CRUST IN VARIOUS
TECTONIC PROVINCES OF SOUTHERN AFRICA

Jan S. V. van Zijl

Geophysics Division, National Physical Research Laboratory
South African Council for Scientific and Industrial Research
Pretoria 0001, South Africa

Abstract. The resistivity structure of the South African crust and upper mantle is being studied by deep and ultradeep electrical soundings (Schlumberger array). Deep electrical soundings are carried out by laying out cable for emission lines (maximum current electrode spacing AB of $\sim$40 km), while ultradeep soundings are made by using overhead telephone or power lines (maximum AB of $\sim$1000 km). Sounding results have been obtained on a variety of tectonic provinces that are given here with the age of the main metamorphic event shown in parentheses: the Rhodesian and Kaapvaal cratons (>2600 m.y.), the Limpopo Mobile Belt (>2700 m.y.), the Namaqua Mobile Belt (1000 m.y.), and the Damara Orogen (500 m.y.). The results show that the cratons are characterized by a highly resistive zone extending downward from the surface to a depth of 5-10 km depending on the locality. The average resistivity of this zone is about 100,000 ohm m with values varying between 30,000 and 400,000 ohm m. In contrast, the mobile belts invariably indicate the presence of a thick but only moderately resistive zone with a resistivity ranging from 2000 to 10,000 ohm m (average, 5000 ohm m) extending from the surface to a depth of 25-30 km. An important feature, deduced from the electrical results, is that this zone continues laterally into the cratons, where it underlies the highly resistive zone. Correlation of these electrical zones with geological information implies that first, the highly resistive zone in cratonic type terrain is associated with vast amounts of massive, strongly consolidated unfractured granitoid material in the form of granitic intrusions, gneisses, and migmatites, second, the moderate resistivities in mobile belts are due to the presence of intensely deformed, fractured, and highly metamorphosed rocks, and third, the lower crust under both cratons and mobile belts is metamorphic and fractured, consisting mainly of a refractory residuum which has been depleted of lighter granitic components. A surprising feature of the electrical sounding results is that a major portion of the upper crust (in mobile belts) and the entire lower crust to a depth of 25-30 km is only moderately resistive, and this provides strong evidence for the presence of free water in these zones. Calculations based on measurements of subsurface water resistivities and the electrical results have provided some crude constraints on the distribution of porosity in the crust. The highly resistive zone of the cratonic terrains has a porosity of less than 0.5%, but there is a severalfold increase in porosity at the transition to the moderately resistive zone which occurs at a depth ranging from 5-10 km. In the mobile belts the porosity varies from about 3% near the surface to 1% at a depth of 25 km. If water is present in the lower crust as is suggested by the electrical results, it may provide a means of explaining the origin of the conductive zone of variable conductance that has been recognized in the depth range 25-40 km. The preferred explanation at this stage is that the conductive zone forms by hydration, and it is shown that serpentinite (hydrated mantle rock) could account for its electrical properties. It is also tentatively suggested that water, by preferential movement, may determine the heat flow pattern observed in cratons and mobile belts. Finally, it is shown that the uppermost mantle is characterized by a thick, highly resistive zone which becomes more conductive as the depth and temperature increase. Its electrical resistivity suggests that it is dry.

Introduction

This paper deals with the results of direct current electrical soundings carried out in southern Africa to study the electrical structure of the crust and upper mantle. The principles of electrical sounding are well known [e.g., Kunetz, 1966; Zohdy et al., 1974; Keller and Frischknecht, 1966] and will not be discussed here. However, for the benefit of the non-specialist some characteristics of the method will be mentioned briefly. The electrical sounding technique used for crustal studies in southern Africa is the Schlumberger array, AMNB, where ideally all four electrodes are spaced symmetrically about a central point and kept in a straight line when a sounding is being carried out. An electrical sounding is carried out by progressively increasing the current electrode separation AB between measurements. At each length AB, current is passed into the earth via the current electrodes A and B, and the resulting electrical field is determined in the central part of the array by measuring the potential difference between the closely spaced measuring electrodes M and N. For each measurement an apparent resistivity ρ_a is calculated. The set of values of ρ_a as a function of the distance AB/2 on a bilogarithmic scale is called a sounding curve and forms the basic document for interpretation.

The great advantage of the Schlumberger technique is that it provides sounding curves of much better quality in heterogeneous terrain than does either the dipole or the Wenner technique [Kunetz, 1966; Orellana, 1972].

Interpretation is carried out in two steps. The first process, physical interpretation, involves the determination of a resistivity-depth model which fits the sounding data. The purpose of the second phase of interpretation is to relate the subsurface resistivity distribution obtained to the geology and tectonic structure of the study area.

As the distance between the current electrodes is increased, the total volume of earth included in the measurement also increases, both vertically and laterally. For a given center position of the array these increasing volumes overlap, and unless the lateral variations in resistivity are too large, successive results will be related strictly to variation in depth. Thus resistivities measured are actually average values for large volumes of earth, and this average is taken over larger and larger volumes as the depth of investigation increases.

When geoelectrical horizons are thick and occur near the surface, their thicknesses and resistivities can be determined uniquely from sounding curves; however, when they become thin with respect to their depth of burial, it is generally no longer possible to determine their thicknesses and resistivities with precision. This ambiguity is reflected in the well-known principles of equivalence and suppression [Kunetz, 1966].

The principle of equivalence applies to a layer whose resistivity is either higher or lower than that of the layers above and below it. The presence of a resistive bed situated between two conductive beds is manifested in a sounding curve by its transverse resistance T, which is the product of its thickness and its resistivity. On the other hand, a conductive bed which lies between two resistive beds shows its presence on a sounding curve by its longitudinal conductance S, which is the ratio of its thickness and resistivity. When equivalence applies, it becomes impossible to distinguish between two resistive beds of different thicknesses and resistivities if they have the same T. Similarly, if two conductive layers are subject to equivalence, they cannot be distinguished if they have the same S value. The principle of equivalence can apply even to relatively thick layers if the resistivity contrast between the layer subject to equivalence and its surroundings is great.

In the case of T equivalence one can determine only the minimum resistivity of the resistive layer (and hence its maximum thickness), while when S equivalence applies, one can determine only the maximum resistivity of the conductive layer (and hence its maximum thickness).

The principle of suppression relates to a layer which has a resistivity intermediate between the layers above and below it. Unless such a layer is relatively thick, its effect cannot be seen clearly on a sounding curve, and its presence may be overlooked in the interpretation.

It is seldom possible to determine the absolute thicknesses and true resistivities of deeply buried layers from the data of one sounding or a few isolated soundings because of the principles of equivalence and suppression. It is only through a comparative study of the common characteristics and progressive changes in form of a series of electrical soundings and by taking all available geological information into account that restrictions can be placed on the values of resistivities and thicknesses imposed by equivalence and suppression. This is one of the reasons for complementing ultradeep soundings with deep soundings, as will be discussed later.

The direct current electrical sounding (ES) method (e.g. Schlumberger) differs in several important respects from the magnetotelluric (MT) sounding and geomagnetic depth sounding (GDS) methods normally used to investigate the electrical properties of the deep crust. The MT technique, which like the ES method is used for the investigation of electrical structure in a vertical sense, is not sensitive to the presence of resistive zones in the crust. The ES method, on the other hand, is sensitive to resistive zones, but, as was noted earlier, it generally responds to their transverse resistance T rather than to their resistivities and thicknesses separately. Nevertheless, the direct current electrical method is the only method capable of providing information about the resistivities of resistive zones, although it should be remembered that the resistivities obtained are minimum values. The MT method, in contrast, is only capable of measuring changes in the thickness of a resistive layer. As has been stated, where a resistive layer becomes thick in comparison with its depth of burial, its true resistivity and its thickness can be determined separately by electrical sounding.

The GDS method is very sensitive to large lateral changes in resistivity and is therefore used as a tool to map highly conductive structures at lithospheric depths. The GDS method has in common with the ES and MT methods the fact that conductive zones are subject to S equivalence. The GDS method differs from the ES method, apart from its use as a tool strictly for lateral exploration, in its lack of ability to detect conductive zones for which the contrast in conductivity is less than a factor of about 100 [Gough, 1974]. This implies that moderately conductive zones within the crust will not be detected by GDS studies.

Deep electrical soundings are thus especially suited to provide information on the distribution of resistive and moderately conductive zones within the crust and uppermost mantle. This work describes the ES results that have been obtained in determining the electrical profile through the crust in most of the major tectonic provinces of southern Africa. As a result of the increasing number of data available for interpretation it has been possible to arrive at a more reliable resistivity model for the crust; consequently, this has necessitated a reinterpretation of previous results [Van Zijl, 1969; Van Zijl and Joubert, 1975].

Sounding Strategy

To probe through the crust, large current electrode spacings AB normally in excess of 100 km are required, and hence it becomes impractical to lay out cable. Instead, existing overhead conductors such as telephone and uncommissioned power lines are used to serve as emission lines. Because of the great lengths of emission lines and the large volumes of earth involved in the measurements, it is extremely important to guard against distortions of the sounding curves due to large lateral variations in resistivity. Lateral effects are studied by the use of multiple MN measuring stations distributed areally in the central zone of an ultradeep sounding. At each MN station the electrical field is recorded in two orthogonal directions so as to determine its configuration. Both MN stations and A and B electrode sites are chosen with great care, all the available geological information of the area to be investigated being taken into account.

As was noted earlier, the value of a single ultradeep electrical sounding can further be greatly enhanced by the execution of additional deep electrical soundings in the area covered by the main sounding. These complementary soundings have a maximum AB spacing of only 40 km. The purpose of these soundings, for which cable is laid to serve as the emission line, is to gain more detailed information about the resistivity distribution in the upper crust. This information facilitates quantitative interpretation and the delineation of tectonic provinces and is also useful in further assessing the likelihood of lateral effects.

Another factor which must be taken into account when crustal direct

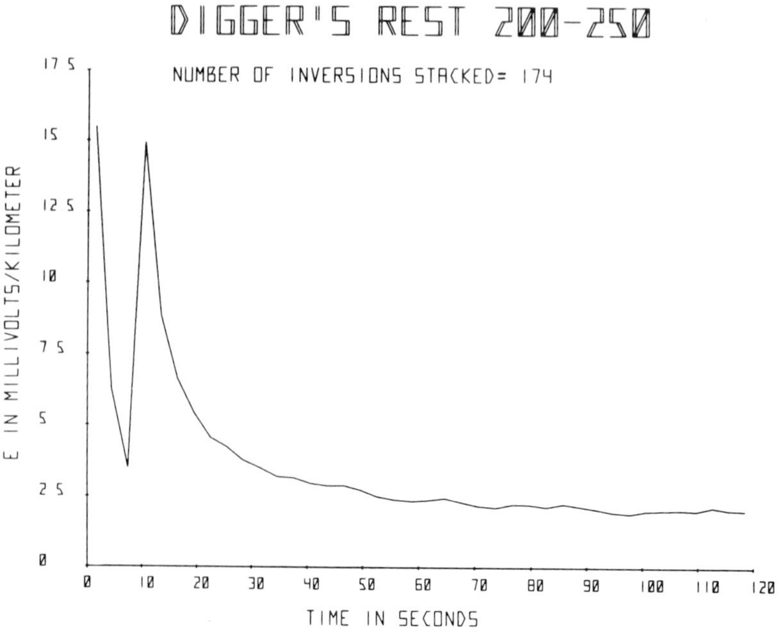

DIGGER'S REST 200-250

NUMBER OF INVERSIONS STACKED = 174

Fig. 1. A stacked electrical field recording at an MN station showing the transient response of earth structure (skin effect) to a square wave emission current. The recording was made at an electrode spacing AB = 450 km with the MN station near the center of AB. MN = 2 km and is parallel to AB. The time between current inversions is 2 min, and the current intensity is 72A. The electrical field has been stacked over 174 current inversions [after Van Zijl and Joubert, 1975].

current electrical soundings are being carried out is the time which the current takes, after being switched on, to reach its steady state value. This transient response of the earth (skin effect) depends on the conductivity distribution with depth in the area and the length of AB, and when these parameters are large, it may take a relatively long time for the transient to vanish. The skin effect is manifested by an overshoot and subsequent exponential decay with time of the electrical field recorded at an MN station towards the steady state value. Figure 1 shows an example of a field measurement over an MN length of 2 km parallel to AB. The AB spacing in this instance was 450 km, and the current 72 A. The electrical field displayed is a stacked version of 174 repeated measurements to minimize noise. It is clearly seen that there is an appreciable amount of skin effect left 1 min after the current has been switched on. The emission current is in the form of a square wave, the half period of which is selected to exceed the duration of the skin effect. Further details of the sounding technique are discussed by Van Zijl [1969] and by Van Zijl and Joubert [1975].

Tectonic Provinces of Southern Africa

Africa has been almost exclusively continental since early Paleozoic times and hence offers an extensive record of early crustal history. The

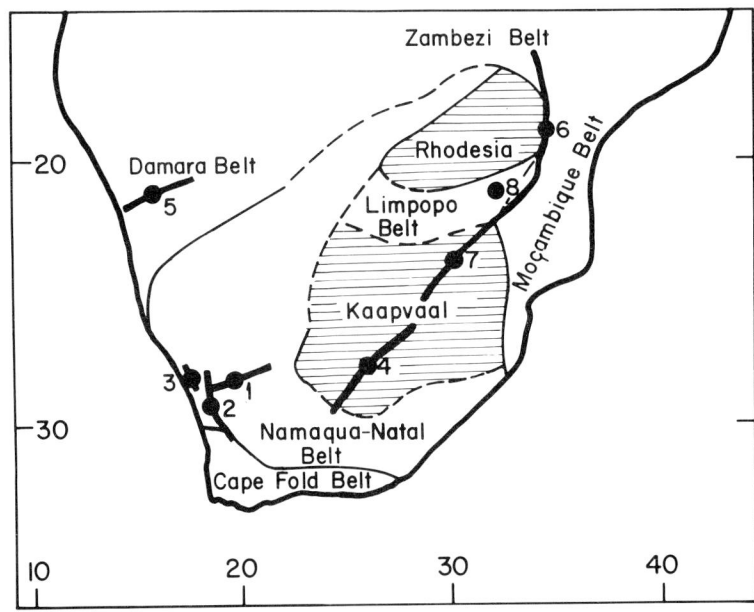

Fig. 2. Map of southern Africa showing the locations of all the ultradeep electrical soundings carried out thus far. Each sounding is numbered and indicates the position of the emission line (heavy lines) and site of the MN station(s) (dots). For details of the maximum length of emission line for each sounding, see Table 2. The sounding locations are shown with respect to the major tectonic units of the subcontinent. Cratons are indicated by the hatched areas. The mobile belts are identified by name. (See Table 1 for correlation between southern African and North American Precambrian tectonic provinces.)

oldest segments of the southern African crust are the cratons which are surrounded by mobile belts of various ages (see Figure 2). In this paper, cratons are defined as terrains which evolved in the Archaean (>2600 m.y.). The cratons are characterized by the intrusion of vast amounts of granitoid material and by minor occurrences of fairly low grade metamorphic rocks. The cratons of southern Africa are thus regarded as ancient, relatively undeformed stable crustal remnants of predominantly granitoid composition which have survived tectonic events since the Archaean [Kröner, 1976]. In contrast, the mobile belts are zones of intense deformation often showing evidence of more than one period of metamorphism. In the older belts, high-grade granulite facies metamorphic rocks are present [Kröner, 1976; Anhaeusser, 1973; Hunter, 1974; Clifford, 1974]. In general, the mobile belts are considered to be ensialic, involving either the rejuvenation of continental floor rocks or deformation of geosynclinal sediments [Clifford, 1970]. Evidence for subduction involving oceanic crust has not yet been forthcoming from the older mobile belts. On the other hand, the origin of the 'young' Cape Fold Belt of Paleozoic age which occurs near the southern boundary of the continent (Figure 2) has been explained in terms of plate tectonics [De Beer et al., 1974; Martini, 1974].

Ultradeep electrical soundings have been carried out on most of the

TABLE 1. Correlation of Major Precambrian Tectonic Provinces in Southern Africa and North America

Southern Africa		North America	
Province	Age m.y.	Province	Age m.y.
Archaean			
Limpopo Mobile Belt	>2700		
Kaapvaal Craton	>2600		
Rhodesian Craton	>2600		
		Superior	2600
Proterozoic			
		Churchill	1300
Namaqua Mobile Belt	1000	Grenville	1000
Damara Orogen	500		

major tectonic provinces (shown in Figure 2). These are (the approximate age of the major orogenic event is shown in parentheses) the Rhodesian and Kaapvaal cratons (>2600 m.y.), the Limpopo Mobile Belt (>2700 m.y.), the Namaqua Mobile Belt (1000 m.y.), and the Damara Orogen (500 m.y.). In view of the comparatively great age of the tectonic events, it is unlikely that residual temperature plays any significant role in causing geophysical anomalies.

To facilitate comparison, the correlation of the major tectonic events in the Precambrian (>500 m.y.) in southern Africa and North America is given in Table 1. Table 2 shows the maximum current electrode separation attained for each ultradeep electrical sounding. Because the sounding curves of ES 3 and ES 4 (Figure 2) were affected by the ocean and by conductive sediments, respectively, they will not be discussed in the text.

Electrical Sounding Investigation of the Rhodesian and Kaapvaal Cratons and the Crosscutting Limpopo Mobile Belt

Tectonic setting. The Rhodesian and Kaapvaal cratons are separated by the ENE striking Limpopo Mobile Belt, which is considered to be a reactivated craton [Mason, 1973]. Its configuration relative to the adjoining cratons is shown in Figures 2 and 3. The belt underwent granulite facies metamorphism earlier than 2700 m.y. ago and was subsequently reheated under retrogressive amphibolite metamorphic conditions at 200 m.y. B.P. [Van Breemen and Dodson, 1972]. The zone of weakness along which the mobile belt evolved was formed before its inception {Mason, 1973}, and the belt is characterized by fairly intense ENE striking fractures especially along the northern and southern marginal zones. Fracturing extended over a long period, brittle fracturing occurring in the later stages of deformation. Between 2000 and 200 m.y. ago, hydrothermal activity took place

TABLE 2. Details of Ultradeep Soundings Carried out in Southern Africa

Sounding	Name	Maximum AB Separation Attained, km	Tectonic Province	Remarks
6	Mavonde	450	Rhodesian Craton	
7	Pietersburg	450	Kaapvaal Craton	
4	Dealesville	600	Kaapvaal Craton	Affected by conductive Karroo sediments
8	Chiredzi	1200	Limpopo Mobile Belt	
1	Pofadder	270	Namaqua Mobile Belt	
2	Kamieskroon	350	Namaqua Mobile Belt	
3	Port Nolloth	170	not certain	Affected by ocean
5	Okasisi	350	Damara Orogen	For AB>200 km, measurements are below noise level

[Jacobsen et al., 1975]. At present the area is slightly active seismically, and a few thermal springs issue from fault planes [Mason, 1973].

It can be seen in Figures 2 and 3 that another mobile belt is situated to the east of the Rhodesian and Kaapvaal cratons. This is the 500 m.y. Moçambique Mobile Belt, which forms part of the Pan-African orogenic event. The Damara Orogen, which will be discussed later, also belongs to this episode.

Electrical results. In Figure 2 the position of the power line which was used as emission line for three ultradeep soundings, ES 6, ES 7, and ES 8, is shown. Electrical field recording sites (MN stations) for the soundings are also indicated. A considerable part of this project involved the combined participation of South African, German, Rhodesian, and Portuguese scientists. A joint paper giving a detailed account of the investigation will be published later. The present discussion will be confined to only a part of the South African effort, namely, that part which has a specific bearing on crustal structure in the rest of the subcontinent.

MN stations for both ultradeep and deep electrical soundings are shown in Figure 3. A careful analysis of all the sounding curves in the region has revealed the presence of four electrical zones (the weathered surface layer being ignored). In Table 3 the deduced resistivity range and average value for each of the zones are given. The electrical characteristics of each zone will be discussed by referring to the two ultradeep sounding curves shown in Figure 4. These two curves, ES 7, obtained on the Kaapvaal Craton, and ES 8, obtained on the Limpopo Mobile Belt, illustrate most of the distinctive features which characterize the individual electrical zones. The sounding curves in Figure 4 are best fitting theoretical curves through the data points for the layered models shown at the bottom of the figure.

Zone 1. As can be seen in Figure 3 the very high resistivities which

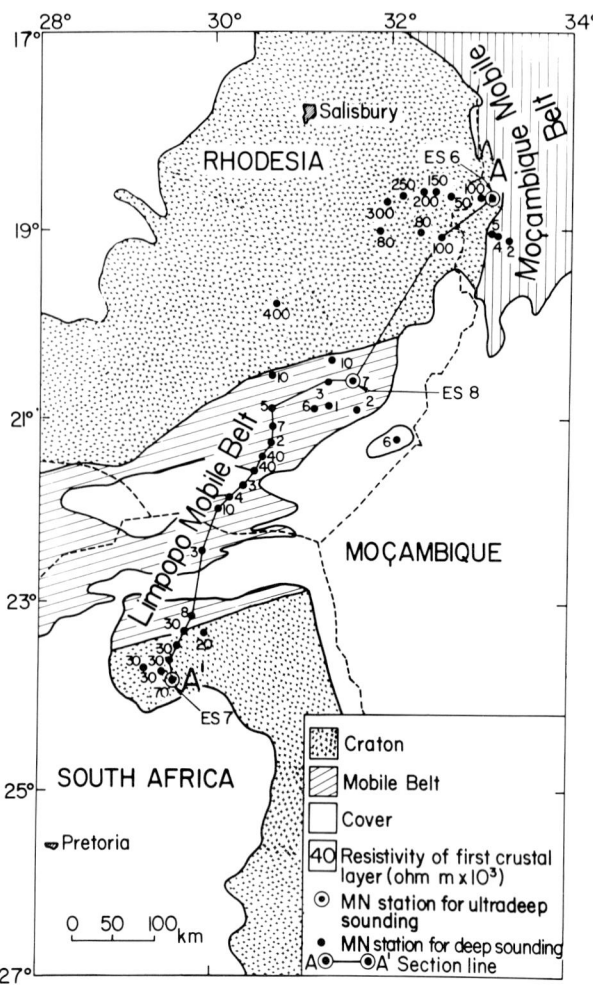

Fig. 3. Map of the eastern part of the Rhodesian Craton, the northeastern part of the Kaapvaal Craton, and the crosscutting Limpopo Mobile Belt in between showing resistivity values of the first crustal layer. Sites of MN stations for both ultradeep electrical soundings (numbered as they are in Figure 2 for orientation) and deep electrical soundings are shown with respect to the geology.

characterize zone 1 are found only on the cratons. Zone 1 occurs immediately below the weathered surface layer. The presence of zone 1 is indicated on the sounding curve of ES 7 by the steep ascending branch between points C and D (Figure 4). Zone 1 is generally subject to T equivalence, so, as was noted earlier, one can only determine T, the transverse resistance (thickness-resistivity product), a minimum resistivity, and a maximum thickness for this zone from each sounding curve which shows its presence. On this basis it has been established that the minimum resistivity of zone 1 in the Kaapvaal and Rhodesian

TABLE 3. Interpreted Resistivity Characteristics of the Electrical Zones Identified From Sounding Results in the Kaapvaal Craton, the Rhodesian Craton, and the Crosscutting Limpopo Mobile Belt

Zone	Range of Resistivities, ohm m	Average Resistivity, ohm m
1	30,000-400,000	100,000
2	2,000-10,000	5,000
3	10-50	unsure - data limited
4	20,000-30,000	unsure - data limited

cratons varies between 30,000 and 400,000 ohm m with an average value of about 100,000 ohm m, while its maximum thickness varies between 5 and 10 km but does not exceed 10 km.

Zone 2. The moderately resistive zone 2 occurs in the Limpopo Mobile Belt immediately underlying the weathered surface layer. Its presence is strikingly illustrated in Figure 3 by the moderate resistivity values of the mobile belt, which are roughly an order of magnitude lower than those on the adjoining cratons. There is a

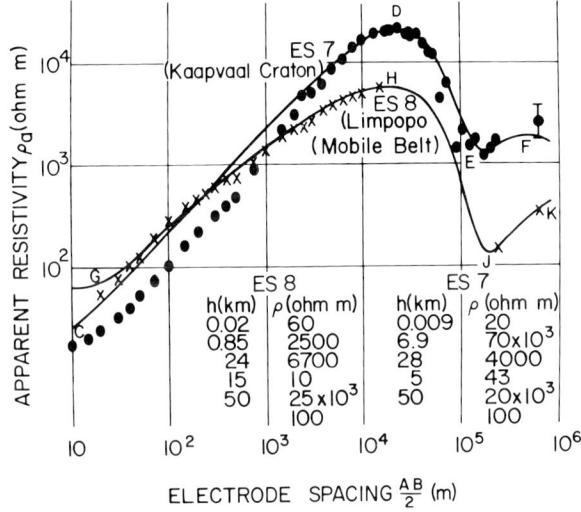

Fig. 4. Ultradeep ES data points and best fitting theoretical sounding curves (solid lines) for soundings ES 7 (Kaapvaal Craton) and ES 8 (Limpopo Mobile Belt). Details of the relevant electrical models to which the theoretical curves apply are shown at the bottom of the figure. The thickness h and resistivity ρ of each electrical zone are given in the models. The apparent resistivity ρ_a value with the large error bar at an electrode spacing AB/2 = 600 km shown with the sounding data for sounding ES 7 was obtained from the results of horizontal profiling at AB = 1200 km (see text).

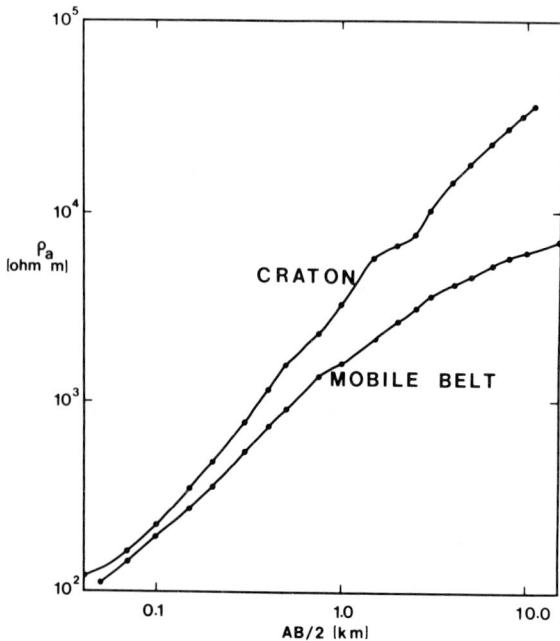

Fig. 5. Two deep ES curves illustrating the dissimilarity in geoelectrical character of the upper crust between a craton and a mobile belt. (In this specific case the curves refer to the Rhodesian Craton and the Limpopo Mobile Belt.)

similar large decrease in resistivity as one passes from the Rhodesian Craton into the Moçambique Mobile Belt. The dissimilarity between the two deep sounding curves in Figure 5 typifies this transition from craton to mobile belt. The presence of zone 2 is shown by the gradual rise of the relevant sounding curve. This is illustrated in Figure 4 by the curve segment GH of the ultradeep sounding curve for ES 8. Zone 2 is not subject to T equivalence, and in the mobile belt its resistivity and thickness can be determined accurately. From the available sounding data it is evident that the resistivity of zone 2 varies between 2000 and 10,000 ohm m with an average value of about 5000 ohm m. Its thickness, on the other hand, remains fairly constant and varies only between 25 and 30 km. An often observed peculiarity of zone 2 is the tendency for its resistivity to increase from its upper boundary, just below the weathered surface layer, to a depth not exceeding 5 km.

One of the rewarding aspects of carrying out a series of closely spaced deep electrical soundings across the mobile belt and into the cratons has been the identification of the lateral continuation of zone 2 into the cratons where it underlies the highly resistive zone 1. Because the resistivity of zone 2 is lower than that of the overlying highly resistive zone 1 but higher than that of the underlying conductive zone 3 (to be discussed below), zone 2 is subject to suppression (explained in the introduction), and its presence becomes very difficult to recognize from isolated ultradeep soundings obtained on the craton. This is illustrated by the ultradeep sounding curve of ES 7 in Figure 4, where it is noted that there is no distinctive feature along the curve segment DE to mark the presence of zone 2. This emphasizes the remark made earlier that a realistic interpretation of crustal soundings can only be made when the results of ultradeep soundings are considered jointly with additional information obtained from complementary deep electrical soundings and other relevant geological information.

Zone 3. The presence of the conductive zone 3 is shown in Figure 4 by the descending segments of the ultradeep sounding curves, from point D to point E on the curve for ES 7 and from point H to point J on the curve for ES 8. Zone 3 is subject to S equivalence, so, as was noted in the introduction, it is only possible to determine its longitudinal conductance S (thickness-resistivity ratio) accurately for those sounding curves which show its presence. There is a large increase in S from the cratons to the Limpopo Mobile Belt as is illustrated in Figure 4 by the relative positions of the rising portions of the curves, segments EF and JK. This downward shift of the rising portions of the curves indicates that S increases from about 100 S on the Kaapvaal Craton (ES 7) to 1500 S on the Limpopo Mobile Belt (ES 8), an increase by a factor of 15. (The SI unit of conductance is siemens (S), where 1 S = 1 mho.) A similar increase has been established for the transition from the Rhodesian Craton to the mobile belt. From the S values it is deduced that zone 3 is weakly represented in the cratons and reaches its maximum development in the mobile belt. In order to delineate more accurately the areas where the conductive zone 3 is well developed, electrical field recordings and apparent resistivity (ρ_a) determinations were made at mobile MN stations distributed over a wide area with a constant electrode separation of AB = 1200 km (AB/2 = 600 km). This is known as horizontal profiling. As can be deduced from the curves in Figure 4, there is an indirect correlation between a ρ_a value measured with an electrode spacing of AB/2 = 600 km and S. The results of the horizontal profiling are shown in the sections of Figure 6. The upper portion of the figure depicts the total electrical field and the lower portion shows apparent resistivity values calculated for the field component parallel to the emission line AB. In spite of the large error bounds (±50%) and insufficient station density it is evident that the values in the northern part of the mobile belt are lower than they are elsewhere, and this indicates that in this area zone 3 reaches its maximum development. Although the conductive zone 3 and lateral variations in its conductance are clearly detected by ultradeep soundings, no sign of its presence was obtained from GDS data furnished by a few magnetic variometers which were installed for the duration of the field program. This finding confirms the limited increase (factor of 15) of the conductance of zone 3 in the mobile belt.

Because zone 3 is subject to S equivalance, it is not possible to allocate realistic thicknesses and resistivities to this zone without further information. This information was provided by the results from the mobile belts to be dealt with later, and will be described in the relevant sections of the paper.

Zone 4. The presence of zone 4, which is highly resistive, is shown

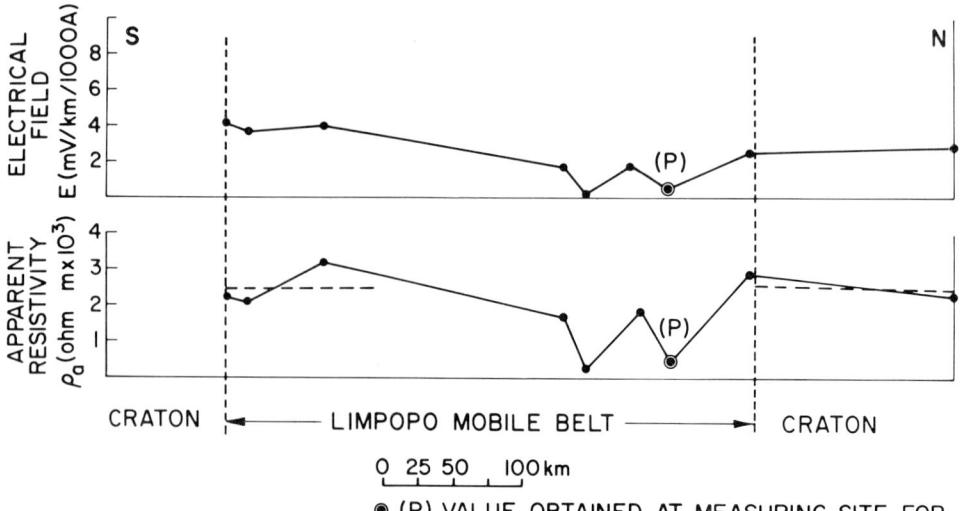

Fig. 6. Results of horizontal profiling across the Limpopo Mobile Belt with AB = 1200 km. The top section shows the variation of the total electrical field, and the bottom section displays apparent resistivities ρ_a calculated from the component of the electrical field parallel to AB. The dashed lines indicate average ρ_a values (see text for further details).

in Figure 4 by the ascending segments of the ultradeep sounding curves: segment EF of the curve for ES7 and segment JK of the curve for ES 8. In the case of ES8 the ascending segment of the curve is based on only two experimental points. These values were determined from very accurate electrical field measurements at various MN stations and are thus considered to be completely reliable. In the case of ES 7 the ascending branch of the curve is very dependent on the last experimental point plotted with the large error bar ($\pm 50\%$). This value was not obtained during the original sounding ES 7, where the maximum current electrode spacing AB was only 450 km (AB/2 = 225 km). It is based on several electrical field measurements carried out with a constant current electrode separation of AB = 1200 km (AB/2 = 600 km) at various MN stations which were confined to those regions where the conductive zone 3 is only weakly developed. These circumstances are similar to those at ES 7 and constitute the basis for using these measurements to obtain an additional experimental point on this sounding curve. Obviously, the MN stations used for the determination of the average apparent resistivity value for this additional experimental point at an electrode spacing of AB/2 = 600 km do not lie near the center of AB as is required by Schlumberger theory. The MN stations have, however, been restricted to the middle third of the AB spacing, and this permits their use in a Schlumberger sounding curve especially if the experimentally determined point lies on the ascending portion of a Schlumberger curve [Kunetz, 1966]. An average value of 2500 ohm m has been obtained for this last point on sounding curve ES 7. In Figure 6 this average value is shown by the dashed lines.

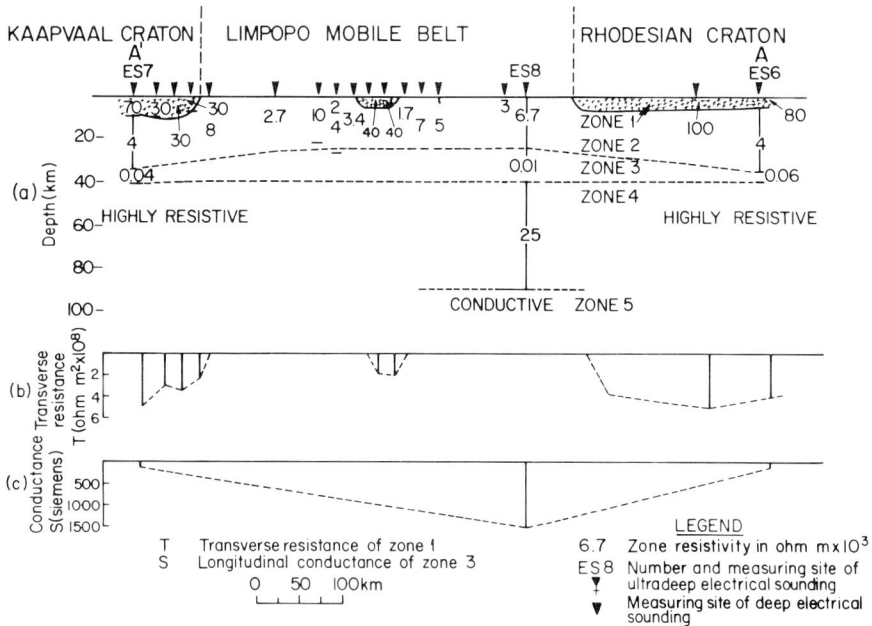

Fig. 7. Geoelectrical sections across the Limpopo Mobile Belt and into the adjoining cratons. (Refer to Figure 3 for the position of the section line.) (a) Resistivity-depth section and identification of electrical zones. (b) Transverse resistance T (thickness-resistivity product) of highly resistive zone 1. (c) Longitudinal conductance S (thickness-resistivity ratio) of conductive zone 3.

The conclusion is thus reached that both sounding ES 7 and sounding ES 8 show the presence of the highly resistive zone 4. Neither its resistivity nor its thickness can be determined with any amount of reliability. Its minimum resistivity is estimated at 20,000 ohm m. Its thickness is probably greater than 50 km.

In the interpretation of the ultradeep sounding curves in Figure 4 a conductive zone (zone 5 of Figure 7a) has been included as the basal semi-infinite layer. Although the presence of this zone has not been observed in the region under discussion, it has been detected in the Namaqua Mobile Belt. Zone 5 has been included in Figure 4 solely to show that its inclusion can be accommodated by the models.

The electrical sounding results as discussed above are shown in the form of a cross section in Figure 7a along the section line A-A' indicated in the map of Figure 3. In Figure 7b the lateral variation of T, the transverse resistance of the highly resistive zone 1, is shown. It is noted that a small remnant of zone 1 remains intact in the mobile belt. The tectonic significance of this will be discussed later. In Figure 7c the lateral variation of S, the longitudinal conductance of the conductive zone 3, is shown. The increased S value in the mobile belt can clearly be seen. The thicknesses and resistivities assigned to zone 3 in Figure 7a will be discussed when the tectonic implications of the electrical results are dealt with.

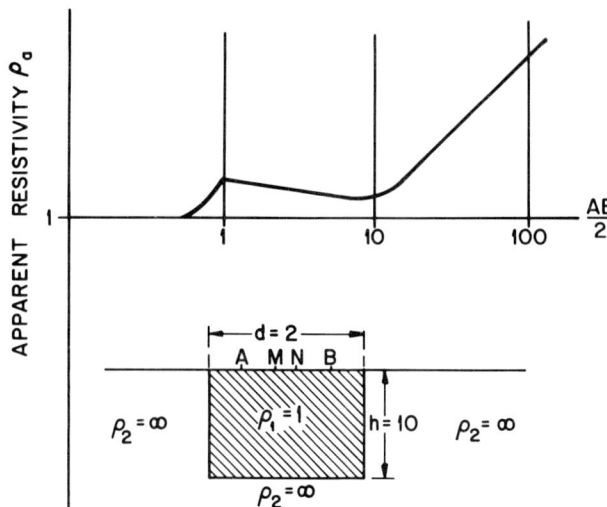

Fig. 8. A theoretical ES curve showing the effect of the edges and base of a two-dimensional conducting slab imbedded in an infinitely resistive medium [after Alpin et al., 1966].

Figure 8 shows a theoretical Schlumberger sounding curve with its midpoint on the center line of a two-dimensional conductive slab. The direction of current electrode expansion is perpendicular to the strike of the conductor which has a thickness of 10 and a width of 2 and is bounded on both sides and at the bottom by an infinitely resistive medium [Alpin et al., 1966]. The situation depicted by the model is more extreme than the actual case, since it does not take into account the presence of the moderately resistive zone 2, which continues under the highly resistive zone 1 associated with the cratons. In spite of this the final ascent of the sounding curve can be correlated with a conductor underlain at depth by a semi-infinite resistive layer. (This curve will rise at $45°$ for the two-layer case described.) It is thus highly unlikely that the final ascent of sounding curve ES 8 is due to a lateral effect caused by the edge of the craton.

Electrical Sounding Investigation of the Namaqua Mobile Belt

Tectonic setting. The Namaqua Mobile Belt occurs to the west and southwest of the Kaapvaal Craton (Figures 2 and 9) and represents a zone of reactivated older basement reconstructed during an intense metamorphic event approximately 1000 m.y. ago [Nicolaysen and Burger, 1965]. There is also evidence of an earlier metamorphic episode which according to Kröner [1976] can be correlated in time with the main phase of deformation in the Limpopo Mobile Belt. This early metamorphic phase also affected the rocks older than 2600 m.y. of the Kheis System and thus gave rise to the Kheis Domain shown in Figure 9. The Namaqua

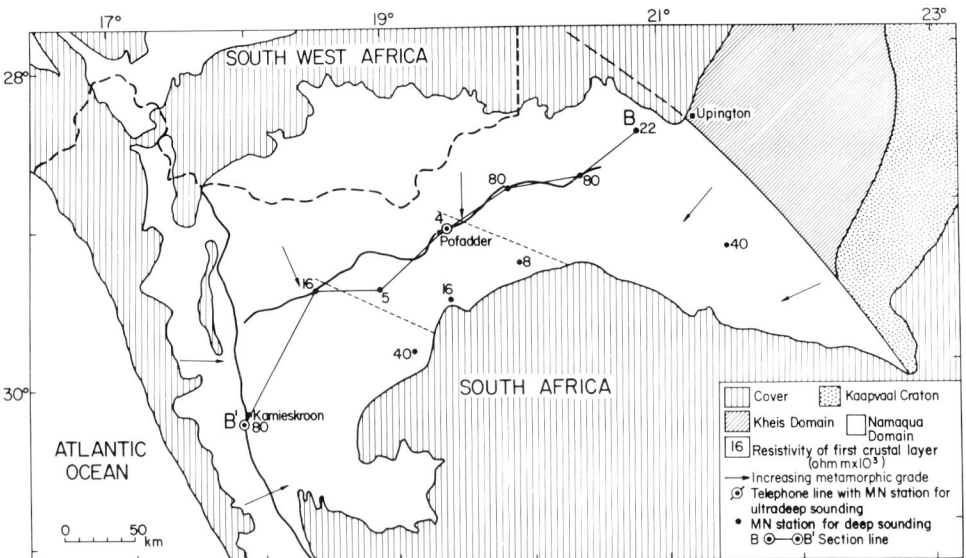

Fig. 9. Map showing the extreme western part of the Kaapvaal Craton and the adjoining Namaqua Mobile Belt. The sites of MN stations for both ultradeep electrical soundings (numbered as they are in Figure 2 for orientation) and deep electrical soundings are shown with respect to the geology.

Mobile Belt differs from the Limpopo Mobile Belt both in age and in its metamorphic character. Granulite facies metamorphism represents the culmination of prograde regional metamorphism [Clifford, 1974] in the Namaqua Mobile Belt, whereas granulite metamorphism in the Limpopo Mobile Belt was followed by a long period of retrograde metamorphism. The central part of the Namaqua Mobile Belt is considered by Joubert [1971] to have been the site of a major regional thermal dome culminating in granulite facies rock assemblages. The postulated thermal dome can be deduced from Figure 9, where the directions of increasing metamorphic grade have been plotted [after Joubert, 1971]. This intense metamorphic event was associated with strong northwest shearing in parts of the mobile belt and along its borders [Blignault et al., 1974; Kröner, 1976]. The rocks of the Namaqua Mobile Belt consist of granite-gneiss and remnants of metavolcanic and metasedimentary rocks of the Kheis System.

Electrical results. Figure 9 shows the location of the telephone lines used as emission lines for the two ultradeep electrical surroundings ES 1 (Pofadder) and ES 2 (Kamieskroon) carried out on the metamorphosed Namaqua Mobile Belt. The measuring stations for these soundings and for AB = 40 km soundings are also shown. At each sounding site the interpreted resistivity of the first crustal layer beneath the weathered surface layer has been indicated in 10^3 ohm m. It is noted that the resistivities are about 1 order of magnitude lower within the central zone of high-grade metamorphism than they are around it. These results are very similar to those obtained in the Limpopo

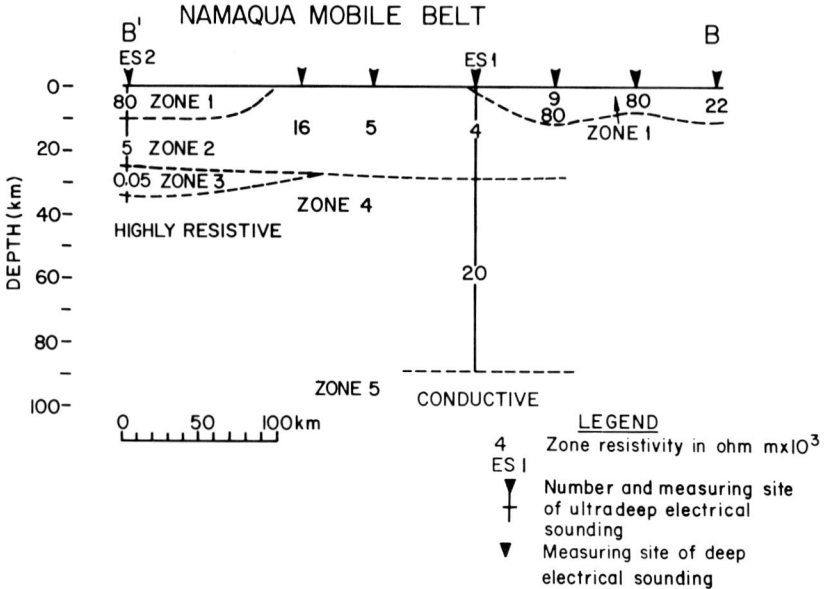

Fig. 10. Resistivity-depth section across the Namaqua Mobile Belt. (Refer to Figure 9 for the position of the section line.)

Mobile Belt area, where the cratons are more resistive than those in the highly metamorphosed mobile belt. The transition from high to lower resistivities is also illustrated in the section of Figure 10 which was deduced from an analysis of all the sounding data in the region. The minimum resistivity of the highly resistive zone varies from 22,000 to 80,000 ohm m and thus broadly corresponds to results obtained over the cratons. Its T value, which ranges from about 4 to 7 x 10^8 ohm m^2, is also typical of the cratons. The similarity of the sounding curve obtained from the Kamieskroon ultradeep sounding, ES 2 (see Figure 11), and the ultradeep sounding curves situated on the Rhodesian and Kaapvaal cratons (soundings ES 6 and ES 7) was noted previously [Van Zijl and Joubert, 1975]. The electrical results thus suggest that except for an inner core of intensely deformed and metamorphosed rocks the Namaqua Mobile Belt consists of rocks which, in a physical sense, are similar to cratons. The electrical zones in Figure 10 are thus correlated with those in Figure 7a. However, more work is needed to define these areas accurately. As occurs in the case of the cratons, the thickness of the highly resistive zone 1 does not exceed 10 km.

The resistivity of the moderately resistive zone 2 associated with the high-grade metamorphic central zone varies between 4000 and 8000 ohm m, and its maximum thickness is 26 km. It is interpreted as continuing under the highly resistive zone 1 as it does in the case of the Limpopo Mobile Belt.

A striking difference between the sections across the Namaqua Mobile Belt (Figure 10) and the Limpopo Mobile Belt (Figure 7a) concerns the development of the conductive zone 3. In the Namaqua Mobile Belt,

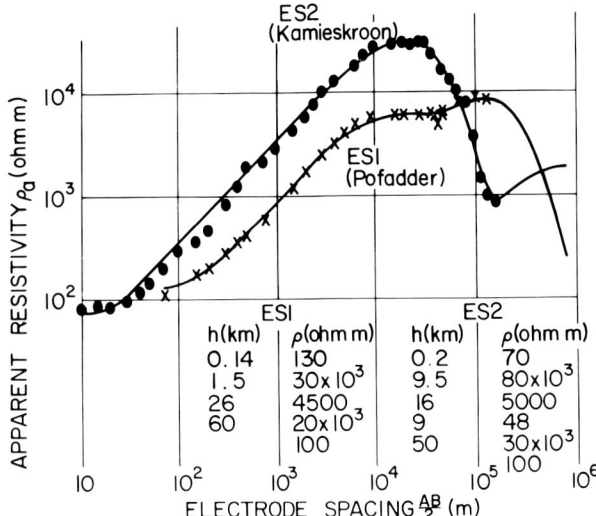

Fig. 11. Ultra-deep ES data points and best fitting theoretical sounding curves (solid lines) for soundings ES 1 (Pofadder) and ES 2 (Kamieskroon). Details of the relevant electrical models to which the theoretical curves apply are shown at the bottom of the figure. The thickness h and resistivity ρ of each electrical zone are given in the models.

zone 3 is absent immediately under the area where zone 2 appears at the surface, whereas in the same tectonic situation in the Limpopo Belt, zone 3 reaches its maximum development. The possible tectonic implication of this difference will be discussed later.

Finally, the section across the Namaqua Mobile Belt shows the presence of a thick resistive zone 4, which has a T value of 1.2×10^9 ohm m^2, followed by a conductive zone 5. Evidence for zones 4 and 5 is obtained from the Pofadder sounding curve, ES 1, shown in Figure 11. The detection of these zones was made possible by the absence of the conductive zone 3 near the base of the crust, which increased the depth of investigation in a very substantial way.

A comparison of the two ultradeep soundings across the Namaqua Mobile Belt can be made by referring to Figure 11, which shows the experimental points and the best fitting theoretical curves for the models at the base of the figure.

In conclusion, it may be mentioned that a large GDS array study, which included this area, did not show the presence of a conductor associated with the Namaqua Mobile Belt [Gough et al., 1973].

Electrical Sounding Investigation of the Damara Orogen

Tectonic setting. The Damara Orogen of South-West Africa is part of the Pan-African tectonic episode which took place about 500 m.y. ago (see Figure 2). The northeasterly striking orogen is approximately 400 km wide and consists of deformed mafic volcanics and pelitic sediments deposited in a southern eugeosynclinal basin and relatively unde-

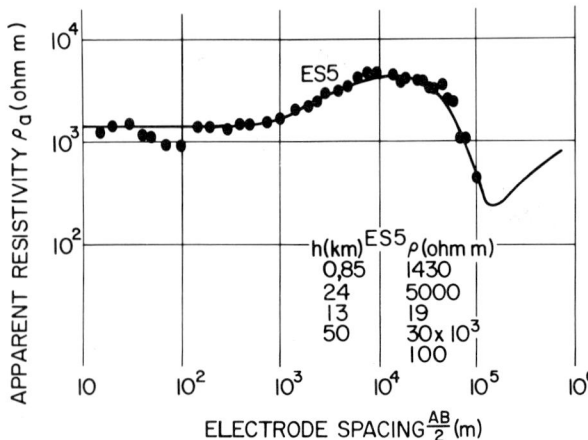

Fig. 12. Ultradeep ES data points and best fitting theoretical sounding curve (solid line) for sounding ES 5 carried out on the Damara Orogen. Details of the relevant electrical model to which the theoretical curve applies are shown at the bottom of the figure. The thickness h and resistivity ρ of each electrical zone are given in the model. Note that the presence of zones 4 and 5 is hypothetical.

formed dolomitic rocks in a northern miogeosynclinal basin. The eugeosyncline over which the ultradeep sounding was carried out was metamorphosed to the amphibolite facies during the Pan-African orogenic event. Granitic masses, probably formed by anatexis, occur extensively [Martin, 1969].

Electrical results. A single ultradeep electrical sounding, ES 5, has been carried out on the Damara Orogen. The location of the telephone line which served as emission line can be seen in Figure 2. The AB direction was chosen parallel to the tectonic grain of the mobile belt. The Damara sounding data and the best fitting theoretical curve are shown in Figure 12 for the model at the bottom of the figure. It is noted that there is no sign of the highly resistive zone 1 that could be associated with cratonic material. There is also no T equivalence, so both the thickness and the resistivity of the moderately resistive zone can be determined with maximum accuracy. Its thickness is 24 km, and its resistivity is 5000 ohm m. These parameters are typical of those found for the Limpopo and Namaqua mobile belts, and hence the moderately resistive zone can be correlated with zone 2.

As occurs in the case of the Limpopo Mobile Belt, the resistivity of zone 2 tends to increase with depth, and it is underlain by the conductive zone 3 at a depth of 24 km. Unfortunately, it has not been possible to determine the longitudinal conductance S of the conductive zone owing to loss of signal strength at large AB separations. In the interpretation of the sounding curve a minimum S has been evaluated and has been used together with the data on both the thick resistive zone 4 and the deep conductive zone 5 obtained from the previously discussed soundings on the Limpopo and/or Namaqua mobile belts. As far as the sounding ES 5 on the Damara Orogen is concerned, however,

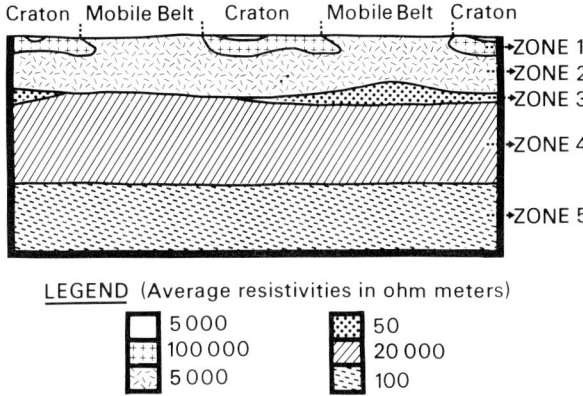

Fig. 13. Schematic generalized model for the crust and uppermost mantle based on the electrical sounding results showing the five zones which have been detected.

the two last mentioned zones are as yet hypothetical. Nevertheless, for the part of the curve where data points are available, there is a strong resemblance between this sounding and sounding ES 8 (Figure 4) carried out on the Limpopo Mobile Belt.

A large GDS array study which included the western part of the Damara Orogen detected an east-west striking anomaly along the border between the miogeosyncline and the eugeosyncline which is thought to be an old zone of weakness [De Beer et al., 1976]. This highly conductive zone lies well to the north of the electrical sounding.

Discussion of Electrical Sounding Results in Terms of Crustal Structure

The generalized electrical model that has been deduced from the ES studies is shown schematically in Figure 13.

Correlation of the relevant electrical zones with the results of geological investigations [Anhaeusser, 1973; Wilson, 1973; Mason, 1973; Joubert, 1971; Martin, 1969; Kröner, 1976] indicates certain relationships. In the following discussion the overburden or cover rocks of the various tectonic provinces have not been considered.

The discontinuous thin uppermost zone consisting of synform structures at the top of zone 1 represents fragments of metamorphosed Archaean metasediments and volcanics which, as was mentioned earlier, occur sporadically in the cratons. These rocks have a resistivity which varies between 2000 ohm m and 8000 ohm m with an average value of about 5000 ohm m (J. S. V. van Zijl, unpublished results, 1976). The highly resistive zone 1 in cratonic terrains is associated with vast amounts of granitoid material in the form of granitic intrusion, gneisses, and migmatites. The moderately resistive zone 2 of the mobile belts can be attributed to the presence of intensely deformed and highly metamorphosed rocks exhibiting features of upper amphibolite and granulite facies metamorphism. The lower crust is metamorphic as deduced from the continuation of zone 2 under the highly resistive zone 1 associated with the cratons.

The view that the stable continental crust of cratonic areas is made up of a lower crust of high-grade metamorphic rocks underlying a more granitic upper crust which also contains remnants of supracrustal metamorphics is supported by numerous studies [e.g., Wegmann, 1935; Lambert and Heier, 1967; Ringwood and Green, 1966; Berckhemer, 1969; Smithson and Decker, 1974]. Geophysical models based particularly on heat flow [Smithson and Decker, 1974] and seismic [Christensen and Fountain, 1975] and gravity [Bott and Smithson, 1967] investigations are compatible with this view. In essence the lower crust is considered to comprise a refractory residuum owing to the depletion of the lighter granitic materials, volatiles, and radioactive elements. This metamorphic zone grades upward into a zone in the upper crust consisting predominantly of granitic material in the form of intrusions, gneisses, and migmatites which were mainly mobilized from melts originating from the lower crust. It has often been noted that rocks of this upper crustal zone both intrude and form the basement to supracrustal metamorphic rocks [Anhaeusser, 1973; Berckhemer, 1969]. The extremely high resistivities associated with the granitoids indicate that the rocks are solid, strongly consolidated, and largely unfractured.

The suggested correlation of the highly resistive zone 1 with cratonic upper crust is significant, since it provides a means of determining the extent of cratonic material even into adjoining mobile belts provided the grade of metamorphism remains low enough to prevent the migration of the granitoid material to higher tectonic levels [Wegmann, 1935]. The results obtained indicate that the distribution of the ancient cratonic material was indeed extensive, and the impression is gained that this layer was broken up by the mobile belts studied. The Limpopo and Namaqua-Natal mobile belts, where highly resistive cratonic material is found on both sides of a core of moderately resistive highly metamorphosed rocks, are good examples. The remnant of highly resistive cratonic material found within the Limpopo Mobile Belt proper (see Figure 7a) is also significant. The view held by many geologists [e.g., Martin, 1969; Kröner, 1976; Shackleton, 1976] that the southern African mobile belts represent reworked and reconstituted cratonic crust is thus supported by the electrical studies. Further evidence comes from complementary electrical information derived from the results of GDS studies [Gough et al., 1973; De Beer et al., 1976] which show that the aforementioned mobile belts are not underlain by a highly conductive lithosphere as would be expected if they represented the traces of subduction zones. The resistivity-derived model indicates that the entire crust in the mobile belts is intensely deformed and is mainly composed of metamorphic rocks which are probably similar to deep crustal rocks in the cratons. In accordance with the views of Kröner [1976] and Shackleton [1976] it seems that the crust has become decratonized with time and is weakened, not stiffened, by events of deformation and metamorphism which have often been polycyclic.

The electrical model can be extended when the absolute values of resistivity obtained from the deep sounding results are taken into account.

The difference in crustal structure between a craton and a mobile belt is best typified by the results obtained across the Limpopo Belt

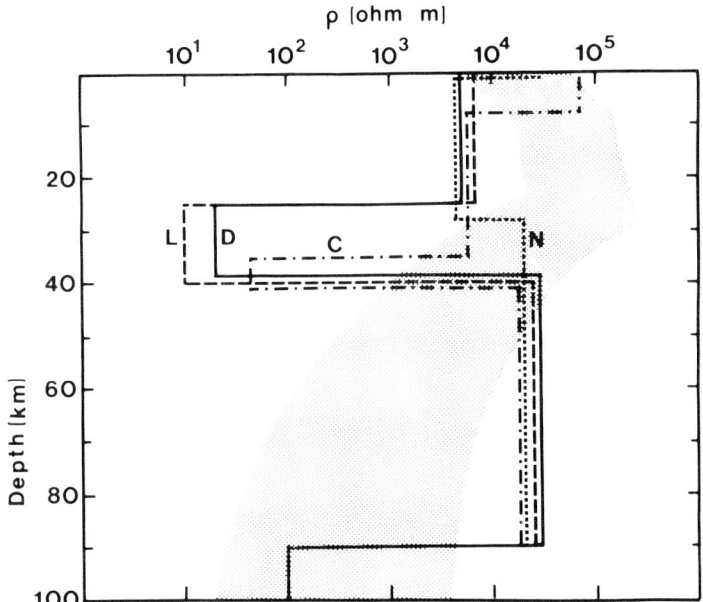

Fig. 14. Resistivity models for the Limpopo (L), Namaqua (N), and Damara (D) mobile belts and the Kaapvaal Craton (C) compared with the laboratory model (shaded) of Brace [1971], which has been extended to 100 km by using the Ringwood continental geotherm [Ringwood, 1969].

and into the adjoining cratons. The results of sounding ES 7 on the Kaapvaal Craton and sounding ES 8 on the Limpopo Mobile Belt (Figure 4) are depicted as resistivity-depth profiles in Figure 14. Also included in the figure is a resistivity model according to Brace [1971] for the stable 'normal continental' heat flow province of the eastern United States. In this model, Brace includes the effect of free water. The effect of temperature on the resistivity of dry rock only becomes dominant at large depths and is manifested by the last curved segment of the model. Brace's original model [Brace, 1971], to a depth of 40 km, has been extended downward to 100 km by using the Ringwood geotherm for shield areas [Ringwood, 1969]. Both lower and upper bounds of the Brace model are shown to allow a more realistic comparison with the sounding results.

It is seen that the Brace model predicts the interpreted resistivity of the highly resistive zone 1, as depicted by the results of sounding ES 7, quite well. However, it should again be emphasized that resistivity values obtained for zone 1 are generally minimum values because of T equivalence. Also, as has been mentioned earlier, the resistivity of zone 1 varies between 30,000 ohm m and 400,000 ohm m depending on its location within a craton. Consequently, the Brace model only applies to those areas where the resistivity of zone 1 tends to be below the average; in general, the model tends to underestimate the resistivity of zone 1.

Zone 2 has a resistivity which is appreciably lower than that predicted by the Brace model even when the spread of resistivities (2000-10,000 ohm m) for zone 2 is taken into account.

The general lack of agreement between the Brace model and the resistivity results for zones 1 and 2, that is, to a depth of about 25-30 km, is mainly due to Brace's assumptions of the porosity-depth relationship. From available temperature gradient estimates [Carte and Van Rooyen, 1969] which vary from $13°C/km$ to $20°C/km$, it can safely be assumed that the temperature to a depth of 25-30 km is too low for conduction through minerals to be important [Brace, 1972]. The crustal rocks conduct electricity by virtue of interconnected cracks filled with water [Brace, 1972]. The space occupied by water can be calculated if the rock resistivity and the water resistivity are known [Brace, 1971]. While the latter parameter cannot be measured directly at great depths, it can be estimated by the extrapolation of water resistivities measured near the surface. Measurements on water samples from boreholes and at a depth of 1200 m in a mine in the study area gave an average value of 15 ohm m and 2 ohm m, respectively. Results from mines in other areas gave values varying from 1 to 4 ohm m. As a first approximation a value of 1 ohm m has been assumed for water at depth. On this basis the porosity of zone 1 over its entire range of resistivities is less than 0.5% which substantiates its basically solid and unfractured nature as noted earlier. In zone 2 the calculated porosities are considerably larger, varying from 3% near the surface, where zone 2 is often more conductive, to about 1% at a depth of 25 km, and this implies that these rocks are cracked and fractured and consequently weakened. These porosity estimates are based on accurate resistivity determinations for zone 2 not subject to equivalence. As was noted previously, zone 2 continues laterally into the cratons. Various studies have indicated that pore pressure is high during metamorphism and that a high degree of crack space is required [e.g., Brace, 1972; Yoder, 1955]. These observations, considered jointly with the intensely deformed nature of these rocks as seen at the surface, may explain the origin of the cracks. The interconnectivity of cracks, which come right up to the surface in mobile belts, implies hydrostatic pore pressure when viewed over geological periods [Healy, 1971]. Although this view may not be generally accepted [Richter and Simmons, 1977] it is supported by the resistivity results over mobile belts which, as was noted above, indicate that the resistivity of zone 2 increases with depth. The origin for the water in the crust may well be meteoric as studies on magmatic bodies have shown [Friedman et al., 1974].

The finding that terrains containing high-grade metamorphic rocks are only moderately resistive is surprising. The explanation that the moderate resistivities reflect an appreciable free water content interconnected through cracks may seem even more surprising, since it would be expected that any free water entering the high-grade metamorphic systems would be consumed in hydration processes of anhydrous minerals. The remarks of Watson [1973] concerning the continued occurrence of granulites in areas where retrograde metamorphism and the introduction of water have been extensive may be appropriate.

The severalfold increase of the porosity in cratonic terrains at the boundary of the highly resistive zone 1 and the underlying zone 2 may account for the low-velocity layer found at this depth in the Canadian Shield [Berry and Fuchs, 1973], a region geologically similar to the areas studied in southern Africa.

The suggestion that free water is available at all levels in the crust has many implications regarding physical processes on one hand

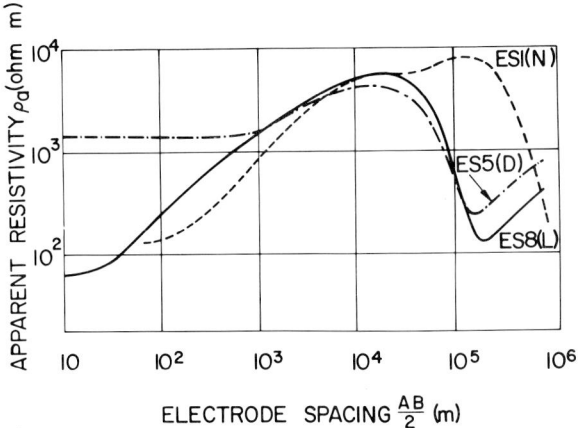

Fig. 15. Best fitting theoretical sounding curves for ultradeep soundings ES 8, ES 1, and ES 5 on the Limpopo (L), Namaqua (N), and Damara (D) mobile belts.

and geochemical effects on the other. The variation of heat flux and the nature of the crust-mantle boundary are only two examples of which brief mention will be made.

It has been observed that thermal gradients and heat flow estimates [Carte and Van Rooyen, 1969; A. E. Carte personal communication, 1976] from mobile belts appear to be anomalously high when compared to those from cratonic terrains, especially when the lower radioactive heat production of highly metamorphosed rocks is taken into account [Smithson and Decker, 1974]. This paradox may possibly be explained by the transport of heat by fluid convection through the cracked metamorphic zone and its concentration in mobile belts.

Regarding the crust-mantle transition zone, the electrical sounding results of the Limpopo and Damara mobile belts show the presence of the conductive zone 3 in the lower crust commencing at a depth of about 25 km and probably extending to the present-day Moho. The main characteristic of zone 3 is its variability in S. This is well illustrated by the relative positions of the final ascending branches of the three sounding curves obtained over the mobile belts shown in Figure 15. This change in S is also illustrated by the corresponding conductivity models in Figure 14. The S value of zone 3 increases from zero in the Pofadder area of the Namaqua Mobile Belt to at least 700 S in the Damara Orogen and reaches a maximum of 1500 S for the Limpopo Mobile Belt. Zone 3 is also weakly represented under the cratons. As shown in Figure 14, the high conductivity of zone 3 cannot be explained by the Brace model even when one takes into consideration the bounds imposed by S equivalence whereby only the maximum resistivity of zone 3 can be obtained. Since the geothermal gradients mentioned earlier also rule out the effect of temperature, it seems much more likely that the high conductivity of zone 3 may be associated in some way with the availability of water near the base of the crust. In each of the mobile belts studies, there is an inverse correlation between the grade of the last metamorphic event and the development of the conductive zone.

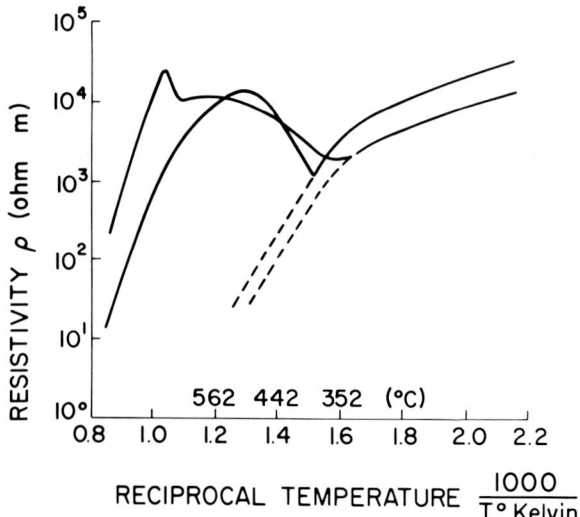

Fig. 16. Resistivity-temperature curves for two samples of serpentinite [after Zablocki, 1964]. The dashed lines show the inferred resistivity-temperature behavior when the water of hydration does not escape.

Sounding ES 1 (Pofadder sounding), which is located within the postulated thermal dome where the last metamorphic event as seen at the surface reached the granulite stage of metamorphism, does not show the presence of the conductive zone 3. On the other hand, zone 3 is well developed in the Damara and Limpopo mobile belts, where the last metamorphism as seen at the surface was less intense. Because of the higher tectonic level of exposure of the last metamorphic event as indicated by its lesser intensity in these mobile belts, the physical effects of this episode of regional metamorphism, such as the required high pore pressure and accompanying high degree of crack space, would continue to correspondingly greater depths than they would in the Namaqua Mobile Belt. It is conceivable that these physical effects could have continued into the uppermost mantle and thus that the conductive layer may have evolved by hydration processes.

The only known hydrated rock with the necessary stability and a low enough resistivity to account for the conductive layer near the crust-mantle transition is serpentinite. The resistivity-temperature behavior of two samples of this rock is shown in Figure 16 [Zablocki, 1964]. It is seen that provided the water of hydration remains in the rock, as would be expected at lower crustal depths, the conductivity of serpentinite is of the right order required by the presence of the conductive layer. This interpretation equates the conductive layer with hydrated mantle, a hypothesis originally put forward for layer 3 of the oceanic crust and for upper-mantle peridotite in continents by Hess in 1959 [Wyllie, 1971]. Parkhomenko et al. [1973] have also recently remarked on the possibility that conductive zones at crustal and upper-mantle depths are due to the presence of serpentinite. Further discussion of the question must, however, await additional resistivity sounding data still to be assembled. The section in Figure 7a shows the bottom of zone 3 as being horizontal and terminating at the present-day Moho, but this is pure speculation. It is interesting to note that detailed interpretation of North American and European seismic results have revealed the presence of a thick transition zone between the crust and the mantle with intermediate P wave velocities ranging from 7 to 8 km/s [Prodehl, 1970; Prodehl, 1977] the position of which corresponds closely with that of the conductive zone 3.

The thick resistive zone 4 in the uppermost mantle (Figure 13 and 14) is observed only in those ultradeep sounding curves with the largest depths of investigation. The high resistivity of zone 4 indicates that it is dry. It may be mentioned that the dry rock resistivities used by Brace for his model are probably too low, since the role of oxygen fugacity was not taken into consideration (A. Duba, personal communication, 1976).

The evidence for the final conductive zones is not very strong; it has only been obtained from the Pofadder sounding curves, ES 1, at very long current electrode spacings. Nevertheless, its presence is adequately predicted by increasing temperature with depth.

Correlation of Sounding Results From Southern Africa With Other Countries

The moderately resistive zone (zone 2) associated with mobile belts was detected in a deep electrical sounding (AB = 22 km) on granitic terrain of the Alpine Orogenic Belt in Germany by Blohm [1972]. It occurs close to the surface, and has an estimated thickness of 24 km and a resistivity of 7000 ohm m.

Both the moderately resistive mobile belt zone and the high-resistivity dry upper mantle are revealed by the deep sounding investigation of Meunier [1975] in the Massif Central, France. Here the moderately resistive zone has a thickness of 30 km and a resistivity of about 5000 ohm m. Its base has been correlated with the Moho by seismic results. The resistive upper mantle is thought to be 50 km thick with a resistivity in the vicinity of 50,000 ohm m. This area was deformed by intense metamorphism in Hercynian times [Holmes, 1965].

The deep sounding results in the United States [Keller et al., 1966] show the presence of a highly resistive zone in some areas and a moderately resistive zone in others, but the pattern is as yet unclear.

Conclusions

Deep and ultradeep electrical sounding results obtained over two cratons and three of the major mobile belts of southern Africa have made it possible for the first time to determine the resistivity profile through the crust and into the upper mantle for those provinces in some detail. These data suggest that specific electrical zones often characterize certain tectonic situations in the different tectonic provinces. The tectonic provinces in southern Africa have been stable since the Precambrian; hence it is very unlikely that residual temperature plays any significant role in determining the resistivity profile in the crust. The geoelectrical investigations thus respond to differences in physical nature between the different tectonic provinces. The physical nature or constitution of a tectonic unit is the end product of its tectonic history, and in this context it should be remembered that the geoelectrical methods record the total effect of past and present conditions. The correlation between tectonics and resistivity has made it possible to interpret the electrical results in terms of crustal structure:

1. There is no continuous high-resistivity zone in the crust.
2. The cratons are characterized by a highly resistive zone extending downward from the surface to a depth of 5-10 km depending on the locality. The average resistivity of this zone is about 100,000 ohm m with values varying between 30,000 and 400,000 ohm m. The rocks are thought to be cold, massive, and largely unfractured, consisting of large volumes of granitoid material in the form of granitic intrusions, gneisses, and migmatites. High resistivities have also been recorded outside the intensely deformed core of the Namaqua Mobile Belt, and this suggests that here too cratonic material exists.
3. The Limpopo Mobile Belt, the central core of the Namaqua Mobile Belt, and the Damara Orogen (eugeosynclinal basin) are characterized by a thick, moderately resistive zone with a resistivity ranging from 2000 to 10,000 ohm m (average, 5000 ohm m) extending from the surface to a depth of 25-30 km. An important feature deduced from the electrical results is that this zone continues laterally into the cratons, where it underlies the highly resistive zone. The mobile belts are made up of intensely deformed, fractured, and highly metamorphosed rocks, the intensity of fracturing increasing toward shallower depths. The lower crust under both cratons and mobile belts is interpreted as being metamorphic and fractured, consisting mainly of a refractory residuum which has been depleted of lighter granitic materials.
4. A surprising finding is that a major portion of the upper crust (in mobile belts) and the entire lower crust to a depth of 25-30 km is only moderately resistive, a situation that provides strong evidence for the presence of free water in these localities. Calculations based on measurements of subsurface water resistivities and the electrical results have provided some crude constraints on the distribution of porosity in the crust. The highly resistive zone of the cratonic terrains has a porosity of less than 0.5% but there is a serveralfold increase in porosity at the transition to the moderately resistive zone in the depth range 5-10 km. This transition may account for the low-velocity layer found at this depth in the Canadian Shield [Berry and Fuchs, 1973], a region geologically similar to the areas studied in southern Africa. In the mobile belts the porosity varies from about 3% near the surface to 1% at a depth of 25 km.
5. If water is present in the lower crust as is suggested by the electrical results, it may provide a means of explaining the origin of the conductive zone of variable conductance that has been recognized in the depth range 25-40 km. The preferred explanation at this stage is that the conductive zone forms by hydration, and it is shown that serpentinite (hydrated mantle rock) could account for its electrical properties. It is interesting to note that detailed interpretation of North American and European seismic results has revealed the presence of a thick transition zone between the crust and the mantle with P wave velocities ranging from 7 to 8 km/s [Prodehl, 1970; Prodehl, 1977].
6. The uppermost mantle is characterized by a thick, highly resistive zone which becomes more conductive as the depth and temperature increase. Its electrical resistivity suggests that it is dry.

7. The electrical results indicate that there is a significant difference in the porosity distribution between cratons and mobile belts. If the heat flow pattern in South Africa is taken as an example, it is suggested that the general tendency for heat flow to be higher over the mobile belts may be due to the preferential movement of water in these zones.

Acknowledgments. The author would like to thank R. M. J. Huyssen, J. H. de Beer, S. J. Joubert, P. L. V. Hugo, R. Meyer, D. Barlow, and the other past and present members of the Geophysics Division for their aid throughout the years in carrying out the fieldwork and reducing the results. J. H. de Beer and S. J. Joubert deserve special thanks for their part in the interpretation of the results and critical review of the manuscript. Without the support of the Department of Posts and Telegraphs, the Electricity Supply Commission, the Gabinete do Plano do Zambeze (Moçambique), the Geological Surveys of Moçambique and Rhodesia, and the Railways Department of Moçambique, the deep sounding projects would not have been poysible. Special thanks are also due to E. Raynham, J. H. Harden, R. B. McAinsh, Crispim de Sousa, Rui Santos Garcia, J. W. Wiles, A. da Costa, and F. Pastor.

References

Alpin, L. M., M. N. Berdichevskii, G. A. Vedrintsev, and A. M. Zagarmistr, Dipole Methods for Measuring Earth Conductivity, Consultants Bureau, New York, 1966.

Anhaeusser, C. R., The evolution of the early Precambrian crust of southern Africa, Phil. Trans. Roy. Soc. London, Ser. A. $\underline{273}$, 359-388, 1973.

Berckhemer, H., Direct evidence for the composition of the lower crust and the Moho, Tectonophysics $\underline{8}$, 97-105, 1969.

Berry, M. J., and K. Fuchs, Crustal structure of the Superior and Grenville provinces of the northeastern Canadian Shield, Bull. seismol. Soc. Amer., $\underline{63}$, 1393-1432, 1973.

Blignault, H. J., M. P. A. Jackson, G. J. Beukes, and D. J. Toogood, The Namaqua tectonic province in South West Africa, Bull. 15, pp. 29-48, Precambrian Res. Unit. Univ. of Cape Town, Cape Town, South Africa, 1974.

Blohm, E. K., Die Methode der geoelektrischen Tiefensondierungen mit grossen Elektrodenentfernungen, Ph.D. thesis, Tech. Univ. of Clausthal, Clausthal, West Germany, 1972.

Bott, M. H. P., and S. B. Smithson, Gravity investigations of granite batholiths, Geol. Soc. Amer. Bull., $\underline{78}$, 859-878, 1967.

Brace, W. F., Resistivity of saturated crustal rocks to 40 km based on laboratory results, in The Structure and Physical Properties of the Earth's Crust, Geophys. Monogr. Ser., Vol 14, edited by J. G. Heacock, pp. 243-255, AGU, Washington, D.C., 1971.

Brace, W. F., Pore pressure in geophysics, in Flow and Fracture of Rocks, Geophys. Monogr. Ser., Vol. 16, edited by H. C. Heard, I. Y. Borg, N. L. Carter, and C. B. Raleigh, pp. 265-274, AGU, Washington, D.C., 1972.

Carte, A. E., and A. I. M. van Rooyen, Further measurements of heat flow in South Africa, Upper Mantle Project, Geol. Soc. S. Afr., Spec. Publ., 2, 445-448, 1969.

Christensen, N. I., and D.M. Fountain, Constitution of the lower continental crust based on experimental studies of seismic velocities in granulite, Geol. Soc. Amer. Bull,, 86, 227-236, 1975.

Clifford, T. N., The structural framework of Africa, in African Magmatism and Tectonics, edited by T. N. Clifford and I. G. Gass, pp. 1-26, Oliver and Boyd, Edinburgh, 1970.

Clifford, T. N., Review of African granulites and related rocks, Geol. Soc. Amer. Spec. Pap., 156, 1-49, 1974.

De Beer, J. H., J. S. V. van Zijl and F. K. Bahnemann, Plate tectonic origin for the Cape Fold Belt, Nature, 252, 675-676, 1974.

De Beer, J. H., J. S. V. van Zijl, R. M. J. Huyssen, P. L. V. Hugo, S. J. Joubert, and R. Meyer, A magnetometer array study in South West Africa, Botswana and Rhodesia, Geophys. J. Roy. Astron. Soc., 45, 1-17, 1976.

Friedman, I., P. W. Lipman, J. D. Obradovich, J. D. Gleason, and R. L. Christiansen, Meteoric water in magmas, Science, 184, 1069-1072, 1974.

Gough, D. I., Electrical conductivity under western North America in relation to heat flow, seismology, and structure, J. Geomagn. Geoelec., 26, 105-123, 1974.

Gough, D. I., J. H. de Beer, and J. S. V. van Zijl, A magnetometer array study in southern Africa, Geophys. J. Roy. Astron. Soc., 34, 421-433, 1973.

Healy, J. H. A comment on the evidence for a worldwide zone of low seismic velocity at shallow depths in the earth's crust, in The Structure and Physical Properties of the Earth's Crust, Geophys. Monogr. Ser., Vol. 14, edited by J. G. Heacock, pp. 35-40, AGU, Washington, D.C., 1971.

Holmes, A., Principles of Physical Geology, Thomas Nelson, London, 1965.

Hunter, D. R., Crustal development in the Kaapvaal Craton, 1, The Archaean, Precambrian Res.,1, 259-294, 1974.

Jacobsen, J. B. E., D. C. Rex, and W. J. Sevenster, K-Ar ages of some mafic dykes from the Messina district, Transvaal, and their bearing on the age of copper mineralization, Trans. Geol. Soc. S. Afr., 78, 359-360, 1975.

Joubert, P., The regional tectonism of the gneisses of part of Namaqualand, Bull. 10, pp. 1-220, Precambrian Res. Unit, Univ. of Cape Town, Cape Town, South Africa, 1971.

Keller, G. V., L. A. Anderson, and J. I. Pritchard, Geological survey investigations of the electrical properties of the crust and upper mantle, Geophysics, 31, 6, 1078, 1966.

Keller, G. V. and F. C. Frischknecht, Electrical Methods in Geophysical Prospecting, Pergamon, New York, 1966.

Kröner, A., Proterozoic crustal evolution in parts of southern Africa and evidence for extensive sialic crust since the end of the Archaean, Phil. Trans. Roy. Soc. London, Ser. A, 280, 541-553, 1976.

Kunetz, G., Principles of Direct Current Resistivity Prospecting, Gebrüder Borntraeger, Berlin, 1966.

Lambert, I. B., and K. S. Heier, The vertical distribution of uranium, thorium and potassium in the continental crust, Geochim. Cosmochim. Acta, 31, 377-390, 1967.

Martin, H., Problems of age relations and structure in some metamorphic belts of southern Africa, Geol. Ass. Can. Spec. Pap., 5, 17-26, 1969.

Martini, J. E. J., The presence of ash beds and volcanic fragments in the greywackes of the Karroo System in the southern Cape Province, South Africa, Trans. Geol. Soc. S. Afr., 77, 113-116, 1974.

Mason, R., The Limpopo Mobile Belt - southern Africa, Phil. Trans. Roy. Soc. London, Ser. A, 273, 463-486, 1973.

Meunier, J., L'apport des sondages à courant continu dans l'étude de la résistivité des couches profondes de la terre - comparaison avec des résultats magnéto-telluriques, Ph.D. thesis, Univ. Louis Pasteur, Strasbourg, France, 1975.

Nicolaysen, L. O., and A. J. Burger, Note on an extensive zone of 1000 million year old metamorphic and igneous rocks in southern Africa, Sci. Terre, 10, 500-518, 1965.

Orellana, E., Prospeccion Geoelectrica en Corrienta Continua, Paraninfo, Madrid, Spain, 1972.

Parkhomenko, E. I., B. P. Belikov, and E. Dvorzhak, Influence of serpentinization upon the elastic and electrical properties of rocks, Izv. Akad. Nauk. SSSR, Fiz. Zemli, 8, 101-108, 1973. (Izv. Akad. Sci. USSR, Phys. Solid Earth, Engl. Transl. 8, 551-556, 1973.)

Prodehl, C., Seismic refraction study of crustal structure in the western United States, Geol. Soc. Amer. Bull., 81, 2629-2646, 1970.

Prodehl, C., The structure of the crust-mantle boundary beneath different tectonic areas of North America and Europe as derived from explosion seismology, in The Earth's Crust, Geophys. Monogr. Ser., vol. 20, edited by J. G. Heacock, AGU, Washington, D.C., this volume, 1977.

Richter, D., and G. Simmons, Microcracks in crustal igneous rocks : Microscopy, in The Earth's Crust, Geophys. Monogr. Ser., vol 20, edited by J. G. Heacock, AGU, Washington, D.C., this volume, 1977.

Ringwood, A. E., Composition and evolution of the upper mantle, in The Earth's Crust and Upper Mantle, Geophys. Monogr. Ser., vol. 13, edited by P. J. Hart, pp. 1-17, AGU, Washington, D.C., 1969.

Ringwood, A. E., and D. H. Green, Petrological nature of the stable continental crust, in The Earth Beneath the Continents, Geophys. Monogr. Ser., vol 10, edited by J. S. Steinhart and T. J. Smith, pp. 611-619, AGU, Washington, D.C., 1966.

Shackleton, R. M., Pan-African structures, Phil. Trans. Roy. Soc. London, Ser. A, 280, 491-498, 1976.

Smithson, S. B. and E. R. Decker, A continental crustal model and its geothermal implications, Earth Planet. Sci. Lett., 5, 1-12, 1974.

Van Breemen, O., and M. H. Dodson, Metamorphic chronology of the Limpopo Belt, southern Africa, Geol. Soc. Amer. Bull. 8, 2005-2018, 1972.

Van Zijl, J. S. V., A deep Schlumberger sounding to investigate the electrical structure of the crust and upper mantle in South Africa, Geophysics, 34, 450-462, 1969.

Van Zijl, J. S. V., and S. J. Joubert, A crustal geoelectrical model for South African Precambrian granitic terrains based on deep Schlumberger soundings, Geophysics, 40, 657-663, 1975.

Watson, J. V. Effects of reworking on high grade gneiss complexes, Phil. Trans. Roy. Soc. London, Ser. A, 273, 443-456, 1973.

Wegmann, E., Zur Deutung der Migmatite, Geol. Rundsch., 26, 305-350, 1935.

Wilson, J. F., The Rhodesian Archaean Craton - An essay in cratonic evolution, Phil. Trans. Roy. Soc. London, Ser. A, 273, 389-412, 1973.

Wyllie, P. J., The Dynamic Earth: Textbook in Geosciences, John Wiley, New York, 1971.

Yoder, H. S., Role of water in metamorphism, Geol. Soc. Amer. Spec. Pap., 62, 505-524, 1955.

Zablocki, C. J., Electrical properties of serpentinite from Mayaguez, Puerto Rico, in A Study of Serpentinite of the Amsoc Core Hole Near Mayaguez, p. 1188, National Academy of Sciences, National Research Center, Washington, D.C., 1964.

Zohdy, A. A. R., G. P. Eaton, and D. R. Maby, Application of surface geophysics to ground-water investigations, chapter D1, Techniques of Water-Resources Investigations of the USGS, pp. 116, U. S. Geological Survey, Reston, Va., 1974.

FLAMBEAU ANOMALY: A HIGH-CONDUCTIVITY ANOMALY IN THE SOUTHERN EXTENSION OF THE CANADIAN SHIELD

Ben K. Sternberg[1] and C. S. Clay

Department of Geology and Geophysics, University of Wisconsin
Madison, Wisconsin 53706

Abstract. Electrical resistivity measurements in northern Wisconsin show the presence of a large high-conductivity anomaly in the crust. The anomaly, which we refer to as the Flambeau Anomaly, is approximately 20 km wide, lies roughly along 46°N latitude, and extends east of 91°W longitude for more than 100 km. Detailed airborne electromagnetic surveys over parts of the anomaly show that the anomalous region consists of four to six roughly parallel conductive formations. Core samples from these formations indicate that graphite, sulfides, and iron formation are the principle conductive materials causing the anomaly. Numerical analysis of resistivity soundings surrounding the Flambeau Anomaly indicates the following layered earth crustal structure: (1) a surficial cover of 0 to 100 m thickness has a resistivity of a few hundred ohm meters; (2) a second layer has a resistivity of a few thousand ohm meters and a thickness of 5 to 10 km; (3) a third layer has a resistivity of greater than 10^5 ohm m; (4) at depths of 15 to 25 km the resistivity decreases to the order of 1000 ohm m. Our model of the Flambeau Anomaly consists of five thin vertical half plane conductors. Numerical modeling indicates that the conductors extend to a depth of the order of 10 km. The Flambeau Anomaly appears to be a major structural feature in the southern extension of the Canadian Shield.

Introduction

Our purpose is to report the nature of an extensive low-resistivity anomaly in the southern extension of the Canadian Shield. North and south of the anomaly (Figure 1) the resistivities of the shield rocks are generally very high.

A great many resistivity soundings were made in northern Wisconsin in connection with site surveys for the Navy's Sanguine, Wisconsin Test Facility, communication antennas. The measurements included 68 dipole-dipole dc resistivity soundings. The locations of the

---

[1] Now at Continental Oil Company, Geophysics Research, Ponca City, Oklahoma 74601.

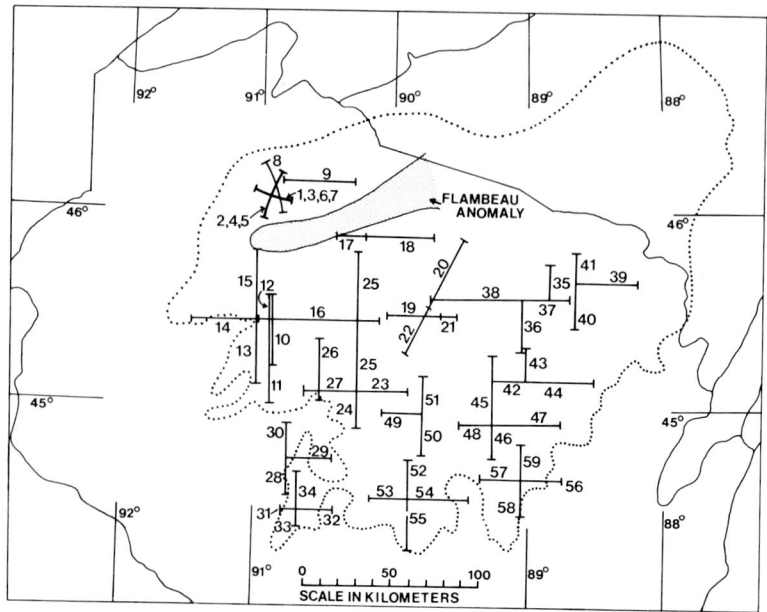

Fig. 1. Southern extension of the Canadian Shield. The region within the dotted line consists of crystalline Precambrian rocks covered by a thin (0 to 100 m) layer of glacial till. Surrounding this region, the Precambrian shield rocks are covered by increasing thicknesses of relatively young, sedimentary rocks. The shaded region labeled Flambeau Anomaly is the high-conductivity anomaly discussed in this paper. The numbered lines are the locations of the dc dipole-dipole resistivity soundings shown in Figure 2.

soundings are plotted in Figure 1. The data are from the following reports: Naval Electronic Systems Command [1972], Davidson et al. [1974], DECO [1965, 1966, 1967, 1968], GTE Sylvania [1972], and Lahman and Nelson [1964]. Some of these reports are classified, but the conductivity data were released and are on file at the Geology and Geophysics Department of the University of Wisconsin.

A compilation of all the available sounding curves is shown in Figure 2. The apparent resistivity is plotted at an effective spacing (effective spacing equals one half of the actual transmitter-receiver separation for in-line or polar dipoles and is equal to the actual separation for broadside or equatorially displaced dipoles).

All the sounding curves indicate very high apparent resistivities (greater than 10^4 ohm m) at 10- to 20-km spacing. In other words, generally high resistivity rocks occur between the transmitter and receiver dipoles. Sternberg [1974] interpreted the soundings made near Clam Lake (soundings 1-9 in Figures 1 and 2) using the following plane-layered model: a surficial layer about 10 m thick with a resistivity of a few hundred ohm meters, a layer with a resistivity

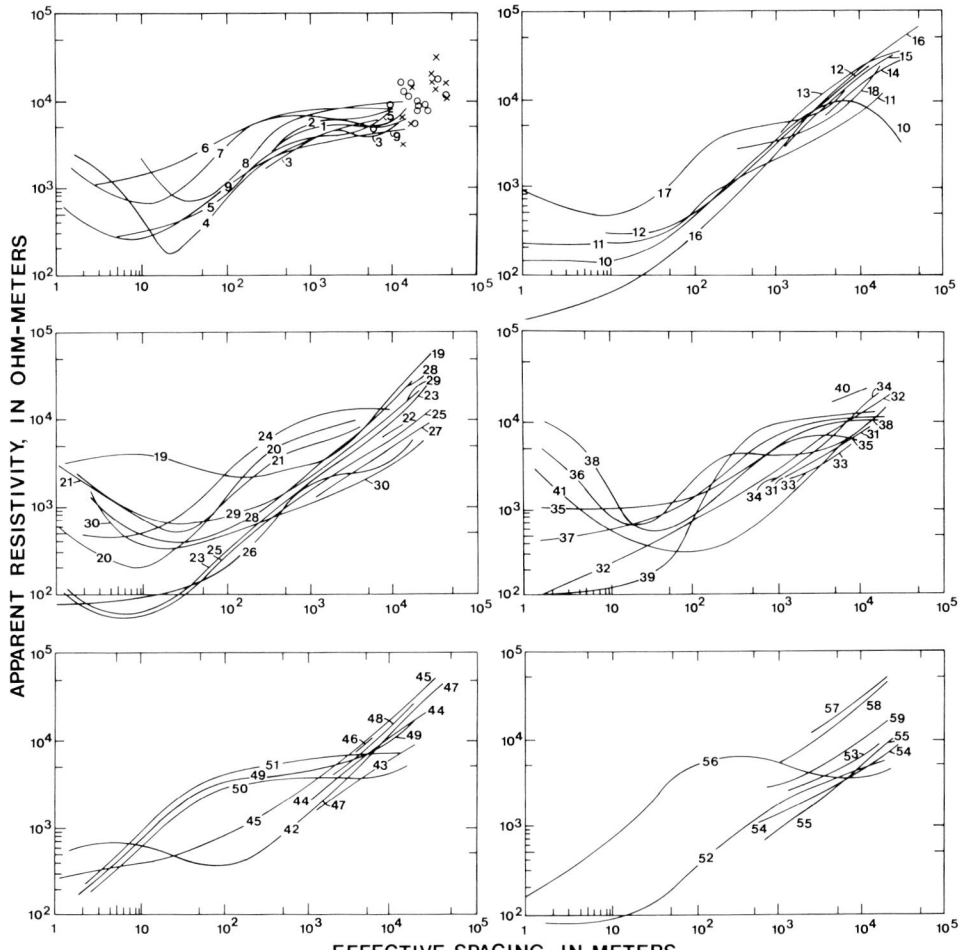

Fig. 2. Direct current resistivity soundings. The numbers on the sounding lines correspond to the numbers on the location map in Figure 1. Soundings 1-3 are from GTE Sylvania [1972] and Davidson et al. [1974]. The crosses in the upper left-hand panel are from the University of Texas-Austin, 1974, and the circles are from the University of Wisconsin, 1973, 1974, and 1975 (see Figure 3). Soundings 23 and 27 are from Lahman and Nelson [1964]. Soundings 13-16, 19-22, 31-34, 39-48, and 52-54 are from DECO [1965]; 23-25, 27-30, and 49-51, from DECO [1966]; 8-12, 17, 18, 26, and 35-38, from DECO [1967]; and 4-7, from DECO [1968].

of a few thousand ohm meters down to about 7 km; and a layer with a resistivity of greater than 10,000 ohm m below 7 km. The limited spacing of the dc soundings prevents detection of any deeper layers. The soundings south of the region labeled Flambeau Anomaly in Figure 1 have sounding curves similar to those near Clam Lake. Most of the variations in apparent resistivity for electrode spacings of less than 1 km can be attributed to different thicknesses of the surficial sediments.

Several of the sounding curves in Figure 1 extend just to the boundary of the low-resistivity anomaly (Flambeau Anomaly) which we discuss in this paper. None of the reported soundings crossed this feature.

Dowling's [1970] magnetotelluric (MT) soundings on the southern extension of the Canadian Shield also show high-resistivity rocks (1000 to 12,000 ohm m) in the upper crust. These rocks overlie less resistive rocks at 15- to 25-km depth.

The dc resistivity measurements and magnetotelluric measurements cover a broad section of the southern extension of the Canadian Shield and show that the shield consists typically of very high resistivity rocks to depths of the order of 10 km.

In 1973 we began a series of measurements to determine in more detail variations in resistivity of the shield, both laterally and vertically. The Navy's Sanguine test antennas near Clam Lake, Wisconsin, were used as a source to extend the dc resistivity soundings (Figure 2a) to a range of 41 km. Plane-layered interpretations of electromagnetic transient data from these sites indicated a conductive layer at depths of the order of 20 km [Sternberg, 1974; Kan, 1975; Johnston, 1975]. Sites at larger range showed extremely low signals and apparent resistivities. These sites were south of the northern edge of the area labeled Flambeau Anomaly in Figure 1. The abrupt drop of apparent resistivity precluded a layered model interpretation, and we believed that these sites were over a very conductive feature [Sternberg, 1974].

In the summer of 1974 we entered a joint study of the crust with groups from the University of Texas-Austin, the University of Texas-Dallas, and the Colorado School of Mines. The Texas groups and our group used transmissions from the Sanguine antennas. Parties from the Colorado School of Mines made dipole-dipole measurements along an east-west line and south of what we now call the Flambeau Anomaly.

The combined results from this field season confirmed the presence of an extensive low-resistivity anomaly in the crust (the Flambeau Anomaly). The results of the joint study also indicate that the crust surrounding the Flambeau Anomaly probably has a maximum resistivity of greater than 10^6 ohm m [Keller and Furgerson, 1977]. The magnetotelluric soundings by Bostick et al. [1977] show the continuity of the lower crustal conductive layer in the area surrounding the Flambeau Anomaly.

In 1975 we studied in greater detail the nature of the low resistivity anomaly. These measurements included audiomagnetotelluric (AMT) soundings and long- and short-range dipole-dipole soundings.

In this paper we present some of the geophysical-geological results for this low-resistivity anomaly.

Long-Range Dipole-Dipole Transmissions

The transmissions were from the Navy's Sanguine test antennas to widely spaced receiver sites. The locations of the transmitter and receiver sites are shown in Figure 3. Either the north-south

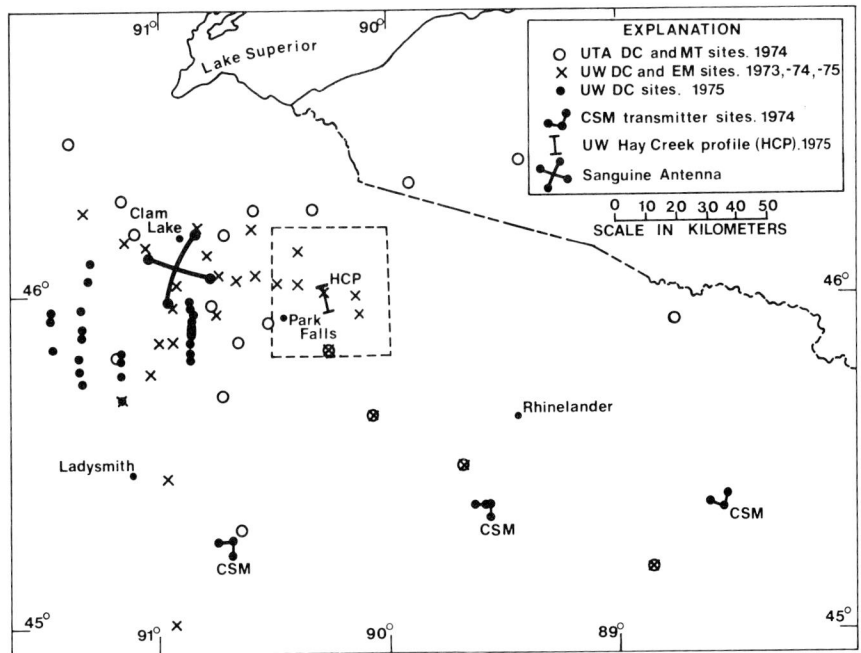

Fig. 3. Location of transmitter and receiver sites. University of Texas-Austin (UTA) recorded dc signals from the Sanguine antennas and natural source magnetotelluric (MT) data. University of Wisconsin (UW) recorded dc and electromagnetic (EM) transient signals from the Sanguine antennas. Colorado School of Mines received dc signals at 309 sites, at ranges up to 60 km, from three transmitter locations labeled CSM. The dashed line box near 90°W, 46°N is the region of the INPUT survey (Figure 5).

or the east-west antennas (24 and 22 km long) were energized by square wave pulses of dc current. The periods and amplitudes of the square waves were 2 s and 70 A for transmissions north of the Flambeau Anomaly and 10 s and 140 A for transmissions to sites south of the anomaly. The electric field signals were received with a pair of orthogonal grounded wire dipoles having spacings of 30 to 300 m.

The received signal level is typically well below the level of naturally occurring noise at the distant sites. We used a PDP 8/E computer at the field sites denoted by crosses in Figure 3 to average (or stack) a number of transmitted pulses in order to improve the signal to noise ratio. The signal to noise improvement for N transmissions is proportional to $N^{1/2}$. We digitized and recorded the entire transient wave form, but we will discuss only the dc signals (fundamental of the square wave pulse) in this paper. At the field sites denoted by solid circles in Figure 3 the dc signal to noise ratio was large enough that we could use just an amplifier and a strip chart recorder. Details of the experimental procedures are given by Sternberg [1977b].

The received signals have an initial transient that is associated with the change of current at the antenna, and then the signal tends to its dc value. North of the Flambeau Anomaly and at ranges up to 50 km, received voltages at 2-s periods differed by less than 10% from those at 10-s periods for both the N-S and E-W antenna transmissions; so the signals were approximately dc. South of the anomaly, and at the extreme ranges, the broadside (equatorial) dipole received voltages at 10-s and 40-s periods differed by less than 10%; so the signals were approximately dc. We could not verify that the in-line (polar) dipole results had actually reached dc at 10-s periods south of the Flambeau Anomaly. We tried 40-s in-line transmissions, but we could not detect this low-level signal because of the relatively large natural noise levels at these periods. We used apparent resistivities calculated from the 10-s transmissions, and we believe they are probably close to dc.

We removed the effect of the source-receiver geometry by calculating the resistivity of a uniform half space that would give the same field as was actually observed at the receiving site. This resistivity is called the apparent resistivity (ρ_a) and is calculated as follows [Keller et al., 1975]:

$$\rho_a = 2\pi R_1^2 \frac{E_t}{I} \left[1 + \left(\frac{R_1}{R_2}\right)^4 - 2\left(\frac{R_1}{R_2}\right)^2 \cos D \right]^{-1/2} \quad (1)$$

where R_1 and R_2 are distances along a line from the ends of the transmitter dipole to the receiver site location, D is the angle between lines R1 and R2, E_T is the received vector sum electric field, and I is the current in the transmitter dipole. Qualitatively, apparent resistivity represents a combination of all the rock resistivities between the transmitter and receiver, including those rocks at depths comparable to the transmitter-receiver spacing and to a lesser extent including the resistivities of rocks off to the side and off the ends of the transmitter-receiver line.

The local fields at the receiving dipoles are often distorted by local anomalies, and an appreciable broadside component is observed for an in-line transmission, and vice versa. The apparent resistivities for each transmission are calculated from the vector sum of orthogonal components of the received electric field.

A map of the apparent resistivities for transmissions from the Sanguine antennas to each receiver location is shown in Figure 4. The value of ρ_a is plotted at each receiver location. Qualitatively, the region indicated by the 10^4-ohm-m contour line has high-resistivity rocks between the transmitter and receiver. The region enclosed by the 10-ohm-m contour is a low-resistivity region, or there are low-resistivity rocks between the transmitter and receiver. At large range (greater than 100 km) in the south and east directions the apparent resistivity increases again to over 10^3 ohm m. This indicates that the crustal rocks in this region have a high resistivity.

The measurements taken along lines south and southeast from the antenna indicate a large E-W trending low-resistivity feature, as shown in Figure 1. F. X. Bostick and H. W. Smith suggested the name

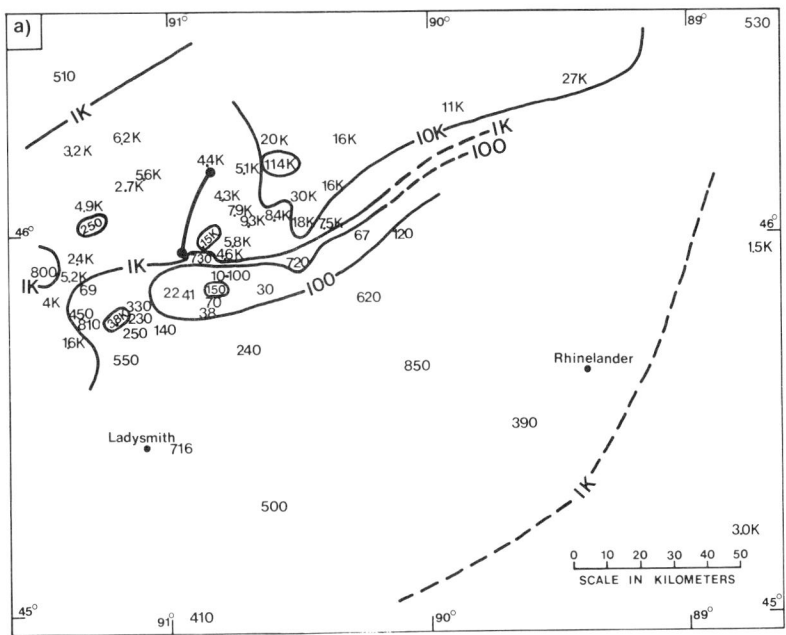

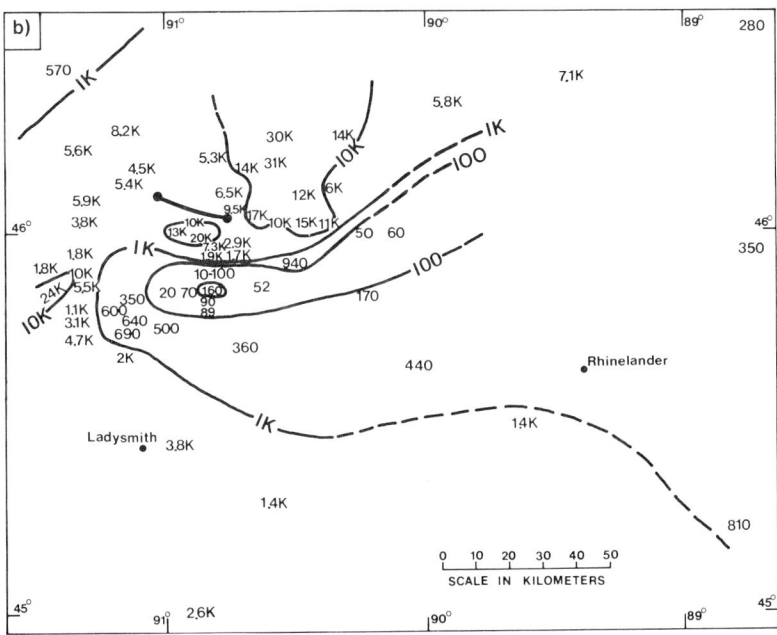

Fig. 4. Direct current apparent resistivity map. (a) North-south Sanguine antenna transmissions. Apparent resistivities in ohm meters are plotted at the receiver position. The data include UTA and UW measurements. (b) East-west Sanguine antenna transmissions.

Flambeau Anomaly because this feature is roughly coincident with the Flambeau River.

Near-Surface Characteristics of the Flambeau Anomaly

We sought help from commercial exploration groups concerning the near-surface nature of the Flambeau Anomaly. Kennecott Copper Corporation furnished the map of an INPUT MARK V survey shown in Figure 5.

INPUT (induced pulse transient) is an airborne electromagnetic surveying system developed by Barringer Research Ltd. (INPUT is a registered trademark of Barringer Research Ltd., Rexdale, Ontario, Canada.) A half-sine wave pulse lasting 1.0 ms was transmitted from a loop encircling the airplane. The secondary field induced in the earth after turnoff of the primary pulse is sensed with a small receiver coil in a 'bird' trailing from the plane. The secondary field signal is sampled at six times after the pulse turnoff (0.3, 0.5, 0.7, 1.1, 1.5, and 1.9 ms). The alternating polarity half-sine wave pulse is repeated every 2.97 ms. Each of the six sample amplitudes is integrated with a time constant of 3 s and displayed continuously on a multichannel strip chart recorder. A large secondary field response from the earth indicates highly conductive rock beneath the airplane. The depth of penetration depends upon the resistivity of the surface layer and is of the order of 10 to a few hundred meters. The INPUT MARK V system is described in more detail by Boniwell [1967] and Becker et al. [1972].

The shaded areas in Figure 5 represent regions where the second-channel secondary field deviated a prescribed amount from the background noise level. Most of these shaded regions correspond to a large secondary field on all six channels, indicating very low resistivity material.

The high density of data from this survey shows the detailed distribution of the low-resistivity material within the Flambeau Anomaly region. The anomaly appears to be a complicated band of interfingering low-resistivity features or conductors.

Glacial till covers nearly the entire region of the airborne INPUT survey, and outcrops are scarce. Kennecott Copper Corporation gave us cores from two drill holes near Parker Lake, 1 km east of Butternut. Calumet and Hecla, Universal Oil Products supplied us with cores and survey information from an area 4 km southwest of Park Falls. Jones and Laughlin Steel Corporation gave the Wisconsin Geological and Natural History Survey drill hole information and cross sections for an area about 1 km west of Butternut.

The Kennecott drill hole intersected a 10- to 15-m-wide zone of 'pencil quality' graphite and up to 20% pyrite. The conductive sheet is steeply dipping and is covered by about 40 to 50 m of glacial till. Surrounding the conductor is chlorite schist, actinolite schist, lapilli tuff, meta-arkose, metagraywacke, sheared pink granite, plus disseminated pyrite and graphite. We measured the resistivities of the most conductive sections of the core and found that they averaged about 0.03 ohm m. The conductive material occurs in patches and veins, and the average bulk resistivity might be considerably higher than this.

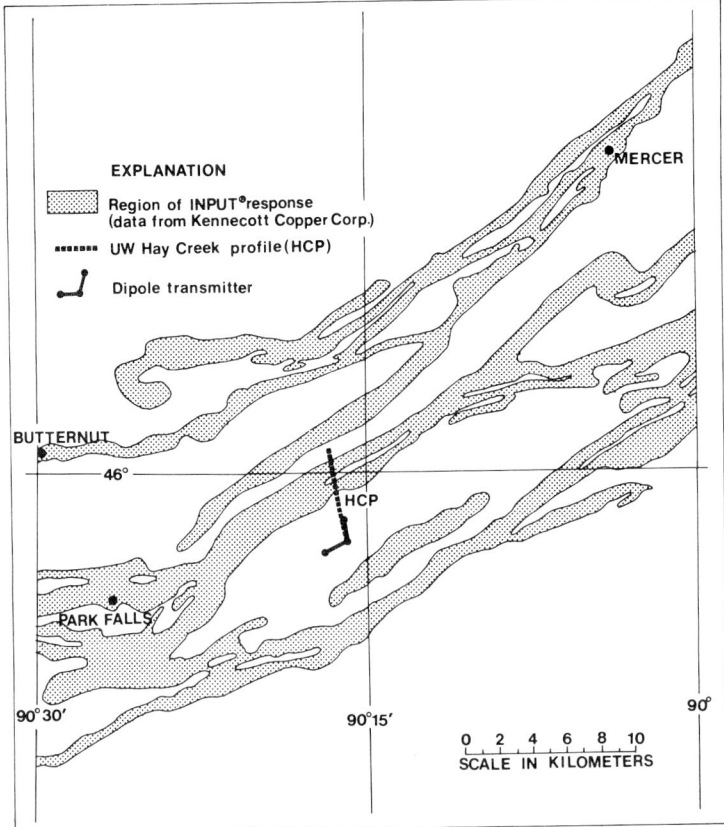

Fig. 5. Airborne INPUT survey conductivity anomalies (supplied by Kennecott Copper Corporation, 1974). Regions of large INPUT response (due to secondary EM fields induced in highly conductive rocks) are shaded. The actual width of the highly conductive rock is usually less than the width of the INPUT anomaly.

The Jones and Laughlin cores from just west of Butternut are from the same electrical and magnetic anomaly as the Kennecott cores, which are from just east of Butternut. The Jones and Laughlin cores contain magnetic iron formation. Evidently, facies changes can occur over short distances along a single conductor (stratigraphic unit). Dutton [1975] has analyzed the cores from this property and found fine chalcopyrite associated with this magnetic cherty iron formation. The iron formation is contained in three to five roughly vertical sheets about 10 to 30 m wide. The surrounding rocks are generally granite and schist.

The Calumet and Hecla cores indicated a number of steeply dipping graphite and disseminated pyrite-pyrrhotite conductors. Fine veins of graphite and scattered grains of pyrite occur throughout the 150 m of the core. Traces of chalcopyrite were also reported. Abundant graphite and pyrite occur in 1- to 5-m-wide bands at several locations along the drill holes. The rocks surrounding the conductors are amphibolite, pegmatite, and schist. We measured the electrical resistivity of samples from the most conductive parts of the core and found that they averaged about 0.1 ohm m.

Figure 6 represents our estimate of the areal extent of the near-surface expression of the Flambeau Anomaly. Since this map is derived from a combination of different measurements having different spatial resolutions, we indicate the bounds given by each of these methods. The INPUT survey has the highest resolution. The INPUT anomalies from Figure 5 are shown as dark bands between longitude 90°30'W and 90°W on Figure 6. Rocks having very low resistivities occur somewhere within each of the plotted bands. From Figure 4 we find very low apparent resistivities between 90°30'W and 91°W for the long-range Sanguine antenna transmissions. West of 91°W, apparent resistivities increase to greater than 10^3 ohm m, indicating that the conductors terminate around 91°W. The northern edge of the anomaly is well determined wherever a line of sites crosses the first conductor. The western and southern edges are not as well determined. On Figure 6 we use dark shading to indicate areas where we believe that there are conductors and light shading to indicate where conductors may exist. We do not know how far east of 90°W the Flambeau Anomaly extends; however, preliminary results from field work in progress indicate that the anomaly extends beyond 89°W.

We can see from Figure 6 that the Flambeau Anomaly is a large feature which dominates the electrical properties of the crust across a section of northern Wisconsin. We now turn our attention to the characteristics of this anomaly at depth, first for a single conductor and then for the anomaly as a whole.

Investigation of a Single Conductive Feature Within the Flambeau Anomaly: Hay Creek Profile

The Hay Creek profile crosses a conductive feature about 15 km northeast of Park Falls. We will compare airborne INPUT data, airborne earth magnetic field data, ground audiomagnetotelluric (AMT) soundings, and ground dipole-dipole transmissions (Figure 8). The locations of the dipole-dipole and AMT sites are shown in Figure 7. The airborne data are along line A-B. We will use the AMT and dipole-dipole measurements to model the resistivity structure quantitatively.

The INPUT profile and magnetic field measurements in Figure 8a are from the airborne surveys of Kennecott Copper Corporation. Qualitatively, an increase in the INPUT response corresponds to the presence of conductors beneath the sensing system. The INPUT profile indicates that there are two subsurface conductors, one at 1.8 and one at 2.6 km south of A. The earth's magnetic field channel shows a 1400-γ anomaly over the conductor 1.8 km south of A, but does not show an anomaly over the conductor 2.6 km south of A.

The AMT surveying method is described by Strangway et al. [1973]. Briefly, the AMT method uses naturally occurring signals (primarily thunderstorm activity) as a source. The electric and magnetic

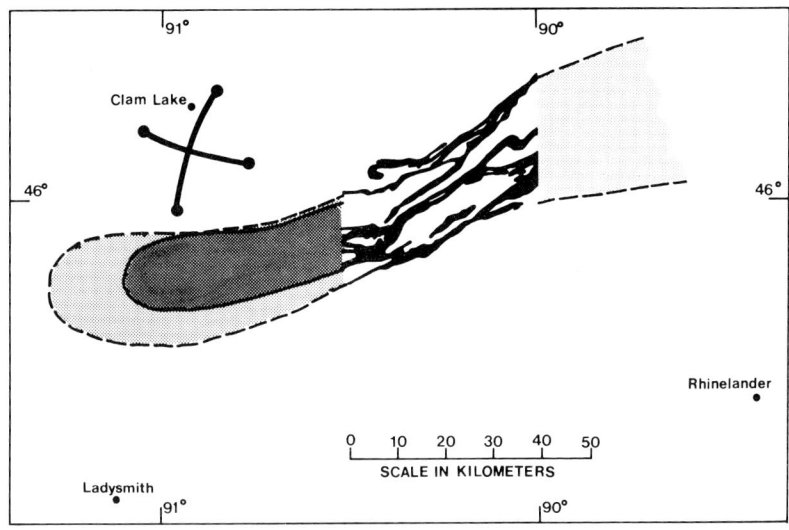

Fig. 6. Estimated extent of the Flambeau Anomaly. The black bands between 90°30'W and 90°W are the INPUT conductivity anomalies from Figure 5. The dark shaded area west of the INPUT survey contains an unknown number of conductors and is based on the results of the Sanguine antenna transmissions (Figure 4) which have relatively low spatial resolution. The light shaded area surrounding the dark shaded region represents the outer limit of where there may be conductors. The light shaded area east of the INPUT survey has conductors and is a probable extension of the anomaly (surveys in progress). The region further east is unknown.

fields are measured at the receiver site in orthogonal directions, and an apparent resistivity (ρ_a) is calculated from

$$\rho_a = \frac{0.2}{f} \left(\frac{|Ex|}{|Hy|} \right)^2 \qquad (2)$$

where f is the frequency of the signal being considered, E_x is the electric field in mV/km, and H_y is the magnetic field in gammas. Because of the skin depth effect, high-frequency signals have shallow depth penetration, and low-frequency signals have large penetration.

The AMT measurements were made with equipment loaned to us by Kennecott Copper Corporation. The equipment was similar to that described by Strangway et al. [1973]. Both the electric and the magnetic field channels had a set of 10 narrow band filters which cover the frequency range from 14 to 10,000 Hz. At each site we made scalar apparent resistivity measurements with the electric dipole (E) parallel to our estimate of the strike of the conductor and the magnetic field sensor (H) perpendicular to the strike. Then both the E and the H sensors were rotated 90° to obtain measurements of

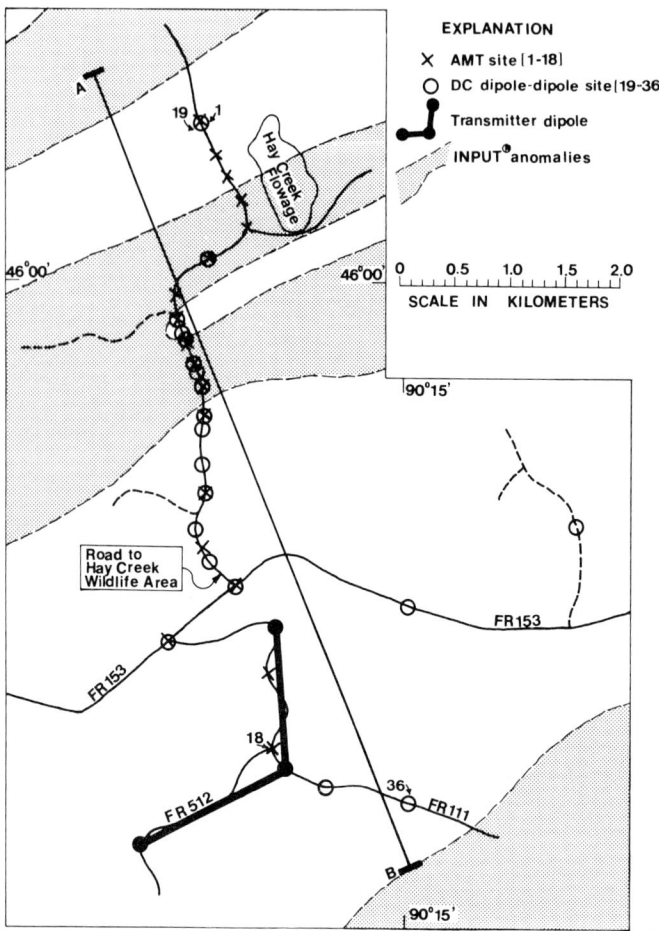

Fig. 7. Locations of the dipole-dipole and AMT sites for the Hay Creek profile.

the scalar apparent resistivities for E perpendicular and H parallel. We estimated the strike of the conductor from the INPUT survey (Figure 5). We believe that the error in our estimate of the strike direction at any given site is less than 20°. Some sites were occupied twice, and the measurements agreed to within a factor of 2. We believe that the lack of agreement between results on different days is due to variations in the direction and characteristics of the natural source.

We compared our AMT measurements, which were made with Kennecott equipment, with AMT soundings made by the University of Texas-Austin (UTA) at the same location. The UTA equipment, which was built by them, measured tensor impedences over the frequency range (3×10^{-2}) to 1000 Hz. Above 100 Hz the apparent resistivities calculated from the two sets of equipment agreed to within a factor of 2. As the frequency decreased below 100 Hz, the Kennecott

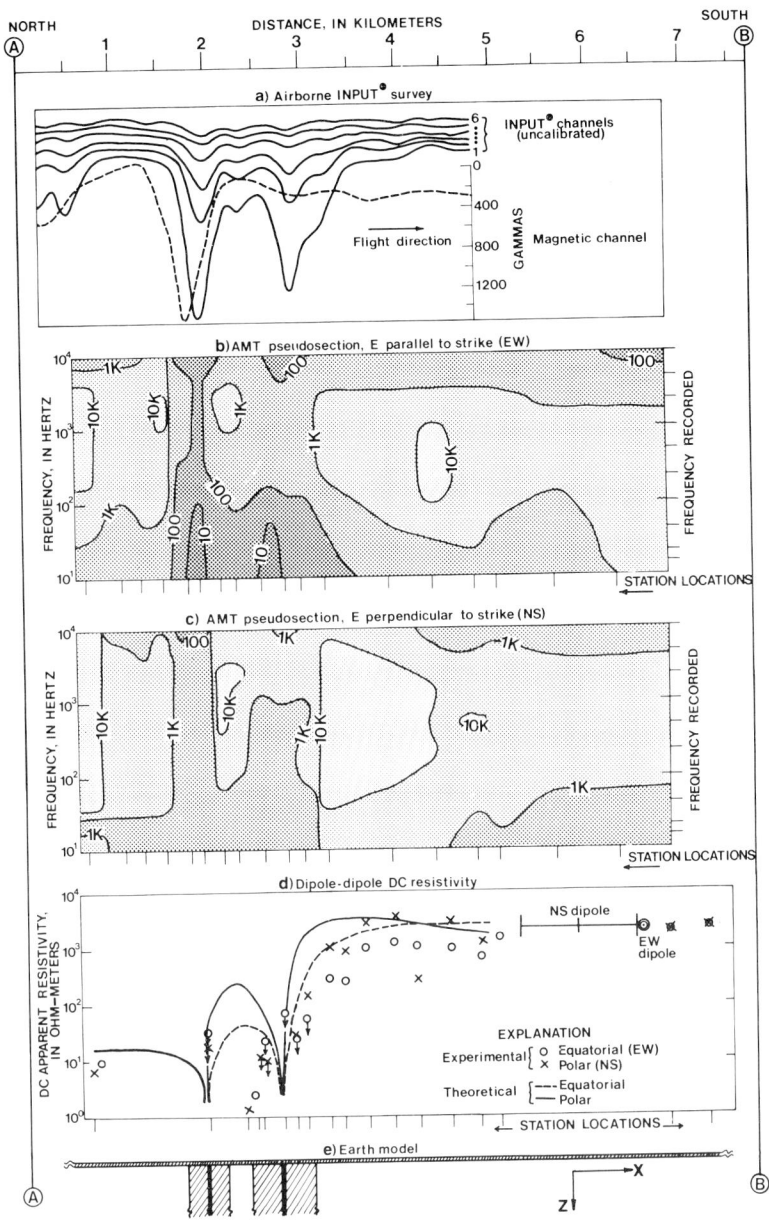

Fig. 8. Measurements along the Hay Creek profile. (a) Airborne INPUT survey and earth magnetic field (furnished by Kennecott Copper Corporation). The INPUT response is displaced about 125 m south due to the distance the receiver bird is behind the airplane.
(b) AMT pseudosection showing electric (E) field parallel to strike. Contours are apparent resistivity in ohm meters. (c) AMT pseudosection showing electric (E) field perpendicular to strike.
(d) Dipole-dipole apparent resistivity. (e) Earth model. The numerical values for the model are in the caption for Figure 9.

equipment apparent resistivity measurements decreased in relation to the UTA measurements. We have attributed the decrease in the Kennecott equipment apparent resistivities to low signal to noise ratios in the magnetic field channel below 100 Hz.

Our measurements of the AMT apparent resistivities at Hay Creek are shown as pseudosections in Figures 8b and 8c. The station locations are shown in Figure 7 and are indicated by short vertical lines at the bottom of Figures 8b and 8c. These figures are called pseudosections because high frequencies (shallow penetration) are plotted at the top and low frequencies (deeper penetration) are plotted at the bottom. Qualitatively, the lowest apparent resistivities along the profile, i.e., the 10-ohm-m contours, locate the positions of the conductors. The contours are approximately symmetric on each side of the anomaly, and this indicates that the features are also roughly symmetric.

The dipole-dipole measurements were similar to the Sanguine antenna dipole-dipole transmissions described earlier except that we set up a small dipole transmitter about 4 km south of the conductors, as shown in Figure 7. The transmitting antennas were 1.3 km long in the north-south direction and 1.5 km long in the east-west direction. Ten-second-period square waves at about 5 A of current were transmitted. On the transmitter side of the conductors the receiving dipoles were 30 m long. The signal to noise voltage ratios were large enough for the signals to be recorded on a chart recorder after amplification. Stations beyond the conductive anomaly at sites 19 and 21 had very low signals because the current was short-circuited in the conductors. Here we used 200-m dipoles and a computer to stack repeated transmissions of the signals. The transmissions were converted to apparent resistivities and are plotted in Figure 8d as values for each receiver site.

At some of the recording sites the signal was too far below the noise to calculate reliably the dc level. For these sites we calculated an upper bound on the apparent resistivity and indicated this in Figure 8d by a plotted point with a downward pointing arrow.

Numerical Computations for the AMT and dc Dipole-Dipole Profiles

Our interpretations are made by computing fields for an assumed earth model, comparing these with the observed fields, and adjusting the model until we get a match between the calculated and observed fields. We make the assumption that the conductive inhomogeneities in the Flambeau Anomaly can be approximated with a model which has constant resistivity in the strike direction (y) and variable resistivity in the other two coordinate directions (z downward and x perpendicular to strike); i.e., the resistivity is $\rho(x, z)$. This simplification greatly reduces computer costs and is generally adequate to determine the gross characteristics of the long, linear conductors in the Flambeau Anomaly.

The AMT models were calculated with a computer program published by Jones and Pasco [1971]. We also used refinements to this program given by Pascoe and Jones [1972], Williamson et al. [1974], and Jones and Thomson [1974].

The resistivity structure $\rho(x, z)$ is divided into a grid with 40 x 40 cells. Within each cell we have constant resistivity, but we can assign any arbitrary resistivity to each cell in the grid. Jones and Pascoe use a finite difference approximation to the wave equation. Application of the finite difference equation to each cell in the grid, along with appropriate boundary conditions, yields a set of simultaneous difference equations to be solved for the surface electric and magnetic fields. For the two-dimensional dependence of $\rho(x, z)$ the solution can be separated into two independent cases: electric (E) field perpendicular to strike, and E field parallel to strike. Details of the calculations are given by Jones and Pascoe [1971] and Pascoe and Jones [1972].

The size of each cell in the grid depends on the frequency (f) and the resistivity (ρ) within that cell. For accuracy in the finite difference approximation we always kept each dimension of the cell less than one third of the skin depth (where skin depth in kilometers is approximately equal to $(1/2)(\rho/f)^{1/2}$). When we cross a boundary from a conductive to a more resistive region, we avoid a sudden jump in the size of the cells by gradually increasing the dimensions of each cell. The edges of the grid were made greater than three skin depths distant from an inhomogeneity. The fields at the sides of the grid are set equal to the fields of a matching horizontally layered earth. The fields along the bottom are set equal to the fields of a matching half space. Computations of E, H, and apparent resistivity along the profile are made for 10, 100, 1000, and 10,000 Hz. We use this apparent resistivity to plot pseudosections.

The theoretical dipole-dipole transmissions require a different procedure because the electrodes are point sources and the fields are three dimensional even though $\rho(x, z)$ is dependent upon two dimensions. Madden [1971] uses a spatial Fourier transform to represent the field variation in the strike direction and derives a set of two-dimensional differential equations which describe the potential fields. By making an analogy between two-dimensional transmission line equations and the two-dimensional potential equations, Madden replaces the potential equations with a set of lumped-circuit resistivity networks. Potentials at the surface are found by Kirchoff's circuit laws, and the final fields are given by an inverse Fourier transformation. The details of this calculation are given by Madden [1971, 1972]. We use a computer program that was based on Madden's resistor network formulation and furnished by Geotronics Incorporated, Austin, Texas.

We divide the resistivity cross section $\rho(x, z)$ into a grid of cells. The size of the grid used for the dc dipole-dipole calculations was variable, up to a maximum of 22 cells deep by 71 cells across. For each type of resistivity model we made the grid denser and larger, until the change in apparent resistivity from one grid calculation to another was less than 10%.

Although we give a model which approximately matches the AMT and dc dipole-dipole measurements, the model is not unique. We shall attempt to find those parameters which are well determined and find a range of values for the poorly determined parameters.

Interpretation of AMT Data: Hay Creek Profile

Recognizing that the actual resistivity structure along the Hay Creek profile is probably very complicated, we determine a simplified structure by trial and error comparisons of computations and data. We try models until the fit to the AMT pseudosection and the dc dipole-dipole transmissions is close and further improvement of the fit depends upon minor details in the resistivity structure.

Our model for the conductors along the Hay Creek profile is shown in Figures 8e and 9c. This model includes those parameters resolved by the AMT soundings and the dc dipole-dipole transmissions.

The AMT computations are sensitive to (1) the location of the very low resistivity conducting sheets, (2) the dip of the sheets, (3) the resistivity ρ and thickness t of the sheets in the form t/ρ (t/ρ is referred to as the conductance or conductivity thickness product), and (4) the moderately conducting region on each side of the highly conducting sheets. We have superimposed the observed and calculated results in Figure 9.

We used the 10-ohm-m contour in Figure 8b to locate the positions of the conducting sheets. Our tests of effect of dip for a single sheet showed that dips shallower than 30° from vertical gave noticeable asymmetry. Since asymmetry is not clearly evident in the data (Figures 8b and 8c), we assume that the half planes are nearly vertical.

A glacial till layer 30 m thick with a 250-ohm-m resistivity overlies the section. The electrical properties of the glacial till layer were determined by a short-range (1 m to 1 km) dc dipole-dipole sounding located along Forest Road 153 and near the 'Road to Hay Creek Wildlife Area' (Figure 7). This 250-ohm-m glacial till layer is also consistent with a layered model interpretation of the AMT high-frequency data. The 3000-ohm-m bedrock half space resistivity provides a reasonable fit to the low-frequency AMT data and the large spacing (>1 km) dipole-dipole data away from the conductors.

The E parallel AMT results show a broad dip in apparent resistivity to about 10 ohm m at 10 Hz over the conductors. The width and minimum of the E parallel anomaly are determined primarily by the product of conductivity and thickness (= t/ρ) of the highly conductive half planes in the model. Conductivity-thickness products of 40 and 60 mhos provide the best fit. Conductivity-thickness products of 18 and 12 mhos lead to apparent resistivity minima roughly 10 times higher than observed. Conductivity-thickness products of 180 and 120 mhos lead to apparent resistivity minima roughly 3 times lower than observed. Increasing the conductivity-thickness product beyond 1000 mhos does not appreciably change the minima or the width of the anomaly; so there is a saturation effect which limits our ability to resolve extremely high conductivity, thin, half plane models.

The E perpendicular AMT results show a sudden drop in apparent resistivity from 3000 ohm m to several hundred ohm meters in a region surrounding each conductive half plane. This can be accounted for by including a zone in the model of 300-ohm-m material, which surrounds each highly conductive half plane.

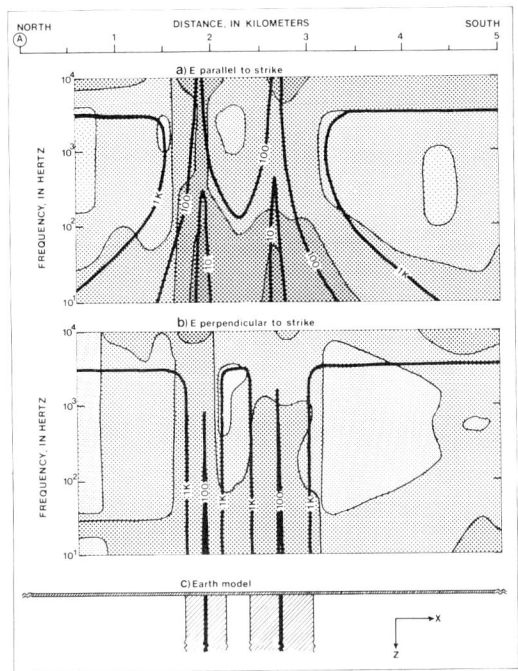

Fig. 9. Comparison of measurements and model calculations for Hay Creek AMT data. The calculated model results are shown as heavy contour lines superimposed on the data from Figures 8b and 8c. (a) E field parallel to strike. (b) E field perpendicular to strike. (c) Earth model. Starting at the left the resistivities and widths for the model are 3000 ohm m and infinite to north; 300 ohm m and 200 m; 0.5 ohm m and 20 m (conductivity-thickness product equal to 40 mhos); 300 ohm m and 200 m; 3000 ohm m and 250 m; 300 ohm m and 300 m; 0.5 ohm m and 30 m (conductivity-thickness product equal to 60 mhos); 300 ohm m and 300 m; and 3000 ohm m infinite to south.

This region is presumably due to disseminated conductive minerals surrounding the massive conductor.

We cannot separately resolve the conductivity and thickness of the half planes using E parallel AMT results. We find that the theoretical E perpendicular results drop close to the true resistivity of the conductor directly over the conductive anomaly. With our 200-m sampling interval we did not happen to have a site directly over a conductor. It would require far more recording sites in order to use the E perpendicular results to determine separately the conductivity and thickness. We can say that the width of the conductors must be less than 200 m. We chose 30 m and 20 m for the widths and 0.5 ohm m for the resistivity of the model conductors, which is consistant with what we found from drill cores at Butternut and Park Falls.

Neither the E perpendicular nor the E parallel results (in the frequency range 10 Hz to 10 KHz) are very sensitive to the depth extent of a vertical half plane conductor when it extends greater than about 1 km below the surface. Depth extents shallower than 1 km are not compatible with dc dipole-dipole data.

Interpretation of dc Dipole-Dipole Data: Hay Creek Profile

The dc dipole-dipole data are sensitive to the conductivity-thickness product of the half plane conductor and to the depth extent of the half plane. We have superimposed the field data points and calculated curves in Figure 8d.

As we approach the conductive anomaly on the transmitter side of the conductor, the apparent resistivity drops from about 3000 ohm m to less than 10 ohm m. Within the highly conductive region we have too few receiver sites to reliably determine the rather complex structure. The single site on the far side of the anomaly is used to model the depth extent and conductivity-thickness products of the vertical half planes.

The two half planes with conductivity-thickness products of 40 and 60 mhos which fit the AMT data also fit the dc dipole-dipole data. The transmission across the anomaly is quite sensitive to the conductivity-thickness product of the sheets. For example, an increase of the conductivity-thickness products to 120 and 180 mhos gives an apparent resistivity of approximately 1 ohm m at the furthest site from the transmitter. A decrease of the conductances to 12 and 18 mhos gives about 10^2 ohm m at the same position.

We found that when the depth extent is less than about 5 km, the calculated polar apparent resistivity is more than a factor of 3 larger than the equatorial apparent resistivity. Since the observed polar and equatorial apparent resistivities are nearly equal, we conclude that the half plane must extend to a depth greater than about 5 to 7 km. To find out how much deeper it extends, in the next section we interpret data taken with a larger transmitter-receiver separation and consequently deeper depth resolution.

We cannot determine the dip or separately determine the conductivity and thickness of the half plane using dc data from a single transmitter site. Had we been able to set up many more transmitter sites, including transmitters directly over the conductors, we could resolve these quantities.

We have not included the effect of conductors south of the transmitter and north of the furthest receiver site in our dipole-dipole modeling. A conductor with a conductivity-thickness product of 60 mhos roughly 1 km from either the transmitter or receiver, but not between the transmitter and receiver, changes the calculated apparent resistivity by at most 30%.

Interpretation of the Long-Range dc Dipole-Dipole Transmissions

Our interpretation of the Flambeau conductivity anomaly includes the structure north and south of the anomaly. The details of our interpretation of the electrical structure north of the Flambeau Anomaly is given in another paper [Sternberg, 1977b]. Briefly, we found that the following layered earth model is appropriate for this region: (1) The near-surface layer has variable thickness (0 to 100 m) and a resistivity of a few hundred ohm meters. This layer is too thin to affect the long-range soundings. (2) A second layer has a resistivity of a few thousand ohm meters to at most a few ten thousand ohm meters and a thickness of about 5 to 10 km. (3) Below this is a layer with a resistivity greater than 10^5 ohm m. (4) At depths of 15 to 25 km the resistivity drops to the order of 1000 ohm m. South of the Flambeau Anomaly the results of Bostick et al. [1977] and Keller and Furgerson [1977] and our interpretations of the Navy contractor conductivity data (Figure 2) indicate a similar structure. We have found that the long-range dc resistivity calculations are not very sensitive to the precise values used in the layered earth model. We have used the following layered earth model for all our resistivity calculations: (1) 4000-ohm-m resistivity, 8500-m thickness; (2) 100,000-ohm-m resistivity, 8500-m thickness; and (3) 1000-ohm-m resistivity extending to the bottom of the grid.

We use Madden's resistor network formulation (see section on numerical computations) to calculate the dc fields when the resistivity structure is a function only of x and z. We investigate in the next section the validity of using a model with infinite extent along the y axis to approximate the Flambeau Anomaly. Briefly, we find that transmissions in the southeasterly direction pass through the anomaly, are shielded from currents leaking around the end, and can be treated as if the conductors in the Flambeau Anomaly were infinite along the y direction. Transmissions to the south would require three-dimensional models of the resistivity structure for interpretation.

A map showing our model for transmissions along the southeast direction appears in Figure 10. We use five vertical sheet conductors to represent the conductive anomalies inferred from the INPUT data. The conductors extend to infinity along the strike direction. Cross sections perpendicular to the strike of the models are shown in Figure 11. The effects of having horizontal interconnections between the vertical sheet conductors as shown in Figure 11b were tested for a few cases. The interconnections change the apparent resistivity by less than 10%, and Figure 11a was used as the standard model. We vary the conductivity-thickness product and the depth extent of the conductors to obtain a fit to the data. For each model calculation the five vertical conductors have the same conductivity-thickness product and depth extent.

The long-range measurements are too sparse to model the complex structure within the anomalous region. Instead, we model the transmissions across the anomaly and use this to determine the gross resistivity structure of the anomaly.

Theoretical model calculations and experimental apparent resistivities are shown in Figures 12a, 12b, and 12c. The measurements are taken from Figure 4 and are shown as a function of range from the center of the Sanguine antenna. The measurements at 60, 80, and 190 km are consistent in that the equatorial (broadside)

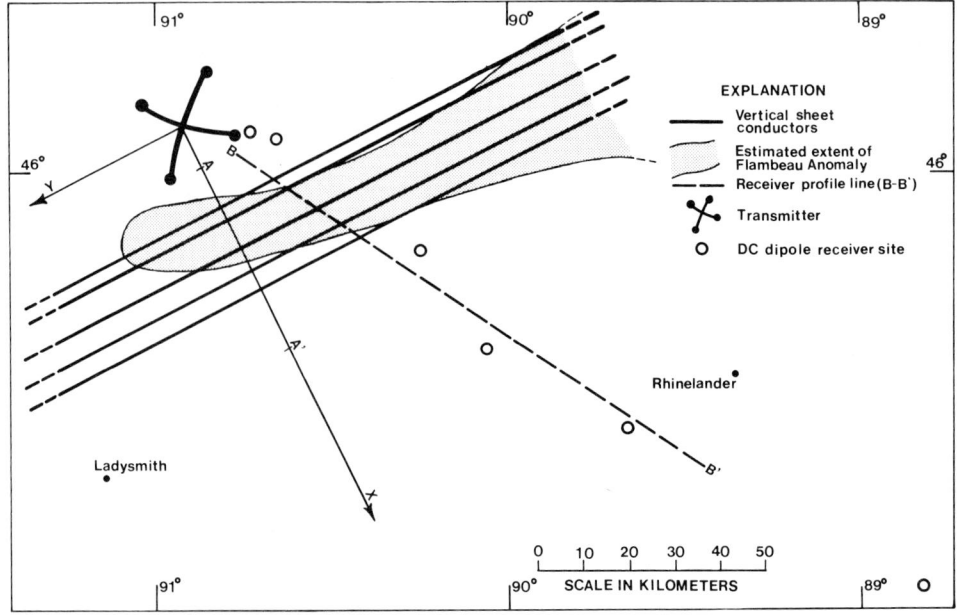

Fig. 10. Map view of the Flambeau Anomaly and model approximation. The five parallel lines coincident with the estimated extent of the Flambeau Anomaly are the locations of vertical sheet conductors which were used to model the anomaly conductors. The resistivity profile is calculated along the southeasterly dashed line (B-B').

dipole apparent resistivities are larger than the polar dipole or in-line apparent resistivities. The polar apparent resistivities are larger than the equatorial at 120 km, presumably owing to some local conductivity inhomogeneity. Although it is not shown, the averages of the polar and equatorial apparent resistivities fall on a smooth curve. Keller and Furgerson [1977] suggest that using

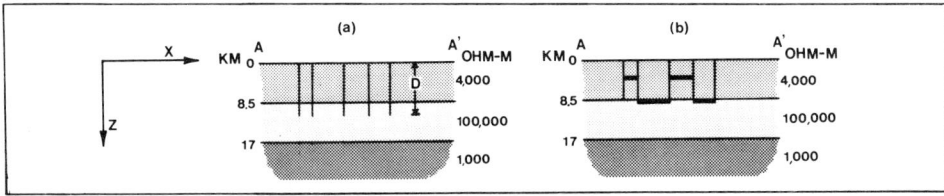

Fig. 11. Model cross sections. The cross section is along A-A' (Figure 10). (a) The five vertical lines represent sheet conductors which extend to various depths (D). (b) Connections between the vertical conductors were simulated. The width of each vertical conductor was 85 m, and they had variable resistivity. The thickness of the connecting segments was 708 m, and they had the same resistivity as the vertical conductors.

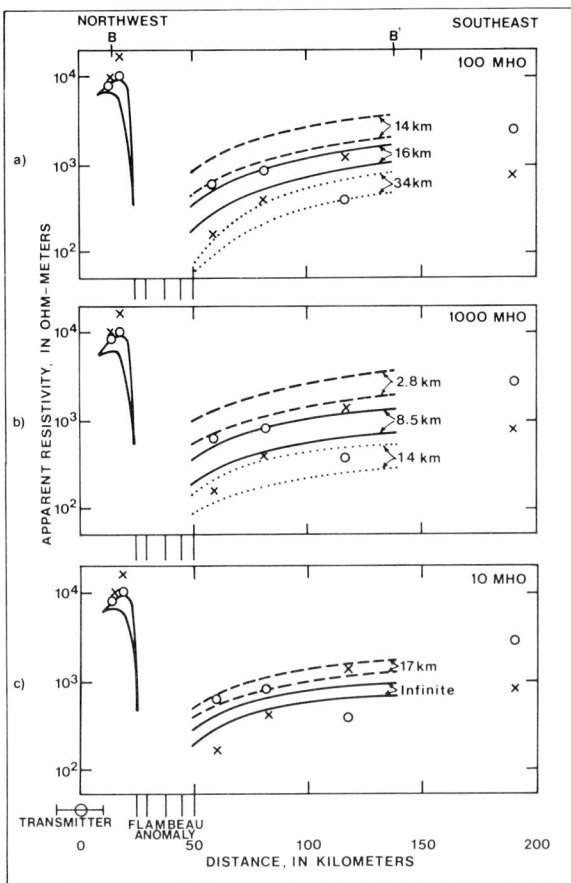

Fig. 12. Comparison of long-range dc dipole-dipole data and model calculations. The model is shown in Figures 10 and 11. The conductivity-thickness product for each of the five conductors is assumed to be (a) 100 mho, (b) 1000 mho, and (c) 10 mho. The assumed depth extents of the vertical conductors are listed to the right of each pair of calculated polar and equatorial curves. Between 0 and 25 km, the curves for all the depth extents overlap. The calculated equatorial curve is always higher than the polar curve between 50 and 135 km. Polar is higher than equatorial between 0 and 25 km. The observed polar data are indicated by crosses, and equatorial data by circles.

the average of the maximum and minimum apparent resistivities reduces the effect of geologic noise. Since the amplitude of the polar components that we measured may not exactly equal the dc amplitude, we have chosen to display both polarizations rather than average them.

We find that we cannot separately determine the conductivity-thickness product (t/ρ) of the conductors and their depth extent from a single profile line of dc resistivity sites. We need an independent estimate of the conductivity-thickness products of the conductors. At Hay Creek we found the conductivity-thickness products of two adjacent conductive sheets to be about 40 and 60 mhos. Analysis of the cores from Park Falls and Butternut indicates the average conductivity-thickness products of these conductors to be less than a few hundred mhos. Mining companies have carried out proprietary airborne and ground electromagnetic surveys over this region and have found conductivity-thickness products of the order of 10 to 100 mhos for the anomalies. Elsewhere in the Canadian Shield, Fraser [1972, 1974] has compiled an extensive set of conductivity-thickness product determinations. He found that the conductivity-thickness products for all the conductors ranged from a few mhos to as high as 1000 mhos, with an average of approximately 50 mhos. To cover a wide range, we will calculate models with conductivity-thickness products of 10, 10^2, and 10^3 mhos for each conductor. We will use these models to estimate the depth extent of the Flambeau Anomaly conductors.

In Figure 12a we assumed a conductivity-thickness product of 100 mhos for each of the five vertical sheet conductors. We calculated theoretical apparent resistivities for conductor depth extents of 14, 16, and 34 km. We found that a 16-km depth extent gives a best fit (of the three depths) to the transmissions across the Flambeau Anomaly. In Figure 12b we used what we consider to be the upper limit of the expected conductivity-thickness product for the conductors--1000 mhos. Calculations for depth extents of 2.8, 8.5, and 14 km have shown 8.5 km to be the best fit. In Figure 12c we used the expected lower limit--10 mhos. We tried depth extents of 17 km and effectively infinite. The 'infinite' depth extent gave the best fit. These results are summarized as follows:

Conductivity-Thickness Product	Depth Extent, km
Average (100 mhos)	16
Upper limit (1000 mhos)	8.5
Lower limit (10 mhos)	'infinite'

Clearly the Flambeau Anomaly conductors must extend deeply into the crust in order to account for the long-range transmissions across the anomaly.

Interpretation of the Effect of Finite Length of the Flambeau Anomaly

In the previous section we assumed that the models have infinite strike length. It would be too expensive to use existing computer techniques to calculate a complete three-dimensional model for the Flambeau Anomaly. We can instead estimate the effect of currents flowing around the end of the Flambeau Anomaly by assuming that the resistivity structure depends upon x and y, i.e., $\rho(x, y)$, and has infinite depth extent in the +z and -z directions. A model which extends to infinity in the +z and -z directions is equivalent to a model extending from zero to plus infinity, with air from zero to minus infinity and with current sources equal to twice those for

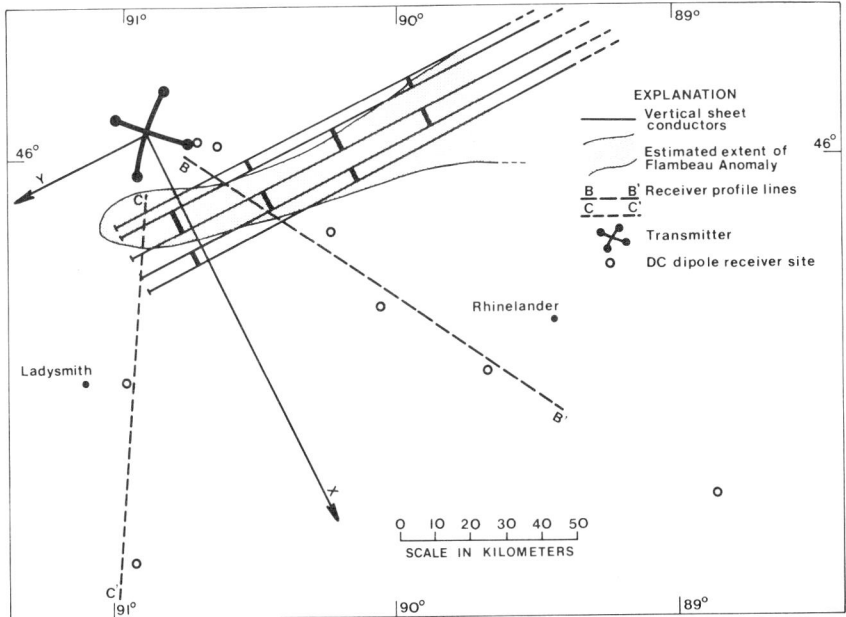

Fig. 13. Map view of the Flambeau Anomaly and terminated conductor model approximation. The five parallel lines coincident with the estimated extent of the anomaly are the locations of vertical sheet conductors which were used to model the anomaly conductors. The heavy connecting lines represent conductive connections between the sheets that were included in some of the model calculations. Fields were calculated along profile lines southeast of B-B' and south of the transmitter C-C'.

the +z and -z model. A plan view of the model is shown in Figure 13. The conductivity-thickness product of each vertical half plane is 10 mhos. The half space resistivity is 10,000 ohm m.

We calculated the apparent resistivities along profile lines south and southeast of the transmitter for two models, and the results are shown in Figure 14. In one model the conductors go to infinity in the +y and -y directions, and in the other the conductors terminate around 91°W. Along the southeast profile the termination of the conductors increases the apparent resistivity less than 30%. Along the south profile the termination of the conductors increases the apparent resistivity about 2 to 3 times. These calculations show that the models used in the previous section (which had infinite length along y) are adequate to determine the fields along the southeast profile. The effect of terminating the conductors around 91°W is much less than the effects of depth extent which we studied in that section.

The INPUT data in Figure 5 show that the conductors intersect. To simulate the intersections, we added short vertical half plane connections between the long conductive sheets (Figure 13). We found that these connections changed the apparent resistivity by

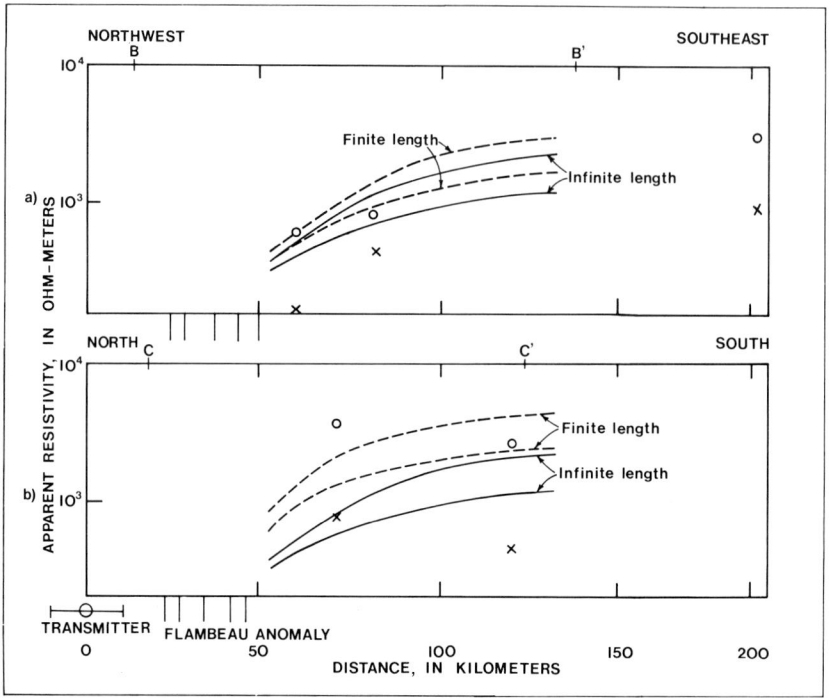

Fig. 14. (a) Apparent resistivities along the southeast profile line B-B' for the finite and infinite strike length models. (b) South profile line C-C'. For each pair of calculated curves the equatorial is always higher than the polar. In each graph we superimpose the observed polar data (crosses) and the equatorial data (circles).

less than 5% for the terminated conductor model and may safely be ignored.

Other Geophysical and Geological Information and the Flambeau Anomaly

Comparison of the Flambeau (electrical) anomaly with other geophysical and geological information is difficult because the available data are often too sparse for the complexity of the region. From our interpretation of data at Hay Creek and from drill cores at Park Falls and Butternut we see that the conductors comprising the Flambeau Anomaly are generally just a few tens of meters wide. It is difficult to collect geologic samples at close spacings in this region because most of the bedrock is covered by glacial till. Any type of ground survey is difficult because there are few access roads and off-road areas have dense trees, brush, and swamps.

Figure 15 shows the Flambeau Anomaly superimposed on a Bouguer gravity map. The anomaly is in a region of generally high gravity values (about -30 mGal). The values north and south are about -40 to -50 mGal. Koo [1976] gives the results of gravity surveys in the

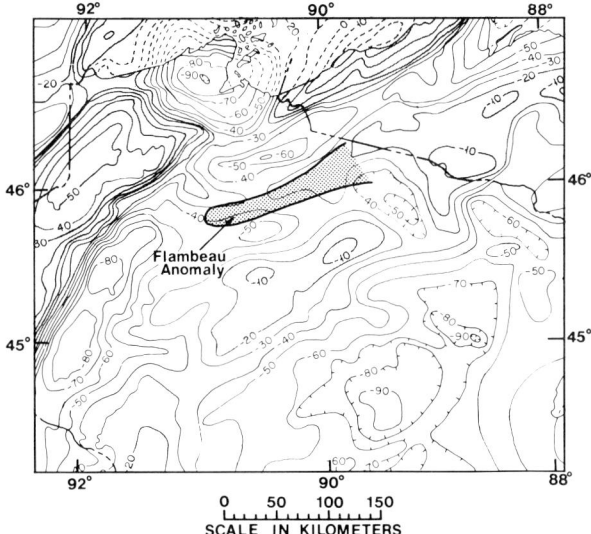

Fig. 15. Bouguer gravity map [from Craddock, 1972]. The Flambeau Anomaly (shaded area) occurs in a region of high gravity values.

area. Assuming an anomalous block of material roughly 15 km deep, he finds that a density contrast of 0.05 to 0.1 g/cm^3 is sufficient to account for the gravity high. Many other combinations of depth extent and density contrasts also fit the gravity data.

The Wisconsin Geological and Natural History Survey has recently completed new aeromagnetic maps of the northern third of the state. Our comparisons of the Flambeau Anomaly with these maps show that the Flambeau Anomaly is generally parallel to the major trends of magnetic anomalies. In the INPUT survey area, roughly one fifth of the conductivity anomalies also have corresponding magnetic anomalies that are greater than several hundred gammas.

The Flambeau Anomaly is superimposed on a generalized geologic map in Figure 16. The trend of the anomaly is closely parallel to the trend of a region of middle Precambrian, dominantly metasedimentary and metavolcanic rocks. Detailed geologic mapping of the Flambeau Anomaly area is in progress (R. Black, C. V. Guidotti, M. G. Mudrey, Jr., and P. K. Sims, personal communications, 1977). Preliminary results indicate that the Flambeau Anomaly occurs at the boundary between middle Precambrian metasedimentary and metavolcanic rocks and lower Precambrian gneisses. Much more geologic work will be required before the relationship between Flambeau Anomaly conductors and local geology is understood.

The major constituent of the cores from Park Falls and Butternut is graphite. Graphite may result from the metamorphism of inorganic carbon (carbonates) or organic carbon (life-forms). In order to provide some information on the possible origin of the conductive material in the Flambeau Anomaly, S. N. Ahmad and E. C. Perry, Jr., (manuscript in preparation, 1977) measured carbon isotope ($^{13}C/^{12}C$) ratios for these samples. They found isotope ratios (relative to

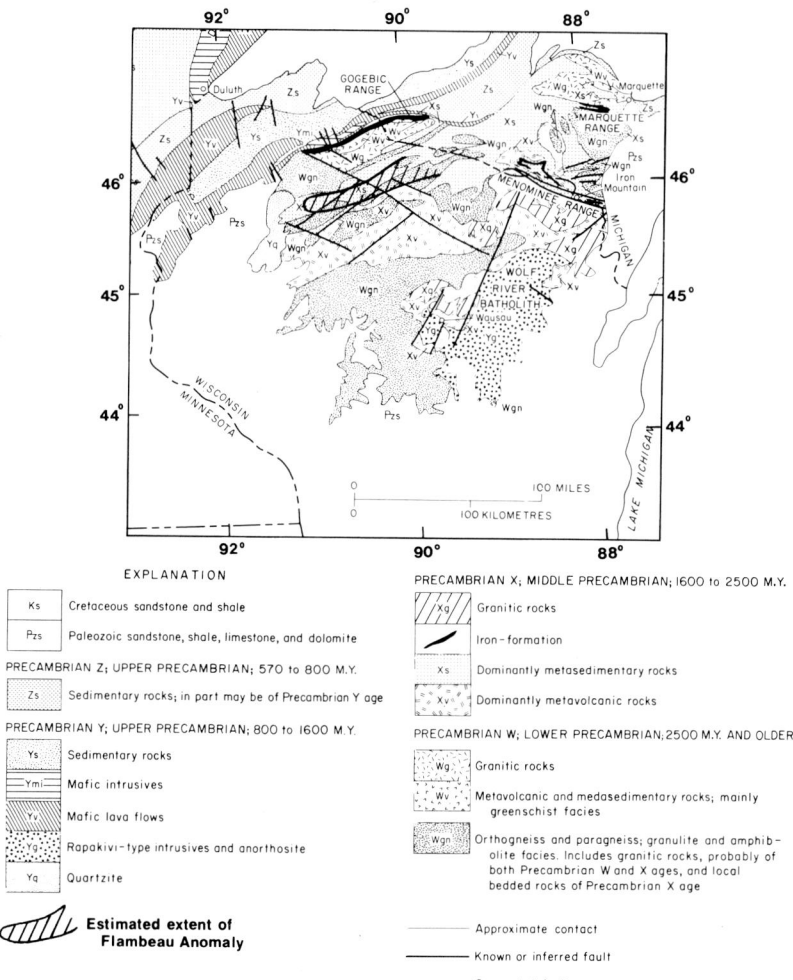

Fig. 16. Generalized geologic map of Precambrian rocks [from Sims, 1976]. The Flambeau Anomaly (hatched area) occurs in a region between gneisses and dominantly metasedimentary and metavolcanic rocks.

the PDB standard) of -23.5 ppt for the Park Falls samples and -20.0 ppt for the Butternut samples. Eichmann and Schidlowski [1975] discuss the carbon isotope analysis method and have compiled the results of 58 analyses made on Precambrian samples of coexisting inorganic and organic carbon. They found an average isotope ratio for organic carbon of -24.7 ± 6.0 ppt and for inorganic carbon of $+0.9 \pm 2.7$ ppt. The close correspondence between the isotope ratios from the Flambeau Anomaly cores and the results of Eichmann and Schidlowski for organic carbon strongly suggests that the Flambeau Anomaly graphite was derived from organic carbon sediments.

Dutton [1975] reported chalcopyrite associated with iron formation just west of Butternut. This occurs along the strike of one of the Flambeau Anomaly conductors. Very fine chalcopyrite was found associated with graphite in cores from the Flambeau Anomaly near Park Falls. We do not know whether economical concentrations of base metals occur within the Flambeau Anomaly. Elsewhere in the Canadian Shield, important copper bodies and other base metal bodies are frequently associated with iron formation, graphite, or other noneconomic conductors [Paterson, 1969].

Gough and Camfield [1972] have recently reported a large magnetic variation anomaly in the vicinity of the Black Hills of South Dakota. On the basis of limited outcrop and borehole data they suggest that the cause of this anomaly is a long linear belt of metamorphic rocks which includes highly conductive graphitic schists. Information is not available on the detailed near-surface characteristics of the conductors nor on their depth extent (except that they are reported to be at 'crustal depths'). The scale of the conductors and the association with graphitic conductors bear a striking resemblance to the Flambeau Anomaly. In a subsequent paper, Alibi et al. [1975] postulate an extension of this conductive region into Saskatchewan for a total conductor length of greater than 1000 km.

Conclusions

We have described a large region of anomalously high electrical conductivity within the resistive crust. On the basis of our modest spatial samples it is estimated to be 15 km wide and more than 100 km long. The anomalous region consists of narrow zones or sheets of highly conductive material such as graphite, sulfides, and iron foundation. We have used numerical models to estimate the depth extent and conductivity-thickness products of the conductive features. The range of values which fit the data is as follows: 10 mhos and essentially infinite depth, 100 mhos and 16-km depth, and 1000 mhos and 8.5-km depth. North and south of the Flambeau low-resistivity anomaly the upper crustal rocks contain small-scale conductive inhomogeneities but generally have resistivities greater than 10^5 ohm m.

Analysis of carbon isotope ratios suggests that the graphite in the conductors was derived from organic carbon sediments. In analogy with modern day conditions we postulate that the conductive materials in the Flambeau Anomaly accumulated as organic sediments in a basin which had restricted circulation and a reducing environment. The resulting black mud deposits were metamorphosed to yield graphite. At various locations within the basin, concentrations of sulfides and iron formation were deposited. These sediments have subsequently been folded to produce the present band of steeply dipping beds. The presence of sheared granite in the drill hole at Butternut indicates that very large shearing motions may have occurred within the anomaly.

As more electrical surveys are carried out, we may expect to find other areas where there are large, extremely high conductivity zones within the resistive continental crust.

Acknowledgments. We have received generous help from many people
and organizations. In Madison, C. R. Bentley, C. E. Dutton,
C. V. Guidotti, L. G. Medaris, R. P. Meyer, M. G. Mudrey, Jr.,
H. F. Wang, and C. T. Young have helped during various phases of the
research and preparation of this paper. F. X. Bostick, H. W. Smith,
and J. E. Boehl, University of Texas-Austin; G. V. Keller and
R. B. Furgerson, Colorado School of Mines; and S. N. Ahmad and
E. C. Perry, Jr., Northern Illinois University have given us advance
copies of their results. C. M. Swift assisted us with the AMT
equipment and interpretation of the INPUT data. T. R. Madden
assisted with the two-dimensional dc models. Geotronics Incorporated,
Austin, Texas, furnished the two-dimensional dc modeling program.
Kennecott Copper Corporation gave us the INPUT data and cores and
loaned us a set of AMT equipment. The Minerals Division of Calumet
and Hecla, Universal Oil Products furnished cores, logs, and the
results of electromagnetic surveys. We thank the following
reviewers for their helpful comments: J. G. Heacock, B. R. Lienert,
and J. S. V. Van Zijl. We thank the Navy's Project Sanguine office
for allowing us to use the Wisconsin Test Facility antennas and
for releasing the conductivity survey data. Many students at the
University of Wisconsin have helped to obtain these data, including
R. Daneshvar, S. Johnston, M. Ko, J. Koo, D. Kositzke, R. Kust,
M. Peterson, P. Probert, and B. Spevack. This work was supported
by the following grants: National Science Foundation GA 37171,
DES75-04879, EAR 75-04879 A01; Office of Naval Research contract
N00014-67-A-0128-0026; and University of Wisconsin Industrial Research
grant 160706. Contribution 334 of the Geophysical and Polar Research
Center, Department of Geology and Geophysics, University of
Wisconsin-Madison.

References

Alabi, A. O., P. A. Camfield, and D. I. Gough, The North American
 Central Plains conductivity anomaly, Geophys. J. Roy. Astron. Soc.,
 43, 815-833, 1975.
Becker, A., C. Gauvreau, and L. S. Collett, Scale model study of time
 domain electromagnetic response of tabular conductors, Can. Inst.
 Mining Trans., 75(725), 90-95, 1972.
Boniwell, J. B., Some recent results with the INPUT airborne EM
 system, Can. Inst. Mining Bull., 60, 325-332, 1967.
Bostick, F. X., H. W. Smith, and J. E. Boehl, Magnetotelluric and
 dipole-dipole soundings in northern Wisconsin, in The Earth's
 Crust, Geophys. Monogr. Ser., vol. 20, edited by J. G. Heacock,
 AGU, Washington, D. C., this volume, 1977.
Craddock, C., Bouguer gravity map showing midcontinent gravity high,
 in Geology of Minnesota: A Centennial Volume, edited by P. K. Sims
 and G. B. Morey, Minnesota Geological Survey, St. Paul, Minnesota,
 1972.
Davidson, D., D. N. Macklin, and K. Vozoff, Resistivity surveying
 as an aid in Sanguine site selection, IEEE Trans. Commun. Technol.,
 22(4), 389, 1974.
DECO, 1964 conductivity measurements in Wisconsin, Minnesota, and

Michigan (SECRET NOFORN), Rep. 30-P-9, Westinghouse Geores. Lab, Boulder, Colo., 1965.

DECO, 1965 conductivity survey in Wisconsin, compiled by J. J. Jacobson, Rep. 30-P-13, Westinghouse Geores. Lab, Boulder, Colo., available from DDC, #AD-487 458, 1966.

DECO, 1967 conductivity surveys in Wisconsin (SECRET), Westinghouse Geores. Lab, Boulder, Colo., Dec. 1967.

DECO, 1968 conductivity survey testbed phase I, compiled by C. J. Wideman, final report, Westinghouse Geores. Lab, Boulder, Colo., Aug. 1968.

Dowling, F. L., Magnetotelluric measurements across the Wisconsin arch, J. Geophys. Res. 75(14), 2683, 1970.

Dutton, C. E., Chalcopyrite in rocks associated with magnetic iron formation in Wisconsin, open file report, U. S. Geol. Surv. and Univ. of Wis., Madison, Geol. and Natur. Hist. Surv., Madison, Wisconsin, 1975.

Eichmann, R., and M. Schidlowski, Isotopic fractionation between coexisting organic carbon-carbonate pairs in Precambrian sediments, Geochim. Cosmochim. Acta, 39, 585-595, 1975.

Fraser, D. C., A new multicoil aerial electromagnetic prospecting system, Geophysics, 37(3), 518, 1972.

Fraser, D. C., Survey experience with the DIGHEM AEM system, Can. Mining Met. Bull., 67, 97-103, 1974.

Gough, D. I., and P. A. Camfield, Convergent geophysical evidence of a metamorphic belt through the Black Hills of South Dakota, J. Geophys. Res., 77, 3168-3170, 1972.

GTE Sylvania, Sanguine site survey, final report, parts II, III, IV, Needham, Mass., 1972.

Johnston, S. C., Frequency domain analysis of ac dipole-dipole electromagnetic soundings, M.S. thesis, Univ. of Wis., Madison, 1975.

Jones, F. W., and L. J. Pascoe, A general computer program to determine the perturbation of alternating electric currents in a two-dimensional model of a region of uniform conductivity with an embedded inhomogeneity, Geophys. J. Roy. Astron. Soc., 24, 3-30, 1971.

Jones, F. W., and D. J. Thomson, A discussion of the finite difference method in computer modeling of electrical conductivity structures, A reply to the discussion by Williamson, Hewlett and Tammemagi, Geophys. J. Roy. Astron. Soc., 37, 537-544, 1974.

Kan, T., Ray theory approximation in geoelectromagnetic probing, Ph.D. thesis, Univ. of Wis., Madison, 1975.

Keller, G. V., and R. B. Furgerson, Determining the resistivity of a resistant layer in the crust, in The Earth's Crust, Geophys. Monogr. Ser., vol. 20, edited by J. G. Heacock, AGU, Washington, D. C., this volume, 1977.

Keller, G. V., R. B. Furgerson, D. Y. Lee, N. Harthill, and J. J. Jacobson, The dipole mapping method, Geophysics, 40(3), 451, 1975.

Koo, J., A gravity survey of the Flambeau Anomaly, Wisconsin, M.S. thesis, Univ. of Wis., Madison, 1976.

Lahman, H., and P. Nelson, Deep resistivity results from North

Carolina, Virginia, Pennsylvania, Wisconsin and Missouri, Rep. 5, available from DDC, #AD-612 609, 1964.

Madden, T. R., The resolving power of geoelectric measurements for delineating resistive zones within the crust, in The Structure and Physical Properties of the Earth's Crust, Geophys. Monogr. Ser., vol. 14, edited by J. G. Heacock, p. 95, AGU, Washington, D. C., 1971.

Madden, T. R., Transmission systems and network analogies to geophysical forward and inverse problems, Tech. Rep. 72-3, Mass. Inst. of Technol., Cambridge, 1972.

Naval Electronic Systems Command, Navy conductivity data summary, attachment T to proposal, Spec. Commun. Proj. Office, Sanguine Div., Washington, D. C., 1972.

Pascoe, L. D., and F. W. Jones, Boundary conditions and calculation of surface values for the general two-dimensional electromagnetic induction problem, Geophys. J. Roy. Astron. Soc., 27, 179, 1972.

Paterson, N. R., Exploration for massive sulphides in the Canadian Shield, Mining and Groundwater Geophysics 1967, Econ. Geol. Rep. 26, pp. 275-289, Geol. Surv. of Can., Ottowa, Canada, 1969.

Sims, P. K., Precambrian tectonics and mineral deposits, Lake Superior region, Econ. Geol., 71, 1092-1118, 1976.

Sternberg, B. K., Controlled source electromagnetic soundings of the crust in northern Wisconsin, M.S. thesis, Univ. of Wis., Madison, 1974.

Sternberg, B. K., Electrical resistivity structure of the crust in the southern extension of the Canadian Shield, Ph.D. thesis, Univ. of Wis., Madison, 1977a.

Sternberg, B. K., Electrical resistivity structure of the crust in the southern extension of the Canadian Shield--layered earth models, submitted to J. Geophys. Res., 1977b.

Strangway, D. W., C. M. Swift, and R. C. Holmer, The application of audiofrequency magnetotellurics (AMT) to mineral exploration, Geophysics, 38, 1159, 1973.

Williamson, K., C. Hewlett, and H. Y. Tammemagi, Computer modeling of electrical conductivity structures, Geophys. J. Roy. Astron. Soc., 37, 533-536, 1974.

HIGH ELECTRICAL CONDUCTIVITIES IN THE LOWER CRUST OF THE NORTHWESTERN
BASIN AND RANGE: AN APPLICATION OF INVERSE THEORY TO A CONTROLLED-
SOURCE DEEP-MAGNETIC-SOUNDING EXPERIMENT

Barry R. Lienert[1] and David J. Bennett[2]

Institute for Geological Sciences, the University of Texas at Dallas,
Richardson, Texas 75080

Abstract. Data from a unique controlled-source deep-magnetic sounding experiment suggest the existence of a zone of high electrical conductivity in the lower crust beneath the region immediately to the north of Walker Lake, Nevada. This region is located in the northwestern portion of the Basin and Range province, where a significant induction anomaly had previously been observed from geomagnetic deep-sounding data. Amplitudes and phases of the vertical and horizontal magnetic fields observed at 14 variometer sites spaced at 5 km intervals and yielding frequencies as low as 6 cph, suggest that the electrical conductivity structure is reasonably one-dimensional throughout the region. Generalised linear inverse theory has been used to obtain best fitting one-dimensional three-layered models for each site. The results suggest a highly conducting (> 0.2 Ω^{-1} m^{-1}) zone at a depth of about 20 km which has a thickness of at least 10 km. This 20-km-deep high-electrical-conductivity zone may correspond to a zone of low seismic velocity inferred both in this area and further to the east at similar depths. Zones of partial melt in the crust resulting from higher than normal temperatures would account for the observed electrical conductivities.

Introduction

A large amount of geomagnetic deep-sounding data has now been accumulated from large-scale magnetometer array studies (for a review see Lilley [1975]). An intriguing question which many of these studies have posed is, 'At what depth in the crust or upper mantle are the conductive zones that such studies delineate really situated?' Much debate has centered around this question and how it relates to possible zones of low seismic velocity in the lower crust such as those proposed by Landisman et al. [1971]. Unfortunately, the interpretations based on the geomagnetic deep-sounding data are usually unable to resolve the depth of conductive structures unambiguously, as Garland [1972] has

---

[1] Now at Astrochemistry Branch, NASA/Goddard Space Flight Center, Greenbelt, Maryland 20771
[2] Now at Physics Department, Victoria University of Wellington, Wellington, New Zealand

recently observed. An ideal method for determining the depth of such structures is the use of nonuniform controlled-source magnetic fields. The problem with this approach is that the magnitudes of the electric currents required to produce magnetic fields which are capable of penetrating to depths greater than 10 km are very large: of the order of hundreds or even thousands of amperes.

During the fall of 1971 a unique opportunity arose to use a high-voltage dc transmission line for a study of the electrical conductivity of the crust using the nonuniform controlled-source magnetic field technique. In cooperation with the Bonneville Power Administration and the Los Angeles Department of Water and Power, the Northwest-Southwest Intertie was made available for a series of deep-magnetic-sounding experiments. An array of 30 magnetic variometers of the type described by Gough and Reitzel [1967] were deployed in selected areas on either side of the transmission line. The locations of the transmission line and the variometer sites from which data were obtained for the present investigation are shown in Figure 1. The southern end of the power line was grounded in Santa Monica Bay to ensure a remote path for any return currents, while the northern end was grounded in the vicinity of The Dalles, Oregon. Current was pulsed through the line at a base frequency of 3 cph for 7 hours (1800-0100 hours local time) and 6 cph for the last hour of the experiment. The maximum currents generated were approximately 300 A. The square wave format used for the switching cycles during the last hour was found to give satisfactory signal to noise ratios at the first four odd harmonic frequencies (6, 18, 30, and 42 cph).

A number of the profile lines in the experiment indicated zones of high conductivity in the crust. However, the northernmost (A-B) line (Figure 1) was unique in showing both significant crustal electrical conductivity and reasonable one dimensionality. This is indicated in Figure 2, which consists of plots of the horizontal (resolved perpendicular to the power line) to vertical magnetic field ratios for the four basic frequencies obtained from Fourier analysis of the data for the final hour. It is easily shown from simple image theory that for an infinitely conducting half space whose surface is at a depth h^* below an infinite line current source flowing in the y direction, the horizontal (B_x) to vertical (B_z) magnetic field ratio at a horizontal distance x from the line is given [Keller and Frischknecht, 1967] by

$$x/2h^* = B_x/B_z \qquad (1)$$

Values of h^* calculated for least squares straight lines through the observed data points are also shown in Figure 2. The small errors in the fitted slopes indicate that the data are consistent with the above type of one-dimensional modeling approach at each of the four frequencies shown.

This simple approach can be extended to conductors of finite order by using a 'complex image' approach [Weaver, 1971]. Instead of an image current at a depth h^* below the surface of an infinite conductor we have a current placed at a complex depth $h + \delta(1-i)$ below the surface of a conductor with conductivity σ. The total depth to the image current from the source is now $2h + \delta(1 - i)$ instead of $2h^*$, which gives

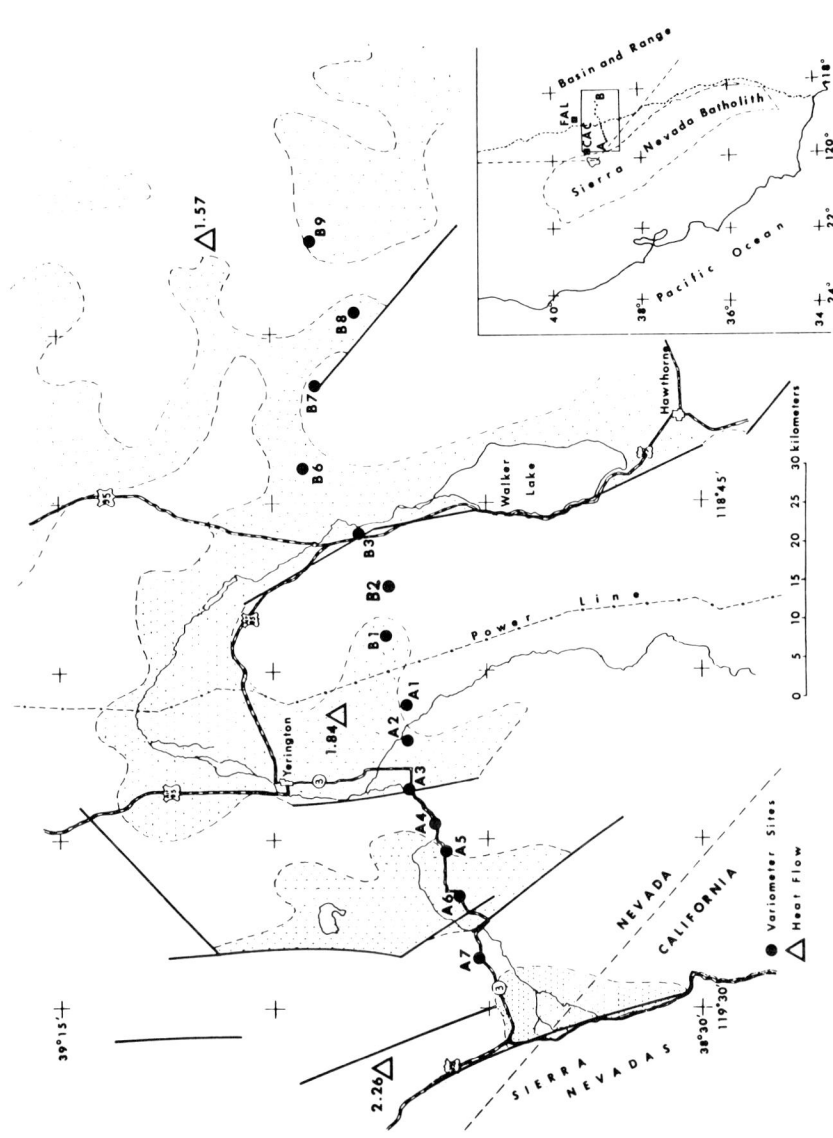

Fig. 1. Location map showing the power line and variometer sites from which magnetic field data were obtained. Heavy lines are major faults, while stippled areas represent alluvial valleys. The heat flow values are from Sass et al. [1971]. Stations CAC and FAL (inset) are sites at which geomagnetic variation data were obtained by Schmucker [1970].

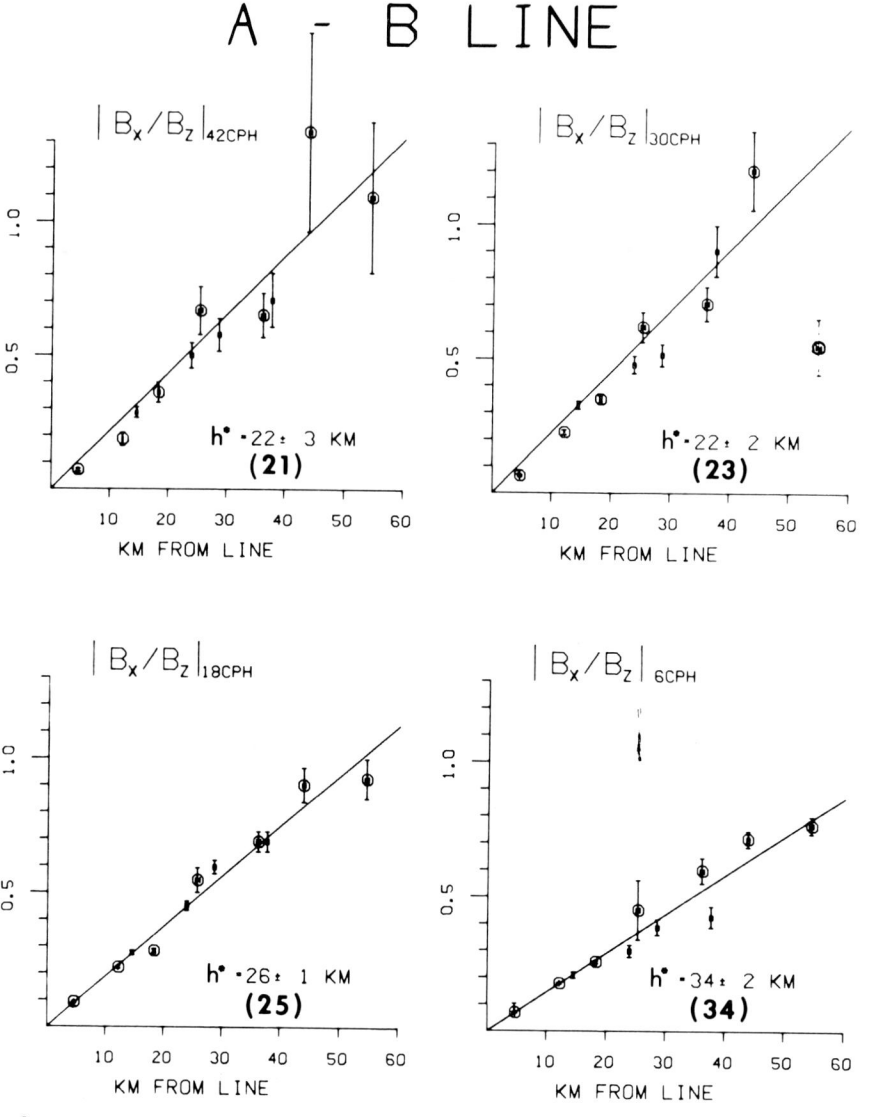

Fig. 2. Horizontal to vertical magnetic field amplitude ratios at the first four harmonic frequencies. The circled points are stations to the east of the power line, and other points are western stations. The h* values represent effective depths to an infinite conductor at each frequency. The values beneath the h* values were calculated for a 0.1 Ω^{-1} m^{-1} conductive zone at a depth of 14 km using complex image theory.

$$h^* = h + (\delta/2)(1-i) \qquad (2)$$

where δ is the well-known skin depth parameter given by

$$\delta = 30.2(f\sigma)^{-\frac{1}{2}} \qquad (3)$$

[Schmucker, 1970], f is the inducing field frequency in cycles per hour, σ is the electrical conductivity in mhos per meter, and δ is in kilometers. Substituting (2) into (1) and taking the modulus gives

$$B_x/B_z = x/2[(h + \delta/2)^2 + \delta^2/4]^{\frac{1}{2}} \tag{4}$$

The 'apparent' slopes should therefore increase with increasing frequency. An estimate of h, the depth to the finite conductor, and its conductivity σ, can be obtained by least squares fitting of the depths h^*, obtained from (1) at each frequency, to the four approximate equations

$$|h^*| \simeq h + \delta/2 = h + 15.1(\sigma f)^{-\frac{1}{2}} \tag{5}$$

for corresponding values of f. The least squares solution to this equation is considerably simpler than that to the exact expression. Although the error in (5) is 30% for h = δ, it decreases to 10% for h = δ/2, which was the smallest value of h obtained in this study.

Weaver [1971] has also shown that this image approximation is a good one when the image current flows at more than one skin depth from the point of observation, a condition that is also reasonably well satisfied in our case. The values obtained for the least squares fit are h ≃ 14 km and σ ≃ 0.1 Ω^{-1} m^{-1}, which when substituted into (2) give predicted values of h^* of 34, 25, 23 and 21 km, in good agreement with the values of h^* shown in Figure 2.

Although the results of this simple modeling approach are similar to the observed data, the question of uniqueness, particularly the possible effect of highly conducting layers near the surface, is an important one to consider. The observed values of h^* cannot be satisfied by moving the entire conductive zone upward, but it may be possible to have a model with high surface conductivities coupled with a conductivity increase at a greater depth which also satisfies the data. Additionally, the data set also contains phases of the horizontal and vertical fields which have been ignored in the analysis thus far. These phases are easy to determine in this experiment, since the times at which the current was turned on (or off) can be measured to within 1 s on each of the records. These times can also be compared with timing marks generated within each variometer, the latter being synchronized to within a few seconds of standard radio time signals.

The use of a general method of determining which of several classes of models best fit the data in a least squares sense and what resolution and errors these data imply is therefore desirable. Such a method, generalized linear inverse theory, was introduced to geophysical problems by Backus and Gilbert [1967,1968,1970] and has recently been reviewed by Wiggins [1972]. In the following section we shall discuss how generalized linear inverse theory was applied to the magnetic field data in order to obtain the electrical conductivity structure of the crust in the vicinity of Walker Lake, Nevada.

The Inverse Problem

In the matrix formalism of Wiggins [1972] the general linear inverse problem is to solve the system of linear equations

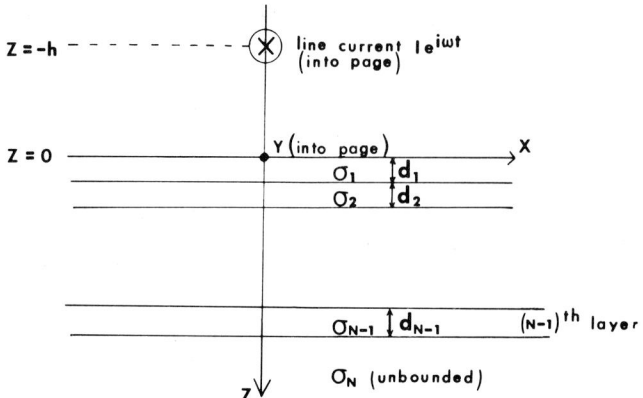

Fig. 3. The N-layered one-dimensional conductivity model.

$$A \cdot \Delta p = \Delta c \qquad (6)$$

where Δp is a 1 x n column matrix representing changes in n parameters in an assumed model which is being fitted to the data, Δc is a 1 x m column matrix representing the differences between m observed data values and the corresponding values predicted by the model, and A is the m x n matrix of partial derivatives of the predicted data values with respect to the model parameters. In this study we shall take an N-layered conducting half space beneath an infinite line current (Figure 3) as the model to be inverted. The data values will be the amplitudes and phases of the horizontal and vertical magnetic fields observed at 6, 18, 30, and 42 cph for a single station. The model parameters will therefore be the thicknesses d_1, d_2, d_3, ..., d_{N-1} and the conductivities σ_1, σ_2, ..., σ_N, as defined in Figure 3.

The solution to the system of equations (6) has been discussed by Wiggins [1972] and Jackson [1972]. The approach adopted in this paper is essentially the same as that used by both of these authors. The major steps involved are summarized in Figure 4. However, some aspects of the process relevant to this paper are as follows.

1. The covariance matrix of the observations, S, is assumed to be diagonal. This in effect says that errors in the observations are independent [Kaula, 1966].

2. The weighting matrix, W, for the parameters [Jackson, 1972] is also taken to be diagonal. Its elements are set equal to the current values of the model parameters which are taken to have a constant *fractional* variance. This also transforms the parameters into a logarithmic space, removing their dimensionality.

3. Rather than limit each parameter change separately to a maximum value [Glenn et al., 1973], we choose to limit the value of Δp^* which is related to the Δp by

$$\Delta p = \sum_{i=1}^{p} V_i \, \Delta p_i^* \qquad (7)$$

[Wiggins, 1972] where V_i is the ith eigenvector of $A^T A$ and p is the number of degrees of freedom. The eigenvectors V_i and their eigenvalues λ_i were obtained by using the Givens-Householder method [Ortega, 1967].

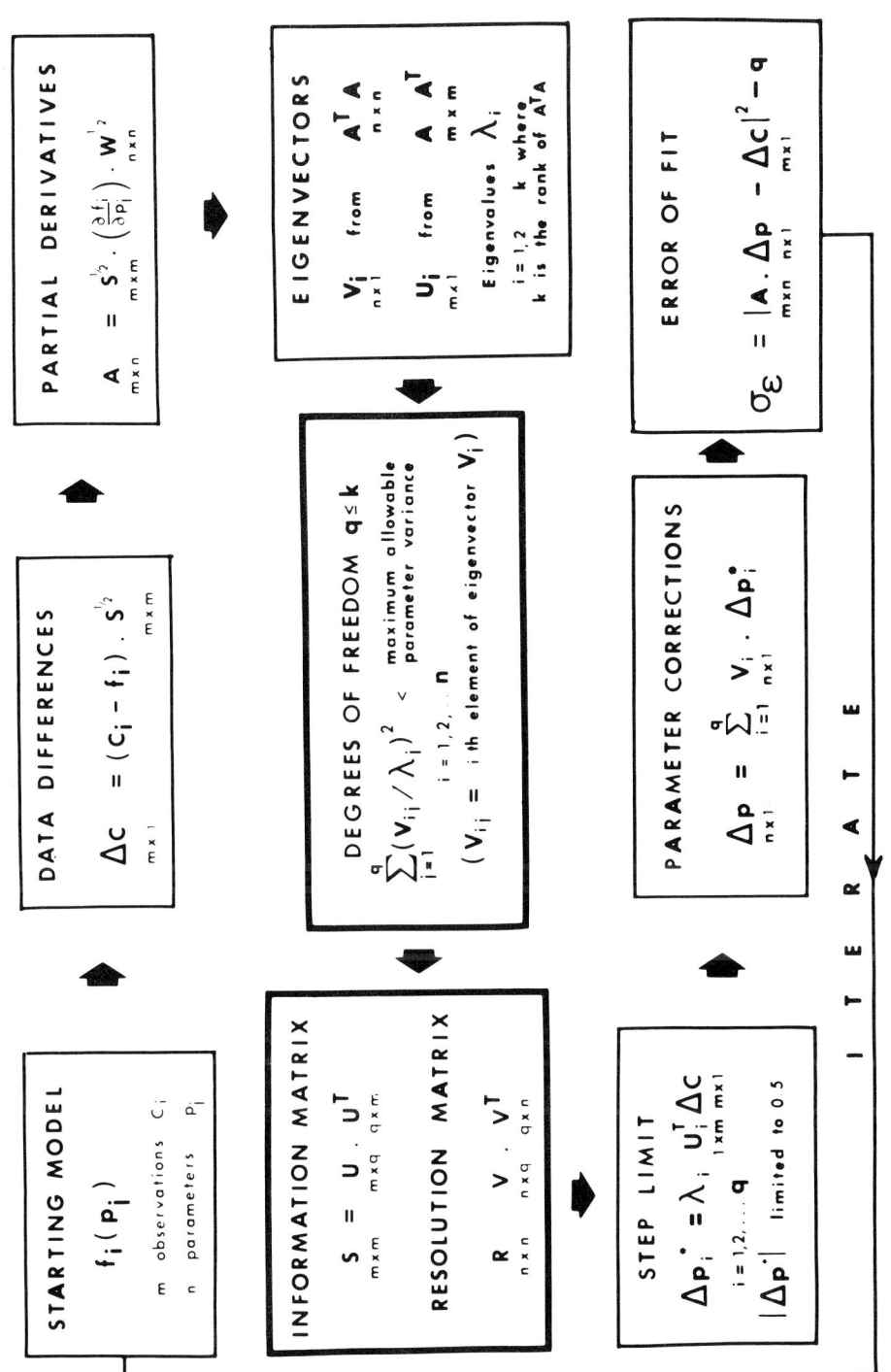

Fig. 4. Flow diagram of inversion procedure. The notation is similar to that used by Jackson [1972] and Wiggins [1972].

The limiting value of Δp^* was taken as 0.5. This is a good compromise between speed of convergence and stability of the solution for this particular data set.

One aspect of the overall process that is rarely discussed is the calculation of errors using

$$\text{var } \{\Delta p_i\} = \sum_{j=1}^{p} (v_{ij}/\lambda_j)^2 \tag{8}$$

where v_{ij} is the ith element of the jth eigenvector of $A^T A$ and λ_j is the corresponding eigenvalue. This gives the variance in each parameter *correction* at each stage of the iteration [Wiggins, 1972]. However, the variance in the finally determined parameters will be affected by the variances occurring at each iterative step. Are these errors cumulative or are the variances in the final corrections the only significant ones to consider? This clearly relates to the question of uniqueness. At each stage in the iteration the parameter correction variances give a range over which the least squares criteria can be satisfied. By simply applying the calculated corrections before reiterating, we are choosing the midpoint of the range of solutions at each iterative step. Although this is a reasonable enough choice, it still neglects many 'directions' in parameter space in which the solution could proceed.

In the present study we shall merely present the errors in the final parameter corrections. This in effect says that the errors are those in fitting the data to the final model to which the scheme converged. But we must accept that the manner in which this final model was obtained is in itself nonunique. When parameters are reduced to the point that they are no longer resolvable, their final values are upper or lower limits only, the variances given by (8) having little or no significance.

Calculation of the Partial Derivatives

In order to solve the system of equations defined by (6) it is necessary to obtain the matrix of partial derivatives of the magnetic field values measured at the surface of an N-layered half space below an infinite periodic line current. Summers and Weaver [1973] have described a method by which these field values can be obtained in terms of sine and cosine integrals, namely

$$B_x/B_{z,p} = 2hx/(x^2 + h^2) - x\int_0^\infty g_0(\xi, 0) \cos(\xi x) e^{-h\xi} d\xi \tag{9}$$

$$B_z/B_{z,p} = -x\int_0^\infty g_0(\xi, 0) \sin(\xi x) e^{-h\xi} d\xi \tag{10}$$

where $B_{z,p}$ is the primary vertical magnetic field, B_x is the horizontal magnetic field, and B_z is the total vertical magnetic field. The kernel g_0 is defined by

$$g_0(\xi, 0) = e^{\xi z} Q_0 \tag{11}$$

and x, h, and z are defined as shown in Figure 3. Q_0 is a parameter which is a function of the frequency of the inducing field as well as the conductivities and thicknesses of the individual layers. Q_0 can be obtained by applying the usual boundary conditions at each of the layer interfaces which results in a recursion relation (see appendix).

The required partial derivatives can now be obtained by differentiating (9) and (10) with respect to the parameter values in each layer. Since the functions under the integral signs are all continuous, the order of integration and differentiation can be reversed. The problem is then to determine the derivatives of the function $g_0(\xi, 0)$ and hence Q_0 with respect to each of the layer parameters. These can be obtained by differentiating the recursion relation for Q_0 which leads to a system of recursion relations which are described in the appendix. Numerical tecniques such as the fast Fourier transform must be used to integrate the resulting kernels according to (9) and (10) in order to obtain the magnetic field values and their derivatives. Details of these tecniques are also presented in the appendix.

An example of the derivatives calculated by using these methods for a three-layered model similar to the model obtained from image theory is shown in Figure 5. A number of interesting features are apparent in these curves. Most of the derivatives increase with increasing distance from the line implying better resolution of the corresponding parameters at large distances. The derivatives with respect to σ_1, the conductivity of the uppermost layer, are very small. This is a consequence of the extremely large skin depth (about 170 km at 6 cph) in a layer having a conductivity of 0.005 Ω^{-1} m^{-1}. The derivatives with respect to the conductivity of the lowermost layer, σ_3, are also small, implying that this parameter will not be very well resolved. The best resolved parameter is d_1, the thickness of the uppermost layer. This parameter has large derivative values for both horizontal and vertical field phases and amplitudes.

Results

The results of an inversion of data collected at the easternmost station using a three-layered model, are presented in Figure 6. The limiting value of the variance in (8) was taken as 50%, which in this case gave three degrees of freedom in the final solution. The resolution matrices for the initial model and the model to which the scheme converged are also shown. The magnitudes of the diagonal elements of these resolution matrices give an indication of the degree of resolution for each parameter [Jackson, 1972]. If a parameter can be resolved independently of the other model parameters, its corresponding diagonal element will be unity. Otherwise, the resolution will be distributed among the other parameters to an extent indicated by the magnitudes of the corresponding elements in the same row (or column). For example, the resolution matrix for the initial model in Figure 6 indicates that the conductivity of the deepest layer, σ_3, is the most well resolved parameter whereas only linear combinations of the remaining parameters can be resolved.

The fractional changes in the parameters indicate that convergence is obtained for this station after eight iterations. The average magnitude of the changes reflects the limiting value used for Δp^*, which in this case was 0.5. The final model obtained is very similar to the one obtained earlier from image theory with a low-conductivity layer overlying a highly conductive zone at a depth of 17±2 km. The large standard deviations in some of the parameters reflect the compromise made by allowing large variances in the parameter corrections in order to

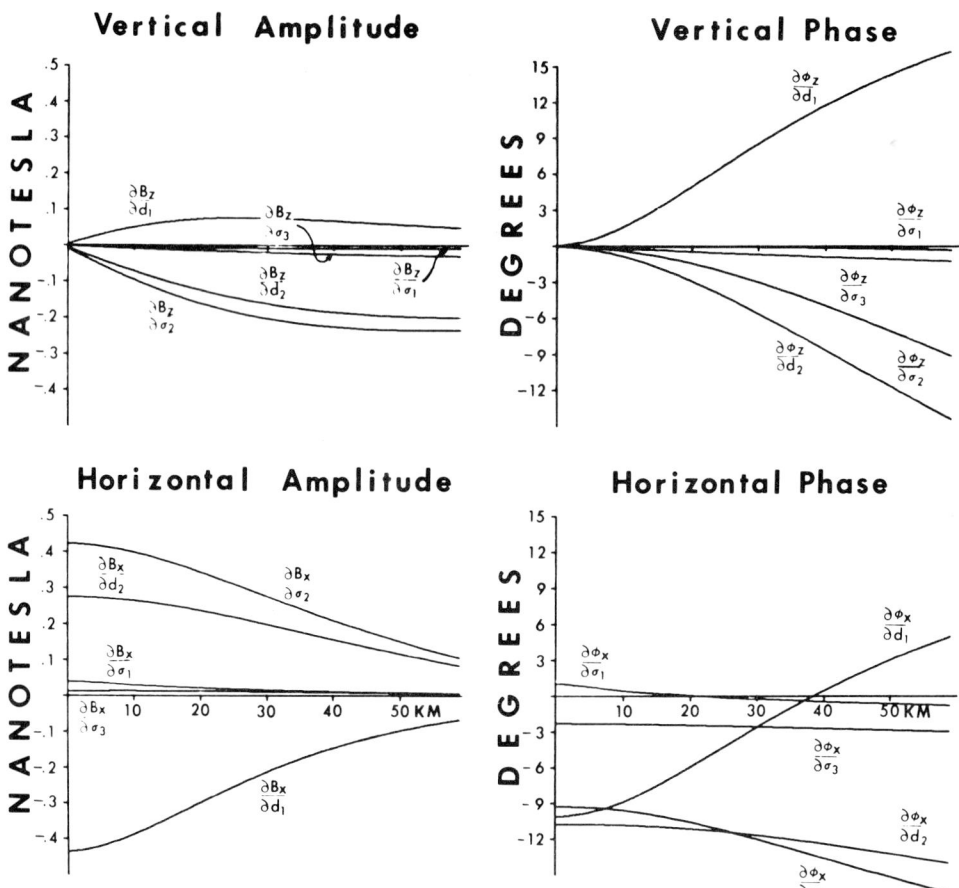

Fig. 5. Numerically calculated partial derivatives with respect to fractional parameter changes for a three-layered model plotted as a function of distance from the power line. The parameter values for this particular set of partial derivatives are $\sigma_1 = 0.005 \; \Omega^{-1} \; m^{-1}$, $d_1 = 14$ km, $\sigma_2 = 0.2 \; \Omega^{-1} \; m^{-1}$, at a frequency of 6 cph.

obtain better resolution. These variances could have been reduced by allowing only 2 degrees of freedom in the solution. However, this would reduce the the number of parameters that can be independently resolved. The limiting value of 50% for variances in (8) resulted in a reasonable 'trade-off' for this particular data set. The resolution matrix for the final model indicates that resolution is now confined mainly to the parameters d_1, σ_2, and d_2. This is to be expected, since the skin depth at the highest frequency (42 cph) in the first layer will still be about 66 km, which is much larger than this layer's thickness (17 km).

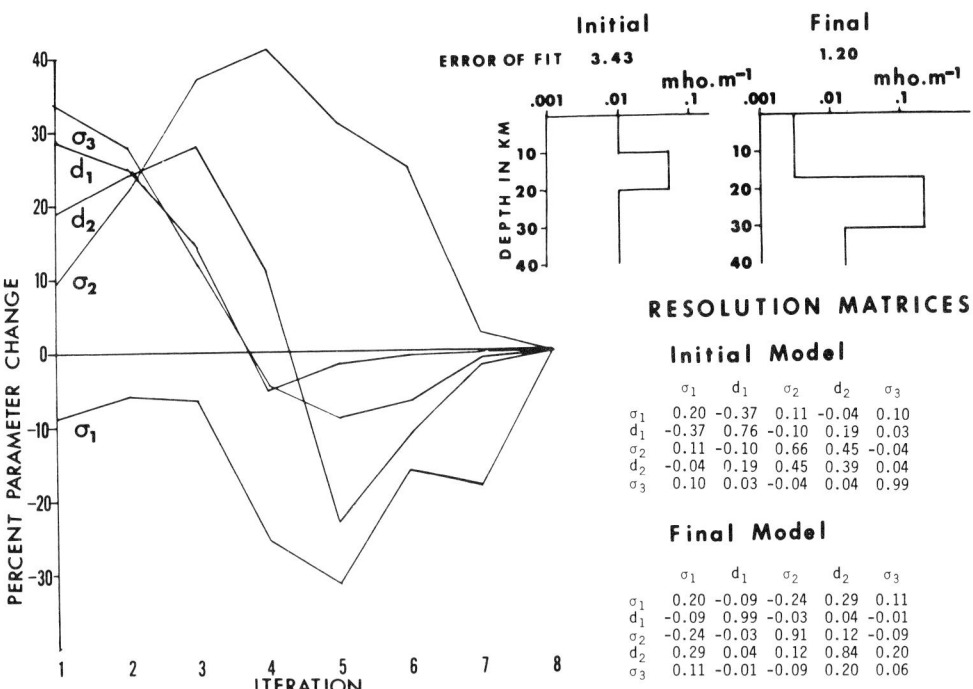

Fig. 6. Inversion result for station A7, 37.8 km from the power line. The fractional parameter changes for each iterative step are shown, along with the initial and final models and their resolution matrices.

Its conductivity, σ_1 will therefore have little effect on the solution. The conductivity of the lowest layer, σ_3, will also have little effect due to the 'screening' effect of the highly conductive second layer.

Plots of the relative sizes of the diagonal elements of the information matrices, defined in Figure 4 and calculated for this same station, are shown in Figure 7. The plots indicate that for both the initial and final models, most of the information resides in the horizontal phases at 6 and 18 cph and the vertical phases at 18 and 42 cph. This is partly due to the larger errors in the amplitude data which included absolute calibration errors. However, the partial derivatives plotted in Figure 5 also show that the phases are more sensitive than the amplitudes to changes in a lot of the parameters.

Results of similar inversions using the same starting model are presented in Table 1 for all the stations in the profile line. In this table, parameters for which the diagonal element of the resolution matrix is less than 0.1 are not assigned a variance. Instead, an upper or lower limit is given, depending on which direction the parameter is being changed before it becomes unresolvable. Where this direction was not clear, the final parameter value is footnoted in the table. The horizontal data alone are used to obtain inversion results for the four stations closest to the power line. Convergence is not initially obtained for these stations owing to the anomalous behaviour of their

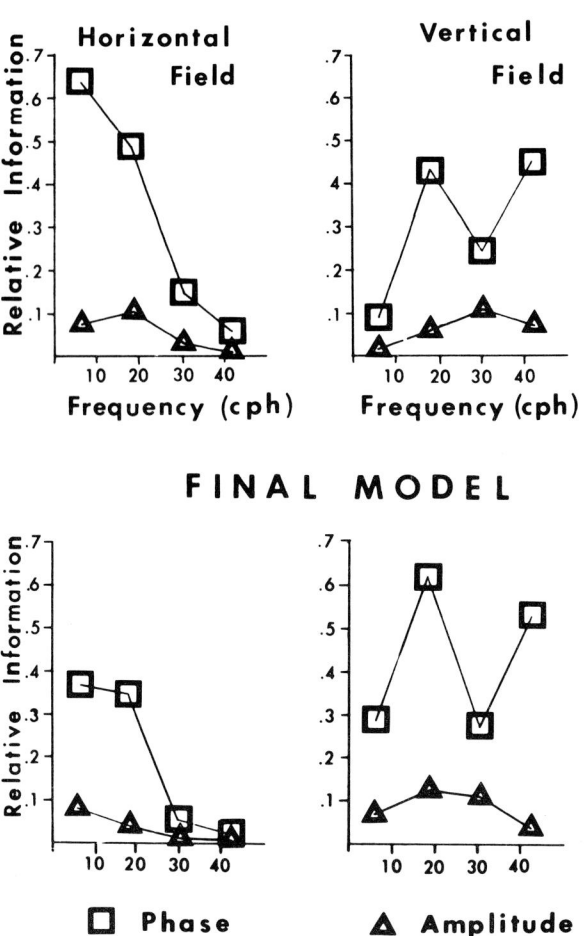

Fig. 7. Magnitudes of the diagonal elements of the information matrices $S = U \cdot U^T$ (Figure 4) for the initial and final models obtained for station A7 (shown in Figure 6).

vertical field phases. These lead rather than lag the phases of the primary field, which is inconsistent with the assumption of one-dimensionality. The source of these anomalous effects is high conductivities at shallow depths near stations A3 to the west, and B3 to the east. These high conductivities may be due to sediments filling the alluvial valleys in which these two stations were located. They may also be due to the presence of conductive fluids within faults which underlie both of these stations (Figure 1). Modeling of these two-dimensional effects will not be attempted in this paper. Numerical solutions to this type of problem have been obtained by Hohmann [1971]. His results indicate that a conductive zone having a limited lateral extent and a thickness of 0.02 skin depths will have a very small effect on the phases of horizontal magnetic fields produced by an

TABLE 1. Inversion Results for the A-B Line.

Station	Latitude	Longitude	x, km	σ_1, Ω^{-1} m^{-1}	d_1, km	σ_2, Ω^{-1} m^{-1}	d_2, km	σ_3, Ω^{-1} m^{-1}	q	σ_ϵ
A7	38°46'	119°26'	37.8	0.003 ± 44%	17 ± 2	0.2 ± 42%	14 ± 8	0.01†	3	1.2
A6	38°47'	119°20'	28.7	0.008 ± 41%	19 ± 3	0.1 ± 11%	20 ± 3	0.03 ± 14%	3	1.7
A5	38°48'	119°16'	24.0	<0.002	24 ± 2	0.09 ± 6%	14 ± 1	<0.003	2	2.6
A4*	38°49'	119°14'	19.5	<0.008	34 ± 2	0.3 ± 12%	12 ± 1	0.01†	2	1.5
A3	38°50'	119°11'	14.5	0.01 ± 13%	27 ± 3	0.1 ± 25%	22 ± 2	>0.06	3	3.1
A2	38°50'	119°06'	8.7	<0.001	19 ± 1	0.4 ± 5%	8 ± 1	>0.04	2	2.3
A1*	38°51'	119°03'	4.6	<0.001	23 ± 3	>1.3	>24	0.01†	1	0.9
B1*	38°52'	118°57'	-4.7	<0.007	23 ± 1	0.9 ± 8%	8 ± 1	<0.004	2	3.0
B2*	38°51'	118°51'	-12.2	<0.002	19 ± 2	0.5 ± 5%	9 ± 5	0.01†	3	0.6
B3*	38°54'	118°48'	-18.3	0.01 ± 21%	7 ± 4	0.08 ± 42%	12 ± 2	0.09 ± 21%	3	1.5
B6	38°58'	118°42'	-25.4	<0.006	28 ± 2	0.7 ± 19%	10 ± 2	0.01†	2	3.3
B7	38°57'	118°35'	-36.2	<0.001	24 ± 2	0.2 ± 7%	13 ± 1	<0.005	2	1.8
B8	38°54'	118°29'	-43.9	<0.002	20 ± 1	0.2 ± 4%	14 ± 1	<0.004	2	2.8
B9	38°57'	118°22'	-54.5	<0.001	24 ± 1	0.2 ± 4%	10 ± 1	>0.03	2	2.7

The parameter values σ_1, σ_2, and σ_3 are the conductivities, and d_1 and d_2 the thicknesses, obtained by inverting a three-layered model for each station; σ_ϵ is the error of fit, and q the number of degrees of freedom allowed in the solution (Figure 4); x is the horizontal distance of each station from the power line. Vertical data are not used for the stations marked with asterisks, either for reasons discussed in the text or, in the case of station A4, because data were not recorded.
* Horizontal data only.
† Not resolvable.

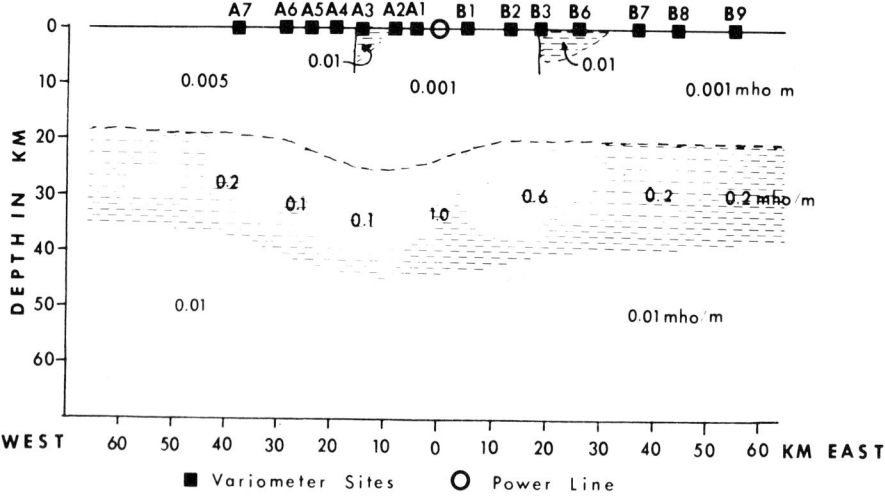

Fig. 8. Cross section of proposed conductivity structure beneath the Walker Lake, Nevada profile line. This shows the conductive zone at 20 km and areas of high surface conductivities inferred from the inversion results.

infinite line current source, even for conductivity contrasts as high as 100. In a surrounding medium having a conductivity of 0.005 $\Omega^{-1}m^{-1}$ at a frequency of 6 cph, 0.02 skin depths is about 3.5 km which is a reasonable figure for the thickness of sediment in parts of the Walker Lake area. It therefore seems reasonable to interpret the horizontal data alone at these four stations as giving an indication of the deep electrical conductivity structure. As is apparent from Table 1 these data then give models that are reasonably consistent with those obtained for the more extreme stations.

The results are summarized pictorially in Figure 8. The most anomalous result is that for station B3, where the value of d_1 is 7 km. It is rather uncertain as to whether the increase in σ_2 for the stations nearest the power line represents a real variation in the deep conductivity or merely an attempt by the inversion process to satisfy features of the data that are due to the shallow effects discussed above. Local increases in the electrical conductivities at depth and also near the surface may explain the very narrow (<100 km) conductivity anomaly which Schmucker [1970] observed in the same area. Schmucker [1970] obtained a reversal in the anomalous vertical geomagnetic variation fields observed at Carson City, to the east, and Fallon, to the west, (see Figure 1) at periods as low as 2 cph. His interpretation of the anomaly was a semicylindrical 'hump' on an infinite conductor at a depth of 80 km, with a radius of 40 km. However, more detailed array studies in 1974 (D.J. Bennett, in preparation) performed in this same area have suggested that the lateral changes in electrical conductivity are somewhat more diffuse. Two-dimensional modeling of this array data in conjunction with the power line data may resolve the question of the extent of the lateral variation in the electrical conductivity.

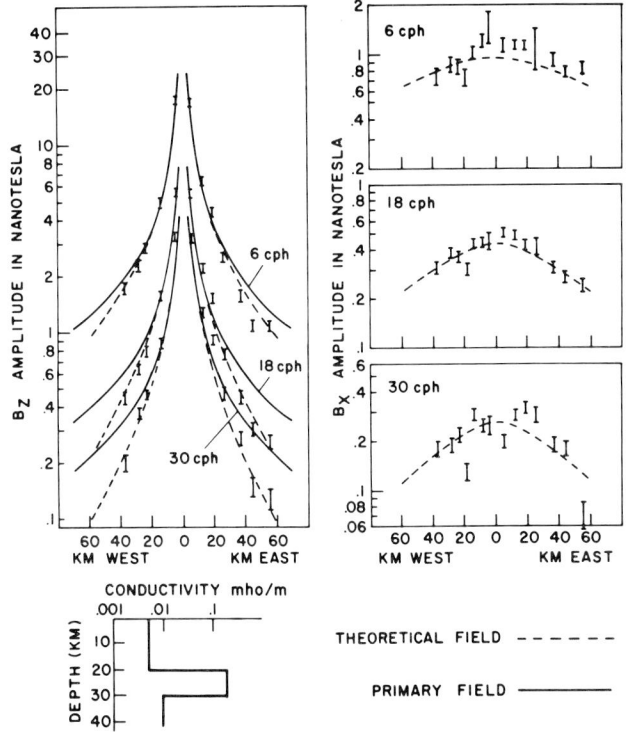

Fig. 9. Amplitudes of the horizontal (B_x) and vertical (B_z) magnetic fields at the first three harmonic frequencies. The solid curve represents the primary vertical field, while the dashed curve represents theoretical values for the indicated model.

The results presented in Table 1 nevertheless provide confirmation of the type of model obtained by using image theory, a substantial conductive zone at about 20 km being obtained for most of the stations. The reason for the smaller depth (14 km), obtained from image theory modeling is partly the better resolution obtained by using the phase data, and partly the effect of giving the conductive layer a finite thickness. The overall fit of the data at the first three frequencies is shown in Figures 9 and 10, which show the amplitude and phase data. The dashed curves are for the 'average' model which is shown in Figure 9. As is expected, the fit of the data to this model is very good at the lowest frequency but deteriorates at the higher frequencies. The anomalous effects of the two shallow conductive zones discussed earlier are particularly apparent in the vertical phase data as well as in the horizontal amplitudes. The overall impression is still one of shallow effects disturbing the main effect of deep underlying structure, since the anomalous features of the data increase markedly at the higher frequencies.

Discussion

The application of generalized linear inverse theory has provided a good degree of confidence in the type of model initially obtained from

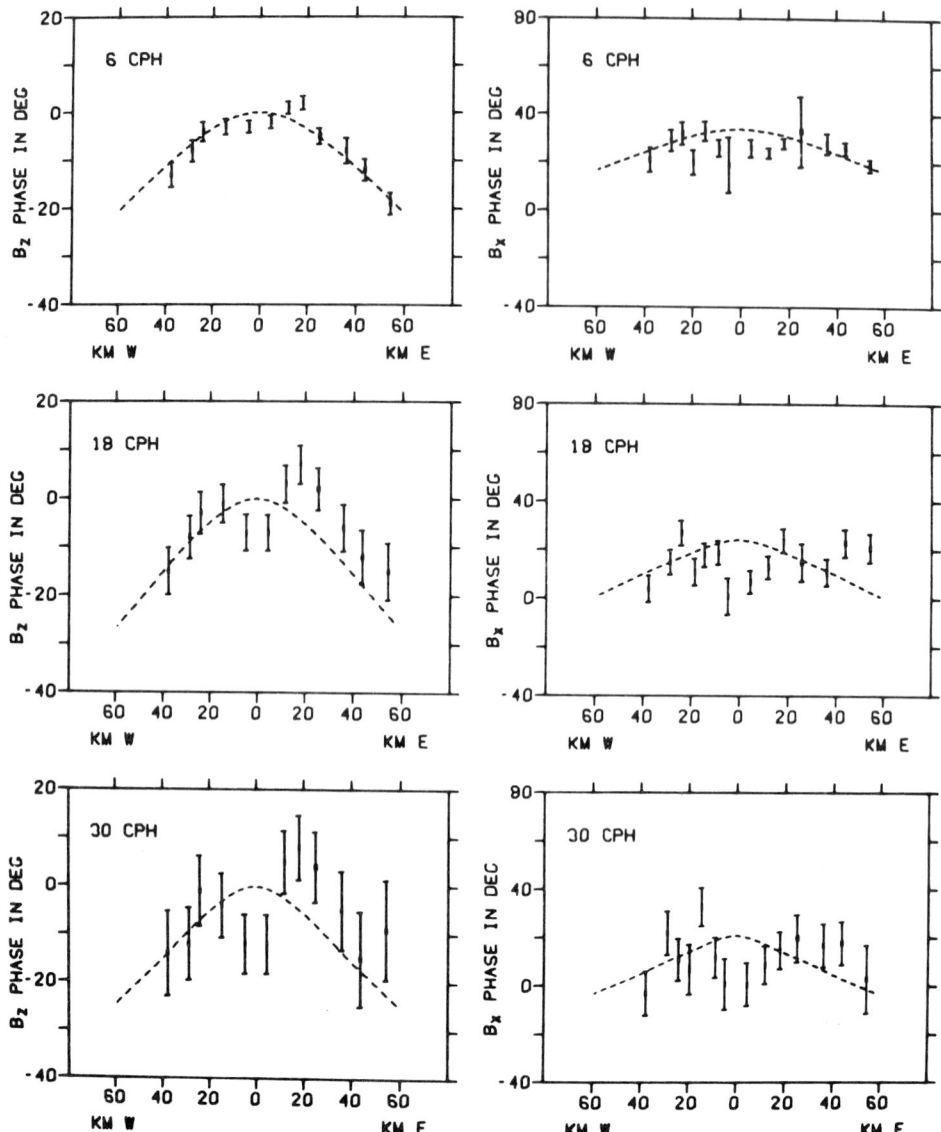

Fig. 10. Phases of the horizontal and vertical magnetic fields. Dashed curves are theoretical values for the model shown in Figure 9.

image theory. Although the use of inverse theory has demonstated the existence of two-dimensional structures, it has also shown that the effect of this type of structure on the data may be confined to portions which can be selectively removed to yield a more reliable one-dimensional model. The number of degrees of freedom allowed in the solution at each station is fairly small, but resolution is very good for the parameters of interest, especially the depth, d_1, to the conductive zone. The rapid convergence initially obtained for most of the stations to a final model which is markedly different from the starting model, has conclusively demonstrated that the data cannot be satisfied solely by one-dimensional models having high conductivities at shallow depths.

A zone of high electrical conductivity at a depth of about 20 km has interesting implications for crustal structure. Sass et al. [1971] obtained high heat flow values ranging from 1.5 to 2.3 HFU (1 HFU = 1 $\mu cal\ cm^{-2}\ s^{-1}$) at three stations in this area (Figure 1) which are typical of values for the Basin and Range province. The geotherm of Roy et al. [1968] for the Basin and Range province suggests that the temperatures at depths of 20-25 km are 500°-600°. Brace [1971] used this geotherm to infer an increase in electrical conductivity at similar depths; however, the values he obtained for the conductivities at such depths are still fairly low - about $10^{-4}\ \Omega^{-1}\ m^{-1}$. This estimate of the conductivity could easily be increased 2-4 orders of magnitude if partial melting occurred [Presnall et al., 1972; Waff, 1974]. Whether partial melting is possible at temperatures as low as 600°C is still a matter of some debate. However, 20 km is considered to be the minimum depth at which crustal melting could occur [Wyllie, 1971].

It is not clear whether this zone of high electrical conductivity can be correlated with the low seismic velocity observed in the Basin and Range province by Mitchell and Landisman [1971], Prodehl [1970], and more recently by Braile et al. [1974]. Landisman et al. [1971] have proposed that a crustal zone of low seismic velocity and high electrical conductivity is a fairly general feature of the continental crust. As Healy [1971] has observed, the seismic evidence is in many cases equivocal. However, the results of this study support the concept of a zone of high electrical conductivity and low seismic velocity in the lower crust beneath Walker Lake, Nevada.

Conclusions

1. Both simple image theory and generalized linear inversions for three-layered models provide results that are consistent with a large increase in electrical conductivity at a depth of about 20 km in the vicinity of Walker Lake, Nevada.
2. The anomaly observed by Schmucker [1970] is undoubtedly due to this zone of high conductivity. However, the lateral changes in the electrical conductivity in this area are probably more diffuse than those he initially proposed.
3. The extremely high conductivities of 0.2-$0.3\ \Omega^{-1}\ m^{-1}$ inferred for this zone strongly suggest partial melting in the lower crust.

Appendix: Calculation of the Field Values and Their Partial Derivatives for an Infinite Line Current Above an N-layered Half Space

The function $g_0(\xi, z)$ is defined by

$$g_0(\xi, z) = Q_0 e^{\xi z} \tag{A1}$$

(A1) is Summers and Weavers' [1973] equation (80) with $\rho = \xi$, where their notation, coordinate system, and units have been followed. The function Q_0 depends on the conductivities and thicknesses of the layers and can be obtained by using the recursion relation

$$Q_{n-1} = u_{n-1}/(v_{n-1} + v_n - u_n Q_n) \tag{A2}$$

(A2) is Summers and Weavers [1973] equation (39) with μ_n, μ_{n-1}, ..., μ_1 set equal to μ_0 while u_n and v_n are given by

$$u_n = \nu_n \operatorname{csch}(\nu_n d_n) \tag{A3}$$

$$v_n = \nu_n \coth(\nu_n d_n) \tag{A4}$$

and

$$\nu_n = (\xi^2 + \omega\sigma_n)^{\frac{1}{2}} \tag{A5}$$

d_n and σ_n are the thickness and conductivity, respectively, of the nth layer, as shown in Figure 3. The recursion commences with

$$Q_{N-1} = u_{N-1}/(v_{N-1} + v_N) \tag{A6}$$

(A2) being used to iterate down to Q_1. Q_0 is then obtained by using

$$Q_0 = 2\xi/(\xi + v_1 - u_1 Q_1) \tag{A7}$$

The derivatives that we require are

$$\partial Q_0/\partial d_n \qquad n = 1, 2, \ldots, N - 1$$

and

$$\partial Q_0/\partial d_n = (i\omega/\nu_n)\partial Q_n/\partial \sigma_n \qquad n = 1, 2, \ldots, N - 1 \tag{A8}$$

The problem now is to obtain recursion relations for $\partial Q_m/\partial d_n$ and $\partial Q_m/\partial \nu_n$ where $m = N - 1, \ldots, 2, 1, 0$. These partial derivatives can be obtained by differentiating (A2), (A6), and (A7) with respect to the appropriate parameters for each layer. The form of these recursion relations will clearly depend on the values of m and n in each layer. The different cases and their appropriate recursion relations appear below.

Case 1: $m = N - 1$

$$u_m \partial Q_m/\partial \nu_n = -Q_m^2 \qquad n = N \tag{A9}$$

$$\partial Q_m/\partial \nu_n = 0 \qquad n = N - 1, \ldots, 2, 1 \tag{A10}$$

$$\partial Q_m/\partial d_n = 0 \qquad n = N - 1, \ldots, 2, 1 \tag{A11}$$

Case 2: $N - 1 > m > n$

$$\partial Q_m/\partial d_n = 0 \qquad m = N - 2, N - 1, \ldots, n + 1 \tag{A12}$$

$$\partial Q_m/\partial \nu_n = 0 \qquad m = N - 2, N - 1, \ldots, n + 1 \tag{A13}$$

Case 3: $m = n$

$$u_m \partial Q_m/\partial d_n = -Q_m^2[v_m^2 + v_m(v_{m+1} - u_{m+1}Q_{m+1})] \tag{A14}$$

$$u_m \partial Q_m/\partial \nu_n = -(Q_m^2/\nu_m)[d_m v_m^2 + (d_m v_m + 1)(v_{m+1} - u_{m+1}Q_{m+1})] \tag{A15}$$

Case 4: $m = n - 1$

$$u_m \partial Q_m/\partial d_n = Q_m^2(u_{m+1}^2 - u_{m+1}Q_{m+1} - \partial Q_{m+1}/\partial d_n) \tag{A16}$$

$$u_m \partial Q_m/\partial \nu_n = (Q_m^2/\nu_n)[(u_{m+1}^2 - u_{m+1}v_{m+1}Q_{m+1})d_n$$
$$- (v_{m+1} - u_{m+1}Q_{m+1}) + u_{m+1}v_{m+1}\partial Q_{m+1}/\partial \nu_n] \tag{A17}$$

Case 5: $0 < m < n - 1$

$$u_m \partial Q_m/\partial \nu_n = Q_m^2 u_{m+1} \partial Q_{m+1}/\partial \nu_n \tag{A18}$$

$$u_m \partial Q_m/\partial d_n = Q_m^2 u_{m+1} \partial Q_{m+1}/\partial d_n \tag{A19}$$

Case 6a: $m = 0$, $n = 1$

$$\partial Q_0/\partial \nu_1 = (Q_0^2/2\xi \nu_1)[(u_1^2 - u_1 v_1 Q_1)d_1$$
$$- (v_1 - u_1 Q_1) + u_1 v_1 \partial Q_1/\partial \nu_1] \tag{A20}$$

$$\partial Q_0/\partial d_1 = (Q_0^2/2\xi)(u_1^2 - u_1 v_1 Q_1 + u_1 \partial Q_1/\partial d_1) \tag{A21}$$

Case 6b: $m = 0$, $1 < n < N$

$$\partial Q_0/\partial \nu_n = (Q_0^2/2\xi \nu_n) u_1 \partial Q_1/\partial \nu_n \tag{A22}$$

$$\partial Q_0/\partial d_n = (Q_0^2/2\xi) u_1 \partial Q_1/\partial d_n \tag{A23}$$

Case 6c: $m = 0$, $n = N$

$$\partial Q_0/\partial \nu_n = -(Q_0^2/2\xi u_1) u_1 \partial Q_1/\partial \nu_n \tag{A24}$$

We have now obtained recursion relations in $u_m \partial Q_m/\partial d_n$ and $u_m \partial Q_m/\partial \nu_n$ from which $\partial Q_0/\partial d_n$, $\partial Q_0/\partial \nu_n$ (and therefore $\partial Q_0/\partial \sigma_n$) can be obtained for all values of n, commencing with $m = N - 1$ and iterating down to $m = 0$, for each partial derivative. A numerical problem arises in evaluating the quantity $Q_m^* = v_m - u_m Q_m$ when the values of ν or d are very small since v_m and u_m then become very large (see (A3) and (A4)). This problem can be overcome by defining a separate recursion relation for Q_m^*, namely,

$$Q_{m+1}^* = [(v_m^2/u_m) + Q_m^*]/(1 + Q_m^*/v_m) \tag{A25}$$

This can readily be obtained by rearranging (A2) and using the fact that

$$u_m^2 - v_m^2 = \nu_m^2 \tag{A26}$$

(A25) is well behaved as v_m becomes large, Q_{m+1}^* tending to the value of Q_m^* when d_m or v_m become small.

The remaining problem is to evaluate integrals of the type

$$\partial B_z/\partial d_n \big|_{z=h=0} = x B_{z,p} \int_0^\infty \partial Q_0/\partial d_n \sin(\xi x) \, d\xi \tag{A27}$$

For distances from the line x, that are much greater than one skin depth, the fast Fourier transform (FFT) is a convenient numerical technique. However, if closely spaced values in the range $0 < x < \delta$ are required, the FFT technique is not very efficient. This is because a large number of equally spaced values of ξ must be used to obtain the same number of coefficients, and only a small number of these coefficients are used at one end of their range. In this case it is more efficient to evaluate the integrals by using either Gaussian quadrature techniques, as done by Frischknecht [1967], or Simpson's rule. It is then necessary to evaluate finite sums such as

$$\partial B_z/\partial d_n \simeq x \, B_{z,p} \sum_0^{N_{max}} \partial Q_0/\partial d_n \sin(n\Delta\xi) \, \Delta\xi \tag{A28}$$

where N_{max} is chosen to make the last term of the series negligibly small compared to the overall sum. The values of $\Delta\xi$ and N_{max} were determined experimentally by starting with small values of $\Delta\xi$ and large values of N_{max}, then comparing the calculated values with those of Price [1950] for a uniform conducting half space. The values obtained for the various derivatives were checked by varying the respective parameters by small amounts and determining the changes in the numerically calculated field values. The field values could be determined to within 5% by using $N_{max} = 49$ and $\Delta\xi = 0.008$. Although some of the derivatives were then only accurate to about 10%, this was found to have a negligible effect on the inversion procedure. The derivatives shown in Figure 5 were calculated by using Simpson's rule with $N_{max} = 199$ and $\Delta\xi = 0.002$ and are accurate to better than 1%.

Acknowledgements. This experiment was made possible by Dr. Anton Hales, who arranged the tests with the Bonneville Power Administration and the Los Angeles Department of Water and Power in conjunction with Dr. George Keller of the Colorado School of Mines. We gratefully acknowledge the help of Charlie Simmons, who performed the field work, and White Chaipayungpun, who provided many of the necessary computer software routines. Research supported by the Earth Sciences Section, National Science Foundation, NSF Grant DES74-22950. Contribution no 329 Institute for Geological Sciences, University of Texas at Dallas.

References

Backus, G. E., and J. F. Gilbert, Numerical application of a formalism for geophysical inverse problems, Geophys. J. Roy. Astron. Soc., 13, 247-276, 1967.

Backus, G. E., and J. F. Gilbert, The resolving power of gross earth data, Geophys. J. Roy. Astron. Soc., 16, 169-205, 1968.

ackus, G. E., and J. F. Gilbert, Uniqueness in the inversion of gross earth data, Phil. Trans. Roy. Soc. London, Ser. A, 266, 123-192, 1970.

Brace, W. F., Resistivity of saturated crustal rocks to 40 km based on laboratory studies, in The Structure and Physical Properties of the Earth's Crust, Geophys. Monogr. Ser., vol. 14, edited by J. G. Heacock, 243-256, AGU, Washington, D. C., 1971.

Braile, L. W., R. B. Smith, G. R. Keller, R. M. Welch, and R. P. Meyer, Crustal structure across the Wasatch Front from detailed seismic refraction studies, J. Geophys. Res., 79, 2669-2677, 1974.

Frischknecht, F. C., Fields about an oscillating magnetic dipole over a two layer earth, Colo. Sch. Mines Quart., 62(1), 1-81, 1967.

Garland, G. D., Electrical conductivity anomalies - Mantle or crust, Comments Earth Sci. Geophys., 3, 167-172, 1972.

Glenn, W. E., J. Ryu, S. H. Ward, W. J. Peeples, and R. J. Phillips, The inversion of vertical magnetic dipole sounding data, Geophysics, 38, 1109-1129, 1973.

Gough, D. I., and J. S. Reitzel, A portable three-component magnetic variometer, J. Geomagn. Geoelec., 19(3), 203-215, 1967.

Healy, J. H., A comment on the evidence for a worldwide zone of low seismic velocity at shallow depths in the earth's crust, in The Structure and Physical Properties of the Earth's Crust, Geophys. Monogr. Ser., vol 14, edited by J. G. Heacock, 35-40, AGU, Washington, D. C., 1971.

Hohmann, G. W., Electromagnetic scattering by conductors in the earth near a line source of current, Geophysics, 36(1), 101-131, 1971.

Jackson, D. D., Interpretation of inaccurate, insufficient, and inconsistent data, Geophys. J. Roy. Astron. Soc., 28, 97-109, 1972.

Kaula, W. M., Theory of Satellite Geodesy, Blaisdell, Waltham, Mass., 1966.

Keller, G. V., and F. C. Frischknecht, Electrical Methods in Geophysical Prospecting, Pergamon, New York, 1967.

Landisman, M., S. Mueller, and B. J. Mitchell, Review of evidence for velocity inversions in the continental crust, in The Structure and Physical Properties of the Earth's Crust, Geophys. Monogr. Ser., vol. 14, edited by J. G. Heacock, 11-34, AGU, Washington, D. C., 1971.

Lilley, F. E. M., Magnetometer array studies: A review of the interpretation of observed fields, Phys. Earth Planet. Interiors, 10(3), 231-239, 1975.

Mitchell, B. J., and M. Landisman, Geophysical measurements in the southern great plains, in The Structure and Physical Properties of the Earth's Crust, Geophys. Monogr. Ser., vol 14, edited by J. G. Heacock, 77-93, AGU, Washington, D. C., 1971.

Ortega, J., The Givens-Householder method for symetric matrices, in Mathematical Methods for Digital Computers, John Wiley, New York, 1967.

Presnall, D. C., C. L. Simmons, and H. Porath, Changes in electrical conductivity of a synthetic basalt during melting, J. Geophys. Res., 77, 5665-5672, 1972.

Price, A. T., Electromagnetic induction in a semi-infinite conductor with a plane boundary, Quart. J. Mech. Appl. Math., 3(4), 385-410, 1950.

Prodehl, C., Seismic refraction study of crustal structure in the western United State, Geol. Soc. Amer. Bull., 81, 2629-2646, 1970.

Roy, R. F., D. D. Blackwell, and F. Birch, Heat generation of plutonic rocks and continental heat flow provinces, Earth Planet. Sci. Lett., 5, 1-12, 1968.

Sass, J. H., A. H. Lachenbruch, R. J. Munroe, G. W. Green, and T. H. Moses, Heat flow in the western United States, J. Geophys. Res., 76, 6376-6413, 1971.

Schmucker, U., Anomalies of geomagnetic variations in the western United States, Bull. Scripps Inst. Oceanogr., 13, 1970.

Summers, D. M., and J. T. Weaver, Electromagnetic induction in a stratified conducting half-space by an arbitrary periodic source, Can. J. Phys., 51(10), 1064-1074, 1973.

Waff, H. S., Theoretical considerations of electrical conductivity in a partially molten mantle and implications for geothermometry, J. Geophys. Res., 79, 4003-4010, 1974.

Weaver, J. T., Image theory for an arbitrary quasi-static field in the presence of a conducting half-space, Radio Sci., 6(6), 647-653, 1971.

Wiggins, R., The general linear inverse problem: Implications of surface waves and free oscillations for earth structure, Rev. Geophys. Space Phys., 10(1), 251-285, 1972.

Wyllie, P. J., Experimental limits for melting in the earth's crust and upper mantle, in The Structure and Physical Properties of the Earth's Crust, Geophs. Monogr. Ser., vol. 14, edited by J. G. Heacock, 279-301, AGU, Washington, D. C., 1971.

CRUST AND UPPER MANTLE NEAR THE WESTERN EDGE OF THE GREAT PLAINS

W. Chaipayungpun and M. Landisman

Geoscience Program, University of Texas at Dallas
Richardson, Texas 75080

Abstract. Magnetotelluric (MT) data were recorded digitally at six sites in the High Plains of eastern Colorado. These sites were selected to be near a U.S. Geological Survey seismic refraction profile, in order to obtain contiguous estimates for both resistivity and velocity in the crust and upper mantle. Quantitative interpretation of the MT apparent resistivities was effected via their transformation into an orthogonal asymptotic coordinate system, consisting of the conductance S, the integral of the conductivity thickness product, and the depth H. The derived S versus H function is associated with a simple, fast inversion procedure which enhances the attractiveness of the MT method. Inferred resistivity distributions for the shallower layers are comparable to results derived from nearby well logs. Our preferred interpretation contains a crustal low resistivity zone at depths near 20 km which may be related to low values of compressional velocity at similar depths. The present MT data as interpreted in the S versus H coordinate system also suggest a relatively conducting region in the upper mantle at depths near 120 km, which may correspond to the base of the lithosphere. Our individual 'spot' determinations of conductivity, and thus temperatures in the lower lithosphere for eastern Colorado, are concordant with heat flow studies which show the geotherms rising westward toward the Rocky Mountain front. The westward upwarp of the mantle geotherms in the eastern Colorado transition belt may be related to regional Laramide tectonism, which produced a relatively shallow partial melt zone beneath the Rocky Mountains and other portions of western North America. Similar or even more marked upwarping of the geotherms can be seen to the south in New Mexico, where this transition belt merges with the Rio Grande rift zone, located in an area of Quaternary and Recent volcanism.

Introduction

This paper presents results derived from magnetotelluric (MT) studies in the High Plains of eastern Colorado which provide estimates of subsurface electrical resistivity distributions for the crust and upper mantle. Resistivities in the lower crust and upper mantle are strongly dependent upon temperature. Spatial changes in resistivity can be correlated with tectonic features and other geophysical parameters, such as seismic wave velocity, attenuation, and density, which are also affected by temperature [Swift, 1967; Uyeda and Rikitake, 1970; Gough, 1973; Schmucker, 1973; Garland, 1975]. The lower temperatures and pressures at shallower depths are coupled with a different electrical conductivity regime. These resistivities are primarily dependent upon small values of effective porosity and the ionic conductivities of included interstitial fluids [Archie, 1942] as well as the chemical composition of the rocks [Parkhomenko, 1967; Shankland, 1975].

Recent electrical studies in several continental regions have provided evidence for an anomalous zone of low resistivity in the upper and middle parts of the earth's crust [e.g., Migaux et al., 1960; Keller et al., 1966; Van Zijl, 1969, 1977; Berdichevskii et al., 1969, 1972,

1973; Dowling, 1970]. Mitchell and Landisman [1970, 1971a, b] noted the correlation between the zone of low electrical resistivity in west Texas and zones of low seismic velocity in Oklahoma and eastern New Mexico at depths of the order of 20 km. They suggest the possibility that interstitial fluid is a common causal mechanism for the reduction of resistivity and the reduction of seismic velocity. Mitchell [1973] also reported that crustal surface wave attenuation measurements provide evidence at midcrustal depths for a significant increase in Q, which was discussed in terms of a transition to a dryer more basic regime beneath the anomalous zone.

Pakiser and Zietz [1965], from transcontinental seismic, aeromagnetic, and gravity surveys combined with geological information, suggest that the continental United States is divided by the Rocky Mountain system into a western superprovince with widespread Cenozoic diastrophism and volcanism, and an eastern superprovince with almost no Cenozoic tectonism. The transition belt between these two superprovinces is situated along the eastern front of the Rocky Mountains at the western edge of the North American craton, which has been relatively stable since Cambrian time. This craton is characterized by moderately shallow basins and broadly warped arches [Ham and Wilson, 1967]. The High Plains of eastern Colorado, near the western margin of the craton, is thus an excellent locale for studying intraplate transitional properties of the crust and upper mantle.

MT observations were recorded at six sites near a U.S. Geological Survey (USGS) seismic refraction profile between Lamar and Sterling in eastern Colorado [Jackson et al., 1963]. These MT experiments constitute a northward extension of the crustal studies reported by Mitchell and Landisman [1970, 1971a, b] in the southern Great Plains.

The inferred resistivity distributions were obtained via an asymptotic transformation of the frequency dependent apparent resistivity into a corresponding function of the conductance S, the integral of the conductivity thickness product, and the depth H. The log S versus log H coordinates provide a natural orthogonal coordinate system for the interpretation of MT data [Bostick, 1977, also private communication, 1974]. A simple and fast inversion procedure utilizing the above asymptotic transformation is described in the section on S and H inversion. This procedure easily permits the introduction of geological and geophysical information into the interpretation process.

Data Analysis

Magnetotelluric data from the six recording sites shown in Figure 1, near Anton, Joes, Idalia, Akron, Yuma, and Wray in the High Plains of eastern Colorado, provided the observational material used in this study. The data were recorded by a gain-ranged digital field system that had a frequency range of roughly five decades, from about several thousand seconds to 0.03 s (30 Hz). This spectral interval was covered by separately recording four different data subsets: Band 1, from the longest periods to 10 s, sampled once per second; Band 2, 100 to 1 s, sampled 10 times per second; Band 3, 10 to 0.1 s (10 Hz), sampled 100 times per second; and Band 4, 3 to 0.03 s (30 Hz), sampled 333 times per second. Special care was taken to insure that the system noise remained far smaller than the recorded natural signals. Further information about the data acquisition and reduction system can be found in a report by Landisman and Chaipayungpun [1977]. The data recorded at each site usually consisted of at least two Band 1 observation subsets, with durations of about 9 hours (2^{15} points) and of nearly 18 hours (2^{16} points). At least three Band 2 subsets, and four to six runs for Bands 3 and 4, each containing either 2^{13}, 2^{14}, or 2^{15} points, were also recorded. These latter short runs were recorded before, between, and after the long ones in order to monitor as many different source polarizations as possible. Only a brief summary of the data reduction sequence will be reviewed in the following paragraphs.

The magnetotelluric impedance tensor Z which relates the measured horizontal components of the electric field E and magnetic field H is defined by

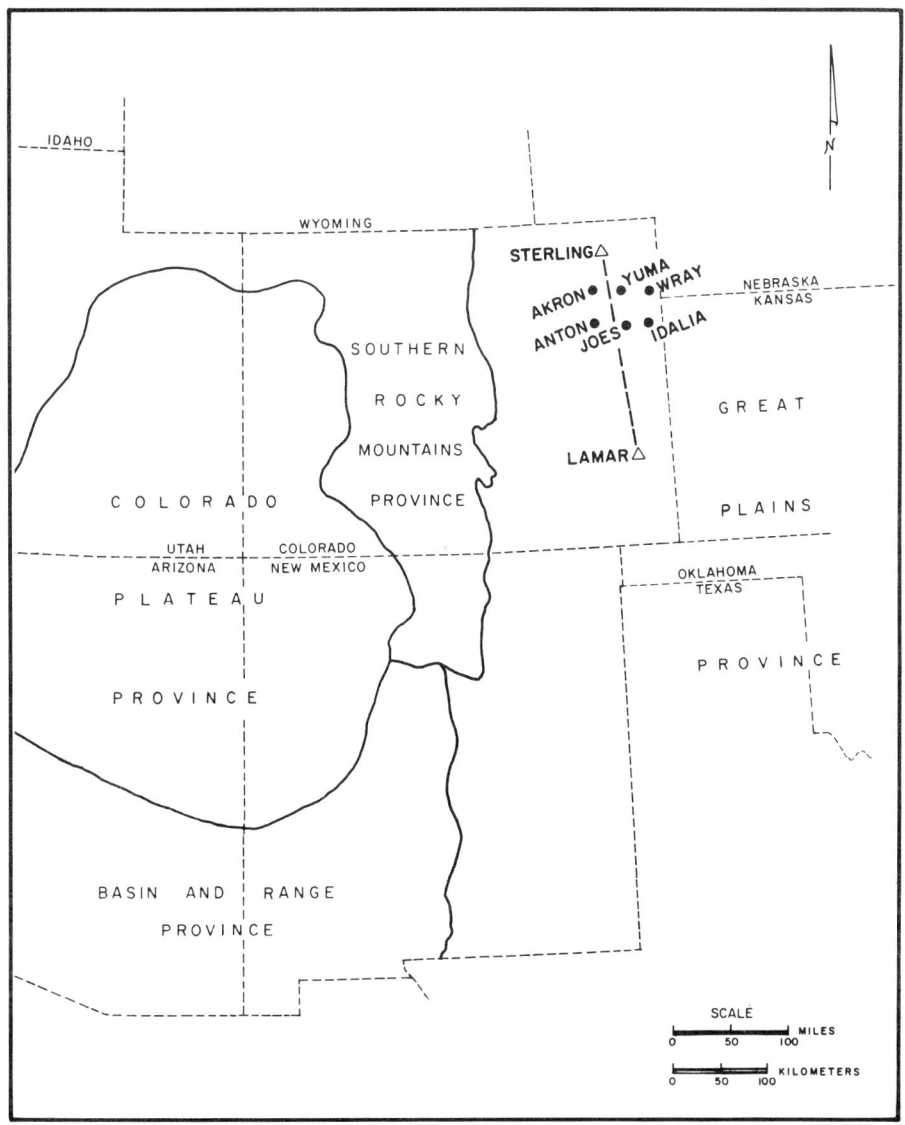

Fig. 1. Map showing locations of six magnetotelluric stations and USGS Lamar-Sterling seismic refraction profile in the High Plains of eastern Colorado.

$$\begin{vmatrix} E_x \\ E_y \end{vmatrix} = \begin{vmatrix} Z_{xx} & Z_{xy} \\ Z_{yx} & Z_{yy} \end{vmatrix} \begin{vmatrix} H_x \\ H_y \end{vmatrix} \qquad (1)$$

Various methods have been proposed for obtaining estimates of the magnetotelluric impedance tensor [e.g., Cantwell, 1960; Bostick and Smith, 1962; Madden and Nelson, 1964; Mann, 1965; Swift, 1967; Rankin and Reddy, 1969; Sims and Bostick, 1969; Hermance, 1973]. The method described by Swift [1967] and reviewed by Vozoff [1972] was adopted in this study. It provides stable least squares estimates for the elements of the impedance tensor from the power and cross-power spectra, which were deduced from multiple subsets selected from the observed E and H field time series [Cooley and Tukey, 1965; Welch, 1967]. The frequency dependent amplitudes and phases of the elements of the apparent resistivity 'tensor' were calculated from

$$\rho_{ij} = (\omega\mu)^{-1} |Z_{ij}^2| \qquad\qquad i, j = x \text{ or } y \qquad (2a)$$

$$\phi_{ij} = \tan^{-1}(\text{Im } Z_{ij}/\text{Re } Z_{ij}) \qquad\qquad i, j = x \text{ or } y \qquad (2b)$$

Coherency estimates between $E_i^{Predicted}$ and $E_i^{Observed}$ were computed with values of $E_i^{Predicted}$ derived from (1), according to the method described by Swift [1967]. The resulting coherency estimates provided a useful criterion for the acceptance or rejection of individual data points; usable data were distinguished by coherency values of at least 0.9.

The most striking characteristic of the results from all these sites is related to the fact that for most periods the magnitudes of the diagonal elements of the apparent resistivity 'tensor' are one or two orders of magnitude smaller than those of the off-diagonal elements. These relationships are present in the data in Figure 2, which were observed near the settlement of Anton. The diagonal elements ρ_{xx} and ρ_{yy} displayed in Figures 2a and 2d are about two orders of magnitude smaller than the off-diagonal elements ρ_{xy} and ρ_{yx} displayed in Figures 2b and 2c. The significantly smaller diagonal apparent resistivity 'tensor' elements observed near Anton, and the similar magnitudes of the off-diagonal 'tensor' elements, suggest that the subsurface resistivities in this portion of the High Plains are characterized by a high degree of lateral homogeneity. This suggestion of lateral homogeneity is supported by the small amplitudes for the observed ratio, typically 12% or less, for the vertical component Z in comparison with the horizontal component H of the ambient magnetic field. Chaize and Lavergne [1968], after a fairly extensive evaluation of MT observations, report that observed Z/H ratios of 15% or less are usually associated with high quality MT results. Such data are nearly devoid of the complications usually associated with multidimensional conductivity distributions. Additional confirmation comes from the small lateral gradients in the vertical induction field, which were observed during a Deep Geomagnetic Sounding (DGS) array experiment in this same general area [Reitzel et al., 1970]. Both of these observational results indicate that the lateral conductivity gradients are unusually small in this portion of the High Plains.

Such a high degree of lateral homogeneity permits corresponding amplitudes and phases of the two off-diagonal elements of the 'tensor' apparent resistivity to be practically indistinguishable for these High Plains recording sites. A simple, fast, and efficient inversion procedure for such data is indicated in Figure 3 and will be described in the next section. The results for our six stations are presented in Figures 4a to 4f. Values for the xy elements are indicated by plus symbols, while those of the yx elements are denoted by squares. Only the yx values are given for Akron, where the xy data are unavailable because of instrumental malfunction. We believe this to be the first set of observations for which one can superimpose the xy and yx elements of the 'tensor' apparent resistivity for both amplitude and phase; it is virtually impossible to distinguish the difference between them over almost the entire frequency range.

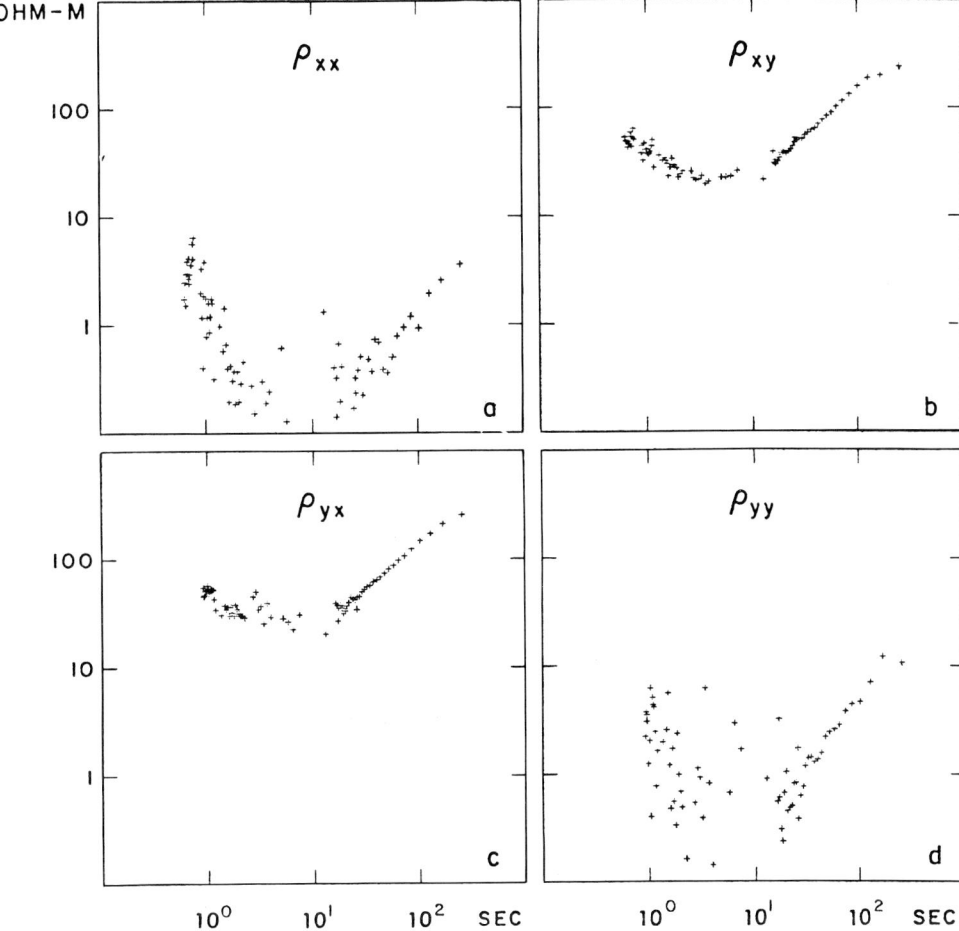

Fig. 2. Elements of the apparent resistivity 'tensor' as observed near Anton, Colorado. Off-diagonal elements ρ_{xy} in part b, and ρ_{yx} in part c, have nearly identical magnitudes which greatly exceed those of the diagonal elements ρ_{xx} in part a and ρ_{yy} in part d. See text for further evidence of lateral homogeneity in this area.

S and H Inversion of Magnetotelluric Observations

Cagniard [1953] and Berdichevskii [1968] were among the first to discuss the asymptotic properties of the MT apparent resistivity ρ_{app} (in ohm-m) as a function of the period T (in seconds). The long period 45° asymptotic rise of the log ρ_{app} versus log T function can be associated with a constant value of the sum of the conductance S (in mhos) for models overlying a perfectly insulating half space. Similarly, Berdichevskii [1968] noted that the log ρ_{app} versus log T curve asymptotically approaches a 45° descent for long periods when the underlying half space is a perfect conductor. The descending asymptote in this case corresponds to the depth to the conductive half space H (in meters). These asymptotic limiting cases of a perfectly insulating half space or a perfectly conducting half space, when overlain by logarithmically increasing values of conductance S or thickness H, respectively, span the

$$\rho_{app} = \frac{1}{\omega\mu S^2}$$

$$\rho_{app} = \omega\mu H^2$$

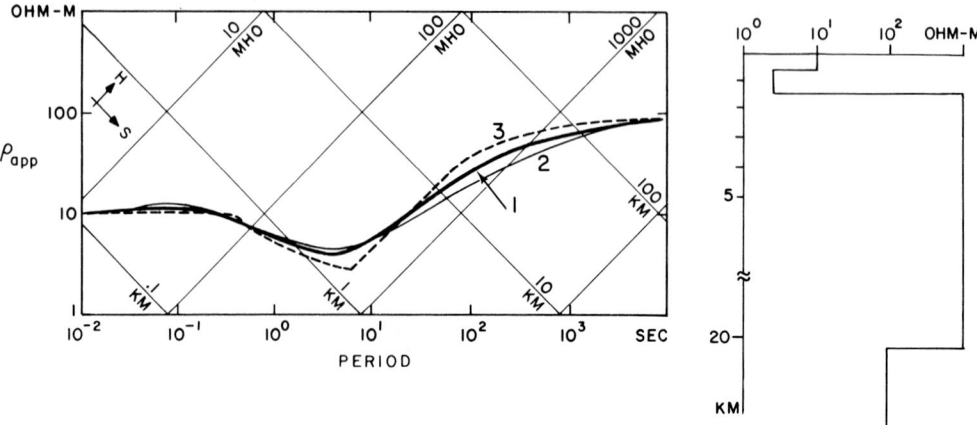

Fig. 3. S and H inversion [after Bostick, 1977, also private communication, 1974], integral S (in mhos) and thickness H (in kilometers), are shown in diagonal coordinates. Observed apparent resistivity (ρ_{app} versus T) is indicated by curve 1, and computed first iterate by curve 2. Stable deduced S and H model (must be interpreted in S and H coordinates) after several iterations is indicated by curve 3. For discussion, see text.

log ρ_{app} versus log T data plane with a second, 'mutually orthogonal' set of coordinates. These latter orthogonal coordinates directly correspond to the numerical values of the physical properties of the set of models just as the original coordinates correspond to the numerical values of the data set.

The mutually orthogonal log S versus log H model-related coordinate system can thus be used as a set of natural coordinates for MT apparent resistivity interpretation, as is displayed in Figure 3 [Bostick, 1977, also private communication, 1974]. The corresponding, mutually orthogonal log ρ_{app} versus log T coordinates for the observed and calculated MT results are also included in Figure 3. Note that in the alternative log ρ_{app} versus log $T^{1/2}$ coordinate system used, for example, by Cagniard [1953], the asymptotic slopes become 63.4°; the log S and log H lines then do not constitute an orthogonal coordinate pair. The presentation of MT apparent resistivity data (ρ_{app} versus T) in the bilogarithmic S - H coordinate system (shown in Figure 3) provides a rapid, simple, and easily used means for estimating the S versus H characteristics of the resistivity distribution beneath an experimental site. In Figure 3, curve 1 represents the 'observed' apparent resistivity which here corresponds to the model on the right. Curve 2 will be discussed later in the section. Curve 3 represents an S versus H version of the model. The similarity between the curve 3 and curve 1 shows the close relationship between the data and the model in their respective coordinate systems. The role played by these S versus H coordinates, intermediate between the observational data and the deduced model, is closely equivalent to the similar function performed by the Dar Zarrouk curve in DC Vertical Electrical Sounding (VES) investigations [Maillet, 1947; Kunetz and Rocroi, 1970].

The transformation from ρ_{app} versus T coordinates to S versus H coordinates can be

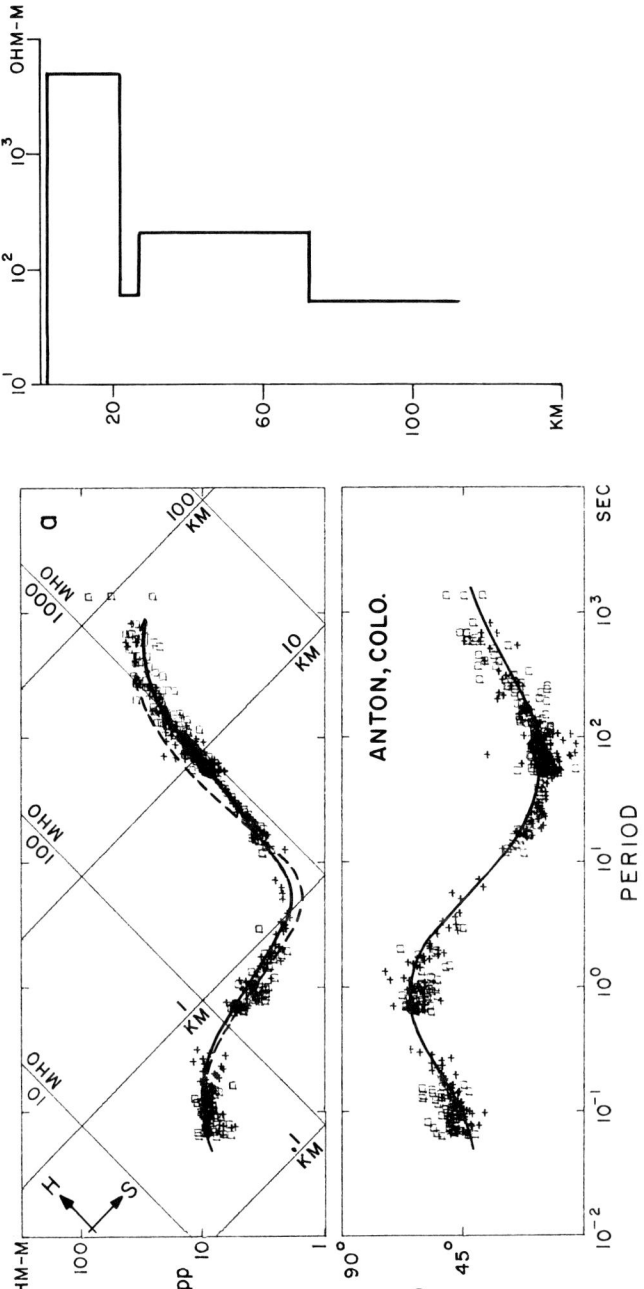

Fig. 4. Apparent resistivity (ρ_{app} versus T) and E minus H phase shift (ϕ versus T) observed at (a) Anton, (b) Joes, (c) Idalia, (d) Akron, (e) Yuma, and (f) Wray in the High Plains of eastern Colorado. Elements ρ_{xy} and ρ_{yx} of the apparent resistivity 'tensor' are indicated by pluses and squares, respectively. Values of the conductivity-thickness integral S (in mhos) and thickness H (in kilometers) are shown for diagonal coordinates. Deduced model, dashed line in S and H coordinates, produces calculated solid lines, ρ_{app} versus T and ϕ versus T. Open circles in parts e and f indicate long period apparent resistivities derived from Deep Geomagnetic Sounding array data for the Great Plains [Kuckes, 1973]. Subsedimentary resistivity depth function is shown at right.

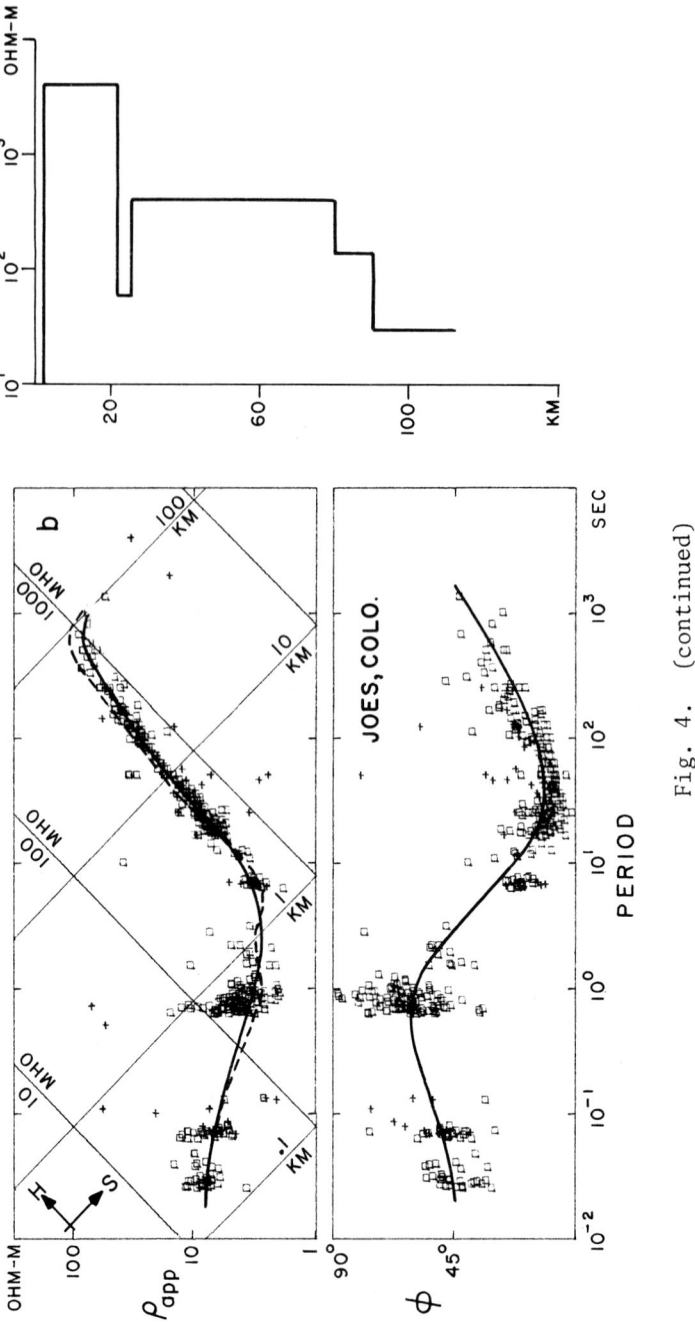

Fig. 4. (continued)

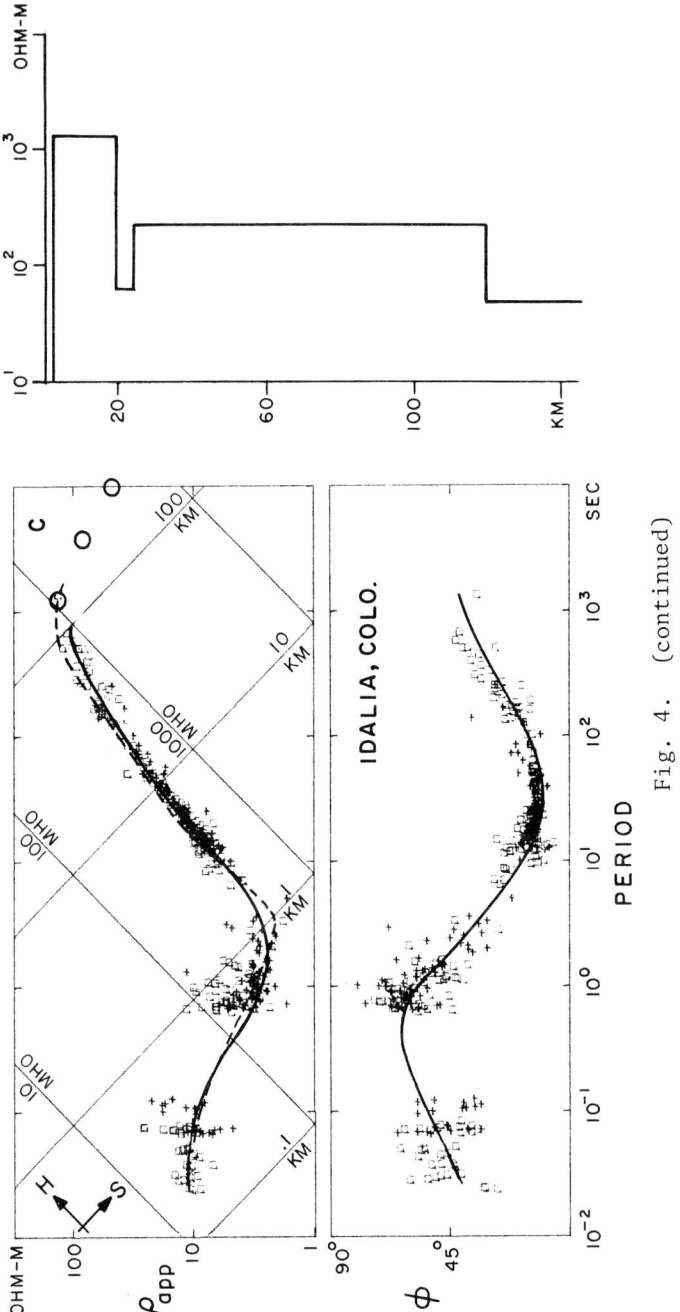

Fig. 4. (continued)

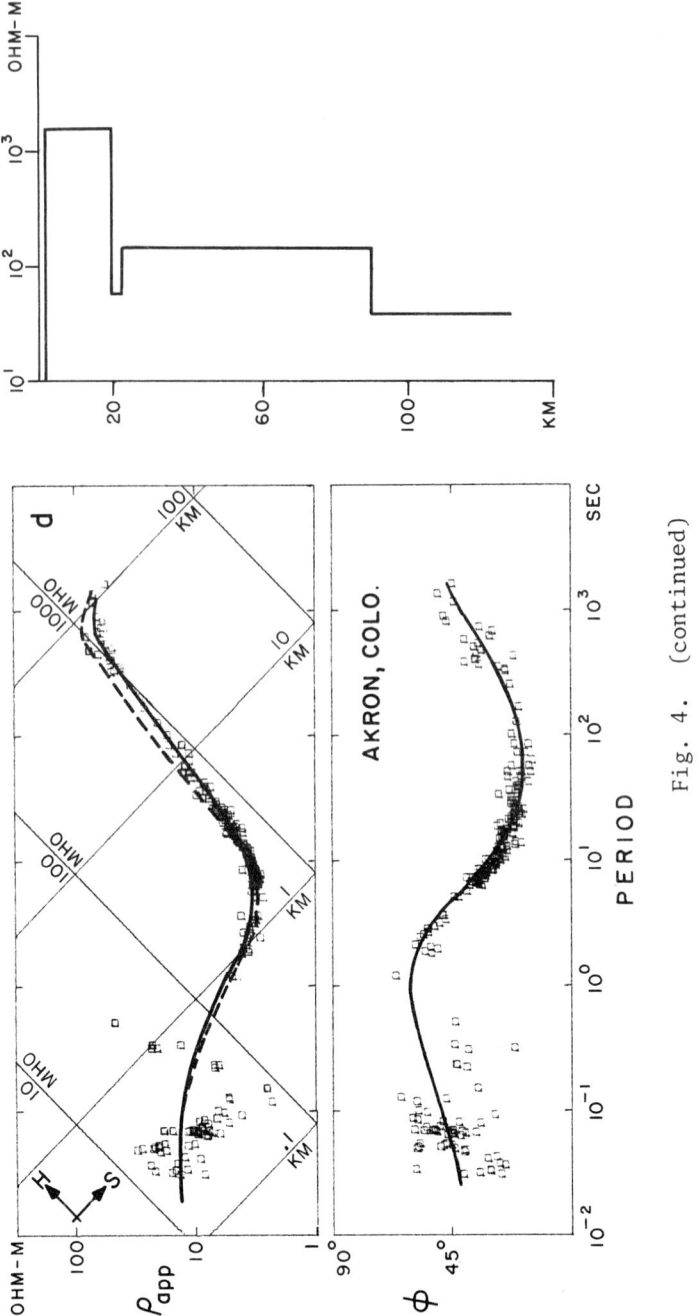

Fig. 4. (continued)

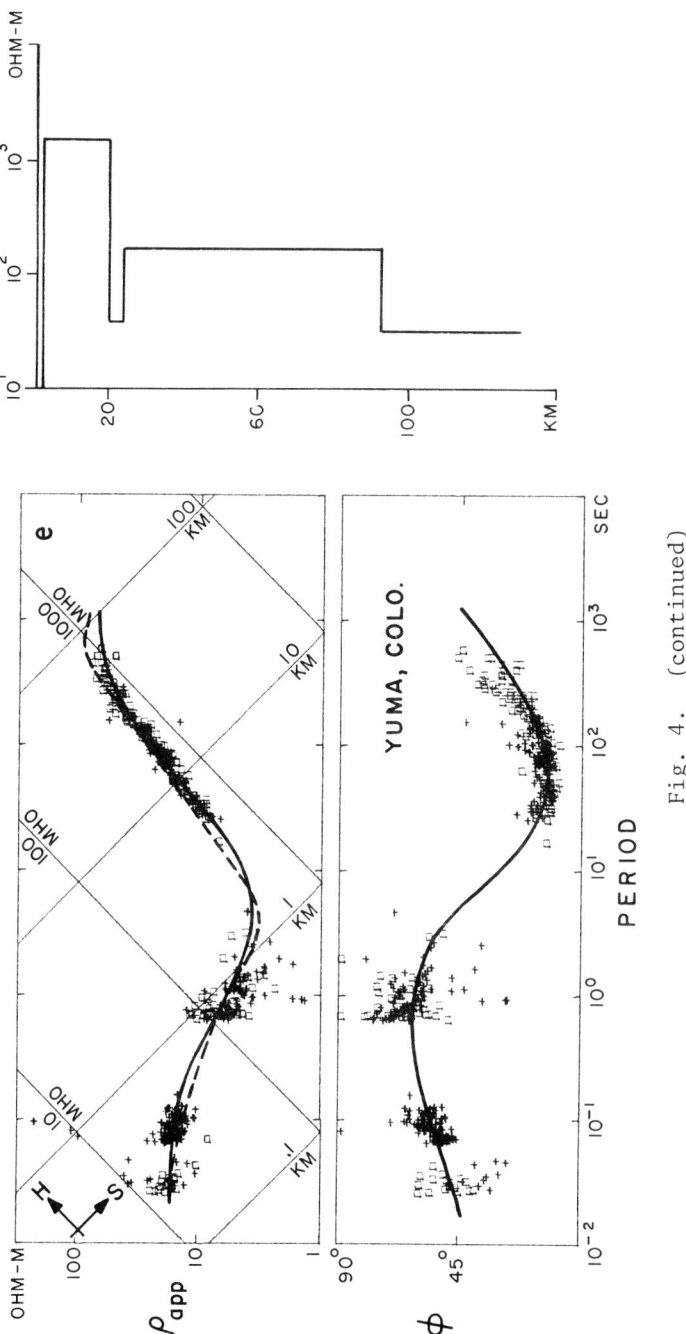

Fig. 4. (continued)

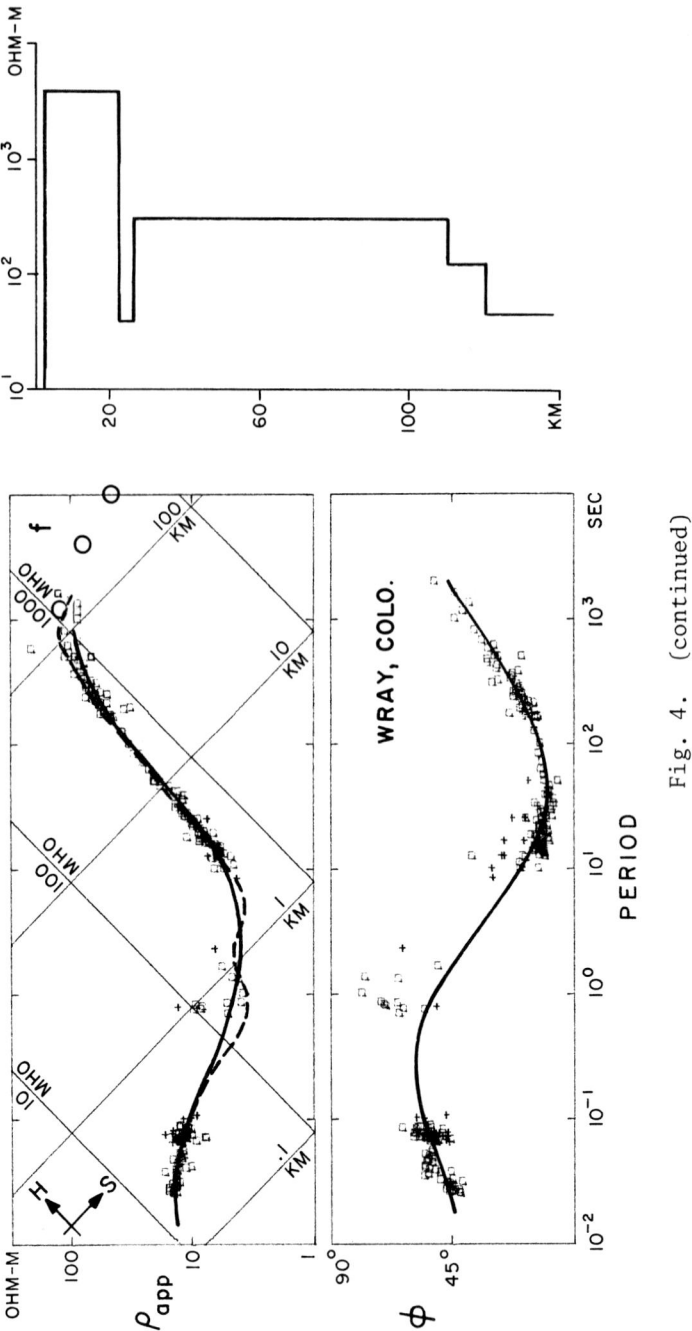

Fig. 4. (continued)

accomplished by using the following asymptotic expressions [Keller and Frischknecht, 1966; Berdichevskii, 1968]:

$$\rho_{app} = (\omega\mu S^2)^{-1} \quad (3)$$

$$\rho_{app} = \omega\mu H^2 \quad (4)$$

These two expressions are easily derived for a single homogeneous, isotropic layer overlying either a perfectly insulating half space (3) or a perfectly conducting half space (4) [e.g., Keller and Frischknecht, 1966]. S is then the conductivity-thickness product for the surficial layer, and H is its thickness.

More generally, these values correspond to the total values accumulated for the quantities S and H after summing the contributions for each of the layers above the half space. The transformation from the log ρ_{app} versus log T to the log S versus log H coordinate system is simply effected by a mere rotation and translation of the data in the respective bilogarithmic coordinate spaces presented in Figure 3. A continuous log S versus log H curvilinear function can then be approximated by a sequence of short straight-line segments, each of which can be conceptually viewed as corresponding to an individual layer. The resistivity and thickness values for each of these layers can be determined from the 'differences' in the S and H values at successive intersections of straight line segments. The representation of a set of observed MT apparent resistivity values by a smooth continuous curve or alternatively by a sequence of line segments, obviously involves a certain amount of judgment. As the lengths of the straight-line segments used to approximate the data become shorter, the one-to-one correspondence of one segment per layer causes the number of layers in the resulting model to increase accordingly.

A simple and fast inversion procedure derived from the above relation of MT observations to the S and H properties of a model will now be described by using the model and curves in Figure 3 as an illustrative example. This inversion procedure utilizes the above asymptotic expressions, (3) and (4), to transform the observed apparent resistivity data, represented by a multiple line-segment approximation to the observational points, into a multilayered horizontally stratified (i.e., one dimensional) isotropic model. The accumulating sums of conductance (S, in mhos) and depth (H, in meters) values for the collection of layers which represent this model correspond closely to the observed values of ρ_{app} versus T. The procedure can be summarized in the following steps

a) Approximate the observed apparent resistivity curve (curve 1 in Figure 3) with a sequence of straight line segments. This approximation can be accomplished by sampling the observed data at a specific log periodic interval. The interval is made small enough to insure adequate representation of each data subset without the loss of essential information.

b) Transform each of the resulting ρ_{app} versus T points into S and H coordinates by using (3) and (4).

c) Convert all of the corresponding S versus H points into individual resistivity estimates ρ_i and thickness estimates t_i. Start downward through the model, proceed from the shortest period observations, which are sensitive only to the top of the model, and skip 'noisy' data points which appear to violate the physical requirement that both resistivities and thicknesses must be intrinsically positive for each of the layers. (This latter requirement, in turn, implies that the slopes between any pair of data points on the log ρ_{app} versus log T scale 'cannot exceed $\pm 45°$,' which slopes correspond to the asymptotic limits for a perfectly insulating or a perfectly conducting half space [Cagniard, 1953; Berdichevskii, 1968].)

d) Calculate the MT apparent resistivities ρ_{app} at each sampled period T for this derived one-dimensional model, shown as curve 2 in Figure 3.

e) Project into S and H coordinates the differences between the observed apparent resistivities ρ_{app} and the values computed at each sampled period, as is done in step d).

f) Use these individual increments of S and H to adjust and correct the S and H values for each of the layers of the initial model derived in step c).

g) Repeat steps d), e), and f) until a close correspondence between the observed and computed apparent resistivity curves is achieved for all periods, i.e., until the sum of the squared ρ_{app} errors is minimized. The final S and H model, whose number of layers equals the number of sampled data points, constitutes the so-called 'comprehensive' approximation to the S versus H curve (according to the discussion of Kunetz and Rocroi [1970]). This is shown as curve 3 of Figure 3.

The advantage of this inversion method is that it provides a 'good' reasonably close starting model. The convergence of this process is conveyed by the sequence of models, which moves toward a stable configuration at the same time as the sequence of computed apparent resistivities moves toward the sampled data points and becomes a smoothed representation of the MT observations. When this iterative process has been completed, usually after only three or four attempts, a simplification and consolidation of the multilayered model is necessary. This overly detailed model, which contains one layer per line segment, is analogous to the 'comprehensive' model derived from Schlumberger DC Vertical Electrical Sounding data by Kunetz and Rocroi [1970]. The Dar Zarrouk function first proposed by Maillet [1947] has been modified by Zohdy [1974] in order to develop an alternative computational inversion method for VES observations. An initial overly detailed model is again consolidated with the help of pertinent external information. The simplification and consolidation can be achieved by replacing the multiply segmented model with an equivalent but simpler description having a smaller number of segments.

Additional geological and geophysical information can also be introduced at this stage to provide assistance for the task of selecting the location of each of the vertices of the new simplified set of line segments. The S and H values of the new model can be checked and compared against the initially derived values, since they serve as a constraint upon the new simplified model. The computed apparent resistivity values for the consolidated model will be a close approximation to the observed values as long as the corresponding new set of line segments is a good approximation to the 'comprehensive' S and H values. Recalculation of MT apparent resistivities for the consolidated model will produce values that do not differ significantly from those deduced initially. The process of consolidation of the model necessarily involves much of the same subjectivity that is involved in choosing the starting model for trial and error or other types of inversion procedures. This delimitation produces a class or type of model which, on the basis of other information, is considered to be 'geologically reasonable.'

Interpretation

The MT apparent resistivity data ρ_{app} versus T, for each of the six experimental sites, were sampled six times per decade on a log periodic scale ranging from about 0.03 s (30 Hz) to approximately 3000 s. Each of these sampling points represents an intersection between adjoining straight-line segments that approximate the observed apparent resistivities. The log sampling interval then determines the sizes of the incremental S and H values that correspond to each of the individual layers in the starting model. This sampling interval therefore should be small enough that the segmented approximation accurately represents the data without introducing an excessive number of layers. Under these constraints, six points per decade appears to be optimal for the present set of data. The sampled log ρ_{app} versus log T data values were then used to generate a multilayered S and H model by using the transformation (3) and (4) from ρ_{app} versus T to S versus H values, and thereafter from the S and H increments to individual layer resistivities and thicknesses.

Iterative S and H Inversion

A computer program was written to perform the iterative S and H inversion procedure. The forward calculation of the MT complex impedance tensor is similar to that described by Patrick and Bostick [1969]. The S and H inversion process permitted the sums of the squared error differences between the observed and computed apparent resistivities to be minimized after only three or four iterations. The multilayered, horizontally stratified models produced by this process are denoted in Figures 4a to 4f by the 'dashed lines,' which 'must be interpreted in the S and H coordinate system.' The 'calculated' apparent resistivities (ρ_{app} versus T) for these S and H models are within 2% of those corresponding to the average values of 'observed' apparent resistivity at each of the sampled periods. These sampled values, representing the average of the observed apparent resistivities at each site, are indicated by the solid lines in Figures 4a to 4f. Subsedimentary resistivity depth distributions are also shown at the right hand side in these figures.

Simplification and Consolidation of the S and H Models

Geological and geophysical information from water well drilling logs (Division of Water Resources, State of Colorado), exploration well logs (Geological Information Library of Dallas), electric logs [Keller, 1968], an isopach map of the entire sedimentary column for the High Plains of eastern Colorado [Rocky Mountain Association of Geologists, 1972], and the seismic P wave velocity distribution [Landisman et al., 1974; Landisman and Chaipayungpun, 1977] were used to simplify and consolidate the 'comprehensive' multilayered model for the resistivity distribution as follows:

Sedimentary layer. At all of the sites shown in Figure 1, there is a relatively strong upward flexure of the S versus H function, denoted by the dashed line at conductances of several hundred mhos and total thicknesses of about 1 to 1.6 km. This flexure indicates a sharp reduction in conductance at greater depths. These reduced contributions to the conductance are indicative of the sediment-basement interface and can be considered to be clear evidence for a relatively conductive sedimentary sequence overlying a highly resistive basement. For these six sites in eastern Colorado the average depth to basement is approximately 1500 m, as can be estimated from the S versus H functions. These estimates of total sedimentary thickness do not differ from values read from a map of basement depths [Rocky Mountain Association of Geologists, 1972] by more than ±100 m. The resistivity of the two shallowest layers varies between 5 and 20 ohm-m, and the sum of their thicknesses ranges from about 200 to 600 m, as determined by the high frequency ρ_{app} asymptotes and the S and H values. These resistivities are similar to the values reported by Harthill [1967] from Audio MT (AMT) surveys in the same area. These two units may correspond to the Tertiary Ogallala formation plus the Cretaceous Arapahoe and Fox Hills formations which are widely observed in local well logs [e.g., Jackson, 1962; Harthill, 1967; Keller, 1968]. The major sedimentary unit below these is the Pierre shale, with resistivities of 1.2 to 2.5 ohm-m and thicknesses of 750 to 1250 m. A tabulation of the MT results for the sedimentary layers interpreted for the station at Anton, Colorado is included in an earlier report [Landisman and Chaipayungpun, 1977].

Crustal Layers. The steep nearly asymptotic rise of the MT apparent resistivity values between 10 and 200 s shows the typical effects of the resistive basement beneath the conductive sediments. On the basis of the S and H values derived from the apparent resistivity data, several resistivity distributions can be inferred for the crust and upper mantle. Of these S and H conformable models, three of the more likely yet distinctly different ones can be described as follows: (1) a one-layered model with a resistivity of the order of 200 ohm-m; (2) a two-layered model with a resistive 2000 ohm-m upper crustal layer overlying a 240 ohm-m lower crust which begins at a depth of about 20 km; and (3) a three-layered model, similar to model 2, but distinguished from it by the inclusion of a conductive layer of 50 ohm-m between depths of 20 and 25 km.

The phase differences between corresponding horizontal components of the electric and magnetic fields were also considered for each of these models in relation to the observed phase differences as functions of period T. Recent reinterpretation of the USGS seismic refraction profile from Lamar to Sterling in eastern Colorado [Landisman et al., 1974; Landisman and Chaipayungpun, 1977] suggests the possibility of a zone of reduced compressional velocity at depths of about 20+ km. Model 3 is the preferred interpretation. Its conductive channel in the mid-crust is not only thoroughly compatible with the present MT data for both apparent resistivity and phase differences, but is also consonant with the seismic interpretation.

Upper-mantle conductive zone. The long-period data from all six MT observation sites require a relatively conductive region in the upper mantle, at a depth of about 100 km, as is indicated by the relatively sharp change in the slope of the apparent resistivity curve at periods near 600 to 1000 s. The depth of the conductive zone was estimated at each site from the corresponding long period corner in the S versus H curve, plus the consideration of long-period data for the nearby Great Plains. The long-period trends of the present MT data are concordant with a set of apparent resistivities derived from results reported by Kuckes [1973], after his elegant analysis of Deep Geomagnetic Sounding data for the western Great Plains [Porath and Dziewonski, 1971]. This set of apparent resistivities is indicated by the open circles, which accompany the MT data for Idalia and Wray in Figures 4c and 4f. The resulting collection of values for the depth to the upper-mantle conductive zone at these six sites can be interpreted in terms of an interface which rises toward the west. A possible correlation of this conductive zone with the main low-velocity zone (LVZ) in the upper mantle is discussed in the following section.

Discussion

The interpretation presented for the sedimentary columns corresponds closely with the two shallowest sedimentary units reported by Keller [1968] after analysis of resistivity logs from deep wells in the Denver Basin and surrounding areas. The first of these two units can be recognized on the basis of its average electrical resistivity of about 5 to 20 ohm-m and thickness of about 300 m. The underlying Pierre shale has a resistivity of 1.2 to 2.5 ohm-m, and a thickness of about 700 to 1200 m.

There are other more resistive units beneath the Pierre shale [Jackson, 1962; Harthill, 1967; Keller, 1968] that can be recognized in well log analyses for depths between roughly 1000 and 1500 m. Some of these units have been economically important to the petroleum industry. The MT response to these porous units tends to be suppressed between the signatures of the rather conductive Pierre shale and the deeper highly resistive basement. This suppression of the response for units having intermediate values of resistivity constitutes an example of the Principle of Suppression, first postulated for DC Vertical Electrical Soundings by Maillet [1947]. Nevertheless, the approximate depth to this exploration target can be inferred from the MT depth estimate for the interface between the Pierre shale unit and the resistive basement, which can be of assistance prior to costly drilling operations.

Crustal models containing an electrically conductive zone, preferably located about 20+ km beneath the earth's surface, represent an interpretation that is in accord with the entire body of High Plains data presented in this report. Reserving for a later study the discussion of topics usually associated with depth resolution, we note that the models described above are simultaneously compatible with our MT apparent resistivity and phase difference observations, and with data from the USGS Lamar-Sterling seismic refraction experiment. They are also consonant with seismic refraction and MT models containing a similar zone of low resistivity and low compressional velocity at corresponding depths previously interpreted from other data sets recorded in neighboring portions of the southern Great Plains [Mitchell and Landisman, 1970, 1971a, b; Landisman et al., 1971]. The present discussion now turns from sedimentary and crustal features to the interpretation of those at greater depths.

Discussion of results inferred from the present MT data set necessarily must carefully consider the relationship of the data from these six High Plains observational sites to the local and regional distributions of electrical conductivity. As is noted in the earlier section on data analysis, the present MT data, over a broad spectral range, display three important and distinctly different criteria signifying lateral homogeneity in the conductivity distribution. The first criterion is exhibited by the ratios between the observed vertical and horizontal magnetic field components and the two other criteria are evident in the observed elements of the horizontal impedance tensor, as illustrated in Figure 2. These criteria can be considered primary evidence for lateral homogeneity in the sediments and crust near each of our recording sites.

The effects of the generally westward rise in the mantle conductive zone beneath the High Plains thus are not discernible in the body of MT data from these experimental locations, except at extremely long periods. For models resembling those in Figures 4a to 4f, varying contributions to the conductance S at site dependent cumulative thicknesses (or depths) of 70 km or more simply do not produce effective perturbations of apparent resistivity unless the periods are greater than roughly 600 s. Thus the observed lateral homogeneity of the sediments and crust in the High Plains explains the relative ease with which 'one-dimensional' models are able to replicate the present MT data, as can be seen in Figures 4a to 4f. The two-dimensional models usually reserved for more complicated survey areas are ordinarily characterized by two separate MT apparent resistivity curves, corresponding either to parallelism or perpendicularity of the horizontal electric field with respect to the structural strike direction [e.g., Keller and Frischknecht, 1966; Vozoff, 1972]. Two-dimensional models with 'buried' lateral conductivity anomalies, such as those reported by Patrick and Bostick [1969], are associated with results that differ from those for the more common model types with surficial inhomogeneities. The two MT apparent resistivity curves, termed E parallel and E perpendicular, coverge to a 'single curve' for periods shorter than those which significantly penetrate the anomalous 'buried' conductor (see Figures 4.5 and 4.12 of Patrick and Bostick, [1969]). Similar relations can be expected to prevail in the present case.

It is also important to remark that even though the map in Figure 1 indicates an interstation spacing of about 30 km and an aperture for these MT observations of perhaps 60 km across the six stations, the effective aperture represented by our combined data set, indicated in Figures 4c and 4f, is far greater. The MT apparent resistivities in these figures, for the stations at Wray and Idalia near the eastern boundary of Colorado, join smoothly into the long period results derived from the extensive set of DGS array data reported by Porath and Dziewonski [1971]. Their array sites effectively sample broad areas of the Great Plains. The combined data therefore permit comparison between the present MT results and those derived from DGS and other geophysical investigations focused on the contrast between the eastern and the western provinces of the United States. Collectively, the combined data set in Figure 4c for Idalia, Colorado, indicates a depth of roughly 120 km to the Great Plains upper-mantle conductive zone. The data for Wray in Figure 4f yield about the same result.

The shallower estimate of about 70 km, shown in Figure 4a for one of the westernmost stations, at Anton, Colorado, grades into slightly deeper values for the stations at Joes, Akron, and Yuma. The shallow conductive depths for our more westerly MT sites can be compared to other similar geophysical results for the western United States. Swift's [1967] MT study interpreted the anomalous conductive upper mantle at depths of about 50 km beneath the southwestern United States to correlate with a zone of low velocity, high temperature, and partial fusion. Swift's depth for the LVZ is remarkably similar to the 50 km depth reported by Massé, Landisman, and Jenkins [1972] for the same region, a result derived from seismic record sections of Nevada Test Site recordings. Gough [1974] and Garland [1975] have presented excellent reviews of the correlation between conductivity anomalies and other geophysical parameters for western North America and other regions.

The relatively sharp increase in conductivity at depths of about 120 km beneath the Great Plains can also be interpreted in terms of the top of LVZ, which marks the base of the lithosphere, even though seismic attenuation and other studies probably preclude significant concentrations of partial melt materials in this area [Hales and Doyle, 1967; Anderson and

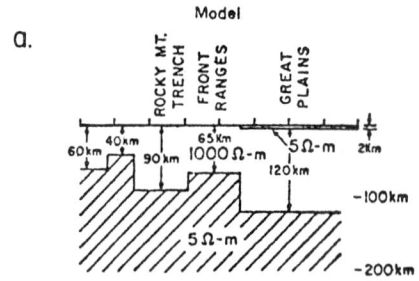

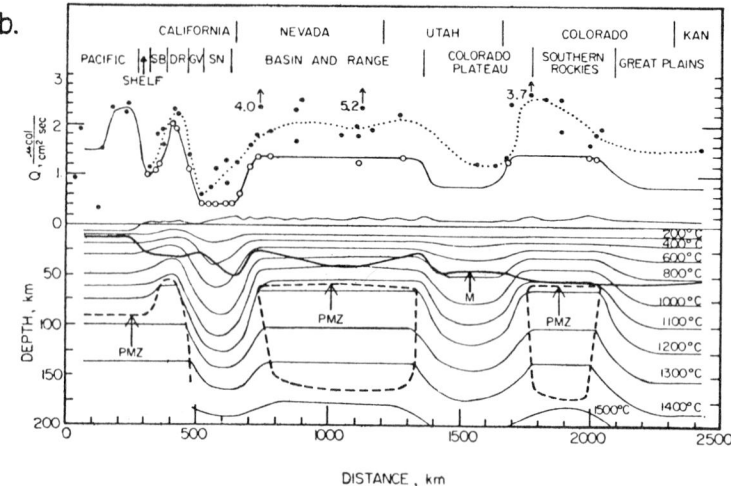

Fig. 5. (a) East-west section shows Deep Geomagnetic Sounding model deduced for Rocky Mountain front near U.S.-Canadian border [after Porath et al., 1971]. (b) East-west profile shows heat flow observations (above) and section (below) shows derived upper mantle model for the western United States. PMZ indicates partial melt zone [after Roy et al., 1972]. Text compares westward rise of 1200°C geotherm, east of Rocky Mountain front to westward rise of mantle conductive zone, Figure 4, for MT stations in Figure 1.

Sammis, 1970; Nur, 1971]. Kurita [1973] reported that long period P and S wave spectra for arrivals from deep earthquakes indicate that depths to the upper mantle LVZ range from about 150 km beneath the Interior Plains of North America to slightly less than 100 km under the continental shelf of the Gulf of Mexico. Massé's [1973] interpretation of mantle travel times also suggests that the LVZ might be found at depths of about 90 to 110 km beneath central North America. These values correlate well with the present estimate of about 120 km for the depth to the conductive region in the upper mantle for the Great Plains.

This lithospheric thickness estimate of 120 km beneath the Great Plains is also consonant with the same depth estimated for a proposed conductor beneath the northern Great Plains along the U.S.-Canadian border. This latter value was deduced from ratios of vertical to horizontal normal magnetic fields observed in the Deep Geomagnetic Sounding experiments of Porath et al. [1971]. Their interpretation is reproduced in Figure 5. This figure also shows a temperature distribution for the crust and upper mantle of the western United States,

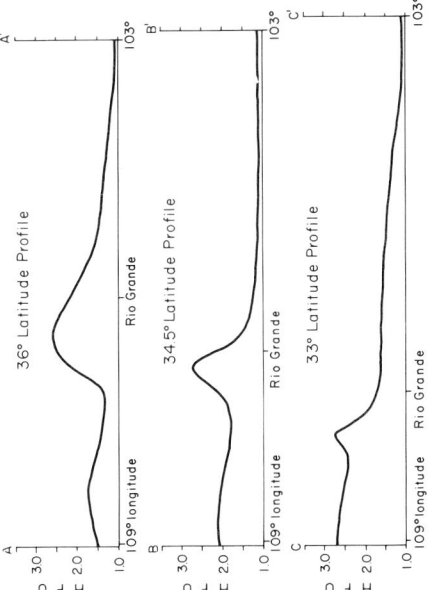

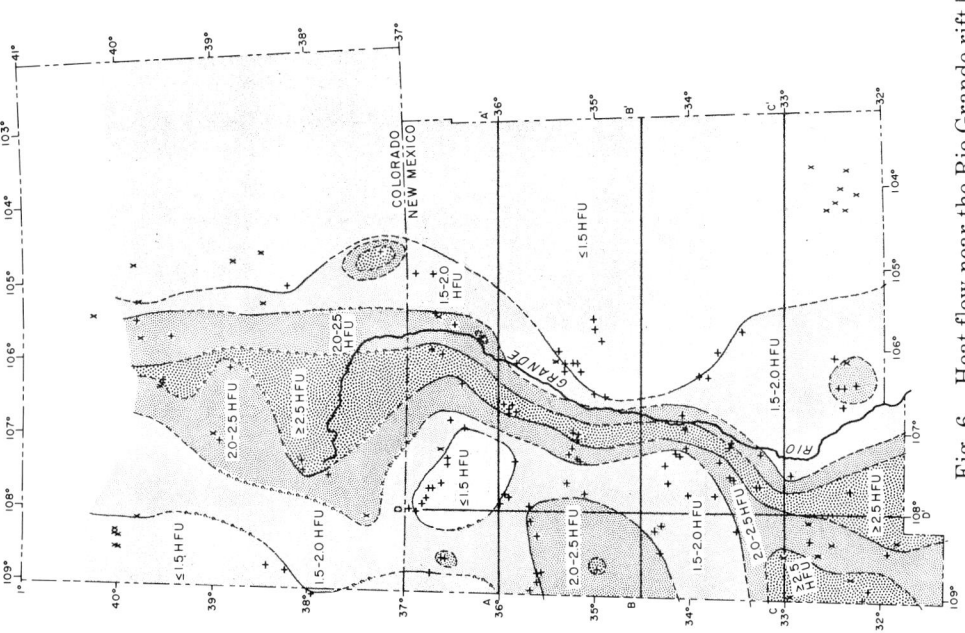

Fig. 6. Heat flow near the Rio Grande rift [after Reiter et al., 1975]. Text compares westward rise of long-wavelength components in heat flow profiles at right to similar heat flow increase near the Rocky Mountain front in Colorado (shown in Figure 5b).

including part of eastern Colorado, as proposed by Roy et al. [1972]. The conductive mantle, at depths of about 65 km under the Rocky Mountain front range in Figure 5a, corresponds closely with the easternmost partial melt zone (PMZ) in Figure 5b. The upwarp of the mantle geotherms in the transition belt between the Great Plains and the southern Rocky Mountains, in Figure 5b, exhibits a remarkable similarity to the presently inferred westward rise of the upper mantle conductive zone, from about 120 km beneath the Great Plains, in Figures 4c and 4f, to about 70 km in Figure 4a for Anton, which is nearly our closest station to the southern Rocky Mountains. The interface defining the top of the conductive region may correspond approximately to the 1200°C geotherm, which marks the top of the PMZ indicated at about 65 km beneath the Rocky Mountains, in Figure 5b.

The westward upwarp of the mantle geotherms in the transition belt in eastern Colorado may be a relic of Laramide tectonism, which provided a relatively shallow PMZ beneath the Rocky Mountains and other regions in western North America. Roy et al. [1972] indicate that the southern Rocky Mountain area has a regional geothermal characteristic similar to that of the Basin and Range province. The Basin and Range and Rocky Mountain provinces are characterized by pronounced negative Bouguer gravity anomalies, which imply mass deficiencies that could be attributed to buoyant zones within a mobile upper mantle [Wollard and Joesting, 1964; Pakiser and Zietz, 1965]. Similar but even more severe local upwarping of the mantle geotherms has been postulated to the south, in New Mexico, where the southern Rocky Mountains merge into the Rio Grande rift zone [Decker and Smithson, 1975]. The rift zone coincides with an area of Quaternary to Recent volcanism, as shown in Figure 6 [after Reiter et al., 1975]. This feature is associated with a major geothermal anomaly area whose heat flow values are greater than 2.5 HFU (μ cal cm^{-2} s^{-1}). Decker and Smithson [1975] also indicate a thinning of the crustal layers, in association with shallow occurrences of mobile mantle material, which unite to produce a relative Bouguer gravity high along the Rio Grande rift.

The MT data presented here are consonant with previous studies of the upper mantle by Archambeau et al. [1968, 1969], Helmberger and Wiggins [1971], and others which indicate that the mantle LVZ is much shallower in the western United States. The rise of the LVZ, whose appearance signals the base of the lithosphere, favors marked westward thinning of the lithosphere, which overlies a mobile tectonically active upper mantle beneath the western United States.

Acknowledgments. The authors thank Francis X. Bostick, Jr., of the University of Texas at Austin for his valuable suggestion concerning the inversion procedure and for his generous and helpful counsel on MT methods. We are grateful to Victor Ditter, John C. Holtorf, Lawrence Grauel, Leo Korf, and Alvin J. Cox for their hospitality and generosity in permitting us access to their property in northeastern Colorado in order to observe the data discussed in this report. Acknowledgments are due to various members of the engineering and scientific staff, who helped to prepare, test, and operate the field equipment and to develop the data reduction software. Computing facilities were furnished by the Institute for Geoscience, University of Texas at Dallas (UTD). A portion of the data reduction was performed at the Energy Research and Development Administration Computing Center, Courant Institute of Mathematical Sciences, New York University. We are grateful to the Office of Naval Research and the National Science Foundation for many years of continuous support without which this study could not have been achieved. Supplemental funds were also furnished by the Extra-Mural Geothermal Program of the U.S. Geological Survey and by the Institute for Geoscience, UTD. It is a pleasure to thank Ruth Ricamore for her help with the figures. This is contribution 341 of the Geoscience Program, UTD.

References

Anderson, D. L., and C. Sammis, Partial melting in the upper mantle, Phys. Earth Planet. Interiors, 3, 41-50, 1970.

Archambeau, C. B., R. F. Roy, D. D. Blackwell, D. L. Anderson, L. R. Johnson, and B. Julian, A geophysical study of continental structure (abstract), Eos, Trans. AGU, 49, 328, 1968.

Archambeau, C. B., E. A. Flinn, and P. G. Lambert, Fine structure of the upper mantle, J. Geophys. Res., 74, 5835-5866, 1969.

Archie, G. E., The electrical resistivity log as an aid in determining some reservoir characteristics, Trans. AIME, 146, 54, 1942.

Berdichevskii, M. N., Electrical prospecting by the magnetotelluric profiling method, Nedra, Moscow, Russia, 1968.

Berdichevskii, M. N., V. P. Borisova, L. L. Vanyan, I. S. Feldman, and I. Yakovlev, Anomaly of the electrical conductivity of the earth's crust in Yakutiya, Izv. Acad. Sci. USSR Phys. Solid Earth, 10, 633-637, 1969.

Berdichevskii, M. N., L. L. Vanyan, I. S. Feldman, and G. Porstendorfer, Conducting layers in the earth's crust and upper mantle, Gerlands Beitr. Geophys., 81, 197-198, 1972.

Berdichevskii, M. N., V. P. Borisova, L. L. Vanyan, V. P. Golovkov, V. I. Dmitriyev, V. G. Dubroskiy, G. I. Kolomiytseva, N. M. Rotanova, E. B. Faynberg, and I. S. Feldman, Deep electromagnetic sounding, Izv. Acad. Sci. USSR Phys. Solid Earth, 7, 454-462, 1973.

Bostick, F. X., Jr., and H. W. Smith, Investigation of large scale inhomogeneity in the earth by the magnetotelluric method, Proc. IRE 40(11), 2339-2346, 1962.

Bostick, F. X., Jr., A simple almost exact method of MT analysis, Appendix to Workshop Report on Electrical Methods in Geothermal Exploration, pp. 174-183, Dept of Geol. and Geophys., Univ. of Utah, Salt Lake City, Utah, 1977.

Cagniard, L., Basic theory of the magnetotelluric method of geophysical prospecting, Geophysics 18, 605-635, 1953.

Cantwell, T., Detection and analysis of low frequency electromagnetic signals, Ph.D. dissertation, Dept. of Geol. and Geophys., Mass. Inst. of Technol., Cambridge, Mass., 1960.

Chaize, L., and M. Lavergne, Signal et bruit en magnetotellurique, Geophys. Prospect., 18, 64-87, 1968.

Cooley, J. W., and J. W. Tukey, An algorithm for the machine calculation of complex Fourier series, J. Math. Comput., 19, 297-301, 1965.

Decker, E. R., and S. B. Smithson, Heat flow and gravity interpretation across the Rio Grande Rift in southern New Mexico and west Texas, J. Geophys. Res., 80, 2542-2552, 1975.

Dowling, F. L., Magnetotelluric measurements across the Wisconsin Arch, J. Geophys. Res., 75, 2683-2698, 1970.

Garland, G. D., Correlation between electrical conductivity and other geophysical parameters, Phys. Earth Planet. Interiors, 10, 220-230, 1975.

Gough, D. I., The geophysical significance of geomagnetic variation anomalies, Phys. Earth Planet. Interiors, 7, 379-388, 1973.

Gough, D. I., Electrical conductivity under western North America in relation to heat flow, seismology, and structure. J. Geomag. Geoelec., 26, 105-123, 1974.

Hales, A. L., and H. A. Doyle, P and S travel time anomalies and their interpretation, Geophys. J. Roy. Astron. Soc., 13, 403-415, 1967.

Ham, W. E., and J. L. Wilson, Paleozoic epeirogeny and orogeny in the central United States, Amer. J. Sci., 265, 332-407, 1967.

Harthill, N., An evaluation of the audio magnetotelluric method, M.S. thesis, Colo. Sch. of Mines, Golden, Colo., 1967.

Helmberger, D., and R. A. Wiggins, Upper mantle structure of midwestern United States, J. Geophys. Res., 76, 3229-3245, 1971.

Hermance, J. F., Processing of magnetotelluric data, Phys. Earth Planet. Interiors, 7, 349-364, 1973.

Jackson, D. B., Electrical properties of the sedimentary section in the High Plains area: Denver, Colorado, U.S. Geol. Surv. Tech. Lett. Rep. ARPA contract 193-61, U.S. Geol. Surv., Denver, Colo., 1962.

Jackson, W. H., S. W. Stewart, and L. C. Pakiser, Crustal structure in eastern Colorado from seismic refraction measurements, J. Geophys. Res., 68, 5762-5776, 1963.

Keller, G. V., Electrical prospecting for oil, Quart. Colo. Sch. Mines, 63(2), 268 pp, 1968.

Keller, G. V., L. Anderson, and J. Pritchard, Geophysical survey investigations of the electrical properties of the crust and upper mantle, Geophysics, 31, 1078-1087, 1966.

Keller, G. V., and F. C. Frischknecht, Electrical methods of geophysical prospecting, 519 pp., Pergamon, New York, 1966.

Kuckes, A. F., Correspondence between the magnetotelluric and field penetration depth analyses for measuring electrical conductivity, Geophys. J. Roy. Astron. Soc., 32, 381-385, 1973.

Kunetz, G., and J. P. Rocroi, Automatic processing of electrical soundings, Geophys. Propect., 18, 157-198, 1970.

Kurita, T., Upper mantle structure in the central United States from P and S wave spectra, Phys. Earth Planet. Interiors, 8, 177-201, 1973.

Landisman, M., and W. Chaipayungpun, First results from electrical and seismic studies of low resistivity, low velocity material beneath eastern Colorado, Geophysics, 42, 804-810, 1977.

Landisman, M., W. Chaipayungpun, D. Loewenthal, and F. Abramovici, Progress report on electrical and seismic crustal studies in the southwestern Great Plains (abstract), Eos Trans. AGU, 55, 288, 1974.

Landisman, M., St. Mueller, and B. J. Mitchell, Review of evidence for velocity inversions in the continental crust, in The Structure and Physical Properties of the Earth's Crust, Geophys. Monogr. Ser., vol. 14, edited by J. G. Heacock, pp. 11-34, AGU, Washington, D.C., 1971.

Madden, T., and P. Nelson, A defense of Cagniard's magnetotelluric method, Office Nav. Res. Rep., Proj. NR-371-401, Geophys. Lab., Mass. Inst. Of Technol., Cambridge, Mass., 1964.

Maillet, R., The fundamental equations of electrical prospecting, Geophysics, 12, 529-556, 1947.

Mann, J. E., Jr., The importance of anisotropic conductivity in magnetotelluric interpretation, J. Geophys. Res., 70, 2940-2942, 1965.

Massé, R. P., Compressional velocity distribution beneath central and eastern North America, Bull. Seismol. Soc. Amer., 63, 911-936, 1973.

Massé, R. P., M. Landisman, and J. B. Jenkins, An investigation of the upper mantle compressional velocity distributions beneath the Basin and Range Province, Geophys. J. Roy. Astron. Soc., 30, 19-36, 1972.

Migaux, L., J. Astier, and P. Revol, Essai de determination experimental de la resistivite electrique des couches profondes de l'ecorce terrestre, C. R. Acad. Sci., 251, 567-569, 1960.

Mitchell, B. J., Surface wave attenuation and crustal anelasticity in central North America, Bull. Seismol. Soc. Amer., 63, 1057-1071, 1973.

Mitchell, B. J., and M. Landisman, A detailed investigation of a crustal section across Oklahoma, Geol. Soc. Amer. Bull., 81, 2647-2656, 1970.

Mitchell, B. J., and M. Landisman, Electrical and seismic properties of the earth's crust in the southwestern Great Plains, Geophysics, 36, 363-381, 1971a.

Mitchell, B. J., and M. Landisman, Geophysical measurements in the southern Great Plains, in The Structure and Physical Properties of the Earth's Crust, Geophys. Monogr. Ser., vol. 14, edited by J. G. Heacock, pp. 77-93, AGU, Washington, D.C., 1971b.

Nur, A., Viscous phase in rocks and the low-velocity zone, J. Geophys. Res., 76, 1270-1277, 1971.

Parkhomenko, E. I., Electrical properties of rocks, 314 pp., Plenum, New York, 1967.

Pakiser, L. C., and I. Zietz, Transcontinental crustal and upper mantle structure, Rev. Geophys. Space Phys., 3, 505-520, 1965.

Patrick, F. W., and F. X. Bostick, Jr., Magnetotelluric modeling techniques, Tech. Rep. 59, 100 pp., Electron. Res. Center, Univ. of Tex., Austin, 1969.

Porath, H., and A. Dziewonski, Crustal electrical conductivity anomalies in the Great Plains province of the United States, Geophysics, 36, 382-395, 1971.

Porath, H., D. I. Gough, and P. A. Camfield, Conductive structures in the northwestern United States and southwest Canada, Geophys. J. Roy. Astron. Soc., 23, 387-398, 1971.

Rankin, D., and I. Reddy, A magnetotelluric study of resistivity anisotropy, Geophysics, 34, 438-449, 1969.

Reiter, M., C. L. Edwards, H. Hartman, and C. Weidman, Terrestrial heat flow along the Rio Grande Rift, New Mexico and southern Colorado, Geol. Soc. Amer. Bull., 86, 811-818, 1975.

Reitzel, J.S., D. I. Gough, H. Porath, and C. W. Anderson III, Geomagnetic deep sounding in the western United States, Geophys. J. Roy. Astron. Soc., 19, 213-235, 1970.

Rocky Mountain Association of Geologists, Geologic atlas of Rocky Mountain region, 331 pp., Denver, Colo., 1972.

Roy, R. F., D. D. Blackwell, and E. R. Decker, Continental heat flow, in The Nature of the Solid Earth, edited by E. C. Robertson, pp. 506-544, McGraw-Hill, New York, 1972.

Schmucker, U., Regional induction studies: a review of methods and results, Phys. Earth Planet. Interiors, 7, 365-378, 1973.

Shankland, T. J., Electrical conduction in rocks and minerals: parameters for interpretation, Phys. Earth Planet. Interiors, 10, 209-219, 1975.

Sims, W. E., and F. X. Bostick, Jr., Methods of magnetotelluric analysis, Tech. Rep. 58, Electron. Res. Lab., Univ. of Tex., Austin, 1969.

Swift, C. M., Jr., A magnetotelluric investigation of an electrical conductivity anomaly in the southwestern United States, Ph.D. dissertation, 211 pp., Mass. Inst. of Technol., Cambridge, Mass., 1967.

Uyeda, S., and T. Rikitake, Electrical conductivity anomaly and terrestrial heat flow, J. Geomagn. Geoelec., 22, 75-90, 1970.

Van Zijl, J. S. V., A deep Schlumberger sounding to investigate the electrical structure of the crust and upper mantle in South Africa (resistivity), Geophysics, 34, 450-462, 1969.

Van Zijl, J. S. V., Electrical studies of the deep crust in various tectonic provinces of southern Africa, in The Earth's Crust: Its Nature and Physical Properties, Geophys. Monogr. Ser., vol. 20, edited by J. G. Heacock, AGU, Washington, D.C., 1977.

Vozoff, K., The magnetotelluric method in the exploration of sedimentary basins, Geophysics, 37, 98-141, 1972.

Welch, P. D., The use of the fast Fourier transform for the estimation of power spectra: a method based on time averaging over short, modified periodograms, IEEE Trans. Audio Electroacoustics, AU-15(2), 70-73, 1967.

Woollard, G. P., and H. R. Joesting, Bouguer gravity anomaly map of the United States, scale 1:2,500,000, U.S. Geol. Surv., Washington, D.C., 1964.

Zohdy, A. A. R., Use of Dar Zarrouk curves in the interpretation of vertical electrical sounding data, U.S. Geol. Surv. Bull. 1313-D, 41 pp., 1974.

CRUSTAL STRESS IN THE CONTINENTAL UNITED STATES AS DERIVED FROM HYDROFRACTURING TESTS

Bezalel C. Haimson

Department of Metallurgical and Mineral Engineering, University of Wisconsin, Madison, Wisconsin 53706

Abstract. In the last 10 years, hydrofracturing has emerged as the most important technique for measuring stress in the earth's crust. Laboratory confirmation of the predicted hydrofracture orientation and of the theoretical stress-hydrofracturing pressure relationships, combined with very encouraging initial testing in oil fields, led to our first major scientific measurement, at Rangely, Colorado. The success of this test as part of an earthquake control experiment led to a number of additional hydrofracturings throughout the United States. To date, such tests have been conducted in nearly 20 states from California to South Carolina and from Idaho to Louisiana, within a depth range of 30-5100m. A pattern of principal stress direction and magnitude profile is beginning to emerge. It appears that throughout the United States all three principal stresses are compressive, with the major horizontal principal compression typically oriented in the northeastern quadrant. Near the San Andreas fault the direction of the major horizontal principal stress is N15°E, in accord with the fault strike - slip characteristic. In the Sierra Nevada Mountains it becomes N25°E, gradually rotating to N35°E-N45°E in Nevada, and averaging N60°E east of the Rockies. At shallow depths (0 - 600m), both horizontal principal compressions are often larger than the vertical, notably in the Midwest and the Appalachians. At greater depths the predominant stress regime throughout the United States is one in which the vertical component is intermediate in magnitude. Only toward the lower end of the range of depths tested does the vertical component appear to approach the magnitude of the largest horizontal compressive stress. Linear regressions of all the hydrofracturing stress results throughout the United States with respect to depth yield high correlation coefficients and the following relationships: $\sigma_{Hmin} = 20 + 0.16D$, $\sigma_{Hmax} = 75 + 0.24D$, and $\sigma_V = 0.25D$, where σ_{Hmin}, σ_{Hmax} and σ_V are the least horizontal, largest horizontal, and vertical compressive stresses, respectively, measured in bars, and D is depth in meters (limited to 0 - 5000m). We compare hydrofracturing results with known overcoring stress measurements and with focal mechanism solutions.

Introduction

The magnitude and direction of regional stresses are intimately related to plate tectonics, igneous intrusions, faulting, and earthquake triggering. Rock stress determination is essential in such practical applications as earthquake control, design of underground excavations, and extraction of minerals and geothermal heat by in situ methods.

Several methods of in situ stress measurement have been developed in the last 30 years. They require a borehole drilled from the surface to the depth of interest. However, with the exception of hydrofracturing, all the methods are limited to very short holes of 50m or less. Hydrofracturing, on the other hand, has been successfully used in boreholes reaching 5000m. Unlike other methods, it does not require overcoring or knowledge of rock elastic parameters, it estimates average stresses over large areas, and it employs simple and rugged equipment which is commercially available.

The hydrofracturing stress-measuring technique consists of sealing off a section of a borehole at the required depth by means of two inflatable rubber packers and pressurizing the isolated segment by using a hydraulic fluid such as water. At some critical (also called breakdown) pressure the rock at the borehole bursts and develops a tensile fracture. This fracture is extended away from the hole by continued pumping. When pumping is stopped with the hydraulic circuit kept closed, a 'shut-in' pressure is recorded. This is the pressure necessary to keep the fracture open. The breakdown and shut-in pressures, which are carefully monitored during the test, can be directly related to the prevailing in situ stresses. An impression packer or other borehole and surface geophysical instruments can be used to determine the orientation and inclination of the hydrofracture which develops along a plane perpendicular to the direction of the least principal compressive stress. In this manner, both the magnitudes and the directions of the principal stresses can be evaluated.

The theory of hydrofracturing has been detailed elsewhere [Hubbert and Willis, 1957; Kehle, 1964; Haimson, 1968, 1974a, 1976a] and only tne major results will be briefly outlined here. Typically a vertical borehole is subparallel to one of the principal stresses. Provided the tested section is not prefractured, the hydrofracture initiates along a plane perpendicular to the direction of the least horizontal compressive stress (σ_{Hmin}). The shut-in pressure (P_s) needed to keep the hydrofracture open when pumping is stopped is approximately equal to the compressive stress across the fracture plane:

$$\sigma_{Hmin} = P_s \quad (1)$$

If the vertical stress (σ_V) is not the least compressive stress it is calculated from the weight of the overlying rock:

$$\sigma_V = \gamma D \quad (2)$$

where γ is the rock pressure gradient and D is the depth. The gradient can be established from rock density measurements along the extracted core or from gravity logs. If σ_V is the least principal compressive stress, a vertical fracture will nonetheless initiate at first, yielding the first shut-in pressure (P_{s1}). Soon the fracture will turn into horizontal, seeking the path of least resistance, and a second shut-in pressure (P_{s2}) will be recorded. Clearly $P_{s1} > P_{s2}$ and:

$$\sigma_{Hmin} = P_{s1} \tag{3}$$

$$\sigma_V = P_{s2} \tag{4}$$

In this situation, both the horizontal principal stresses and the vertical stress will be directly determined by hydrofracturing pressures [Haimson, 1976a, 1976b; Zoback et al., 1977].

The magnitude of the largest horizontal compressive stress (σ_{Hmax}), ignoring possible effects of fluid flow into the rock, is obtained from:

$$\sigma_{Hmax} = T + 3\sigma_{Hmin} - P_c - P_o \tag{5}$$

where P_o is the rock pore pressure at the tested depth; T is the hydrofracturing tensile strength, which can be determined in the laboratory as well as in the field.

The directions of the two horizontal principal stresses are determined from the orientation of the vertical hydrofracture which coincides with the orientation of σ_{Hmax}. This coincidence of directions as well as the other theoretical results have been verified numerous times in the laboratory [Haimson, 1968, 1974a; Haimson and Avasthi, 1975; Zoback et al., 1976].

The first indications of the potential of hydrofracturing as a stress-measuring technique came from calculations reported by Scheidegger [1962] and Kehle [1964] based on routine oil well hydraulic fracturing jobs run for production stimulation purposes. Haimson and Stahl [1970] reported on three series of oil field hydrofracturing jobs (in the states of New York, Illinois, and Ohio) in open holes, where impression packers and bottom hole pressure transducers were used to determine crack orientation and fracturing pressures, respectively. The results within each group of tests are very consistent with respect to critical and shut-in pressures and fracture directions. The consistency of the results strongly suggests that they are closely related to local in situ stresses. Stress calculations based on these hydrofracturing jobs were subsequently published by Haimson [1974a]. Following the laboratory study and the initial field results, all indications were that hydrofracturing had a great potential as a deep hole stress-measuring technique.

The first major use of hydrofracturing as a deep borehole stress-measuring technique was in connection with the earthquake control experiment at Rangely, Colorado [Haimson, 1972, 1973; Raleigh et al., 1972, 1976]. Intense seismic activity centered in and around the Rangely oil field in the vicinity of a strike-slip fault had been recorded in the area for some years. A primary objective of the research program was to determine whether a correlation exists between earthquake triggering and formation pore pressure. The latter had been artificially raised through water flooding of the field. A quantitative solution relating pore pressure to slip initiation along the fault, which in turn could have triggered the earthquakes, requires knowledge of the in situ stress condition in or below the oil-bearing strata. Hydrofracturing was selected as the stress-determining method. We tested the bottom section of a newly drilled oil well located at the southern boundary of the oil field, approximately 1000m west of the fault, near the earthquake prone region. The results that we obtained

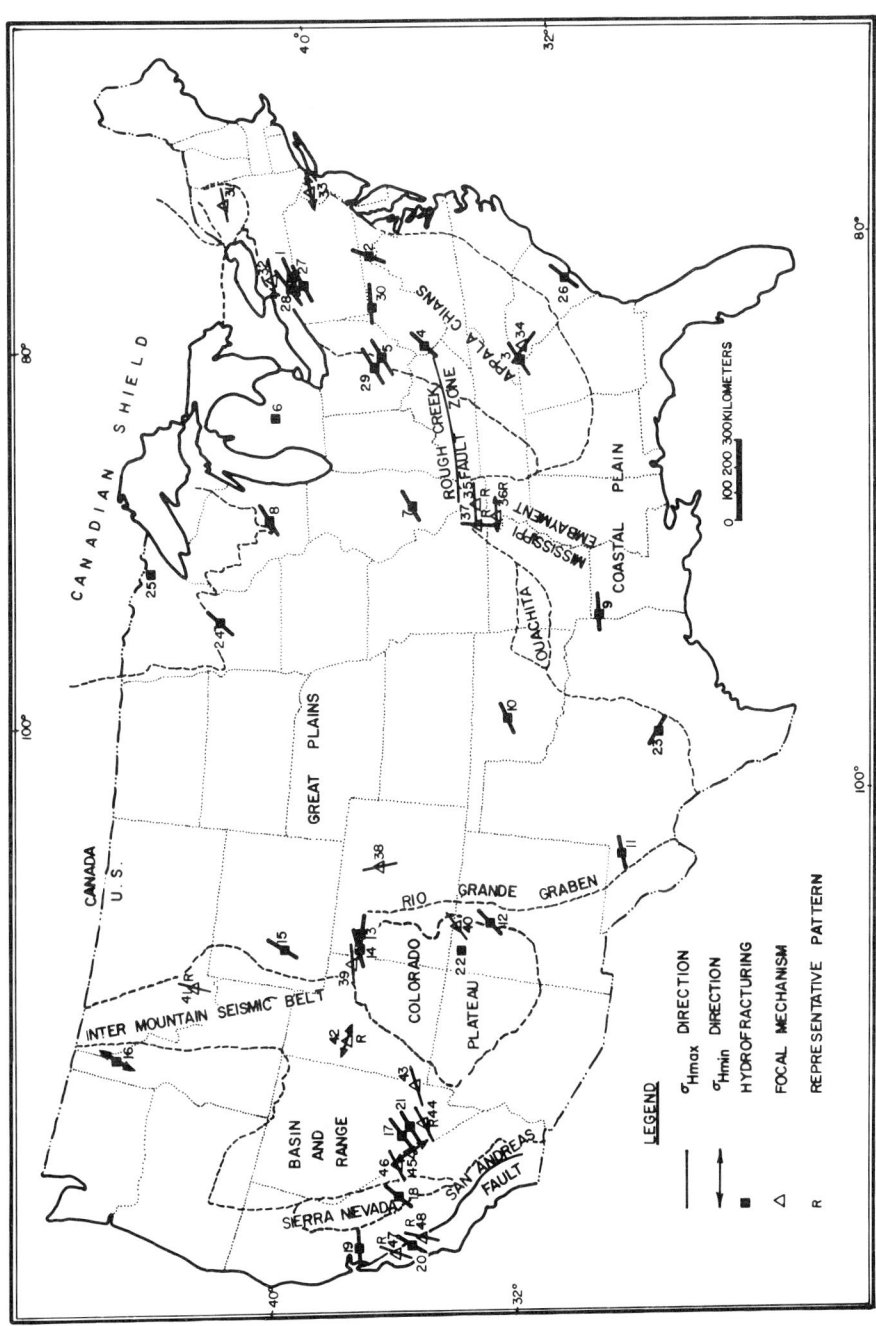

at that depth (1900m below the surface) were σ_V = 435 bars (vertical), σ_{Hmin} = 315 bars (horizontal at N20°W), and σ_{Hmax} = 590 bars (horizontal at N70°E). The hydrofracturing test showed that the vertical principal stress was intermediate in magnitude, strongly indicating strike-slip faulting. The horizontal principal stress directions are in accord with both the N50°E strike of the fault and its right lateral slip direction. The magnitudes of the measured principal stresses were used to predict a critical pore pressure of 240 bars for fault slip to occur. This value is within 10% of the earthquake-related threshold pressure monitored at the site. The Rangely stress measurement not only demonstrated the reliability of hydrofracturing and its importance to future earthquake control studies but also showed the effectiveness of the method in studying crustal stress.

The success of the Rangely experiment led to more hydrofracturing measurements sponsored by both government and private industry. The results obtained to date are hardly sufficient to describe in detail the state of stress in North America. However, measurements in some 20 states do provide an emerging pattern of stress magnitudes and directions. Table 1 lists the measurements and Figure 1 is a map of the United States showing both the location of known hydrofracturing tests and the respective maximum horizontal stress directions observed. For comparison, representative focal mechanism solutions in areas of hydrofracturing tests are also shown. Four clusters of measurements are of particular interest, and they will be discussed separately in the following sections.

California - Nevada

Hydrofracturing stress measurements have been conducted in each of the three tectonic provinces: Coast Range (Livermore and San Ardo), Sierra Nevada (Helms pumped storage project near Dinkey Creek) and Basin and Range (Rainier Mesa, Nevada test site). The latter is the most thoroughly tested location, with two sets of hydrofracturing measurements [Haimson et al., 1974; Tyler and Vollendorf, 1975], a series of overcoring tests [Hooker et al., 1971], and numerous focal mechanism solutions [Hamilton and Healy, 1969; Lindh et al., 1973]. The overwhelming evidence from the hydrofracturing measurements is that within the depth range tested the state of stress favors strike-slip faulting with the major horizontal compressive stress at N35°E - N45°E (sites 17 and 21, Table 1). Both the overcoring tests and the focal mechanism solutions corroborated these results.

Fig. 1. Map of the continental United States showing locations of known hydrofracturing stress measurements and the directions of the largest horizontal compressive stress as obtained in these tests. For comparison, focal mechanism solutions in the areas of hydrofracturing measurements are also shown. Where many local solutions are available bearing a consistent trend only representative patterns are displayed. Numbers next to hydrofracturing tests refer to sites in Table 1. Focal mechanism results were taken from Sbar and Sykes [1973] (sites 31-33), Talwani [1977] (34), Street et al. [1974] (35-37), Smith and Sbar [1974] (40-46), Raleigh [1974] (38, 39, 47, 48).(Page 579.)

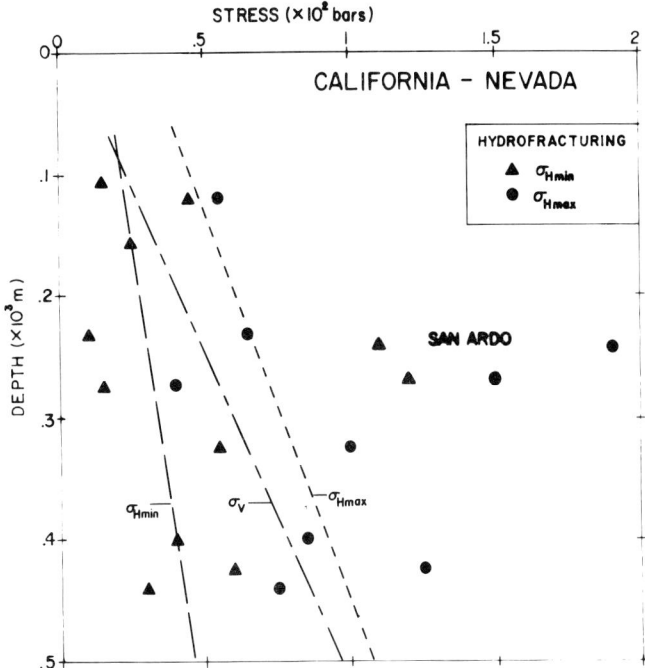

Fig. 2. Variation of horizontal and vertical stresses with depth in California - Nevada, based on hydrofracturing tests. Shaded area covers the San Ardo test results.

Measurements in the Sierra Nevada Mountains (site 18, Table 1) between 160-325m depth yielded a strike-slip faulting stress regime with the maximum compression at N25°E [Haimson, 1976b]. Common to both sites (17 and 18) is a linear increase with depth of all three principal stresses.

Hydrofracturing at Livermore, California [Zoback et al., 1977], between 100-150m yielded σ_{Hmin} values within the range expected on the basis of our previous measurements (site 19, Table 1). The direction of σ_{Hmax} (the magnitude of which was not determined) was found to be N68°E.

At San Ardo, California, within 3 km from the San Andreas fault, Zoback et al. [1977] measured unusually high horizontal stresses, rendering σ_V the least compressive stress (site 20, Table 1). The horizontal stress values, plotted in Figure 2, are significantly out of the range of values obtained in other tests. This discrepancy remains unexplained. We speculate, however, that the close vicinity of the fault and the shallow depths of the tests are reflected in the San Ardo results. The direction of σ_{Hmax} is N15°E, in agreement with both the strike and the right lateral slip of the fault, as well as with focal mechanism solutions as shown in Figure 1 [also see Raleigh, 1974].

Generalizing from just a few tests it is remarkable that the direction of σ_{Hmax} appears to gradually rotate as one moves eastward, from N15°E at San Ardo to N25°E in the Sierra Nevada, to N35°E - N45°E at the Nevada test site. The great majority of results east of the Basin and Range show a N60°E direction of σ_{Hmax} (Figure 1).

Figure 2 presents all the test results in California - Nevada (based on Table 1) as a function of depth (100-500m). There is a definite increase in stress magnitudes with depth, and σ_V is predominantly intermediate in value. Linear regressions of the principal stresses, excluding those determined at San Ardo, have only fair correlation coefficients (0.5-0.6) but yield the following relationships:

$$\sigma_{Hmax} = 25 + 0.17D; \qquad \sigma_{Hmin} = 15 + 0.07D; \qquad \sigma_V = 5 + 0.18D$$

where stresses are in bars and D is depth in meters. The trend of σ_V is to become the largest compression at greater depths. Indeed both strike-slip and normal fault conditions are encountered in the Sierra Nevada and the Basin and Range as evidenced by some focal mechanism solutions (Figure 1).

Wyoming - Colorado - New Mexico

The hydrofracturing measurements in these states have generally been conducted on the periphery of the Colorado Plateau. The first measurement was at Rangely, Colorado (site 14, Table 1), the results of which are detailed in the Introduction of this paper. The measured stresses were corroborated not only by the type of existing fault and the monitored earthquake triggering pore pressures, but also by near-surface overcoring tests, focal mechanism solutions and local geological structures [Raleigh et al., 1976]. Additional shallow measurements by Bredehoeft et al., [1976] some 50 km east of Rangely (site 13, Table 1) yielded a nearly hydrostatic stress regime with σ_{Hmax} direction averaging N90°E.

Significant hydrofracturing tests have been conducted near the Valles Caldera, New Mexico (site 12, Table 1) where preliminary results indicate that at depths of 750-3000m the largest horizontal compressive stress is oriented at N35°E with σ_{Hmin} roughly one half of σ_V [L. Aamodt. personal communication, 1975]. Additional hydrofracturing results near Farmington, New Mexico, and in the Green River Basin, Wyoming (sites 15 and 22, Table 1) have been published by Power et al., [1976]. In the Wyoming site several methods of fracture delineation yielded a direction of N25°E for σ_{Hmax}.

By plotting the measurements from Table 1 as a function of depth (Figure 3) a striking linear relationship is obtained for each principal stress. Within the range of depths of 500-3000m the stress depth relationships are:

$$\sigma_{Hmax} = -30 + 0.31D; \qquad \sigma_{Hmin} = 10 + 0.16D; \qquad \sigma_V = -10 + 0.25D$$

with a correlation coefficient of 0.97 or better. The vertical stress is the intermediate component throughout. Principal stress directions are not too consistent in the area, although they are limited to the northeastern quadrant for σ_{Hmax} (Figure 1). Focal mechanism solutions near Rangely and in northern New Mexico are in agreement with local hydrofracturing directions.

TABLE 1. Known Hydrofracturing In Situ Stress Measurements In The Continental United States

Site-Reference No.	Location	Rock Type	Depth** m	σ_v # bars	σ_{Hmin} bars	σ_{Hmax} bars	σ_{Hmax} direction
1	Allegany Co., New York	sandstone	510	140	140	195	N75°E
2	Berkeley Co., West Virginia	shale	25	7	10	125	N25°E
			135	35	70	240	
3	Salem, South Carolina	gneiss	120	30	70	90	N60°E
			255	65	180	260	
4	Wayne, West Virginia	shale	835	210	180	400	N50°E
5	Hocking Co., Ohio	sandstone	810	185	155	285	N65°E
6	Ithaca, Michigan	limestone	1230	320	295	505	--
		dolomite	2805	730	400	560	
		sandstone	3660	950	645	885	
		shale	5110	1325	960	1470	
7	Illinois§	carbonate	100	25	25	75	N60°E
8	Montello, Wisconsin	granite	75	20	70	160	N65°E
			190	50	70	160	
9	Caddo-Pine Island, Louisiana§	chalk	425	100	75	--	E-W
10	Kingsfisher Co., Oklahoma§	--	--	--	--	--	N65°E
11	West Texas§	--	--	--	--	--	N75°E
12	Valles Caldera, New Mexico	granite	760	200	145	--	N35°E
			1980	515	345	--	
			2925	760	390	--	
13	Meeker, Colorado	oil shale	245	55	35	65	E-W
			470	105	85	120	
14	Rangely, Colorado	sandstone	1900	435	315	590	N70°E
15	Green River Basin, Wyoming§	--	2775	630	520	815	N25°E
16	Kellogg, Idaho	quartzite	2285	620	260	430	N75°W
17	Rainier Mesa, Nevada	tuff	230	40	10	65	N35°E
			410	75	40	85	
18	Dinkey Creek, California	granite	160	45	45	55	N25°E
			325	90	55	100	

TABLE 1. (continued)

Site-Reference No.[*]	Location	Rock Type	Depth[**] m	σ_V[#] bars	σ_{Hmin} bars	σ_{Hmax} bars	σ_{Hmax} direction
19	Livermore, California	gravel, sand	110	20	15	--	N70°E
			155	30	25	--	
20	San Ardo, California	sandstone	240	55	110	225	N15°E
			270	60	120	160	
21	Rainier Mesa, Nevada	tuff	275	50	15	40	N45°E
			440	80	30	75	
			415	75	60	125	
22	Farmington, New Mexico[§]	--	930	210	145	210	--
23	Marble Falls, Texas	granite	255	65	35	160	N60°W
			305	75	45	240	
24	St. Cloud, Minnesota	granite	15	5	70	200	N40°E
25	Ely, Minnesota	--	300	70	100	185	--
26	Summerville, South Carolina	shale	200	45	30	50	N50°E
27	McKean Co., Pennsylvania	sandstone	--	--	--	--	N70°E
28	Allegany Co., New York	sandstone	--	--	--	--	N60°E
29	Hocking Co., Ohio	sandstone	--	--	--	--	N65°E
30	Coal Mine in West Virginia[§]	shale	220	50	60	145	E-W

[*] References: 1, 5, 7 - Haimson [1974a]; 2 - Unpublished; 3, 8, and 18 - Haimson [1976b]; 4 - Haimson [1977a]; 6 - Haimson [1977b]; 9 - Strubhar et al. [1975]; 10 - von Schonfeldt et al. [1973]; 11 - Fraser and Pettitt [1962]; 12 - L. Aamodt, personal communication, 1975; 13 - Bredehoeft et al. [1976]; 14 - Haimson [1973]; 15, 22 - Power et al. [1976]; 16 - Haimson [1974b]; 17 - Haimson et al. [1974]; 19, 20 - Zoback et al. [1977]; 21 - Tyler and Vollendorf [1975]; 23 - Roegiers [1974]; 24, 25 - von Schonfeldt and Rough [1971]; 30 - Parsons and Dahl [1972].

[**] For sites where a range of depths under 500m was tested, yielding a linear stress-depth relationship, only the results at the upper and lower points are given. Otherwise average values (one stress result per location) or distinct values obtained at different depths are given.

[#] The value of σ_V was calculated from known or assumed rock density and depth of testing, except for sites, 2, 3, 8, 20 where it was obtained from hydrofracturing results.

[§] Precise location unknown. Abbrevation: Co. - County.

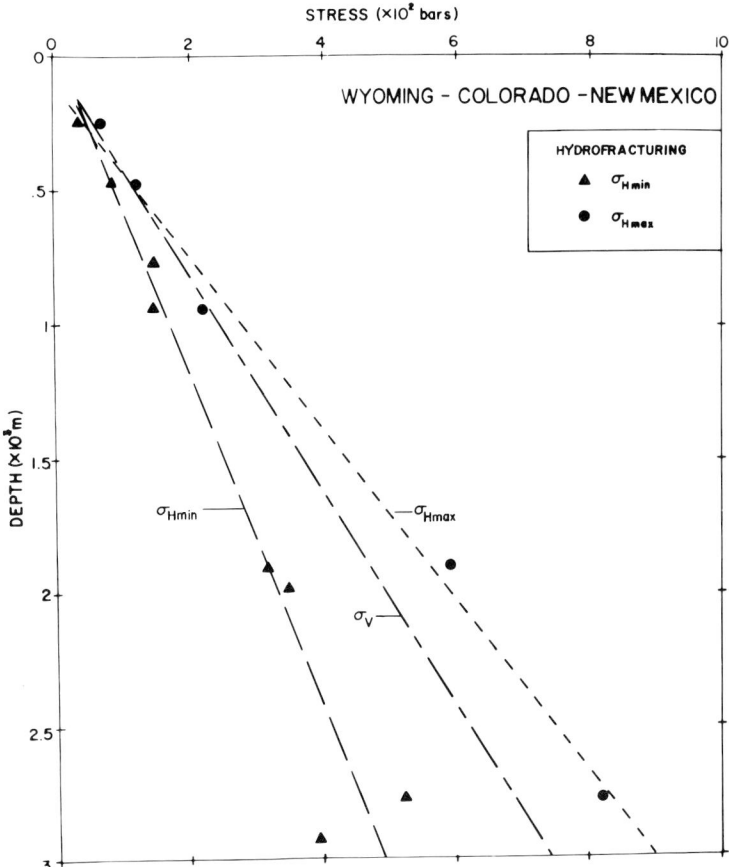

Fig. 3. Varation of horizontal and vertical stresses with depth in the Wyoming-Colorado-New Mexico area, based on hydrofracturing tests.

The Midwest

Measurements have been conducted in the Canadian Shield (Montello, Wisconsin; St. Cloud and Ely, Minnesota), the Michigan Basin (Ithaca, Michigan), the Illinois Basin, and in the Appalachian Plateau (Hocking County, Ohio).

The significant result of the Canadian Shield tests (sites 8, 24 and 25, Table 1) is perhaps, that the magnitude of the vertical component is in all cases the least compressive stress. The relatively high horizontal stresses at shallow depths are not unusual in basement rock of ancient shields [Ranalli and Chandler, 1975]. The direction of σ_{Hmax} is N65°E at Montello and N40°E at Ely.

A series of tests was conducted near the axis of the Michigan Basin (site 6, Table 1) at unprecedented depths ranging from 1200-5100m, resulting in a tentative stress profile in formations from Devonian to Precambrian [Haimson, 1977b]. The shallowest test at 1230m yielded a vertical stress approximately equal to the least horizontal compres-

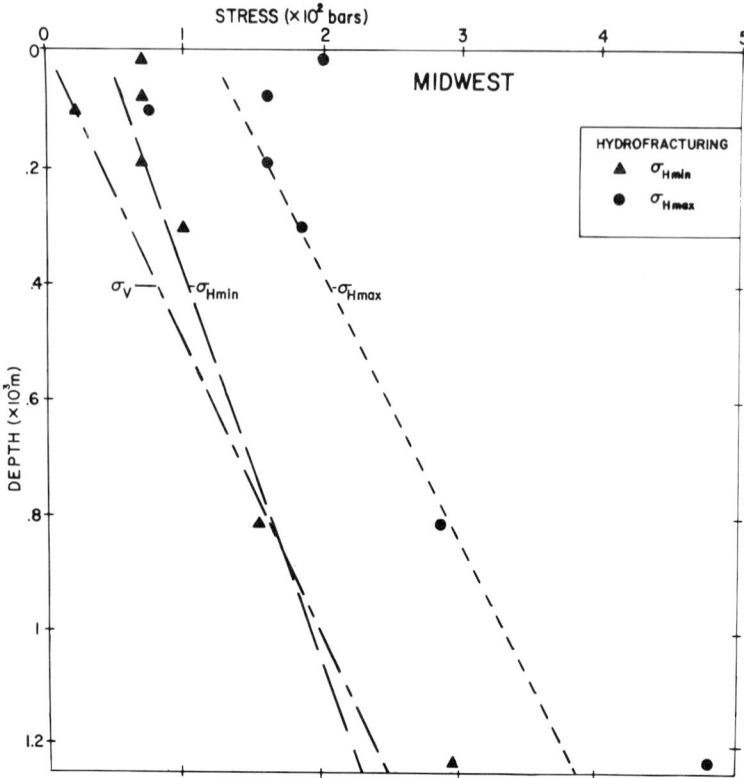

Fig. 4. Variation of horizontal and vertical stresses with depth between 100 and 1500m in the Midwest, based on hydrofracturing tests.

sion. Only the deeper measurements resulted in a σ_V increasingly higher than σ_{Hmin}, trending toward the largest principal stress. Tests in Ohio and Illinois (sites 5 and 7, Table 1) at shallow depths of 810m and 100m, respectively, also indicated that the least horizontal compression is approximately equal to the vertical component.

Figure 4 depicts the generalized stress regime in the top 1200m below surface in the Midwest. Within 0-800m the vertical stress is the least principal compression. Below 800m it gradually becomes the intermediate component. Linear regression analysis for 0-1200m yields:

$$\sigma_{Hmax} = 115 + 0.28D; \quad \sigma_{Hmin} = 40 + 0.19D; \quad \sigma_V = 0.25D$$

with very high correlation coefficients (0.92 or better).

We were not able to determine stress direction in the Michigan testhole. The other measurements in the Midwest indicate a predominant direction of N60°E (Figure 1).

The Midwest is largely a stable, inactive area and no focal mechanism solutions are known in the vicinity of the test sites. However, bordering on the south of the Midwestern states is the Mississippi Embayment, where numerous earthquakes occur, the focal mechanism solutions of which indicate a considerably more complex stress pattern

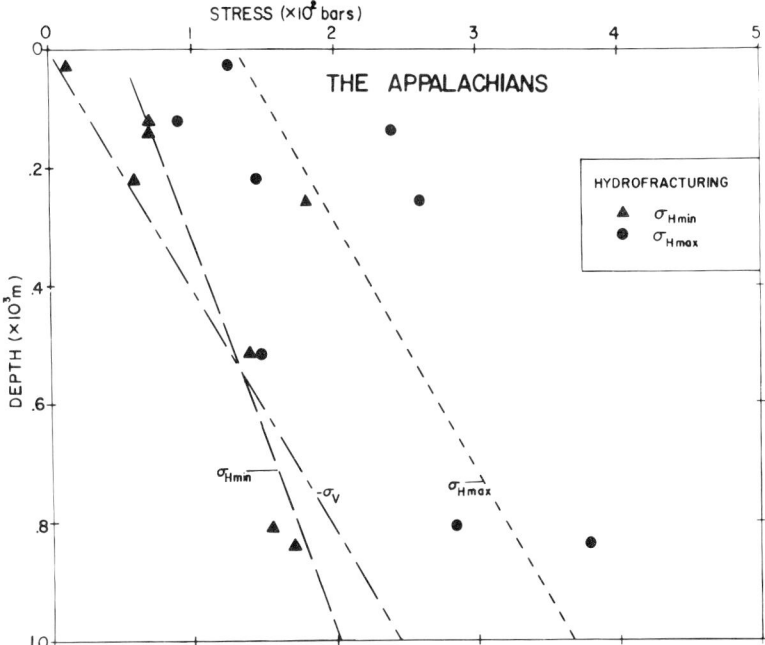

Fig. 5. Variation of horizontal and vertical stresses with depth in the Appalachians, based on hydrofracturing tests.

(Figure 1), at least locally, than the simple ENE compression observed elsewhere [Street et al., 1974]. No in situ stress measurements have been conducted in the area although the need for such tests is obvious.

The Appalachians

Included in this region are all the states that form the Appalachian system, as shown in Figure 1. Measurements have been conducted in Allegany County, New York (sites 1 and 28, Table 1), McKean County, Pennsylvania (site 27), Hocking County, Ohio (sites 5 and 29), West Virginia (sites 2, 4 and 30), and near Salem, South Carolina (site 3).

The results of the hydrofracturing tests conducted between 0-850m indicate thrust fault conditions in the upper 600m, with both horizontal stresses higher than the vertical. However, the two tests conducted at about 800m (sites 4, 5, Table 1) indicate that σ_V is intermediate at that depth. Figure 5 presents the stress results as a function of depth. Linear approximations yield:

$$\sigma_{Hmax} = 125 + 0.25D; \quad \sigma_{Hmin} = 50 + 0.16D; \quad \sigma_V = 0.24D$$

with fairly good correlation coefficients (0.78 or better). An interesting characteristic of these relationships is that throughout the depths tested σ_{Hmin} is approximately one half of σ_{Hmax}.

The directions of the principal stresses are remarkably consistent

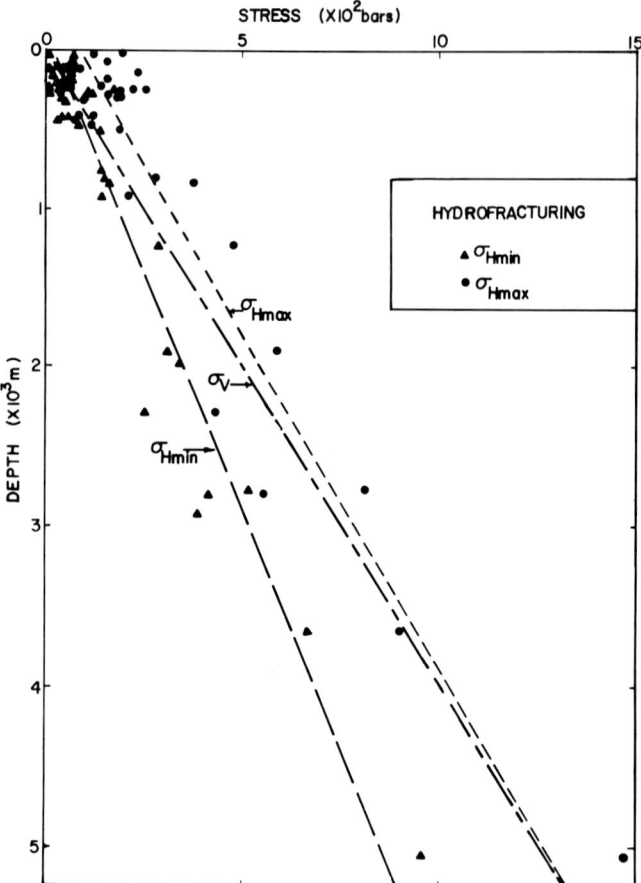

Fig. 6. Variation of horizontal and vertical stresses with depth in the continental United States, based on all the known hydrofracturing test results (see Table 1 for data and references).

with a mean σ_{Hmax} orientation of N65°E. Focal mechanism solutions are generally in agreement, indicating thrust fault conditions with σ_{Hmax} in a ENE direction (sites 31 and 32, Figure 1). An interesting exception is the focal mechanism solution at Lake Jocassee (site 34, Figure 1) showing σ_{Hmax} direction to be at N45°W [Talwani, 1977]. This is almost 90° from σ_{Hmax} direction determined nearby (site 3) by hydrofracturing.

Discussion and Conclusions

The accumulating data on crustal stresses in the United States, based on recent hydrofracturing measurements, have a common denominator in the direction of the largest horizontal stress. This direction is remarkably consistent between northeast and east-northeast (Figure 1). Similar stress directions have also been determined by other methods. Particularly, overcoring tests in the eastern part of the country

[Sbar and Sykes, 1973] and fault plane solutions in the West [Raleigh, 1974] have supported the hydrofracturing results. The direction of the major horizontal stress in the United States does not appear to be clearly related to ancient tectonic events (Appalachians, southern Canadian Shield) but does reflect the mechanism of active faulting (Rangely, San Andreas). The relationship of the stress directions to plate tectonics has been discussed by Sbar and Sykes [1973] and Raleigh [1974]. Despite the generally consistent directions, local deviations of up to $\pm 45°$ do exist and require individual measurement of stress orientations to evaluate their significance.

The ability to use hydrofracturing to measure stress at depths never attempted before is beginning to provide us with a more complete picture of the stress profile. In Figure 6 we have combined all the known hydrofracturing results in the United States and have obtained an overall variation of horizontal and vertical stresses with depth down to 5 km. Similar profiles have been obtained before on the basis of other stress measurement methods [Hast, 1973; Herget, 1974; Gay, 1975; Ranalli and Chandler, 1975]. These measurements, however, were conducted near the surface or from underground openings, a condition which could have affected the results, and were limited in overall depth.

The hydrofracturing results have been fitted with linear curves given by the following relationships:

$$\sigma_{Hmax} = 75 + 0.24D; \quad \sigma_{Hmin} = 20 + 0.16D; \quad \sigma_V = 0.25D$$

where stresses are in bars and depths (D) are in meters. The vertical stress is the result of both overburden weight calculations and hydrofracturing measurements. The general stress regime appears to be one in which thrust faulting conditions dominate at 0-500m depth, followed by strike-slip faulting at 500-5000m, and by normal faulting at greater depths. Focal mechanism solutions have shown that such changes in the state of stress with depth can exist locally [Talwani, 1977]. The relationships show that at shallow depths (0-1000m) the average horizontal stress, given by $\sigma_{Havg} = 50 + 0.20D$, tends to be higher than the vertical stress, as was previously noted by Hast [1973], Gay [1975], and others. The phenomenon of high horizontal stresses at shallow depths is particularly evident in the Canadian Shield, and the southern Appalachians. Beneath 1000m the average horizontal stress gradually drops and becomes approximately equal to $0.8\sigma_V$ at 5000m. The trend is similar to that observed by Gay [1975] in southern Africa. It is to be remembered that most of the hydrofracturing measurements beneath 500m were taken in sedimentary rock formations associated with geologically stable areas of little seismic activity. The results to date indicate a tendency toward a hydrostatic state of stress with increasing depth. This can be established by calculating the ratios between the principal deviatoric stresses and the average stress. The ratio tends to diminish with depth for each of the stresses. It is also interesting to note that the gradients of σ_{Hmin} and σ_{Hmax} are consistent with those determined by von Schonfeldt et al., [1973], using data from hydraulic fracturing stimulation treatments of some 3000 oil wells.

We emphasize here that Figures 2-6 are preliminary in nature and many more measurements are needed before the stress-depth relationships can be unequivocally established. We also stress that our confidence in the accuracy of the σ_{Hmin} values is somewhat higher than that in the σ_{Hmax} magnitudes, which are also more widely scattered with respect to the averaging curve. Moreover, Figures 2-6 are not intended for use in lieu of measurements. Local stresses could significantly vary from the plotted average, as can be verified from the results of measurements described in this paper.

In conclusion, the emergence of the hydrofracturing technique has provided us with a preliminary picture of crustal stress distribution in the United States. Many more measurements are required to complete the picture. Large areas in the Great Plains, the Coastal Plain, and the Columbia Plateau have never been tested. Only a few shallow measurements have thus far been conducted in the earthquake prone region along the San Andreas fault. In addition to covering new terrain, emphasis in future tests should also be placed on selected deep measurements (5-10 km), which could be the key to establishing reliable crustal stress gradients.

References

Bredehoeft, J. D., R. G., Wolff, W. S. Keys, and E. Shutter, Hydraulic fracturing to determine the regional in situ stress field, Piceance Basin, Colorado, Geol. Soc. Amer. Bull., 87, 250-258, 1976.

Fraser, C. D. and B. E. Pettitt, Results of a field test to determine the type of orientation of a hydraulically induced formation fracture, J. Petrol. Tech., 14, 463-466, 1962.

Gay, N. C., In-situ stress measurements in southern Africa, Tectonophysics, 29, 447-459, 1975.

Haimson, B. C., Hydraulic fracturing in porous and nonporous rock and its potential for determining in-situ stresses at great depth, Ph.D. thesis, Univ. of Minn., Minneapolis, 1968.

Haimson, B. C., Stress measurements in the Weber sandstone at Rangely, Colorado (abstract), Eos Trans. AGU, 53, 524, 1972.

Haimson, B. C., Earthquake related stresses at Rangely, Colorado, in New Horizons in Rock Mechanics, edited by H. Hardy and R. Stefanko, pp. 689-708, American Society of Civil Engineers, 1973.

Haimson, B. C., A simple method for estimating in-situ stresses at great depths, Field Testing and Instrumentation of Rock, ASTM Spec. Tech. Publ. 554, pp. 156-182, 1974a.

Haimson, B. C., Stress measurements in faults and their vicinities, semiannual report, grant 14-08-0001-G118, U.S. Geol. Survey, 1974b.

Haimson, B. C., The hydrofracturing stress measuring technique - method and recent field results in the United States, in Proceedings of the International Society for Rock Mechanics Symposium on Investigation of Stress in Rock - Advances in Stress Measurement, pp. 23-30, The Institution of Engineers, Sydney, Australia, 1976a.

Haimson, B. C., Preexcavation deep-hole stress measurements for design of underground chambers - case histories, in Proceedings of the 1976 Rapid Excavation and Tunneling Conference, edited by R. J. Robbins and R. J. Conlon, pp. 699-714, Soc. Mining Engineers, New York, N.Y., 1976b.

Haimson, B. C., A stress measurement in West Virginia and the state of stress in the southern Appalachians (abstract), Eos Trans. AGU, 58, 493, 1977a.

Haimson, B. C., Crustal stress in the Michigan Basin, J. Geophys. Res., 82, in press, 1977b.

Haimson, B. C., and J. M. Avasthi, Stress measurements in anisotropic rock by hydraulic fracturing, in Applications of Rock Mechanics, edited by E. R. Hoskins, Jr., pp. 135-156, American Society of Civil Engineers, 1975.

Haimson, B. C., J. Lacomb, A. H. Jones, and S. J. Green, Deep stress measurements in tuff at the Nevada test site, in Advances in Rock Mechanics, vol. IIa, pp. 557-561, Nat. Acad. of Sci., Washington, D.C., 1974.

Haimson, B. C., and E. Stahl, Hydraulic fracturing and the extraction of minerals through wells, in Proceedings of 3rd Symposium on Salt, pp. 421-432, Northern Ohio Geological Society, Cleveland, Ohio, 1970.

Hamilton, R. M., and J. H. Healy, Aftershocks of the Benham nuclear explosion, Geol. Soc. Amer. Bull., 59, 2271-2281, 1969.

Hast, N., Global measurements of absolute stress, Phil. Trans. Roy. Soc. London, 274, 409-419, 1973.

Herget, G., Ground stress determination in Canada, Rock Mech., 6, 53-64, 1974.

Hooker, V. E., T. R. Aggson, and D. C. Bickel, In-Situ determination of stresses in Rainier Mesa, Nevada test site, report, U.S. Bur. of Mines, 1971.

Hubbert, M. K. and D. G. Willis, Mechanics of hydraulic fracturing, Trans. AIME, 210, 153-160, 1957.

Kehle, R. O., Determination of tectonic stresses through analysis of hydraulic well fracturing, J. Geophys. Res., 69, 259-266, 1964.

Lindh, A. G., F. G. Fisher and A. M. Pitt, Nevada focal mechanisms and regional stress fields (abstract), Eos Trans. AGU, 54, 1133, 1973.

Overbey, W. K., Jr., and R. L. Rough, Surface studies predict orientation of induced formation fractures, Producers Monthly, 16-19, June 1968.

Overbey, W. K. Jr., and R. L. Rough, Prediction of oil and gas bearing rock fracture from surface structural freatures, U.S. Bur. of Mines report of investigation 7500, 1971.

Parsons, R. C., and H. D. Dahl, A study of the cause of roof instability in the Pittsburgh coal seam, in Proceedings of the Seventh Canadian Rock Mechanics Symposium, pp. 79-78, Mines Branch Department of Energy, Mines and Resources, Ottawa, Canada, 1972.

Power, D.V., C. L. Schuster, R. Hay and J. Twombly, Detection of hydraulic fracture orientation and dimensions in cased wells, J. Petrol. Tech., 28, 1116-1124, 1976.

Raleigh, C. B., Crustal stress and global tectonics, in Advances in Rock Mechanics, vol. Ia, pp. 593-597, Nat. Acad. of Sci., Washington, D.C., 1974.

Raleigh, C. B., J. H. Healy, and J. D. Bredehoeft, Faulting and crustal stress at Rangely, Colorado, in Flow and Fracture of Rocks, Geophys. Monogr. Ser., vol. 16, edited by H. C. Heard, I. Y. Borg, N. L. Carter, and C. B. Raleigh, pp. 275-284, AGU, Washington, D.C., 1972.

Raleigh, C. B., J. H. Healy, and J. D. Bredehoeft, An experiment in earthquake control at Rangely, Colorado, Science, 191, 1230-1237, 1976.

Ranalli, G., and T. E. Chandler, The stress field in the upper crust as determined from in situ measurements, Geol. Rundsch., 64, 653-674, 1975.

Roegiers, J., The development and evaluation of a field method for in situ stress determination using hydraulic fracturing, Ph.D. thesis, Univ. of Minn., Minneapolis, 1974.

Sbar, M. L., and L. R. Sykes, Contemporary compressive stress and seismicity in eastern North America: An example of intra-plate tectonics, Geol. Soc. Amer. Bull., 84, 1861-1882, 1973.

Scheidegger, A. E., Stresses in earth's crust as determined from hydraulic fracturing data, Geol. Bauw., 27, 45-50, 1962.

Smith, R. B. and M. L. Sbar, Contemporary tectonics and seismicity of the western United States with emphasis on the Intermountain Seismic Belt, Geol. Soc. Am. Bull., 85, 1205-1218, 1974.

Street, R. L., R. B. Herrmann and O. W. Nuttli, Earthquake mechanics in the central United States, Science, 184, 1285-1287, 1974.

Strubhar, M. K., J. L. Fitch, and E. E. Glenn, Jr., Multiple vertical fractures from an inclined wellbore--A field experiment, J. Petrol. Tech., 27, 641-647, 1975.

Talwani, P., Stress distribution near Lake Jocassee, South Carolina, Pure and Applied Geophys., 115, in press, 1977.

Tyler, L. D. and W. C. Vollendorf, Physical observations and mapping of cracks resulting from hydraulic fracturing in situ stress measurements, paper presented at the Annual Meeting of the Society of Petroleum Engineers of AIME, Dallas, Texas, 1975.

von Schonfeldt, H. A., An experimental study of open-hole hydraulic fracturing as a stress measurement method with particular emphasis on field tests, Ph.D. thesis, Univ. of Minn., Minneapolis, 1970.

von Schonfeldt, H. A., R. O. Kehle, and K. E. Gray, Mapping of stress field in the upper earth's crust of the U.S., final technical report, grant 14-08-0001-122278, U.S. Geol. Surv., 1973.

Zoback, M. D., F. Rummel, R. Jung, Alheid, H. J., and Raleigh, C. B., Rate controlled hydraulic fracturing experiments in intact and pre-fractured rock, Int. J. Rock Mech. Mining Sic., in press, 1976.

Zoback, M. D., J. H. Healy, and J. C. Roller, Preliminary stress measurements in Central California using the hydrofrac technique, Pure and Applied Geophys., 115, in press, 1977.

Zoback, M. D. and J. H. Healy, In situ stress measurements near Charleston, South Carolina (abstract), Eos Trans. AGU, 58, 493, 1977.

HIGH-ACCURACY DETERMINATION OF TEMPORAL VARIATIONS OF CRUSTAL RESISTIVITY

H. Frank Morrison, Robert F. Corwin, and Mark Chang

Engineering Geoscience, University of California
Berkeley, California 94720

Abstract. A network of large-scale dipole-dipole arrays has been used since early 1973 to monitor temporal changes of electrical resistivity along 50 km of the San Andreas fault south of Hollister, California. Transmitter-receiver separations of up to 20 km are used, and a synchronous detection system is used to obtain accurate signal voltage amplitudes in the presence of telluric current noise. Error bars (95% confidence interval on the mean) on weekly resistivity readings range from a few tenths of a percent of the mean for receivers with high signal-to-noise ratios to a few percent at noisy remote receivers; week-to-week resistivity variations generally fall within the same range. Three significant resistivity variations have been observed: a previously reported 24% variation which appeared to be related to an earthquake of magnitude 4.2 which occurred within the array, a long-term 13% variation between a single transmitter-receiver pair, and a 10% variation seen at three stations from one transmitter.

An analysis of the effects of surface layer variations and of inhomogeneities near the dipoles suggests that the major variations are not the result of near-surface changes brought about by rainfall or water table fluctuations. Models based on isotropic changes in porosity, with attendant resistivity changes predicted by Archie's law, have serious limitations in explaining the variations because they result in unreasonable crustal uplifts. A simple anisotropic vertical slab model can explain the observed variations, but its relationship to physical processes in the fault zone is not clear.

Introduction

Brace and Orange [1968] observed strong changes in resistivity of rocks prior to fracture under stress and suggested that the buildup of stress prior to an earthquake might be detected by changes in earth resistivity. Field experiments by Barsukov [1970] showed apparent resistivity variations related to earthquake activity, and high-accuracy continuous resistivity measurements by Yamazaki [1967] showed steplike changes apparently associated with strain steps. It appeared from these results that at least for some rock types, small strains (changes in porosity) might result in significant changes in bulk resistivity which could be measured from the surface of the earth. To study this phenomenon, a portion of the San Andreas Fault south of Hollister, California, (Figure 1) was chosen on the basis of a history of consistent moderate seismicity [Bolt and Miller, 1975]. The geology of the area consists basically of granites to the west of the fault zone and Franciscan sed-

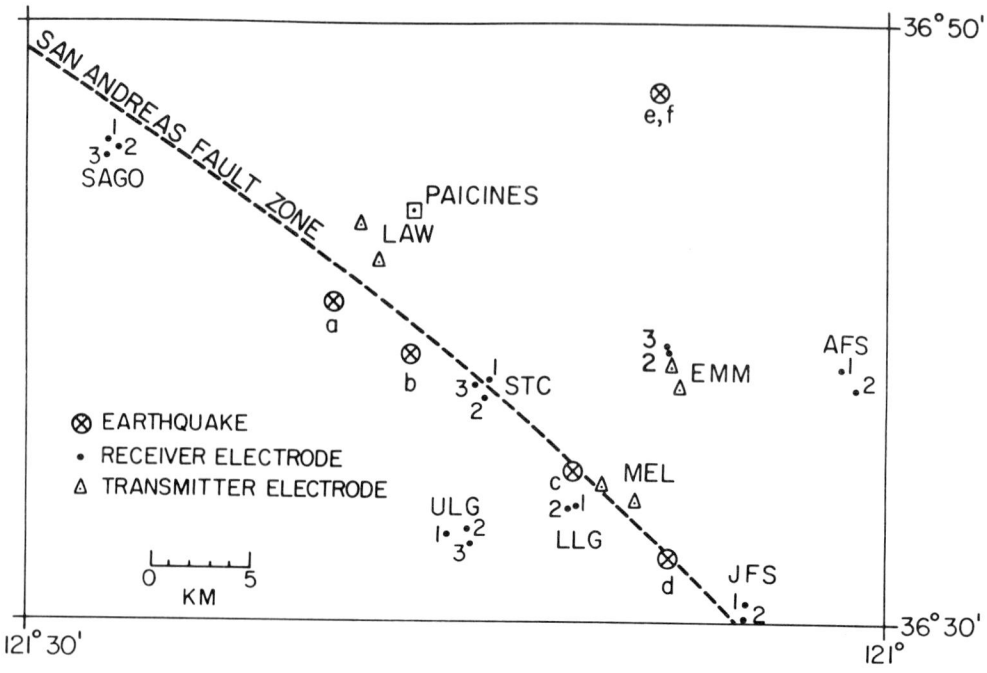

Fig. 1. Resistivity array. Data for earthquakes are given in Table 2.

iments to the east [Wilson, 1943]. A preliminary dipole-dipole survey was carried out over parts of the area [Mazzella, 1976], and resistivity monitoring began between a transmitter to the east of the fault and two receivers to the west of the fault in early 1973. A resistivity variation was seen beginning in April 1973, followed in June of that year by an earthquake of magnitude 4.2 located, rather fortuitously, directly between the transmitter and the receivers [Mazzella and Morrison, 1974]. On the basis of this early encouraging result the resistivity array has been expanded to cover about 45 km of the fault zone, and the measuring equipment has been improved to the point where most of the apparent resistivity values measured are accurate within about 1-3%. This paper describes the resistivity network, instrumentation, and data and briefly discusses some possible mechanisms for temporal variations of the measured resistivity of the San Andreas fault zone.

Resistivity Network

The electrical resistivity of the earth is measured by injecting current into the earth through a pair of transmitter electrodes and measuring the potential generated across a pair of receiving electrodes. The ratio of the received voltage V to the transmitted current I is proportional to the overall resistivity of the earth beneath the array. The apparent resistivity ρ_a is the value calculated for a particular transmitter-receiver pair when it is assumed that the earth is of uniform resistivity. The Hollister resistivity network, shown in Figure 1, includes three current transmitters (located at Law Ranch, LAW; Emmet School, EMM; and Melendy Ranch, MEL) and seven voltage receivers (located at San Andreas Geophysical Observatory, SAGO; Stone Canyon, STC; Emmet School, EMM; Antelope Fire Station, AFS; Jefferson School, JFS; Lower La Gloria Road, LLG; and Upper La Gloria, ULG). The receivers at ULG, STC, and SAGO are three-electrode arrays; the others are single dipoles. Weekly measurements are made from the EMM transmitter to ULG, LLG, STC, JFS, AFS, and a local receiver at EMM; from MEL to LLG, ULG, STC, and JFS (the nearby LLG dipole serves as a local receiver for MEL); and from LAW to STC, ULG, and SAGO. Readings are made in the evening to take advantage of the reduced telluric noise level. Table 1 shows transmitter-receiver separations, nominal V/I ratios, and apparent resistivities for each transmitter-receiver combination. Table 2 lists all earthquakes with magnitudes greater than 3 that have occurred within the network since it was established.

Instrumentation

The EMM transmitter is powered by a 100-KW ac motor-generator set. The output is rectified and reversed in polarity every 5 s to produce a 10-s period square wave of 160 A peak to peak at 450 V. A period of 10 s was chosen for all the transmitters to avoid the electromagnetic coupling that occurs at higher frequencies and interference from the high level of telluric activity in the 20- to 100-s period band. The MEL and LAW transmitters use three-phase line power at 480 V and after rectification supply approximately 200 A peak to peak at 650 V. In order to bring the current electrode resistance down to 5-10 ohms, three 1.2 m x 2.4 m steel plates, buried just below the surface, are used at all sites. (Steel was selected after laboratory tests showed it to have a lower rate of weight loss per ampere hour than either lead or aluminum.) Connection to the EMM transmitter electrodes is made through #4 AWG copper cable. Connection to the plates at MEL is through cable containing four strands of #12 AWG copper wire, and at LAW through #1 AWG copper wire. All transmitter cables are enclosed in plastic pipe and buried in trenches about 0.5-1.0 m deep. They are periodically checked for high voltage current leaks by disconnecting the plates from the cables and running the transmitter into the resulting open circuit. Copper-copper sulfate electrodes, buried at a depth of 1 m, are used at all receiving stations except STC and SAGO, where the electrodes are of lead. The electrodes at STC are buried at a depth of about 100 m [Bufe, 1973], and those at SAGO just below the surface.

Initial resistivity measurements were made by using strip chart recorders to measure the transmitted current and received signal amplitudes from which the V/I ratio is calculated. Because the signal-to-noise ratio varies from about 1 at noisy stations to 20 at quiet stations, data reduction was tedious. Often, when the strip chart records were used, many hours were required to obtain a V/I ratio with an error bar of less than $\pm$ 10% of the mean value [Mazzella, 1976]. (The error bar used in this work is defined as two standard errors about the mean [Ostle, 1963]. Thus assuming the measured values to be normally distributed about the true mean, one is 95% confident that the true mean V/I ratio lies within the error bar. The error bar is expressed as a percentage of the mean value.)

TABLE 1. Transmitter-Receiver Separations, Nominal V/I Ratios, and Apparent Resistivities

TRANSMITTERS

Receiver dipole	EMM			MEL			LAW		
	S, km	V/I, ohms	ρ_a, ohm m	S, km	V/I, ohms	ρ_a, ohm m	S, km	V/I, ohms	ρ_a, ohm m
SAGO 1-2	32			34	1.00×10^{-6}	73	15	5.08×10^{-6}	202
SAGO 1-3	32			34	1.04×10^{-6}	110	15	4.93×10^{-6}	71*
SAGO 2-3	32			34	2.17×10^{-5}	7	15	7.23×10^{-7}	6
STC 1-2	10	2.26×10^{-6}	14	10	1.24×10^{-5}	30*	11	1.40×10^{-6}	4
STC 1-3	10	2.18×10^{-6}	18	10	3.50×10^{-5}	5	11	2.26×10^{-6}	4
STC 2-3	10	4.45×10^{-6}	28	10	2.37×10^{-5}	3	11		
LLG 1-2	10	6.12×10^{-7}	2	2.5	2.80×10^{-5}	21	19	4.47×10^{-6}	220
ULG 1-2	15	1.06×10^{-5}	132	9	2.43×10^{-5}	21	19	3.65×10^{-6}	0*
ULG 1-3	15	1.18×10^{-5}	370	9	3.75×10^{-6}	10*	19	1.22×10^{-7}	2*
ULG 2-3	15			9	1.79×10^{-6}	3	19	1.9×10^{-7}	10
JFS 1-2	15	1.34×10^{-6}	10	10	6.95×10^{-6}	60	30		
AFS 1-2	9	4.46×10^{-5}	280	14			25		
EMM 2-3	1.3	4.68×10^{-4}	11						
Transmitters as receivers									
EMM				7.7	2.34×10^{-6}	5			
MEL	7.7	2.34×10^{-6}	5				20	1.25×10^{-6}	8
LAW				20	1.25×10^{-6}	8			

S is approximate transmitter-receiver separation; ρ_a is apparent resistivity. Because of the large effect of small errors in electrode location, the values of ρ_a are approximate. Blank spaces indicate that values were not measured.

* Data are unreliable because of possible large geometric factor error.

TABLE 2. Earthquakes With Magnitude Greater Than 3 Within the Resistivity Network From January 1, 1973, to July 30, 1976

Event	Date	Magnitude	Depth, km
a	June 14, 1975	3.0	3.8
b	March 26, 1975	3.1	3.8
c	June 22, 1973	4.2	9.0
d	December 9, 1975	3.0	8.0
e	December 29, 1975	3.4	9.0
f	March 19, 1976	4.2	8.6

Data are from University of California, Berkeley Seismographic Station. Locations are given in Figure 1.

To alleviate the problem of obtaining accurate data in the presence of high telluric noise, receivers employing a synchronous detection principle now are used [Morrison, 1975]. Crystal oscillator clocks in the transmitter and receiver are synchronized at the beginning of a measuring interval, and then the 10-s square wave from the transmitter oscillator is used to reverse the polarity of the direct current entering the ground. No further connection between the transmitter and receiver is required. The receiver operates by first converting the incoming signal to a frequency by using a voltage-to-frequency converter (VFC) with a center frequency of 5 kHz. The oscillator in the receiver unit is used to drive an up-down counter which adds cycles from the VFC during the positive half of the transmitter cycle and subtracts cycles during the negative half. Thus at the end of 10 s the net count is proportional to the averaged amplitude of the incoming square wave signal, with noise at periods greater or less than 10 s greatly reduced. The effects of initial overshoot or damping or of imperfect synchronization are minimized by not sampling the initial 1.88 s and final 0.63 s of the 5-s half cycle (i.e., only half of each half cycle is sampled, beginning after three eights of the half cycle has elapsed). A similar scheme is used to determine the transmitted current by measuring the voltage across a 100 µ ohm manganin shunt. The resistance of the shunt is accurate to ±0.1%, and its temperature coefficient is about 20 ppm/°C.

In practice, 10 of these 10-s measurements are averaged by the detector, and then the averaged zero-to-peak millivolt amplitude of the signal for the 100-s interval is shown on a liquid crystal display and manually recorded by the operator. From 5 to 36 of these 100-s readings, depending on the signal-to-noise level at the receiver, are

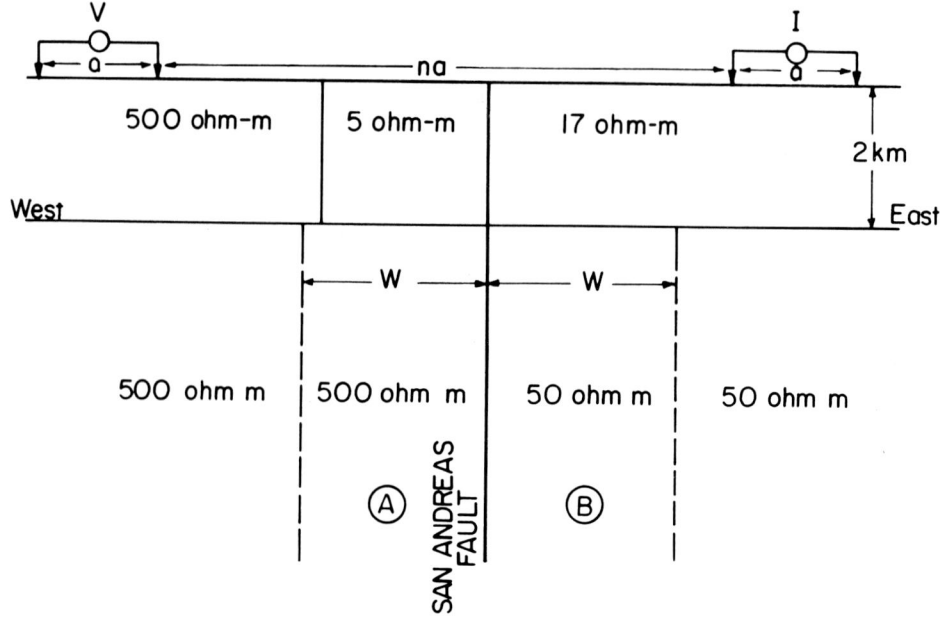

Fig. 2. Resistivity cross section between EMM transmitter and ULG re- based on data from Mazzella 1976. W is the width of the zones of varying resistivity.

averaged to obtain the final V/I ratio and error bar. The receivers are calibrated with an internal reference voltage at the beginning and end of each run and are checked every 3 months against a laboratory standard. Use of this system has resulted in a considerable improvement in data quality, as may be seen by comparing the error bars of the data taken in early 1973 (Figure 3) with those from mid-1974 and later, when the synchronous detectors were put into regular use.

Data

The resistivity survey across the fault zone [Mazzella, 1976] showed that at depths greater than about 2 km the resistivity of the granite to the west of the fault is about 500 ohm m, and that of the Franciscan sediments to the east about 50 ohm m. Apparent resistivities for transmitter-receiver pairs lying within or close to the fault zone (for example, LAW-STC, MEL-STC, and MEL-JFS) are much lower, about 4-10 ohm m (Table 1). A zone of very low resistivity (about 2-5 ohm m) underlies the LLG receiver, which is somewhat to the west of the active fault trace but within an area of frequent small earthquakes. A composite resistivity cross section between EMM and ULG based on these data is shown in Figure 2.

The resistivity data from the EMM, MEL, and LAW transmitters, plotted as percent deviations from a constant reference V/I ratio, are shown in Figures 3, 4, and 5, respectively. The numbers following the receiver station codes refer to apparent resistivities calculated from voltages observed between those electrodes (for example, STC 12 is between electrodes 1 and 2). Week-to-week variations of the resistivity readings are generally of the same order as the error bars, about 1 or 2% for strong receivers (signal-to-noise ratio of about 10 or greater) to 4 or 5% for weak receivers on noisy nights (signal-to-noise ratio of about 1 or less). Variations that are significantly larger than the error bars are discussed below.

Two significant variations appear in the EMM transmitter data (Figure 3). First, there was an anomaly of about 24% peak to peak at the ULG and LLG receivers from April through June 1973. Second, the resistivity at the AFS receiver has varied continuously since measurements began in January 1975, the total variation being about 13%. The 1973 variation was followed by an earthquake of magnitude 4.2 at a depth of 9 km (event c in Table 2 and on Figure 1) located between the EMM transmitter and the ULG receiver and about 2 km northwest of the LLG receiver. All other earthquakes of magnitude greater than 3 that have occurred within the network are shown in Figure 1 and listed in Table 2.

The only significant resistivity variation recorded from the MEL transmitter (Figure 4) was a single-point negative deviation in early December 1974 seen at all receivers and ranging up to 10%. This variation was coincident with a large creep event which was recorded over a length of the fault including the area between MEL and STC. The readings were made during moderate rainfall, but similar amounts of rain during the preceding 2 months did not seem to affect the data. However, a receiver malfunction noted shortly after the readings throws some doubt on the quality of the data. Several large single-point variations (up to 7%) have been recorded from the LAW transmitter (Figure 5), but because of the large error bars characterizing these data, these variations do not appear to be significant.

Discussion

The observed resistivity variations might be caused by (1) instrumental or measuring errors, (2) shallow resistivity variations (caused, for example, by surface moisture changes), or (3) changes in the resistivity of the rocks at depth caused by tectonic and hydrologic processes.

Experimental Errors

The measurement errors could arise from malfunction of the synchronous detectors or transmitter electronics or from leakages from the current or receiving cables. Transmitter current leakage might occur through insulation breakdown in wet ground and would result in a drop in the measured apparent resistivity. This error is minimized by high-voltage open circuit tests of the transmitter, especially at times when anomalous resistivities are noted. Accidental grounding of the receiver electrode cables also would result in a drop in apparent resistivity, and this possible source of error is checked by measurement of the contact resistance and electrode potential difference before each run.

The synchronous detectors are internally calibrated before and after each reading and are externally calibrated in the laboratory every 3

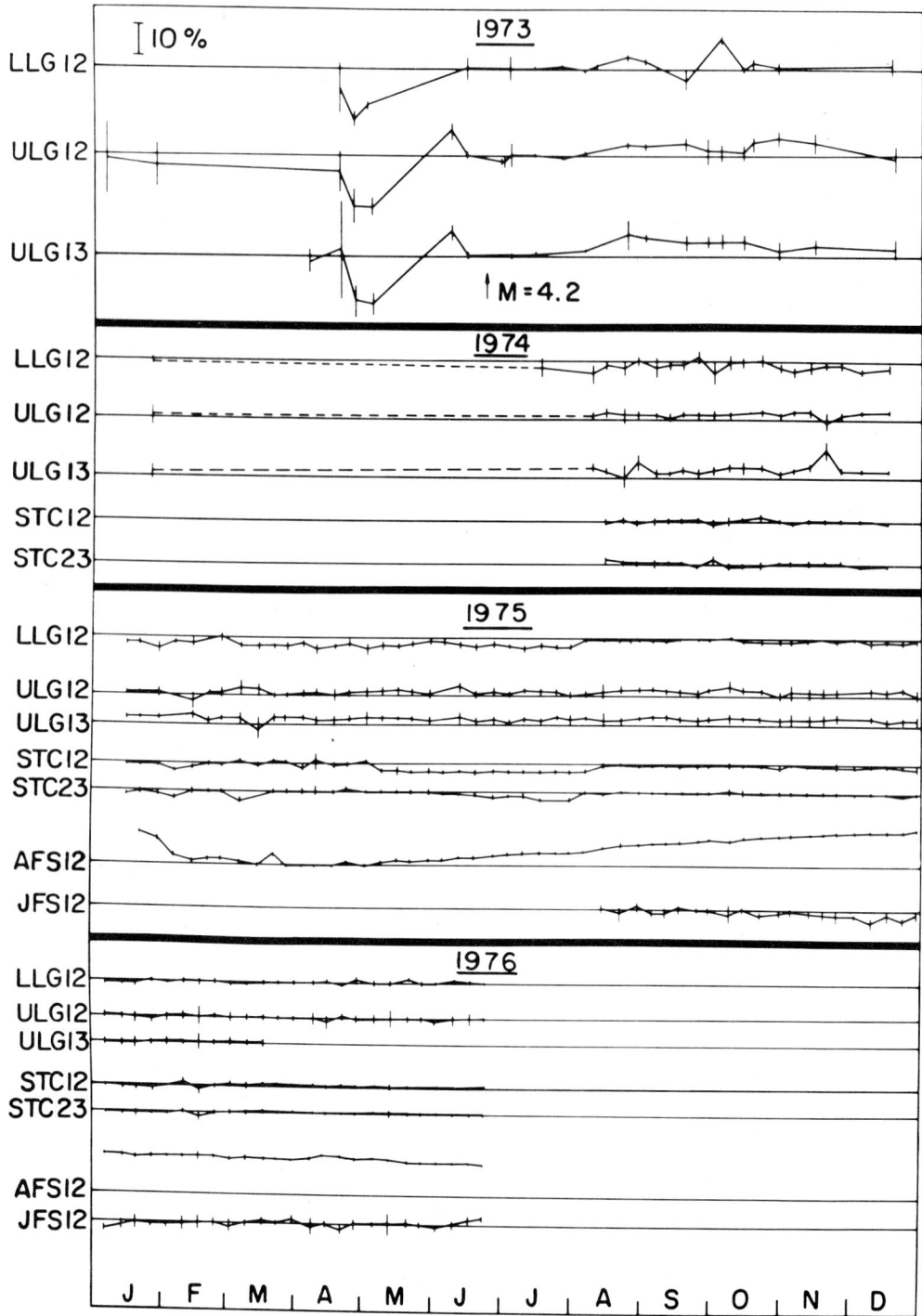

Fig. 3. Resistivity data from EMM transmitter.

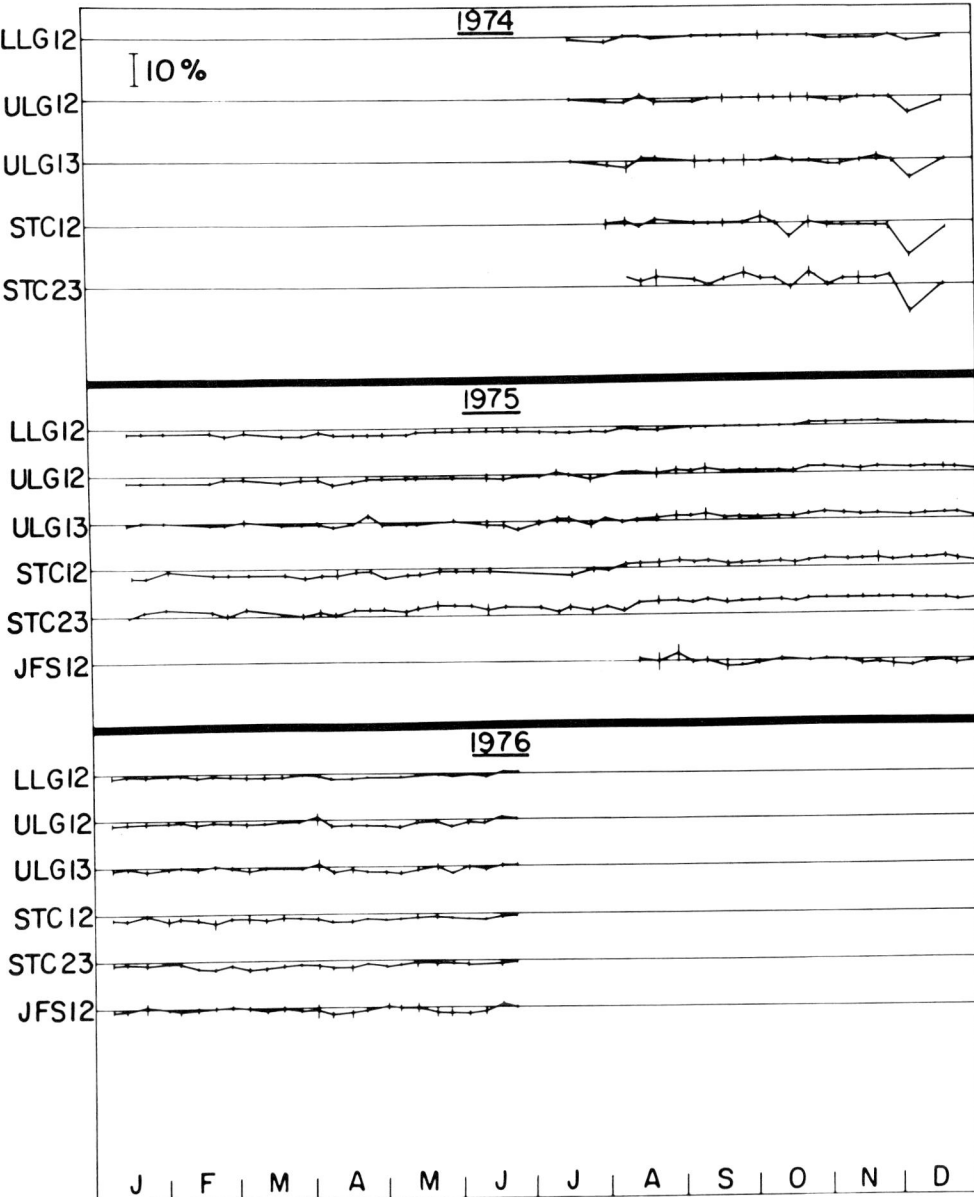

Fig. 4. Resistivity data from MEL transmitter.

months. The same instrument is used to take readings at several different receiver sites (for example, AFS, STC, and ULG), so the effects of long-term drift of a single instrument would be apparent. The synchronization of the transmitter and all the receivers is checked after each run, and the reading is repeated if synchronization is not within

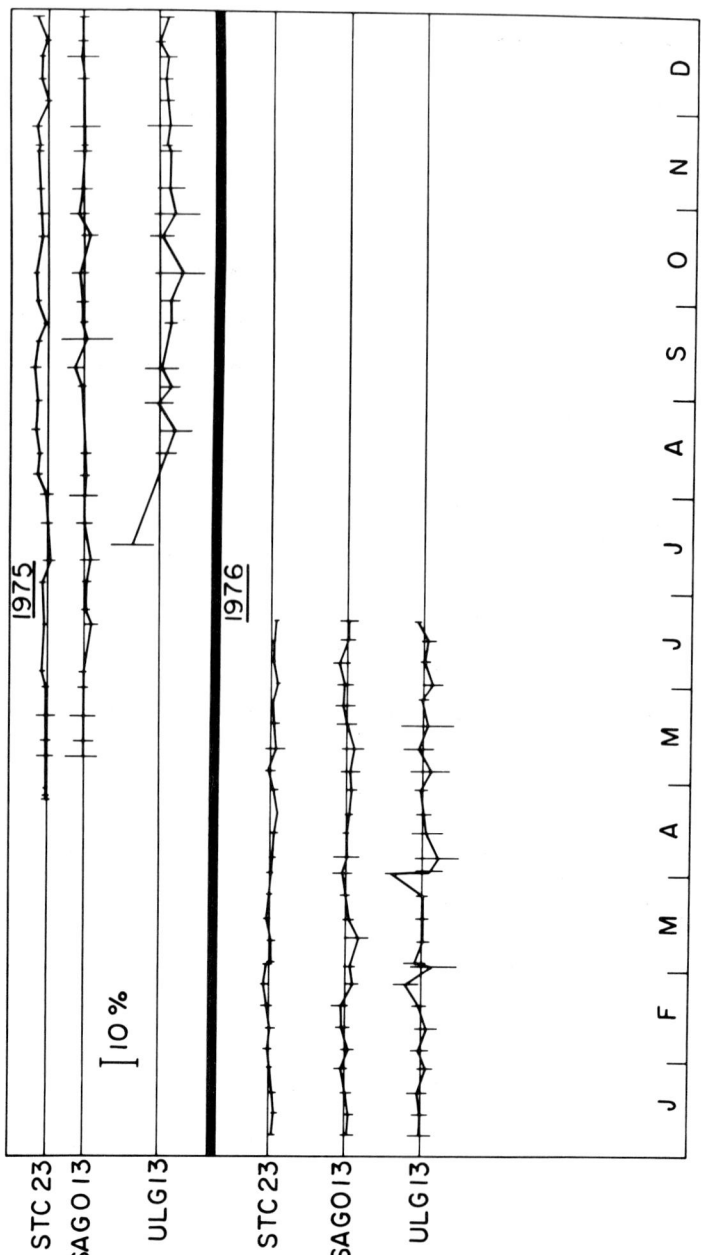

Fig. 5. Resistivity data from LAW transmitter.

specifications. With all these precautions the possibility of erroneous readings is small. If such errors have occurred, they are limited to occasional single points and would not affect long-term variations.

The large error bars characteristic of many of the 1973 data reflect the difficulty of obtaining a sufficient number of usable signal amplitudes from strip chart records rather than any inherent error in the data. [Mazzella 1976] describes the exhaustive methods used to obtain reliable data during this period. Since the synchronous detectors were put into use in mid-1974, the ease of obtaining data has allowed quick rechecks of any apparently anomalous readings for possible errors. Subtle changes in the wave form of the received signal also might affect the resistivity readings. We have not noticed such changes, but they might be difficult to detect.

Near-Surface Resistivity Variations

The possibility that shallow resistivity variations may affect the apparent resistivity of large arrays demands careful consideration. In this context, shallow refers to variations that might be caused by changing near-surface water saturation, changes in the water table by rainfall, or changes in resistivity caused by the diurnal or seasonal temperature wave. Few observational data are available on long-term variations in near-surface resistivity. Yamazaki [1967] reported long-term changes of about 10% in the resistivity of a small (5 m) array installed in a cave and concluded that the changes may have been influenced by rainfall. Barsukov and Sorokin [1973] found that the resistivity of a surficial soil layer (1-5 m) seemed to show a seasonal variation of 50%. Bufe [1973] described a possible seasonal resistivity cycle of about 10% for an 800-m array located at STC. Fitterman [1974] used a highly accurate resistivity variometer with 200-m spacing near MEL and saw virtually no seasonal variation within his 0.1% accuracy despite the occurrence of heavy seasonal rains.

In areas of complex, and certainly three-dimensional, geology the evaluation of the effects of these shallow variations is difficult. However, bounds on these influences can be obtained from horizontally layered models and from simple two-dimensional models.

Fitterman [1974] has presented an analysis of the sensitivity of a Schlumberger resistivity array to changes in the upper, or surface, layer of a simple two-layer model. He found that surface variations have more effect for conductive surface layers than for resistive surface layers. We have made a similar analysis for the dipole-dipole arrays used in this study, and a summary of this analysis is presented in Figure 6. We have chosen a surface layer of thickness equal to the dipole length a and arbitrarily consider the case where the electrode separation is 10 times the dipole length (n = 10). Percentage changes in the apparent resistivity (positive and negative) are plotted as a function of percentage change in the first-layer resistivity for several cases of resistivity contrast between the first and the second layer (ρ_1/ρ_2). As Fitterman [1974] has shown, moderate changes in a resistive first layer have little effect; a change of ±50% for an initial contrast of 100:1 has less than a 0.01% effect on the overall apparent resistivity. On the other hand, a change of only 5% for a conductive surface layer (ρ_1/ρ_2 = .01) has a 5% effect on the overall resistivity. As the

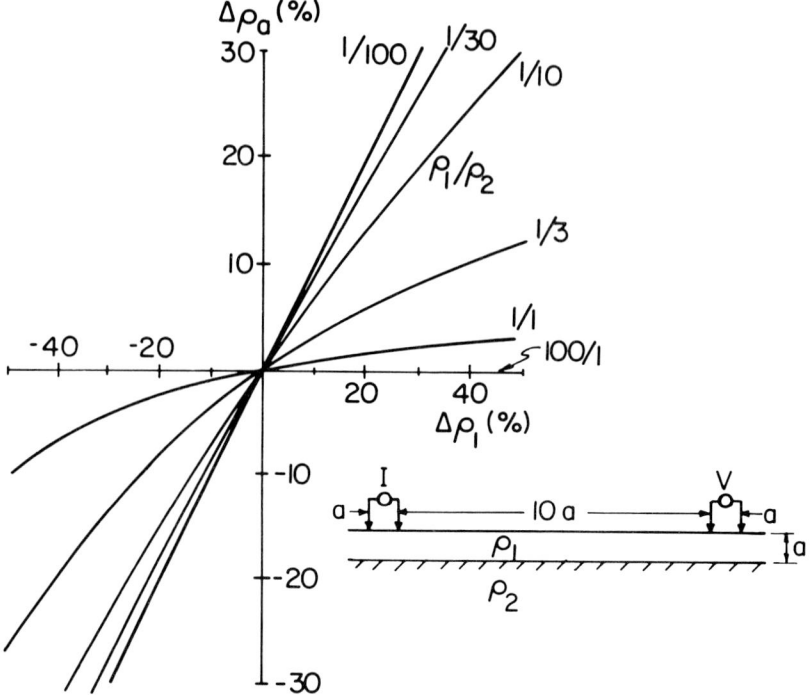

Fig. 6. Effects of surface layer resistivity variations on the apparent resistivity of a dipole-dipole array.

spacing n becomes larger, the effects are reduced. It can readily be appreciated that changes of less than 50% in a layer only 10-100 m thick would have no measurable effect on the apparent resistivities for most of the transmitter-receiver pairs.

Ideally, small resistivity arrays should be installed at each transmitter and receiver site to monitor the resistivity in the first few hundred meters. In practice, our near-surface resistivity monitors have been limited to a small (100 m) dipole located 500 m north of the EMM transmitter (EMM 23 on Figure 1) and the nearby LLG receiver for the MEL transmitter. Resistivity variations at both these sites have been within the measurement accuracy of a few tenths of a percent for the entire measurement period, an indication that near-surface resistivity changes must be confined to depths of a few meters at most. Fitterman [1974] found variations of less than 1% for his monitoring system near MEL. At the EMM 23 receiver the depth of the water table, observed in a nearby well, varies seasonally from 3 to 6 m with no effect on the apparent resistivity measured at EMM 23.

The uniform layer model is certainly an idealized representation of the Hollister area. To investigate the effects of changes in near-surface ground resistivity close to or beneath either dipole, we have studied the effects of a shallow inhomogeneity of rectangular cross section and infinite extent in the direction perpendicular to the line through the electrodes (Figure 7). This inhomogeneity might be a crude

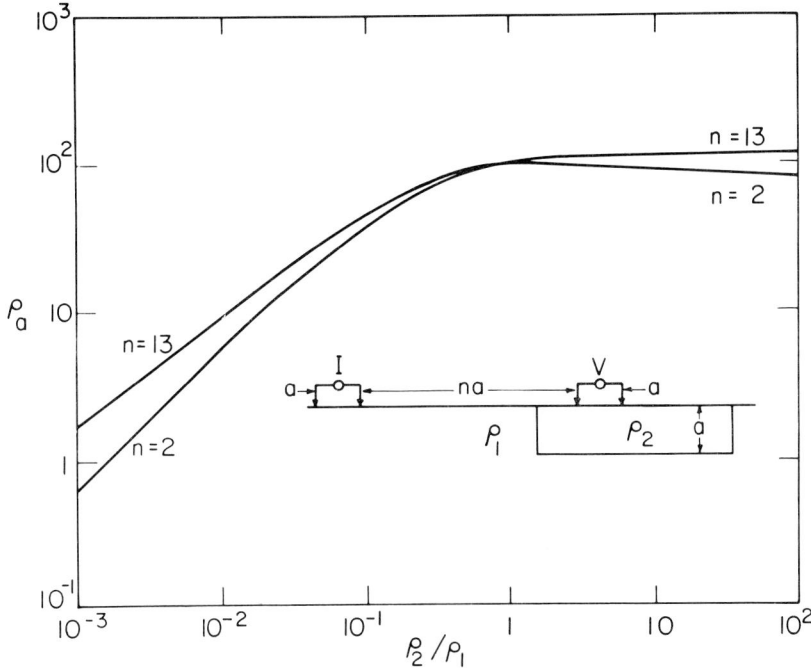

Fig. 7. Effect of resistivity contrast of a shallow inhomogeneity ρ_2 on the apparent resistivity of a dipole-dipole array. The inhomogeneity is infinitely extended in the direction perpendicular to the line joining the electrodes.

representation of a volume of recent sediments, or alluvium, in Franciscan terrane. The resistivity of the block is ρ_2 and the background resistivity is ρ_1. In this model the thickness of the inhomogeneity is set equal to the dipole length a, the width is 5 times the dipole length, and the results are shown for dipole spacings of 2a and 13a. Again the effect of a resistive inhomogeneity is negligible, but a conductive inhomogeneity can dominate the resulting apparent resistivity values even at large spacings. In fact, once the contrast ρ_2/ρ_1 is less than 0.1, a variation of 10% in the resistivity of the inhomogeneity produces roughly the same percentage variation in the overall apparent resistivity. This is a much stronger effect than is observed with the uniform surface layer, especially considering that less volume of ground is affected, and points out the importance of determining the resistivity distribution before attempting to interpret the results. On the basis of the above analysis a site with low local resistivity might be expected to show large apparent resistivity variations caused by local shallow resistivity variations. The LLG receiver shows resistivities considerably below the surrounding values (Table 1), but no seasonal changes in resistivity have been seen at the LLG receiver (Figures 3 and 4). The apparent resistivity between EMM and AFS is unusually high (280 ohm m) for a transmitter-receiver pair located in the normally less resistive Franciscan terrane (Figure 2). A resistivity of approximately 30 ohm m was measured by using a 50-m local transmitter centered between the receiving electrodes at AFS. Thus a large contrast exists in the resistivity section between these two stations, and very large variations at AFS could produce the observed seasonal variations of approximately 10%. We are currently endeavoring to make regular local resistivity measurements at AFS.

Tectonic strain-resistivity variations

The motivation for this experiment was that tectonic processes, particularly those related to possible dilatant effects [Scholz et al., 1973] may be responsible for resistivity variations. Except in rare cases where the rock minerals are themselves electrically conductive, current is carried primarily through the pore fluids. For a very wide variety of rock types the electrical resistivity has been found to be related to the porosity through the following empirical expression first discovered by Archie [1942]:

$$\rho_r = \rho_f \phi^{-m} \qquad (1)$$

where ρ_r is the resistivity of the rock, ρ_f is the resistivity of the pore fluid, ϕ is the fractional porosity, and m is an empirical constant, usually about 2.

Changes in rock porosity might therefore be expected to yield changes in resistivity. Considerable care must be taken, however, in using Archie's law to predict the change in resistivity accompanying a strain in the form of a porosity change. For example, during dilatancy the pore geometry of stress-induced microfractures may be quite different from the normal pore geometry, and the new porosity-resistivity relation might require a change in the exponent m. Fatt [1957] studied the stress-resistivity behavior of a sample that required an exponent of 3.4. Yamazaki [1974] has studied a sample of rock taken from the site of his continuous monitoring resistivity experiment and has observed strain versus resistivity changes that would require an exponent of about 10 in Archie's law if a rock porosity of 10% is assumed. For most rocks a relatively simple sequence of events takes place as a sample is stressed [Brace et al., 1965; Brace and Orange, 1968]. At low confining pressures, cracks close up, and the resistivity increases quite rapidly for a pressure rise of several kilobars. At higher stresses the resistivity remains relatively constant until, at about half the fracture stress, dilatancy begins, the porosity increases, and the resistivity falls dramatically before fracture occurs. There are two regimes where stress-induced resistivities might be expected, one in the upper few kilometers, where stress changes simply open and close joints and fractures, and another at greater depth and higher stress levels, where dilatancy may be taking place. There are convincing arguments based on laboratory experiments and on analysis of field data on joint and fracture distribution [Brace, 1971] that the joint porosity for many rocks (10-1000 ohm m) is simply not a great enough fraction of the normal pore porosity for changes in it to affect the bulk resistivity. Whether this is generally true remains to be determined from careful field measurements.

Constraints on the dilatancy-induced porosity changes have been studied by Hanks [1974]. He has shown that for the dilatancies required to explain seismic velocity anomalies, the strains are such that they require unreasonable crustal uplifts. T.R. Madden (personal communication, 1976) has shown that the work required to dilate a volume large enough to produce observable apparent resistivity changes also is unreasonable.

We have considered several simple resistivity models in which portions of the subsurface undergo changes in intrinsic resistivity due either to simple crack closure mechanisms or to dilatancy. The approach has been empirical in the sense that we have sought resistivity models to explain the observed variation, in this case the 20% variation observed before the magnitude 4.2 earthquake in June 1973. An alternative approach, which has been used by Fitterman [1976], is to develop a model of rock behavior during certain hypothesized stress changes and to predict the associated resistivity changes. As more data emerge to define better the actual resistivity changes and their distribution, these two approaches will become reconciled.

A simplified two-dimensional model of the resistivity distribution near the fault zone is shown in Figure 2. The model is based on a field survey of the area [Mazzella, 1976] and on the measured apparent resistivities from the permanent transmitter-receiver pairs in the area. The transmitter and receiver are both 1.5 km in length and are 14 km apart. The fault zone is located in the middle of the array. A two-dimensional finite element method was used in the calculations of apparent resistivity. In Table 3 we have shown the changes in apparent resistivity effected by varying the resistivity in two zones, A and B, which lie adjacent to the fault in the granite and in the sediments, respectively (Figure 2). The resistivity changes are assumed to be isotropic. Widths W of 2 km and 5 km are considered for each zone, and changes of 25 and 50% in the zone resistivity are used. For the case of a zone 2 km in width the intrinsic resistivity changes of the fault zone produce rather small apparent resistivity changes for the array considered here. For the case of a 5 km width the changes can be as large as 14 or 15%. The problem with models of this sort is that it seems highly unlikely that such large volumes could undergo such dramatic changes in porosity (the ultimate factor controlling resistivity at depth) without causing measurable, if not catastrophic, uplift of the surface.

To illustrate the conflict between uplift and a significant change in resistivity, a buried spherical volume has been selected (Figure 8). This model is presently the only one available to us for analyzing three-dimensional inhomogeneities, and a spherical dilatant zone may not be unreasonable for small earthquakes. We consider here a spherical dilatant region of resistivity ρ_2 centered at the earthquake focus. The background or half-space resistivity is ρ_1. The electric potentials on the surface due to a point current source also on the surface have been calculated by using the quasi-analytic solution formulated by Large [1971]. It is assumed that the resistivities involved are isotropic and that changes in resistivity are the results of changes in porosity and hence of changes in volume of the sphere.

The crustal uplift at the epicenter of a dilated sphere buried in a half space can be calculated by the following equation formulated by Singh and Sabina [1975]:

TABLE 3. Changes in Apparent Resistivity ($\Delta\rho_a/\rho_a$) Caused by Resistivity Variations ($\Delta\rho/\rho$) and Changes of Width W of Zones A and B (Figure 2)

ZONE	$\Delta\rho/\rho$, %	$\Delta\rho_a/\rho_a$, % W = 2 km	W = 5 km
A	-50	-0.7	-1.7
	-25	-0.4	-0.9
B	-50	-5.8	-7.0
	-25	-3.0	-13.7
	+50	+5.7	+14.7
A and B	-50	-4.3	-15.3
	-25	-2.2	-7.8

$$\Delta h = -(4/3)\varepsilon_f (r/d)^2 r(1 + \nu) \quad (2)$$

where Δh is the uplift at the epicenter, $\varepsilon_f = \Delta V/3V$ ($\Delta V/V$ is a free fractional volumetric strain), r is the radius of the sphere, d is the depth to the center of the sphere, and ν is Poisson's ratio of the rock. If $\nu = 0.25$ and $\varepsilon = \Delta V/V$, (2) can be simplified to

$$\Delta h = -0.556\varepsilon (r/d)^2 r \quad (3)$$

The uplift thus is given in terms of the same parameters used in describing the resistivity variations associated with the same dilatant sphere.

If we assume that Archie's law can be used to explain changes in resistivity associated with porosity changes (strain), then the resistivity change can be related to the porosity change via the relationship

$$\Delta\rho_r/\rho_r = m(\Delta\phi/\phi_0) \quad (4)$$

If we further assume that $\Delta\phi$ represents the volumetric strain ε and take a value of 2 for m as suggested by Brace [Brace et al., 1965], this relation can be further reduced to

$$\Delta\rho_r/\rho_r = -2\varepsilon/\phi_0 \quad (5)$$

or

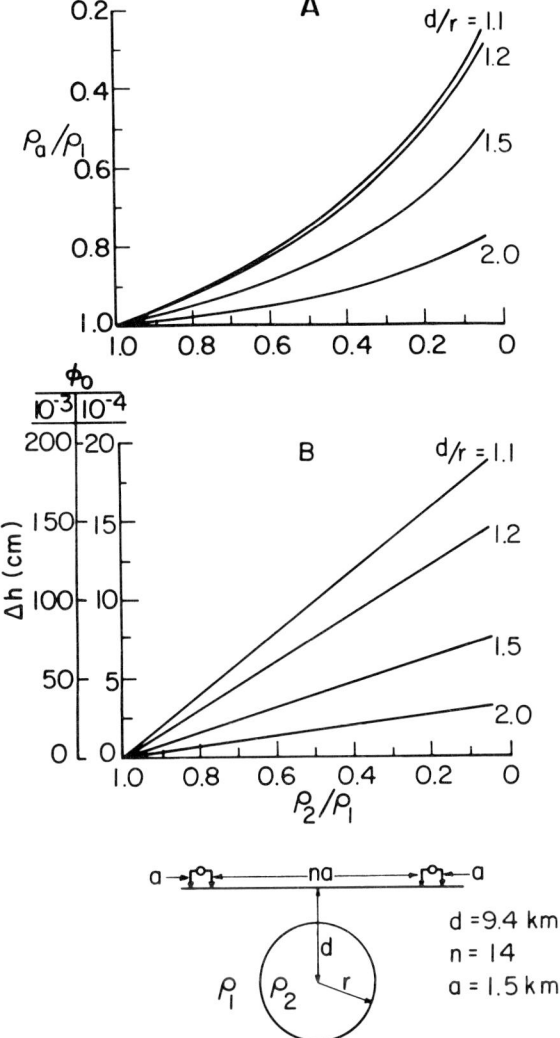

Fig. 8. Dipole-dipole apparent resistivity change ρ_a/ρ_1 and crustal uplift Δh over a dilatant sphere as a function of intrinsic resistivity contrast. Uplifts are calculated for ρ_2/ρ_1 for two values of initial porosity ϕ_0.

$$\varepsilon = -(\phi_0/2)\cdot(\Delta\rho_r/\rho_r) \qquad (6)$$

The size of the dilatant region can be estimated by the aftershock zone [Anderson and Whitcomb, 1973; Rikitake, 1975] via the following relationship:

$$\log R = 3 + M/2 \qquad (7)$$

where M is the magnitude of the main event and R is the radius of the aftershock zone in centimeters. For an earthquake of magnitude 4, the aftershock radius, according to (7), would be around 1 km. The real dilatant zone may be from 2 to 8 times the size of the aftershock zone, i.e., 2 to 8 km [Anderson and Whitcomb, 1973]. Spheres in this size range are considered in the following analysis.

A dipole-dipole resistivity array like the one discussed in the previous section is assumed except that here we consider the dilatant region as a spherical zone centered at a depth d of 9.4 km (the focal depth of the magnitude 4.2 earthquake). The apparent resistivity versus the intrinsic resistivity change of the sphere for different sphere radii r is plotted in Figure 8a. It can be seen that a 20% change in the apparent resistivity ρ_a would require a 45% change in the intrinsic resistivity of a spherical dilatant zone with radius r = 8.5 km (d/r = 1.1) and would require a 90% change in the intrinsic resistivity for a sphere with r = 4.7 km (d/r = 2). The volumetric strain ε for a corresponding intrinsic change in resistivity is calculated from (6) by assuming a value for the initial porosity ϕ_0. The crustal uplifts may then be calculated as a function of the intrinsic resistivity change by substituting (6) in (3). The results showing crustal uplifts for two initial porosities, 10^{-3} and 10^{-4}, are plotted in Figure 8b.

For low initial values of ϕ_0, small absolute changes would result in large percentage changes and thus in large resistivity variations. For an initial rock porosity $\phi_0 = 10^{-4}$, a 45% change in the resistivity of the sphere would imply a crustal uplift of 8 cm over the center of the sphere, a value which is not unreasonable. However, Brace 1975 has pointed out that the pore porosity of crustal rocks cannot be much smaller than 10^{-3}; indeed the low values of apparent resistivity even for the largest arrays bear this out. For $\phi_0 = 10^{-3}$, a crustal uplift of 80 cm would be necessary to accommodate the resistivity variation; for $\phi_0 = 10^{-2}$ (a more reasonable value for the Hollister region) the uplift would be 8 m. Uplifts of this magnitude would be difficult to overlook, and there is no evidence in tilt data from the region that such an uplift preceded the earthquake.

It seems difficult on the basis of the above argument to explain the observed resistivity variation without an unreasonably high crustal uplift. However, it should be realized that resistivity often is strongly anisotropic [Keller and Frischknecht, 1966], as is tectonic strain. The calculations of crustal uplift are based on the assumption of ideal homogeneous elastic volumetric expansion. In the actual field situation for the strike-slip faulting characteristic of the San Andreas fault, the preferential crack orientation would be vertical and parallel to the principal stress direction [Nur et al., 1973]. In this case the crustal uplift might not be as large for the observed resistivity change. Lee [1976] has modeled the fault zone as a transversely anisotropic dike and has found that changes in the coefficient of anisotropy of the dike may result in changes in the apparent resistivity measured across the dike larger than those that would be caused by changes in isotropic resistivity alone.

This study again considered a dipole-dipole array perpendicular to a vertical fault zone with a width of 1 km (Figure 9). The model is a simplified version of the interpreted resistivity cross section shown in Figure 2. The fault zone, which in this model need not correspond to the actual fracture or gouge zone, is assumed to be transversely anisotropic; it is characterized by one resistivity perpendicular to the zone boundaries ρ_t and another parallel to the boundaries ρ_ℓ. The coefficient of anisotropy λ is given by

$$\lambda = (\rho_t/\rho_\ell)^{1/2} \tag{8}$$

and for convenience an equivalent zone resistivity ρ_m is defined by

$$\rho_m = (\rho_t \rho_\ell)^{1/2} \tag{9}$$

The potential measured at the receiver is equivalent to the potential that would have been measured if the fault zone were isotropic and had a scaled resistivity of $\lambda\rho_m$ and a width multiplied by λ. The potential can then be calculated by the formula given by Van Nostrand and Cook, [1966] for the isotropic dike case.

As an illustration of the effects of a zone which can become anisotropic, the apparent resistivities for a 15-km dipole-dipole array across the 1 km wide zone are plotted in Figure 9 for various values of the coefficient of anisotropy. To simplify the analysis the equivalent resistivity ρ_m of the zone is left equal to the initial resistivity ρ_D as the anisotropy is varied. The apparent resistivities ρ_a for the array are normalized by the apparent resistivity ρ_a^0 for the initial isotropic zone.

It can be seen from this figure that if the anisotropic zone is initially resistive, the apparent resistivity variations can be significant for small values of λ. For example, a zone of initial resistivity of 500 ohm m can produce a decrease in apparent resistivity of 10% if the coefficient of anisotropy becomes 1.3.

Very little analysis exists on the anisotropic behavior of rocks under stress. At low stress levels it might be reasonable to assume that the opening and closing of joint or fracture porosity could be responsible for the development of anisotropy. In the simplest case of a rock with a single set of parallel fractures, Parkhomenko [1967] gives a simple formula for ρ_t, the resistivity perpendicular, and ρ_ℓ, parallel, to the fractures:

$$\rho_t = \rho_r - (\phi_f \rho_r - \rho_f) \tag{10}$$

$$(1/\rho_\ell) = (\phi/\rho_f) + (1-\phi_f)/\rho_r \tag{11}$$

Here ρ_r is the resistivity of the nonfractured rock, presumably following a conventional Archie's law, and ρ_f is the resistivity of the fluid

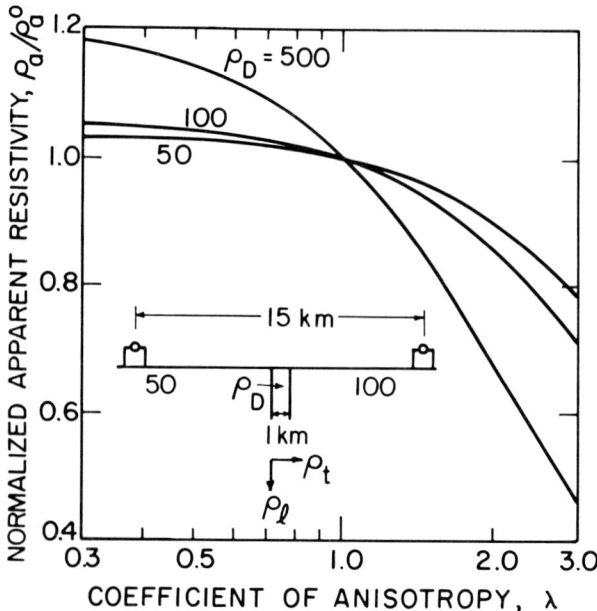

Fig. 9. Normalized apparent resistivity ρ_a/ρ^0 for a dipole-dipole array across an anisotropic zone. The apparent resistivity is normalized by the apparent resistivity ρ_a^0 calculated for the isotropic case ($\lambda = 1$).

in the fractures (and in the normal pores). The fracture porosity ϕ_f is not necessarily the same as the porosity in the unfractured rock.

We can apply this simple model to a fractured zone in the granite which has a bulk resistivity interpreted from the resistivity survey of 500 ohm m and a pore fluid resistivity of 1 ohm m. If we consider a joint porosity of 10^{-3}, the above expressions for ρ_t and ρ_f combine to yield a coefficient of anisotropy of 1.22.

This kind of modeling is illustrative of the results that could occur in favorable circumstances. The existence of such fracture zones aligning to produce a net anisotropy under stress is conjectural, and the final explanation will require much more analysis of the resistivity of rocks in situ.

Conclusions

Careful measurements of electrical resistivity using a large resistivity array show that accurate measurements of crustal resistivities can be made which may relate to temporal changes in tectonic strain. Variations in the resistivity of near-surface volumes of rock with dimensions comparable to those of the dipoles can affect the overall apparent resistivities if the rocks are more conductive than their surroundings. However, measurements of the resistivities at several of the dipole sites has shown that the shallow resistivities vary by less than 1% even during rainy seasons. Unfortunately, no large earthquakes have occurred within the array since high accuracy measurement began, but an excellent base line series of measurements has been established so that future variations associated with earthquakes will have a high degree of statistical reliability.

Acknowledgments. This research was supported by U.S. Geological Survey Grant 14-08-0001-G-15259.

We gratefully acknowledge the support of The Pacific Gas and Electric Co. for installation of line operated transmitters.

References

Anderson, D. L., and J. H. Whitcomb, The dilatancy-diffusion model of earthquake prediction, Proceedings of Conference on Tectonic Problems of the San Andreas Fault System, Stanford Univ. Publ. Univ. Ser. Geol. Sci., 13, 417-425, 1973.

Archie, G. E., The electrical resistivity log as an aid in determining some reservoir characteristics, Trans, AIME, 146, 54-61, 1942.

Barsukov, O. M., Relationship between the electrical resistivity of rocks and tectonic processes, (Izv. Acad. Sci. USSR Phys. Solid Earth, Engl. Transl., 1, 55-59, 1970.)

Barsukov, O. M., and O. N. Sorokin, Variations in apparent resistivity of rocks in the seismically active Garm region, (Izv. Acad. Sci. USSR Phys. Solid Earth, Engl. Transl., 10, 685-687, 1973.)

Bolt, B. A., and R. D. Miller, Catalogue of Earthquakes in Northern California and Adjoining Areas, 1 January 1910 - 31 December 1972, Seismographic Stations, University of California, Berkeley, 1975.

Brace, W. F., Resistivity of saturated crustal rocks to 40 km based on laboratory measurements, in The Structure and Physical Properties of the Earth's Crust, Geophys. Monogr. Ser., vol. 14, edited by J. G. Heacock, AGU, Washington, D.C., 1971.

Brace, W. F., Dilatancy-related electrical resistivity changes in rocks, Pure Appl. Geophys., 113, 207-217, 1975.

Brace, W. F., and A. S. Orange, Electrical resistivity changes in saturated rocks during fracture and frictional sliding, J. Geophys. Res., 73 (4), 1433-1445, 1968.

Brace, W. F., A. S. Orange, and T. R. Madden, The effect of pressure on the electrical resistivity of water-saturated crystalline rocks, J. Geophys. Res., 70 (22), 5669-5678, 1965.

Bufe, C. G., Indirect measurement of earth strain-electrical resistivity methods, Tech. Rep, ERL 256-ESL 28, Nat. Oceanic and Atmos. Admin., U. S. Dept. of Commer., 13-16, 1973.

Fatt, I., Effect of overburden and reservoir pressure on electric logging formation factor, Bull. Amer. Assoc. Petrol. Geol., 41 (11), 2456-2466, 1957.

Fitterman, D. V., Electrical resistivity and fault creep behavior along strike-slip fault systems, Ph.D. thesis, Mass. Inst. of Technol., Cambridge, 1974.

Fitterman, D. V., Theoretical resistivity variations along stressed strike-slip faults, J. Geophys. Res., 81(26), 4909-4915, 1976.

Hanks, T. C., Constraints on the dilatancy-diffusion model of the earthquake mechanism, J. Geophys. Res., 79, 3023-3025, 1974.

Keller, G. V., and F. C. Frischknecht, Electrical Methods in Geophysical Prospecting, Pergamon, New York, 1966.

Large, D. B., Electric potential near a spherical body in a conductive half-space, Geophysics, 36(4), 763-767, 1971.

Lee, L. S., A study of anisotropic models in the resistivity method, M. S. thesis, Univ. of Calif., Berkeley, 1976.

Mazzella, A., Deep resistivity study across the San Andreas fault zone, Ph.D. thesis, Univ. of Calif., Berkeley, 1976.

Mazzella, A., and H. F. Morrison, Electrical resistivity variations associated with earthquakes on the San Andreas fault, Science, 185 (4154), 855-857, 1974.

Morrison, H. F., 1975 technical progress report on U.S.G.S. grant no. 14-08-0001-G-124, Feb. 1, 1975.

Nur, A., M. L. Bell, and P. Talwani, Fluid flow and faulting, 1, A detailed study of the dilatancy mechanism and premonitory velocity changes, Proceedings of Conference on Tectonic Problems of the San Andreas Fault System, Stanford Univ. Publ. Univ. Ser. Geol. Sci., 13, 391-404, 1973.

Ostle, B., Statistics in Research, 2nd ed., Iowa State University Press, Ames, 1963.

Parkhomenko, E. I., Electrical Properties of Rocks, Monogr. in Geosci., Plenum, New York, 1967.

Rikitake, T., Dilatancy model and empirical formula for an earthquake area, Pure Appl. Geophys., 113, 141-147, 1975.

Scholz, C. H., L. R. Sykes, and Y. P. Aggarwal, Earthquake prediction: A physical basis, Science, 181(4102), 803-810, 1973.

Singh, S. K., and F. Sabina, Epicentral deformation based on the dilatancy-fluid diffusion model, Bull. Seis. Soc. Amer., 65(4), 845-854, 1975.

Van Nostrand, R. G., and K. L. Cook, Interpretation of resistivity data, U.S. Geol. Surv. Prof. Pap. 499, 1966.

Wilson, I. F., Geology of the San Benito Quadrangle, California, Calif. J. Mines Geol., 39(2), 183-270, 1943.

Yamazaki, Y., Electrical conductivity of strained rocks, 3, A resistivity variometer, Bull. Earthquake Res. Inst. Tokyo Univ., 45, 849-860, 1967.

Yamazaki, Y., Coseismic resistivity steps, Tectonophysics, 22, 159-171, 1974.

KINEMATICS AND DYNAMICS OF SLIDING IN SAW-CUT ROCK

H. M. J. Illfelder and H. F. Wang

Department of Geology and Geophysics, University of Wisconsin
Madison, Wisconsin 53706

Abstract. Stress and displacement were recorded at a 500-Hz or 1-kHz sampling rate during slip events on saw-cut granite and sandstone. At confining pressures between 3 and 15 megapascals (MPa), two types of slip events could be distinguished for both rock types on the basis of sharpness and size of the stress drop. The type A event is transitional between stable sliding and stick-slip and is attributed to the fact that the stiffness (spring constant) of the testing machine is lower than the decline of frictional force with displacement. The type B event is inherently unstable because it takes place by self-sustaining brittle fracture of asperities. A gradation of events between type A and type B exists. These observations are interpreted in terms of a general stiffness model applicable to both laboratory and earthquake mechanism studies.

Introduction

Friction behavior observed in the laboratory on faulted or saw-cut rock has also been interpreted to occur on shallow faults in the earth's crust [Brace and Byerlee, 1966; Byerlee and Brace, 1968; Scholz et al., 1972]. Stick-slip instability is analogous to a fault that is locked until the shear stress reaches a level sufficient to create a sudden slip event, and stable frictional sliding is analogous to fault creep [Byerlee and Brace, 1968].

Stable sliding occurs at very low normal stresses with a gradual transition to stick-slip as the normal stress increases [Scholz et al., 1972]. As normal stress increases and slip events occur, the experimenter has a qualitative idea that the nature of the instability changes. Initially, the events are barely noticeable except for slightly more rapid pen motion on an XY or strip chart recorder. Eventually, the events become sharply audible. The purpose of our experiments is to describe these observations more accurately by high-speed recording of the vertical stress and displacement across saw-cut rock specimens. We interpret the results in terms of a general stiffness model for faulting applicable to both laboratory and earthquake mechanism studies [Byerlee, 1970; Nur and Schultz, 1973].

Experimental Procedure

Observations of fault stability depend upon the characteristics of the loading system as well as the intrinsic frictional properties of the fault [Walsh, 1971; Hudson et al., 1972; Nur and Schultz, 1973].

TABLE 1. Comparison of Preliminary and Main Series of Experiments

	Preliminary	Main
Number of recorded events	40	500
Rock specimens	Barre granite	Barre granite or Jordan sandstone
Sample diameter, cm	2.54	5.08
Sample length, cm	7.3	12.7
Sample jacket material	Tygon	Gum rubber
Saw-cut smoothness	150-grit	240-grit
Load frame capacity, N	0.5×10^6	1.2×10^6
Loading system stiffness, N/m	9×10^7	16×10^7
Load cell capacity, N	0.5×10^6	1.1×10^6 or 0.1×10^6
Maximum travel of displacement transducer, cm	±0.64	±2.54
Data acquisition	FM tape recorder or digital oscilloscope	PDP-11/03

Two series of experiments were done in which different loading systems, pressure vessels, and data acquisition methods were used. The general procedure is described below, and specific details summarizing each experimental series are given in Table 1.

Loading

The ends of cylindrical specimens are ground parallel within 0.005 cm and are saw cut at an angle of 30° to the long axis. The saw-cut faces are lapped to produce a uniform initial sliding surface. The samples are dried for 1 day at 80°C in a vacuum oven before being jacketed in flexible tubing. Hydrostatic confining pressure is applied before superimposing a vertical differential load. The servo-controlled loading system is programed to increase the axial load at a rate that is small in relation to the duration of stress drop events.

Signal Conditioning

A strain gage is the sensor in the load cell, and a magnetic core passing through a coil winding is the sensor in the displacement transducer. Each variable is measured in the vertical direction as a voltage with a known calibration factor. The displacement transducer is excited by a high-frequency oscillator which introduces noise of several kilohertz. The displacement transducer contains an internal low-pass filter with a half-power point of 115 Hz.

Data Acquisition

In the preliminary series of experiments the primary recording technique was to accumulate the load and displacement outputs for the entire experiment with a tape recorder. After an experiment the tape output is monitored with an XY recorder to locate slip events. The load and displacement signals are passed through a low-pass filter with a half-power point of 300 Hz before being digitized at 1 kHz and saved on a mass storage device. A secondary recording technique was to use the midsignal trigger feature of a digital oscilloscope to directly capture slip events. The digital oscilloscope is useful for reconnaissance because it gives an immediate visual picture of an event.

In the main series of experiments the load cell and displacement transducer outputs are low-pass-filtered and then are continuously digitized at a rate of 500 Hz by a PDP-11/03 computer. The data are also monitored on an XY recorder. If an event is detected, either because it is audible or because the XY recorder shows sudden motion, then the experimenter can save the previous 1.5 s of information by manually striking a keyboard character. About 25 s are needed to transmit the PDP memory contents to a larger computer with a mass storage device. Although the PDP is programed to have two sets of buffers, not every one of several closely spaced events can be saved. Events can also be missed by badly timed interruptions. Overall, about two thirds of the actual events were saved on disk. For any given event the data collected by this system are equivalent to the data digitized from the earlier tape recorder method.

Experimental Results

Aluminum Specimen

In the two series of experiments load and displacement are measured by transducers external to the triaxial cell. We present evidence that the recorded time behavior of load and displacement applies to the saw-cut sample within the cell, even though the true sample load and displacement are lower than the measured sample load and displacement because of seal friction and strain in the loading frame and ram. Our concern is that high-frequency components of the slip events may be modified mechanically by O ring seals and the mass of the loading piston and electronically by the low-pass filters.

To investigate these effects, an aluminum sample without a saw cut was substituted for a rock sample in the triaxial cell used for the main series of experiments. First, a confining pressure of 7 megapascals (MPa) was applied to the aluminum sample, and then it was rapidly loaded and unloaded to simulate fast stress changes during a slip event in saw-cut rock. Because aluminum stock is a high-Q material, we expect that load and displacement will be in phase. The closely matching shapes of the load and displacement curves of Figure 1 show that there is no significant time phase error introduced from externally located load and displacement transducers.

The displacement recorded during a stress drop event includes not only the sliding along the saw cut of a rock specimen but also the relative lengthening of the loading system. We calculate from Figure 1, using the known Young's modulus for aluminum, that this loading system deflection is about twice the noise level in the displacement recordings.

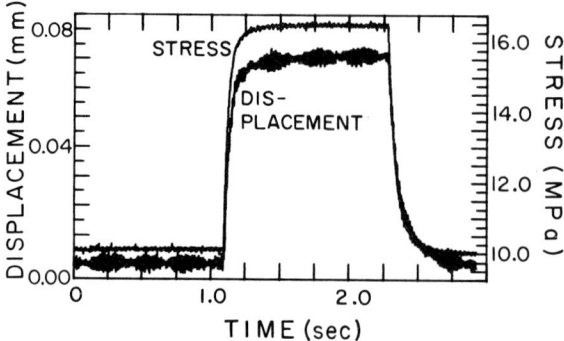

Fig. 1. Axial displacement and axial stress versus time for an aluminum sample at a confining pressure of 7 MPa. The axial stress was programed for a square wave variation.

Event Types

The slip events range from fairly gradual to very abrupt and sharp (Figures 2-7). Although a spectrum of types exists, we find it convenient to call the more gradual stress drop events type A and the sharper stress drop events type B. We did not detect differences in slip behavior between the Barre granite and Jordan sandstone specimens.

Figures 2-6 represent slip events during a single run as the axial stress is increased. The confining pressure increases from 5.5 to 6.5 MPa as the piston displaces confining fluid. During plotting the displacement scale is reset to a new zero for each event. We classify Figures 2 and 3 as type A events, whereas Figures 4, 5, and 6 are type B events. Figure 7 comes from another run and represents a sharper type B event than that in Figure 6. In Figures 5 and 6, precursory sliding can be identified by a slight upward slope of the displacement recording. There is also a downward curvature in the stress recording just prior

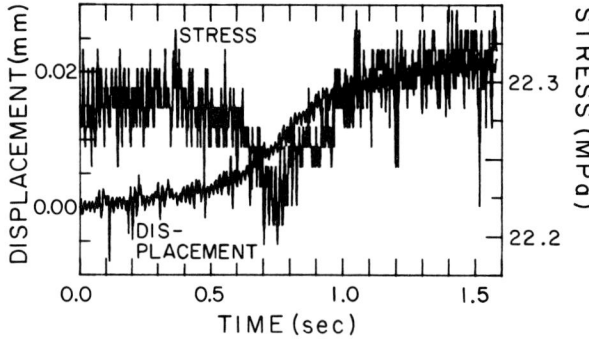

Fig. 2. Axial displacement and axial stress versus time for a type A event in saw-cut Barre granite. The event begins at about 0.4 s. The noise level is high because the event is small. The confining pressure is 5.5 MPa.

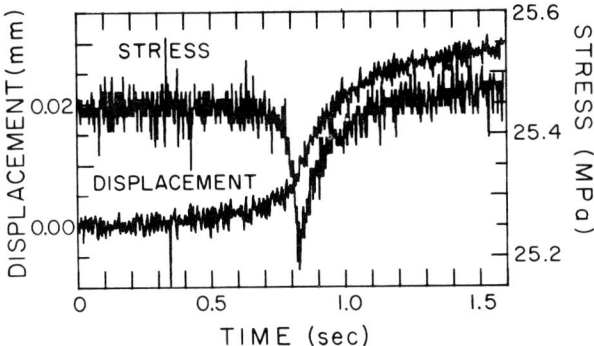

Fig. 3. Axial displacement and axial stress versus time for a type A event. The event begins at about 0.7 s. The confining pressure is 5.7 MPa.

to the main drop, indicating a somewhat gradual approach to the instability that is not seen in Figure 7.

In the type A event the stress drop occurs gradually, and a fair fraction of the total displacement occurs before the stress minimum. In the type B event, most of the stress drop is quite sharp, so that the major portion of the displacement commences with the stress minimum. Type B events are sharply audible when they occur. Type A events are quieter and may not be heard at all but are observed by the more rapid pen motion on an XY recorder. Stress drops in the main series of tests for type A events are 0.6 MPa or smaller; stress drops for events transitional between type A and type B range between 0.6 and 1.5 MPa; stress drops for type B events overlap these transitional events with a lower limit of 0.6 MPa but are usually in the 1.5- to 3.0-MPa range. Thus the sharper character of type B events correlates with larger stress drops.

How much of the total displacement is associated with the initial stress drop and how much with the servo-controlled stress recovery is difficult to assess. The recovery portion of an event is a complex mix of rock and testing machine characteristics. Because of the fast

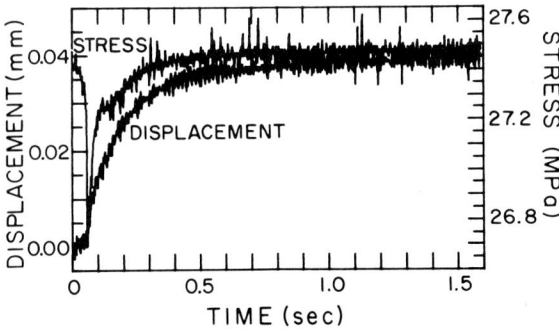

Fig. 4. Axial displacement and axial stress versus time for an early type B event. The confining pressure is 5.9 MPa.

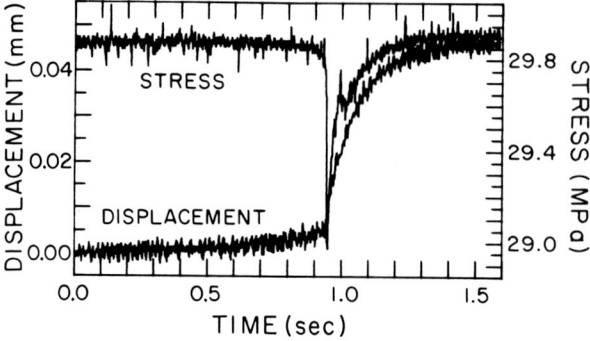

Fig. 5. Axial displacement and axial stress versus time for a type B event at 0.9 s. Note the precursory sliding and the secondary event at 0.97 s. The confining pressure is 6.3 MPa.

response of servo load (Figure 1) we believe that the slow return of stress is a measure of continued fault weakness until the displacement achieves a new position of stability. The total stress drop and total displacement are roughly proportional to each other (Figure 8).

Surface History

A fine powder of fault gouge is found on the saw-cut faces after a test. Occasionally, striations are also found. Their occurrence corresponds to the largest stress drop events.

Some experiments were repeated at the same initial confining pressure without resurfacing the saw-cut faces in order to examine the effect of gouge buildup on event type. The initial event of the repeat experiment could be the same as that in the preceding run, or it could be of the opposite type. The presence of fine gouge does not affect event type.

Confining Pressure

Confining pressure affects the type of event produced. Only type B events are observed for confining pressures of 7 MPa or higher, independent of the previous sliding history of the rock specimen. Both type A and type B events can occur at confining pressures below 7 MPa, but the likelihood of type A events increases at the lower end of confining pressures used. Initial type A events grade to type B within a run because confining pressure increases due to piston displacement as the axial load increases. If the initial event is type B, all subsequent events in the run are also type B.

Loading System

The overall stiffness of a loading system is the ratio of change in applied force to the displacement in the system [e.g., Hudson et al., 1972]. We obtained the elastic displacement for an applied force by summing frame deflection, load cell strain, and piston strain. The

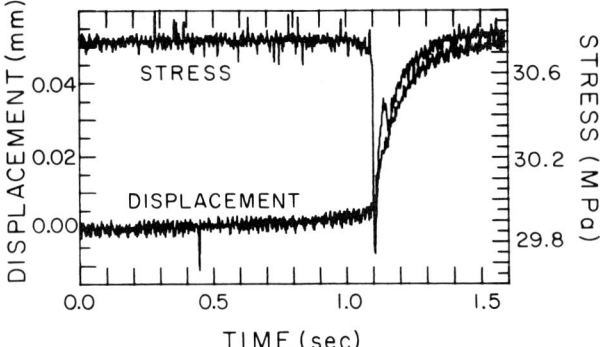

Fig. 6. Axial displacement and axial stress versus time for a type B event. Note the secondary event at 1.15 s. Ignore the glitch in the displacement at 0.45 s. The confining pressure is 6.5 MPa.

frame used in the preliminary series of experiments has a stiffness which is smaller than that used in the main series of experiments (Table 1). For type A events, displacements and stress drops are several times larger in the preliminary series of experiments. For type B events, displacements are several times larger in the preliminary series, but stress drops are only slightly larger.

Discussion

Stable Sliding to Stick-Slip Transition

Scholz et al. [1972] and Johnson [1975] conducted frictional sliding tests on granite and dunite at low normal stresses comparable to those in our tests. They used large rectangular specimens with a saw cut inclined at 30° across the major faces. The samples were loaded biaxially by hydraulic rams at right angles. The vertical ram was advanced by pumping in fluid at a constant rate. In their experiments, Scholz et al. [1972] and Johnson [1975] identified what they call episodic stable sliding: friction behavior transitional between continuous stable sliding and stick-slip. Episodic stable sliding takes place at a rate much slower than that of stick-slip, is inaudible, and can be readily followed by a strip chart recorder.

We believe that a sequence of episodic slides may be combined and appear as a single type A event such as that of Figure 2. The very gradual increase of displacement in Figure 2 suggests that the response of a strip chart recorder may have been sufficient to produce our record. The slip in Figure 2 is about 20 microns. Displacement measured by Scholz et al. [1972] for a sequence of episodic slides is a few tens of microns, the displacement for each episode being a few microns. The stress recovery is much slower in the Scholz et al. [1972] biaxial machine than in our servo-controlled machine. The servo-controlled recovery is so fast that the sliding of several episodes may be superimposed and not separated in time to produce a type A event.

At 0.97 s in Figure 5 and at 1.15 s in Figure 6, just after the large type B stick-slip in each figure, we note a small event with a stress

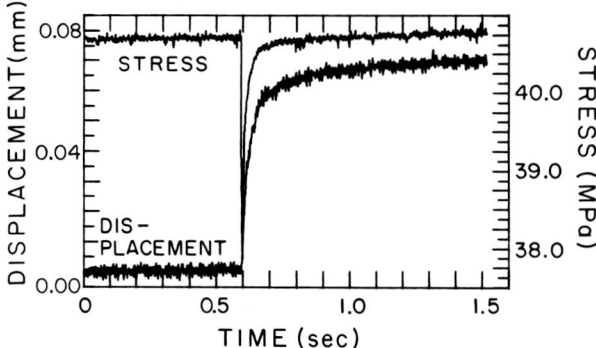

Fig. 7. Axial displacement and axial stress versus time for a very sharp type B event. The confining pressure is 10.2 MPa.

drop of 0.1-0.2 MPa. This small event appears to be type A and may also represent a complete sequence of episodic sliding. This interpretation can be correlated with the Scholz et al. [1972] observation that episodic sliding follows initial stick-slip events after sufficient reloading has occurred.

Sliding Mechanism

From experiments on rough faulted surfaces, Byerlee [1967] attributed sliding at very low normal stresses to the surfaces lifting over the interlocking irregularities until the normal stress reaches a value at which it is easier for the surfaces to slide by breaking through steeply inclined asperities. We believe that this description can also explain the gradual transition between type A and type B events. At normal stresses in the region of type A events, slight instabilities occur as the force required for the surfaces to climb over an irregularity exceeds the friction as the surfaces slide down. The sharper type B events represent brittle fracture of asperities. In the transition between the two sliding mechanisms, both effects are probably intermixed. Thus some precursory sliding takes place in the type B events of Figures 5 and 6 but does not take place in Figure 7.

Stiffness Model

Byerlee [1970] interpreted stick-slip instability to be due to fluctuations in friction force with displacement along a fault. In his model, one end of a spring is connected to a rider on a horizontal flat surface. A tangential force is applied to the rider by pulling the other end of the spring at a constant velocity. Nur and Schultz [1973] called Byerlee's model the stiffness model because the stiffness of the loading system (spring) relative to that of the fault (rider) can explain stable creep, stick-slip, fault locking, and their relationship to one another.

When the slope of the rising portion of the force-displacement curve (Figure 9) is positive, the fault slides in a stable manner if the loading force exceeds the friction force. When the resisting friction

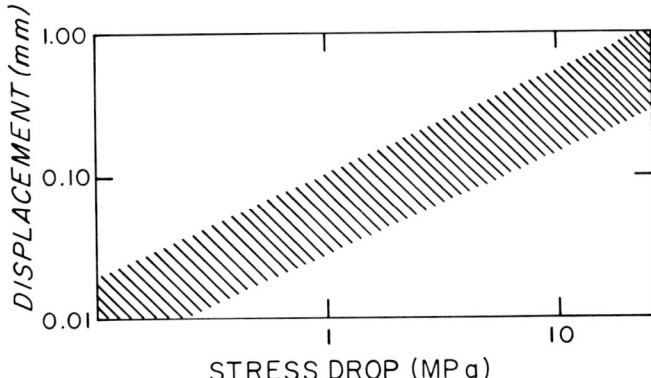

Fig. 8. Total axial displacement versus total axial stress drop for events in the main test series. The band represents about 90% of the observed events.

force drops more rapidly with displacement than the stiffness (spring constant) of the loading system, an instability occurs because of a force imbalance. In Figure 9 the effect of loading system stiffness is compared at two relative maxima of the fault's friction-displacement curve. At the left-hand instability the specimen would be unstable in a soft loading system (shallowly dipping dashed line), whereas it would be stable in a stiff system (steeply dipping dashed line). At the right-hand instability the descending portion of the fault's friction-displacement curve has a positive slope, and an instability exists for even an infinitely stiff loading system.

The two types of instabilities shown in Figure 9 have been found in studies of the postfailure region in the fracturing of intact rock [Wawersik and Fairhurst, 1970]. The left-hand relative maximum corresponds to class I failure, and the right-hand relative maximum corresponds to class II failure. In the postfailure region, energy must still be supplied to class I rocks to continue fracture propagation, whereas fracture propagation is self-sustaining in class II rocks. Class II failure is associated with more brittle rocks.

We believe that type A events are analogous to class I failure and that type B events are analogous to class II failure. Type A events may be controllable as stable sliding in a machine of sufficient stiffness, but type B events are inherently unstable because the brittle fracture of asperities is self-sustaining once a critical load level is reached.

The smaller displacements and stress drops for type A events in the stiffer testing machine can be understood schematically from the left-hand relative maximum of Figure 9. The force imbalance is restored to equilibrium sooner for a stiffer machine. For type B events, however, the right-hand relative maximum applies. Stress drop is less dependent upon machine stiffness, but displacement is smaller for higher stiffnesses. We described this observation between our two loading systems in the experimental results section. Also, Byerlee and Brace [1968] found that stress drop was independent of machine stiffness for stick-slip occurring at a confining pressure of 185 MPa.

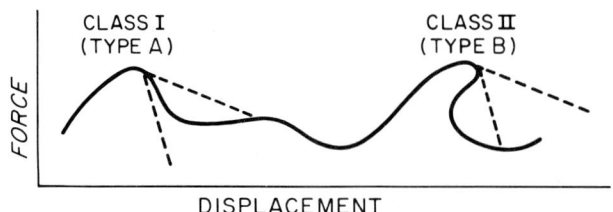

Fig. 9. Schematic force-displacement curve illustrating type A (class I) and type B (class II) behavior. Solid curve represents frictional force versus displacement. Dashed lines represent force-displacement behavior of loading systems with different stiffnesses.

Tectonic Loading

There are obvious differences between smooth rock surfaces at low normal stresses and shallow faults at greater normal stresses in the earth. Nonetheless, Scholz et al. [1972] point out that much of the creeping zone of the San Andreas fault appears to be near the transition between stable sliding and stick-slip. In this sense the laboratory experiments reflect in a general way frictional behavior of sliding surfaces.

The earth has a low stiffness relative to that of laboratory testing machines [Walsh, 1971]. Observed behavior on shallow faults will include the combined effects of the earth as a loading machine in applying tectonic stresses. Self-sustaining instabilities (type B) will exist regardless of loading system stiffness, although we might expect an event to be accentuated by a low-stiffness system. But some earthquakes may be type A and are instabilities because of the low stiffness of the earth. For this type of event the distinction between earth stiffness and fault stiffness should be kept in mind in interpreting temporal seismicity patterns.

Conclusion

The stiffness model provides a schematic framework in which to interpret our observations of type A and type B events. Type A events represent an instability only insofar as the loading system is not stiff enough to prevent the occurrence of a force imbalance. Sliding in type A events probably takes place because the surfaces ride over asperities. The sharper and larger type B events represent an inherent instability independent of the stiffness of the loading system. At higher confining pressures, only type B events occur because sliding takes place by brittle fracturing of asperities. More generally, the stiffness model provides a means for understanding observations on tectonic faults in terms of laboratory results. The relation of loading system stiffness to fault stiffness can be especially important for observed friction behavior in a transition region between continuous stable sliding and stick-slip.

Acknowledgments. Our preliminary experiments were performed mostly with borrowed equipment. We thank B. I. Sandor, B. C. Haimson,

C. S. Clay, and R. P. Meyer for the loans. The latter three persons also provided useful suggestions during the course of the experiments. We also benefited from suggestions by W. F. Brace, M. Feves, J. G. Heacock, H. Spetzler, and H. Mizutani. We received much help in the laboratory from D. Kositzke and E. Roeloffs. Financial support came from research funds administered by the Graduate School of the University of Wisconsin and by National Science Foundation grants DES75-04290 and EAR76-13363.

References

Brace, W. F., and J. D. Byerlee, Stick-slip as a mechanism for earthquakes, Science, 153, 990-992, 1966.

Byerlee, J. D., Theory of friction based on brittle fracture, J. Appl. Phys., 38, 2928-2934, 1967.

Byerlee, J. D., The mechanics of stick-slip, Tectonophysics, 9, 475-486, 1970.

Byerlee, J. D., and W. F. Brace, Stick-slip, stable sliding, and earthquakes - Effect of rock type, pressure, strain rate, and stiffness, J. Geophys. Res., 73, 6031-6037, 1968.

Hudson, J. A., S. L. Crouch, and C. Fairhurst, Soft, stiff and servo-controlled testing machines: a review with reference to rock failure, Eng. Geol., 6, 155-189, 1972.

Johnson, T. L., A comparison of frictional sliding on granite and dunite surfaces, J. Geophys. Res., 80, 2600-2605, 1975.

Nur, A., and P. Schultz, Fluid flow and faulting, 2, A stiffness model for seismicity, in Proceedings of the Conference on Tectonic Problems of the San Andreas Fault System, edited by R. L. Kovach and A. Nur, pp. 405-416, Stanford University Press, Palo Alto, Calif., 1973.

Scholz, C., P. Molnar, and T. Johnson, Detailed studies of frictional sliding of granite and implications for the earthquake mechanism, J. Geophys. Res., 77, 6392-6406, 1972.

Walsh, J. B., Stiffness in faulting and in friction experiments, J. Geophys. Res., 76, 8597-8598, 1971.

Wawersik, W. R., and C. Fairhurst, A study of brittle rock fracture in laboratory compression experiments, Int. J. Rock Mech. Min. Sci., 7, 561-575, 1970.

HEAT FLOW IN THE UNITED STATES AND THE THERMAL REGIME OF THE CRUST

Arthur H. Lachenbruch and J. H. Sass

U.S. Geological Survey
Menlo Park, California 94025

Abstract. A contour map of heat flow based on 625 observations now available in the conterminous United States shows new detail. Subprovinces of exceptionally high heat flow (>2.5 HFU (1 HFU = 10^{-6} cal/cm^2 s)) in the western states are beginning to emerge as regional features, but their boundaries are still largely unknown. The 'Battle Mountain High,' previously described in north central Nevada, probably extends northeastward to Utah and Idaho and westward almost to California. With the eastern Snake River Plain, a region that probably has large convective loss, it could form a zone of exceptionally high heat loss that extends almost continuously for 1000 km from the vicinity of Steamboat Springs near Reno, Nevada, to Yellowstone Park in Wyoming and possibly northward into the Idaho batholith. A sinuous high heat flow subprovince of comparable length is emerging in the Rio Grande Trough in New Mexico and southern Colorado. The linear relation between surface heat flow and radioactive heat production, so successful in the Sierra Nevada and eastern United States provinces, does not apply in the Basin and Range province. There the variations in heat flow caused by hydrothermal and magmatic convection are probably greater by a factor of 3 or 4 than those caused by crustal radioactivity, and heat flux into the lower crust is not uniform; it is probably controlled by the mass flux of intruding magma. Regional variations in this mass flux, probably associated with crustal spreading, can account for the high heat flow subprovinces, and more local anomalies and silicic volcanic centers as well. Although convective processes cause a large dispersion of heat flow in the Basin and Range province, modal values of reduced heat flow can be used to construct generalized crustal temperature profiles for comparison with profiles for more stable areas and with melting relations for crustal rocks. Theoretical profiles are consistent with the widespread magmatic manifestations observed in the Basin and Range province. Laterally extensive silicic partial melts are possible at midcrustal levels in 'hot' subprovinces like the Battle Mountain High. Effects of hydrologic convection (whether driven by thermal density differences or regional piezometric conditions) are important to an understanding of regional heat flow, especially in tectonically active areas. The 'Eureka Low,' a conspicuous subprovince ($\sim 3 \times 10^4$ km^2) with anomalously low heat flow in southern Nevada, is probably caused by interbasin flow in deep aquifers fed by downward percolation of a small fraction of the annual precipitation. Heat flow observations in such areas provide useful information on regional hydrologic patterns.

Introduction

The net outward flow of heat across the earth's surface is a fundamental term in the energy balance of processes within the earth. Consequently, measurements of this quantity not only contain information about the state of the earth but also about processes associated with the generation, transport, and storage of heat within it. The number of heat flow data has increased tenfold in the last decade, and this has led to a more complete understanding of regional variation of heat flow and its causes. The observations often can be interpreted in terms of very simple, internally consistent models that give useful insights into processes of the lithosphere beneath both oceans and continents [see e.g., Birch et al., 1968; Roy et al., 1968a, b, 1972; Lachenbruch, 1970; Sclater and Francheteau, 1970]. This paper summarizes the data available on regional heat flow in the conterminous United States and discusses some of their implications for the thermal regime and processes within the crust.

A Heat Flow Map

The heat flow data available to us as of June 1976 are presented as coded symbols in Figure 1, and a contoured interpretation of them is shown as Figure 2. The heat flow unit (HFU) is 10^{-6} cal/cm^2 s (41.8 mW/m^2). For continuity, we have included in Figure 1 heat flows measured in the Pacific Ocean near the coast; however, we have made no attempt to contour them. The 625 points from the conterminous United States include published results from many laboratories and 130 or so recent determinations in preparation for formal publication by the U.S. Geological Survey (USGS) (see Diment et al. [1975] and Sass et al. [1976a] for a complete bibliography of published values). Most of the published data are supported by tabulations of thermal gradient and conductivity, but a few of the points have been taken only from published graphs or maps. The quality of the data is quite variable, as many of the determinations were made in holes drilled for purposes other than heat flow measurement; some were made in shallow holes drilled primarily for geothermal energy prospecting, and some in holes (usually in crystalline rock) drilled for scientific studies of regional heat flow. The best and the worst determinations are generally from holes drilled for other purposes; the best because such holes may be drilled to much greater depths than can be justified by limited research budgets and the worst because such holes are sometimes sited poorly for heat flow measurements and they may be sampled inadequately to characterize thermal conductivity. Few determinations have been made in holes drilled for petroleum exploration because of the difficulty of obtaining adequate conductivity samples and undisturbed temperature measurements. In prospecting for geothermal energy, large local anomalies, often in surficial sediments, are the targets, and substantial uncertainties can be tolerated. Such holes, commonly drilled to depths of about 50 m, can, however, give valuable heat flows under favorable hydrologic and topographic conditions. Even in holes drilled for regional heat flow in crystalline rock, depths greater than 250 m rarely can be justified, and uncertainties regarding the regional significance of an individual determination can be substantial.

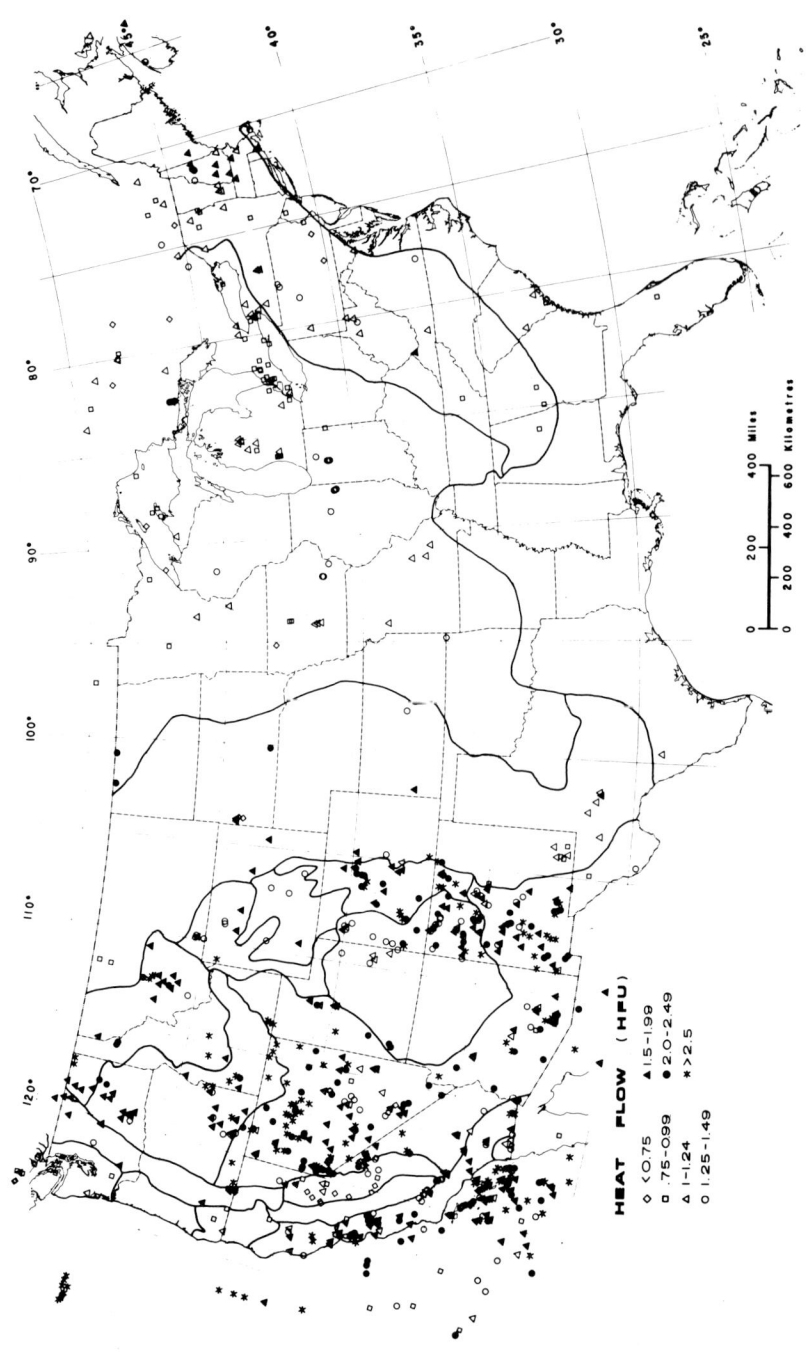

Fig. 1. Observed heat flow in the conterminous United States and some peripheral regions. Physiographic boundaries generalized from Fenneman [1946]. (1 HFU = 10^{-6} cal/cm^2 s = 41.8 mW/m^2).

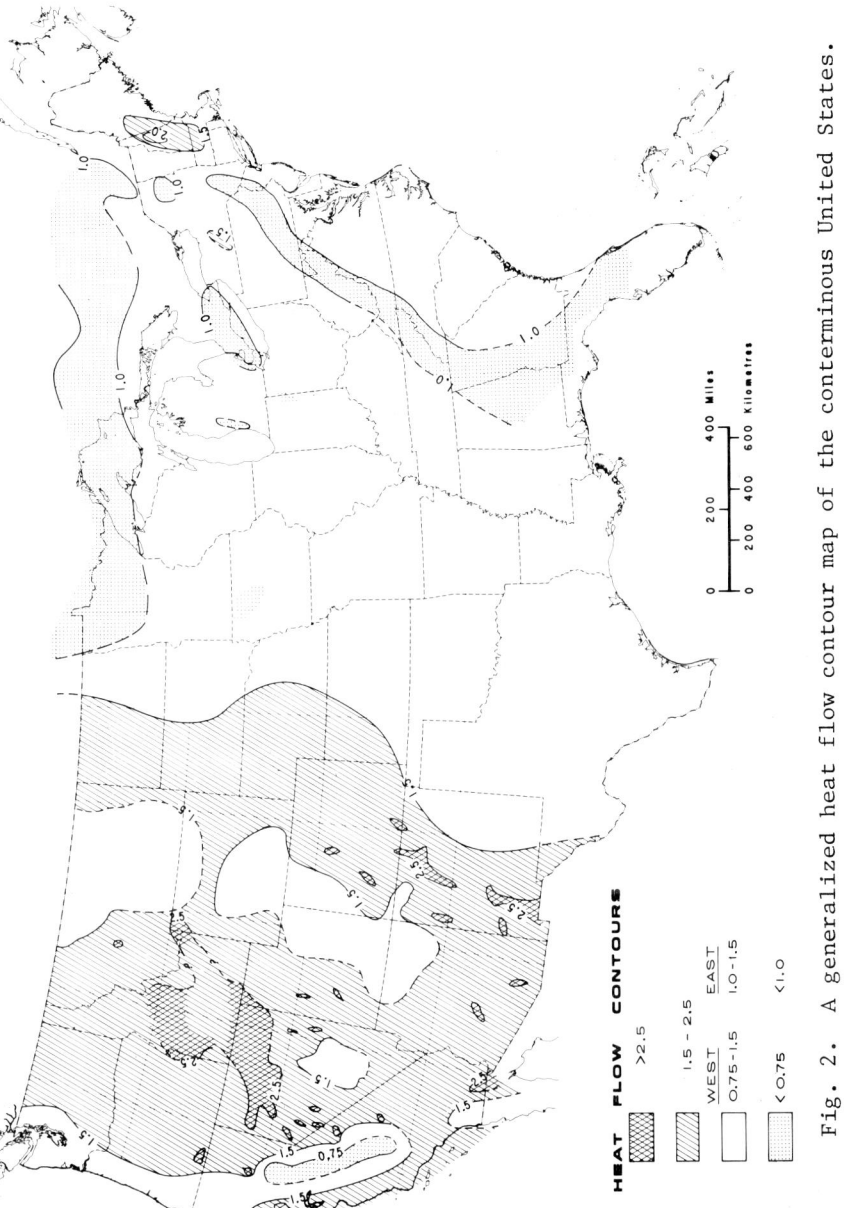

Fig. 2. A generalized heat flow contour map of the conterminous United States.

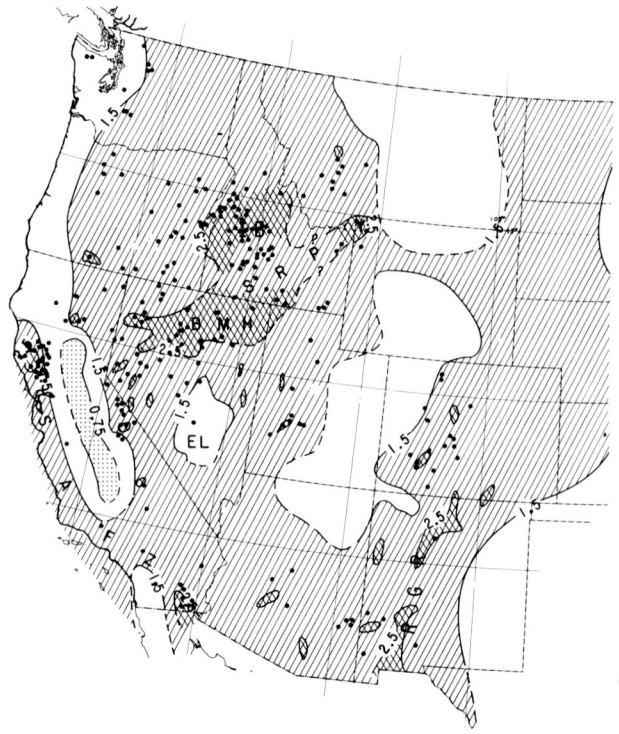

Fig. 3. Regional heat flow and distribution of hydrothermal systems. Dots show locations of hydrothermal systems in the conterminous United States with estimated reservoir temperatures greater than 90°C [Renner et al., 1975]. Abbreviations are BMH for Battle Mountain High, EL for Eureka Low, IB for Idaho batholith, SRP for eastern and central Snake River Plain, Y for Yellowstone thermal area, RGR for Rio Grande Rift, and SAFZ for San Andreas Fault zone.

In studies of regional heat flow it has been customary in the past to avoid regions of hot springs because of their local complexity. However, current interest in volcanic processes and the origin of geothermal energy resources requires that these hot spring areas be understood in relation to their regional thermal and tectonic settings. Figure 3 shows locations of the hotter known hydrothermal systems in the United States. Extending regional heat flow studies into these areas poses problems; the conductive flux at the surface can vary from zero to several hundred heat flow units over distances of a few kilometers, and substantial amounts of heat may be discharged convectively by lateral underflow in shallow aquifers into streams and lakes or at the surface by springs and fumaroles. Under these conditions there is some question about what we should define and map as 'heat flow.' In many regions we cannot feel confident, without hydrologic information, that heat transport in the upper few kilometers is exclusively conductive. However, hydrologic details are generally unknown; the heat flows selected for presentation in Figure 1 represent the upward conductive flux determined

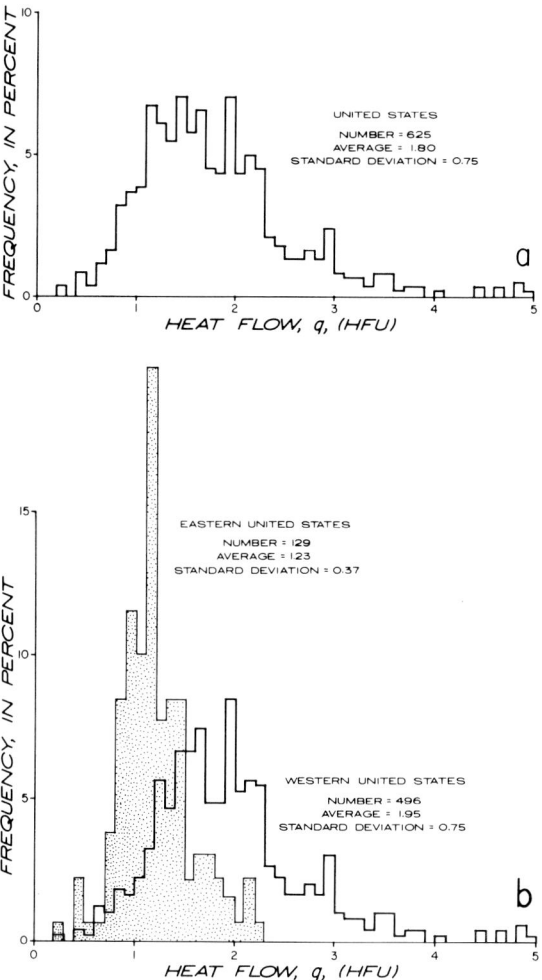

Fig. 4. Histograms of heat flow (a) for the conterminous United States as a single population and (b) for the portion east of the Great Plains (stippled) and the remainder of the conterminous United States (unstippled), treated separately. (For location of the Great Plains, see Figures 1 and 14.)

in holes (usually to depths of at least 100 m) at all sites not obviously disturbed by local water movements. Figure 2 indicates the generalized distribution of heat flow we should expect for observations under these conditions. The data are summarized in histograms in Figure 4.

The gross features shown in earlier regional heat flow maps [Roy et al., 1972; Diment et al., 1972; Blackwell, 1971; Sass et al., 1971; Diment et al., 1975] persist in Figure 2. These are a generally low-to-normal heat flow in the eastern United States (1.5 HFU is approximately the world average) and a generally high heat flow in the west, with zones of lower heat flow in the Colorado Plateau and near the Pacific Coast and very low values in the Sierra Nevada (for identification of the provinces, see Figure 14). The contours in the east are essentially unchanged from the map of Diment et al. [1972]. However, in the west, more information is now available from the work of Combs and Simmons [1973] in the Great Plains, Reiter et al. [1975] and Decker and Smithson [1975] in the Rio Grande Trough and Rocky Mountains, and by D. D. Blackwell, R. G. Bowen, and their associates [Blackwell, 1974; Bowen, 1973; Bowen et al., 1976; Brott et al., 1976] in the Pacific Northwest. Many new data from the USGS, largely in Nevada and the Pacific states, are also available [Lachenbruch and Sass, 1973; Diment et al., 1975].

In the map by Sass et al. [1971] a subprovince of exceptionally high heat flow, the 'Battle Mountain High,' was identified in the northern Great Basin, and one of normal and low heat flow, the 'Eureka Low,' was found in the southern Great Basin. The subsequent measurements tend to confirm the existence of both of these features. It now appears that the Battle Mountain High may be considerably larger than originally indicated, extending across northeastern Nevada at least to the Idaho and Utah borders. It is separated by the Snake River Plain from other regions of exceptionally high heat flow and hydrothermal activity in the Yellowstone, Wyoming, area to the northeast and the Idaho batholith to the north (Figure 3) [Blackwell, 1971; Urban and Diment, 1975; Morgan et al., 1977]. Recent volcanism and hot springs suggest that much of the Snake River Plain (particularly in the east) might have exceptionally large heat loss, although the surface heat flow is complicated by hydrologic conditions [Brott et al., 1976]. Hence a region of exceptionally high heat loss might extend northeastward almost continuously for 1000 km from the vicinity of Steamboat Springs near Reno, Nevada, to Yellowstone Park in Wyoming (see Figure 3). Another feature of the new map is the accumulating evidence for high heat flow in western California throughout a broad band that encloses the San Andreas Fault system [Lachenbruch and Sass, 1973].

The control is still poor in the western Colorado Plateau and eastern Great Basin (Figures 1 and 2); the boundaries drawn there are influenced by the distribution of hot springs and other geophysical data. Although heat flow data are accumulating in some volcanic regions such as parts of the Pacific Northwest [Bowen et al., 1976; Sass et al., 1976b; Brott et al., 1976], it is difficult to evaluate the regional significance of some of them because the rocks are often very permeable with temperatures influenced by hydrologic convection. In the Great Plains and much of the central and eastern United States the control is so poor that contours are rather arbitrarily drawn and a few new measurements could change them over large areas. Work is continuing at heat flow laboratories across the country, and better representations of the geographic distribution of heat flow are expected in the near future; good discussions of the major regional features have already been presented [Roy et al., 1972; Blackwell, 1971; Sass et al., 1971]. Therefore in this paper we shall focus on the general interpretation of heat flow within provinces that seem to have distinct regimes, rather than on the significance of the geographic distribution of the provinces themselves. In particular, we shall discuss the Sierra Nevada, the stable eastern and central United States, and the Basin and Range with its emerging subprovinces. The first represents the coldest crust in the United

States, and the Basin and Range subprovinces are probably among the hottest. Much of the discussion is general and applies to regions outside of these provinces where the heat flow is not as well known. In the San Andreas Fault zone of western California the crustal thermal regime may be unique; it has been discussed elsewhere [Lachenbruch and Sass, 1973; Brune et al., 1969; Henyey and Wasserburg, 1971].

In the next section we present some general background material as context for the discussions to follow. We hope it will make the paper more useful to readers not specialized in the study of terrestrial heat flow.

The symbols most frequently used in the text are as follows:

Θ temperature, °C;
t time, s;
K thermal conductivity, cal/°C s cm;
ρ, ρ' density of static and moving material, respectively, g/cm^3;
c, c' heat capacity of static and moving material, respectively, cal/g °C;
$\alpha = K/\rho c$, thermal diffusivity, cm^2/s;
q vertical conductive heat flux, HFU (10^{-6} cal/cm^2 s);
q^* intercept value from heat flow–heat production curve, HFU;
q_c vertical combined heat flux (convective plus conductive), HFU;
q_r reduced heat flow, HFU;
A heat generation, HGU (10^{-13} cal/cm^3 s);
A_o radioactive heat production in surface rock, HGU;
v vertical (seepage) velocity or volume flux of water or magma, cm^3/cm^2 s;
$s = k/\rho'c'v$, characteristic length for groundwater convection, cm;
k permeability, cm^2;
$\tau(z)$ conductive time constant for distance z;
$\ell(t)$ conduction length for time t;
D characteristic depth for radioactivity or slope of heat flow–heat production curve, km;
Γ thermal gradient for pure conduction, °C/km;
z depth;
h depth of circulation in hydrothermal system;
H depth to top of magma
Θ_m magma temperature, °C.

Some General Considerations and Rules of Thumb

General

In discussing the geothermal regime it is useful to refer to the following general equation which relates the temperature and the processes that generate, transport, and store heat in the crust.

$$-\nabla \cdot \hat{q} \equiv \nabla \cdot (K\nabla\Theta) \tag{1a}$$

$$-\nabla \cdot \hat{q} = -A + \rho'c'\hat{v} \cdot \nabla\Theta + \rho c \frac{\partial \Theta}{\partial t} \tag{1b}$$

Here $\hat{q}$ is the conductive flux vector, Θ is the temperature, and K is the thermal conductivity. The rate of heat generation per unit volume is denoted by A; it could represent the effects of radioactive decay, frictional heating, phase change, or chemical reaction. The values ρ and c are the density and heat capacity of material at any point, and ρ' and c' are the corresponding properties for material (usually water or magma) moving with velocity $\hat{v}$. If the movement is through pores or fractures in a fixed framework, $\hat{v}$ represents the volume flux ('seepage velocity') and not particle velocity. In general, all of the parameters in (1), including $\hat{v}$, are functions of x, y, and z, and some can have significant dependence on temperature or pressure.

Although three-dimensional effects must be kept in mind, useful simple interpretations generally involve quasi-one-dimensional models in which all of the quantities in (1) vary only with depth beneath the surface z. It is customary in geophysics to define the 'heat flow' q as the upward component of conductive flux and to reverse the sign convention; i.e.,

$$q \equiv K \frac{\partial \Theta}{\partial z} \qquad (2)$$

Unless otherwise specified, q will represent the conductive heat flow near the earth's surface z = 0. It is convenient also, in the one-dimensional models, to take the upward velocity v to be positive, although it is in the direction of negative z. With these conventions, (1) for the one-dimensional case reduces to

$$\frac{\partial q}{\partial z} = -A - \rho'c'v \frac{\partial \Theta}{\partial z} + \rho c \frac{\partial \Theta}{\partial t} \qquad (3)$$

where q is the upward conductive heat flux, v is the upward volume flux of material with volumetric heat capacity $\rho'c'$, and ρc is the corresponding quantity in any stationary element.

Interpretation of regional heat flow in terms of crustal regime

Heat flow is determined from measurements of K and $\partial \Theta / \partial z$ (equation (2)) in holes typically drilled to depths of 100-300 m. Whether the value is typical of conditions in the underlying crust must be judged on the basis of internal consistency of observations within each hole and among neighboring ones. Regionally significant heat flows are expected to vary only smoothly over lateral distances much less than a crustal thickness, and as a minimum condition there should be a substantial interval in each hole throughout which the measured heat flow is redundant, i.e., over which the right side of (3) vanishes (after correction, if necessary, for laterally variable temperature and topography at the earth's surface; see, e.g., Blackwell [1973], Birch [1950] and Lachenbruch et al. [1962]). As the earth's surface is a source of temperature change ($\partial \Theta / \partial t \neq 0$) and hydrologic disturbance ($v \neq 0$) we generally have greater confidence in results from deeper holes. The internal consistency of measurements within and among holes in uniform granitic rocks at stations 30 km apart in the Sierra Nevada (Figure 5) provides some confidence that a regional condition is being measured there. By contrast, determination of the heat discharge and its regional significance in hydrothermal areas such as Long Valley, California (Figure 6) poses special problems.

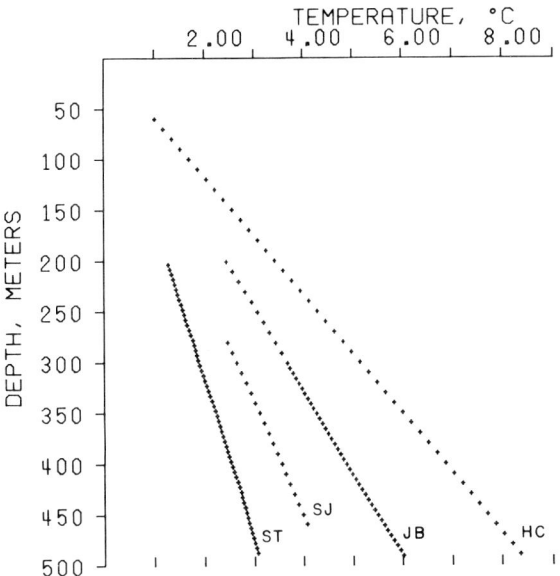

Fig. 5. Temperature measurements in granitic rocks of the Sierra Nevada province adjusted to a common temperature origin at the surface. Stations are about 30 km apart on a line from the western margin (ST) to the range crest (HC) [Lachenbruch, 1968].

The heat flow, normally measured in the upper 1% of the crust, provides only a boundary condition from which we should like to determine the thermal regime of the crust, i.e., to determine $q(z)$ throughout the entire crust. This requires a knowledge of how the terms on the right side of (3) vary throughout the crust. To provide meaningful constraints on these terms we must obtain insight into the physical processes that they represent. Interpretations of the crustal regime generally represent attempts to integrate (3) with simplifications believed to be appropriate for specific regions.

In our discussion, A will represent heat generated by radioactive decay of U, Th, and ^{40}K, elements present in minute amounts in crustal rocks. The process goes on steadily, irrespective of what else might be happening in the earth. The second term on the right in (3) represents effects of relative vertical movement of crustal (and upper mantle) masses; they may be solid blocks moving along faults or magmatic and aqueous fluids generally moving through fractures created by faulting or through pore spaces. As these movements are generally intermittent or short-lived, they generate transient disturbances represented by the last term in (3). Surface indicators of these mass movements are earthquakes, young volcanic rocks, and hot springs, which are shown with the heat flow distribution in Figures 3, 7, and 8. These manifestations are generally concentrated in the western regions of anomalously high heat flow, suggesting that the anomalies there are due primarily to convection and associated transients (i.e., to effects of the last two terms in (3)). The manifestations are rare in regions of low heat flow in the Sierra Nevada and the eastern half of the United States, and the crustal

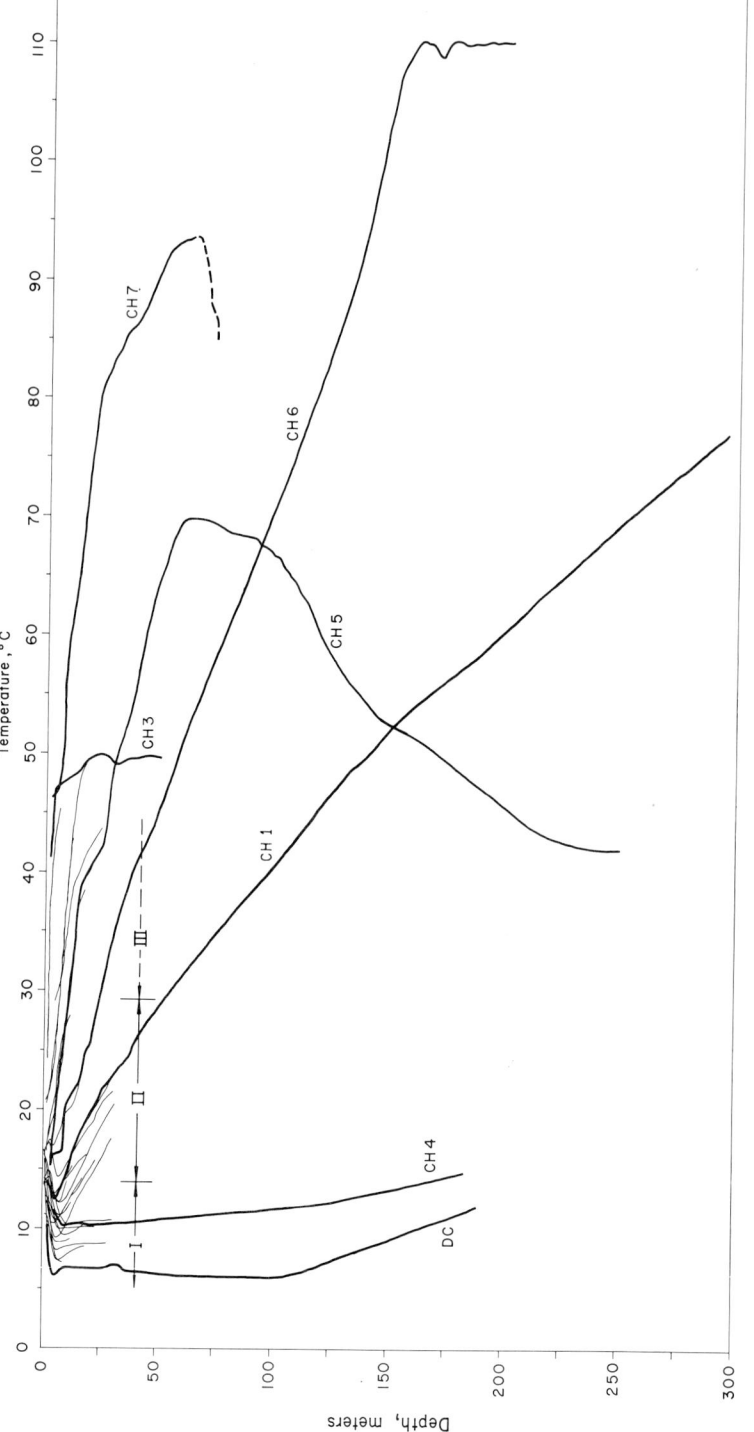

Fig. 6. Temperature measurements in the Long Valley caldera in California, a region with dimensions of about 20 km x 30 km [Lachenbruch et al., 1976b].

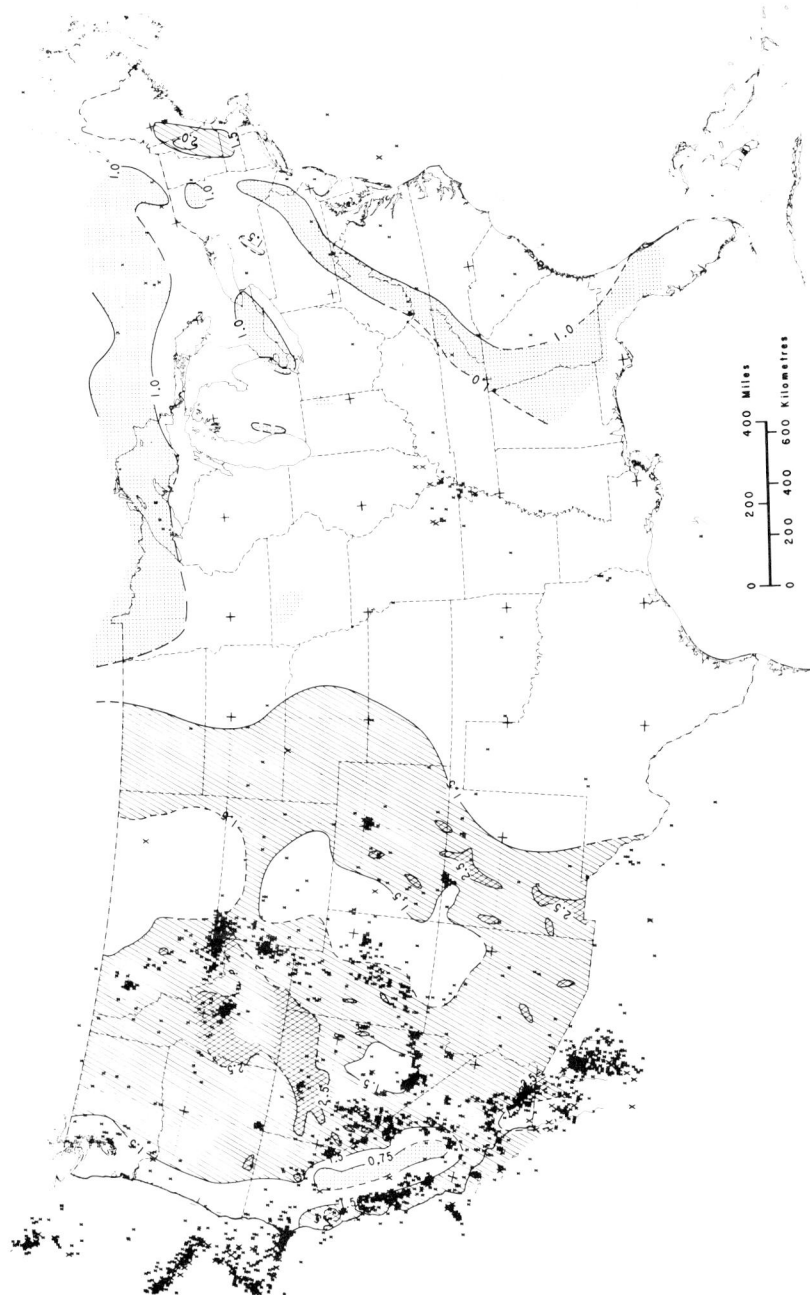

Fig. 7. Regional heat flow and distribution of seismic epicenters for the period 1961-1970. Dots represent earthquakes of magnitude about 3 to 5, crosses greater than 5. National Oceanographic and Atmospheric Administration epicenters have been replotted by J. C. Lahr and P. R. Stevenson of the USGS (personal communication, 1976).

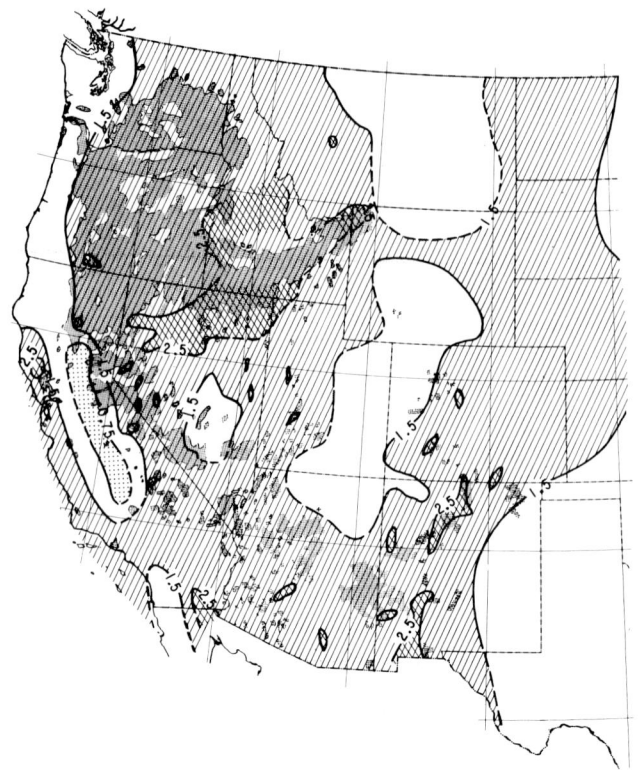

Fig. 8. Regional heat flow and the distribution of volcanic rocks erupted in the last 17 m.y. in the conterminous United States. Distribution of volcanic rocks adapted from Plate II of Stewart and Carlson [1977].

regime in these regions seems to be dominated by radioactivity (first term on the right in (3)).

Heat production and conductive transients

To estimate the relative importance of the terms in (3) we consider the contribution Δq to the surface heat flow that each term might make in a layer of thickness $\Delta z = z_2 - z_1$:

$$\Delta q = \int_{z_2}^{z_1} \frac{\partial q}{\partial z} dz \qquad (4a)$$

$$\Delta q = A\Delta z + \rho' c' v \Delta \theta - \rho c \frac{\partial \theta}{\partial t} \Delta z \qquad (4b)$$

where the parameters in (4b) are taken as appropriate average values and $\Delta \theta$ is the temperature difference across the layer.

If Δz represents a 30-km crust, then the contribution of the first term on the right can be written in dimensional form as follows:

$$\Delta q \text{ (HFU)} = 0.3A \text{ (HGU)} \qquad (5)$$

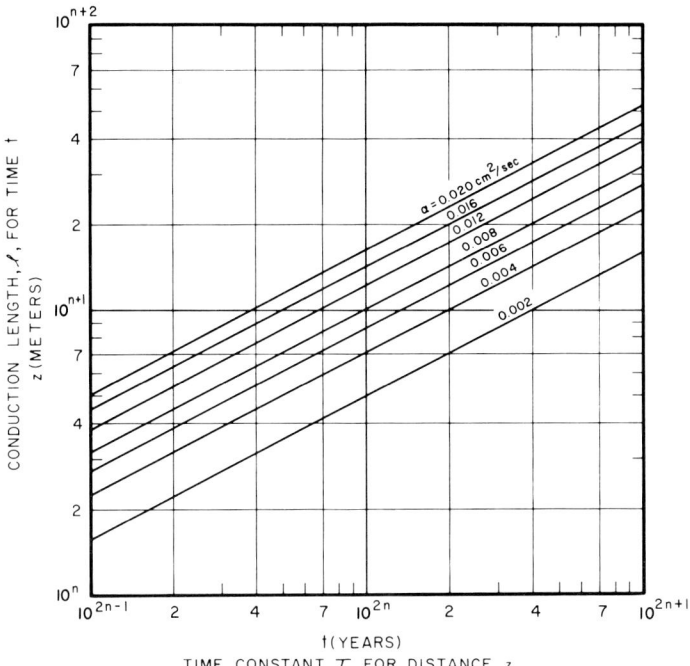

Fig. 9 Relation between conduction length ℓ and time t, or between time constant τ and distance z, for the range of thermal diffusivities (α) of natural earth materials. The scale parameter n may be assigned any convenient value. (For example, if $\alpha = 0.008$ cm^2/s and t = 1 m.y., set n = 3 to obtain ℓ = 10 km. Conversely, given z = 10 km, then τ = 1 m.y.).

Radioactive decay results in generation of heat at the rate of 1-10 HGU in most crustal rocks exposed at the surface. According to (5), if such rocks were distributed throughout the crust, they could account for much or all of the observed surface heat flow. Hence the distribution A(z) is important to an understanding of the crustal regime, and it has been the subject of considerable study.

Skipping to the last term in (4b), we set $\rho c = 0.6$ cal/cm^3 °C and $\Delta z = 30$ km to represent the crust. The contribution of temperature change can be written

$$\Delta q \text{ (HFU)} \sim -0.6 \times 10^5 \times \frac{\partial \theta}{\partial t} \text{ (°C/yr)} \qquad (6)$$

Thus cooling of the entire crust at rates $\sim 10^{-5}$ °C/yr could contribute significantly to measured heat flow; a total of 20°C or 30°C of crustal cooling could account for heat flow at the observed rates for a million years with no other contributions. However, the rate at which the crust can cool is controlled by the mode of heat transfer within it. Hypothetical circulatory convection systems can be contrived to remove heat from the crust at almost any desired rate, but if the heat transfer is by conduction, the rate at which heat may be discharged or stored is severely constrained by the low conductivity and substantial volumetric heat capacity of crustal materials. These constraints can be discussed in terms of simple limiting heat conduction models that can be reduced to convenient rules of thumb for the purpose of general discussion.

The solutions to many simple conduction problems can be expressed in terms of the single dimensionless ratio $z/\sqrt{4\alpha t}$. Normally, t is the time since some sort of disturbance occurred, z is a distance, usually from the source of disturbance, and α represents thermal diffusivity $(k/\rho c)$. We define a characteristic 'conduction length' ℓ and a characteristic 'conduction time constant' τ as follows:

$$\ell(t) \equiv \sqrt{4\alpha t} \tag{7a}$$

$$\tau(z) \equiv z^2/4\alpha \tag{7b}$$

Hence

$$\frac{z^2}{4\alpha t} \equiv \left(\frac{z}{\ell(t)}\right)^2 = \frac{\tau(z)}{t} \tag{7c}$$

and the approach to steady conditions can be expressed in terms of how large t is in relation to τ or how large ℓ is in relation to z, depending on which variables are known. For convenient reference the conduction length ℓ can be found for any t, and the time constant τ can be found for any z for the range of diffusivities for natural earth materials in Figure 9. A reasonable average diffusivity for the entire crust may be around 0.01 cm^2/s.

Some convenient rules of thumb relevant to the measurement of heat flow or the interpretations to be presented are as follows:

1. A periodic temperature change with amplitude B and period P will have a negligible effect (a few percent of B) at depth $\ell(P)$. Thus diurnal fluctuations (P $\sim$ 3 x 10^{-3} years) are negligible in sediments ($\alpha \sim$ 0.002-0.008 cm^2/s) at depths of 30-50 cm (Figure 9, n = -1). The annual wave (Figure 9, P = 1 year, n = 0) penetrates about 5-10 m in poorly conducting sediments and 15 m in crystalline rock (α = 0.014 cm^2/s). The temperature pulsations due to repeated intrusion of a deep crustal layer with a period of 10^5 years would be negligible a few kilometers above (and below) the layer (Figure 9, n = 3). The process would therefore have the same effect at the surface as a continuous intrusion with a uniform (time averaged) temperature.

2. A rapid change in surface temperature at time t = 0 in the amount B perturbs the gradient at the surface by about $B/\ell(t)$, and its effect on temperature may be appreciable ($\sim$15% of B) to depth ℓ but is completely negligible (<1% of B) beneath 2ℓ. Thus a 5°C post-Pleistocene warming 10,000 years ago (Figure 9, n = 2) could disturb the surface gradient of the order of 5°C/km in crystalline rocks; temperature effects would be appreciable in holes to several hundred meters. The gradient disturbance would be greater in sediments, but because of the role of thermal conductivity the effect on heat flow would be less. Temperatures below an intruded layer maintained at constant temperature can be treated by these rules.

3. A rapid change in temperature at depth z will not be detectable in surface heat flow until times approaching 1/2 $\tau(z)$, and the surface heat flow will be almost completely adjusted to the change for times exceeding $\tau(z)$. Thus a rapid (step) temperature change due to intrusion

at the base of a 30-km crust ($\alpha \sim 0.01$ cm^2/s) will not affect surface heat flow for about 3 or 4 m.y., but the entire crust will have equilibrated to the change in 8 m.y. or so (Figure 9, n = 3).

4. Heating (or cooling) by a constant heat source (or sink) at a depth z from time t = 0 will not affect heat flow at the surface until times approaching $\tau(z)$. The surface heat flow will not approach its steady value until $t \gtrsim 100\tau(z)$, but for $t > 3\tau(z)$ temperatures in the layer above depth z can be estimated (within about 10%) by assuming that conditions observed at the surface are steady, i.e., by assuming that heat flow is independent of depth in the layer. The constant source approximates long-term slow intrusion of a sill in which the melt does not survive between intrusive pulses or the thermal recovery of a layer of thickness z after extinction of a hydrothermal system within it. The constant sink approximates effects of a cooling sill after solidification [see Lachenbruch et al., 1976a].

Convection--general considerations

An understanding of regional heat flow in tectonically active areas requires at least a gross understanding of convective processes in the crust. We distinguish between two main types: convection by magma and convection by groundwater. The large-scale effect of magmatic convection on surface heat flow is caused by the upward transport of fluid at temperatures greater than ambient; it always results in a positive contribution to heat flow. The lateral extent of the positive anomaly will be comparable to that of the region intruded. In the steady state the intensity of the anomaly can be determined in terms of the rate (volume flux) of intrusion v from the second term on the right in (3), making allowance in the factor c' for the heat of crystallization. For present purposes, transient effects of convection by magma can be discussed adequately in terms of the rules of thumb presented in the last paragraph.

Groundwater convection is a more difficult problem; it generally involves some flows that are hotter than ambient and others that are colder, and it is mainly confined to the upper crust, close to the surface where heat flow is measured. With the search for geothermal energy it is becoming more important to understand groundwater convection in a regional context. As this process has generally been treated lightly in discussions of regional heat flow, we shall consider it here in somewhat more detail than the processes just discussed.

The magnitude of hydrothermal effects can be estimated by retaining only the second term on the right in (4b). The volumetric specific heat $\rho'c'$ for hot and cold water (and for melted rock as well) is generally from 0.7 to 1 cal/cm^3 °C. Taking the latter value leads to the following dimensional relation for vertical one-dimensional steady convection

$$\Delta q \text{ (HFU)} = 10^6 v \text{ (cm/s)} \Delta\Theta \text{ (°C)} \qquad (8a)$$

$$\Delta q \text{ (HFU)} \sim v \text{ (ft/yr)} \Delta\Theta \text{ (°C)} \qquad (8b)$$

Equation (8b) is a useful rule of thumb. Seepage velocities of 1 ft/yr (0.3 m/yr) would result from Darcian flow under unit hydraulic gradient in a medium with a permeability of only 1 mdarcy (10^{-11} cm^2); they are not uncommon in hydrothermal areas. Such a flow rising across a layer with 10°C temperature difference would contribute 10 HFU to the surface flux.

A second useful relation for steady vertical flow is obtained by neglecting the first and third terms on the right in (3) and writing it in the form

$$\frac{\partial q}{\partial z} = -\frac{\rho' c'}{K} vq \qquad (9)$$

Integration yields

$$q(z_1)/q(z_2) = e^{\Delta z/s} \qquad (10)$$

where s is a characteristic distance with the sign of v.

$$s = K/\rho' c' v \qquad (11a)$$

It is most easily expressed as a rule of thumb in terms of feet.

$$s \text{ (feet)} \sim \frac{100}{v \text{ (ft/yr)}} \text{ for 'sediment' } (K = 3 \text{ mcal/cm s °C}) \qquad (11b)$$

$$s \text{ (feet)} \sim \frac{200}{v \text{ (ft/yr)}} \text{ for 'rock' } (K = 6 \text{ mcal/cm s °C}) \qquad (11c)$$

According to (10) and (11b), steady vertical flow at 1 ft/yr through a 500-foot layer (Δz) would increase the gradient and conductive heat flow in the direction of water flow by the factor $e^5 \sim 150$. This can lead to very large (and short-lived) local fluxes and large temperature differences (equation (8)) unless the gradient on the inflow side is very small. Hence the flow will generally cause most of the layer to be nearly isothermal if $\Delta z \gg |s|$. Thus for downward flows (s negative) of 1 ft/yr the gradient near the surface and the measured heat flow will generally vanish if the layer is only a few hundred feet thick. Similarly, the surface heat flow will be 'washed out' by downward percolation of only 1 in./yr (a small fraction of annual precipitation) through a few thousand feet of porous rocks. This effect obviates the determination of regional heat flow by conventional means over large areas. Some such areas, mantled by permeable volcanic rocks, are of considerable interest as potential sources of geothermal energy.

Temperature in the layer of thickness Δz (equation (10)) is determined by specifying at least one of the boundary temperatures and the other boundary temperature or one of the boundary heat flows $q(z_1)$ or $q(z_2)$. To be consistent, any physical model must also conserve mass of the flowing water. A useful consistent solution for coupled heat and water flow is obtained by identifying $\theta(z_1)$ with the mean ground surface temperature and $q(z_2)$ with the regional heat flow. In this case, water flows horizontally along z_2 with no change in temperature, providing a source (or sink) for the vertical mass flow to (or from) the surface. The model yields a useful rule of thumb; viz., whether the surface heat flow ($q(z_1)$) is significantly different from the regional heat flow ($q(z_2)$) depends upon whether the depth of vertical water flow Δz is small or large in relation to $|s|$ (equation (10)). A second application of (10) [Bredehoeft and Papadopulos, 1965] assigns both boundary temperatures $\Theta(z_1)$ and $\Theta(z_2)$. However, unless the assigned temperature $\Theta(z_2)$ is the value determined by uniform flux from below, $q(z_2)$, the water flowing horizontally along z_2 must be a source of heat as well as mass, its temperature must change horizontally, and the one-dimensional model is only approximate. It is useful to note, however, that from a transient solution for this case [Nathenson, 1977] it can be shown that the stationary condition described by (10) is generally approached after one conductive time constant $(\tau(\Delta z))$ for slow water flow between depths held at constant temperature and sooner if the vertical water flow is vigorous ($|s| \ll \Delta z$).

Equations (8), (10), and (11) give an indication of the enormous effects that hydrologic conditions can have on measured heat flow (see, e.g., Figure 6). In natural systems these effects can be extremely complex, involving variable upward and downward flows (with temperature and pressure dependent properties) in fractures and pore spaces [see, e.g., Sorey, 1975]. These systems may be in delicate balance, vulnerable to the effects of earthquakes or the drilling of wells. The pattern assumed by these flows depends upon the conditions that drive them. There are two distinct cases: (1) the flow is forced by the configuration of fractures and permeable formations and by regional piezometric conditions controlled by precipitation, evaporation, and topography, or (2) the flow results from the instability of groundwater heated from below in a permeable layer. (In general, elements of both driving mechanisms are present.) The second case tends to produce circulating cells with an aspect ratio close to unity [e.g., Sorey, 1975]; it should result in heat flow anomalies that change in sign over lateral distances of the order of the depth of circulation. No such condition applies to the first case, which in extreme circumstances could produce persistent anomalies in surface heat flow on a regional scale (e.g., possibly, the Eureka Low, to be discussed). The foregoing results can be applied to order of magnitude calculations for the first case. We shall now consider the second case, which we shall refer to as hydrothermal systems.

Hydrothermal systems

These systems are initiated when the Rayleigh number R exceeds a critical value [e.g., Lapwood, 1948]. Other things being equal, R increases linearly with the depth of the permeable zone, the permeability of the zone, and the temperature difference across it; increasing any of these could initiate a hydrothermal convection system. To investigate gross relations between these systems and the regional heat flow or magmatism that supplies their heat, we shall discuss some highly idealized quasi-one-dimensional models.

Consider a region near the earth's surface (Figure 10a, z = 0) in which the heat transfer is initially conductive and the temperature profile is linear with gradient Γ (Figure 10b, curve I). The heat flow q is then $K\Gamma$. Suppose that in a region of high heat flow at time t = 0, fractures open in a layer extending from the earth's surface to depth h and that this increases the permeability so much that the groundwater becomes unstable and starts to circulate. Above regions where the water is moving downward the surface heat flow will diminish, and above regions where it is moving upward the surface heat flow will increase.

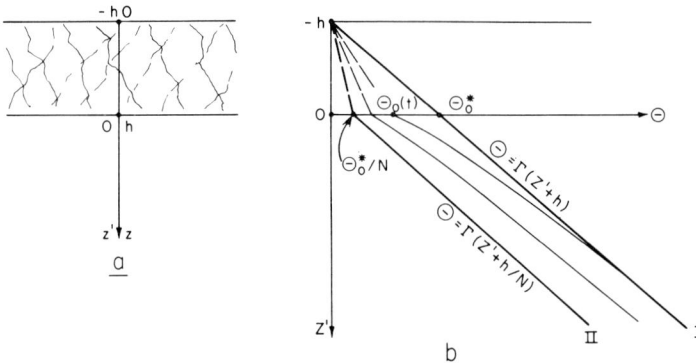

Fig. 10. Idealized one-dimensional model for hydrothermal convection in a surficial layer of thickness h; the heat is supplied by steady regional heat flow (see text).

However, the total heat transport across the layer (integrated over the surface above the fractured region) must increase, as the initial conductive transport is now supplemented by convective transport. For simple one-dimensional order of magnitude calculations we assume that the net effect of convection is to increase the mean heat flux through the layer ($z < h$) by the factor $N > 1$ (e.g., for the layer with uniform vertical flow, upward over half the area and downward over the other half, (10) yields $N = [\exp(h/|s|) + \exp(-h/|s|)]/2$. When the system is in a stationary state and the lower and upper boundaries are held at constant temperatures, N would represent the Nusselt number; in real systems, neither surface would be at uniform temperature. We let $\Theta_o(t)$ be the average temperature at $z = h$ at time t and assume that it is uniform. The surface temperature is assumed to be uniform at the average ambient value taken as zero. This condition would be violated locally by hot springs, the convective discharge of which is included in the factor N. We neglect the change of N that would occur as the system evolves. (A useful discussion of the relation between heat flow and coupled heat and water flow in porous layers has been given by Donaldson [1962] and in isolated fractures by Lowell [1975] and Bodvarsson [1969].)

In this model the mean upward flux from $x = h$ will (initially) exceed the mean flux into $x = h$ by the factor N. After time t the basal temperature will drop from Θ_o^* to some value $\Theta_o(t)$, and the heat flux through the convecting layer will drop to $NK\Theta_o/h$. The convecting system will continue to mine heat from the earth until the flux through it is equal to the regional conductive flux $K\Gamma$ supplied from below, i.e., equilibrium will be established when

$$NK\Theta_o/h = K\Gamma \equiv K\Theta_o^*/h \qquad (12a)$$

or

$$\Theta_o = (\Theta_o^*/N) \qquad t \to \infty \qquad (12b)$$

To estimate the time required for stabilization, we first neglect the heat capacity of the convecting layer $z < h$ and consider the underlying region $z > h$, which we denote by $z' > 0$ (Figure 10); heat loss from the surface $z' = 0$ is proportional to the temperature $\Theta_o(t)$. The differential equation and conditions are

$$\frac{\partial^2 \Theta}{\partial z'^2} = \frac{1}{\alpha} \frac{\partial \Theta}{\partial t} \qquad z' \geq 0 \qquad (13a)$$

$$\Theta = \Theta_o^* + \Gamma z' \qquad t = 0 \qquad z' \geq 0 \qquad (13b)$$

$$\frac{\partial \Theta}{\partial z'} = \frac{N}{h} \Theta_o(t) \qquad z' = 0 \qquad (13c)$$

The solution (modified from Carslaw and Jaeger [1959, p. 71]) for temperature Θ_o at the base of the slab is

$$\Theta_o(t) = \Theta_o^* - \Theta_o^*(1 - 1/N)[1 - e^{-\beta^2} \text{erfc } \beta] \qquad (14)$$

where $\beta^2 = (N^2 \alpha/h^2)t$. Hence $\Theta_o(t)$ approaches its equilibrium value Θ_o^*/N after a time

$$t \sim h^2/N^2 \alpha \qquad (15)$$

which can be read from Figure 9. For $N = 2$ it is the same as the conductive time constant for the slab of thickness h, and for $N = 6$ it is an order of magnitude less.

The sensible heat lost by the slab as its base cools from Θ_o^* to $(1/N)\Theta_o^*$ (neglected in the above calculation) is roughly $\Theta_o^*(1 - 1/N)h\rho c/2$. This heat could sustain the mean anomalous flux of $NK(\Theta_o^*/h)(1 - 1/N)/2$ for a time

$$t \sim h^2/N\alpha \qquad (16)$$

The actual time for stabilization of the slab depends upon the complex behavior of coupled temperature and velocity fields; it will generally result in changing N and perhaps in an increase in h, if thermal contraction results in deepening fractures. However, for larger N (say, >2) likely to be of interest we judge from the above that processes internal to the slab (16), not those beneath it (15), will be controlling and that a steady state is likely to be approached in times of the order of $\tau(h)$ (Figure 9) or less. For larger times the surface flux would still have extreme local variations, but perturbations would integrate to zero, and the average combined flux would equal the regional heat flow q. Stabilization times vary from 1,000 years for $h \sim 400$ m to 100,000 years for $h \sim 4$ km (Figure 9). More active systems (large N) probably stabilize more quickly. (For a layer in which circulation is confined to fractures separated by distance λ, the stabilization time will probably be controlled by $\tau(\lambda)$ if $\lambda > h$ [see, e.g., Bodvarsson and Lowell, 1972; Carslaw and Jaeger, 1959, Figure 12].)

Restoration of the steady regional heat flux at the surface after extinction of a hydrothermal system is a very slow process (governed by the conduction rule of thumb, number 4 above). It can be viewed crudely as the conductive return of curve II, now the initial condition, to curve I, the final condition (Figure 10b). The heat flow anomaly Δq at the surface can be shown to be

$$\Delta q = -q(1 - 1/N) \text{ erf } \frac{h}{\sqrt{4\alpha t}} \qquad (17)$$

where q is the steady regional heat flow. According to (17) the anomaly would be half its initial value when $t \sim h^2/\alpha$ (or $4\tau(h)$) and 10% of its initial value when $\tau \sim 110\tau(h)$.

We illustrate these results with a highly idealized numerical example. A 'one-dimensional' hydrothermal system with depth h = 2 km develops in a region with steady regional flux q of 2.5 HFU. Assume that N = 5 and (perhaps unrealistically) that this value and the depth h persist as the system ages. In early stages the average heat flow from the system will be Nq, i.e., 12 or 13 HFU. After some 25,000 years ($\tau(2 \text{ km})$) or less the average flux will fall to the regional value with 80% (i.e., 1 - 1/N) or 2 HFU being supplied by convective transfer. If the circulation suddenly stopped (e.g., from earthquakes or sealing of fractures), the mean flux would fall to 0.5 HFU, producing a mean local anomaly of -2 HFU; negative anomalies of 0.5 and 0.2 HFU would still persist 1/2 and 3 m.y. after circulation stopped, respectively. Although the example is extreme, it serves to illustrate why the heat flow might be extremely variable in tectonically active provinces where hydrothermal convection systems are common; the relation of the anomalies to the systems that produced them may be obscure.

These highly simplified considerations suggest the following generalizations regarding hydrothermal convection systems supported by regional heat flow in permeable surface layers:

1. The heat flow q (and combined flux q_c) will vary over horizontal distances of the order of depth of circulation, h, during all phases.

2. During an initial phase which might last $\sim\tau(h)$ the mean combined flux from the surface will exceed the regional heat flow.

3. If the system survives, it will reach a stationary stable phase in which the mean combined flux will equal the regional heat flow.

4. In a waning or recovery phase, probably longer by a factor of 10^2 than the initial phase, the mean surface flux will be less than regional heat flow.

5. The mean combined flux at the surface integrated over all phases will equal the regional heat flow; if there was convective loss into surface drainage, the integrated conductive flux will be less than the regional value.

If we can view the Basin and Range province (which is $\sim 10^7$ m.y. old) as containing a random sample of such systems (with life cycles of $\lesssim 10^6$ years), the average combined flux will equal the regional heat flow, and the average conductive flux will be less. Similar generalizations apply to systems confined to widely separated fractures as long as they are sustained by regional heat flow (or a source of heat at depth that is uniform for times much larger than the stabilization time for the hydrothermal system).

In the foregoing discussion, hydrothermal instability was initiated by introducing fractures which increased the permeability k or depth of the fractured layer, h. We assumed a steady 'regional' conductive flux from great depth. Even in the Battle Mountain High the regional flux

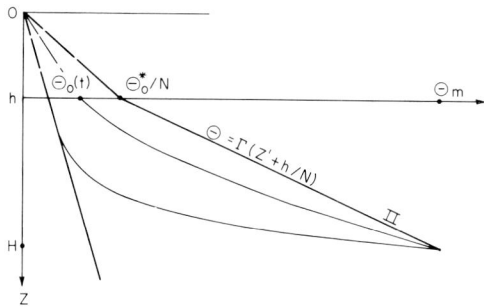

Fig. 11. Idealized one-dimensional model for hydrothermal convection in a surficial layer of thickness h; the heat is supplied by upper crustal magma at depth H (see text).

does not exceed about 3 HFU, and it leads to regional gradients generally in the range of 50°–100°C/km. Intrusions in the upper crust could, of course, produce much higher heat flows locally; if the melts persist long enough, they can generate a steady thermal condition in the overburden, heating the rock all the way to the surface. Under such conditions, hydrothermal instability could be initiated by raising the basal temperature Θ_o of the fractured zone rather than by increasing its permeability. This condition is illustrated in Figure 11 for the simple one-dimensional case of intrusion at depth H with temperature Θ_m. If the magma supply is sufficient to maintain the isothermal condition at z = H during the initial stages, a stationary thermal condition will be approached in the overlying rocks in about $\tau(H)$ (see Figure 9); otherwise, it will take longer [Lachenbruch et al., 1976a]. This stationary condition is represented by curve II in Figure 11, which, of course, is the same (mathematically) as curve II in Figure 10b. In this case, however, the mean surface flux from the developing system may be less than the flux in the steady state, and the steady flux may be much greater than the regional value. In the steady state the mean combined flux q_c from the system can be written

$$q_c = K\Gamma \tag{18a}$$

$$q_c = \overline{K}\Theta_m/H \tag{18b}$$

$$q_c = K\Theta_m/[H - h(N - 1)/N] \tag{18c}$$

where (18c) is obtained by substituting for $\overline{K}$ the harmonic mean conductivity of the overburden (H), using the effective conductivity NK in the fractured zone. The expression in brackets in (18c) could be called the 'effective depth' of magma; it is the depth that would be implied by heat flow observations if convection were absent.

$$H - h\frac{N-1}{N} = K\frac{\Theta_m}{q_c} \tag{19}$$

At the Long Valley caldera in California the mean combined flux at the surface has been estimated by hydrochemical means to be greater than 10

HFU by Sorey and Lewis [1976] and about 16 HFU by White [1965]. The caldera has been a source of volcanism for 2 m.y., and hydrothermal activity has been in progress at least 300,000 years, more intense in the past [Bailey et al., 1976]. If we assume a stationary state and take $q_c \sim 13$ HFU, $\Theta_m = 800°C$, and $K = 5$ mcal/cm s °C, we obtain for the effective depth of magma

$$H - h\frac{N-1}{N} \approx 3 \text{ km}$$

This implies that hydrothermal convection must extend downward (to depth h) within 3 km of the magma with very high N or even closer if the circulation is less vigorous. Structural and seismic evidence [Bailey et al., 1976; Hill, 1976; Steeples and Iyer, 1976] suggests that if magma now exists beneath the caldera, it must be at a depth of at least 6 to 8 km. Thus at least 3 to 5 km is made 'transparent' by hydrothermal convection, and water must be circulating to very great depths.

It has been pointed out that hydrothermal systems supported by regional heat flow probably exhaust the sensible heat in time $\sim \tau(h)$, say, 10^3-10^5 years, and if they survive thereafter, they do not result in anomalous heat loss. In volcanic areas such as Long Valley, hydrothermal systems are evidently supported by upper crustal intrusion, and they can persist for millions of years, discharging heat at an anomalous rate. This behavior imposes severe heat demands on the underlying magmatic system [see Lachenbruch et al., 1976a]. When the effective depth of magma exceeds 10 or 15 km (depending on the choice of K and Θ_m in (18c)), the steady surface flux approaches the regional value characteristic of the Battle Mountain High (~ 3 HFU) and the time constant for the overburden becomes large in relation to the stabilization time for hydrothermal systems. Under these conditions the heat supply for hydrothermal systems would be considered as either regional heat flow or a local magmatic anomaly, depending upon whether or not the magmatic condition were widespread. Both situations probably occur in the Basin and Range province.

Convection and Regional Heat Flow

In general, convection by groundwater in upper crustal rocks poses the greatest obstacle to determining from surface observations the heat flow associated with crustal conditions at depth. Obvious effects can often be eliminated by judicious selection of study sites or by the criterion of internal consistency applied to measurements in deep holes and in neighboring holes. However, because hydrologic effects can be subtle, lingering uncertainties may persist. For regional interpretation it is important to know whether undetected hydrologic anomalies are likely to be the exception or the rule. It has already been pointed out that in some regions of porous volcanic and sedimentary rocks they may be the rule. Thus it is difficult to determine the regional significance of heat flow measurements throughout portions of the Pacific Northwest, including the Cascade volcanos, parts of the Columbia Plateau, and the important Snake River Plain [Brott et al., 1976], and in other recent volcanic areas such as the San Francisco peaks at the edge of the Colorado Plateau. Very detailed hydrologic studies and deep drilling might be required to detect heat from magmatic reservoirs

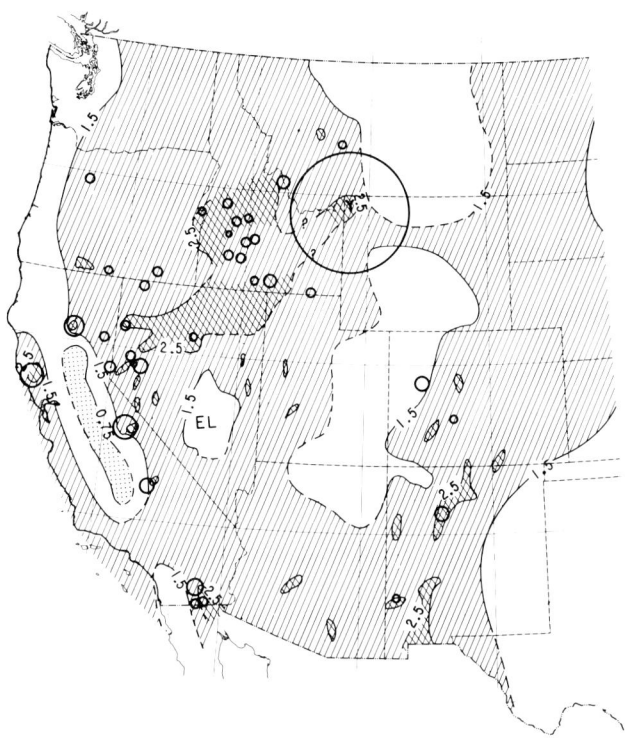

Fig. 12. Regional heat flow and the natural heat discharge of known hydrothermal systems with reservoir temperatures greater than 90°C. Each circle is centered at the location of the system it represents. A heat flow of 1 HFU through the circular area is equivalent to the rate of combined heat discharge estimated for the system. Systems with estimated discharge of less than 3×10^6 cal/s are not shown.

[Smith and Shaw, 1975] likely to underlie some of these regions. A similar problem often occurs in sedimentary basins, and for this reason, sites for regional heat flow studies are often chosen in less permeable rock, despite the more costly drilling.

In interpreting heat flow in the Basin and Range province it will be important to estimate the regional effects of hydrothermal convection systems. Figure 3 shows all the hydrothermal systems with estimated reservoir temperatures greater than 90°C that Renner et al. [1975] were able to identify in the conterminous United States in a recent study. From the data presented by them and from supplementary studies, chiefly those of Olmsted et al. [1975], Mariner et al. [1974], Bowen and Peterson [1970], and Fournier et al. [1976], we have attempted to summarize the natural heat discharge from each of these systems. This discharge comes in varying proportions from conductive loss from the reservoir and from conductive and convective losses from fluid discharged from the reservoir into shallow aquifers or surface drainage. Methods of estimating these discharges have been discussed by Olmsted et al. [1975], White [1965, 1968], Sorey and Lewis [1976], Fournier et al. [1976], and

Morgan et al. [1977]; the methods vary depending upon hydrologic conditions, and, of course, the estimates are subject to substantial uncertainty. Of the 255 systems listed by Renner et al. [1975], we judged that about three dozen had total combined natural discharges greater than about 3×10^6 cal/s; they are shown as circles in Figure 12. The area enclosed by each circle is the area through which an anomalous flux of 1 HFU would be equivalent to the total rate of combined (conductive and convective) heat discharge for the system. Typically, the anomalous regions have an area that is an order of magnitude smaller than the circles in which they are centered. The purpose of this representation is only to place some of the better known systems in a regional perspective. More detailed studies of these and other systems (some perhaps with high mass discharge at lower temperature) will surely change the picture.

Insofar as the 255 systems located in Figure 3 are concerned, the cumulative anomalous discharge is small in relation to the integrated regional flux; with the exception of the Yellowstone system, their effects on the thermal balance of the crust would be local. For the anomalous discharge to equal the integrated regional flux the circles (Figure 12) would have to overlap once on the average throughout most of the western United States. For those systems that might be stabilized above upper crustal intrusions the circle indicates the rate at which heat must be supplied by magmatic convection. For those systems that have stabilized and are supported by regional heat flow the circle represents the area over which a negative regional anomaly of 1 HFU would be sufficient to complete the heat balance.

These results offer some hope that we might be able to find a characteristic regional flux to identify with the crustal regime over large areas of the Basin and Range and similar regions; the most reasonable choice would be the most frequently occurring (or modal) value of the conductive flux. The mean would be biased toward large values by effects of undetected upper crustal intrusions, although the shallower ones would probably be identified by their hot springs and avoided in regional studies. In large regions of high heat flow, most of the local anomalies of unidentified origin are likely to come from hydrothermal convection supported by regional heat flow and modified by the forcing effects of variable topography, permeability, and precipitation. Although the combined anomalous flux from these systems might integrate to zero, the mean conductive flux would be biased toward lower values if there were appreciable convective discharge into surface drainage. (The mode and mean are not appreciably different for the Basin and Range data, possibly because internal drainage minimizes convective loss from shallow aquifers.) If the local convection systems were common, the dispersion of heat flow would be large, and the mode would be poorly defined, making the regional flux more difficult to identify. This is evidently the case in the Basin and Range (to be discussed further below), where highly fractured rocks are difficult to avoid and high heat flow and locally variable topography and precipitation favor small-scale convection systems [Olmsted et al., 1975].

A hydrologic anomaly on a regional scale seems to provide the most reasonable explanation of the Eureka Low subprovince of the Basin and Range (Figure 12). The average of 13 heat flows in this 30,000 km^2 region of south central Nevada is about 1.1 HFU, roughly 1 HFU less than the heat flow believed characteristic of the surrounding Basin and Range province. The deeper holes generally showed thermal evidence of downward moving water; in the deepest hole this evidence persisted to depths greater than 3 km [Sass et al., 1971]. In a careful hydrologic synthesis, Winograd and Thordarson [1975] have shown that an 11,000 km^2 region, straddling the southern boundary of the Eureka Low, is hydrologically integrated into one groundwater basin, although the region contains 10 topographic basins. The interbasin flow occurs to depths up to 1 1/2 km beneath the surface in permeable fractured carbonate rocks underlying the region; discharge is concentrated along a fault line in the Armogosa Desert on the southwestern margin of the system. Eakin [1966] has described a similar system in a 20,000 km^2 region including much of the eastern portion of the Eureka Low, and Dinwiddie and Schroder [1971] report evidence for interbasin flows to depths greater than 2 km in valleys of the central portion of the Eureka Low. A general discussion of the problem has been given by Mifflin [1968], who summarizes evidence for large-scale interbasin flows in regions underlain by fractured carbonate rocks in southeastern Nevada. The observation of interbasin flow systems in this region makes it likely that the entire Eureka Low is caused by such systems and, in fact, that heat flow might be a useful means of studying them. As the heat flow is still poorly known in the Eureka Low, it is likely that the pattern is much more complex than indicated by the single contour that delineates it in Figure 12. Nevertheless, it is useful to make a very simple steady state order of magnitude calculation. Suppose water percolated downward uniformly in the Eureka Low at the average rate of 1 cm/yr, some 5-10% of the local annual precipitation. Then s $\sim$ -2 km for 'rock' (equation (11c)). If the average depth of the interbasin conduit were $\sim$1.4 km, according to (10), the surface heat flow would be roughly half the regional heat flow as required. If the recharge velocity (-v) were cut to 5 mm/yr, the regional depth of water flow would be 2.8 km. In the system studied by Winograd and Thordarson the average recharge rate required to supply the estimated annual discharge is about 2 mm/yr; for the system studied by Eakin it is about 5 mm/yr. These values seem consistent with the foregoing calculations, especially since the hydrologic systems studied each overlap the Eureka Low and may have somewhat higher mean heat flows. If the average heat flow anomaly in the Eureka Low is indeed about -1 HFU, the discharge required to complete the thermal balance can be compared to that of the hydrothermal systems by comparing the circles in Figure 12 to the mapped size of the Eureka Low. Next to the Yellowstone system the Eureka Low would have the greatest heat discharge. However, the temperature of the flow would be low, 30°-60°C above surface ambient according to (10) (when -v = 1 cm/yr and 5 mm/yr are used). Much of this heat is probably discharged convectively by warm springs; if it were not, it could cause a substantial positive heat flow anomaly. The possibility of interbasin flows on the scale suggested by the present configuration of the Eureka Low requires further investigation in connection with proposals for underground storage of nuclear wastes in such areas [see also Hunt and Robinson, 1960].

We shall return to the problem of convective transport in a later section; the next section considers the simpler crustal regime characteristic of regions where convection is unimportant.

Sierra Nevada and Eastern United States: Effects of Radioactivity

It has been pointed out (5) that most crustal rocks seen at the surface contain enough radioactive uranium, thorium, and potassium to contribute appreciably to surface heat flow if such rocks were distributed uniformly through the crust. The cumulative contribution of crustal radioactivity must be known in order to determine from the surface heat flow the flux from the top of the mantle. Additionally, a knowledge of how crustal radionuclides are distributed vertically should lead to a better understanding of the thermal regime and geochemical evolution of the crust. Substantial progress has been made on these problems by studying the relation between surface heat flow and radioactive heat production of plutonic and highly metamorphosed crystalline rock exposed at the surface. Such rocks are the ones most likely to be related geochemically to the inaccessible material on which they rest.

Figure 13 is a plot of measured heat flow q versus radioactive heat production A_o of crystalline drill core or outcrop material sampled at the heat flow site. The 150 or so points include all published results from the conterminous United States and adjacent Mexico and many new points of our own (heat productions were determined by our colleague, Carl Bunker, in Denver). Locations from which the data of Figure 13 were obtained are shown in Figure 14, to be discussed further below. Birch et al. [1968] discovered that a graph of heat flow versus heat production (q, A_o) for sites in New England yields a straight line, and Roy et al. [1968a] showed that the same line accommodated additional observations in the stable central region of the United States. Their line is shown in Figure 13 and labeled 'Eastern U.S.'; data from all of the locations east of the Great Plains (shown in Figure 14) lie close to this line (solid circles, Figure 13). Two other heat flow provinces were defined by (q, A_o) lines presented by Roy et al. [1968a], one for the Sierra Nevada and one for the Basin and Range province. Both lines are shown in Figure 13. The line for the Sierra Nevada province was confirmed by independent studies [Lachenbruch, 1968], and further confirmation has come from a value published more recently [Lachenbruch et al., 1976a]. Ten of the eleven points (crossed circles, Figure 13) interior to the Sierra Nevada physiographic province lie close to the Sierra line. The eleventh point for the Sierra Nevada (DP, near the top of Figure 13) lies far above the Sierra line, as was expected; the site is only 3 km from the Long Valley volcanic center. Most of the other points in Figure 13 are not close to any of these province lines, and they will be discussed later. However, it will be useful first to outline a simple interpretation of the linear relation for the Sierra Nevada and eastern United States; it provides general insight into conditions in the crust and mantle [Birch et al., 1968; Roy et al., 1968a; Lachenbruch, 1968, 1970].

The simplicity of the linear relation suggests a simple model. We assume one-dimensional steady state, nonconvective transfer and retain only the term for heat production $A(z)$ on the right side of (3).

$$\frac{dq}{dz} = -A(z) \qquad (20)$$

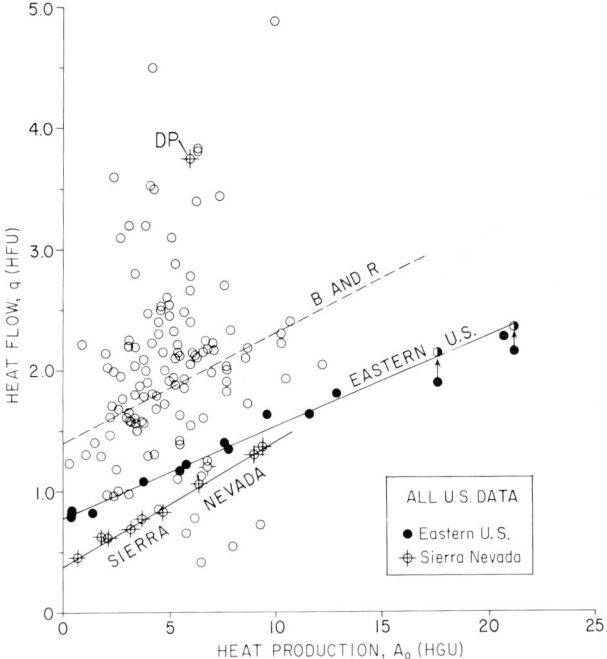

Fig. 13. Observations of heat flow q and radioactive heat production A_o from crystalline rocks in the conterminous United States; linear regression lines are from Roy et al. [1968a] for the Basin and Range (dashed), eastern United States, and Sierra Nevada provinces. Solid circles represent points east of the Great Plains, and crossed circles represent points interior to the Sierra Nevada physiographic province. Vertical arrows represent corrections for finite size of plutons [Roy et al., 1968a]. Three of the open circles on the eastern United States curve at about 1 HFU are from the Klamath Mountains in northern California. The crossed circle slightly above the Sierra Nevada line (q = 1.1) has an uncertain heat production. DP is adjacent to the Long Valley volcanic center.

The linear relation for either province may be written

$$q = q^* + DA_o \qquad (21)$$

where q and A_o are heat flow and heat production measured near the surface z = 0, and q^* and D are the intercept and slope parameters that define the heat flow province. Rocks at sites satisfying (21) vary greatly in age and have different histories of uplift and erosion; unless the relation (21) is an accident of the present, it should remain valid after erosion by an arbitrary amount z at any location. Thus if at any site a layer of thickness z is eroded away, A_o will take on a new value $A(z)$ depending on how radioactivity is distributed with depth, and q will take on a new value $q(z)$, but (21) should still apply. Hence we let q be a function of z and replace A_o by $A(z)$ in (21). Then substi-

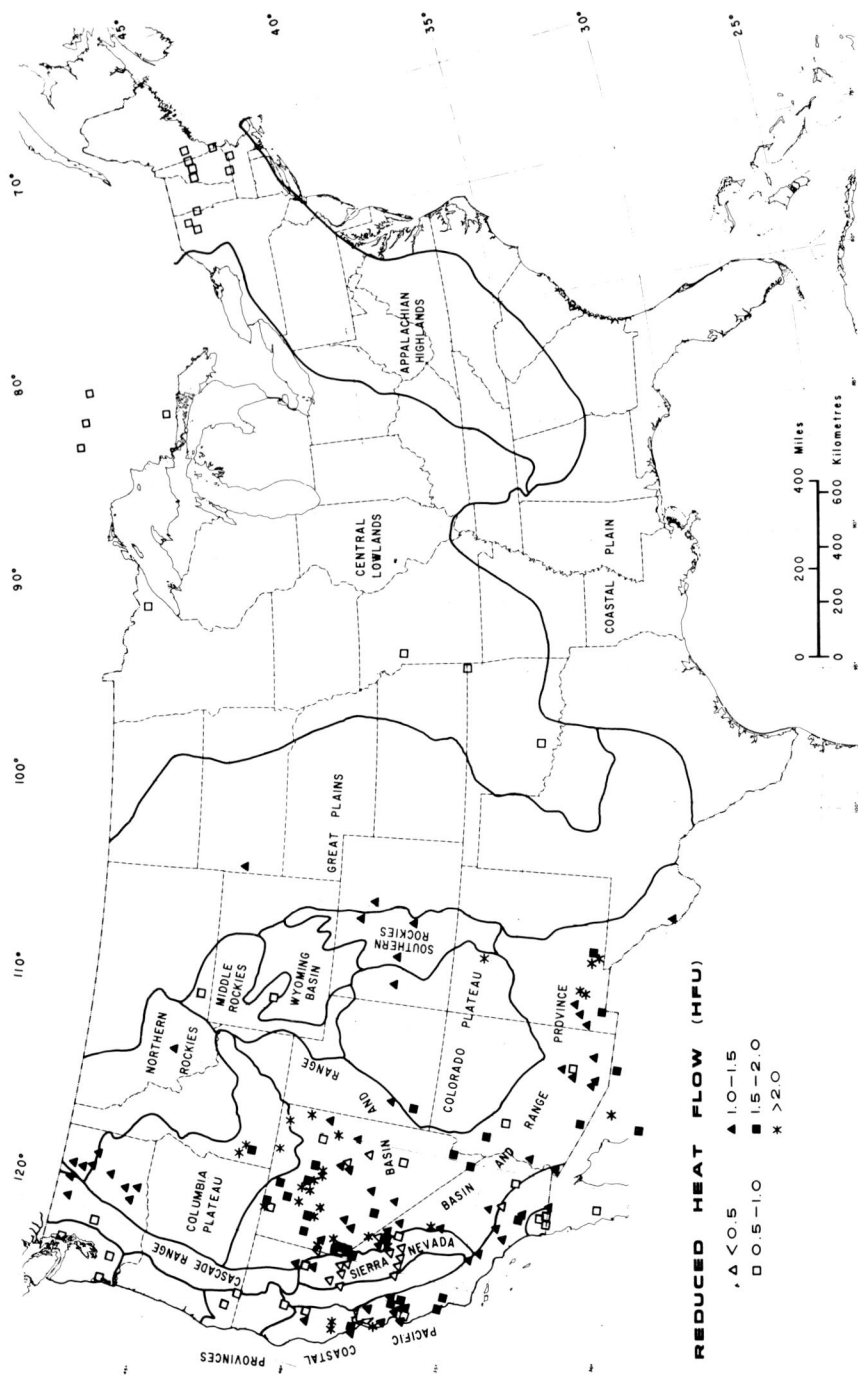

Fig. 14. Reduced heat flows and physiographic provinces (modified from Fenneman [1946].) Abbreviations are KM for Klamath Mountains, SAFZ for San Andreas Fault zone, and GV for Great Valley of California.

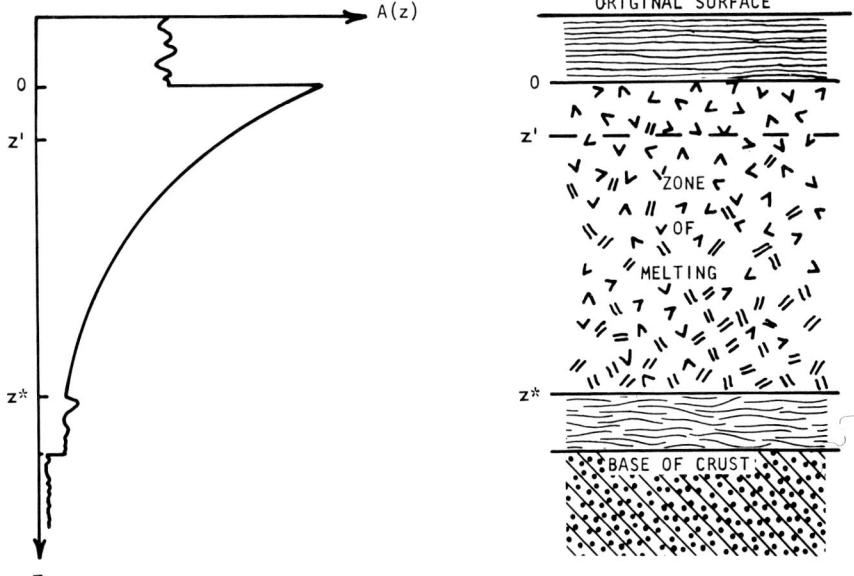

Fig. 15. Conceptual model for the distribution of heat production A(z) in the continental crust (see text).

tuting (21) in (20) yields a unique result for heat production as a function of depth

$$A(z) = A_o e^{-z/D} \qquad (22)$$

where A_o is the value on the presently exposed surface, $z = 0$. Suppose that this distribution extends to some depth z^* and that $q(z^*)$ is the heat flow at that depth. Then by using (22) the surface heat flow q is

$$q = q(z^*) + \int_{z^*}^{0} A(z) \, dz \qquad (23a)$$

$$q = [q(z^*) - DA_o e^{-z^*/D}] + DA_o \qquad (23b)$$

The expression in brackets represents the empirically determined intercept q* (equation (21)). The slope parameter D is about 7.5 km for the eastern United States and 10 km for the Sierra. Hence if the exponentially fractionated layer z^* extends throughout all or most of the crust ($z^* \gg D$), the exponential term in (23b) is small, and $q(z^*)$ corresponds approximately to q*, which will probably be approximately the same as the mantle contribution. (For a more complete discussion, see Lachenbruch [1970].)

This simple model is illustrated conceptually in Figure 15. In the plutonic and highly metamorphosed rocks thought to make up most of the crust, U, Th, and K are fractionated upward exponentially, presumably by some process taking place during stages of partial melting or migration of metamorphic fluids [e.g., Lambert and Heier, 1967; Albarede, 1975].

The characteristic depth D is a parameter characterizing the fractionation process. If we assume that after equilibrium is established, heat flow into the lower crust q* becomes uniform throughout the province, then subsequent measurements in exposed crystalline rocks would generally follow the linear relation observed (equation (21)).

Other assumptions regarding radioactivity of the lower crust are possible, and if the constraint imposed by differential erosion is set aside, source distributions in the upper crust other than (22) are permissible [see, e.g., Roy et al., 1968a; Lachenbruch, 1970; Blackwell, 1971]. Any simple source distribution model is, of course, approximate, as large variations are known to occur on all observable scales. Although direct observational evidence on the form of A(z) is weak, statistical studies of heat production in deep boreholes [Lachenbruch, 1971; Lachenbruch and Bunker, 1971] and geologic studies of differentially eroded plutons [Swanberg, 1972] provide some support for the exponential model. In any case, as first pointed out by Birch et al. [1968], in provinces where the linear relation applies, local variations in heat flow are probably caused by variations in radioactivity strongly concentrated in the upper crust, and heat flow through the lower crust is evidently uniform. Important corollaries are that (1) convective heat transfer is probably insignificant in the crust in these provinces, for otherwise it would have to be uniform throughout each province, and (2) if transient conditions occur, they must be uniform throughout the province. This suggests a deep mantle origin for such transients; e.g., a cool subducted slab deep beneath the Sierra Nevada has been suggested to explain the very low q* there [Roy et al., 1972; Blackwell, 1971].

For the exponential model described, crustal temperatures are given by [Lachenbruch, 1970]

$$\Theta(z) = [q^*z + D^2 A_o (1 - e^{-z/D})]/K \qquad (24)$$

where K is thermal conductivity. Crustal temperatures for the appropriate province parameters (q*, D) and for the ranges of heat production observed in each province are given for an assumed uniform conductivity (K = 6 mcal/cm s °C) in Figure 16. Similar curves, some with different assumptions, can be found elsewhere [Roy et al., 1968a; Lachenbruch, 1968, 1970; Roy et al., 1972; Diment et al., 1975; Lachenbruch et al., 1976a; Blackwell, 1971]. However, in provinces where the linear relation is found to apply, crustal temperatures are severely constrained, and models more specialized than this simple one (equation (24)) do not give significantly different results. (The problem has been discussed by Blackwell [1971].) The chief uncertainty is in the choice of conductivity, which might be adjusted by ±15% with a proportional effect on the temperatures shown. For reference we have shown in Figure 16 the curves for the beginning of melting for granodiorite with pore water present (GSS) and without it (GDS); they are discussed further in the next section.

In the highly radioactive rocks of New England (shaded on map, Figure 2) crustal temperatures probably lie between the central and upper (eastern United States) curves (A_o = 10 and 20 HGU) of Figure 16, whereas for most of the eastern United States, temperatures are expected to lie in the lower half of the range shown (between A_o = 0 and A_o = 10 HGU). For the Sierra Nevada curves, temperatures in the upper range

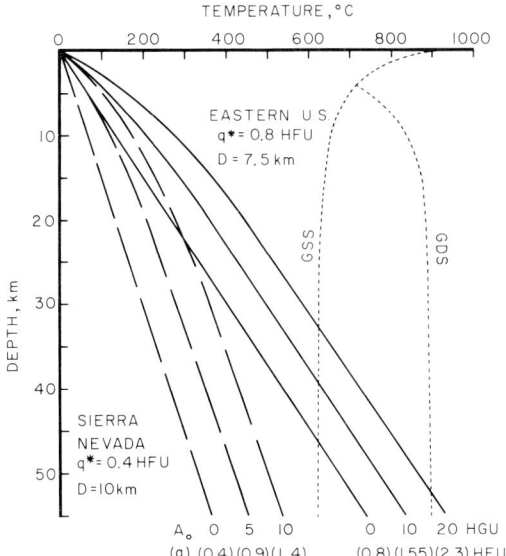

Fig. 16. Steady crustal temperature profiles based on the linear heat flow-heat production relation (equation (24)) for the Sierra Nevada (long-dashed curves) and the central and eastern United States (solid curves) for the range of heat production (A_o) observed in each province. Corresponding values of heat flow (q) are shown in parentheses. Assumed thermal conductivity is 6 mcal/cm s °C. Short-dashed curves GSS (granodiorite saturated solidus) and GDS (granodiorite dry solidus) from Wyllie [1971] show the beginning of melting for rock of intermediate composition.

(A_o = 5 to 10 HGU, Figure 16) are characteristic of the younger (Upper Cretaceous) plutons near the crest of the mountain system, and in the lower range (A_o = 0 to 5 HGU) they generally represent conditions in the western foothills. For the same heat flow, temperatures are lower in the Sierra Nevada, where a smaller fraction of the heat originates at depth ($q^* \simeq 0.4$ HFU in the Sierra and $q^* \simeq 0.8$ HFU in the eastern United States).

The Basin and Range Province: Effects of Radioactivity and Convection

The third heat flow province defined by Roy et al. [1968a] is the Basin and Range province, for which all current (q, A_o) data are shown in Figure 17. The locations from which these data were obtained are indicated by the points within the Basin and Range province boundary in Figure 14 and those points in Mexico south of Arizona [from Smith, 1974]. The regression line (dashed, Figure 17) determined by Roy et al. [1968a] from 12 of 15 available data pairs had a slope D of 9.4 ± 1.3 km and intercept q^* of 1.4 ± 0.09 HFU. It is clear from the scatter in Figure 17 that if the Basin and Range province is in some sense represented by this line, it is not in the same sense that the

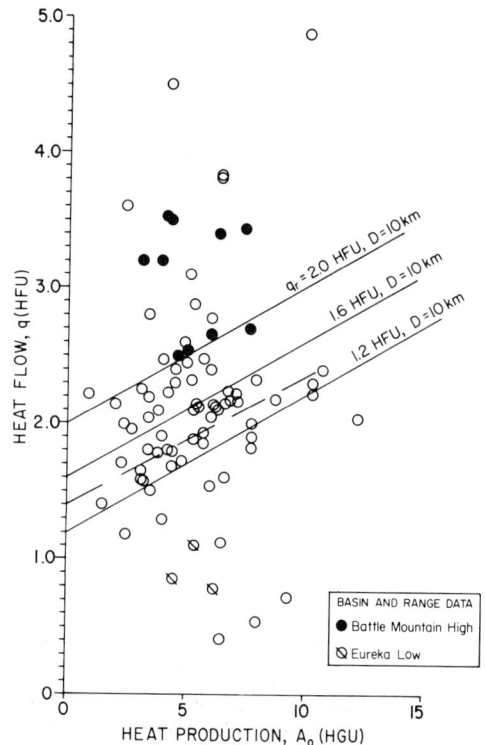

Fig. 17. Observations of heat flow q and radioactive heat production A_o from crystalline rock of the Basin and Range province. Regression line from earlier studies is dashed.

Sierra Nevada and eastern United States provinces are represented by their lines (Figure 13). As the Basin and Range province now has by far the most observations, it could be argued that the linear regression lines will lose their significance in the other provinces, too, as more data are acquired. However, we consider this unlikely, as the density of observation is presently as great in the Sierra Nevada as it is in the Basin and Range, and it was recognized at the outset by Roy et al. [1968a] that the regression analysis was least significant in the Basin and Range province. A more discriminating use of the variable quality data shown in Figure 17 might provide justification for a linear relation in the Basin and Range or parts of it, and this is under study. For the present, however, it appears that the linear regression of q on A_o has little significance in the Basin and Range province, and the question arises whether insights from the foregoing simple model of the linear relation can be applied usefully there.

According to the simple model the steady state contribution from crustal radioactivity is DA_o, where D is a geochemical characteristic of the crust describing its fractionation. This physical interpretation does not depend upon the constancy of lower crustal heat flow (q^*) and is reasonable whether or not the linear relation applies. On this basis, we might subtract the term for crustal radioactivity from the

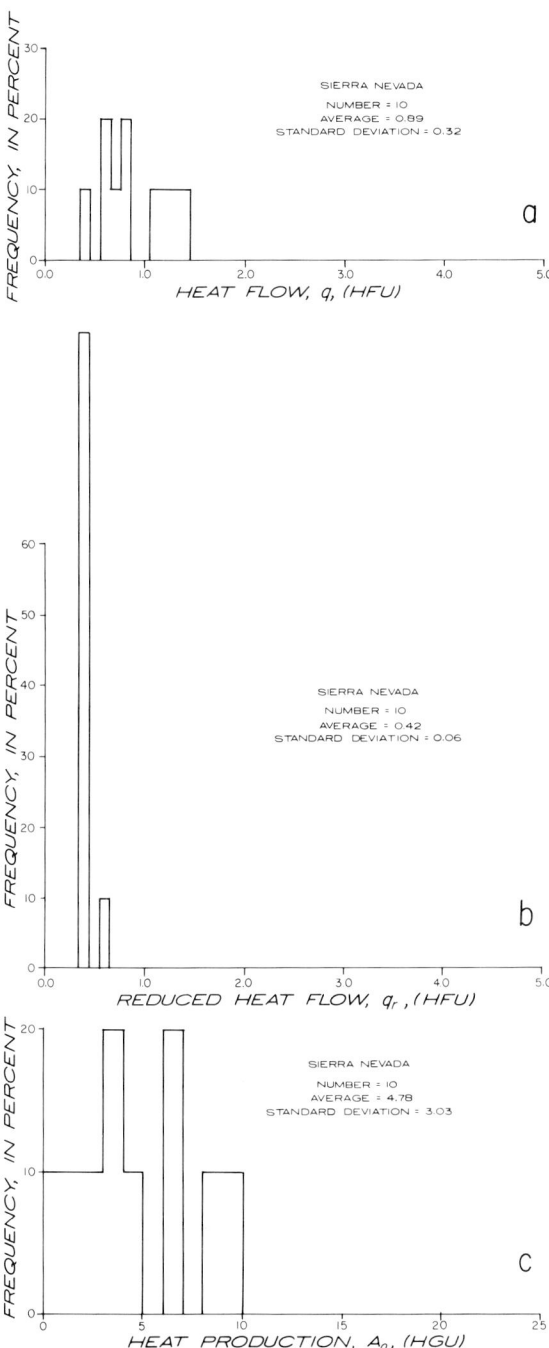

Fig. 18. Histograms of (a) heat flow, (b) reduced heat flow, and (c) radioactive heat production for stations at which both q and A_o were measured in the Sierra Nevada province.

observed heat flow to obtain the 'reduced heat flow' q_r employed by Roy et al. [1972].

$$q_r = q - DA_o \qquad (25)$$

Although D is not known directly for the Basin and Range, we might reasonably assume it to have the constant value of 10 km, firmly established in contiguous rocks of the adjacent Sierra Nevada province (granitic rocks in the western Basin and Range province are part of the Sierra Nevada batholith).

Histograms of the three variables in (25), heat flow q, reduced heat flow q_r, and heat production A_o, are shown in parts a, b, and c, respectively, of Figures 18-20 for the three provinces. For a province in which the linear relation applied in a deterministic sense, q_r would be identical to q*, a constant (equation (21)). It is seen from Figures 18 and 19 that this is nearly the case for the Sierra Nevada and eastern United States. In such provinces the simple interpretive model gives q_r a clear-cut physical meaning; it is the uniform flux q* from the mantle or at least the lower crust. No such interpretation is possible for the Basin and Range, where q_r is widely dispersed (Figure 20b). A second distinctive feature of the Basin and Range data (Figure 20b) is the large mean value for q_r, about 1.6 HFU or twice the value for the stable eastern United States usually considered as normal. On the basis of an earlier discussion we believe that the large dispersion results primarily from convection by groundwater, and the large mean, from convection by magma. Before discussing implications of these two inferences we shall consider how widespread the conditions represented by the sample in Figure 20 might be.

In brackets in Figures 20a and 20b, statistics are given for the Basin and Range province excluding determinations in the anomalous subprovinces (10 values from the Battle Mountain High and 3 values from the Eureka Low). For comparison, we show in Figure 21 the corresponding results for the complete population of heat flow measurements in the Basin and Range province; again the brackets enclose statistics determined with the two subprovinces excluded (20 values from the Battle Mountain High and 13 values from the Eureka Low). The fact that the principal mode, the mean, and the standard deviation are essentially the same for data in brackets in Figures 20a and 21 adds a note of generality to the analysis of data in Figure 20. However, without a more careful study of the individual sites than we have yet undertaken, more refined statistical treatment is not warranted. Reduced heat flows reported from the northern and southern Rocky Mountains and the Columbia Plateau (Figure 14) seem commonly to fall in the modal range (1.2-1.6 HFU) characteristic of the Basin and Range data, and those provinces are expected to have similar crustal regimes [see Blackwell, 1969, 1971; Roy et al., 1972]. We have focused on the Basin and Range province because it seems to represent a fairly continuous tectonic unit, and the heat flow coverage there is relatively dense.

The reduction for crustal radioactivity in the Basin and Range province does not significantly reduce the dispersion, and we should be little better off in estimating the heat flow with a knowledge of local radioactivity than without it. This does not mean that the reduction (25) for crustal radioactivity in the Basin and Range is not valid. The

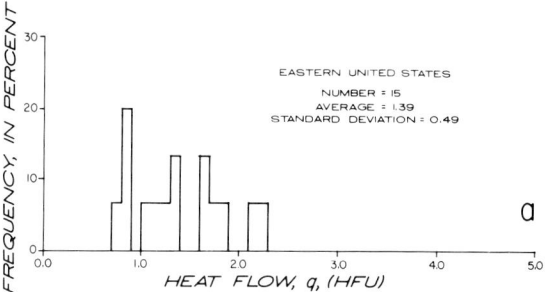
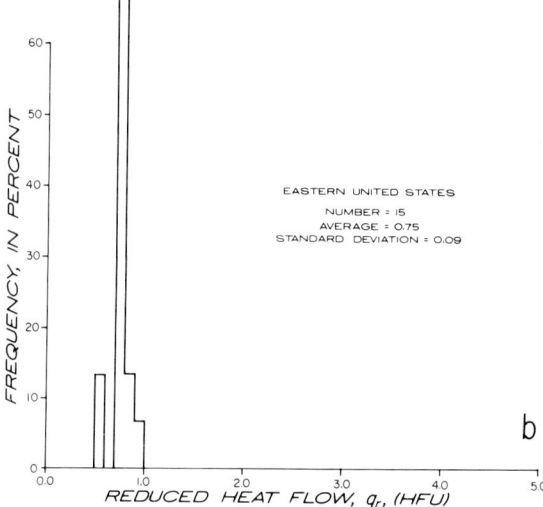
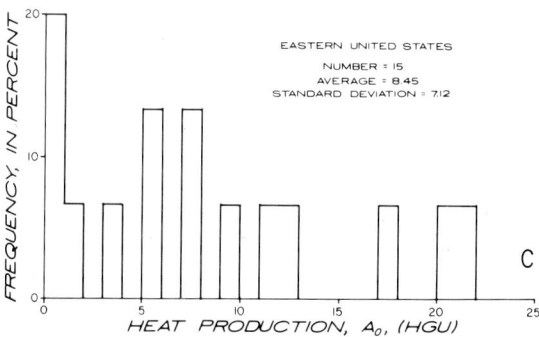

Fig. 19. Histograms of (a) heat flow, (b) reduced heat flow, and (c) radioactive heat production for stations at which both q and A_0 were measured in the eastern United States province (i.e., east of the Great Plains).

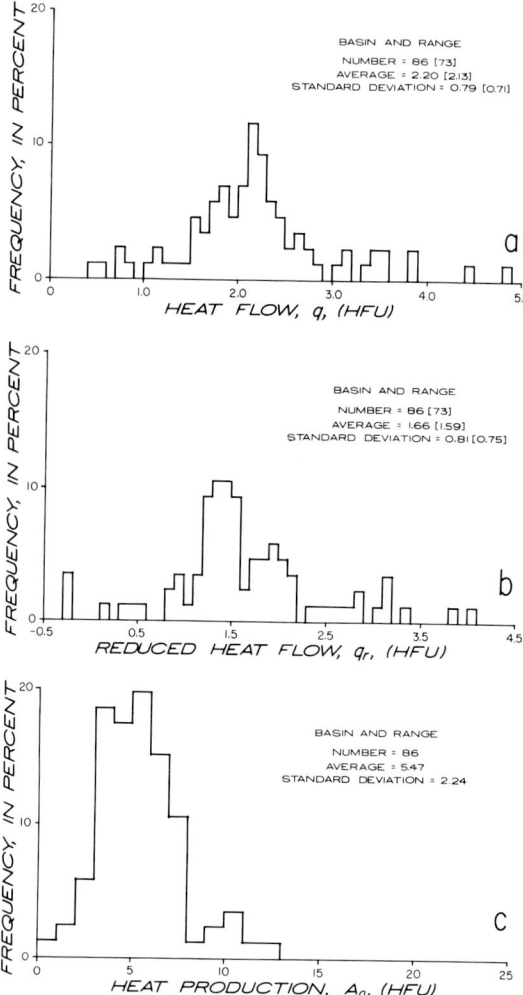

Fig. 20. Histograms of (a) heat flow, (b) reduced heat flow, and (c) radioactive heat production for stations at which both q and A_o were measured in the Basin and Range province. Statistics in brackets in (a) and (b) were determined with data from Battle Mountain High and Eureka Low deleted.

standard deviation of the crustal correction (DA_o with $D = 10$ km, Figure 20c) is only 0.22 HFU, about one fourth of the standard deviations for both q and q_r. If q and A_o were normally distributed, the reduction from q to q_r (equation (25)) would have an insignificant effect on the standard deviation (~ 0.03 HFU). The statistics do suggest that three-dimensional effects, thermal transients, and convection, neglected in the simple theory (equation (20)), are substantially greater (generally by a factor of 3 or 4) than the effects of variable crustal radioactivity. As three-dimensional effects are evidently unimportant in the other provinces, it is likely that the dispersion of

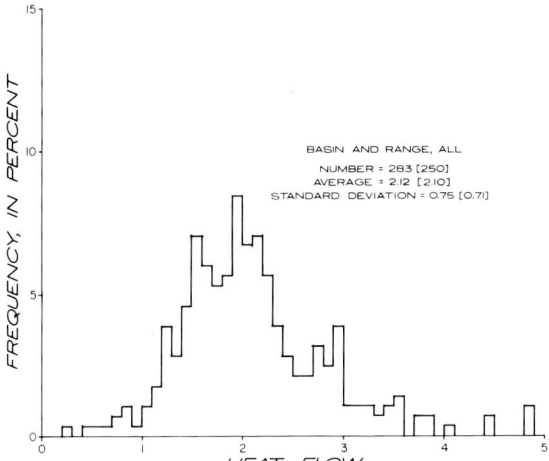

Fig. 21. Histogram of all heat flow data from the Basin and Range province. Statistics in brackets were determined with data from the Battle Mountain High and Eureka Low deleted.

reduced heat flow in the Basin and Range is due primarily to hydrothermal and (to a lesser extent) magmatic convection, including, of course, their time dependent effects, as discussed in an earlier section.

We have mentioned that in the Basin and Range province the mean conductive flux reduced for crustal radioactivity is about twice as large as would be expected in stable regions. We naturally associate this large and variable reduced heat flow with the present pattern of extensional deformation, magmatism, and hot spring activity that has probably characterized the province for the past 17 m.y. [Thompson and Burke, 1974]. It has been shown that convective processes operating solely within the crust over this time (twice the conductive time constant for the crust) would probably reduce the mean conductive flux, not increase it. Hence the excess heat is probably supplied convectively by magma rising across (and possibly beneath) the base of the crust. Regional variations in the intensity of this magmatic upflux are probably responsible for high heat flow subprovinces like the Battle Mountain High and the Rio Grande Trough, and for local silicic volcanic centers like the Long Valley caldera as well. It is likely that these variations are, in turn, caused by local variations in the rate of crustal extension [Lachenbruch et al., 1976a] (A. H. Lachenbruch and J. H. Sass, unpublished models).

It is useful to assume that over large areas the crust is in a quasi-steady state, receiving as much heat by conduction and convection across its base as it gives off by conduction and convection at its surface. As q_r is normally estimated from the conductive heat flow, the actual mantle contribution would be larger than q_r by the amount of convective loss from hot and warm springs and volcanos. However, the heat delivered by post-Oligocene extrusive rocks in the Great Basin is negligible in relation to the conductive heat flow, and judging from Figure 12, the net effect of the hotter known springs might not be important on a regional scale. Nevertheless details of the total hydrologic heat loss are poorly known, and we shall neglect them although they could be significant. In regions where the convective flux is known, it can be accommodated in the computation of q_r (by using combined surface flux q_c instead of q in equation (25)) if the steady state assumption seems to warrant it.

If there is a characteristic flux into the lower crust of the Basin and Range, it is likely that it is represented by the modal value of q_r, with the nonmodal values generally representing superimposed anomalous convective effects. Excluding the two subprovinces, almost half the values of q_r fall in the modal range 1.2-1.6 HFU (Figure 20b). (Interestingly, the value of q* originally determined from linear regression by Roy et al. [1968a] was 1.4 HFU.) The Battle Mountain High, a positive anomaly with lateral dimensions of many crustal thicknesses, has been defined as a region with q > 2.5, but it is essentially unchanged if defined as the region q_r > 2. Figure 22 shows crustal profiles (equation (24)) for a steady conductive mantle flux q_r of 1.2 and 1.6, which might bracket the 'characteristic' conditions in the Basin and Range province, and of 2.0 and 2.5, which are intended to represent (lower) limiting and typical conditions in the Battle Mountain High. (The mean for the 20 heat flows in the Battle Mountain High is 3.0 ± 0.4 standard deviation; the mean for the 10 reduced heat flows is 2.5 ± 0.4 standard deviation.) As such large variations in q_r (from 1.2 to 2.5) have a far greater effect on temperature than variations in radioactivity, the curves are shown only for the near-average A_o of 5 HGU (Figure 20c). Likely variations in A_o (of ±3 HGU, Figure 20c) would change the deep crustal temperatures by only ±50°C. Variations in thermal conductivity of ±15% from the assumed value of 6 mcal/cm s °C would change the temperatures in Figure 22 by ±15%. Shown also in Figure 22 is the curve for the Sierra Nevada ($\overline{q^*}$ = 0.4 HFU) and the curve for q* = 0.8 HFU, which is similar to that for the stable eastern United States except that it is drawn for D = 10 km (instead of 7.5 km) for consistency. The latter is a useful (if somewhat arbitrary) reference for conditions one might expect in the Basin and Range crust if it were underlain by a stable mantle; it differs from the corresponding curve (q* = 0.8 HFU, D = 7.5 km) for the stable eastern United States by less than 45°C.

Although the curves in Figure 22 are drawn as if all of the mantle flux q_r were conducted from the base of the crust, this condition is not required nor is it expected to apply in the hotter regimes. Insofar as our observations at the surface are concerned, the anomalous source may be produced by repeated intrusion at any depth; if the source persists long enough to establish a steady state, the temperatures above it will fall on the appropriate curve of Figure 22. The time τ(z) required for a sill-like constant temperature source to equilibrate the overburden is shown on the depth axis in Figure 22. (A sill-like source whose strength does not change with time takes about three times as long. During the slow cooling following solidification the average gradient anomaly in the overburden will remain rather close to that measured at the surface, provided that the original source persisted for many τ.) Thus a continuing intrusive process that maintained the 20-km temperature at 900°C for more than 3 m.y. would cause a conductive regime above it as indicated by curve F (Figure 22); thereafter, downward extrapolation from surface observations would correctly identify the 20-km temperature.

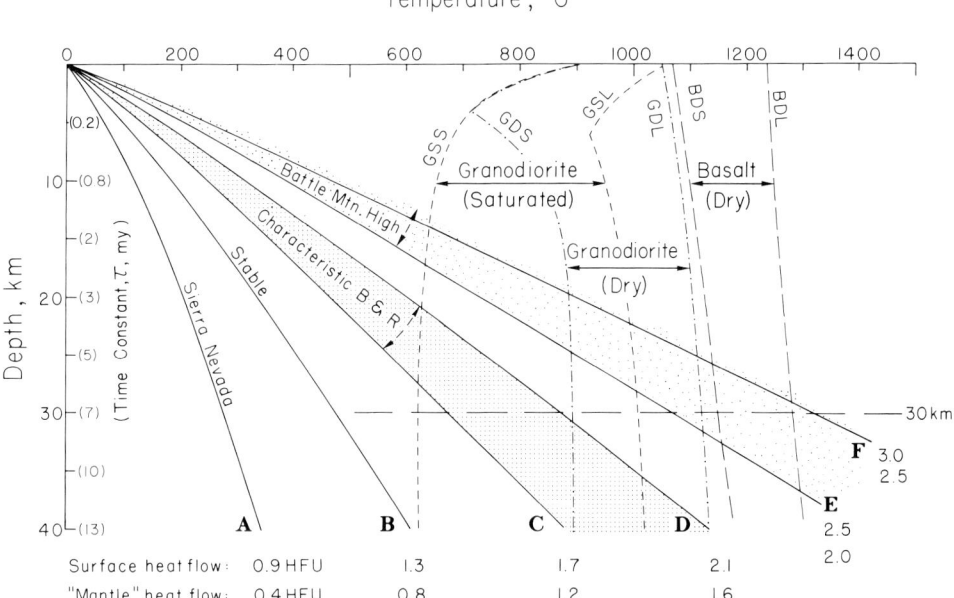

Fig. 22. Generalized conductive temperature profiles for the Sierra Nevada crust (A), a stable reference crust (see text) (B), the characteristic Basin and Range crust (C to D), and lower limiting (E) and typical (F) conditions in the crust of the Battle Mountain High. All curves are drawn for a surface heat production (A_o) of 5 HGU and thermal conductivity (K) of 6 mcal/cm s °C (equation (24)). Corresponding surface heat flow and reduced or 'mantle' heat flow are shown at the bottom of each curve. Melting relations [after Wyllie, 1971] are shown for intermediate crustal rock by the curves GSS (granodiorite saturated solidus), GSL (granodiorite saturated liquidus), GDS (granodiorite dry solidus), and GDL (granodiorite dry liquidus) and for basalt by BDS (basalt dry solidus), and BDL (basalt dry liquidus). In parentheses on the depth axis is shown the time required for the overburden to approach thermal equilibrium after intrusion by a sill maintained at constant temperature.

Above the intrusion we would measure a surface heat flow of 3 HFU and a reduced heat flow of 2.5, and (relative to curve B, Figure 22) the contribution of anomalous convected flux would be 1.7 HFU. The actual anomalous flux into the lithosphere at the time might be greater or less than 1.7 HFU, depending upon whether the lower portion was absorbing or releasing heat, i.e., whether its temperature was approaching the new stationary state from below or above. Convective transfer by magma rising in the lower crust (below 20 km in this example) would cause mean temperatures there to be less than indicated by the conductive curve F (A. H. Lachenbruch and J. H. Sass, unpublished models). Convection in this lower region might involve no more than the upward movement of basalt in narrow conduits en route to the 20-km depth, or it could involve complex networks of basaltic intrusion and secondary diapheric movements of silicic melts and adjustments of solid rock.

In Figure 22 we have shown [from Wyllie, 1971] some limiting melting relations for materials likely to be involved in convective heat transport to and through an anomalously hot crust. The most probable source of anomalous heat is upward migration of basalt (or possibly other mantle-derived magma) beneath or into the base of the crust. For reasonable water contents, most of the crystallization and hence most of the latent heat release will have occurred by the time the basalt cools past its dry solidus (BDS, Figure 22) [Wyllie, 1971; Peck et al., 1966]. Heat introduced by the basalt could raise the temperature and melt fractions of the indigenous crustal material, assumed to have an intermediate (granodiorite) composition. In the presence of excess water ('vapor present condition'), such rocks would begin to melt along the curve GSS (granodiorite saturated solidus) (Figure 22), and melting would be complete at GSL (granodiorite saturated liquidus). If no water occurred in the crustal rock except that bound in hydrous minerals, melting would not begin until temperatures exceeded GDS (granodiorite dry solidus), and it would not be complete until they reached GDL (granodiorite dry liquidus). If a trace of pore water were present, it would dissolve preferentially in a melt (of rhyolite composition) that would begin to form along GSS. With further increase of temperature the increasing melted fraction would become more and more undersaturated, making the residual crystals more difficult to melt; complete melting would not occur until temperatures approached GDL. As the radioactive elements, like the water, move preferentially into the melt, upward migration of the melt might produce the condition illustrated schematically in Figure 15 and at the same time dehydrate the lower crust. Wyllie [1971] has pointed out that unless the lower crust were somehow rehydrated, a second cycle of lower crustal melting would be more difficult, as it would require temperatures in excess of GDS. If repeated cycles did occur, however, we might expect more complete upward fractionation of radioelements. This would appear in surface observations as a decrease in the value of the characteristic depth D (equation (21)).

It is seen from Figure 22 that at the base of a 30-km crust in the Battle Mountain High, basalt melt could be stable, and dry intermediate crustal rock could be completely melted. Hence some heat is probably transferred by magmatic intrusion in the lower crust of the Battle Mountain High, for if it were not, the thermal regime would be conductive, and the base would be nearly or quite all melted. Substantial amounts of melting of most crustal rocks in the presence of pore water could occur near the base of a 30-km crust in the 'characteristic Basin and Range' regimes. Hence 'first cycle' (wet) crustal fractionation could be initiated under the Basin and Range regimes, and 'second cycle' (dry) fractionation could occur under the Battle Mountain High regime. Convection by quasi-steady upward migration of a melted fraction could, of course, convert one regime to the other.

Typically, the upper 20 km of the Basin and Range crust has a seismic velocity ($V_p \simeq 6.0$ km/s) characteristic of silicic rocks, including granite, and the lower 10 km or so has a higher velocity ($V_p \simeq 6.7$ km/s) characteristic of denser (and presumably more refractory) materials, including basalt [e.g., Hill and Pakiser, 1966]. As temperatures in the Battle Mountain High and similar regions can be in the range 700°–900°C at depths of 15-20 km (curve F, Figure 22), laterally extensive partial silicic melts could occur at midcrustal levels in such regions. Convective transfer attending stretching and intrusion of the extending lithosphere could, however, reduce these temperature estimates by 100°C or so (A. H. Lachenbruch and J. H. Sass, unpublished).

In summary, we have found that the linear relation between heat flow and surface radioactivity does not apply generally in the Basin and Range province. For the linear relation to apply, crustal contributions to heat flow should be exclusively from radioactivity, and the mantle flux should be uniform. We believe that the relation fails in the Basin and Range province because both conditions are violated there. Variations in surface heat flow caused by hydrothermal and magmatic convection overshadow variations caused by crustal radioactivity (they are probably greater by a factor of 3 or 4), and the anomalously large mantle flux is not uniform. Mantle heat flux is probably controlled by magmatic mass flux (into or beneath the base of the crust) which varies in intensity, creating subprovinces like the Battle Mountain High and the Rio Grande Rift, and more local heat flow anomalies and volcanic centers as well (Figure 3). Frequently occurring values of reduced heat flow suggest that the mantle flux throughout much of the province might have characteristic values in the range 1.2-1.6 HFU. These considerations form the basis for construction of generalized crustal temperature profiles which can be discussed in terms of melting relations for crustal rocks. Theoretical temperatures are consistent with the extensive manifestations of magmatic activity observed in the province.

Discussion and Summary

Our knowledge of regional heat flow in the United States has been acquired only recently. In his review of the status of geothermal investigations in 1954, Birch [1954a] was able to cite only three 'reasonably adequate' determinations (0.93 HFU in northern Michigan [Birch, 1954b], 1.7 HFU in the Colorado Front Range [Birch, 1950], and 1.29 HFU in the Central Valley of California [Benfield, 1947]). Although we now know that each of these values is quite representative of its geologic province, little could be deduced about regional patterns from three determinations. In a review about a decade later, Lee and Uyeda [1965] listed heat flow from about 40 distinct sites in the conterminous United States; the data indicated that heat flow was generally higher in the tectonically active western United States than in the more stable eastern and central portions. A few years later, publication of over 100 new values [Roy et al., 1968b; Blackwell, 1969] revealed correlations between heat flow and higher-order tectonic and geologic features, chiefly, high heat flow in New England and in the Basin and Range province, the northern and southern Rockies, and the Columbia Plateau and a band of lower heat flow near the Pacific Coast, features shown in the maps of Roy et al. [1972] and Blackwell [1971]. A map by Diment et al. [1972] showed the systematic variations of heat flow in the Appalachian Mountain region. Sass et al. [1971] presented 100 or so additional values revealing the strong correlation of heat flow with the major N-S trending tectonic provinces of California, including the San Andreas Fault zone. They also defined subprovinces of high and low heat flow in the Basin and Range province, to which another, the Rio Grande Rift, has more recently been added [Decker and Smithson, 1975; Reiter et al., 1975]. Further detail is shown in the map presented here, and we can, of course, expect the trend to continue.

Our understanding of these heat flow observations was increased substantially in 1968 by the discovery of the linear relation between heat flow and surface radioactivity in New England [Birch et al., 1968], its application to other provinces [Roy et al., 1968a], and its independent confirmation for the Sierra Nevada [Lachenbruch, 1968]. Curiously, 8 years later we still do not know how general this relation might be. It has been shown above that in one of the provinces where the relation was formerly thought to apply approximately (the Basin and Range province), it does not apply. Although the relation is now supported by studies in crystalline rocks of Canada [Cermak and Jessop, 1971], Australia [Jaeger, 1970; Sass et al., 1976c], India [Rao et al., 1976], and Norway [Swanberg et al., 1974], the most convincing results remain those from the provinces in which the relation was first discovered, the eastern United States and the Sierra Nevada. As anticipated by Roy et al. [1968a], data from the Canadian Shield [Cermak and Jessop, 1971] and to a lesser extent those from central Australia are accommodated reasonably well by the line for the stable eastern United States, but the parameters determined from these areas independently are somewhat different; in particular, the values $q^* = 0.64$ HFU and $D = 11.1$ km for nine points in central Australia [Sass et al., 1976c] seem significantly different. Several isolated determinations such as the three from the Klamath Mountains in northern California (Figure 13 and Lachenbruch and Sass [1973]) and two from Precambrian rocks in southern India [Rao et al., 1976] fall on the eastern United States curve. The data from southern Norway [Swanberg et al., 1974] lie rather close to the Sierra Nevada line; the linear regression analysis there yielded $q^* = 0.48$ HFU and $D = 8.4$ km. However, nine points from Precambrian rocks of western Australia [Sass et al., 1976c] yield $q^* = 0.63$ and $D = 4.5$ km, a line quite different from those observed elsewhere, and four points from northern and central India [Rao et al., 1976] yield $q^* = 0.92$ HFU and $D = 14.8$ km. Judging from our experience with the Basin and Range province, many more observations will be needed in all of these areas, and others, to establish the general significance of the relation between heat flow and radioactivity in crystalline rocks. As we have remarked, the importance of this relation is the requirement it places on the vertical distribution of crustal radioactivity and to a lesser extent on the total contribution of crustal radioactivity to surface heat flow. These requirements relate to the processes responsible for evolution of the continental crust. In order for the linear relation to obtain, several other special requirements must be met [Lachenbruch, 1970]; i.e., three-dimensional conductive effects and magmatic and hydrothermal convection must be negligible in the crust, and crustal transients and mantle heat flow must be uniform throughout the province. Thus the linear relation can be expected to apply only in more stable regions, and even there, only under rather special circumstances. Nevertheless, the insight obtained from the relation can provide a basis for estimating the contribution of crustal radioactivity elsewhere on the continents. Thus the 'reduced heat flow,' obtained by subtracting the estimated crustal contribution from observed heat flow [Roy et al., 1972], might be used to interpret continental heat flow where the linear relation does not apply. As the reduction depends upon the value of D, which varies around the world by a factor of 3, according to presently available studies, the reduction must be applied with caution. The justification for using D = 10 km in the Basin and Range province is provided by the linear relation observed in the Sierra Nevada and the observation that granitic rocks in the western Basin and Range at least are part of the Sierra Nevada batholith. Reasonably confident reductions can sometimes be made in regions where crystalline rock is not exposed if enough information on crustal composition is available from other sources. Examples are the California coast ranges, where the thickness of the Franciscan formation (of known radioactivity) is estimated from seismic studies [Lachenbruch and Sass, 1973], and the Pacific Northwest coastal provinces, where the basement rock is believed to be largely mafic [Blackwell, 1971; Sass et al., 1971] and hence to have rather low radioactivity. It is seen from Figure 14 that reduced heat flow estimated for sites along the Pacific Coast is generally in the range characteristic of the eastern United States except in the band through western and south central California enclosing the San Andreas Fault, where it is generally similar to values in the Basin and Range province.

A rather complete description of regional heat flow in the United States and a thorough understanding of its implications for thermal state and processes in the crust will probably be needed for a comprehensive assessment of our geothermal energy resource and the formulation of rational plans for exploring it and exploiting it. As we have seen, even in regions where the heat flow is low and the likelihood of an exploitable resource is slight, the understanding obtained from heat flow studies can be important for unraveling the more complex thermal problems in high heat flow areas. It is seen from Figures 1-3 that only now, with more than 600 determinations, are the areas of extremely high heat flow beginning to emerge as regional features; their boundaries are generally unknown, however, and few areas are sampled adequately for an understanding of hydrothermal systems in a crustal context. Large areas of the heat flow control map (Figure 1) are blank. Ironically, many of these are oil-producing areas where every year more than 10,000 holes are drilled and temperatures are measured. It is likely that a knowledge of regional heat flow in these areas would increase our understanding of the widely discussed 'geopressured' energy resources of the Gulf Coast [see, e.g., Jones, 1969; Jones and Wallace, 1974; Papadopulos et al., 1975] and of the general problem of thermal evolution of sedimentary basins and the maturation of hydrocarbons. A comprehensive compilation of temperature gradient from some 25,000 sites in oil-producing areas of North America has recently been published by the American Association of Petroleum Geologists in cooperation with the USGS [American Association of Petroleum Geologists-U.S. Geological Survey, 1976]; it contains much useful regional information. However, as was emphasized by Birch [1954a], the principal variable affecting temperature gradient in the outer layers of the crust is thermal conductivity (and locally, water movement). Hence a compilation of the temperature gradient alone can be expected to tell us much more about the variations in conductivity (and locally, water movement) than about variations in the more fundamental quantity, heat flow. Regional heat flow studies are proceeding at a modest pace (limited by the availability of drill holes) at a handful of research laboratories across the country. In view of the importance of regional studies to the energy industries and the enormous amount of drilling they undertake, more interaction between these groups should hold substantial advantage for all.

Heat convected by moving groundwater requires careful attention in heat flow studies; it can perturb or completely dominate the regional flux associated with the crustal regime at depth. In tectonically active regions, open fractures, permeable volcanic rocks, and high heat flow may result in circulatory convection systems driven primarily by thermal density differences. Such systems are expected to produce perturbations to surface heat flux which change sign over lateral distances of the order of the depth of circulation, probably up to many kilometers. If the system is sustained by upper crustal magmatic intrusion, the combined flux at the surface might be much greater than the regional value for millions of years. If the system is sustained by regional heat flow, the combined flux will probably fall to the regional value after 10^3-10^5 years (depending on circulation depth), but a lingering conduction anomaly will persist long after the circulation stops. Perturbations caused by these systems create large dispersion in conductive flux from tectonically active regions, making it difficult to identify and interpret the regional heat flow without a dense network of observations [see, e.g., Blackwell and Baag, 1973; Blackwell et al., 1975; Brott et al., 1976; Combs, 1975; Lachenbruch et al., 1976b; Sass and Sammel, 1976; Sass et al., 1976b, d]. Hydrothermal convection systems constitute most of the targets under investigation as potential sources of geothermal energy, and much will no doubt be learned about their inner workings in the next few years. We expect that the increased understanding of the dynamics of these systems will help in the interpretation of regional heat flow and that the better understanding of regional heat flow will, in turn, increase our general understanding of the broader crustal conditions that generate hydrothermal systems. In regions like the Battle Mountain High, where the average heat flow is about 3 HFU, the steady gradient in poorly conducting sediments is typically about 100°C/km, and under favorable conditions a commercial energy resource could exist beneath deep sedimentary basins, even in the absence of hydrothermal convection [see, e.g., Diment et al., 1975].

Large disturbances to regional heat flow can also be caused by groundwater circulation forced by the distribution of precipitation, topography, and permeable formations. In some regions mantled by permeable volcanic rocks, downward percolation can completely 'wash out' the conductive flux, making it impossible to study regional heat flow by conventional means. In other regions like the Eureka Low, effects can be more subtle. Evidence there suggests that a negative heat flow anomaly over an area $\sim 3 \times 10^4$ km^2 might be caused by interbasin flow in deep aquifers fed by downward percolation of a small fraction of the annual precipitation. Heat flow results of this sort can provide useful information on regional hydrologic patterns with important implications for underground disposal of radioactive waste.

Acknowledgments. We are grateful to our colleagues, R. L. Christiansen, W. H. Diment, D. R. Mabey, L. J. P. Muffler, M. Nathenson, F. H. Olmsted, M. L. Sorey, D. E. White, and D. L. Williams, for helpful comments on the manuscript and to B. V. Marshall, S. P. Galanis, Jr., and R. J. Munroe for assistance with computations and illustrations.

References

Albarede, F., The heat flow/heat generation relationship: An interaction model of fluids with cooling intrusions, Earth Planet. Sci. Lett., 27, 73-78, 1975.

American Association of Petroleum Geologists-U.S. Geological Survey, Geothermal gradient map of North America, scale 1:5,000,000, U.S. Geol. Surv., Reston, Virginia, 1976.

Bailey, R. A., G. B. Dalrymple, and M. A. Lanphere, Volcanism, structure, and geochronology of Long Valley caldera, Mono County, California, J. Geophys. Res., 81, 725-744, 1976.

Benfield, A. E., A heat flow value for a well in California, Amer. J. Sci., 245, 1-18, 1947.

Birch, F., Flow of heat in the Front Range, Colorado, Geol. Soc. Amer. Bull., 61, 567-630, 1950.

Birch, F., The present state of geothermal investigations, Geophysics, 19, 645-659, 1954a.

Birch, F., Thermal conductivity, climatic variation, and heat flow near Calumet, Michigan, Amer. J. Sci., 252, 1-25, 1954b.

Birch, F., R. F. Roy, and E. R. Decker, Heat flow and thermal history in New England and New York, in Studies of Appalachian Geology: Northern and Maritime, edited by E. Zen, W. S. White, J. B. Hadley, and J. B. Thompson, Jr., pp. 437-451, Interscience, New York, 1968.

Blackwell, D. D., Heat flow determinations in the northwestern United States, J. Geophys. Res., 74, 992-1007, 1969.

Blackwell, D. D., The thermal structure of the continental crust, in The Structure and Physical Properties of the Earth's Crust, Geophys. Monogr. Ser., vol. 14, edited by J. G. Heacock, pp. 169-184, AGU, Washington, D. C., 1971.

Blackwell, D. D., Surface ground temperature variations in mountainous regions and a new topographic correction technique for heat flow measurements (abstract), Eos Trans. AGU, 54, 1207, 1973.

Blackwell, D. D., Terrestrial heat flow and its implications on the location of geothermal reservoirs in Washington, Wash. Div. Mines Geol. Inform. Circ., 50, 21-33, 1974.

Blackwell, D. D., and C. G. Baag, Heat flow in a 'blind' geothermal area near Marysville, Montana, Geophysics, 38, 941-956, 1973.

Blackwell, D. D., M. J. Holdaway, P. Morgan, D. Petefish, T. Rape, J. L. Steele, D. Thorstenson, and A. F. Waibel, Results and analysis of exploration and deep drilling at Marysville geothermal area, in The Marysville, Montana Geothermal Project, Final Report, pp. E.1-E.116, Battelle Pacific Northwest Laboratories, Richland, Wash., 1975.

Bodvarsson, G., On the temperature of water flowing through fractures, J. Geophys. Res., 74, 1987-1992, 1969.

Bodvarsson, G., and R. P. Lowell, Ocean floor heat flow and the circulation of interstitial waters, J. Geophys. Res., 77, 4472-4475, 1972.

Bowen, R. G., Progress report on geothermal measurements in Oregon, Ore Bin, 35, 6-7, 1973.

Bowen, R. G., and N. V. Peterson, Thermal springs and wells, Oreg. Dep. Geol. Miner. Ind. Misc. Pap., 14, 1970.

Bowen, R. G., D. D. Blackwell, D. A. Hull, and N. V. Peterson, Progress report on heat-flow study of the Brothers fault zone, central Oregon, Ore Bin, 38, 39-46, 1976.

Bredehoeft, J. D., and I. S. Papadopulos, Rates of vertical groundwater movement estimated from the earth's thermal profile, Water Resour. Res., 1, 325-328, 1965.

Brott, C. A., D. D. Blackwell, and J. C. Mitchell, Heat flow study of the Snake River Plain region, Idaho, Geothermal Investigations in Idaho, Water Inform. Bull. 30, part 8, 195 pp., Idaho Department of Water Resources, Boise, 1976.

Brune, J. N., T. L. Henyey, and R. F. Roy, Heat flow, stress, and rate of slip along the San Andreas Fault, California, J. Geophys. Res., 74, 3821-3827, 1969.

Carslaw, H. S., and J. C. Jaeger, Conduction of Heat in Solids, 2nd ed., Oxford University Press, New York, 1959.

Cermak, V., and A. M. Jessop, Heat flow, heat generation and crustal temperature in the Kapuskasing area of the Canadian Shield, Tectonophysics, 11, 287-303, 1971.

Combs, J., Heat flow and microearthquake studies, Coso geothermal area, China Lake, California, Final Report, Order Number 2800, contract N00123-74-C-2099, 65 pp., Advan. Res. Proj. Agency, Washington, D. C., 1975.

Combs, J., and G. Simmons, Terrestrial heat flow determinations in the north central United States, J. Geophys. Res., 78, 441-461, 1973.

Decker, E. R., and S. B. Smithson, Heat flow and gravity interpretation across the Rio Grande Rift in southern New Mexico and west Texas, J. Geophys. Res., 80, 2542-2552, 1975.

Diment, W. H., T. C. Urban, and F. A. Revetta, Some geophysical anomalies in the eastern United States, in The Nature of the Solid Earth, edited by E. C. Robertson, pp. 544-572, McGraw-Hill, New York, 1972.

Diment, W. H., T. C. Urban, J. H. Sass, B. V. Marshall, R. J. Munroe, and A. H. Lachenbruch, Temperatures and heat contents based on conductive transport of heat, U.S. Geol. Surv. Circ., 726, 84-103, 1975.

Dinwiddie, G. A., and L. J. Schroder, Summary of hydraulic testing in and chemical analyses of water samples from deep exploratory holes in Little Fish Lake, Monitor, Hot Creek, and Little Smoky Valleys, Nevada, Central Nevada-40, Tech. Lett. 474-90, 70 pp., U.S. Geol. Surv., Denver, Colorado, 1971.

Donaldson, I. G., Temperature gradients in the upper layers of the earth's crust due to convective water flows, J. Geophys. Res., 67, 3449-3459, 1962.

Eakin, T. E., A regional interbasin groundwater system in the White River area, southeastern Nevada, Water Resour. Res., 2, 251-271, 1966.

Fenneman, N. M., Physical divisions of the United States, U.S. Dep. of the Interior, Washington, D. C., 1946

Fournier, R. O., D. E. White, and A. H. Truesdell, Convective heat flow in Yellowstone National Park, in Proceedings of the Second United Nations Symposium on the Development and Use of Geothermal Resources, pp. 731-739, U.S. Government Printing Office, Washington, D. C., 1976.

Henyey, T. L., and G. J. Wasserburg, Heat flow near major strike-slip faults in California, J. Geophys. Res., 76, 7924-7946, 1971.

Hill, D. P., Structure of Long Valley caldera, California, from a seismic refraction experiment, J. Geophys. Res., 81, 745-753, 1976.

Hill, D. P., and L. C. Pakiser, Crustal structure between the Nevada Test Site and Boise, Idaho, from seismic refraction measurements, in The Earth Beneath the Continents, Geophys. Monogr. Ser., vol. 10, edited by J. S. Steinhart and T. J. Smith, pp. 391-419, AGU, Washington, D. C., 1966.

Hunt, C. B., and T. W. Robinson, Possible interbasin circulation of ground water in the southern part of the Great Basin, U.S. Geol. Surv. Prof. Pap., 400-B, B273-B274, 1960.

Jaeger, J. C., Heat flow and radioactivity in Australia, Earth Planet. Sci. Lett., 8, 285-292, 1970.

Jones, P. H., Hydrodynamics of geopressure in the northern Gulf of Mexico basin, J. Petrol. Technol., 21, 803-810, 1969.

Jones, P. H., and R. H. Wallace, Jr., Hydrogeologic aspects of structural deformation in the northern Gulf of Mexico basin, J. Res. U.S. Geol. Surv., 2, 511-517, 1974.

Lachenbruch, A. H., Preliminary geothermal model of the Sierra Nevada, J. Geophys. Res., 73, 6977-6989, 1968.

Lachenbruch, A. H., Crustal temperature and heat production: Implications of the linear heat flow relation, J. Geophys. Res., 75, 3291-3300, 1970.

Lachenbruch, A. H., Vertical gradients of heat production in the continental crust, 1, Theoretical detectability from near-surface measurements, J. Geophys. Res., 76, 3842-3851, 1971.

Lachenbruch, A. H., and C. M. Bunker, Vertical gradients of heat production in the continental crust, 2, Some estimates from borehole data, J. Geophys. Res., 76, 3852-3860, 1971.

Lachenbruch, A. H., and J. H. Sass, Thermo-mechanical aspects of the San Andreas, in Proceedings of the Conference on the Tectonic Problems of the San Andreas Fault, pp. 192-205, Stanford University Press, Stanford, Calif., 1973.

Lachenbruch, A. H., M. C. Brewer, G. W. Greene, and B. V. Marshall, Temperatures in permafrost, in Temperature--Its Measurement and Control in Science and Industry, pp. 791-803, Reinhold, New York, 1962.

Lachenbruch, A. H., J. H. Sass, R. J. Munroe, and T. H. Moses, Jr., Geothermal setting and simple heat conduction models for the Long Valley caldera, J. Geophys. Res., 81, 769-784, 1976a.

Lachenbruch, A. H., M. L. Sorey, R. E. Lewis, and J. H. Sass, The near-surface hydrothermal regime of Long Valley caldera, J. Geophys. Res., 81, 763-768, 1976b.

Lambert, I. B., and K. S. Heier, The vertical distribution of uranium, thorium, and potassium in the continental crust, Geochim. Cosmochim. Acta, 31, 377-390, 1967.

Lapwood, E. R., Convection of a fluid in a porous medium, Proc. Cambridge Phil. Soc., 44, 508-521, 1948.

Lee, W. H. K., and S. Uyeda, Review of heat flow data, in Terrestrial Heat Flow, Geophys. Monogr. Ser., vol. 8, edited by W. H. K. Lee, pp. 87-190, AGU, Washington, D. C., 1965.

Lowell, R. P., Circulation in fractures, hot springs, and convective heat transport on mid-ocean ridge crests, Geophys. J. Roy. Astron. Soc., 40, 351-365, 1975.

Mariner, R. H., J. B. Rapp, L. M. Willey, and T. S. Presser, The chemical composition and estimated minimum thermal reservoir temperatures of selected hot springs in Oregon, open file report, 27 pp., U.S. Geol. Surv., Menlo Park, Calif., 1974.

Mifflin, M. D., Delineation of ground-water flow systems in Nevada, Hydrol. Water Resour. Publ. 4, 111 pp., Desert Res. Inst., Reno, Nevada, 1968.

Morgan, P., D. D. Blackwell, R. E. Spafford, and R. B. Smith, Heat flow measurements in Yellowstone Lake and the thermal structure of the Yellowstone caldera, J. Geophys. Res., in press, 1977.

Nathenson, M., The effects of a step change in water flow on an initially linear profile of temperature, in *Proceedings of the Geothermal Reservoir Engineering Workshop*, vol. 2, Stanford Geothermal Program, Stanford, California, in press, 1977.

Olmsted, F. H., P. A. Glancy, J. R. Harrill, F. E. Rush, and A. S. VanDenburgh, Preliminary hydrogeologic appraisal of selected hydrothermal systems in northern and central Nevada, Open File Rep. 75-56, 267 pp., U.S. Geol. Surv., Menlo Park, Calif., 1975.

Papadopulos, S. S., R. H. Wallace, Jr., J. B. Wesselman, and R. E. Taylor, Assessment of onshore geopressured-geothermal resources in the northern Gulf of Mexico basin, U.S. Geol. Surv. Circ., 726, 125-146, 1975.

Peck, D. L., T. L. Wright, and J. G. Moore, Crystallization of tholeiitic basalt in Alae lava lake, Hawaii, Bull. Volcanol., 29, 629-656, 1966.

Rao, R. U. M., G. V. Rao, and H. Narain, Radioactive heat generation and heat flow in the Indian shield, Earth Planet. Sci. Lett., 30, 57-64, 1976.

Reiter, M., C. L. Edwards, H. Hartman, and C. Weidman, Terrestrial heat flow along the Rio Grande Rift, New Mexico and southern Colorado, Geol. Soc. Amer. Bull., 86, 811-818, 1975.

Renner, J. L., D. E. White, and D. L. Williams, Hydrothermal convection systems, U.S. Geol. Surv. Circ., 726, 5-57, 1975.

Roy, R. F., D. D. Blackwell, and F. Birch, Heat generation of plutonic rocks and continental heat flow provinces, Earth Planet. Sci. Lett., 5, 1-12, 1968a.

Roy, R. F., E. R. Decker, D. D. Blackwell, and F. Birch, Heat flow in the United States, J. Geophys. Res., 73, 5207-5221, 1968b.

Roy, R. F., D. D. Blackwell, and E. R. Decker, Continental heat flow, in *The Nature of the Solid Earth*, edited by E. C. Robertson, pp. 506-543, McGraw-Hill, New York, 1972.

Sass, J. H., and E. A. Sammel, Heat flow data and their relation to observed geothermal phenomena near Klamath Falls, Oregon, J. Geophys. Res., 81, 4863-4868, 1976.

Sass, J. H., A. H. Lachenbruch, R. J. Munroe, G. W. Greene, and T. H. Moses, Jr., Heat flow in the western United States, J. Geophys. Res., 76, 6376-6413, 1971.

Sass, J. H., W. H. Diment, A. H. Lachenbruch, B. V. Marshall, R. J. Munroe, T. H. Moses, Jr., and T. C. Urban, A new heat-flow contour map of the conterminous United States, Open File Rep. 76-756, 24 pp., U.S. Geol. Surv., Menlo Park, Calif., 1976a.

Sass, J. H., S. P. Galanis, Jr., R. J. Munroe, and T. C. Urban, Heat-flow data from southeastern Oregon, Open File Rep. 76-217, 52 pp., U.S. Geol. Surv., Menlo Park, Calif., 1976b.

Sass, J. H., J. C. Jaeger, and R. J. Munroe, Heat flow and near-surface radioactivity in the Australian continental crust, Open File Rep. 76-250, 91 pp., U.S. Geol. Surv., Menlo Park, Calif., 1976c.

Sass, J. H., F. H. Olmsted, M. L. Sorey, H. A. Wollenberg, A. H. Lachenbruch, R. J. Munroe, and S. P. Galanis, Jr., Geothermal data from test wells drilled in Grass Valley and Buffalo Valley, Nevada, Open File Rep. 76-85, 43 pp., U.S. Geol. Surv., Menlo Park, Calif., 1976d.

Sclater, J. G., and J. Francheteau, The implications of terrestrial heat flow observations on current tectonic and geochemical models of the crust and upper mantle of the earth, Geophys. J. Roy. Astron. Soc., 20, 509-542, 1970.

Smith, D. L., Heat flow, radioactive heat generation, and theoretical tectonics for northwestern Mexico, Earth Planet. Sci. Lett., 23, 43-52, 1974.

Smith, R. L., and H. R. Shaw, Igneous-related geothermal systems, U.S. Geol. Surv. Circ., 726, 58-83, 1975.

Sorey, M. L., Numerical modeling of liquid geothermal systems, Open File Rep. 75-613, 66 pp., U.S. Geol. Surv., Menlo Park, Calif., 1975.

Sorey, M. L., and R. E. Lewis, Convective heat flow from hot springs in the Long Valley caldera, Mono County, California, J. Geophys. Res., 81, 785-791, 1976.

Steeples, D. W., and H. M. Iyer, Low-velocity zone under Long Valley as determined from teleseismic events, J. Geophys. Res., 81, 849-860, 1976.

Stewart, J. H., and J. E. Carlson, Generalized maps showing distribution, lithology, and age of Cenozoic igneous rocks in the western United States, scale 1:5,000,000, in Cenozoic Tectonics and Regional Geophysics of the Western Cordillera, Spec. Pap., edited by R. B. Smith and G. P. Eaton, Geological Society of America, Boulder, Colorado, in press, 1977.

Swanberg, C. A., Vertical distribution of heat generation in the Idaho batholith, J. Geophys. Res., 77, 2508-2513, 1972.

Swanberg, C. A., M. D. Chessman, G. Simmons, S. B. Smithson, G. Gronlie, and K. S. Heier, Heat-flow-heat-generation studies in Norway, Tectonophysics, 23, 31-48, 1974.

Thompson, G. A., and D. B. Burke, Regional geophysics of the Basin and Range province, Annu. Rev. Earth Planet. Sci., 2, 213-238, 1974.

Urban, T. C., and W. H. Diment, Heat flow on the south flank of the Snake River Rift (abstract), Geol. Soc. Amer. Abstr. Programs, 7, 648, 1975.

White, D. E., Geothermal energy, U.S. Geol. Surv. Circ., 519, 17 pp., 1965.

White, D. E., Hydrology, activity, and heat flow of the Steamboat Springs thermal system, Washoe County Nevada, U.S. Geol. Surv. Prof. Pap., 458-C, 109 pp., 1968.

Winograd, I. J., and W. Thordarson, Hydrogeologic and hydrochemical framework, South-Central Great Basin, Nevada-California, with special reference to the Nevada Test Site, U.S. Geol. Surv. Prof. Pap., 712-C, 126 pp., 1975.

Wyllie, P. J., Experimental limits for melting in the earth's crust and upper mantle, in The Structure and Physical Properties of the Earth's Crust, Geophys. Monogr. Ser., vol. 14, edited by J. G. Heacock, pp. 279-301, AGU, Washington, D. C., 1971.

CHARACTERISTICS OF SELECTED GEOTHERMAL SYSTEMS IN IDAHO

James K. Applegate and Paul R. Donaldson

Department of Geology and Geophysics, Boise State University
Boise, Idaho   83725

Abstract. Many areas of Idaho exhibit geothermal potential. Large areas of the southern portion of the state have been explored by commercial firms, and numerous areas were identified by White and Williams (1975) as having high geothermal potential. Only limited drilling has been accomplished, and most evaluations have been based on geophysical, geological, and geochemical data. Regional geophysical data supplemented by local surveys suggest the lack of a localized heat source. Instead, geological and geophysical studies suggest that the occurrence of geothermal systems in this area results from heat concentration below an insulating layer. The insulation typically comprises an interbedded sequence of sediments and volcanics. The noncontinuous permeability in such a sequence restricts the development of large-scale convection systems. Otherwise the insulating properties of the sequence would be lost to convective heat transfer. Specific geothermal reservoirs result from structural controls such as fault and fracture related porosity. Initial studies in three areas of Idaho support these concepts.

Introduction

Geothermal exploration has, because of its infancy, been primarily based on geologically simple models with a discrete heat source and a discrete reservoir. This classical model invariably requires geological evidence, generally youthful rhyolitic volcanism, for the existence of the localized heat source. The lack of rhyolitic volcanism has the effect of eliminating many otherwise interesting areas from consideration for geothermal exploration. If acceptable alternative models exist that do not require such evidence, then the previously eliminated areas which may be more representative of the low to intermediate temperature systems prevalent throughout the western United States would again be prospective.

Initial studies in three geothermal areas of Idaho suggest a hypothetical model without a localized heat source. This alternative model is characterized by (1) heat entrapped beneath a thermally insulating cover of interbedded sediments and volcanics, (2) the absence of widespread convection in the insulating cover due to the lack of continuous permeability, and (3) resource localization controlled by structural features. In the following discussion, data from each of the areas will be used to establish the plausibility of the hypothetical model.

Boise Geothermal System

Portions of northeast Boise have been heated by geothermal water since the 1890's. The fluids are produced from two shallow (approximately 120 m) wells and are postulated to represent leakage from a deeper reservoir.

Exploration was initiated in early 1975 in an attempt to define the shallow resource and its relationship to the overall geothermal system. Studies undertaken include microseismic monitoring, electrical resistivity mapping, magnetometer profiling, geochemical thermometry, and general geological mapping utilizing vertical and oblique aerial photography. Additional data available included regional gravity, regional magnetics, and U-2 and satellite imagery.

Geology

The Boise area is on the northern edge of the western portion of the Snake River Plain (Fig 1) and lies in the transition zone between the plain and the Idaho Batholith. The Snake River Plain appears to be a tensional basin filled with over 4 km of sediments and volcanics.

The local geology consists of alluvium overlying Tertiary lacustrine sediments and volcanics which lap onto the Idaho Batholith. The batholith is primarily quartz monzonite of Cretaceous age. The local structure is dominated by the northwest-southeast trending Boise fault [Malde, 1959]. This normal fault, downthrown toward the plain, is located along the topographic break between the plain and the foothills.

The Boise fault and a series of north to northeast trending features extending from the batholith toward the plain are well defined on Landsat imagery (Fig 2). The intersections of these two sets of features present prime exploration targets.

Geophysics

Geophysical studies in the area were undertaken to define further the location and extent of the structures suggested by the linear features. If these features are representative of faults, they may be expected to exhibit microseismic activity and to be characterized by resistivity, magnetic, and gravity anomalies.

Continuous microseismic monitoring for over 18 months has failed to delineate any local (within 30 km) seismic activity. Significant activity has been detected from eastern Oregon and from within the Idaho Batholith, suggesting that the transition zone between the western Snake River Plain and the batholith is aseismic.

Resistivity data were acquired for the area by utilizing a rotating bipole source and roving dipole receivers [Furgerson and Keller, 1974; Donaldson and Applegate, 1975]. This technique is particularly sensitive to lateral resistivity changes, which may be indicative of geological structures.

Several electrical parameters can be calculated from this type of resistivity mapping data. Two commonly used parameters are apparent resistivity and total longitudinal conductance. The apparent resistivity map (Fig 3) and the total conductance map (Fig 4) are contoured from over 200 data points which approximate a 0.75-km grid of receiver stations in the survey area.

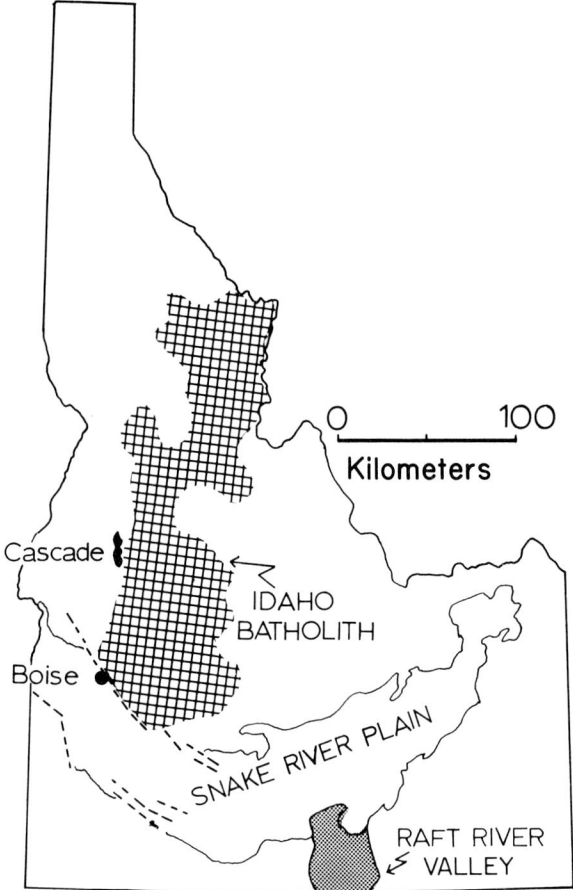

Fig. 1. Index map showing locations of the three (Boise, Cascade, and Raft River) geothermal areas to be discussed and their spatial relationships to the Idaho Batholith and the Snake River Plain.

The apparent resistivity (Fig 3) shows steep gradients probably related to the Boise fault and orthogonal structural features. There is a low apparent resistivity trend associated with hot water producing wells near the old State Penitentiary and a large low to the northeast of the source, which is in an area of former hot spring activity. The total conductance map (Fig 4) shows two linear trends. These two linears appear to correspond to the Boise fault and one of the major north-northeast oriented features. A Schlumberger equivalent equatorial dipole profile shows resistive basement at about 2.5-km.

Magnetic profiles (not shown) and the regional gravity map (Fig 5) provide corroborative evidence for two intersecting fault systems. Additionally, estimates of basement depth on the basin side of the Boise fault from gravity and resistivity data are in substantial agreement. Neither resistivity nor gravity suggests the presence of a localized heat source.

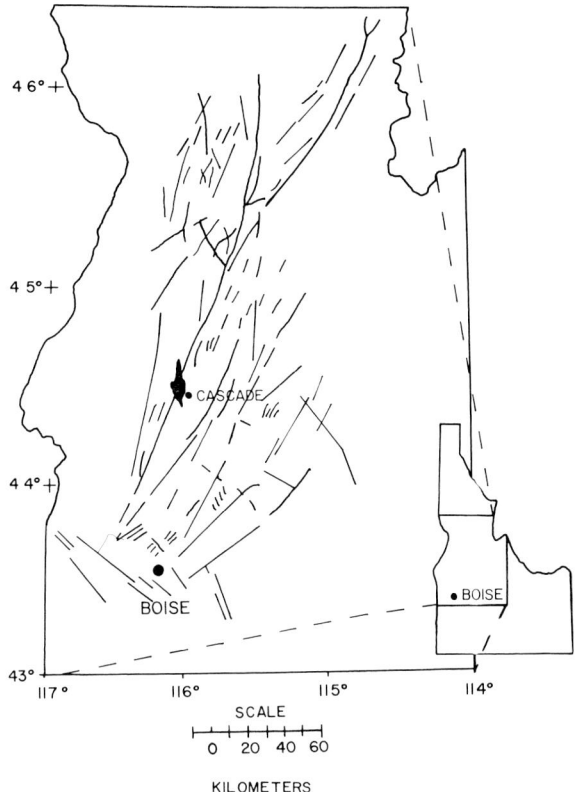

Fig. 2. Map showing major Landsat features in a portion of west central Idaho (modified from Day [1974]).

Discussion

The foregoing data suggest the following geologic model for the Boise geothermal system (1) a fault-controlled basin edge complicated by approximately orthogonally intersecting faults and (2) no indications of a localized heat source. Given the presence of an insulating cover, the existence of a basin boundary provides a mechanism for concentrating the trapped heat. The faults intersecting the basin edge provide the plumbing necessary for introducing the heat transfer medium, meteoric water.

A simplistic calculation [Diment et al., 1975] gives resulting temperatures of approximately 175° to 185°C at the base of a 2.5-km-thick thermal insulator with these properties (1) thermal conductivity of 7.0 mcal/cm s°C for the Idaho Batholith and 3.0 mcal/cm s°C for the sediments and (2) a heat flow of 3.0 μcal/cm^2 s [see Brott et al., 1976]. A silica geothermometry calculation using a mixing model for samples from existing Boise wells predicts an undiluted reservoir temperature of 165°C and 61% dilution (Fig 6). This excellent agreement strongly supports the validity of the model.

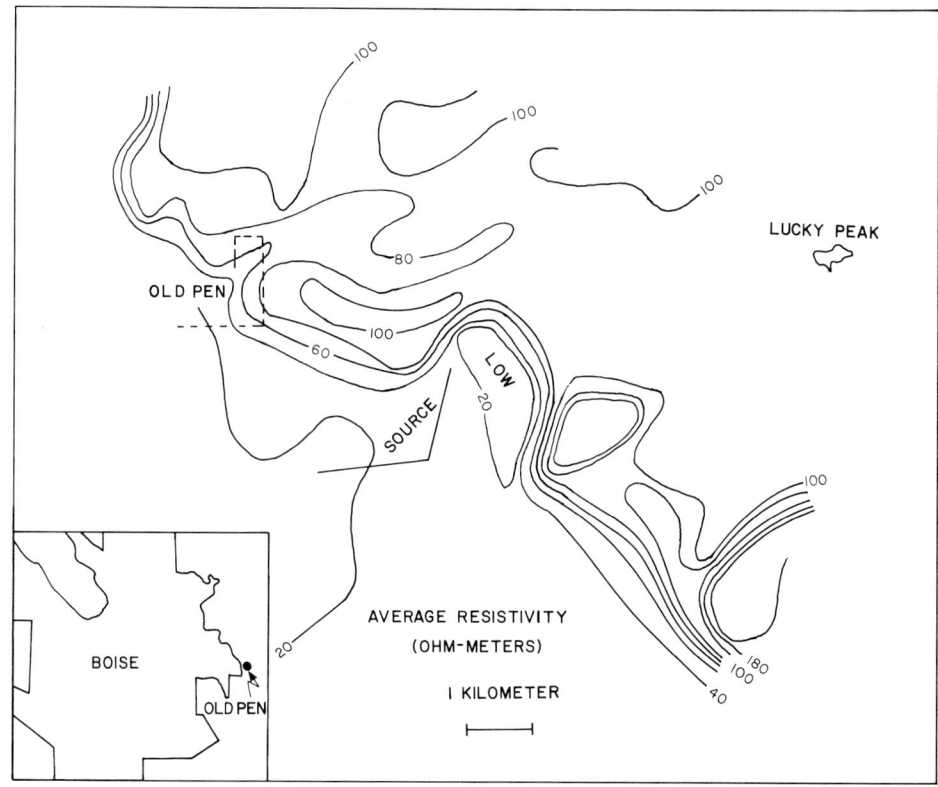

Fig. 3. Average resistivity map for a portion of the Boise area. The average resistivity is obtained by determining the apparent resistivity with the rotating bipole source and roving dipole receiver. A number of structural trends are clearly shown by the contours.

Cascade Geothermal System

The Cascade area lies in an area of extensive hot spring activity [see White and Williams, 1975]. Some local use has been made of the resource for recreation facilities, ice melting, and space heating.

Geological and geophysical studies were begun in 1975 to evaluate the feasibility of further exploration and development [Wilson et al., 1976]. Studies undertaken include microseismic monitoring, geochemical thermometry, aerial reconnaissance, and general geological mapping utilizing vertical and oblique aerial photography. Additional data available included regional and local gravity and satellite imagery.

Geology

The Cascade area is on the western margin of the Idaho Batholith. Three major rock zones in the area are quartz diorite gneiss, quartz

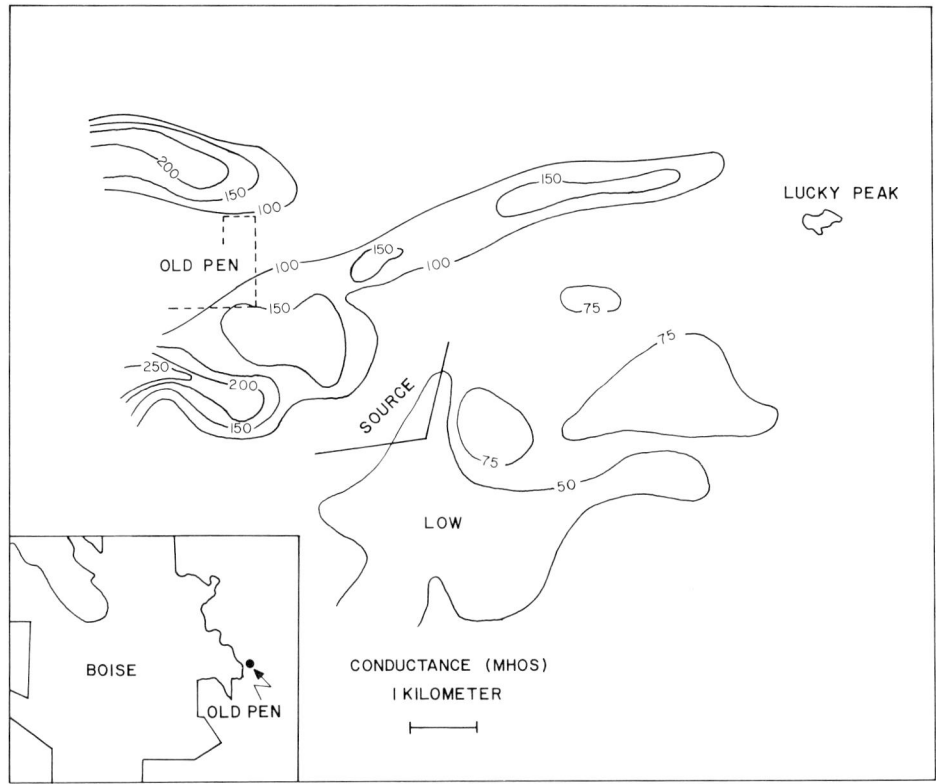

Fig. 4. Total longitudinal conductance map for a portion of the Boise area. The longitudinal conductance assumes cylindrical current spreading. Several major structural trends are prominent.

diorite, and granodiorite. In the western portion of the area, Columbia River basalts of Miocene age overlie the batholithic rocks [Wilson et al., 1976]. The Long Valley downwarp contains Quaternary and Pliocene(?) sediments [Schmidt and Mackin, 1970, p. 11]. A local gravity survey [Kinoshita, 1962] suggests basin fill thicknesses ranging from 600 to 2100 m.

Landsat imagery (Fig 2) shows a significant number of large-scale linear features with lengths of tens of kilometers. The major features have a north-northeast to north trend, while another set has a northeast trend. One of the major features trending north-northeast passes through the Cascade community [Wilson et al., 1976].

Detailed geologic mapping (Fig 7) shows faults with trends similar to the Landsat linears, suggesting that these linears are major fault features. The mapping also indicates a series of east trending faults that are essentially perpendicular to the direction of primary faulting. There are some indications of left lateral movements along these faults [Wilson et al., 1976]. The lack of erosion on these features suggests that there has been recent movement. Work by Cluff and Idriss [1972] indicates evidence for very recent movement along the Cascade-Sweet

Fig. 5 Regional gravity map of a portion of west central Idaho showing the apparent lack of shallow intrusives [Mabey et al. 1974].

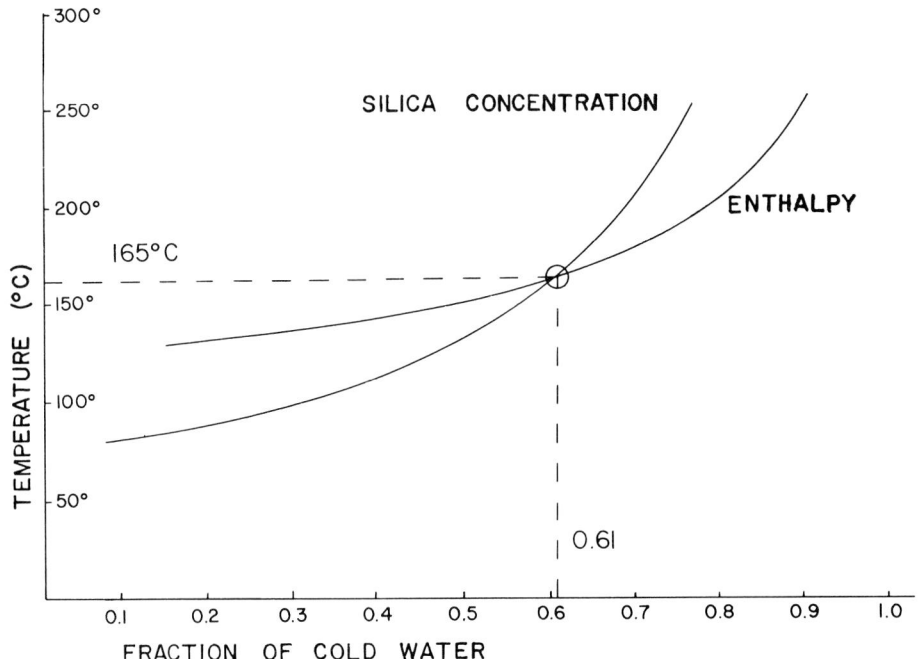

Fig. 6 Graph illustrating a mixing model calculation of a silica predicted temperature for the undiluted reservoir at Boise. The predicted temperature is 165°C with 61% dilution for the water produced.

fault system. This system appears to be related to the major linear through the Cascade area.

Geophysics

Studies of geophysical data from the area were undertaken to define further the location, extent, and activity of the faults indicated by the geological mapping. The earthquake data file from the National Earthquake Information Service indicates 13 locatable events in Valley County, Idaho, and 4 additional events for which 'felt' reports are available. The largest event was a magnitude 6.0 in 1945 located approximately 55-km northeast of Cascade. In 1970 there was a six-event swarm with earthquakes of magnitudes 3.5, 3.7, 3.8, 3.9, 4.3, and 4.3. This swarm occurred about 40-km east of Cascade. The events nearest Cascade were a 1966 quake of magnitude 3.5 20-km north of Cascade and a smaller event (1962 no magnitude given) located 10-km north of Cascade.

Microseismic monitoring was carried out for 8 days during the summer of 1975. The survey detected 24 events with S-P times of less than 4 s. Sixteen of these events occurred in the last 24 hours of monitoring. Locations for seven of these events (Fig 7) suggest activity along the east trending faults through Cascade. The depths for these 7 events fall in the range of 5 to 10 km. The two largest events in this swarm had magnitudes of approximately 0.75. The 17 other events were too

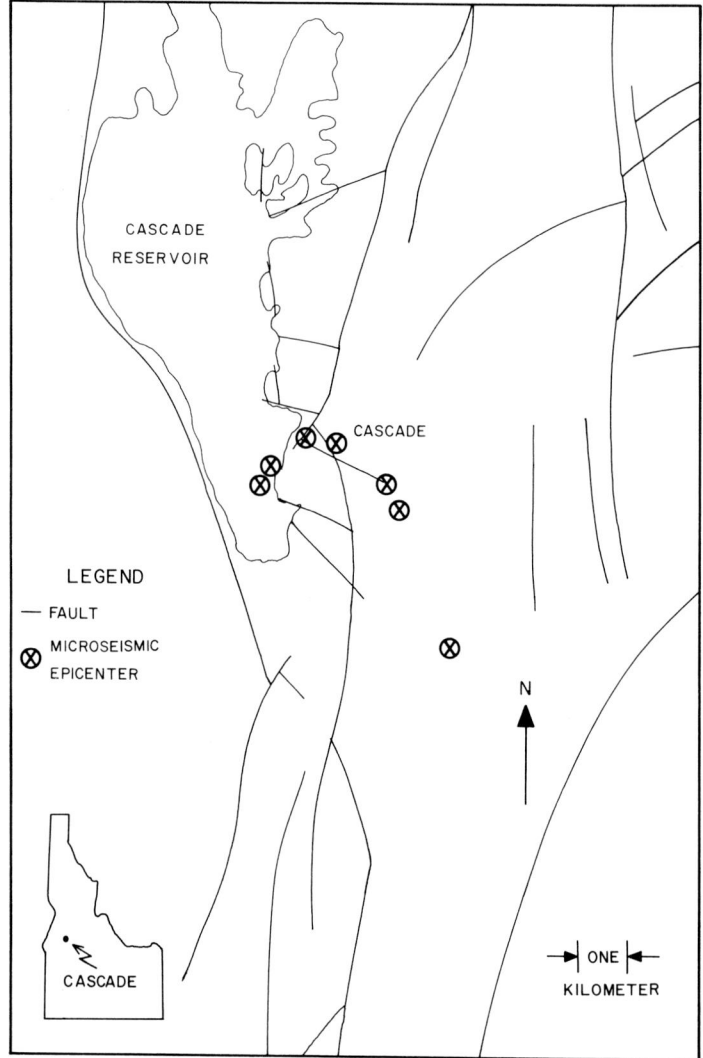

Fig. 7 Map of the Cascade area showing relationship between mapped faults and microseismic events recorded in one 8-day period.

small to be located but were similar in character to the located events [Wilson et al., 1976].

Gravity data [Kinoshita, 1962] (Fig 5) indicate several intermontane basins. The data suggest that these basins are fault bounded.

Discussion

Numerous active hot springs occur in the area and all appear to be fault controlled. The Cascade Reservoir has inundated what are reported to have been the largest and hottest springs. The reported location of

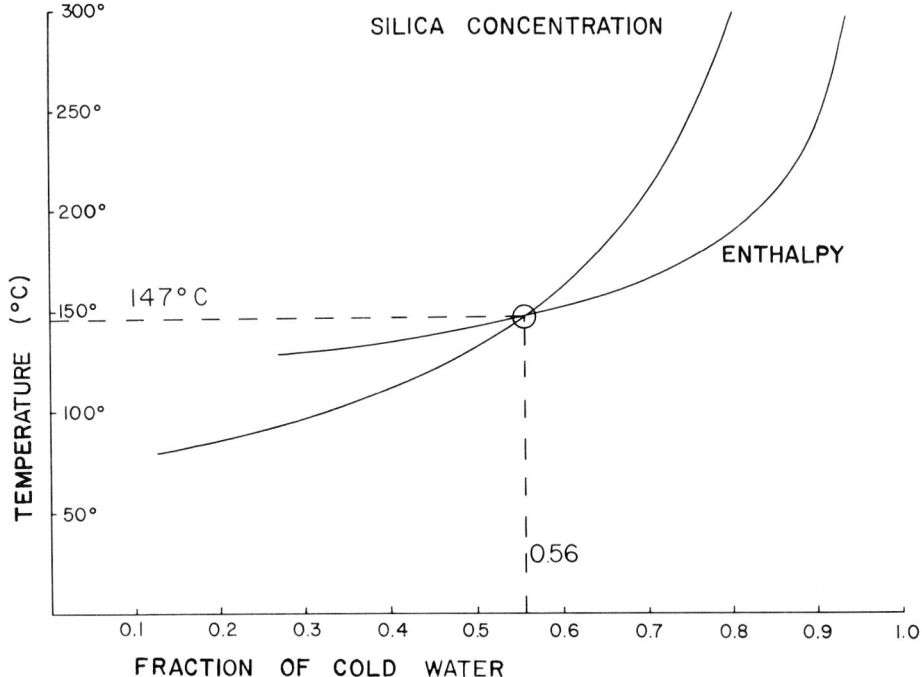

Fig. 8 Graph illustrating a mixing model calculation of a silica predicted temperature for the undiluted reservoir at Cabarton Hot Springs at the south end of Cascade Reservoir. The predicted reservoir temperature is 147°C with 56% cold-water dilution for the spring.

these springs is near the apparent intersection of one of the east trending faults and the major north-northeast trending linear feature. The hottest accessible spring in the area has a surface temperature of 71°C. A silica geothermometry calculation using a mixing model predicts an undiluted source temperature of 147°C and 56% cold-water dilution for this spring (Fig 8). The highest temperature predicted by geothermometry for any of the springs is 179°C [Wilson et al., 1976].

Interestingly, gravity data suggest a greater basin depth beneath springs with higher predicted undiluted temperatures and lesser depths beneath springs with lower predicted temperatures. These observations support the previously postulated model of heat concentration beneath a thermal insulating blanket of sedimentary fill. A thicker insulation would result in greater heat retention and thus higher temperature at the base of the blanket. Furthermore, there is no indication in the gravity data to suggest that near-surface intrusives provide a local heat source.

Raft River Geothermal System

Hot water produced from springs and shallow wells in the Raft River Valley has been used for agricultural purposes for many years. Studies have been undertaken recently to evaluate the commercial potential

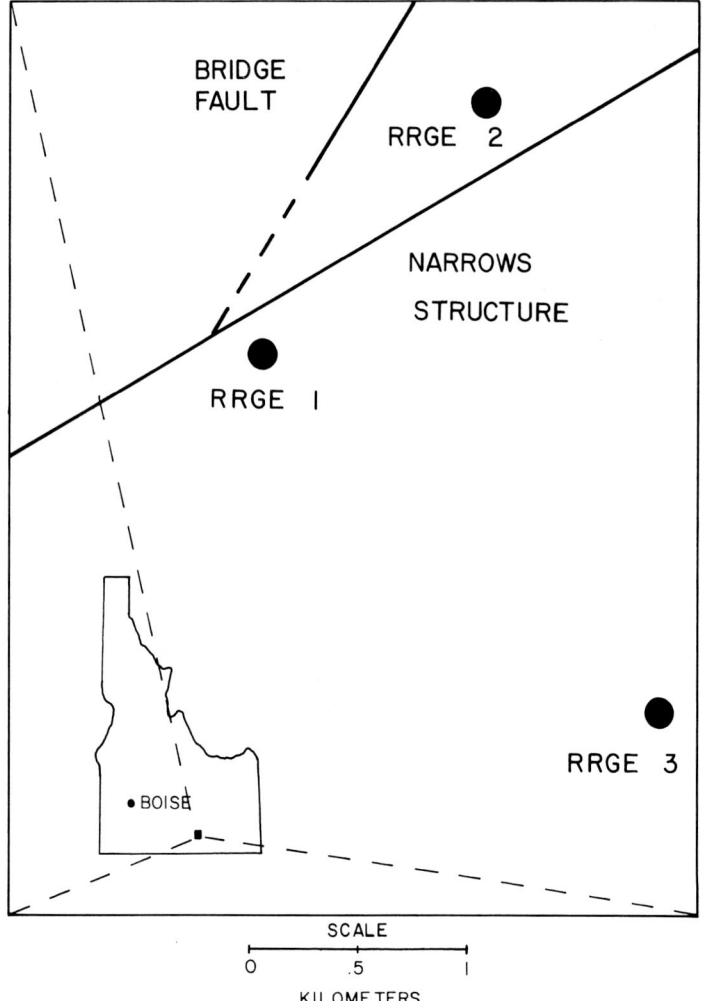

Fig. 9. Schematic map of the Bridge fault, the Narrows structure, and RRGE 1, 2, and 3. The Bridge fault is along the eastern flank of the Jim Sage Mountains, and the Narrows structure separates the Jim Sage Mountains from the Raft River Mountains (Fig. 10).

of the Raft River Valley further [Mabey et al., 1975; Williams et al., 1975; Zohdy et al., 1975; Ackerman, 1975; Nichols and Applegate, 1974]. These studies include a broad spectrum of geophysical and geological investigations. On the basis of these studies, three wells were drilled to depths of approximately 1500 to 1900 m in 1974 and 1975.

Geology

The Raft River Valley, a part of the Basin and Range geomorphic province, is located in south central Idaho, south of the Snake River

Plain and north of the Utah-Idaho state line (Fig 1). The valley is a late Cenozoic structural downwarp bounded by faults on the west, south, and east [Williams et al., 1975]. The downwarp is filled with Tertiary and Paleozoic sediments and volcanics which overlie Precambrian rocks. The Tertiary deposits are composed of (1) 5 to 70 m of Pleistocene and Holocene fan gravels and alluvium, (2) 0 to 200 m of silt and sand, which comprise the Pleistocene Raft Formation, and (3) up to 1800 m of the Pliocene Salt Lake Formation, which consists of tufaceous sediments, volcanics (felsic lava flows and ash flows), and basin fill tufaceous sediments and conglomerates [Williams et al., 1975]. These are underlain by complex Paleozoics which overlie Precambrian rocks.

Two major features appear to be important in the regional structure. The geothermal area appears to be controlled by the intersection of the ENE trending feature through the Narrows (the Narrows structure) and a north trending basin edge feature (the Bridge fault) (Fig 9). The Narrows structure and the Bridge fault are major features on Landsat imagery.

Geophysics

Extensive surface geophysical surveys were conducted by the U.S. Geological Survey. The study includes gravity, magnetic, refraction seismic, resistivity, audio magnetotelluric, self-potential, and telluric current surveys. The geophysical surveys indicate a maximum thickness of about 2-km of Cenozoic sedimentary and volcanic rock, in agreement with the general geologic interpretation [Mabey et al., 1975]. The presence of structural features interpreted from the Landsat imagery is also clearly evident in these geophysical studies (for example, gravity map (Fig 10).

Discussion

Drilling in the Raft River Valley was intended to intersect areas of increased porosity postulated to exist at the intersection of the Narrows structure and the Bridge fault [Nichols and Applegate, 1974; Williams et al., 1975; Mabey et al., 1975]. The first well (RRGE 1) produced water of 145°C from a zone between 1150- and 1370-m deep. Subsequently, two other wells were drilled (RRGE 2 and RRGE 3) to evaluate the reservoir further. These produced equivalent temperature water from zones located between 1170 and 1340 m for RRGE 2 and 1310 and 1580 m for RRGE 3. Figure 9 is a sketch of the locations of RRGE 1, 2, and 3 and of the Narrows structure and the Bridge fault zone.

The very extensive geological and geophysical studies completed in the Raft River Valley quite clearly define the geologic setting of the geothermal system. This setting is in many ways analogous to the Boise and Cascade geothermal systems. As was true in the previously discussed systems, the resource exists in a structurally controlled sediment-filled basin. The approximately 2-km-thick sedimentary cover [Mabey et al., 1975] constitutes sufficient thermal insulation to account for temperatures in the range 145° to 155°C, given the following additional parameters (1) thermal conductivity of 7.0 mcal/cm s°C for basement rocks and 3.0 mcal/cm s°C for the sediments and (2) heat flow of 3.0 μcal/cm^2 s.

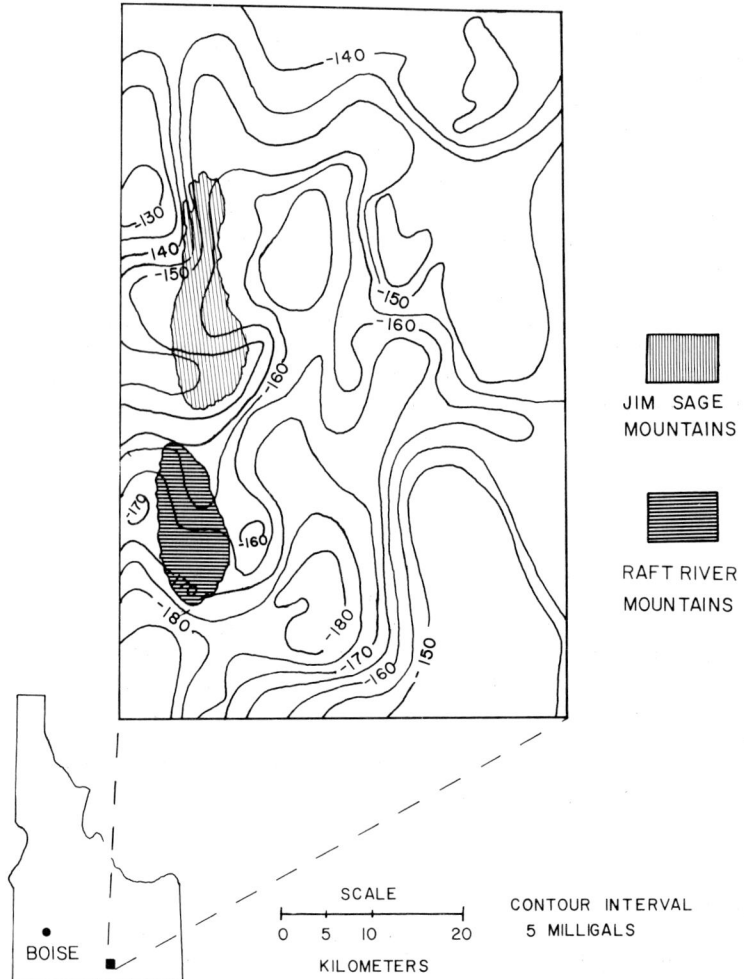

Fig. 10. Gravity map of the Raft River Valley. The Narrows structure separates the Jim Sage Mountains from the Raft River Mountains. The Bridge fault is along the eastern flank of the Jim Sage Mountains. Gravity data is from Mabey et al. [1974].

Once again, the heat which would be trapped at the base of the described column of sediments and volcanics is sufficient to account for temperatures predicted and in this case substantiated by deep drilling. Again, the regional geophysical data do not suggest the presence of a youthful shallow heat source, and indeed none is needed to produce the temperatures observed.

The hypothetical model for this and the previously discussed systems has assumed negligible convective heat transfer due to an absence of continuous connected porosity in the basin fill. This condition has been postulated to exist in the Boise and Cascade geothermal systems

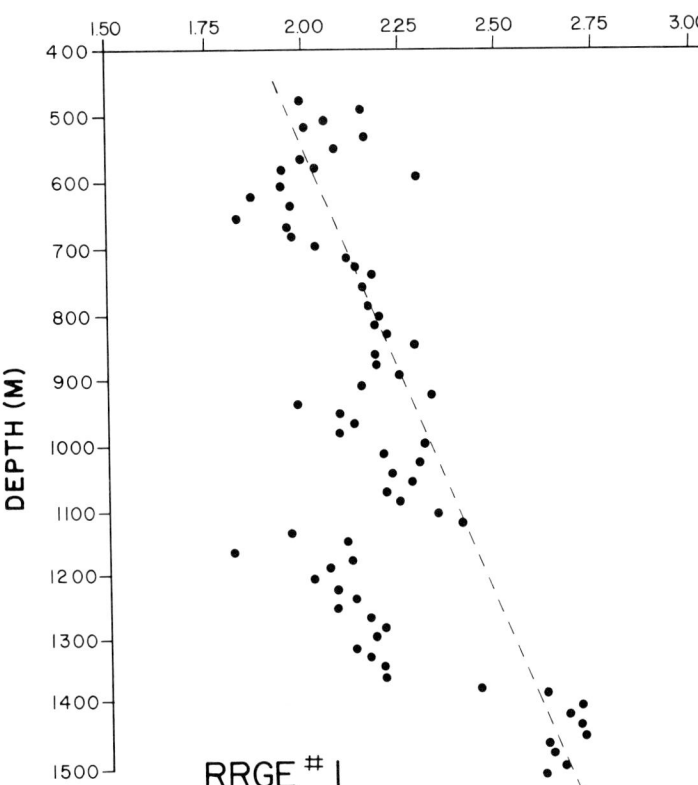

Fig. 11. Plot of bulk density from density log versus depth for RRGE 1. The general trend is that density increases with depth in a sedimentary sequence because of increasing compaction and decrease of porosity. The graph then shows a decrease in porosity with depth until the productive zone (1150 to 1370 m) is encountered [Applegate et al., 1976].

on the basis of the presence of nonpermeable basalt flows interbedded with a variety of permeable and nonpermeable sediments. Geophysical well logs in the Raft River wells provide a picture of widely varying porosity throughout the sedimentary section (Fig 11), which implies the noncontinuous permeability previously postulated.

Summary

Geophysical and geological studies of the Boise, Cascade, and Raft River geothermal systems suggest that the classical model requiring a shallow molten or recently molten mass as a heat source is not valid. These arguments can be extended to account for many geothermal systems in the western United States.

Therefore it is necessary to consider geothermal models which do not

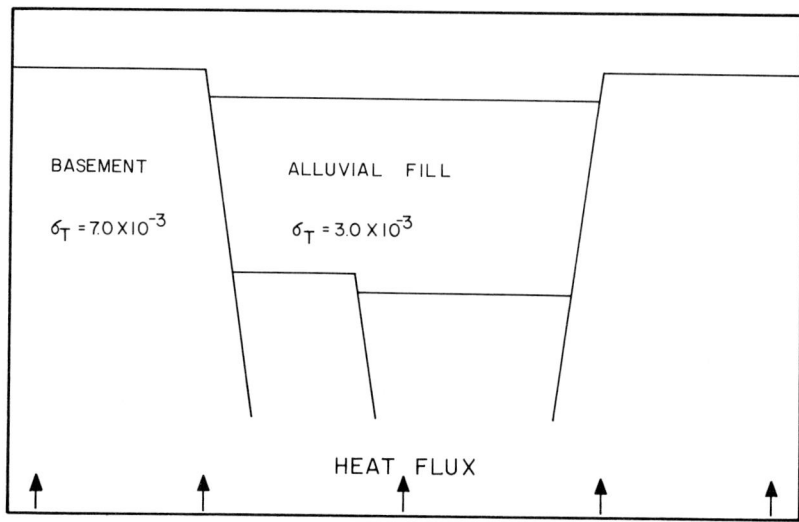

Fig. 12. Schematic diagram of the proposed geological model for the Boise, Cascade, and Raft River geothermal systems. Thermal conductivities (σ) of 3.0×10^{-3} and 7.0×10^{-3} cal/cm s °C were used for the sediments and basement rock, respectively, to calculate the temperatures at the base of the sedimentary section.

require a localized heat source. In this case the ultimate source of heat must be the lower crust or upper mantle. There are several possible explanations for the deep heat source. Whatever the nature of the deep-seated heat source, the result is increased heat flow into the upper crust. If there is increased heat flow into the upper crust, then specific geologic conditions must exist which result in heat entrapment and hence a shallow geothermal system.

The foregoing discussions present a set of conditions sufficient for establishing such a geothermal system. Other authors [Keller, 1975; Diment et al., 1975; Crewdson, 1976] have also suggested similar models to account for geothermal systems without shallow localized heat sources.

The model presented (Fig 12) is compatible with geological and geophysical observations from all three areas considered in Idaho. While other models can be postulated to account for these systems, the parameters specified in previous arguments for this model best fit the observed conditions.

References

Ackerman, H. D., Seismic refraction study in the Raft River geothermal area, Idaho (abstract), Geophysics, 41 (2), 336, 1975.

Applegate, J. K., P. R. Donaldson, D. L. Hinkley, and T. L. Wallace, Borehole geophysics evaluation of the Raft River geothermal reservoir, Idaho (abstract), Geophysics, 42 (1), 138, 1976.

Brott, C., D. D. Blackwell, J. Mitchell, Geothermal investigations in Idaho, 8, Heat flow in the Snake River Plain, Idaho Dep. Reclam. Water Inform. Bull., 30, 1976.

Cluff, L. S., and I. M. Idriss, Investigation and evaluation of the potential activity of the Boise fault, The Veterans Administration Hospital Report, V.A. Proj. 11-5045, Woodward-Lundgren & Ass. Oakland, Calif. 1972.

Crewdson, R. Geophysical work in the Black Rock Desert area, paper presented at the Conference on Exploration for the Geothermal Reservoir, Colo. Sch. of Mines and Nat. Sci. Found., Golden, Colo., May 4-5, 1976.

Day, N. F., Linears map of Idaho, band 5, MSS-ERTS, Idaho Bur. of Mines and Geol., Moscow, 1974.

Diment, W. H., T. C. Urban, J. H. Sass, B. V. Marshall, R. J. Munroe, and A. H. Lachenbruch, Temperatures and heat contents based on conductive transport of heat, U. S. Geol. Surv. Circ., 726, 84-103, 1975.

Donaldson, P. R., and J. K. Applegate, Evaluation of the geothermal potential of the Boise Front, Idaho (abstract), Geophysics, 41 (2), 350, 1975.

Fournier, R. O., and A. H. Truesdell, Geochemical indications of subsurface temperature, 1, Basic assumptions, J. Res. U. S. Geol. Surv., 2, 263-269, 1974.

Furgerson, R., and G. V. Keller, Rotating dipole methods for measuring earth resistivity (abstract), Geophysics, 40 (1), 148, 1974.

Keller, G. V., Rational models for geothermal systems (abstract), Geol. Soc. Amer. Abstr. Programs, 7 (5), 616, 1975.

Kinoshita, W. T., A gravity survey of part of the Long Valley district, Idaho, open-file report, U. S. Geol. Surv. Denver, Colo., 1962.

Mabey, D. R., D. L. Peterson, and C. W. Wilson, Preliminary gravity map of southern Idaho, open-file report, U. S. Geol. Surv., Denver, Colo., 1974.

Mabey, D. R., H. Ackerman, A. A. R. Zohdy, D. B. Hoover, D. B. Jackson, and J. O'Donnell Geophysical studies of a geothermal area in the southern Raft River Valley, Idaho (abstract), Geol. Soc. Amer. Abstr. Programs, 7 (5), 624, 1975.

Malde, H. E., Fault zone along northern boundary of western Snake River Plain, Idaho, Science, 130 (3370), 272, 1959.

Nichols, C. R., and J. K. Applegate, Geologic and geophysical aspects of site selection for a geothermal demonstration project, Raft River, Idaho, Idaho Geothermal R & D Project Report for Dec. 16, 1973 to Mar. 15, 1974, Rep. ANCR-1155, UC-13, pp. 24-34, At. Energy Comm., Idaho Falls, Idaho, 1974.

Schmidt, D. L., and J. H. Mackin, Quaternary geology of Long and Bear Valleys, west-central Idaho, U. S. Geol. Surv. Bull., 1311-A, 1970.

White, D. F., and D. L. Williams, (Eds.), Assessment of geothermal resources of the United States--1975, U. S. Geol. Surv. Circ., 726, 1975.

Williams, P. L., K. L. Pierce, D. H. McIntyre, H. R. Covington, and P. W. Schmidt, Geologic setting of the Raft River geothermal area, Idaho (abstract), Geol. Soc. Amer. Abstr. Programs, 7 (5), 652, 1975.

Wilson, M. D., J. K. Applegate, S. L. Chapman, and P. R. Donaldson, Geothermal investigation of the Cascade, Idaho area, Repo. I, Dep. of Geol. and Geophys., Boise State Univ., Boise, Idaho, 1976.

Zohdy, A. A. R., D. B. Jackson, and R. J. Bisdorf, Exploring the Raft River geothermal area, Idaho (abstract), Geophysics, 41 (2), 382, 1975.

FLUID CIRCULATION IN THE EARTH'S CRUST

Denis Norton

Department of Geosciences, University of Arizona
Tucson, Arizona 85721

Abstract. Numerical simulation of thermally driven fluid flow caused by igneous intrusives in the upper crust indicates that fluid circulation is an inevitable consequence of lateral density gradients in pore fluids characteristic of these environments. Thermal perturbations associated with igneous plutons are predicted to be sufficiently large to generate hydrothermal systems in which the magnitude of convective heat transport exceeds that of conductive heat transport for rock permeabilities greater than 10^{-18} m^2 [Norton and Knight, 1977]. Furthermore, the style of the heat transfer is significantly different from systems in which conduction is the dominant heat transfer mechanism, particularly when the transport and thermodynamic properties of the fluid phase are taken into account. As a consequence of the critical end point which exists in the H_2O and related systems, the region above plutons is predicted to contain extensive vertical zones of nearly constant temperature. These first-order approximations of fluid circulation reveal two points relevant to predicting the thermal regime of the crust: (1) thermal gradients above convection-dominated systems are very nonlinear and cannot uniquely predict subsurface temperatures within our present scope of knowledge and data and (2) since fluid circulation may extend through a considerable portion of the upper crust in tectonically active regions, the thermal regime of these crustal regions is poorly understood.

Introduction

Temperature conditions in the earth's crust are normally predicted on the basis of extrapolated temperature-gradient data, petrologic arguments, and numerical approximations of conductive heat transfer processes in which various thermal energy sources, as well as rock properties, are considered. Analyses of thermal convection have usually indicated fluid circulation to be an important heat transport process, at least in geothermal areas [Elder, 1965; Ribando et al., 1976; Lister, 1974; Lowell, 1975]. However, the consequences of fluid circulation on the thermal conditions in the crust have only recently been analyzed for situations in which (1) the transport and thermodynamic properties of the fluid phase are allowed to vary with temperature and pressure changes and (2) an igneous intrusive body is present in the upper crust.

The unique characteristics of fluid systems for which H_2O is a principal component suggest that the properties of these types of fluids should contribute significantly to the heat transport process in convection-dominated systems [Norton and Knight, 1977]. Enthalpy, density, and viscosity of phases in the pure H_2O system and in salt-H_2O systems are very dependent on temperature and pressure in a temperature-pressure region which starts at the critical end point and extends to higher pressures. As a consequence of this dependence, natural systems are predicted to have thermal characteristics distinctly different from those predicted on the basis of constant fluid properties or even those predicted on the basis of properties approximated by simplistic equations of state. Most equations of state that have been previously used merely predict fluid properties along the two-phase surface in the H_2O or salt-H_2O systems.

Crustal environments which contain hot igneous bodies inevitably cause fluid circulation, and if the intrinsic host rock permeability is greater than 10^{-18} m^2, heat transfer by convection accounts for at least 10% of the total heat transfer, and at permeabilities greater than 10^{-18} m^2, convection greatly predominates over conduction [Norton and Knight, 1977]. Fluid-driving forces are generated as a natural consequence of the near-vertical side contact of intrusives, which cause lateral perturbations in the density of pore fluids. Instability of the fluids and the onset of convection is therefore instantaneous in these systems; the magnitude of the initial fluid flux depends principally on permeability of host intrusive rocks.

The purpose of this communication is to review the nature of fluid circulation related to transient thermal anomalies in the crust and to consider the consequences this fluid circulation might have on our concepts of the thermal environment in the crust.

Numerical Simulation of Fluid Circulation

Lateral density perturbations in fluids contained in the flow porosity of rocks cause fluid flow. The magnitude of this flow in natural systems can be determined by Darcy's law:

$$\bar{q} = \frac{k}{\nu} (\nabla P + \rho \bar{g}) \tag{1}$$

where the mass flux $\bar{q}$ is a function of intrinsic rock permeability k, fluid viscosity ν, density ρ, the gradient in pressure ∇P, and the gravitational vector $\bar{g}$. Density gradients, which may be the result of concentration as well as thermal gradients on the fluid, give rise to $\nabla_x P$ and $\nabla_y P$ terms in the horizontal plane. Although both types of density gradients are ubiquitous in the crust, only those resulting from thermal anomalies are included in the computations.

Thermal anomalies often cause and are coincident with extensive fracture zones and therefore probably represent the most significant contribution to fluid circulation in the crust, except in sedimentary basin environments where large concentration gradients are common. The inferred association of permeable fractured rocks with thermal anomalies in the upper 10-15 km of the crust suggests that fluid circulation is a characteristic feature in these environments.

Fluid flow caused by thermal anomalies related to igneous plutons are effectively scaled and represented in two dimensions by partial differential equations which describe the conservation of mass, momentum, and energy for the fluid-rock system [Norton and Knight, 1977]:

$$\gamma \frac{\partial T}{\partial t} + q\nabla H = \nabla \cdot \kappa \nabla T \qquad (2)$$

and

$$\frac{\nu}{k}\nabla \cdot \frac{\nu}{k} \nabla \Psi = R \frac{\partial \rho}{\partial y} \qquad (3)$$

where T is the temperature; Ψ the stream function; q the fluid flux; t the time; H, ρ, and ν the enthalpy, density, and viscosity of the fluid; k the permeability of the rock; κ the thermal conductivity; γ the volumetric heat capacity of the fluid saturated media; R the Rayleigh number; t the time; ∇ the gradient operator; and y the horizontal coordinate in the two-dimensional section to which these equations apply.

The physical meaning of (2) and (3) is apparent if one considers that the fluid density gradients on the right-hand side of (3), which result from a thermal anomaly, cause fluid circulation. That is, they define gradient values of the stream function and therefore fluid flux, since $q_z = -\partial \Psi/\partial y$ and $q_y = \partial \Psi/\partial z$. The fluid flux q in turn transports heat away from the thermal anomaly, second term on left of (2); at the same time, thermal energy is conducted away from the thermal anomaly, right-hand side of (2). Both of these processes give rise to a decrease in temperature with respect to time and therefore decrease the horizontal fluid density gradients. And, consequently, the thermal anomaly is decreased by combined convective and conductive heat transfer. Equations (2) and (3) are approximated by finite difference numerical equations which permit computation of the values of the dependent variables at discrete points in the domain from initial and boundary values specified for the system. The numerical analysis provides the option to include variable transport properties of the fluid (H_2O system) and rock, general boundary and initial conditions, and radioactive and volumetric heat sources in a two-dimensional domain. The transport processes related to the transient thermal anomaly are approximated by a time sequence of steady state numerical solutions to (2) and (3), computed at explicitly stable time intervals. An alternating-direction-implicit finite difference method is used to approximate the spatial derivatives at discrete intervals of the order of 0.1 to 0.5 of the system height. Fluid pressure in the system is computed at each steady state step by integration of (2), in which the fluid properties, viscosity and density, are expressed as a function of temperature and pressure. The following discussion relies on computations and analyses using these methods.

The Nature of Fluid Circulation Related to Thermal Anomalies

The style of fluid circulation in the upper 10 km of crust and the nature of pluton cooling has been simulated [Norton and Knight, 1977] for a variety of host rock permeability values and geometries.

Convection dominates heat transfer in hot igneous pluton environments if rock permeabilities are of the order of 10^{-18} m^2 or greater, resulting in a spatial redistribution of thermal energy significantly different than that in similar environments in which conductive heat

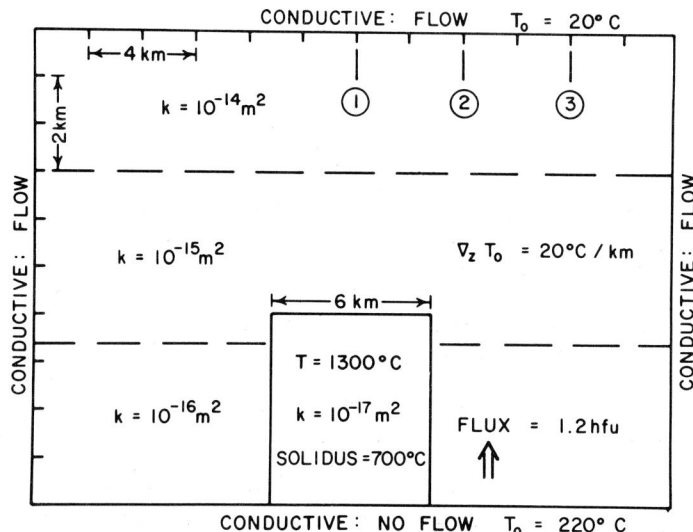

Fig. 1. Two-dimensional cross section of a crust 10 km deep and 24 km wide containing an igneous pluton 4 km high and 6 km wide. Initial value and boundary conditions are for a magma body emplaced instantaneously at 6 km below the surface. Pluton permeability is effectively zero until the temperature of discrete points in the body decreases to 700°C; then permeability at those points is set to 10^{-17} m^2. Regional heat flux is set at 1.2 HFU for the duration of the system. The initial thermal gradient is 20 C/km, whereas the magma is homogeneous and is connected to a magma reservoir below the base of the pluton at T = 1300°C for the initial 50,000 years. Thermal conductivity is constant at 0.6 cal/m sec °C; the circulating fluid is pure H$_2$O and does not react with the enclosing rocks.

transfer predominates. Although the time duration of convection-dominated thermal anomalies is similar to that of conduction-dominated systems when the pluton itself is impermeable, the cooling time is significantly shortened by increases in permeability such as might accompany extensive fracturing of the pluton. The direct application of these modeling results to actual systems must be made with caution since the in situ values of rock permeability are virtually unknown. However, analogies drawn between permeability values of rocks for which permeability data are available and estimates of permeability suggest that permeability values exceeding the 10^{-18} m^2 minimum may characterize a substantial portion of the upper crust [Norton and Knapp, 1977].

A numerical model of a system which illustrates the convective transfer of heat around igneous plutons is presented. A basaltic magma at ∼1300°C is presumed to be emplaced relatively rapidly, with respect to the rate of heat transfer away from the magma body, into host rocks whose permeability increases upward from 10^{-16} m^2 to 10^{-14} m^2 (Figure 1). The relatively rapid intrusion rate only requires magma flow velocities on the order of a few centimeters per year, a value which seems to be reasonable. Since cooling of magmas is normally accompanied by fracture development resulting from reactions that increase or decrease the pluton volume, the pluton permeability is changed from effectively zero to 10^{-17} m^2 as the temperature of discrete grid points in the pluton decreases to <700°C, thereby simulating fracture development and permitting fluid circulation through the pluton.

Boundary conditions selected for this system are analogous to natural systems where thermal energy is conducted through all the boundaries. The bottom and top boundary temperatures are set to 220°C and 20°C, respectively. Thermal conductivity of the domain is assumed to be constant, 0.6 cal/ms °C, and since the bottom boundary is conductive, the domain has a constant regional flux of 1.2 μcal/m^2 sec (HFU) and the host rocks have an initial background thermal gradient of 20°C/km. The relative permeability values within the domain are set to simulate the decrease in continuous fractures with depth, and the magnitude of the permeability is set to illustrate the effects of fluid circulation. The side and top boundaries are permeable, but the base is impermeable in order to further simulate the decrease in permeability with depth. The permeable top boundary condition does not, however, permit convection of thermal energy out of the system. This latter condition simulates natural systems which do not have hot springs emerging at the top boundary, e.g., the fluids flow through and thermally equilibrate with the rocks at the top boundary. This system was then simulated by using a spatial discretization of 160 points, which results in a 0.1 vertical increment and a 0.06 horizontal increment. The numerical approximations represent the partial differential equations to within a truncation error of the order of 0.05 times the value of the dependent variable. Discrete time increments are computed on the basis of stability criteria, which results in convergence errors of the order of 0.005 times the dependent variable.

The thermal anomaly, introduced by the pluton, causes pore fluids in the host rocks to circulate from the sides and top boundaries of the domain toward the pluton then upward along its side margins and out the top of the domain (Figures 2-4). This circulation pattern significantly increases the heat flux over the pluton top with respect to a purely conductive process. As a consequence of the convective heat transfer, thermal gradients in the domain directly over the pluton are relatively steep near the surface, i.e., 0-0.5 km, decrease sharply and remain constant over several kilometers, then gradually increase toward the pluton top (Figures 5a-5c). The convective heat flux at the surface directly above the pluton varies from 1.2 HFU at the initial time to 15 HFU at 8×10^4 years, whereas the vertical component of convective flux at 0.5 km depth ranges from 0.5 HFU at 2×10^4 years to 20 HFU at 8×10^4 years and then gradually decreases to 10 HFU at 1.5×10^5 years.

The caveat about these values at the surface is that they are arbitrary to the extent that they are a function of the numerical discretization. However, the relative comparison between the values in the same system at various times is a reasonable approximation of what can be expected in nature. Finer discretization merely results in a nonlinear thermal gradient and predicts it to better precision. Progressive fracturing of the pluton contributes to the persistence of

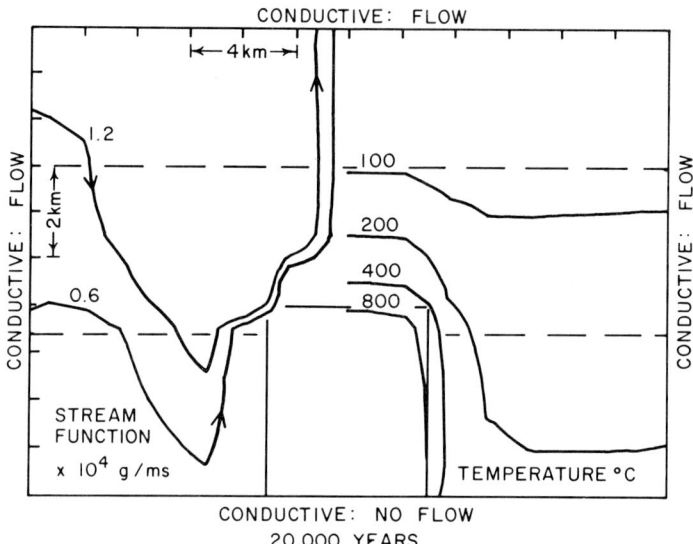

Fig. 2. Scalar stream function (g/m sec) and temperature (degrees Celsius) distributions after 2×10^4 years elapsed time, illustrating steady state fluid circulation and temperature, respectively. Vertical fluid fluxes of the order of 5×10^{-5} g/m^2 sec are realized 3 km directly above the pluton, which, together with conductive heat transfer, cause the 100°C isotherm to migrate upward at about 0.05 m/yr.

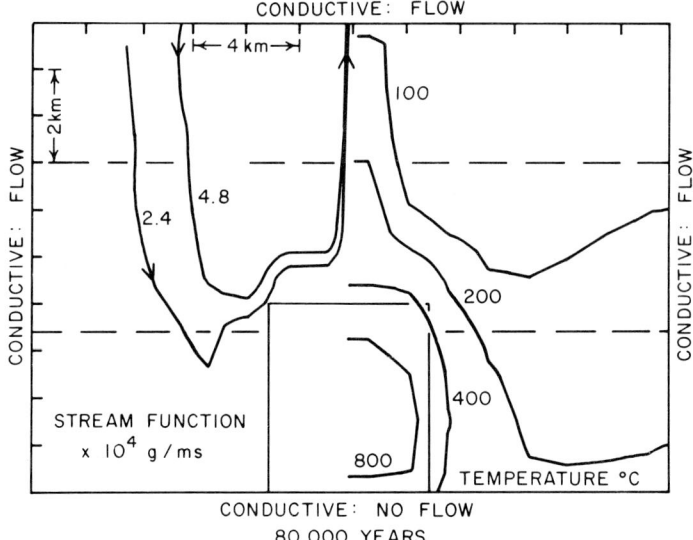

Fig. 3. Scalar stream function and isotherm distributions after 8×10^4 years elapsed time. Vertical fluid fluxes of the order of 10^{-3} g/m^2 sec are realized 3 km directly above the pluton, and 8 km laterally away from this upflow zone, downward fluid fluxes, $\sim 10^{-7}$, g/m^2 sec, occur.

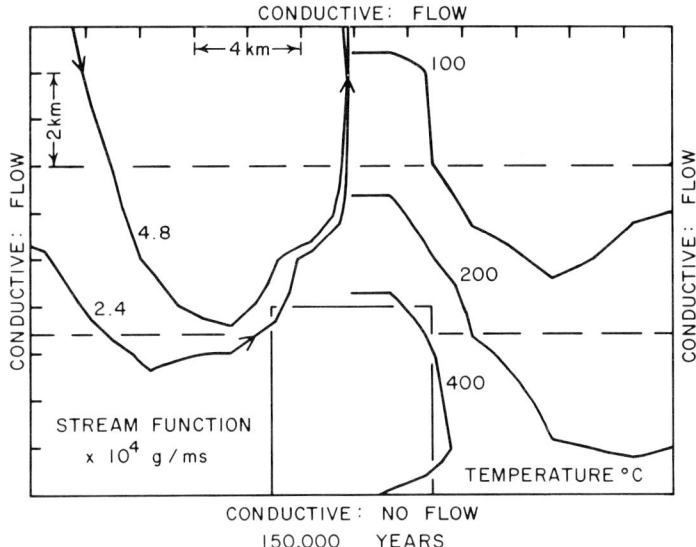

Fig. 4. Scalar stream function and iostherm distributions after 1.5×10^5 years elapsed time. Vertical fluid fluxes at comparable positions in previous times have decreased to about 50% of the fluxes at 8×10^4 years. The 100°C and 200°C isotherms have moved to slightly deeper portions in response to the decreasing convective flux. The outer 1.5 km of the upper 2 km of the pluton fractured at 10^5 years, thereby increasing the cooling rate of the body. The average pluton temperature is 800°C at this time.

large convective heat fluxes over a long time period. The estimated time duration for which convective fluxes will be greater than the regional heat flux in the upper 2 km of the system is about 5×10^5 years.

Laterally away from the pluton, thermal gradients in the fluid downflow zone are depressed below the regional gradients as a result of the convective heat flux of -3 HFU. In these regions, at cooling times $\sim 1.2 \times 10^5$ years, the isotherms are depressed downward with respect to their regional position, cf. 200°C isotherm. The portion of the anomaly, at temperatures >100°C, in the upper 3 km is dispersed over an area equivalent to the pluton top.

The several-kilometer vertical extent of relatively constant thermal gradients in the host rocks overlying the pluton and the corresponding temperature values, 100-400°C, are characteristic of convection-dominated systems which we have analyzed [Norton and Knight, 1977]. Transport and thermodynamic properties of supercritical fluid in the H_2O system and salt-H_2O systems are characterized by extremes which contribute to these thermal gradient features (Figure 6). In the region which extends from the critical end point, $\sim$ 375°C and 220 bars for the H_2O system, derivatives of fluid density and enthalpy with respect to temperature at constant pressure are maximums, and fluid viscosity is a minimum. Therefore thermal perturbations at these conditions result in the largest fluid density gradients which togeth-

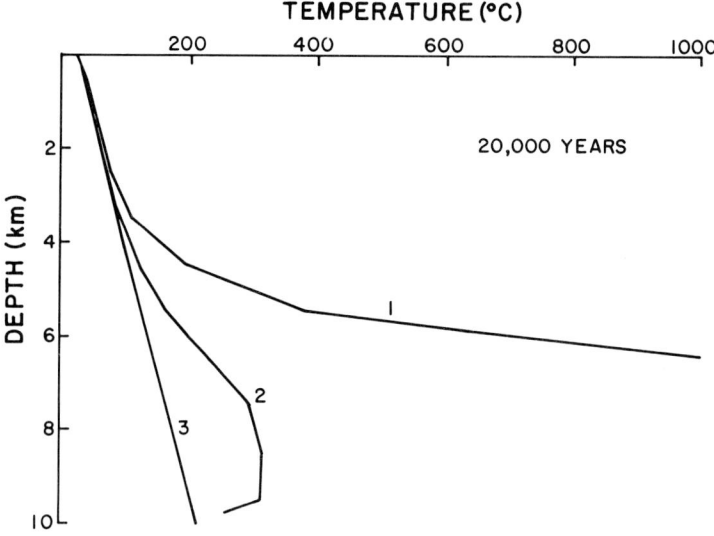

Fig. 5a. Vertical thermal gradients from the surface to the base of the system at elapsed time of 2×10^4 years. Vertical sections are located along the center line, line 1, of the pluton; 1 km away from the side wall of the pluton, line 2; and 5 km away from the side wall of the pluton, line 3 (cf. Figure 1 for positions).

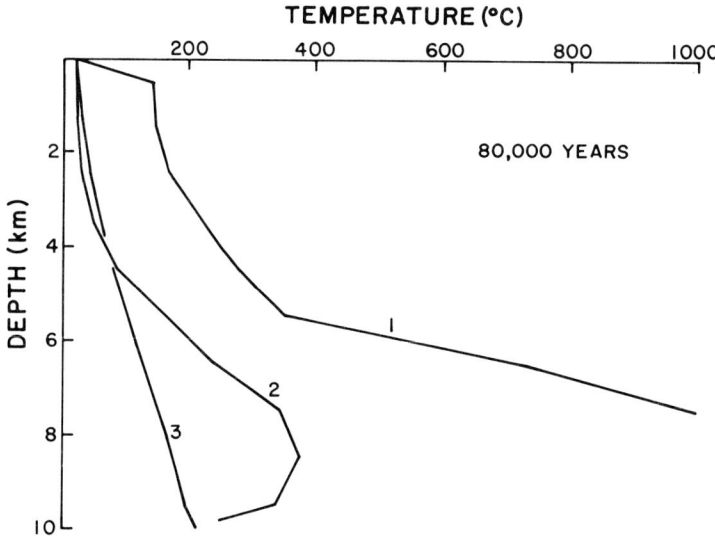

Fig. 5b. Vertical thermal gradients from the surface to the base of the system at elapsed time of 8×10^4 years. Vertical sections are located along the center line, line 1, of the pluton; 1 km away from the side wall of the pluton, line 2; and 5 km away from the side wall of the pluton, line 3 (cf. Figure 1 for positions).

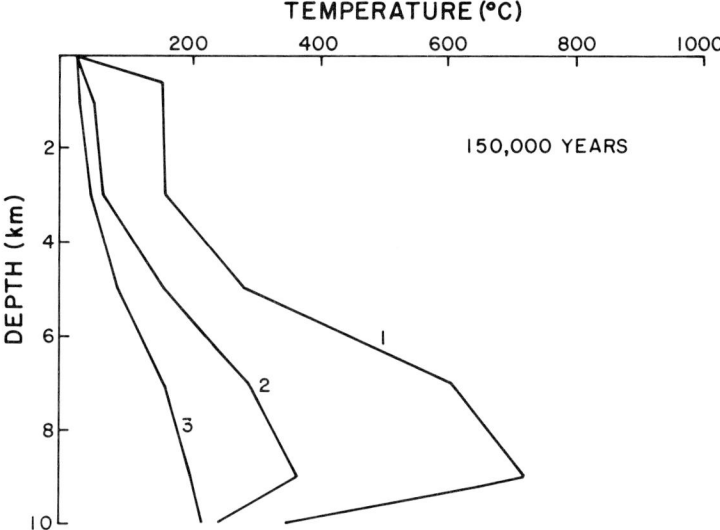

Fig. 5c. Vertical thermal gradients from the surface to the base of the system at elapsed time of 1.5×10^5 years. Vertical sections are located along the center line, line 1, of the pluton; 1 km away from the side wall of the pluton, line 2; and 5 km away from the side wall of the pluton, line 3 (cf. Figure 1 for positions).

er with the minimum in the fluid viscosity tend to maximize the fluid fluxes. The heat capacity of the fluid is also a maximum under these conditions, and hence the convective heat transport is maximized in this temperature-pressure region. As a point of interest these extremes tend to decrease in magnitude from the region near the critical end point to lesser extremes at higher pressures. The critical end point and related extremes in fluid properties are displaced to higher temperatures and pressures for salt-H_2O systems (Figure 6). Fluids which contain dissolved components equivalent to a 3 m NaCl solution have a critical end point at 590°C and 850 bars, and the extremes in fluid properties extend into a region analogous in position to the pure H_2O system (Figure 6).

The nature of the thermal gradients within permeable rocks overlying thermal anomalies in a natural system is clearly predicated on the values of fluid properties and permeability. Since the fluid properties, at least for the H_2O system, are relatively well known, one can reasonably assume that these thermal gradients will at least be characteristic of environments where rock permeabilities are $\geq 10^{-18}$ m^2 and anomaly temperatures are $\geq 375°C$ at depths where pressures are greater than 220 bars, i.e., ~ 2.2 km. In natural environments where dissolved components are relatively more concentrated, these effects will be realized at progressively greater depths and slightly higher temperatures.

The example system discussed above contains relatively high values of rock permeabilities with respect to our current best guesses of conditions in the crust. However, fluid circulation effects have been

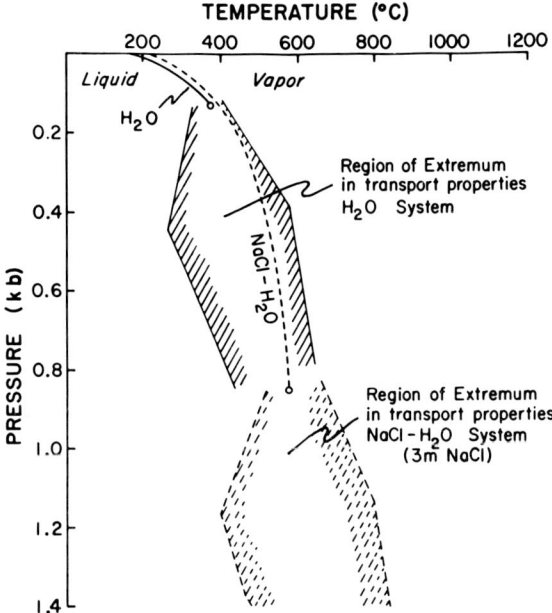

Fig. 6. Temperature-pressure sections through the NaCl-H₂O systems depicting the two-phase surface, liquid-vapor, and critical end point for 0 m and 3 m NaCl concentrations. The approximate region of anomalous extreme in transport properties of supercritical fluid is depicted for 0 m and 3 m solutions. Note the shift of the critical end point and associated anomalous regions to higher temperatures and pressures as a result of adding NaCl to the system.

observed in models where the pluton tops are 12-km deep, within low-permeability rocks (10^{-20} m²), but are overlain by higher permeability (10^{-18} m²) zones which simulate vertical fractures. Thermal gradients are more linear in these systems than in the system discussed above, but only a few percent contribution to the heat flux by convection may have a significant effect on our interpretations of the thermal environment in the crust. The most significant feature of the simulated system is that vertical thermal gradients in systems where convective heat transfer occurs do not provide a unique set of data with which subsurface temperatures can be predicted.

Fluid Circulation in the Crust

Fluid circulation in the upper crust is predicted to be more extensive than was previously thought; its magnitude may be large enough to contribute significantly to the redistribution of thermal energy in this environment. The magnitude of the contribution of fluid circulation to heat transport depends entirely on the magnitude of permeability in crustal rocks and the distribution, with respect to thermal anomalies, of fluid-saturated rocks with permeabilities $\geq 10^{-18}$ m². The minimum permeability value is realized in rocks which have continuously open-planar fractures spaced 0.1 km apart with an effective aperture of ∼20 µm [Norton and Knapp, 1977; Snow, 1970]. This abundance of continuous fractures is easily realized in tectonically active regions and in pluton environments [Villas, 1975; Villas and Norton, 1977], but apertures and continuity of fractures with respect to depth are unknown. In tectonically quiescent regions neither abundance nor aperture of fractures have been documented. However, indirect evidence suggests that permeabilities sufficient to permit significant heat transfer by convection may be realized in the upper crust. First, in tectonically active regions, continuous fractures develop to considerable depths, as is indicated by earthquake hypocenter data. Second, igneous intrusive processes contribute to development of fracture sets in the rocks they intrude. The extent and magnitude of the permeability resulting from combined tectonic and igneous intrusive events is clearly conducive to extensive fluid circulation, as is evidenced by eroded equivalents to these environments which show abundant mineral alteration, as well as by large gains and losses of chemical components in and adjacent to fractures. Transport of thermal energy into the crust by magma or simple conduction also produces fractures due to the differential thermal expansion of pore fluids and rocks [Knapp and Knight, 1977]. In tectonically less active regions, permeability values can be inferred from electrical and, perhaps, seismic properties, empirical relationships between pore continuity, and electrical resistivity [Brace, 1971] or variations in seismic wave velocity [Nur and Simmons, 1969]. These indirect lines of evidence suggest that crustal rocks contain a fluid phase, which may be relatively concentrated in dissolved components, and that they are sufficiently permeable to warrant further efforts toward quantitative determination of bulk rock permeability.

Analyses of transport phenomena in permeable media suggest that fluid circulation through fractured rocks may contribute significantly to heat transfer through the crust, at least to depths of 10-15 km. As a consequence of fluid circulation, several effects may be realized in nature: lower than normal thermal gradients over several-kilometer vertical distances in the upper crust, abnormally low conductive thermal values coincident with fluid downflow zones, and gross errors in predicting subsurface temperatures by downward extrapolation of thermal gradients. These effects are undoubtedly present in active geothermal systems and can be predicted, with reasonable confidence, to occur in the vicinity of virtually all igneous bodies emplaced into the upper crust. The more widespread realization of the effects in more normal crust is mere speculation at this time, and many questions remain that will require more precise numerical models and data acquisition. However, this first approximation suggests that the nature of the upper crustal environment may indeed be the result of dynamic fluid systems, the extent of which is unknown.

Acknowledgments. This work was supported by funds provided by ERDA, contract E-11-1-2763, and NSF grant EAR74-03515 A01 for which the author is grateful. Discussions with Professor H. C. Helgeson, R. Knapp, J. Knight, and R. N. Villas have contributed to the author's ideas. I am also grateful for discussions with W. F. Brace and for his published work, which provided the incentive to consider the problem of permeability distribution in the crust, and to L. McLean for her editorial assistance in preparing the manuscript.

References

Brace, W. F., Resistivity of saturated crustal rocks to 40 km based on laboratory measurements, in The Structure and Physical Properties of the Earth's Crust, Geophys. Monogr. Ser., vol. 14, edited by J. G. Heacock, pp. 243-255, AGU, Washington, D. C., 1971.

Elder, J. W., Physical processes in geothermal areas, in Terrestrial Heat Flow, Geophys. Monogr. Ser., vol. 8, edited by W. H. K. Lee, pp. 211-239, AGU, Washington, D. C., 1965.

Knapp, R., and J. Knight, Differential thermal expansion of pore fluids, fracture propagation, and microearthquake production in hot pluton environments, J. Geophys. Res., 82, in press, 1977.

Lister, C. R. B., On the penetration of water into hot rock, Geophys. J. Roy. Astron. Soc., 39, 465-509, 1974.

Lowell, R. P., Circulation in fractures, hot springs, and convective heat transport on mid-ocean ridge crests, Geophys. J. Roy. Astron. Soc., 40, 351-365, 1975.

Norton, D., and R. Knapp, Transport phenomena in hydrothermal systems: The nature of porosity, Amer. J. Sci., 277, in press, 1977.

Norton, D., and J. Knight, Transport phenomena in hydrothermal systems: Cooling plutons, Amer. J. Sci., 277, in press, 1977.

Nur, A., and G. Simmons, The effect of saturation on velocity in low porosity rocks, Earth Planet. Sci. Lett., 17, 183-193, 1969.

Ribando, R. J., et al., Numerical models for hydrothermal circulation in the oceanic coast, J. Geophys. Res., 81(17), 3007-3012, 1976.

Snow, D. T., The frequency and aperture of fractures in rocks, J. Rock Mech., 7, 23-40, 1970.

Villas, R. N., Fracture analysis, hydrodynamic properties, and mineral abundances in altered igneous rocks at the Mayflower Mine, Park City District, Utah, Ph.D. dissertation, 254 pp., Univ. of Utah, Salt Lake City, 1975.

Villas, R. N., and D. Norton, Irreversible mass transfer between circulating hydrothermal fluids and the Mayflower stock, Econ. Geol., 72, in press, 1977.

NUMERICAL SOLUTIONS FOR TRANSIENT HEATING AND FLUID WITHDRAWAL IN A LIQUID-DOMINATED GEOTHERMAL RESERVOIR

Ping Cheng and Lall Teckchandani

Department of Mechanical Engineering, University of Hawaii
Honolulu, Hawaii 96822

Abstract. The transient responses in a liquid-dominated geothermal reservoir resulting from a sudden heating and withdrawal of fluids are studied numerically. The idealized two-dimensional rectangular reservoir is confined by cap rock at the top, heated by bedrock from below, and recharged continuously through vertical boundaries from the sides. With appropriate approximations, the governing equations can be combined and reduced to two nonlinear partial differential equations in terms of dimensionless temperature and pressure, having a dimensionless parameter D, which is a measure of the relative importance of convective to conductive heat transfer. The finite difference form of the temperature equation is solved numerically by the alternating direction implicit method, while that of the pressure equation is solved by the Crank-Nicholson implicit method. It is found that for a two-dimensional reservoir 6000 m wide and 1500 m thick, characterized with D = 4000, a steady state condition is reached approximately 7000 years after a sudden heating from below. Oscillatory convection begins noticeably for a reservoir having D > 4000, a higher frequency of oscillations occurring as the value of D is increased. The contraction of isotherms in a reservoir due to fluid withdrawal from both a line and a plane sink is demonstrated.

Introduction

The study of heat and mass transfer in a porous medium can yield considerable insight into the nature of convective heat transfer in geothermal reservoirs. On the basis of a linear stability analysis, Lapwood [1948] shows that for a porous medium bounded by two parallel impermeable surfaces and heated from below, free convection begins when the value of the Rayleigh number of the medium is equal to $4\pi^2$. The Rayleigh number is found to be the governing parameter for free convection in a porous medium, which is defined as $Ra = \rho K g \beta_t \Delta T h/\mu\alpha$; here ΔT and h denote the temperature difference and the distance between the parallel surfaces; ρ, β_t, and μ the density, thermal expansion coefficient, and viscosity of the fluid; K and α the permeability and the equivalent thermal diffusivity of the saturated porous medium; and g the gravitational acceleration. The value of the critical Rayleigh number for the onset of free convection in a porous medium has been verified by experiments [Katto and Masuoka, 1967].

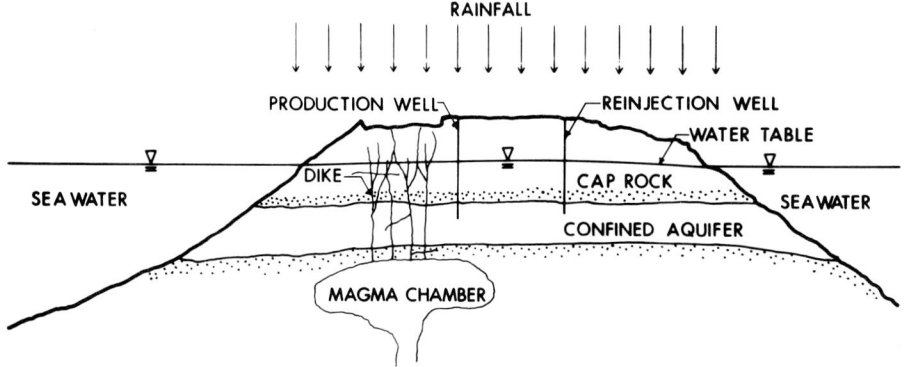

Fig. 1. Island aquifer with geothermal heat sources.

For the study of convection in a porous medium with Rayleigh numbers in excess of the critical value, nonlinear analysis or numerical methods must be used. For example, on the basis of the boundary layer simplifications, Cheng and his co-workers [Cheng, 1977a,b,c; Cheng and Chang, 1976; Cheng and Minkowycz, 1977; Minkowycz and Cheng, 1976] have obtained a number of analytical solutions for the prediction of steady heat transfer rates and the size of the hot-water zone adjacent to hot intrusives. On the other hand, there has been a considerable amount of work on the numerical study of steady, free convection in porous media (see Witherspoon et al. [1975] for a review of the literature).

Comparatively little work has been done on the study of transient convective heat transfer in porous media. It appears that the first numerical solution for the transient free convection in an enclosed porous medium heated nonuniformly from below is due to Elder [1967]. In an experimental investigation of free convection in a porous medium heated from below, Combarnous and Le Fur [1969] observe that oscillatory convection exists when Ra = 280-300. The oscillatory convection was subsequently confirmed numerically by Horne and O'Sullivan [1974a] as well as by Caltagirone [1975]. It was found that the value of the Rayleigh number for which oscillatory convection begins to appear depends on the type of thermal and hydrodynamic boundary conditions as well as on the aspect (or width/length) ratio of the enclosed domain of the porous medium. The physical mechanism for the existence of the oscillatory convection was attributed to the thermal instability of the thermal boundary layers adjacent to the heated or cooled surfaces.

Most recently, the problem of transient, combined, free and forced convection in porous media has also received considerable attention. For example, the problem of withdrawal of fluids from a point sink in a liquid-dominated geothermal reservoir has been studied numerically by Horne and O'Sullivan [1974b] whereas the corresponding problem for a two-phase geothermal reservoir has been studied numericably by Lasseter [1975]. These studies are useful for the prediction of the behavior of geothermal wells under different operating conditions.

In this paper, transient responses in an aquifer resulting from both a sudden heating and withdrawal of fluids are studied numerically. As is shown in Figure 1, the boundary conditions considered in this paper

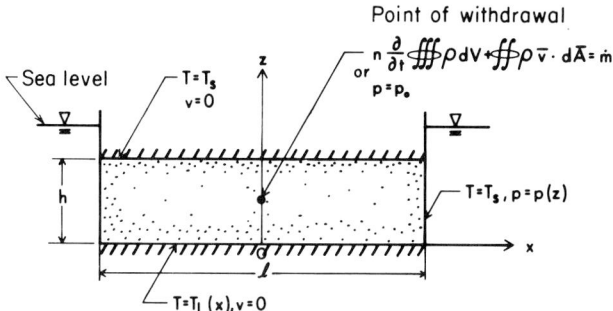

Fig. 2. Vertical cross section of an idealized geothermal reservoir.

differ from those of the previous work in that the reservoir is constantly recharged through the sides, a situation thought to be analogous to the heat transfer mechanism which applies to at least some of the Hawaiian shield volcanoes. The mathematical model is formulated on the basis of Darcy's law together with the conservation of heat and mass in a saturated porous medium with constant properties. With appropriate approximations the governing equations can be combined and reduced to two nonlinear partial differential equations of parabolic type in terms of dimensionless temperature and pressure, with a dimensionless parameter D, defined by $D \equiv \rho K g h / \alpha \mu$, which is related to the Rayleigh number by $Ra = D \Delta T \beta_t$. The temperature equation is solved numerically by the alternating direction implicit method, while the pressure equation is solved by the Crank-Nicholson implicit method. It is found that for a two-dimensional idealized rectangular aquifer (Figure 2) with recharge from the sides through vertical boundaries, oscillatory convection begins approximately at $D = 4000$ (which is equivalent to a Rayleigh number of 200 if $\Delta T = 250°C$ and $\beta_t = 2 \times 10^{-4}/°C$) and the period of oscillation decreases as the value of D is increased. For a reservoir with a depth of 1500 m having $D = 4000$ it is found that isotherms reach a steady state 7000 years after a step function heating from below. The contraction of isotherms in a geothermal reservoir resulting from the withdrawal of fluids for the cases of both a line and a plane sink is also investigated. Isotherms in the reservoir are compared before and after the fluid withdrawal.

Formulation of the Problem

In the mathematical formulation of the problem the following simplifying assumptions are made:
1. The groundwater temperature is everywhere below boiling for the pressure at each depth.
2. The groundwater and the rock formation are in local thermodynamic equilibrium.
3. The rock formation is nondeformable.
4. The physical properties of the groundwater and the rock formation, i.e., the thermal conductivity, specific heat, viscosity, and permeability, are constant.
5. Darcy's law applies.

6. Density variations are considered only in the buoyancy force term of Darcy's law and in the transient term $\partial \rho / \partial t$ in the continuity equation.

With these assumptions the governing equations for convective heat transfer in a geothermal system are

$$n \, \partial \rho / \partial t + \rho_a \, \text{div} \, \bar{v} = 0 \tag{1}$$

$$\bar{v} = -(K/\mu)(\nabla p - \rho \bar{g}) \tag{2}$$

$$\rho = \rho_a [1 - \beta_t (T - T_a) + \beta_p (p - p_a)] \tag{3}$$

and

$$\sigma \, \partial T / \partial t + \bar{v} \cdot \nabla T = \alpha \nabla^2 T \tag{4}$$

where the subscript a denotes some reference condition; n is the porosity of the rock formation; $\bar{v}$ and $\bar{g}$ are the Darcy velocity vector and the gravitational acceleration vector, respectively; β_t and β_p are the coefficient of thermal expansion and the isothermal compressibility, defined by $\beta_t \equiv (1/\rho)(\partial \rho/\partial T)_p$ and $\beta_p \equiv (1/\rho)(\partial \rho/\partial p)_T$; p and T are the pressure and temperature; α is the equivalent thermal diffusivity, defined by $\alpha = k_m/(\rho_a C)_f$ with $k_m = (1 - n)k_r + nk_f$, where k is the thermal conductivity and the subscripts m, r, and f denote the saturated porous medium, the rock formation, and the fluid, respectively; σ is defined by $\sigma \equiv (\rho C)_m/(\rho_a C)_f$ with $(\rho C)_m = (1 - n)(\rho C)_r + n(\rho_a C)_f$, where ρ and C are the density and the specific heat; μ is the viscosity of the fluid; and K is the permeability of the rock formation.

Instead of Figure 1 we shall consider its corresponding idealized model, shown in Figure 2, where h and ℓ are the depth and the width of the vertical cross section of the reservoir, the origin of the coordinates being located at the center of the bedrock. We shall now consider the boundary conditions and initial conditions for the following two problems.

<u>Formation of a geothermal reservoir</u>. The idealized two-dimensional reservoir, shown in Figure 2, is assumed to be initially motionless and isothermal at T_s and is suddenly heated by intrusives such that the bedrock has a temperature distribution $T_L(x)$. Accordingly, the initial conditions of the problem are

$$T(x, z, 0) = T_s \tag{5}$$

and

$$p(x, z, 0) = p_a + p_s(z) \tag{6}$$

where $p_s(z) \equiv \{\exp[\beta_p \rho_a g(h - z)] - 1\}/\beta_p$ and p_a is the hydrostatic pressure at z = h in the sea, which is assumed to be an isothermal compressible fluid. Boundary conditions for the problem are:

$$T(\pm \ell/2, z, t) = T_s \qquad p(\pm \ell/2, z, t) = p_a + p_s(z) \tag{7}$$

$$T(x, h, t) = T_a = T_s \qquad v(x, h, t) = 0 \qquad (8)$$

$$T(x, 0, t) = T_L(x) \qquad v(x, 0, t) = 0 \qquad (9)$$

where $T_L(x)$ is prescribed.

Withdrawal of fluids from a geothermal reservoir. We shall study the contraction of isotherms resulting from the withdrawal of fluids from a line or a plane sink for the two-dimensional reservoir shown in Figure 2. The initial conditions for the problem will be the prevailing pressure and temperature distribution in the reservoir at the time of withdrawal. Mathematically, we have

$$p(x, z, 0) = p_i(x, z) \qquad (10)$$

and

$$T(x, z, 0) = T_i(x, z) \qquad (11)$$

where $p_i(x, z)$ and $T_i(x, z)$ are the prescribed initial pressure and temperature. Boundary conditions for the problem are the same given by (7)-(9). In addition, we need to specify either the pressure at the points of withdrawal, i.e.,

$$p(x_o, z_o, t) = p_o \qquad (12)$$

or the withdrawal rate. Furthermore, we note that (1) is not applicable at the points of withdrawal, where conservation of mass requires that

$$n \frac{\partial}{\partial t} \oiiint \rho dV + \rho_a \oiint \bar{v} \cdot d\bar{A} = \dot{m} = \rho_a Q^* \qquad (13)$$

where m and Q^* are the mass rate and volume rate at the point of withdrawal and the integrations must enclose the point of withdrawal. Equation (13) is the relation for specifying the withdrawal rate. It is worth noting that (13) reduces to (1) for $\dot{m} = 0$ or $Q^* = 0$.

We now express the governing equations and the boundary and initial conditions in dimensionless form. To this end we introduce the following dimensionless variables:

$$P \equiv (p - p_a)/\rho_a gh \qquad (14a)$$

$$\theta \equiv (T - T_a)/(T_m - T_a) \qquad (14b)$$

$$U \equiv u\mu/\rho_a gK$$

$$V \equiv v\mu/\rho_a gK$$

$$X \equiv x/h \tag{14c}$$

$$Z \equiv z/h \tag{14d}$$

$$\tau \equiv \alpha t/\sigma h^2 \tag{14e}$$

where T_m is the maximum temperature of the bedrock. In terms of the dimensionless variables, (1)-(4) can be rewritten as

$$D\left[U\frac{\partial\theta}{\partial X} + V\frac{\partial\theta}{\partial Z}\right] + \frac{\partial\theta}{\partial\tau} = \nabla^2\theta \tag{15}$$

$$D\left[\nabla^2 P + \varepsilon_p \frac{\partial P}{\partial Z} - \varepsilon_t \frac{\partial\theta}{\partial Z}\right] = n\left[\varepsilon_p \frac{\partial P}{\partial\tau} - \varepsilon_t \frac{\partial\theta}{\partial\tau}\right] \tag{16}$$

where the dimensionless parameters ε_t, ε_p, and D are given by

$$\varepsilon_t \equiv \beta_t(T_m - T_a) \tag{17a}$$

$$\varepsilon_p \equiv \beta_p \rho_a g h \tag{17b}$$

$$D \equiv \rho_a K g h/\alpha\mu = Ra/\varepsilon_t \tag{17c}$$

where $Ra \equiv \rho_a Kgh\beta_t(T_m - T_a)/\alpha\mu$ is the Rayleigh number.

The dimensionless velocity components can be computed from

$$U = -\frac{\partial P}{\partial X} \tag{18}$$

$$V = -\left(\frac{\partial P}{\partial Z} + 1 + \varepsilon_p P - \varepsilon_t \theta\right) \tag{19}$$

For the problem of the formation of a geothermal reservoir the initial and boundary conditions (5)-(9) in dimensionless form are

$$\theta(X, Z, 0) = 0 \tag{20}$$

$$P(X, Z, 0) = \left\{\exp[\varepsilon_p(1-Z)] - 1\right\}/\varepsilon_p \tag{21}$$

$$\theta(\pm L/2, Z, \tau) = 0 \qquad P(\pm L/2, Z, \tau) = \left\{\exp[\varepsilon_p(1-Z)] - 1\right\}/\varepsilon_p \tag{22}$$

$$\theta(X, 1, \tau) = 0 \qquad V(X, 1, \tau) = -\left(\frac{\partial P}{\partial Z} + 1 + \varepsilon_p P - \varepsilon_t \theta\right) = 0 \tag{23}$$

$$\theta(X, 0, \tau) = \theta_L(X) \qquad V(X, 0, \tau) = -\left(\frac{\partial P}{\partial Z} + 1 + \varepsilon_p P - \varepsilon_t \theta\right) = 0 \tag{24}$$

For the problem of withdrawal of fluids the dimensionless boundary conditions are also given by (22)-(24). Furthermore, we must specify either the pressure or the rate of withdrawal at the line or plane sinks. Thus (12) and (13) in dimensionless form are

$$P(X_o, Z_o, \tau) = P_o \qquad (25)$$

$$n \oiiint \left[\varepsilon_p \frac{\partial p}{\partial \tau} - \varepsilon_t \frac{\partial \theta}{\partial \tau} \right] dv + \oiint \bar{v} \cdot d\bar{s} = Q \qquad (26)$$

where $Q = Q^*/\alpha h$, $v \equiv V/h^2$, and $S \equiv A/h^2$. The initial conditions (10) and (11) in dimensionless form are

$$P(X, Z, 0) = P_i(X, Z) \qquad (27)$$

and

$$\theta(X, Z, 0) = \theta_i(X, Z) \qquad (28)$$

where $P_i(X, Z) \equiv [p_i(x, z) - p_a]/\rho_a g h$ and $\theta_i(X, Z) \equiv [T_i(x, z) - T_a]/(T_m - T_a)$.

It should be noted that if the transient term in (1) is assumed to be negligible, the right-hand side of (16) vanishes, and the resulting equation will be of elliptic rather than parabolic form. With this assumption the first term in (26) also vanishes. Furthermore, (16) is also of elliptic form if $\varepsilon_p = 0$ or $\beta_p = 0$.

Finite Difference Equations

The governing nonlinear equations (15) and (16) are of parabolic type, and therefore exact analytical solutions are not possible. For this reason we shall approximate the set of nonlinear partial differential equations by a set of nonlinear algebraic equations, on the basis of the finite difference method, which can be solved numerically. To this end we divide the problem domain into rectangular grids, the coordinates of the grid points being given by (X_i, Z_j), where $X_i = (i - 1)\Delta X$ (with $i = 1, 2, \ldots, M$) and $Z_j = (j - 1)\Delta Z$ (with $j = 1, 2, \ldots, N$). Similarly, the dimensionless time is divided into discrete time levels which will be indicated by superscripts. Thus $\theta_{i,j}^k$ and $P_{i,j}^k$ are the dimensionless temperature and pressure at the point (X_i, Z_j) and the time level k. The finite difference equations corresponding to (15) with boundary conditions (22a), (23a), and (24a) will be solved on the basis of the alternating direction implicit method (ADI), while the equation corresponding to (16) with mixed boundary conditions (22b), (23b), and (24b) will be solved by the Crank-Nicholson implicit method.

Formation of a geothermal reservoir. We now obtain the finite difference equations corresponding to (15) on the basis of the alternating direction implicit method (ADI) for the problem of the formation of a geothermal reservoir. The method consists of splitting the time step into two halves (k + 1/2) and (k + 1). For the first time level (k + 1/2) the partial derivatives of θ with respect to Z will be treated as being implicit (i.e., evaluated at the unknown time level k + 1/2), whereas the partial derivatives of θ with respect to X and all other unknowns will be treated as being explicit (i.e., evaluated at the known time level k). Next, to achieve a solution for the complete time level (k + 1), the directions are alternated by treating partial derivatives of θ with respect to X as being implicit and all others as being explicit. The resulting algebraic equations for each of the half steps will be of a tridiagonal form that can be solved noniteratively. Thus the finite difference equation corresponding to (15) for any interior node is

$$D\left[U_{i,j}^k\left(\frac{\theta_{i+1,j}^k - \theta_{i-1,j}^k}{2\Delta X}\right) + V_{i,j}^k\left(\frac{\theta_{i,j+1}^{k+1/2} - \theta_{i,j-1}^{k+1/2}}{2\Delta Z}\right)\right] +$$

$$\frac{\theta_{i,j}^{k+1/2} - \theta_{i,j}^k}{\Delta\tau/2} = \frac{\theta_{i+1,j}^k - 2\theta_{i,j}^k + \theta_{i-1,j}^k}{(\Delta X)^2} +$$

$$\frac{\theta_{i,j+1}^{k+1/2} - 2\theta_{i,j}^{k+1/2} + \theta_{i,j-1}^{k+1/2}}{(\Delta Z)^2} \qquad (29)$$

for time level (k + 1/2) and

$$D\left[U_{i,j}^k\left(\frac{\theta_{i+1,j}^{k+1} - \theta_{i-1,j}^{k+1}}{2\Delta X}\right) + V_{i,j}^k\left(\frac{\theta_{i,j+1}^{k+1/2} - \theta_{i,j-1}^{k+1/2}}{2\Delta Z}\right)\right] +$$

$$\frac{\theta_{i,j}^{k+1} - \theta_{i,j}^{k+1/2}}{\Delta\tau/2} = \frac{\theta_{i+1,j}^{k+1} - 2\theta_{i,j}^{k+1} + \theta_{i-1,j}^{k+1}}{(\Delta X)^2} +$$

$$\frac{\theta_{i,j+1}^{k+1/2} - 2\theta_{i,j}^{k+1/2} + \theta_{i,j-1}^{k+1/2}}{(\Delta Z)^2} \qquad (30)$$

for time level (k + 1), where

$$U_{i,j}^k = -\left[\frac{P_{i+1,j}^k - P_{i-1,j}^k}{2\Delta X}\right] \qquad (31a)$$

$$V_{i,j}^k = -\left[\frac{P_{i,j+1}^k - P_{i,j-1}^k}{2\Delta Z} + 1 + \varepsilon_p P_{i,j}^k - \varepsilon_t \theta_{i,j}^k\right] \qquad (31b)$$

Equations (29) and (30) can be rearranged to give

$$(R - 2r_2)\theta_{i,j+1}^{k+1/2} + (4r_2 + 1)\theta_{i,j}^{k+1/2} - (R + 2r_2)\theta_{i,j-1}^{k+1/2} =$$

$$(1 - 4r_1)\theta_{i,j}^k + (2r_1 - S)\theta_{i+1,j}^k + (2r_1 + S)\theta_{i-1,j}^k \qquad (32a)$$

for time level $(k + 1/2)$ and

$$(S - 2r_1)\theta_{i+1,j}^{k+1} + (4r_1 + 1)\theta_{i,j}^{k+1} - (S + 2r_1)\theta_{i-1,j}^{k+1} =$$
$$(1 - 4r_2)\theta_{i,j}^{k+1/2} + (2r_2 - R)\theta_{i,j+1}^{k+1/2} + (2r_2 + R)\theta_{i,j-1}^{k+1/2} \quad (32b)$$

for time level $(k + 1)$, where $r_1 \equiv \Delta\tau/[2(\Delta X)^2]$, $r_2 \equiv \Delta\tau/[2(\Delta Z)^2]$, $R \equiv V_{i,j}{}^k D\Delta\tau/(2\Delta Z)$, and $S \equiv U_{i,j}{}^k D\Delta\tau/(2\Delta X)$. The matrix coefficients of (32a) and (32b) are of a tridiagonal form which can be solved noniteratively.

The finite difference equations corresponding to boundary conditions (22a), (23a), and (24a) valid for all times $(k, k + 1, k + 2, \ldots)$ are

$$\theta_{1,j} = \theta_{Mj} = 0 \qquad j = 1, 2, \ldots, N$$

$$\theta_{i,N} = 0 \qquad i = 1, 2, \ldots, M \qquad (33)$$

$$\theta_{i,1} = \theta_L[(i-1)\Delta X] \qquad i = 1, 2, \ldots, M$$

and the initial condition for temperature is

$$\theta_{i,j} = 0 \qquad i = 1, 2, \ldots, M \qquad j = 1, 2, \ldots, N \qquad (34)$$

The pressure equation (16) with mixed boundary conditions (22b), (23b), and (24b) does not permit the use of the ADI method because of significant round-off errors in solving the resulting tridiagonal matrix. Thus for the pressure equation the Crank-Nicholson implicit method was used instead. The resulting finite difference equation corresponding to (16) for any interior node is

$$D \left\{ \frac{P_{i+1,j}^{k+1} - 2P_{i,j}^{k+1} + P_{i-1,j}^{k+1} + P_{i+1,j}^{k} - 2P_{i,j}^{k} + P_{i-1,j}^{k}}{(2\Delta X)^2} + \right.$$

$$\frac{P_{i,j+1}^{k+1} - P_{i,j}^{k+1} + P_{i,j-1}^{k+1} + P_{i,j+1}^{k} - P_{i,j}^{k} + P_{i,j-1}^{k}}{2(\Delta Z)^2} +$$

$$\varepsilon_p \left[\frac{P_{i,j+1}^{k+1} - P_{i,j-1}^{k+1} + P_{i,j+1}^{k} - P_{i,j-1}^{k}}{4(\Delta Z)} \right] -$$

$$\varepsilon_t \left[\frac{\theta_{i,j+1}^{k+1} + \theta_{i,j-1}^{k+1} + \theta_{i,j+1}^{k} - \theta_{i,j-1}^{k}}{4(\Delta Z)} \right] \right\} =$$

$$n\varepsilon_p \left[\frac{P_{i,j}^{k+1} - P_{i,j}^{k}}{\Delta\tau} \right] - n\varepsilon_t \left[\frac{\theta_{i,j}^{k+1} - \theta_{i,j}^{k}}{\Delta\tau} \right]$$

which can be rewritten as

$$P_{i,j}^{k+1} = \left(\frac{n}{2} + \varepsilon_p + 2C_1 + 2C_2\right)^{-1}\left\{\left(\frac{n}{2} + \varepsilon_p - 2C_1 - 2C_2\right)P_{i,j}^k + \right.$$

$$C_1\left(P_{i+1,j}^{k+1} + P_{i-1,j}^{k+1} + P_{i+1,j}^k + P_{i-1,j}^k\right) +$$

$$\left(C_2 + \frac{D\varepsilon_p\phi_2}{2}\right)\left[P_{i,j+1}^{k+1} + P_{i,j+1}^k\right] +$$

$$\left(C_2 - \frac{D\varepsilon_p\phi_2}{2}\right)\left[P_{i,j-1}^{k+1} + P_{i,j-1}^k\right] -$$

$$\frac{D\varepsilon_t\phi_2}{2}\left[\theta_{i,j+1}^{k+1} - \theta_{i,j-1}^{k+1} + \theta_{i,j+1}^k - \theta_{i,j-1}^k\right] +$$

$$\left.\frac{n\varepsilon_t}{2}\left[\theta_{i,j}^{k+1} - \theta_{i,j}^k\right]\right\} \quad (35)$$

where $C_1 \equiv D\Delta\tau/[2(\Delta X)^2]$, $C_2 \equiv D\Delta\tau/[2(\Delta Z)^2]$, and $\phi_2 \equiv \Delta\tau/(2\Delta Z)$.

Boundary conditions (22b), (23b), and (24b) in finite difference form are

$$P_{1,j} = P_{M,j} = \left\{\exp[\varepsilon_p(1-j\Delta Z)] - 1\right\}/\varepsilon_p \quad j = 1, 2, \ldots, N \quad (36)$$

$$\frac{4P_{i,2} - 3P_{i,1} - P_{i,3}}{(2\Delta Z)} + 1 + \varepsilon_p P_{i,1} - \varepsilon_t \theta_{i,1} = 0 \quad i = 1, 2, \ldots, M \quad (37)$$

$$\frac{3P_{i,N} - 4P_{i,N-1} + P_{i,N-2}}{(2\Delta Z)} + 1 + \varepsilon_p P_{i,N} - \varepsilon_t \theta_{i,N} = 0 \quad (38)$$

$$i = 1, 2, \ldots, M$$

where the initial condition for pressure in finite difference form is

$$P_{i,j} = \left\{\exp[\varepsilon_p(1-j\Delta Z)] - 1\right\}/\varepsilon_p \quad i = 1, 2, \ldots, M \quad j = 1, 2, \ldots, N \quad (39)$$

<u>Withdrawal of fluids from a geothermal reservoir</u>. In deriving the finite difference equations for this problem the mesh will be chosen such that the mesh points coincide with the points of withdrawal. Most of the finite difference equations derived in the previous section also hold for this problem except that the initial conditions (34) and (39) will be replaced by the temperature and pressure distributions at the time of withdrawal. Furthermore, (35) holds for all interior nodes except at the points of withdrawal, where either the pressure or the withdrawal rate must be specified. For a horizontal line sink located

at (I, J) the finite difference equation corresponding to (26) is

$$n\left[\varepsilon_p\left(\frac{P_{I,J}^{k+1} - P_{I,J}^k}{\Delta\tau}\right) - \varepsilon_t\left(\frac{\theta_{I,J}^{k+1} - \theta_{I,J}^k}{\Delta\tau}\right)\right] +$$

$$D\left[\frac{P_{I+1,J}^{k+1} - 2P_{I,J}^{k+1} + P_{I-1,J}^{k+1}}{(\Delta X)^2} + \frac{P_{I,J+1}^{k+1} - 2P_{I,J}^{k+1} + P_{I,J-1}^{k+1}}{(\Delta Z)^2}\right] +$$

$$\varepsilon_p\left(\frac{P_{I,J+1}^{k+1} - P_{I,J-1}^{k+1}}{2\Delta Z}\right) + \varepsilon_t\left(\frac{\theta_{I,J+1}^{k+1} - \theta_{I,J-1}^{k+1}}{2\Delta Z}\right) = \frac{Q}{\Delta X \Delta Z} \qquad (40)$$

Numerical Procedure

Formation of a geothermal reservoir. The numerical procedure is as follows.
1. Select values for D, ε_p, and ε_t, and incorporate the initial conditions of $\theta_{i,j}^k$ and $P_{i,j}^k$ from (34) and (39).
2. Choose a suitable time step.
3. Calculate $U_{i,j}^k$ and $V_{i,j}^k$ from (31a) and (31b).
4. Find a new temperature $\theta_{i,j}^{k+1}$ using (32a) and (32b).
5. Find a new pressure $p_{i,j}^{k+1}$ from (35) by the Gauss-Seidel iteration method. At the end of each iteration the pressure values at the top and lower boundary are recalculated by using (37) and (38). The iteration process will stop if the maximum difference in pressure between two successive iterations is less than 10^{-5}.
6. Find new values of $U_{i,j}^{k+1}$ and $V_{i,j}^{k+1}$ from (31a) and (31b) using the new values of $P_{i,j}^{k+1}$ and $\theta_{i,j}^{k+1}$.
7. Repeat steps (4)-(6) to advance the solution to the next time level.

Withdrawal of fluids from a geothermal reservoir. The numerical procedure is as follows.
1. Select values of D, ε_p, and ε_t, and incorporate suitable initial conditions of pressure and temperature.
2. Choose points of withdrawal, and impose suitable pressures at these points.
3. Execute steps (2)-(5) as outlined in the procedures for solving the problem of the formation of a geothermal reservoir on all points of the mesh except at the point of withdrawal.
4. Calculate the value of Q using (40).
5. Repeat steps (3) and (4) to advance the solution to the next time level.

Numerical Results and Discussion

Computations were carried out with double precision for an aquifer with an aspect ratio of 4, having $\theta_L(X) = \exp[-2(X)^2]$ and $\theta_a = \theta_s = 0$. Other parameters chosen for the computation are $\varepsilon_t = 0.05$ (corresponding to $\beta_t = 2 \times 10^{-4}/°C$ and $(T_m - T_a) = 250°C$), $\varepsilon_p = 0.0025$ (corresponding to $\beta_p = 1.66 \times 10^{-9} m^2/kg$ and $\rho_a gh = 1.5$ kg/m²), $M = 41$ (or $\Delta X = 0.05$, which corresponds to 75 m if $h = 1500$ m), and $N = 81$ (or $\Delta Z = 0.0125$, which corresponds to 18.75 m). The value of D for a viable geothermal reservoir ranges from 2000 to 20,000. A very small time step $\Delta \tau = 2.5 \times 10^{-5}$ (corresponding to 5 months on the real time scale if $\alpha = 2.3 \times 10^{-3} m^2/h$, $h = 1500$ m, and $\sigma = 0.6$) was used for the initial time steps, while larger values up to $\Delta t = 10^{-4}$ (approximately 20 years) were used as the solution advances to later times.

<u>Formation of a geothermal reservoir.</u> When an aquifer, initially isothermal and motionless, is suddenly heated from below, a conduction layer adjacent to the heated surface begins to build up and thus to generate a hydrostatic pressure gradient along the horizontal impermeable surface. When the value of the Rayleigh number based on the thickness of the thermal boundary layer exceeds its critical value, the pressure gradient will induce a convective movement such that the cold water from the ocean starts to move inland from the lower portion of the aquifer. The cold water from the ocean is gradually heated by the bedrock until it reaches the neighborhood of maximum heating, where it rises up as a thermal plume, spreads out under the cap rock, and finally discharges into the ocean. The development of the isotherms for an aquifer with $D = 4000$ is shown in Figure 3, where $\tau = 0.001$, corresponding to 200 years on the real time scale. The convective pattern reaches a steady state at $\tau = 0.035$, which corresponds to approximately 7000 years. For an aquifer with $D > 4000$, oscillatory convection becomes noticeable. Figure 4 shows the development of isotherms in an aquifer having $D = 7000$. The first peak values of isotherms occur at approximately $\tau = 0.02$ (corresponding to 4000 years), after which the isotherms begin to contract until $\tau = 0.04$ (i.e., 8000 years). Starting from $\tau = 0.04$, the isotherms move upward again until approximately $\tau = 0.06$ (i.e., 12,000 years), which is the second peak value of the isotherms. The oscillatory convection appears to be due to the creation and detachment of a mass of hot fluid rising up from the region of maximum heating. Comparison of Figures 3 and 4 shows that the rate of heating of the groundwater depends strongly on the value of D, being faster for higher values of D. Computations are also carried out for other values of D. It is found that the period of oscillation decreases as D increases.

Thus for a reservoir with $D > 4000$ the size of the hot-water zone may expand or contract owing to oscillatory convection. However, as the period of oscillation is of the order of thousands of years, it may be difficult to detect in field observations. The effect of oscillation has no practical consequence for the operation of a geothermal power plant, since the life-span of a power plant is considerably less in comparison with the period of oscillation. The fact that oscillatory convection exists, however, may have some implications for the assessment of geothermal resources and geophysical exploration, since it indicates that at certain times the size of the hot-water zone may be small in spite of the fact that the reservoir is very permeable and a magma chamber may exist at shallow depth.

<u>Withdrawal of fluids from a geothermal reservoir.</u> When a production well is located in a geothermal reservoir, additional pressure gradients

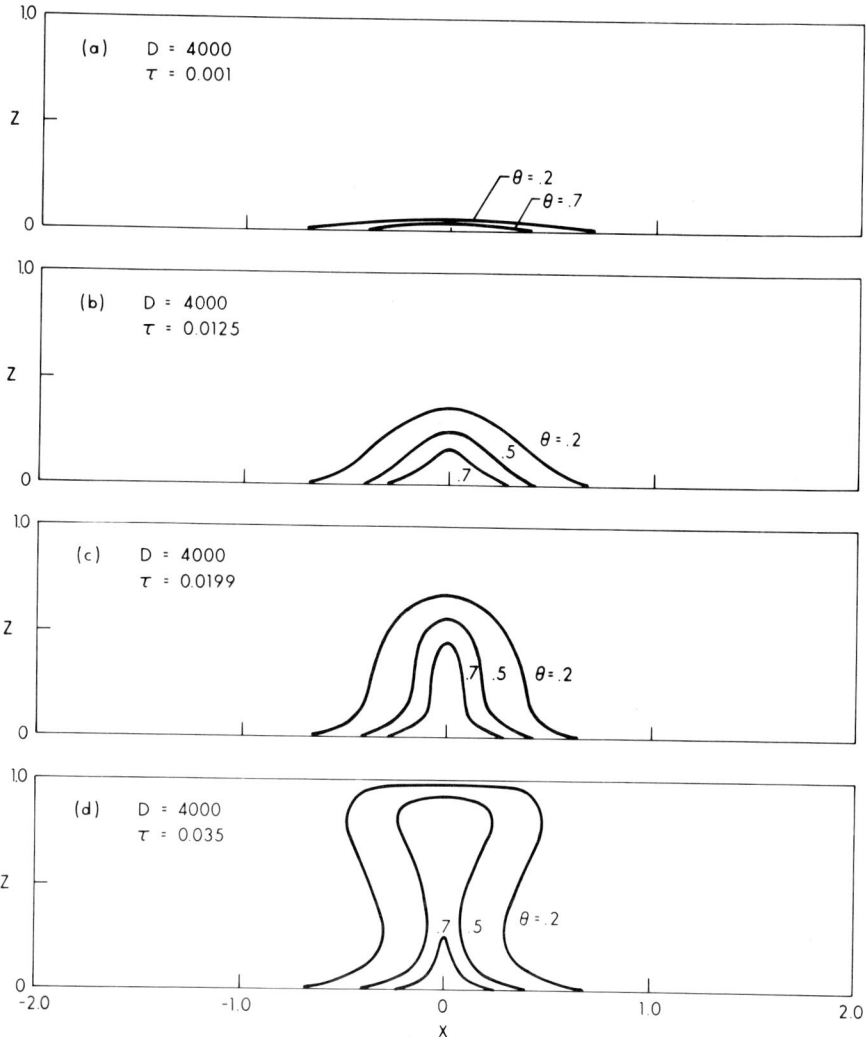

Fig. 3. Development of isotherms in a geothermal reservoir at D = 4000.

will be generated. Thus for the fluid below the well it will experience a favorable pressure gradient which tends to accelerate its vertical motion owing to buoyancy force; consequently, heat convection from the heated surface is enhanced. On the other hand, for the fluid above the production well it will experience an adverse pressure gradient which tends to retard its vertical motion owing to buoyance force. If the negative pressure gradient is sufficiently strong that it overcomes the buoyancy force, the isotherms then begin to contract. The study of the rate of contraction of isotherms in a geothermal reservoir is important, since it will determine the life-span of a geothermal well.

Figure 5 shows the contraction of isotherms resulting from the withdrawal of fluid along a horizontal line sink with $P_0 = 0$ in a reservoir having D = 7000. The horizontal line sink, which appears as a point on

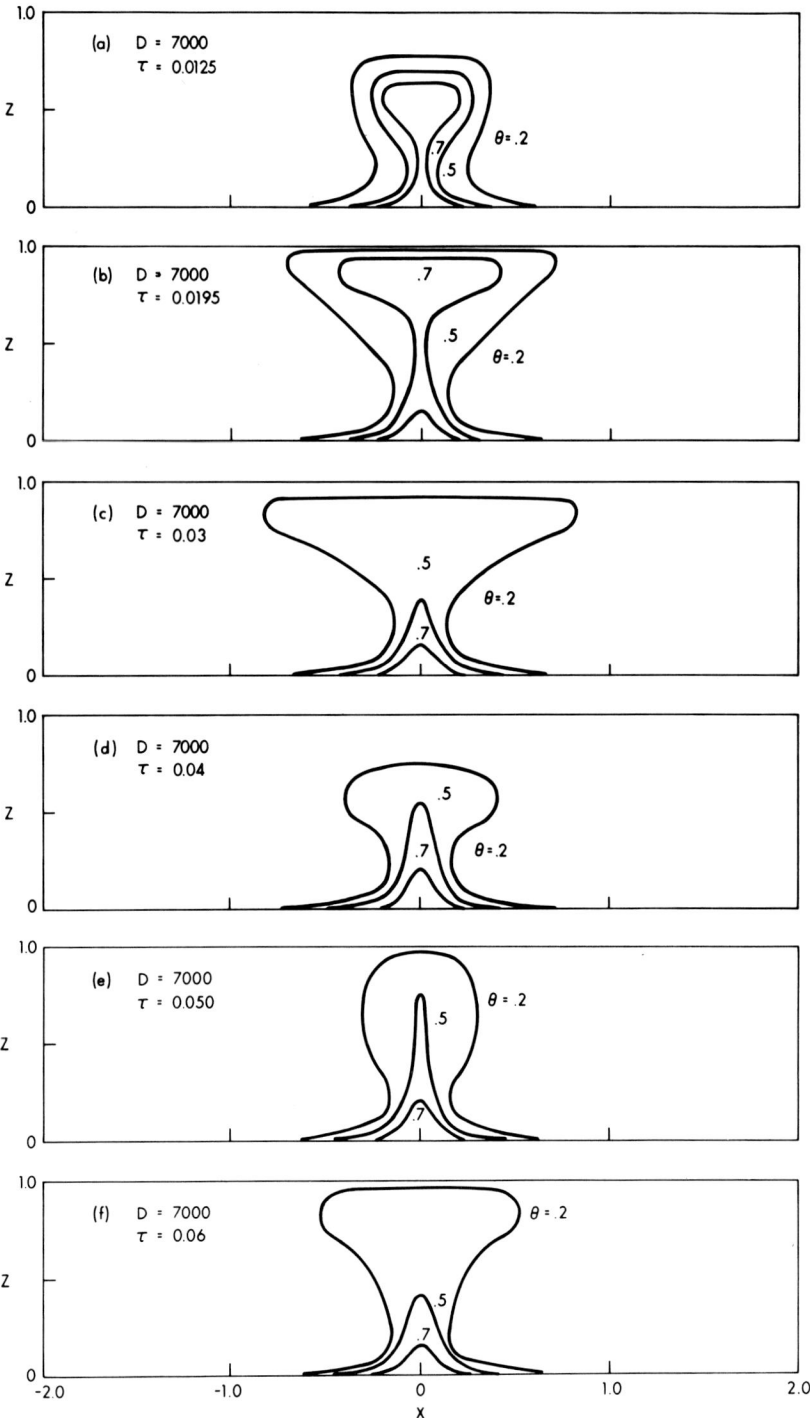

Fig. 4. Development of oscillatory convection in a geothermal reservoir at D = 7000.

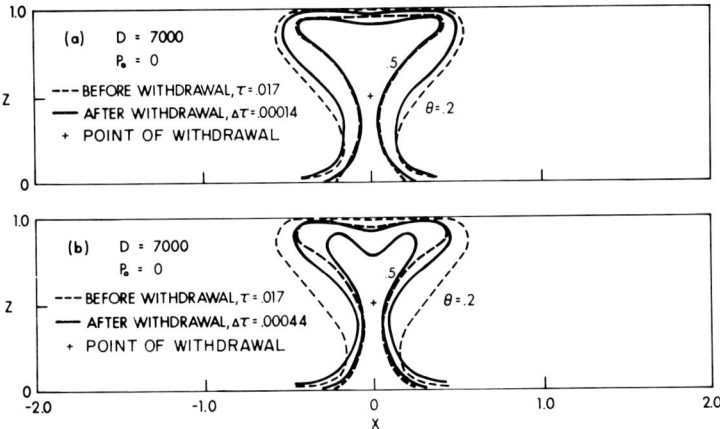

Fig. 5. Contraction of isotherms in a geothermal reservoir resulting from withdrawal of fluids from a horizontal line source.

the vertical plane, is located at X = 0 and Z = 0.5, i.e., directly above the point of maximum heating. The volume rate of withdrawal as calculated from (40) is Q ≅ 2200 or Q* = Qαh ≅ 7.6 m³/h (if α = 2.3 x 10⁻³m²/h and h = 1500 m) per meter of the horizontal line sink. The isotherms before the fluid withdrawal are shown by dashed lines which correspond to the condition at τ = 0.017. The solid lines in Figures 5a and 5b indicate the isotherms after 30 years and 100 years of continuous withdrawal of fluids at Q ≅ 2200. While it is shown in the figure that the isotherms hardly change after 30 years of continuous operation, the temperature of the groundwater above the sink decreases noticeably after 100 years of operation.

Figure 6 shows the contraction of isotherms resulting from the withdrawal of fluid along a vertical plane sink in an aquifer having D = 7000. The vertical plane sink, which appears as a vertical line on the vertical X-Z plane, is located vertically upward from the point (0, 0.5) to the top of the aquifer. The pressure along the sink is such that P = 0 at the point (0, 0.5) and increases linearly upward with height. The isotherms before the fluid withdrawal are for the same condition as those in Figure 5 (i.e., τ = 0.017) and are shown by dashed lines. The solid lines are the isotherms after 30 years of continuous withdrawal of fluid at Q ≅ 5500, i.e., at the rate of 19 m³/h per meter perpendicular to the X-Z plane. At this rate of withdrawal it is shown that the temperature of the groundwater in the upper portion of the reservoir decreases noticeably after 30 years of operation. The contraction of isotherms will decrease the temperature of the hot water withdrawn from the well and consequently the amount of recoverable energy. It should be noted that the rate of contraction of isotherms depends not only on the withdrawal rate but also on the size of the heat source as well as on the location of the well relative to that of the heating surface.

Computations were also carried out for the case of ε_p = 0 (i.e. β_p = 0) as well as for the case for which the transient term in the continuity equation is neglected. It is found that the effects of isothermal compressibility and the term $\partial p/\partial t$ are small. It is therefore concluded that the Boussinesq approximations, usually invoked in

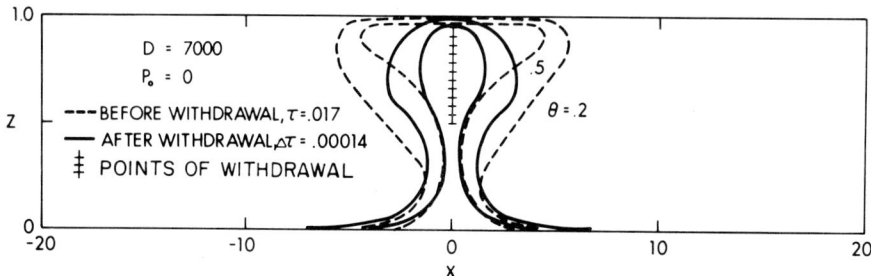

Fig. 6. Contraction of isotherms in a geothermal reservoir resulting from withdrawal of fluids from a vertical plane sink.

classical free convection problems, are also a good approximation for combined free and forced convection in a porous medium.

Concluding Remarks

To study the large scale convection motion in a geothermal reservoir, the reservoir is often idealized as a saturated porous medium. The mathematical formulation of the problem leads to a pair of coupled dimensionless partial differential equations with D (a measure of relative importance of convection to conduction heat transfer) as a parameter. Since exact analytical solutions for the set of nonlinear partial differential equations are difficult (if not impossible) to obtain, numerical solutions are often used.

In the present paper, the effects of transient heating and transient fluid withdrawal in a porous medium, bounded at the top and bottom by impermeable surfaces with recharge through vertical boundaries, are studied numerically. The former problem is applicable to the formation of a volcanic island geothermal reservoir, and the latter has important applications for the prediction of the life span of a geothermal well. It is found that:

1. For a two-dimensional rectangular reservoir, 6000 m wide and 1500 m thick, having D = 4000, a steady state condition is reached approximately 7000 years after a magma chamber is intruded from below. For a reservoir with D > 4000, the isotherms in the reservoir would rise and fall periodically with time and no steady state condition exists. The period of oscillation is of the order of 8000 years for a reservoir with D = 7000. The frequency of oscillation increases as the value of D is increased. The oscillatory convection is owing to the generation and subsequent detachment of hot bulbs rising from the bottom heated surface.

2. For a sufficiently strong withdrawal rate, it is shown that isotherms begin to contract as hot water is sucked into the well. The continuous contraction of the isotherms would eventually decrease the temperature of the withdrawal fluids, and consequently, the amount of recoverable geothermal energy. If the geological and hydrological conditions of a geothermal reservoir are known approximately, the present model can be used to predict the life span of a geothermal well under different operating conditions.

Acknowledgments. This study is part of the Hawaii Geothermal Project, funded in part by the RANN program of the U.S. National Science Foundation (grant GI-38319), by the U.S. Energy Research and Development Administration (contract EY-76-C-03-1093), and by the State and County of Hawaii.

References

Caltagirone, J. P., Thermoconvective instabilities in a horizontal porous layer, J. Fluid Mech., 72, 269-287, 1975.

Cheng, P., Combined free and forced convection flow about inclined surfaces in porous media, Int. J. Heat Mass Transfer, 20, in press, 1977a.

Cheng, P., Similarity solutions for mixed convection from horizontal impermeable surfaces in saturated porous media, Int. J. Heat Mass Transfer, 20, in press, 1977b.

Cheng, P., Integral methods for convective heat transfer in porous layers, Hawaii Geothermal Project Tech. Rept. No. 22, 1977c.

Cheng, P., and I. D. Chang, Buoyancy induced flows in a porous medium adjacent to impermeable horizontal surfaces, Int. J. Heat Mass Transfer, 19, 1267-1272, 1976.

Cheng, P., and W. J. Minkowycz, Free convection about a vertical flat plate embedded in a saturated porous medium with application to heat transfer from a dike, J. Geophys. Res., 82, in press, 1977.

Combarnous, M., and B. Le Fur, Transfer de chaleur par convection naturelle dans une couche poreuse horizontale, C. R. Acad. Sci., Ser. B, 269, 1009-1012, 1969.

Elder, J. W., Transient convection in a porous medium, J. Fluid Mech., 27, 609-623, 1967.

Horne, R. N., and M. J. O'Sullivan, Oscillatory convection in a porous medium heated from below, J. Fluid Mech., 66, 339-352, 1974a.

Horne, R. N., and M. J. O'Sullivan, Oscillatory convection in a porous medium: The effect of throughflow, paper presented at 5th Australasian Conference on Hydraulics and Fluid Mechanics, Univ. of Canterbury, Christchurch, New Zealand, 1974b.

Katto, Y., and T. Masuoka, Criterion for the onset of convective flow in a fluid in a porous medium, Int. J. Heat Mass Transfer, 10, 297-309, 1967.

Lapwood, E. R., Convection of a fluid in a porous medium, Proc. Cambridge Phil. Soc., 44, 508-521, 1948.

Lasseter, T. J., The numerical simulation of heat and mass transfer in multidimensional two-phase geothermal reservoirs, Pap. 75-WA/HT-71, Amer. Soc. of Mech. Eng., Houston, Texas, 1975.

Minkowycz, W. J., and P. Cheng, Free convection about a vertical cylinder embedded in a porous medium, Int. J. Heat Mass Transfer, 19, 805-813, 1976.

Witherspoon, P. A., S. P. Neuman, M. L. Sorey, and M. J. Lippman, Modelling geothermal systems, paper presented at International Meetings on Geothermal Phenomena and Their Applications, Accad. Naz. dei Lincei, Rome, March 3-5, 1975.

NUMERICAL CALCULATION OF TWO-TEMPERATURE THERMAL CONVECTION IN A
PERMEABLE LAYER WITH APPLICATION TO THE STEAMBOAT SPRINGS THERMAL
SYSTEM, NEVADA

D. L. Turcotte

Department of Geological Sciences, Cornell University
Ithaca, New York 14853

R. J. Ribando and K. E. Torrance

Sibley School of Mechanical and Aerospace Engineering,
Cornell University, Ithaca, New York 14853

Abstract. One approach to the study of hydrothermal circulations is to treat the earth's crust as a permeable medium. An object of such studies is to understand hot springs. However, in most studies of flows in permeable media the temperatures of the fluid and matrix are assumed to be equal. Since the surface temperature is prescribed to be the ambient temperature in such a case, hot springs are not possible. In the present paper the mean temperatures of the fluid and matrix are treated as independent variables, two energy equations are written, and heat transport between fluid and matrix is assumed to be proportional to the difference between the local mean temperatures of the fluid and the matrix. Numerical solutions for two-dimensional thermal convection in a layer of permeable material are obtained for the case of a permeable upper boundary. For large values of the Rayleigh number it is found that significant temperature differences between fluid and matrix are restricted to a thin boundary layer near the permeable upper boundary of the convection cell. A boundary layer analysis is presented which allows the modification of isothermal solutions to include two-temperature effects. A capillary model relates the permeability and heat transfer coefficient to capillary diameters and spacing. The results of the numerical calculations are applied to the Steamboat Springs, Nevada, thermal system. The temperature of the hot springs and the measured temperatures in boreholes are used to deduce an applicable Rayleigh number and heat transfer coefficient. The model is reasonably successful in explaining the behavior of this hydrothermal system.

Introduction

Convection of water in the crust is known to play a dominant role in hydrothermal systems. In some cases this convection occurs in granular materials, but more often, it occurs through cracks and fractures. Frequently, circulation patterns are dominated by fault zones. If a large fraction of the circulation is confined to a relatively narrow zone, an aquifer model may be applicable. Lowell [1975] has modeled hydrothermal systems in this way.

If the circulation is relatively broad and uniform, a permeable medium model may be applicable. This type of model has been applied to hydrothermal convection in the oceanic crust by Lister [1972] and by Ribando et al. [1976]. These calculations of thermal convection cells in the oceanic crust have been used to explain the spatial periodicity of the heat flow measurements obtained adjacent to the Galapagos spreading center by Williams et al. [1974]. It was concluded that the mean heat flow near a spreading center can induce thermal convection in the permeable volcanic rocks of the oceanic crust.

On the continents, strong hydrothermal circulations occur in localized regions. Although some hot (warm) springs appear to be generated by the mean continental heat flow (examples are the warm springs of West Virginia and Georgia), most are associated with abnormally high, near-surface geothermal gradients. These high geothermal gradients are usually attributed to the cooling of a near-surface magma body.

The purpose of this paper is to model continental hot springs in terms of thermal convection in a uniform permeable layer heated from below. In previous studies of crustal hydrothermal circulations using permeable media models the temperatures of the fluid and matrix have been taken to be equal [Ribando and Torrance, 1976]. Since the upper boundary of the convecting layer is taken to be the ambient temperature, the fluid leaving the layer is also at the ambient temperature. In order to study hot springs it is necessary to relax the constraint that the temperatures of the fluid and matrix be equal.

A number of studies of flows in permeable media have been carried out which assume that the mean temperatures of the fluid and matrix differ and which include an appropriate heat transfer coefficient between the two. The appropriate equations have been derived by Bland [1954]. They have been applied to problems involving flows through pebble bed heat exchangers [Nelson and Galloway, 1975]. Two-temperature thermal convection in a permeable layer with impermeable boundaries has been studied by Combarnous [1972] and by Combarnous and Bories [1974]. These authors assumed that the heat transfer between fluid and matrix was linearly proportional to the temperature difference. They obtained numerical calculations for the velocity and temperature distributions. However, there appear to have been no previous studies of two-temperature convection with a permeable upper boundary which are directly relevant to crustal hydrothermal circulations.

Formulation of the Problem

Initially, we will consider a macroscopic model for two-temperature thermal convection in a permeable layer. Clearly on a microscale the temperatures of the fluid and matrix vary on the scale of the flow capillaries. We assume that this scale is small compared with the scale of the overall convection (the macroscale) so that it is appropriate to define a local mean fluid temperature T_f and a local mean matrix temperature T_m. We assume that the heat transfer between matrix and fluid is proportional to the temperature difference. The upper boundary of the layer is assumed to be permeable, and the matrix temperature is taken to be the ambient temperature T_1. Fluid entering the layer through the upper boundary is taken to have the ambient temperature. Fluid leaving the layer has a temperature greater than T_1, hot springs thus resulting. The lower boundary of the layer is taken to be impermeable, and we assume a constant heat flux q_o through this boundary.

The appropriate equations for steady flow with the Boussinesq approximation have been given by Combarnous [1972]. The Boussinesq approximation assumes that density variations may be neglected except in the gravitational body force term. The equations are

$$\nabla \cdot \bar{v} = 0 \qquad (1)$$

$$0 = \nabla p - (\mu_f/k)\bar{v} + \alpha_f \rho_f g (T_f - T_1) \bar{\varepsilon}_z \qquad (2)$$

$$\rho_f c_{pf} (\bar{v} \cdot \nabla) T_f = n \lambda_f \nabla^2 T_f + H(T_m - T_f) \qquad (3)$$

$$0 = (1 - n)\lambda_m \nabla^2 T_m + H(T_f - T_m) \qquad (4)$$

where $\bar{v}$ is the Darcian velocity (Darcy's law is assumed to be valid), p the deviation of the pressure from the hydrostatic state, n the flow porosity, k the permeability, μ the viscosity, λ the thermal conductivity, ρ the density, α the coefficient of thermal expansion, c_p the specific heat at constant pressure, H the heat transfer coefficient per unit volume between fluid and matrix (a constant), g the gravitational acceleration, and $\bar{\varepsilon}_z$ the unit vector in the upward direction.

The convection is assumed to take the form of two-dimensional horizontal rolls with horizontal axis in a horizontal layer of depth d. Each convection cell has a width w. Adjacent convection cells are counterrotating. The symmetry conditions on the vertical boundaries between cells require $v_x = 0$ and $\partial T_m/\partial x = \partial T_f/\partial x = 0$. The boundary conditions on the lower boundary $z = -d$ require that $v_z = 0$, $\partial T_f/\partial z = 0$, and $\lambda_m \partial T_m/\partial z = -q_o$. The requirement that the upper boundary be permeable is satisfied if the fluid pressure is a constant. Without loss of generality this constant can be taken to be zero; i.e., $p = 0$ at $z = 0$. The temperature boundary conditions on the upper boundary are $T_m = T_1$ and $T_f = T_1$ if $v_z < 0$ (flow into the layer) and $\partial^2 T_f/\partial z^2 = 0$ if $v_z > 0$ (flow out of the layer).

It is convenient to introduce nondimensional variables $v' = vd/\kappa_r$, $x' = x/d$, $T' = \lambda^*(T - T_1)/q_o d$, and $p' = kp/\mu_f \kappa_r$ and parameters

$$Ra = \frac{k\rho_f g \alpha_f d^2 q_o}{\mu_f \kappa_r \lambda^*} \qquad Ha = \frac{Hd^2}{\lambda^*} \qquad w' = \frac{w}{d}$$

where $\kappa_r = \lambda^*/\rho_f c_{pf}$ and $\lambda^* = n\lambda_m + (1 - n)\lambda_f$. This value for λ^* implies that heat conduction in the matrix and fluid occurs in a parallel mode. In addition to the appropriately defined Rayleigh number Ra, solutions will also depend on the nondimensional heat transfer coefficient between matrix and fluid Ha, the cell aspect ratio w', the porosity n, and the thermal conductivity ratio λ_f/λ_m.

Substitution of the nondimensional variables and parameters into (1)-(4) yields

$$\bar{\nabla}' \cdot \bar{v}' = 0 \tag{5}$$

$$0 = -\nabla'p' - \bar{v}' + \text{Ra} T_f' \bar{\varepsilon}_z \tag{6}$$

$$(\bar{v} \cdot \nabla') T_f' = n(\lambda_f/\lambda^*) \nabla'^2 T_f' + \text{Ha}(T_m' - T_f') \tag{7}$$

$$0 = (1-n)(\lambda_m/\lambda^*) \nabla'^2 T_m' + \text{Ha}(T_f' - T_m') \tag{8}$$

When the nondimensional Stokes stream function ψ' is introduced,

$$\nabla' \times \psi' = \bar{v}' \tag{9}$$

the momentum equation (6) can be written

$$\nabla'^2 \psi' = \text{Ra}\, \partial T_f'/\partial x' \tag{10}$$

In terms of the nondimensional variables the boundary conditions become

at $x' = 0, w'$: $v_x' = 0$, $\partial T_f'/\partial x' = \partial T_m'/\partial x' = 0$

at $z' = 0$: $p' = 0$, $T_m' = 0$,
$T_f' = 0$ if $v_z' < 0$, $\partial^2 T_f'/\partial z'^2 = 0$ if $v_z' > 0$

at $z' = -1$: $v_z' = 0$, $(\lambda_m/\lambda^*)\partial T_m'/\partial z' = -1$, $\partial T_f'/\partial z' = 0$

To study the stability problem for the onset of convection, the above set of equations can be linearized. For the boundary conditions with the fluid and matrix temperatures equal (Ha $\to \infty$) the critical Rayleigh number is 17.7 [Nield, 1968]. At Rayleigh numbers less than this, convection does not occur. In order to determine the amplitude of the convection it is necessary to retain the nonlinear term on the left side of (7); to solve the nonlinear problem, numerical techniques are generally required.

Numerical Solutions

In order to solve the problem posed above, finite differences are used to approximate (7)-(10). Three-point central differences are used for linear space derivatives, and conserving, upwind differences are used for the nonlinear convection terms in the fluid energy equation. Starting with initial data, the stream function is advanced by iteratively solving a difference approximation of (10). One Gauss-Seidel iteration is used, followed by several successive overrelaxation iterations. The temperature fields are then explicitly advanced by including the time-dependent terms $\partial T_f'/\partial t'$ and $\partial T_m'/\partial t'$ in (7) and (8), respectively, which are approximated with two-point forward differences. The procedure is repeated until a steady state is reached.

Dirichlet boundary conditions (boundary conditions on the function) are treated as known boundary data during the iteration procedure. At boundaries with Neumann conditions (boundary conditions on the normal derivative of the function) special forms of the governing equations are required. For boundaries with an adiabatic or prescribed heat flux condition a finite difference energy balance is applied to the volume element adjacent to the boundary. For the calculations reported here a uniform 26 x 26 grid was used. The horizontal cell dimension w' was divided into 25 equal increments and the unit layer thickness was also divided into 25 equal increments. Convergence with mesh size was confirmed by using coarser and finer grids on selected test problems. We estimate that our results are probably correct to about 5% for global quantities.

Computed isotherms and streamlines for Ra = 100 and Ha = 100 are shown in Figure 1. Isotherms for both the matrix and the fluid are indicated. Other parameters for this flow, and for all calculated flows in this paper, are w' = 0.6, n = 0.2, and λ_f/λ_m = 3. A large fraction ($\approx$ 90%) of the fluid flow enters and leaves the permeable layer through the permeable top; only about 10% recirculates within the cell. As the cold fluid circulates through the matrix, its temperature remains below the matrix temperature for some time as it heats up. In the region of ascending flow the matrix temperature drops, and the fluid temperature is higher than the matrix temperature. The temperature of fluid flowing out of the permeable layer at the center of the ascending flow is T_f' = 0.110. This is 19.5% of the maximum matrix temperature in the permeable layer.

As mentioned previously, the aspect ratio for the flow shown in Figure 1 is w' = 0.6. This was the maximum stable aspect ratio for Ra = 100 and Ha = 100. When the aspect ratio for the calculation was increased above 0.6, the flow broke into two cells which eventually reached steady state convection. When the Rayleigh number for the calculation was increased above 100, the flow with aspect ratio of 0.6 remained stable with a single cell.

The maximum velocity of the fluid flowing out of the permeable layer at z' = 0 is a measure of the rate of flow through the layer. This velocity $(v_z')_{max}$, is plotted against the Rayleigh number in Figure 2 for Ha = 200, 500, 1000, and ∞. The last value is the case in which $T_f' = T_m'$ everywhere in the permeable layer. We correlate the maximum velocities with the empirical relation

$$(v_z')_{max} = 0.425[1 + (3/Ha^{1/3})](Ra - Ra_{crit}) \qquad (11)$$

As Ha is decreased, the temperature difference between the ascending and the descending fluid increases, the circulation of fluid through the layer thus being enhanced.

Usually, in calculations of thermal convection in a fluid layer the Nusselt number (the ratio of the heat transferred across the permeable layer with convection to the heat transferred without convection) is determined as a function of Rayleigh number. However, when the heat flux is prescribed, as in our calculations, it is appropriate to obtain the dependence of the maximum nondimensional temperature at the base of the layer as a function of the Rayleigh number. We find that the maximum matrix temperature is insensitive to the dimensionless heat

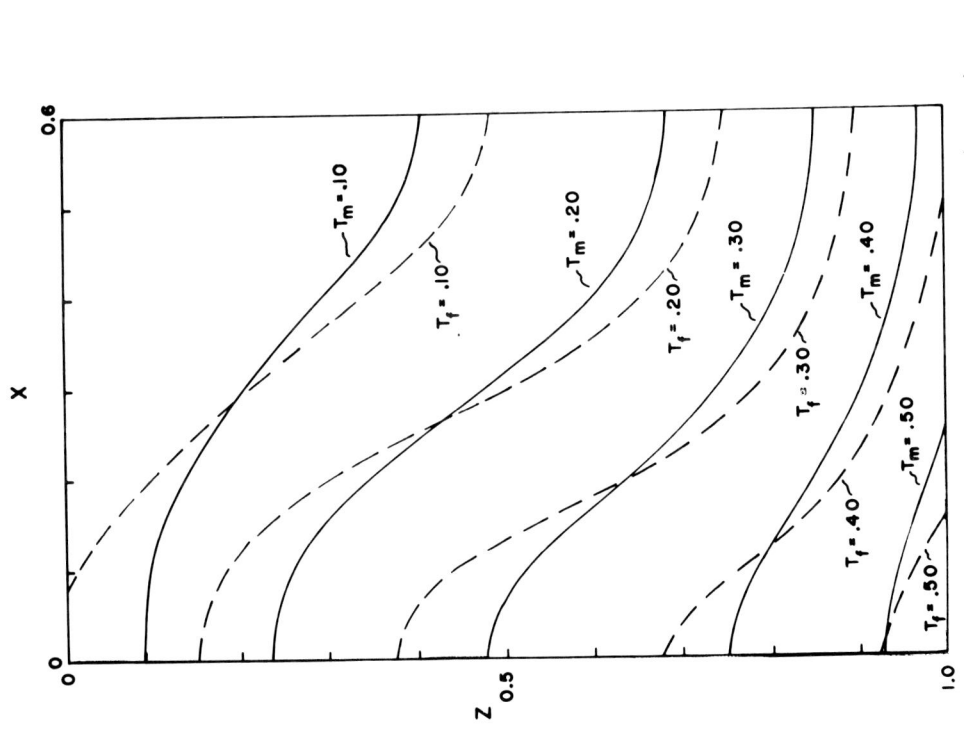

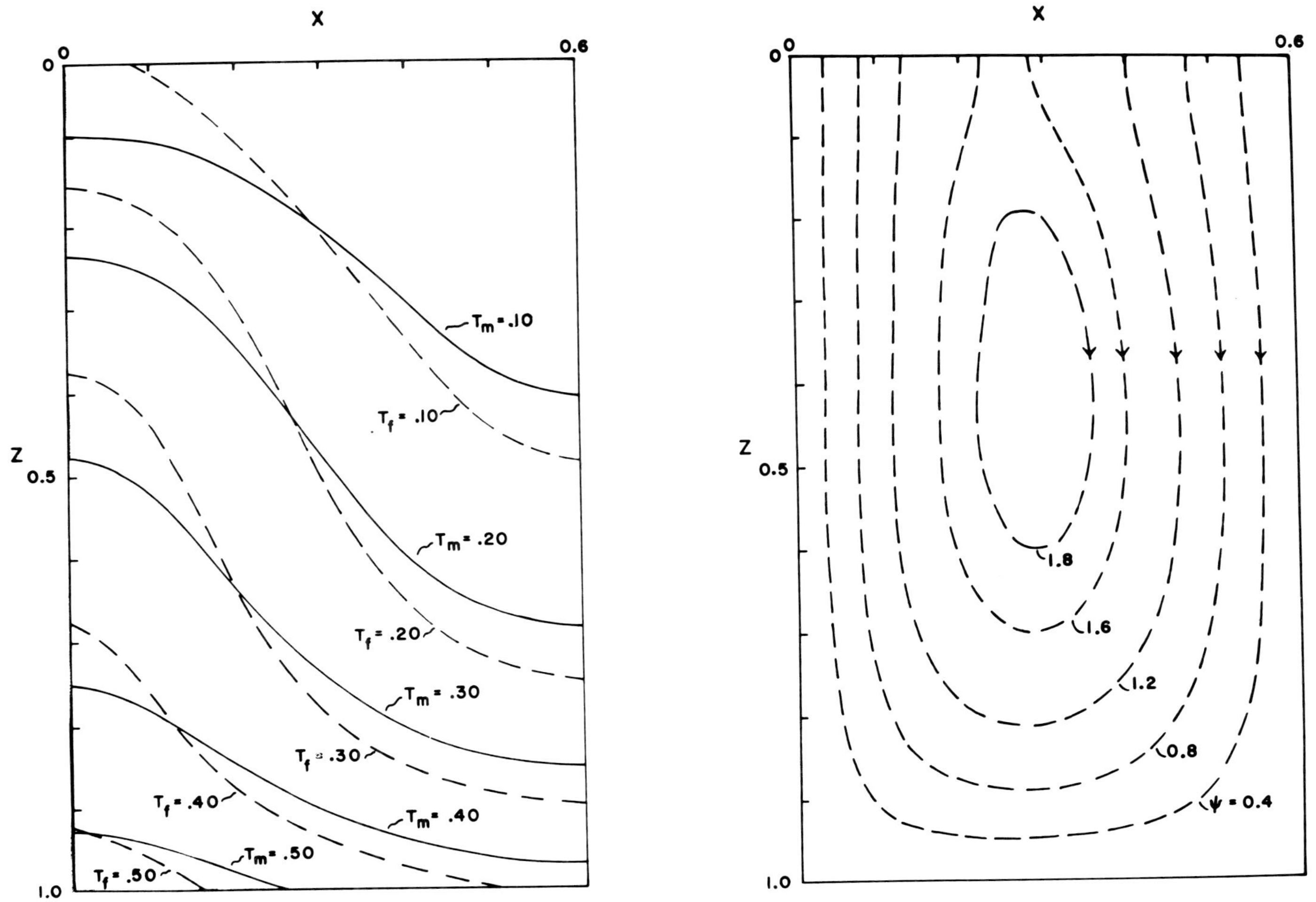

Fig. 1. Isotherms (left) and streamlines (right) for two-dimensional two-temperature thermal convection in a porous layer heated from below; Ra = 100, Ha = 100, and w' = 0.6.

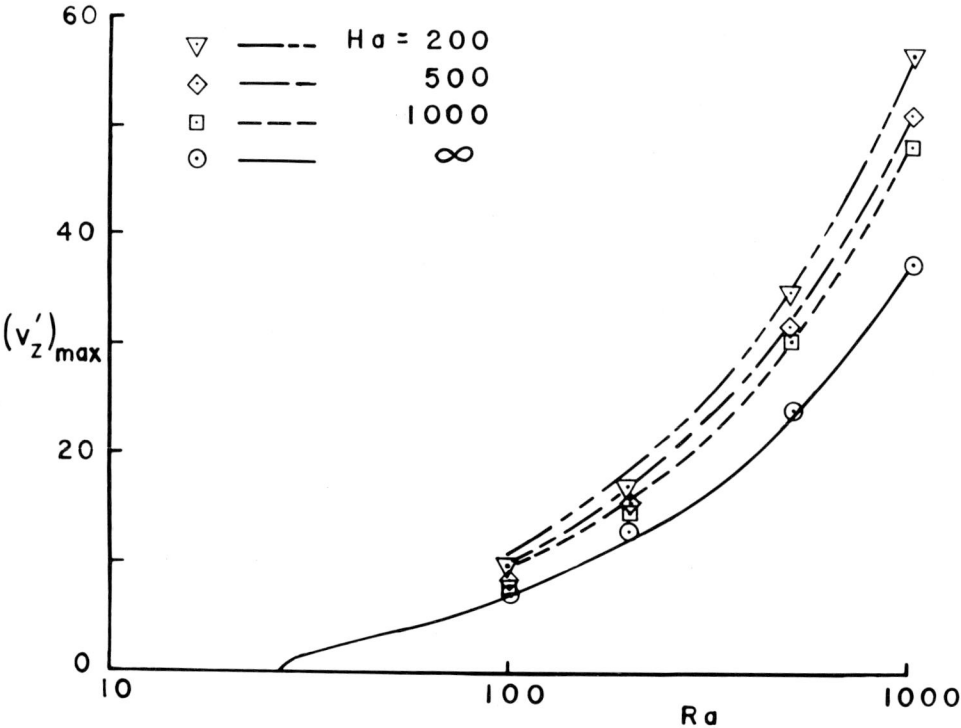

Fig. 2. Dependence of the maximum nondimensional fluid velocity out of the porous layer on the Rayleigh number for various values of Ha. The points are from numerical calculations, and the lines from the empirical correlation given in (11).

transfer coefficient and its dependence on Rayleigh number correlates well with the empirical expression

$$(T_m')_{max} = [1 + 0.0625(Ra - Ra_{crit})^{0.6}]^{-1} \qquad (12)$$

Although the water entering the permeable layer on its upper boundary has the same temperature as the matrix, the water flowing out has an excess temperature. This excess temperature is a maximum at the center line of the ascending flow ($x' = 0$). The numerically determined values for the ratio of the maximum surface fluid temperature to the maximum matrix temperature on the base of the layer are given in Figure 2. It is seen that the fluid temperature at the surface increases with Rayleigh number. As Ra is increased, the flow velocity also increases, and the fluid has less time to equilibrate its temperature with that of the matrix. The surface fluid temperature decreases with Ha as would be expected; however, there is still a substantial temperature excess for Ha = 1000. The numerical calculations show that for large values of Ha and Ra the fluid and matrix temperatures are nearly equal within the permeable layer except for a thin boundary layer near the top, where fluid is leaving the layer.

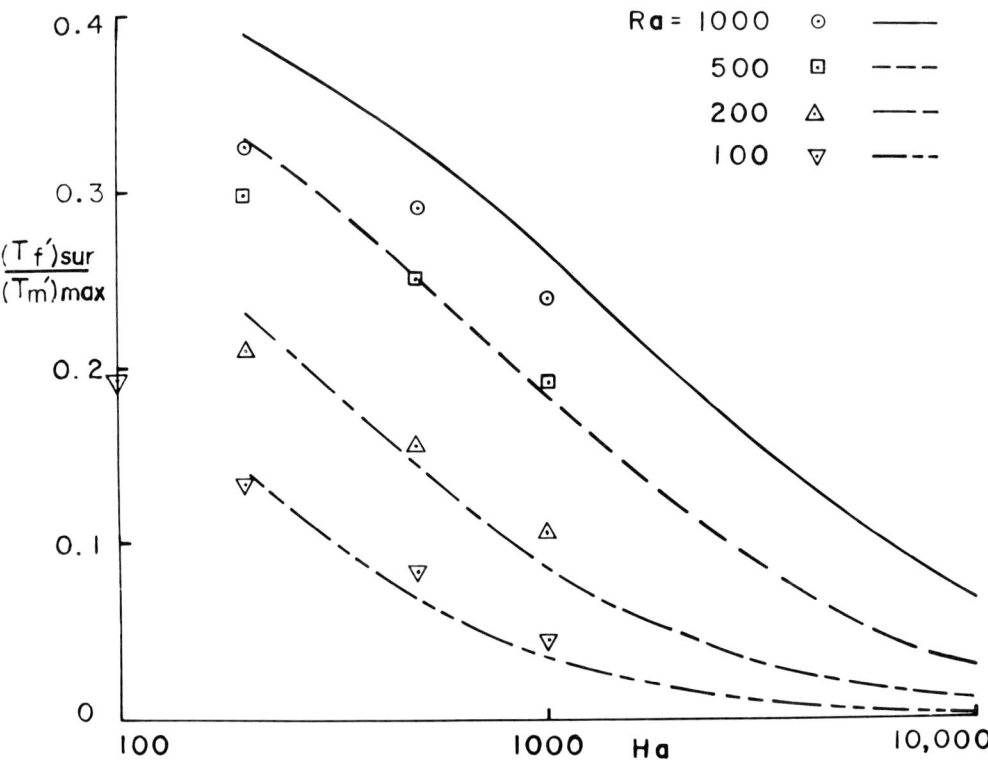

Fig. 3. Ratio of the nondimensional surface fluid temperature at x' = 0 to the maximum nondimensional matrix temperature on the base of the layer as a function of the dimensionless heat transfer coefficient for several Rayleigh numbers. The points are from numerical calculations, and the lines from (17).

Boundary Layer Analysis

If the fluid and matrix temperatures are nearly equal except near the upper boundary of the permeable layer, a boundary layer analysis should be applicable. Within the interior of the permeable layer the single temperature (Ha $\to \infty$) solution should be valid. This two-dimensional solution can be matched to a one-dimensional, two temperature solution valid in the thin boundary layer near the upper boundary. Within the boundary layer the vertical velocity is taken to be a constant equal to its surface value, and the energy equations (7) and (8) reduce to

$$v_z' (dT_f'/dz') = Ha(T_m' = T_f') \qquad (13)$$

$$0 = (d^2T_m'/dz'^2) + Ha(T_f' = T_m') \qquad (14)$$

where it has been assumed that horizontal gradients can be neglected and that the porosity is small, i.e., n = 0.

The boundary conditions are $T_m' = T_f' \to T_\infty'$ as $z' \to -\infty$ and $T_m' = 0$ at $z' = 0$. The solution of (11) and (12) for these boundary conditions is

$$T_m' = T_\infty' (1 - e^{mz'}) \qquad (15)$$

$$T_f' = T_\infty'\{1 - [1 - (m^2/Ha)]e^{mz'}\} \qquad (16)$$

where

$$m = [(Ha^2/4v_z'^2) + Ha]^{1/2} - (Ha/2v_z')$$

Our numerical calculations show that it is appropriate to take $T_\infty' = 1/2\,(T_m')_{max}$. Substitution of this value into (16) gives the temperature of the fluid flowing out of the permeable layer.

$$(T_f')_{sur} = \frac{(T_m')_{max}}{2Ha}\left[\frac{Ha^2}{4v_z'^2} + Ha\right]^{1/2} - \frac{Ha}{2v_z'}^2 \qquad (17)$$

By taking the value of $(v_z')_{max}$ given by (11) this surface temperature is compared with the values obtained from the numerical calculations in Figure 3. Although this boundary layer analysis can only be approximately valid, reasonably good agreement with the numerical calculations is obtained. The agreement improves as the Rayleigh number is increased because this reduces the heat conduction in the matrix below the thermal boundary layer.

Capillary Model for Permeability

In order to evaluate the permeability and the heat transfer coefficient it is necessary to prescribe a microscopic model for the flow through the permeable medium. Many models for permeability have been proposed; these are summarized by Bear [1972]. For the purposes of this paper we propose to use one of the simplest capillary models. We consider a matrix filled with equally spaced circular capillaries as illustrated in Figure 4. Each capillary has a diameter δ, and the distance between capillaries is b. The porosity n is

$$n = \pi\delta^2/4b^2 \qquad (18)$$

We assume $b \gg \delta$.

The permeability k for this model, if slow laminar flow in the capillaries is assumed, is given by [Bear, 1972]

$$k = \pi\delta^4/128\,b^2 \qquad (19)$$

Further approximations are required in order to determine a heat transfer coefficient. Because the diameter of the capillaries is small in comparison with the spacing, the variation in temperature within the capillaries can be neglected in comparison with the temperature

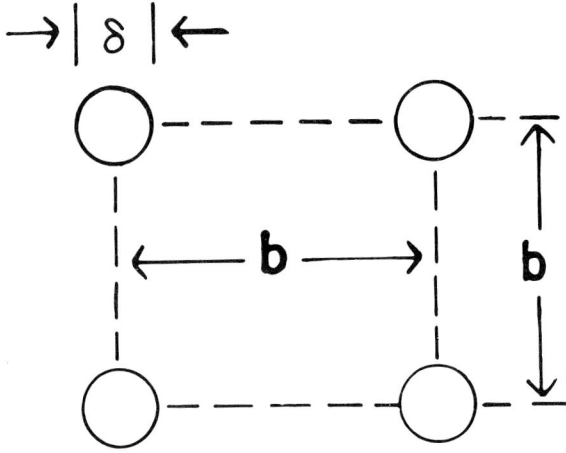

Fig. 4. Illustration of the capillary model.

variation in the matrix. We approximate the temperature distribution in the matrix with

$$\frac{T - T_m}{T_f - T_m} = \frac{\ln(2r/b)}{\ln(\delta/b)} \quad (20)$$

This is the temperature distribution in a cylindrical annulus with $T = T_f$ at $r = \delta/2$ and $T = T_m$ at $r = b/2$. When this temperature distribution is used to evaluate the heat transfer between fluid and matrix, the heat transfer coefficient is given by

$$H = 2\pi\lambda_m/[b^2 \ln(b/\delta)] \quad (21)$$

and the dimensionless heat transfer coefficient is given by

$$Ha = 2\pi d^2/[b^2 \ln(b/\delta)] \quad (22)$$

since $\lambda^* = \lambda_m$ in the limit $n \to 0$. The permeability is strongly dependent on the capillary size, but the heat transfer coefficient is primarily dependent on the capillary spacing. Since the capillary spacing b must be a small fraction of layer depth d for the model to be valid, it is seen from (22) that only relatively large values of Ha (>100) are consistent with the model. Although this model is highly idealized, previous studies [Bear, 1972] show that it should give reasonably accurate estimates for the microscopic structure of the permeable rock.

Steamboat Springs, Nevada, Hydrothermal System

Steamboat Springs in southern Washoe County, Nevada, near the California border, is an area of hot springs with some geysering. Extensive studies of this hydrothermal system have been carried out by Thompson and White [1964], White et al. [1964], and White [1968].

Probably the best data on the extent of the hydrothermal system come from measurements of the groundwater temperature at the water table. A large number of shallow wells have been drilled in the area, and the temperature data have been tabulated by White [1968]. A contour map of the groundwater temperature at the water table is given in Figure 5. A well-defined thermal anomaly exists with a horizontal extent of 5-10 km. Most authors would attribute this thermal anomaly to a magma body at depth. The three areas of boiling hot springs are shown by the darkened regions. The central area is known as the main terrace, and the southern area as the low terrace. The Steamboat Springs fault zone appears to provide structural control for the low and main terraces.

The U.S. Geological Survey drilled a number of drill holes immediately adjacent to the hot springs. The bottom temperatures logged during the drilling of GS3 [White, 1968] are given in Figure 6. This drill hole was located on the western edge of the main terrace about 60 m from the nearest boiling spring.

Geochemical studies indicate that at least 95%, and probably a larger fraction, of the water from the hot springs is meteoric water. Clearly, a thermal convection system exists. We will apply our numerical solution for thermal convection in a permeable layer to this hydrothermal system. Our fluid and matrix temperature profiles at $x = 0$ for Ra = 1000 and Ha = 1000 are compared with the measured values in Figure 6. In order to fit the thermal boundary layer thickness of the data we take d = 500 m. This depth is considerably less than most estimates. White [1957] suggested a depth of circulation of about 3 km on the basis of the solubility requirements of alkali chlorides. A larger Rayleigh number would give a greater depth of circulation. However, the permeability is undoubtedly a strong function of depth, so that our results can only be approximately valid. In order to fit the magnitude of the temperature below the thermal boundary layer we take $q_o d/\lambda_m = 1725°C$. If the mean ambient temperature is $T_1 = 10°C$, the temperature of the water flowing out of the layer is $94.5°C$. This is very near the boiling temperature at this altitude. The data points tend to fall between the predicted fluid and matrix temperatures. This is not surprising, since the rock matrix is expected to have variations in temperature depending on the proximity of flowing water. The predicted temperatures are also in good agreement with values obtained from geothermometry of the springwaters. A temperature of $207°C$ is obtained from the last equilibration of SiO_2, and $226°C$ from the last equilibration of Na-K-Ca [Renner et al., 1975]. Although the choices for Ra and Ha are somewhat arbitrary, the results are in quite good agreement with observations considering the many approximations in the model.

In the remainder of our analysis we assume the following average properties for the fluid and matrix:

$$\mu_f = 1.33 \times 10^{-3} \text{ g/cm s} \qquad \rho_f = 1.14 \text{ g/cm}^3$$

$$\alpha_f = 1.13 \times 10^{-3} \text{ °C}^{-1} \qquad \lambda_m = 8 \times 10^{-3} \text{cal/cm s °C}$$

$$\kappa_r = 7.5 \times 10^{-3} \text{ cm}^2/\text{s}$$

With $q_o d/\lambda_m = 1725°C$, d = 500m, and $\lambda_m = 8 \times 10^{-3}$ cal/cm s °C we find

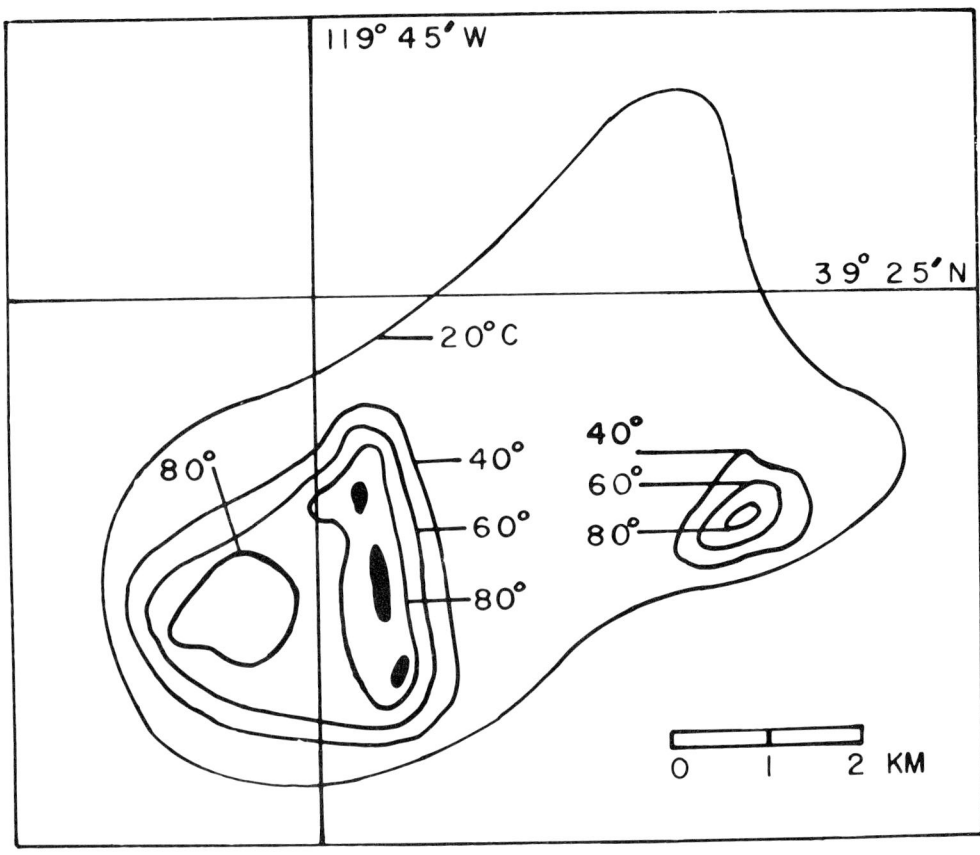

Fig. 5. Distribution of groundwater temperatures in the Steamboat Springs, Nevada, hydrothermal system.

that $q_0 = 2.65 \times 10^{-4}$ cal/cm^{2}s (265 HFU). This is the mean heat flow in the convecting region. From our numerical calculations we find that 37% of the surface heat flux is due to the convection of hot water and 62.3% is due to conduction to the surface through the matrix.

From our numerical calculations the maximum nondimensional Darcian velocity at the surface is $v_{max}' = 48.9$. When the definition of the nondimensional velocity is used, the actual Darcian velocity is $v_{max} = 7.33 \times 10^{-6}$ cm/s. The flux of fluid out of the cell averaged over the cell gives a nondimensional Darcian velocity $\bar{v}' = 12.28$. The corresponding mean velocity is $\bar{v} = 1.84 \times 10^{-6}$ cm/s. The approximately 70 hot springs associated with the Steamboat Springs thermal system discharge about 200 ℓ/min. However, White [1968] concluded that a considerable fraction of the discharge of the thermal system flows directly into Steamboat Creek. On the basis of measurements of the chlorine content of Steamboat Creek, White [1968] estimated that the total discharge of the thermal system is 4275 ℓ/min. Using our mean velocity of 1.84×10^{-6} cm/s we require a discharge area of 3.87 km^2 to provide the estimated total discharge. It is seen from Figure 5

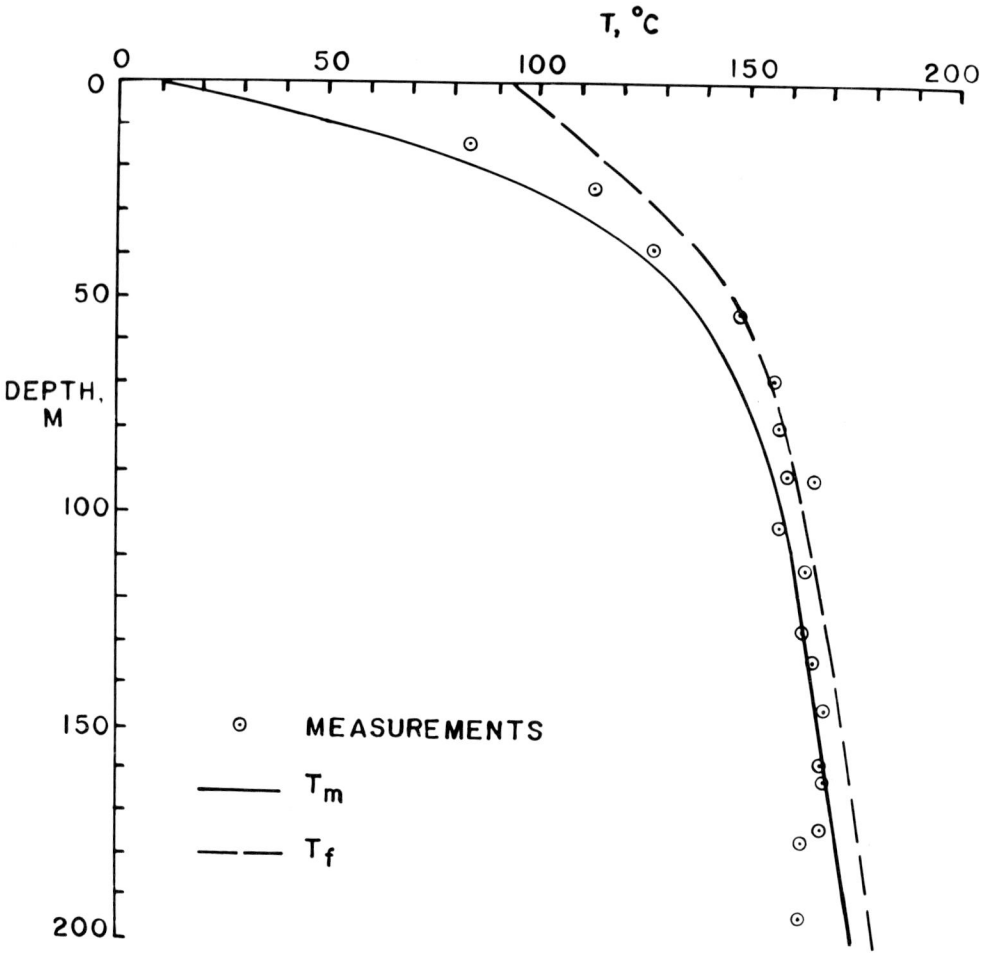

Fig. 6. Bottom temperatures logged during the drilling of GS3 [White, 1968] compared with the calculated fluid and matrix temperatures for Ha = 1000, Ra = 1000, d = 500 m, and $q_0 d/\lambda_m$ = 1725°C.

that this is a reasonable area for the convection system associated with the hot springs. When this area and the mean heat flow q_0 = 2.65 x 10^{-4} cal/cm² s obtained above are used, the total heat loss from the system is Q_0 = 10.3 x 10^6 cal/s. This compares with the value Q_0 = 11.8 x 10^6 cal/s given by White [1968].

The definition of the Rayleigh number can be used to determine the permeability. With Ra = 1000, d = 500 m, q_0 = 2.65 x 10^{-4} cal/cm² s, we find that k = 9.38 x 10^{-11} cm². This appears to be a reasonable value intermediate between values for sandstones and oil rocks.

Using our capillary model for permeability, we can now derive values for the diameter of the capillaries δ and the spacing b. From (19) and (22) with Ha = 1000, d = 500 m, and k = 9.38 x 10^{-11} cm² we find that δ = 0.290 cm and b = 1.36 x 10^3 cm. From (18) the porosity is

$n = 3.6 \times 10^{-8}$. Using this value of the porosity and the mean Darcian velocity $\bar{v} = 1.84 \times 10^{-6}$ cm/s, we find that the mean velocity in the capillaries is 52 cm/s.

The two-temperature permeable layer analysis can explain the presence of hot springs. The flow capillaries must be relatively large ($\simeq$ 0.1-1 cm), the spacing between capillaries must be relatively large ($\simeq$ 10 m), and the flow velocities must be sufficiently high to prevent thermal equilibration ($\simeq$ 0.1-1 m/s).

Conclusions

The use of a two-temperature model for thermal convection in a porous layer is reasonably successful when it is applied to the Steamboat Springs, Nevada, thermal system. The predicted flow rates are in reasonable agreement with observations. The predicted depth of the circulation is probably too shallow. Several explanations for this discrepancy can be given. The Rayleigh number that we have chosen could be too small. However, a more likely explanation is a depth-dependent permeability. As the lithostatic pressure increases, it is expected that cracks and fractures will close. Our previous studies [Ribando and Torrance, 1976; Ribando et al., 1976] have shown that the layer depth is equivalent to the scale depth for an exponentially decreasing permeability.

The microscopic model shows that the spacing between flow capillaries must be relatively large and that the flow velocities in the capillaries must be relatively large in order for significant temperature differences between the fluid and matrix to develop. It should be emphasized that the models used in this paper are highly idealized. Constant permeability has been assumed, constant fluid properties have been assumed, and the circularly capillary model is certainly not representative of rock permeability. Considering the approximations the agreement between theory and observations seems quite reasonable.

Observations indicate the presence of a thin thermal boundary layer (thickness, 50 m) in the region of upwelling convection. Our calculations show that a large Rayleigh number is required to obtain such a thin thermal boundary layer. In order to get the fluid through this boundary layer without temperature equilibration, large velocities in relatively large channels are required. Apparently, the high permeability of the Steamboat Springs fault satisfies these conditions and allows fluid to escape through the thermal boundary layer without thermal equilibration.

Acknowledgments. This research has been supported by the Division of Engineering of the National Science Foundation under grant ENG 75-14596 and by the Earth Sciences Section of the National Science Foundation under grant DES 72-01522. One of the authors (R.J.R.) held a National Science Foundation energy-related traineeship (geothermal).

References

Bear, J., Dynamics of Fluids in Porous Media, Elsevier, New York, 1972.
Bland, D. R., Mathematical theory of the flow of a gas in a porous solid

and of the associated temperature distribution, Proc. Roy. Soc., Ser. A, 221, 1-28, 1954.

Combarnous, M., Description du transfert de chaleur par convection naturelle dans une couche poreuse horizontale à l'aide d'un coefficient de transfert solide-fluide, C. R. Acad. Sci., Ser. A, 275, 1375-1378, 1972.

Combarnous, M., and S. Bories, Modelisation de la convection naturelle au sein d'une couche poreuse horizontale à l'aide d'un coefficient de transfert solide-fluide, Int. J. Heat Mass Transfer, 17, 505-515, 1974.

Lister, C. R. B., On the thermal balance of a mid-ocean ridge, Geophys. J. Roy. Astron. Soc., 26, 515-535, 1972.

Lowell, R. P., Circulation in fractures, hot springs, and convective heat transport on mid-ocean ridge crests, Geophys. J. Roy. Astron. Soc., 40, 351-365, 1975.

Nelson, P. A., and T. R. Galloway, Particle-to-fluid heat transfer in dense systems of fine particles, Chem. Eng. Sci., 30, 1-6, 1975.

Nield, D. A., Onset of thermohaline convection in a porous medium, Water Resour. Res., 4, 553-560, 1968.

Renner, J. L., D. E. White, and D. L. Williams, Hydrothermal convection systems, Assessment of Geothermal Resources of the United States, U.S. Geol. Surv. Circ., 726, 1975.

Ribando, R. J., and K. E. Torrance, Natural convection in a porous medium: Effects of confinement, variable permeability, and thermal boundary conditions, J. Heat Transfer, 98, 42-48, 1976.

Ribando, R. J., K. E. Torrance, and D. L. Turcotte, Numerical models for hydrothermal circulation in the oceanic crust, J. Geophys. Res., 81, 3007-3012, 1976.

Thompson, G. A., and D. E. White, Regional geology of the Steamboat Springs area, Washoe County, Nevada, U.S. Geol. Surv. Prof. Pap., 458A, 1964.

White, D. E., Thermal waters of volcanic origin, Geol. Soc. Amer. Bull., 68, 1637-1658, 1957.

White, D. E., Hydrology, activity, and heat flow of the Steamboat Springs thermal system, Washoe County, Nevada, U.S. Geol. Surv. Prof. Pap., 458C, 1968.

White, D. E., G. A. Thompson, and C. A. Sandberg, Rocks, structure, and geologic history of Steamboat Springs thermal area, Washoe County, Nevada, U.S. Geol. Surv. Prof. Pap., 458B, 1964.

Williams, D. L., R. P. Von Herzen, J. G. Sclater, and R. N. Anderson, The Galápagos spreading center: Lithospheric cooling and hydrothermal circulation, Geophys. J. Roy. Astron. Soc., 38, 587-608, 1974.

APPENDIX. Recommendations for Future Crustal Research

There were 54 participants in the symposium on the Nature and Physical Properties of the Earth's Crust. Papers were presented to the entire group, which on the final day outlined the following General Research Considerations. Subsequently, smaller groups were formed which wrote Research Recommendations for specific areas. This was done while details of the presentations of the week were still fresh in the minds of the participants.

General Research Considerations

1. How does physical or chemical stratigraphy relate to geological stratigraphy? What processes affect these relationships?
2. In what ways does the crust relate to the mantle? In what ways does the lithosphere relate to the asthenosphere?
3. How do scaling effects influence our interpretations? What is the relationship between laboratory and field data?
4. What is the spectrum of crustal inhomogeneities?
5. To what extent and in what sense are different physical properties continuous in the crust? How are these properties affected by tectonic activity? How do they relate to each other?
6. What is typical of 'typical' crustal models? Data should be integrated into models. Broad-scale commonalities should be recognized (despite local inhomogeneities).
7. What are the definitive properties of various geologic provinces? To what degree do shields represent the lower (pre-Cambrian) crust? How can we best identify specific environments and associate physical and geological properties with them? Geologic province types should be studied on a multidisciplinary basis.
8. What are the artifacts of our measurement techniques?
9. What is the dynamic nature of the crust? Such evidence is needed to deduce evolutionary processes.
10. Where can we best establish common areas for conducting multidisciplinary studies? There is a need to agree on test range(s). Since different geophysical techniques respond to different parameters, it will be necessary to resolve differences between the models which result. For example, studies should include (a) seismic reflection, wide-angle reflection, and refraction, (b) deep drilling, and (c) electrical, thermal, geological, gravitational, and magnetic probing, etc., as they are appropriate.

Entire Assembly

Research Recommendations for Geology

The central problem in crustal studies is simply to describe the crust, the Moho, and the relation of the crust to the upper mantle. Until we know what the crust is like from top to base and how the crust varies laterally, hypotheses on crustal genesis and the role of plate tectonics in the development of ancient crust must remain rather speculative.

Studies should begin with the definition of major crustal provinces; that is, we must know what the different kinds of crust are that need to be described. Then the problem is to determine what vertical sections through the various types of crust are like. Vertical crustal sections may be determined in a number of ways with highly varying degrees of detail, certainty, and effort. One of the most satisfactory approaches is to recognize areas where vertical sections or deep portions of the crust are exposed by unusual tectonic activity, for example, the Ivrea zone in the Alps or possibly the Jotunheimen area in Norway. Unfortunately, the number of such occurrences is probably small and biased. Another approach is to investigate inclusions from diatremes which represent a natural boring through the crust and offer a relatively cheap opportunity to study samples of deep crustal material with the restriction that the samples are mixed up. Structural studies in shields across domes and major faults will give some idea of the vertical distribution of rock types. Laboratory petrologic studies may furnish geobarometers and geothermometers that allow us to place exposed rocks at their proper depth. More isotope studies are needed to determine the derivation of crustal material and fractionation patterns. Seismic reflection studies offer a more detailed picture of the crust than has hitherto been possible, especially when they are combined with other information. All these studies should eventually lead to deep drilling, the ultimate approach.

The problem of vertical movements and material transfer in the consolidated crust is still a major question. Vertical movements are commonly recorded in sedimentary rocks which require further study in order to define the problem properly. A mechanism must be found that explains vertical oscillations measured in kilometers and that still maintains isostatic equilibrium. What is the nature of the Moho? Does it move? Or are vertical movements largely accommodated in the upper mantle? What does it mean when deep crustal rocks such as granulites are exposed at the surface with 30 to 40 km of crust beneath them? What causes the formation of deep basins? How does the Moho vary within a tectonic province and from one province to another? All of these questions are fundamental to understanding the nature of the crust and crustal dynamics.

Geologic studies should be conducted with modern tectonic hypotheses in mind so that data collected can be used to attack these problems. However, facts and hypotheses must be kept separate. Complex problems should be studied in areas where exposures are adequate enough to allow their solution.

Many of these problems are so vast and complicated that their solutions may require central planning of coordinated studies. A general attempt must be made to explain conflicting views and conflicting observations and not necessarily to arrive at one interpretation but at least to understand why there is disagreement. Geological and geophysical interpretations need to be reconciled. Data integration workshops to coordinate various individual studies are needed for this purpose.

Scott B. Smithson
Chairman
A. R. Green
M. Holdaway
J. Eichelberger
S. Kaufman
Elaine Padovani

Research Recommendations for Internal Friction Measurements

Results of laboratory internal friction (Q^{-1}) measurements lead to several conclusions: (1) The removal of H_2O from open fractures raises Q, and Q > 2000 has been achieved in olivine basalts after outgassing. (2) Trace amounts of H_2O absorbed into thoroughly outgassed fractures lower the Q dramatically and $Q \simeq 10^2$ is typical for rock containing fractures exposed to laboratory atmosphere, while $Q \simeq 10$ is typical for 'moist' rocks. (3) The effects described in the first and second conclusions are reversible and have been observed over a frequency range from ~25 kHz to 50 Hz, at low confining pressures, and for longitudinal, flexural, and torsional waves. Therefore the effects are thought to be active for seismic waves. (4) The details of the mechanisms for the effects are presently not understood. Detailed application of the effects to the interpretation of seismic Q data requires an understanding of the mechanisms. It is recommended that the sources of the mechanisms be explored. The associated experiments should include confining pressure, pore pressure, temperature, frequency, and crack porosity as parameters. The measurement samples should be well characterized, particularly with regard to the nature and density of fractures. New improved techniques of characterizing fractures developed by the Massachusetts Institute of Technology should be incorporated into the characterization procedure. The selection of samples should also reflect current knowledge of crustal material.

B. R. Tittmann
Gene Simmons

Research Recommendations for the Evaluation of Physical and Mechanical Properties

A knowledge of the physical and mechanical properties of rocks has great practical importance. For example, the exploration, assessment, and development of mineral, oil, and geothermal areas depend upon the availability of laboratory data to predict in situ properties and to interpret geophysical field studies. Relationships among the seismic, electrical, thermal, and pressure sensitive properties of rocks and such quantities as their lithology, density, porosity, fluid content, and strength should be established in the laboratory in order to provide the background upon which to base such interpretations. We recognize several problem areas in the study of physical properties.

Scaling is a fundamental difficulty, and we must resolve the following questions. (1) Are short-term experiments in the laboratory applicable to longer-term behavior in the field? (2) Do microcracks control the physical response of rocks in the field to the same degree as they do in the laboratory? (3) How do joints and large fractures affect physical and mechanical properties? (4) Is there a relationship between microcracks and joints? (5) What are the effects of wavelength versus frequency? Do different seismic or electrical frequencies sample different wavelength effects?

Experiments which will help resolve these questions include (1) laboratory and field measurement of the physical and mechanical properties of a suite of granite samples of various dimensions from the same quarry and comparison of these laboratory measurements with measurements in the quarry, (2) examination of the effects of heterogeneity to determine whether samples which are homogeneous on one scale are homogeneous on a larger scale, and (3) examination of the effects of joints and fractures by carefully mapping them and performing quarry scale experiments.

Other studies central to the understanding of physical and mechanical properties of rocks include the following.

1. <u>Pore pressure</u>. All pressure sensitive properties (e.g., compressibility, seismic velocity, and electrical resistivity) are affected by pore pressure. Yet pore pressure is rarely measured in situ, and we have very little insight concerning effective stress states at depth in the crust. Field measurements of pore pressure should be included when velocity, resistivity, etc., logs are obtained, and these field data should be compared with laboratory results made under controlled conditions of pore and confining pressure.

2. <u>Temperature effects</u>. The effects of temperature on the physical properties of earth materials in situ should be documented in the laboratory. Special emphasis should be placed on the response of fluids and minerals to temperature and the relationships between laboratory and field behavior. These studies would be especially applicable to geothermal energy development.

3. <u>Cycling effects</u>. Stress cycling and temperature cycling to various maximum values and the response of rocks to repeated cycles of fluid injection and drainage should be studied to determine the behavior of rocks as a function of cycling history. Understanding cycling effects is important to such problems as the understanding of

tectonically active zones, secondary oil recovery, the development of geothermal energy in hot dry rocks, and the underground storage of waste material.

4. <u>Interrelationships among physical and mechanical properties</u>. By developing relationships between various mechanical and physical properties (e.g., electrical resistivity versus velocity or velocity versus crack porosity) we will be able to predict the behavior of rocks in situ and to interpret field observations given only one or two measured properties.

Understanding the stress-strain response of rocks is important for the development of earthquake prediction control capabilities and for the exploration of natural resources in tectonically active zones. The following aspects of stress-strain behavior should be examined. (1) The analysis of paleostress and paleotectonics by using structural geology, microcrack history, fluid inclusions, mineralogy, and petrology should be investigated. (2) The determination of present in situ stress by using direct methods of overcoring and hydrofracturing and indirect methods of seismic, electrical and geodetic surveys is important. Temporal effects of stress can be measured by using the in situ indirect methods stated above. (3) Laboratory studies that include the analysis of rock behavior under simulated conditions of stress, confining pressure, pore pressure, and temperature are also important. Limits to in situ stress can be determined from laboratory measurements of the strength of rock under simulated field conditions.

M. Feves
Chairman
M. Batzle
B. Brace
R. Corwin
S. Jones
D. Norton
H. Pratt
D. Turcotte
H. Wang
N. Warren

General Research Recommendations for Seismology

The techniques of seismology provide the most direct probe available to the earth scientist in his study of the earth. Recent advances in field and interpretation techniques have improved the ability to study the physical properties of the earth's crust and thereby to contribute significantly to the study of its nature and tectonic history. In addition to their general scientific importance, such studies are essential to an improved understanding of energy and natural resource exploration and of natural geological hazards and their mitigation.

Problem Areas

Some of the specific characteristics of and questions concerning the earth's crust which seismologists should investigate are the following.
1. Are discontinuities in the earth's crust sharp or transitional? What is their geological and petrological interpretation?
2. Are discontinuities, layers, or zones of similar physical properties continuous or semicontinuous across geological province boundaries?
3. What is the geographical distribution of crustal features such as low velocity, high Poisson's ratio, low Q, and high conductivity zones? Is there a common physical or chemical explanation for these observed properties?
4. What physical and chemical conditions can be inferred from field observations of compressional and shear wave velocities and seismic wave attenuation in the crust?
5. How do crustal structure and the distribution of physical properties of the crust relate to the dynamics of crustal evolution and tectonics?
6. Can we use our knowledge of the history and properties of the oceanic crust to aid in the interpretation of the more complicated continental crust and the study of the relationship and differences between the two?
7. What is the relationship of the crust to the underlying lower lithosphere and asthenosphere? For example, What is the nature of the coupling of the lithospheric plates to the asthenosphere? And how does it influence the crust?

Recommendations

The seismology group recognizes that various geophysical methods provide complementary as well as redundant information on physical properties; thus optimum mapping of the nature of the crust requires a multidisciplinary attack. The following are suggestions for the initial phase of a detailed seismological investigation of the earth's crust.
1. A geophysical study should be undertaken in a single area so that several seismic and other appropriate geophysical techniques can be employed in order to compare results and interpretations of the various methods and to allow an integrated interpretation. A

by COCORP (Committee for Continental Reflection Profiling) has been or will be employed.
specific problem to investigate by using a multimethod approach is the nature of discontinuities in the crust and at the crust mantle boundary. It is suggested that this study might most efficiently be accomplished at a location where deep seismic reflection profiling

2. With the experience gained in the multidisciplinary study described above, geophysicists will be in the position to address many of the important contemporary geological problems. For example, a tectonically active area should be studied with special emphasis on the relationships among earthquake distribution and focal mechanisms (state of stress) and the physical properties of the crust determined by various seismological and other field investigations.

3. Future seismological field measurements (both in oceanic and on continental areas) should include compressional and shear wave velocity analysis and amplitude studies to aid in the interpretation of fine structure and in the determination of anelasticity (Q^{-1}) where possible.

4. We recommend that laboratory studies of compressional and shear wave velocities and Q of a broad suite of rocks, especially metamorphic rocks, be accomplished under a wide range of conditions appropriate to the earth's crust, including variations in confining pressure, temperature, saturation, and pore pressure. Such studies will allow comparison and inference from the laboratory and field investigations.

At present, a study of continental interiors forms one phase of the U.S. Geodynamics Program. The efforts of the Geodynamics Committee have largely been directed toward understanding the structure and evolution of sedimentary basins. The goals cited in this report differ from those of that committee in that we seek information pertaining to a broader range of crustal models and properties. However, the knowledge obtained by the two groups should be complementary and should lead to an enhanced knowledge of the structure and evolution of the crust.

G. Sutton
  Cochairman
L. Braile
  Cochairman
C. Prodehl
S. Mueller
M. Berry
A. Sanford
S. Kaufman
B. Mitchell
J. Orcutt
R. Buffler
D. Jurdy
S. Jones

Research Recommendations for Electrical Surveys

Although electrical surveying techniques have been used by geophysicists to study earth structures since early in this century, the methods are not as technically advanced as they are in some other geophysical disciplines, such as seismology, gravity, or geomagnetism. However, in the past decade there has been a rapid development of the capabilities of electrical surveying methods. Increased interest stems from the application of electrical methods for prospecting for oil and gas and, more recently, for prospecting for geothermal systems. In addition, electrical methods are beginning to find some application in the study of crustal structure. Three techniques in particular are being used; they are known as the direct current resistivity, the magnetotelluric, and the electromagnetic sounding methods. Each method appears to have its own special advantages and disadvantages, and so the methods are not strictly competitive for a specific problem.

Resolving Power

The electrical methods do not yet compete with the seismic method in terms of resolution or reliability. In part this shortcoming is a consequence of a lack of experience, but in part it is also the result of limitations inherent in the physics of electrical methods. The limitations to the precision with which earth structure can be determined are of two types.

One limitation is a lack of sensitivity of electromagnetic fields to earth structures smaller than a given size. For example, commonly, the strength of an electric or an electromagnetic field is measured with an accuracy of a few percent. If an earth structure being studied causes a change in the behavior of the field by less than this amount, it is unlikely that the earth structure can be detected. The solution to this problem is perhaps straightforward: the accuracy with which measurements are made can be increased, so that the effect of the structure being studied becomes larger than the error of measurement.

The other limitation is nonuniqueness, which is a different type of limitation. For example, it should be possible to detect a bed 1 ft (30 cm) thick at a depth of 1000 ft (300 m) if the resistivity of that bed is a thousandfold lower than the resistivity of the overlying rocks. However, precisely the same thing can be said about a bed 2 ft (60 cm) thick with a resistivity 500-fold lower than the resistivity of the overlying beds or a bed 5 ft (1.5 m) thick with a resistivity 200-fold lower, both at a depth of 1000 ft (300 m). In all three cases the effect on the electric or electromagnetic field will be substantially the same, so that the three cases cannot be distinguished. An increase in the precision of measurement does not provide a solution for this problem because the nature of the effect caused by the bed at depth does not change as the accuracy is increased. Auxilliary information must be provided before any distinction can be made between the cases.

RESEARCH RECOMMENDATIONS 745

Accuracy of Measurement

The accuracy with which electrical field surveys are carried out does not strain the current state of the art in instrumentation. It is common practice to measure direct current resistivity values in the field with an accuracy of ±5%. Certainly, equipment for more precise measurements is easily available, but the argument is made that usually errors introduced by small local changes in resistivity contribute a greater scatter to the data than is caused by errors in measurement. In order to detect a structure in the earth, it is probably necessary to have a change in resistivity that amounts to about 20% with present field standards. For a layered medium this means that a bed can be detected in which the contrast in either transverse resistance or longitudinal conductance amounts to about 50% of that in the overlying bed. In order to increase accuracy in such field surveys, it will be necessary to improve the measurement accuracy and to reduce the scatter of data caused by local inhomogeneities in resistivity. The first can be accomplished by routine improvements in field procedure and measurement practice, but the second requires careful thought. It may be that the scatter of field data can be reduced by making highly redundant measurements and performing some sort of averaging process, or it may be that certain configurations of electrodes will have minimal sensitivity to local changes in resistivity.

Magnetotelluric Method

With the magnetotelluric method the following considerations apply. The accuracy with which resistivity can be determined at a given frequency depends on the statistical behavior of the natural electromagnetic noise of the earth. Long observation times provide one with a better opportunity to get reliable estimates of earth resistivity but significantly increase the cost of magnetotelluric surveys. A question to be raised is whether it is more desirable to have a larger number of locations occupied with a lower accuracy so that the effects of lateral changes in resistivity can be evaluated or whether it is better to observe for a long period of time at a single station so that apparent resistivities can be determined with less uncertainty. It may well be that the use of a controlled source to provide frequencies that are not abundant in the natural electromagnetic specyrum will be a desirable adjunct to magnetotelluric sounding. Research should be directed to the question of how effectively the results of a magnetotelluric sounding can be improved by using an artificial source for at least part of the spectrum.

Data Reduction and Interpretation

Interpretive techniques for electrical surveying methods have advanced rapidly in recent years. The behavior of the electric or electromagnetic field in the presence of a layered structure, structures with simple geometric shapes, or structures with arbitrary shapes has been solved. Perhaps the most significant advance of recent years has been the application of mathematical inversion techniques to find earth models which closely simulate the results observed in field surveys. The best success has been obtained with inversion in those cases when the earth can be modeled by a one-dimensional medium, that is, by a sequence of layers in which the resistivity varies only with depth. Here a number of inversion techniques have been developed in which it is possible to simulate a set of field data with an average error of less than 1% by using very limited computer time. Only in this case do the capabilities of the interpretation method exceed the quality of the field data which are provided. For two-dimensional and three-dimensional problems, while the mathematical solutions are available, the numerical techniques are not yet advanced enough to permit inversions to be done with a practical amount of work. It is probable that further research both on numerical techniques for determining the electric and electromagnetic fields about bodies of arbitrary shapes and on numerical methods for doing inversions will be productive areas of research. It is quite important in this research that the door be left open for new approaches. The art of modeling and inversion is still at too early a stage of development for the best lines of attack or the ultimate limitations to be recognized.

Multidisciplinary Approach

The product of further development of techniques for measuring and interpreting electrical structures should be a capability for defining electrical structures with resolutions comparable to or better than those now obtained in gravity and density modeling. An important factor in reaching this goal will probably be our ability to incorporate diverse geophysical data into the interpretation of electrical data. For example, the seismic technique is capable of measuring earth structures with considerable accuracy. Often seismic velocities and electrical resistivities vary together. In porous rocks, both acoustic wave speed and resistivity are determined by the porosity. Within a given rock unit, resistivity and wave speed will correlate with a reasonably high degree of reliability. Therefore if a problem in nonuniqueness arises in the interpretation of electrical data, the earth's structure can be determined from seismic data, while the earth resistivity would be determined with an electrical method. Less work has been done on the interrelation of resistivity and wave speed in nonporous rocks, but it is likely that for restricted ranges of conditions, resistivity and wave speed will correlate in crystalline rocks. Much more research needs to be done on this topic before we can determine to what degree seismic results will assist us in evaluating electrical survey data.

Summary

In summary, while electrical surveying methods are undergoing rapid development at the present time, they must be considered to be relatively far behind the other geophysical methods in their present state of development. When electrical methods are improved to the point that they include the current level of sophistication of modern digital recording and data processing techniques, they should provide

RESEARCH RECOMMENDATIONS

us with an increasingly valuable tool for studying earth structures, including those associated with the accumulation of natural resources and those that may give rise to natural hazards. Even today, electrical methods are quite useful for prospecting for oil and gas, minerals, and geothermal energy. In addition, the use of resistivity measurements to assist in earthquake prediction and volcanic hazard prediction offers considerable promise.

> G. Keller
> Cochairman
> J. van Zijl
> Cochairman
> C. Clay
> B. Sternberg
> N. Harthill
> D. Strangway
> F. Dowling
> C. Skokan
> B. Lienert
> R. Corwin

Research Recommendations For The Thermal State Of The Crust

It is the objective of research on the thermal state of the crust to define the thermal regime (steady state distribution of temperature and fluid pressure) and the thermal evolutionary processes which affect changes in the regime active in the crust.

Measurements of surface heat flow provide the best direct evidence for this evolution. The role of crustal radioactivity is known to be important; and the study of it requires that heat flow measurements and near-surface concentrations of radioactivity be determined simultaneously. The role of deep crustal heat production versus the mantle contribution is poorly understood (and needs to be studied).

Parametric Data

In order to quantify the thermal history of crustal rocks, it is necessary to obtain certain parametric data as defined by transport theory. These include thermal conductivity, heat capacities, coefficients of thermal expansion, compressibilities (of fluids in a water-salt system) and of rocks over temperatures from ambient to 1000°C and for pressures from 1 bar to 10 kbar, permeabilities of rocks over a scale of fracture distributions, and viscosity of salt-water systems.

The study of magmas requires a knowledge of all the above parameters except permeability. It also requires data in the viscoelastic and the plastic-brittle transition regions. Such data would then permit the analysis of hypothetical thermal regimes in the crust by a numerical simulation of transport processes. Improvements are required in the simulation of systems which exist for $10^4 - 10^6$ years and incorporate hundreds of cubic kilometers of crust.

Theory

The theory of heat transport must provide for an efficient and effective simulation of thermal energy transport processes operative in the crust and for the transport properties of various phases. Improvements needed in numerical simulation include more efficient solutions to the partial differential equations. Theoretical developments must be applied to real crustal environments.

Field Surveys

Heat flux from the crust provides a boundary condition required to interpret subsurface conditions. Surveys directed toward this object objective should measure thermal gradients, conductive and convective fluxes, in situ thermal conductivity, permeability (bulk rock and fracture controlled), and fluid compositions.

Hydrothermal Systems

The role of hydrothermal systems is particularly relevant to current problems in the use of geothermal energy, and therefore we suggest the following studies: (1) in situ measurements of temperature, two-phase flow, permeability, fluid composition, host rock composition, mineral content, and characteristics of fluid inclusions, (2) surface measurements of heat flow, gravity, water fluxes and geochemical and electrical properties, (3) laboratory measurements of transport properties of fluids and rocks, (4) modeling that includes numerical simulation and scale models, and (5) measurements of the origin of heat from magma chambers, natural geothermal gradients, and radiogenic sources.

D. L. Turcotte
D. Norton

List of Participants

The following people participated in the symposium: J. K. Applegate, Boise State University; P. C. Badgley, Office of Naval Research; M. Batzle, Massachusetts Institute of Technology; M. Berry, Energy, Mines and Resources, Ottawa; H. Bezdek, Office of Naval Research; W. F. Brace, Massachusetts Institute of Technology; L. W. Braile, Purdue University; R. Buffler, University of Texas at Galveston; J. L. Carter, University of Texas at Dallas; P. Cheng, University of Hawaii; C. S. Clay, University of Wisconsin at Madison; R. Corwin, University of California at Berkeley; F. Dowling, Office of Naval Research; J. Eichelberger, Los Alamos Scientific Laboratory; M. Feves, Massachusetts Institute of Technology; and A. R. Green, Exxon Production Research Company.

Other participants were B. C. Haimson, University of Wisconsin at Madison; N. Harthill, Group Seven, Inc.; J. G. Heacock, Office of Naval Research; M. Holdaway, Southern Methodist University; R. M. Housley, Rockwell International Science Center; S. Jones, Chevron Oil Field Research Company; D. Jurdy, Princeton University; S. Kaufman, Cornell University; G. V. Keller, Colorado School of Mines; A. H. Lachenbruch, U. S. Geological Survey; M. Landisman, University of Texas at Dallas; B. Lienert, University of Texas at Dallas; M. Manghnani, University of Hawaii; B. J. Mitchell, Saint Louis University; W. Muehlberger, University of Texas at Austin; S. Mueller, Institut für Geophysik, Zurich; D. Norton, University of Arizona; J. Oliver, Cornell University; and J. Orcutt, Scripps Institution of Oceanography.

Completing the list of participants are E. Padovani, University of Texas at Dallas; R. Phinney, Princeton University; H. R. Pratt, Terra Tek; C. Prodehl, University of Karlsruhe, Karlsruhe; D. A. Richter, Massachusetts Institute of Technology; A. R. Sanford, New Mexico Tech; C. Skokan, Colorado School of Mines; J. Skokan, Gulf Minerals Resources Company; G. Simmons, Massachusetts Institute of Technology; S. B. Smithson, University of Wyoming; B. K. Sternberg, University of Wisconsin at Madison; D. Strangway, University of Toronto; G. Sutton, University of Hawaii; B. R. Tittmann, Rockwell International Science Center; D. L. Turcotte, Cornell University; J. S. van Zijl, National Physical Research Laboratory, Pretoria; H. Wang, University of Wisconsin at Madison; N. Warren, University of California at Los Angeles; and K. Westhusing, Energy Research and Development Administration.

INDEX

Terms are indexed by the first page of the paper in which they are discussed. All terms were supplied by the authors.

Absence of reflected waves, 243
Almandine garnet, 79
Amplitudes, 427
Anelasticity, 427
Anisotropic resistivity effects, 593
Anisotropy, 289
Apparent resistivity, 440, 501
Asperities, 615
Attenuation, 197, 405, 427
Audiomagnetotelluric (AMT) survey, 501
Basalt-andesite-dacite-rhyolite association, 57
Basin and Range Province, 349, 427
Batholiths, 57
Battle Mountain High, 626
Bouguer anomaly, 385
Bounds (seismic), 371
Brittle fracture, 615
Calc-alkaline rocks, 57
Canadian shield, 319
Capillary models, 722
Caribbean, 271
Cementation, 233
Central Wisconsin, 95
COCORP, 243
Complexity of deep crustal structure, 243
Compliance, 119
Compressibility, 95
Compressional wave velocity, 95, 271
Conductance, 440, 501
Conductive heat transfer, 693
Confining pressure, 197
Conrad discontinuity, 289, 385
Continuity of reflectors, 243

Controlled atmosphere, 181
Controlled source experiment, 531
Convection, 705
Convection in magma, 57
Convective heat transfer, 693
Cordierite granulite, 79
Cordillera, 319
Cracks, 149, 181
Crack porosity distribution, 95
Cratons, 470
Crust-mantle boundary, 349
Crust-mantle transition zone, 289, 349
Crust-Moho transition, 243
Crustal cross section, 349
Crustal evolution, 1
Crustal extension and heat flow, 626
Crustal radioactivity and heat flow, 626
Crustal reflections, 254
Crustal refraction, 553
Crustal stress, 576
Crustal structure, 254, 427
Crustal temperature profiles, 626
Crustal types, 1
Crustal velocities, 254
dc dipole-dipole survey, 501
Deep basement reflectors, 243
Deep electrical soundings, 470
Delay time, 371
Density, 95
Density gradients, 693
Depth of the strongest velocity gradient, 349
Dielectric losses, 181
Differential strain analysis, 95, 233
Diffractions, 243

Dipole array, 440
Dipole mapping method, 440
Dipole moment, 440
Earthquake prediction, 593
East Pacific Rise, 405
Eastern Colorado High Plains, 553
Elastic moduli nonself-consistent calculations, 119
Elastic moduli self-consistent calculations, 119
Electrical conductivity, 181, 531
Electrical resistivity, 95, 470, 593
Electrical soundings, 470
Electrical structure of continental crust, 470
Electrical structure of lithosphere, 470
Ellipticity, 470
Episodic sliding, 615
Equatorial dipole sounding, 440
Eurasia, 405
Eureka Low, 626
Experimental equilibria, 79
Explosion seismology, 349
Extremal bounds, 371
Fault displacement, 615
Fault friction, 615
Fault gouge, 615
Fault plane solution, 385
Fault stability, 615
Fine structure, 289
Finite difference (AMT) model calculations, 501
Flambeau anomaly, 501
Fluid phase, 79
Fluids in rocks, 289
Fracture zones, 693
Fracturing, 233
Frederick diabase, 95
Fused silica, 95, 197
Gasbuggy (nuclear explosion), 385
Geobarometer, 79
Geophysical exploration, 676
Geothermal, 676, 722
Geothermal gradient, 79
Geothermal reservoir, 705
Geothermal systems, 233, 693
Geothermometry, 19, 79
Graben areas, 349
Granite, 215
Granite melting, 79

Graniteville granite, 149
Granitic intrusions, 289
Granodiorite, 57
Granulite facies, 79
Graphite, 501
Gulf of Mexico, 271
Healed microcracks, 149
Heat flow, 626, 676
Heat flow contour map, 626
Heat transfer, 693
Hercynian Europe, 349
Heterogeneity of the basement, 243
Heterogeneous crustal model, 254
Higher-mode surface waves, 405
Hot springs, 722
Hot springs and regional heat flow, 626
Hydration processes, 470
Hydrofracturing, 576
Hydrothermal, 233
Hydrothermal systems, 626
Hydrology and heat flow, 626
Idaho (geothermal), 676
Igneous rocks, 95, 197
In situ stress, 576
In situ tests, 215
Inhomogeneities, 319
Input (airborne EM survey), 501
Internal friction, 197
Interrelationships among physical properties, 95
Inverse theory, 531
Inversion, 371
Inversion of attenuation data, 405
Iron formation, 501
Joint permeability, 215
Laboratory measurements at low frequencies, 197
Laboratory tests, 215
Laccolithic zone of granitic intrusions, 289
Laminated high-velocity layer, 289
Laminated structure, 319
Layer solution, 371
Layering of the crust, 289
Lithosphere, 553
Lithospheric thickness (determination), 553
Loading system stiffness, 615

INDEX 753

Longitudinal conductance, 440
Low-velocity zones, 289, 531
Magma (body), 385
Magma chambers, 57
Magmatism and heat flow, 626
Magnetic fields, 531
Magnetotellurics (MT), 440, 553
Marine geophysics, 271
Mean crustal velocity, 349
Mellen gabbro, 149, 181
Melting in the crust, 57
Metamorphic fronts, 289
Metamorphic lower crust, 470
Microcracks, 95, 149
Microearthquakes, 385
Middlebrook felsite, 149
Mixing of magmas, 57
Mobile belts, 470
Moduli-pressure curve crack populations, 119
Mohorovicic discontinuity, 289, 385
Montello granite, 181
Mountain root, 349
MT and well logs, 553
Multiplicity, 371
New Mexico, 385
Nonuniqueness, 371
North America, 405
Northwestern Atlantic, 271
Numerical simulation, 693
Ocean bottom seismograph, 371
Olivine basalt, 197
Open microcracks, 149
Orogenic belts, 349
Outgassing, 197
PMP phase, 349
Pacific Ocean, 405
Partial melting, 531
Pelitic metamorphic rocks, 79
Permeability, 215, 233
Petrology of andesitic rocks, 57
Physical properties, 233
Poisson's ratio, 385
Polar dipole sounding, 440
Pore aspect ratio, 119
Pore closing pressure, 119
Pore pressure effects, 319
Pore pressures, 215, 289
Pore strain, 119
Pore volume changes, 119
Porosity of crust, 470

Porous media, 722
Precambrian basement, 149, 181
Precambrian shield rocks, 501
Precambrian tectonic provinces, 470
Production well, 705
Pseudoisochron, 57
Q, 405, 427
Q structure model, 197
Red River quartz monzonite, 149
Reduced heat flow, 626
Reflection studies, 319
Reflections, 289, 385
Reflections from Moho, 243
Reflections in Precambrian rocks, 254
Refraction, 371
Refraction studies, 319
Resistivity, 233, 440, 676
Resistivity structure, 501
Resolution, 371
Resolving kernels, 371
Retrograde travel time curve, 349
Rio Grande rift, 385
Rock porosity, 593
Rock resistivity, 593
Rotating dipole survey, 440
S and H inversion, 553
San Andreas fault zone, 593
Sandstone, 215
Sanguine, Wisconsin, test facility antennas, 501
Scanning electron microscopy, 149
Schlumberger array, 440, 470
Sealed microcracks, 149
Sealing, 233
Sedimentary basin evolution, 1
Seismic Q, 197
Seismic reflection, 271
Seismic reflection profiling, 243
Seismic refraction, 427
Seismic structure, 319
Seismicity, 676
Shear wave velocity, 95
Shield areas, 349
Skrainka diabase, 149
Socorro, 385
Southeast Missouri, 95
Southern extension of the Canadian shield, 501
Stable sliding, 615
State of stress, 576

Stick slip, 615
Stiffness model, 615
Stouts River rhyolite, 149
Strain of oblate spheroidal pores, 119
Stress, 215, 576
Stress directions map of United States, 576
Stress drop, 615
Stress variation with depth in the United States, 576
Strontium isotopes, 57
Sulfide mineralization, 501
Surface conduction, 95
Surface waves, 405
Swarms (earthquake), 385
Synthetic seismograms, 427
Tectonic stress, 576
Tectonic units, 349
Temperature, 197
Thermal anomalies, 693
Thermal convection, 722
Thermal gradients, 722
Tigerton gabbro, 149
Total field apparent resistivity, 440
Transition zones, 289, 319
Transmission line analogy model calculations, 501
Transverse resistance, 440
Travel time, 371
Travel time inversion, 371
Uniqueness, 371
Upwarp, 553
Vacuum, 197
Veining, 233
Velocity, 215
Velocity anisotropy, 319
Velocity-depth profiles, 319
Velocity gradients, 349, 427
Velocity inversions, 289, 349
Vibrating bar technique, 197
VIBROSEIS, 243
VIBROSEIS survey, 319
Volcanism, 57
Volume strain, 119
Water content, 119, 289, 470
Wausau granite, 149, 181
Westward upwarp of mantle geotherm, 553
Xenoliths, 19
Xenoliths in andesitic rocks, 57

QE
511
S94
1976

JUN 21 1978